AF750351

Wetlands of the world I

Handbook of vegetation science

FOUNDED BY R. TÜXEN
H. LIETH, EDITOR IN CHIEF

Volume 15/2

Wetlands of the world: Inventory, ecology and management

Volume I

Africa, Australia, Canada and Greenland, Mediterranean, Mexico, Papua New Guinea, South Asia, Tropical South America, United States

Edited by

DENNIS WHIGHAM, DAGMAR DYKYJOVÁ
and SLAVOMIL HEJNÝ

KLUWER ACADEMIC PUBLISHERS
DORDRECHT / BOSTON / LONDON

Library of Congress Cataloging-in-Publication Data

Wetlands of the world I : inventory, ecology, and management / edited
by D.F. Whigham, D. Dykyjová, and S. Hejný.
p. cm. -- (Handbook of vegetation science)
Includes bibliographical references and index.
ISBN 0-7923-1685-1 (HB : alk. paper)
1. Wetland flora. 2. Wetlands. 3. Wetland ecology. 4. Wetlands--Management. I. Whigham, Dennis F. II. Dykyjová, Dagmar.
III. Hejný, Slavomil. IV. Series.
QK911.H3 pt. 15/2
[QK938.M3]
581 s--dc20
[333.91'8] 92-8365

ISBN 0-7923-1685-1

Kluwer Academic Publishers incorporates the publishing programmes of
Dr. W. Junk Publishers, MTP Press, Martinus Nijhoff Publishers, and D. Reidel Publishing Company.

Distributors

for the United States and Canada: Kluwer Academic Publishers, 101 Philip Drive, Norwell, MA 02061, USA
for all other countries: Kluwer Academic Publishers Group, P.O. Box 322, 3300 AH Dordrecht, The Netherlands

PRINTED IN THE NETHERLANDS

Contents

Preface

The impetus for this volume was the 2nd International Wetlands Conference which was held in June, 1984 at Trebon, Czechoslovakia. An overview of the worlds wetlands was one of the themes of the conference and it was decided that a useful follow-up would be a publication on the same topic. The initial goal was to cover as many of the worlds wetlands as possible in one volume and to have an emphasis on wetland ecology, biota, classification, and management. Individuals who made presentations at the Trebon conference were asked to prepare chapters and the editors also solicited other contributions.

For a variety of reasons, the initial goal has been difficult to reach, especially coverage of the entire globe, and it has been necessary to publish the contributions in more than one volume. Volume 1 represents the completion of the first phase of the project and it covers most of the Western Hemisphere, Australia, most of Africa, the Indian subcontinent, the Mediterranean region, and Papua New Guinea. Volume 2 will contain chapters on Western Europe, Northern Europe, Central Europe, most of northern and western Asia, the Middle East, and Indonesia. It is our hope that Volume 2 will appear in the near future and, if possible, a third volume will be published if authors can be secured to cover areas such as the Far East, other parts of the Indo-Pacific region, and New Zealand.

It is our hope that these volumes will be useful to those who seek an overview of the worlds wetlands and an introduction to the literature on their distribution, biota, management, and especially their ecology. The editors also hope that the chapters in these volumes will provide information that supplements earlier publications about the distribution of wetlands (Scott and Carbonell 1986, Carp 1980).

Information about the ecology of wetlands has increased tremendously in recent years. In addition to *Aquatic Botany*, there are now two journals devoted to wetlands (e.g., *Wetlands* published by the Society of Wetland

Scientists, *Wetland Ecology* and *Management* published by SPB Academic Publishing) and a variety of treatises have appeared in recent years (e.g., Davis and Gasse 1988, Denny 1985, Burgis and Symoens 1987, Ellenbroek 1987, Hughes and Hughes 1992, Hook *et al.* 1988, Lugo *et al.* 1990, Mitsch and Gosselink 1986, National Wetlands Working Group 1988, Patten 1990, Rodwell 1991, Sharitz and Gibbons 1989, Verhoeven 1992, Whigham *et al.* 1990, van der Valk 1990. There is still, however, much to be learned before effective management of these valuable resources will be possible.

Wetlands have been studied in great detail in some areas (USA, Canada, Australia) while information for other areas hasn't yet reached the stage of having adequate biotic inventories (e.g., New Guinea). Few natural wetlands exist in many parts of the world (e.g., the Mediterranean region and the Indian subcontinent) and ecological information is very rudimentary for other regions (e.g., Africa, Mexico, South America) even though excellent individual studies have been conducted. It is our hope that the papers presented in these volumes will provide an impetus to encourage additional studies of one of the worlds most important types of ecosystems.

References

Burgis, M. J. and J. J. Symoens (eds.) (1987) African Wetlands and Shallow Water Bodies. Éditions de l'ORSTOM, Institut Français de Recherche Scientifique Pour le Développement en Coopération, Paris, France. 650 pp.

Carp, E. (compilor) (1980) Directory of Wetlands of International Importance in the Western Palearctic. International Union for Conservation of Nature and Natural Resources, Gland, Switzerland. 506 pp.

Davis, B. and Gasse (eds.) (1988) African Wetlands and Shallow Water Bodies. Bibliography. Travaux et Documents No. 211. ORSTROM, Paris, France. 502 pp.

Denny, P. (ed.) (1985) The Ecology and Management of African Wetland Vegetation. Dr. W. Junk Publishers, Dordrecht, The Netherlands. 343 pp.

Ellenbroek, G. A. (1987) The Ecology and Productivity of an African Wetland System: The Kafue Flats, Zambia. Kluwer Academic Publishers, Dordrecht, The Netherlands. 267 pp.

Hook, D. D., McKee, W. H., Jr., Smith, H. K., Gregory, J., Burrell, V. G. Jr., DeVoe, M. R., Sojka, R. E., Gilbert, S., Banks, R., Stolzy, L. H., Brooks, C., Matthews, T. D. and Shear, T. H. (eds.) (1988) The Ecology and Management of Wetlands. Volume 1: Ecology of Wetlands. Timber Press, Portland, Oregon, USA. 592 pp.

Hook, D. D., McKee, W. H., Jr., Smith, H. K., Gregory, J., Burrell, V. G. Jr., DeVoe, M. R., Sojka, R. E., Gilbert, S., Banks, R., Stolzy, L. H., Brooks, C., Matthews, T. D. and Shear, T. H. (eds.) (1988) The Ecology and Management of Wetlands. Volume 2: Management, Use and Value of Wetlands. Timber Press, Portland, Oregon, USA. 394 pp.

Hughes, R. H. and Hughes, J. S. (1992) A Directory of African Wetlands. IUCN. Glands, Switzerland and Cambridge, United Kingdom. 820 pp.

Lugo, A. E., Brinson, M. and Brown, S. (eds.) (1990) Forested Wetlands. Ecosystems of the World 15. Elsevier, Amsterdam, The Netherlands. 527 pp.

Mitsch, W. J. and Gosselink, J. G. (1986) Wetlands. Van Nostrand Reinhold Company, New York, New York, USA. 539 pp.

National Wetlands Working Group (1988) Wetlands of Canada. Ecological Land Classification

Series, No. 24. Sustainable Development Branch, Environment Canada, Ottawa, Ontario, and Polyscience Publications Inc., Montreal, Quebec, Canada. 452 pp.

Patten, B. D. (ed.) (1990) Wetlands and Shallow Continental Water Bodies. Volume 1: Natural and Human Relationships. SPB Academic Publishing bv, The Hague, The Netherlands. 759 pp.

Rodwell, J. S. (ed.) (1991) British Plant Communities. Volume 2. Mires and Heaths. Cambridge University Press. Cambridge, United Kingdom. 628 pp.

Scott, D. A. and Carbonell, M. (compilors) (1986) A Directory of Neotropical Wetlands. IUCN and IWRB Slimbridge, United Kingdom. 684 pp.

Sharitz, R. R. and Gibbons, J. W. (eds.) (1989) Freshwater Wetlands and Wildlife. CONF-8603101, U.S. Department of Energy, National Technical Information Service, Springfield, Virginia, USA. 1265 pp.

van der Valk, A. (ed.) (1989) Northern Prairie Wetlands. Iowa State University Press, Ames, Iowa, USA. 400 pp.

Verhoeven, J. T. A. (ed.) (1992) Fens and Bog in The Netherlands: Vegetation, History, Nutrient Dynamics and Conservation. Kluwer Academic Publishers. Dordrecht, The Netherlands. 490 pp.

Whigham, D. F., Good, R. E. and Kvet, J. (eds.) (1990) Wetland Ecology and Management: Case Studies. Kluwer Academic Publishers, Dordrecht, The Netherlands. 180 pp.

August 29, 1992

DENNIS WHIGHAM
DAGMAR DYKYJOVÁ
SLAVOMIL HEJNÝ

Ralph E. Good
(Feb. 24, 1937 – Dec. 11, 1991)

This volume is dedicated to the memory of our wetlands colleague and friend, Dr. Ralph E. Good. Ralph, Distinguished Professor of Botany at Rutgers University, was a tireless and dedicated ecologist who served the scientific community in a variety of ways. He served on the Governing Board (1980–1982) and Board of Directors (1983–1986) of the American Institute of Biological Sciences. The Ecological Society of America (ESA) benefited in numerous ways from his service over many years. He was a member and/or chair of numerous committees and was the ESA Business Manager (1973–1979) and Vice President (1978–1980). For his numerous efforts to ESA, Ralph received its Distinguished Service Citation in 1989. He was named as a fellow of the American Association for the Advancement of Science in 1990 in recognition of his contributions in the field of ecology. Ralph was also heavily involved in regional and university activities. He served as President of the Philadelphia Botanical Club (1973–74) and the New Jersey Academy of Sciences (1978–1980). At Rutgers, Ralph was Chair (1978–1982) of the Biology Department at Rutgers-Camden and was the Director of its Biology Graduate Program from 1988 until his death in 1991. His service to the university was recognized in 1985 when he received the Rutgers Presidential Award for Distinguished Public Service. His commitment to public service is perhaps best represented by his efforts to establish the Pinelands National Reserve, the first reserve of its type in the USA. Ralph worked with local, state, and federal agencies to help establish the reserve and was instrumental in creating the Rutgers Division of Pinelands Research 1981. With his wife Norma and his many graduate students, Ralph published numerous papers on a variety of terrestrial and wetland topics. His energy and dedication will be missed by all who knew him.

Acknowledgments

The editors and authors would like to acknowledge individuals and organizations who have contributed to completion of this volume.

Editors – The editors would especially like to thank Meridel Jellifer for the numerous hours that she spent at the wordprocessor working on the manuscripts. It was an often difficult and long task and her effort is greatly appreciated. Margaret McWethy drafted figures in the chapters on Africa and Mexico. Mary Bates of the U.S. Fish and Wildlife Service provided valuable assistance with the chapter on U.S. wetlands. DFW would like to thank Dr. David Correll, Director SERC for providing financial assistance and for agreeing that Mrs. Jellifer could work on the project. Additional financial help was provided by UNESCO (Contract SC/RP/204.079.4).

Australia – Research facilities and funding were provided by the University of New England and the Royal Botanic Gardens Sydney.

Papua New Guinea – Partial funding for this chapter was provided by the University of Papua New Guinea Research and Publications Committee.

United States – The chapter was prepared by scientists working for the U.S. Government. As such, the material in this chapter are exempted from copyright rules and regulations. Support for the chapter was provided by the U.S. Fish and Wildlife Service and the following national and state cooperators to the National Wetlands Inventory (NWI): U.S. Army Corps of Engineers, Department of the Navy, U.S. Environmental Protection Agency, U.S. Bureau of Reclamation, Alaska, Colorado, Connecticut, Delaware, Florida, Hawaii, Illinois, Indiana, Kentucky, Maine, Maryland, Michigan, Minnesota, Nevada, New Mexico, Oregon, Pennsylvania, South Carolina, South Dakota,

Tennessee, Utah, Virginia, Washington, and Wyoming. Other NWI contributors are Puerto Rico, North Slope Borough (Alaska), Ducks Unlimited, Bonneville Power Administration, Yukon Pacific Corporation, and Cominco Alaska Exploration Incorporated.

List of contributors

I. Editors

DENNIS F. WHIGHAM
Smithsonian Environmental Research Center
Box 28
Edgewater, Maryland 21037
USA

DAGMAR DYKYJOVÁ
Department of Hydrobotany
Institute of Botany
Czechoslovak Academy of Science
379 82 Třeboň
Czechoslovakia

SLAVOMIL HEJNÝ
Institute of Botany
Czechoslovak Academy of Science
252 43 Průhonice
Czechoslovakia

II. Authors

CHARLES M. BREEN
Institute of Natural Resources
University of Natal
P.O. Box 375

Pietermaritzburg
3200
South Africa

R. H. BRITTON
Station Biologique de la Tour du Valat
Le Sambuc
13200 Arles
France

MARGARET A. BROCK
Biology Department
University of New England
Armidale
New South Wales
Australia, 2351

A. J. CRIVELLI
Station Biologique de la Tour du Valat
Le Sambuc
13200 Arles
France

PATRICK DENNY
English Nature
Northminster House
Peterborough PE1 1UA
United Kingdom

C. M. FINLAYSON[1,2]
[1]Alligator Rivers Region Research Institute
Office of the Supervising Scientist
Private Mail Bag
Jabiru, NT 0886
Australia

[2]Present address:
International Waterfowl and Wetland Research Bureau
Slimbridge
Gloucester GL2 7BX
United Kingdom

V. GLOOSCHENKO[1,2]
[1]Wildlife Branch
Ontario Ministry of Natural Resources
Whitney Block
Toronto, Ontario
Canada M7A 1W3

[2]Present address:
U.S. Fish and Wildlife Service
75 Spring Street, S.W.
Atlanta, GA 30303
USA

W. A. GLOOSCHENKO[1,2]
[1]Lakes Research Branch
National Water Research Institute
P.O. Box 5050
Burlington, Ontario
Canada L7R 4A6

[2]Present address:
KEMRON Environmental Services
2986 Clairmont Rd.
Suite 150
Atlanta, GA 30329
USA

BRIJ GOPAL
Jawaharlal Nehru University
School of Environmental Sciences
New Mehrauli Road
New Delhi
India

JAN HEEG
Department of Zoology
University of Natal
P.O. Box 375
Pietermaritzburg
3200
South Africa

SURREY JACOBS
Royal Botanic Gardens
Sydney
New South Wales
Australia 2000

DAVID M. JOHN
Department of Botany
The Natural History Museum
Cromwell Road
London SW7 5BD
United Kingdom

WOLFGANG J. JUNK
Max-Planck-Institut für Limnologie
Arbeitsgruppe Tropenokologie
Postfach 165
W-2320 Plön
Germany

K. KRISHNAMURTHY
Centre of Advanced Study in Marine Biology
Annemalai University
Parangipettai 608 502
Tamil Nadu
India

CHRISTIAN LÉVÊQUE
Office de la Recherche Scientifique et Technique Outre-Mer
213 Rue Lafayette
75480 Paris Cedex 10
France

LEONARD E. NEWTON
Department of Botany
Kenyatta University
P.O. Box 43844
Nairobi,
Kenya

INGRID OLMSTED
Centro de Investigacion Cientifica de Yucatan, A.C.
Apartado Postal 87
Cordemex, Merida
Yucatan, Mexico

PATRICK L. OSBORNE[1,2]
[1]Biology Department
University of Papua New Guinea
P.O. Box 320
National Capital District
Papua New Guinea

[2]Present address:
Water Research Laboratory
Faculty of Science and Technology
University of Sydney
Richmond
New South Wales
Australia, 2753

MAITLAND SEAMAN
Department of Zoology
University of the Orange Free State
P.O. Box 339
Bloemfontein
9300
South Africa

C. TARNOCAI
Land Resource Research Institute
Agriculture Canada
K.W. Neatby Building
Ottawa, Ontario
Canada K1A 0C6

RALPH W. TINER
U.S. Fish and Wildlife Service
Region 5
Newton Corner, MA 02158
USA

I. VON OERTZEN
CAB International
Wallingford
Oxon OX10 8DE
United Kingdom

WILLIAM WILEN
U.S. Fish and Wildlife Service
Division of Habitat Conservation
400 Arlington Square
1849 C. St., NW
Washington, DC 20240
USA

S. ZOLTAI
Canadian Forestry Service
5320-122nd Street
Edmonton, Alberta
Canada T6H 3S5

Wetlands of Africa: Introduction

P. DENNY

Abstract

This chapter provides an overview of African wetlands divided into three main geographic areas: Eastern, Western and South Africa. The main types of wetland vegetation and their distributions are outlined. The interrelationships of geomorphology, climate, soil types, and water quality are examined in relation to wetland development. The dynamics of specific wetlands in each of the geographical areas are then examined in detail. Included are the Swamps of the Upper Nile, the Rift and high altitude lakes of Eastern Africa, the Niger and its floodplains, the Lower Senegal Valley, Coastal Lagoons of the Ivory Coast, and Lake Chad in West Africa. The vast floodplains of Southern Africa including the Pongolo River floodplains, the Mkuze Wetland System, Nyl River floodplains, and various pans and dambos. The major man-made lakes, their impact on the environment and problems with waterweeds are also considered. African wetlands are fragile ecosystems under serious threat and there are many pressures on their long-term survival. Thus, management strategies for their conservation and sustainable development are discussed in relation to the needs of the people of Africa.

Introduction

In this chapter, although the whole of the African continent, Madagascar, and offshore islands of the eastern South Atlantic and western Indian Oceans come within the brief (Fig. 1), consideration will be confined to the mainland. The reason for this is twofold: the continent occupies by far the largest area and contains the most extensive wetlands; and information on offshore islands is sparse. A factor which must be borne in mind continuously whilst reading the text is the deficiency of knowledge of African vegetation, and wetlands in particular. The sheer size of the continent; the often inhospitable

D.F. Whigham et al. (eds.), Wetlands of the World I, 1–128.

Figure 1. Map of Africa indicating the approximate position of countries. The approximate boundaries of the three parts (Eastern, Western, Southern) of Africa covered in this chapter are indicated. Some names have changed since the preparation of the map.

climate, difficult terrain, limited access and resources for surveys, all contribute to the situation. The number of wetlands that have received more than cursory attention is very few. The reader is thus advised that the majority are still to be studied and any review must be biased. To get things into perspective, consider the following example. In Great Britain some areas of the Norfolk Broads and accompanying wetlands (which cover around 10 km^2) have been scheduled as nature reserves or sites of special scientific interest. In Africa, such small areas would probably not warrant a second glance.

Wetlands such as the Sudd in the Upper Nile (16,300 km^2 permanent swamp + 15,500 km^2 seasonal flooding, (Mefit-Babtie 1983, Howell *et al.* 1988), the Okavango (16,000 km^2, Howard-Williams and Thompson 1985), or the Niger Inland Delta (2,000 km^2, Howard-Williams and Thompson 1985) are only just beginning to attract attention. Many small wetlands on offshore islands or within Africa may provide refuge as important as the Norfolk Broads but, as yet, are unappreciated.

Two publications (White 1983, Denny 1985a) cover much of the material needed in this chapter. White's publication embodies all types of African vegetation, is the most thorough to date, and is likely to be a standard reference for many years to come. Denny's is a pan-African approach to wetlands specifically, and considers fully their ecology, dynamics, and management. Data have been extracted from both texts and the reader is referred to them for detailed information. Broader texts covering African inland waters with chapters on aquatic plants include Beadle (1981) and Symoens *et al.* (1981) whilst a useful source book by Luther and Rzoska (1971) lists information for waters proposed for conservation. Some information can be gleaned from Walter (1971) who discusses African vegetation in the different climatic zones and Lind and Morrison (1974) who consider East African vegetation. A new book by Finlayson and Moser (1991) provides an excellent overview of global wetlands and includes a chapter by Denny on African systems. Davies and Walker (1986) considers wetlands associated with river systems and has some good examples from Africa. A number of books consider the general limnology and ecology of specific sites, e.g. *Flore et Faune Aquatiques de l'Afrique Sahelo-Soudanienne* (Durand and Lévêque 1980), *Lake Chad* (Carmouze *et al.* 1983), *The Nile*, (Rzoska 1976), *The Niger* (Grove 1985), *Lake Sibaya* (Allanson 1979), *The Jonglei Canal* (Howell *et al.* 1988), *Lake Chilwa* (Kalk *et al.* 1979), *The Kafue Flats* (Ellenbrock 1987), and *Lake McIlwaine* (Thornton 1982) but it must be remembered that wetland and floodplain vegetation accounts for only part of the text and sometimes, is superficially described. A list of 794 cross-indexed citations of books, publications and reports on wetlands from various regions of Africa, compiled by Thompson *et al.* (1985) on the other hand, is essential reading for additional references.

Books recently published do much to enhance our present knowledge. These include: *The Evolution of Africa's Rare Animals and Plants* (Kingdon 1990), *Inland Waters of Southern Africa* (Allanson *et al.* 1990), *The Inland Waters of Tropical West Africa* (John 1986), and *Plant Ecology in West Africa: Systems and Processes* (Lawson 1986). A very comprehensive works entitled *African Wetlands and Shallow Waterbodies*, published by ORSTOM, Paris, is composed of three volumes: (1) *Bibliography* edited by B. R. Davis and F. Gasse; 1988, (2) *Directory* edited by M. J. Burgis and J-J. Symoens,

1987, and (3) *Structure, Functioning and Management* edited by C. Breen and C. Lévêque. Finally, the IUCN is also compiling a *Directory of African Wetlands* (Mephan and Mephan in press).

Broad vegetation zones

Africa has the second largest landmass in the world and a more diverse flora than any equivalent area (White 1983). The wetlands are only a small part of the entire vegetation range and before we focus our attention on these it is useful to summarize broad vegetation zones. There are several ways in which this can be done. Traditionally, descriptive terms such as "semi-desert shrublands" and "savanna" etc., which inherently incorporate physiographic, climatic, and edaphic features in vegetation classification has been used to delimit physiognomic types. White (1983) argues that these terms are restrictive as they are a subjective selection of a few features and do not necessarily reflect the flora. He therefore favours ". . . a chorological system based on the patterns of geographical distribution shown by entire floras . . ." He feels that in the first instance vegetation classification should be *without* reference to the physical environment. Thus his classification is in floristic regions termed *Phytochoria* which are based on the richness of their endemic flora at the species level. Physiographic, climatic, and faunistic (including anthropogenic) pressures will moderate the vegetation (hence the physiognomic classification) and there are large areas of similarity in the two classifications but (i) a particular physiognomic type may incorporate two or more distinct floras and (ii) the boundaries of a particular flora may transcend several physiognomic types.

The effect of environment

Environmental characters help us to understand the dynamics of vegetation and a brief mention should be made of the major ones. First consider altitude. The continent can be divided broadly into two parts: Low and High Africa (Fig. 2). Low Africa, which is to the north and west, is composed largely of sedimentary basins and upland plains below about 600 m a.s.l. (e.g. the catchment areas of the Lower Nile, Niger, and Zaire rivers and the Lake Chad basin). Within Low Africa there are mountainous regions such as the Atlas Mountains (4,165 m), Jebel Marra (3,042 m, see Wickens 1976) and the headwaters of the Niger (Guinean Highlands, 1,853 m) but these are relatively few. In contrast, High Africa to the south and east is mainly about 1,000 m. The rifting and faulting of the continent along two major faultlines running approximately NE/SW from Ethiopia to Zimbabwe (and the associated volcanic activity), provide some of the highest mountains (The Rwenzori

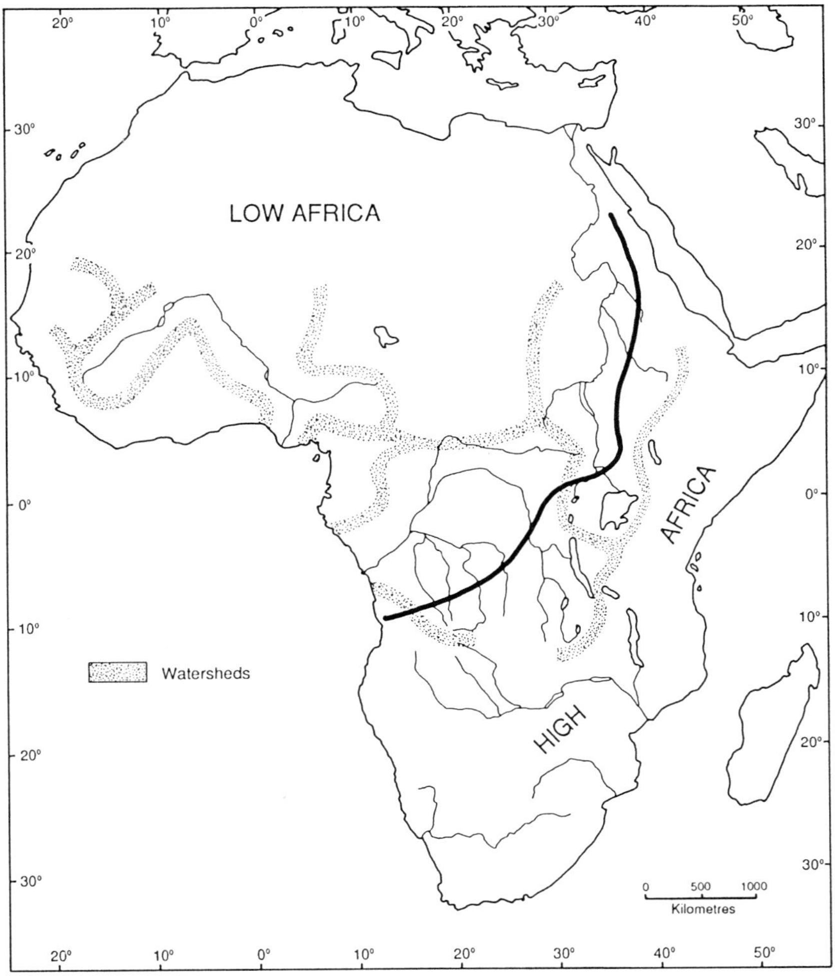

Figure 2. Map showing the main watersheds and the approximate division between High (Normally above 1000 m a.s.l.) and Low (up to 600 m a.s.l.) Africa (after White 1983; Fig. 2).

Mountains, 5,109 m and Mt. Kilimanjaro, 5,895 m) and deepest lakes (The Rift Valley Fault Lakes) as well as high plateaux. Within a plateau is the Basin of Lake Victoria (1,130 m) and some of the richest papyrus swamps.

Inevitably, with a continent the size of Africa, which spans some 60° of latitude and 6,000 m altitude, a wide range of climate will occur. Of the ten principle climatic types of the world proposed by Walter and Lieth (1960–67) in their *Klimadiagramm Weltatlas*, no less than five can be found in Africa (Fig. 3). i.e.

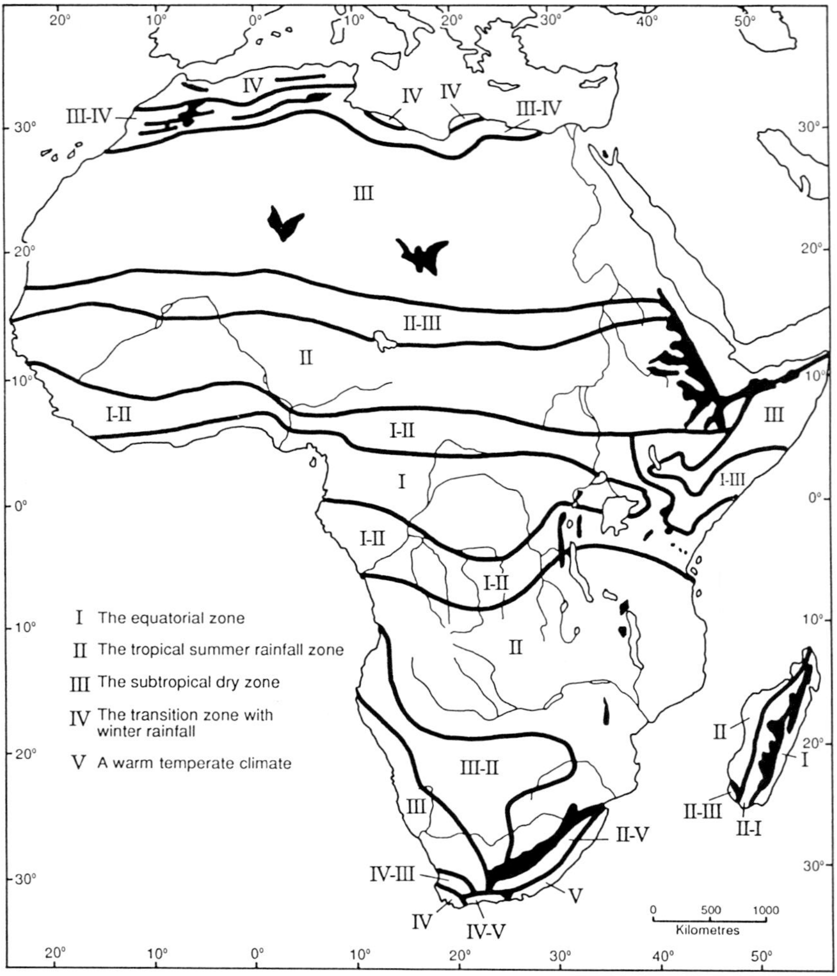

Figure 3. The major climatic zones of Africa (after Walter *et al.* 1975; Map 9).

I. Equatorial: hot, humid or with two rain seasons.
II. Tropical: hot with summer rain.
III. Sub-tropical: hot and arid.
IV. Mediterranean: arid summers and winter rains; rarely frosty.
X. Mountain

Naturally, there will be broad correlations between climatic types and vegetation (compare Figs. 3 and 5) but, as White (1983) points out, the main phytochoria are characterized by a variety of forcing factors of which climate is only one. Therefore, whilst local climatic information provides valuable additions for the understanding of vegetation dynamics, correlations between

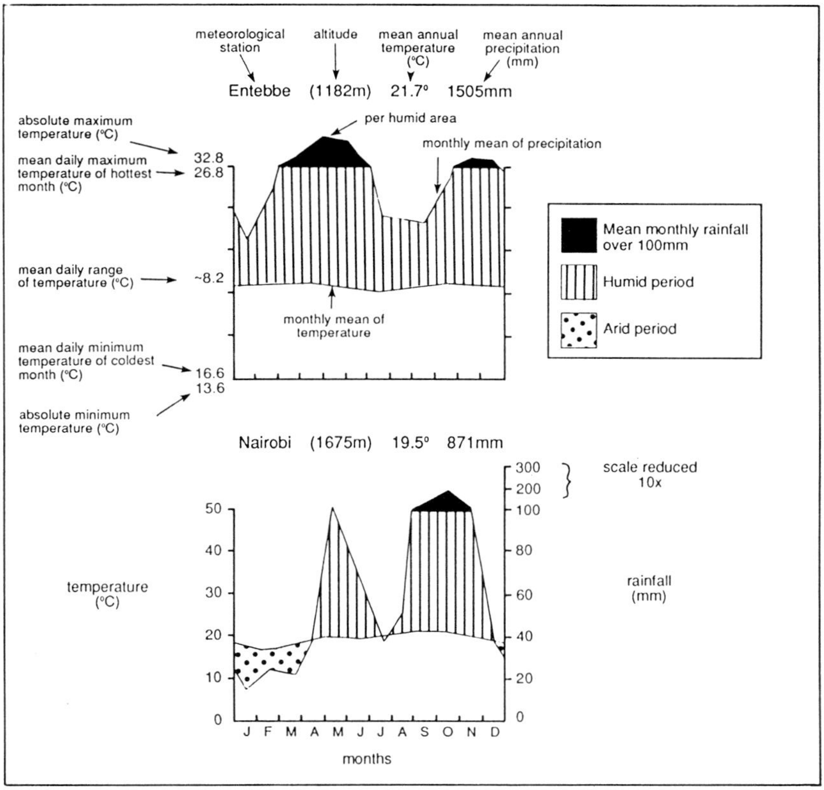

Figure 4. An example of two climatic diagrams with annotations to assist interpretation (from Walter *et al.* 1975; Figs. 6 and 8).

primary phytochoria and major climatic types are of limited value. What are helpful are the climatic diagrams published in Walter and Lieth (1960–67) and Walter *et al.* (1975) which give data from local meteorological stations. These need a little explanation as similar diagrams will be used in the succeeding pages to provide quick reference to regional climate in wetland areas. The climatic diagrams plot mean monthly temperatures and rainfall on the same scale such that 20 mm rainfall is equivalent to 10° C increase in temperature above zero. A rainfall curve above the temperature curve (hatched area) tends to indicate a relatively humid period, and below, (stippled area) a relatively arid period. These data do not correspond directly to potential evaporation but provide a rule-of-thumb reference for waterbalance. Mean monthly rainfalls exceeding 100 mm are also indicated. An example of such a diagram is given in Fig. 4.

Unlike other vegetation types, wetland physiognomy is generally more directly related to the height of the watertable and flooding periodicity than precipitation and evaporation. However, towards the drier end of a flood-plain catena, climate has a more direct effect. Absolute, extreme, and mean temperatures along with daily and seasonal ranges in temperature also can directly affect the wetland flora. *Potamogeton schweinfurthii*, for example, is confined largely to the warm tropics whilst *Potamogeton lucens* is a species of cooler climates (Denny 1985b). Mangroves, likewise, are not frost resistant.

Classification of vegetation

White (1983) classifies a phytochorion which has both more than 50% of its species confined to it and a total of more than 1000 endemic species into the highest rank, which he terms a *Regional Centre of Endemism*. Between regional centres will be found *Transition Zones*, and if these are of the same magnitude as the centres of endemism, they are given a name and are ranked equally. Passing through a transition zone, a flora from one centre is replaced spatially by another. Sometimes, more than one main phytochorion is involved – The Lake Victoria Mosaic is an example – and the vegetation is much more complex forming a mosaic of different physiognomic types with different floristic relationships. The main phytochoria of the African mainland and their distributions are shown in Fig. 5. The major phytochoria will support a variety of vegetation types characterised by their physiognomy. Sixteen main types have been identified (White 1983, Tables 1 and 3), five which follow are of relevance to wetlands:

Forests.	A continuous stand of trees at least 10 m tall, their crowns interlocking.
Grasslands.	Land covered with grasses and other herbs, either without woody plants or the latter not covering more than 10% of the ground.
Mangroves.	Open or closed stands of trees or bushes occurring on shores between high- and low-water mark.
Herbaceous wetlands.	Freshwater, swamp and aquatic vegetation.
Halophytic vegetation.	Saline and brackish swamp.

Forests and *Grasslands* are both very large entities of mainly non-wetland habitat. However, they include important wetland communities such as swamp and riparian forests and seasonal- and permanently-flooded grasslands. Flooded grasslands are often classified in the broader category of edaphic grasslands as the extent of soil waterlogging is a major determinant of the plant community (Vesey-Fitzgerald 1963). The broad distribution of *Forests* and *Grasslands* communities together with *Mangroves*, *Herbaceous wetlands* and *Halophytic vegetation* are shown in Fig. 6 and are described below. As *Herbaceous wetlands* have clearly defined physiognomic communi-

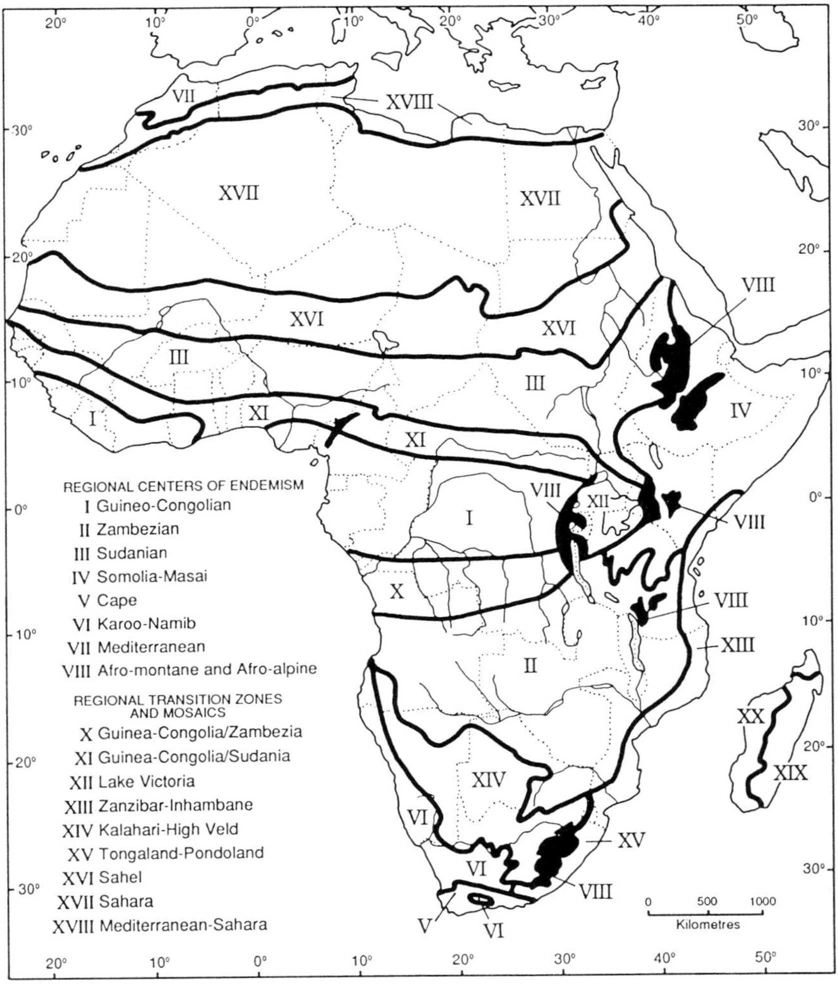

Figure 5. The main phytochoria (phytogeographical areas) in Africa (redrawn from White 1983; Fig. 4).

ties dependent upon water depth (see Table 1) each will be described separately.

Forests

Swamp forests

Additional reading can be found in: Thompson (1985), White (1983), Thompson and Hamilton (1983), Hall and Swaine (1981), and Boughey (1957). Swamp forests contain dense stands of trees from 10–50 m or more in height,

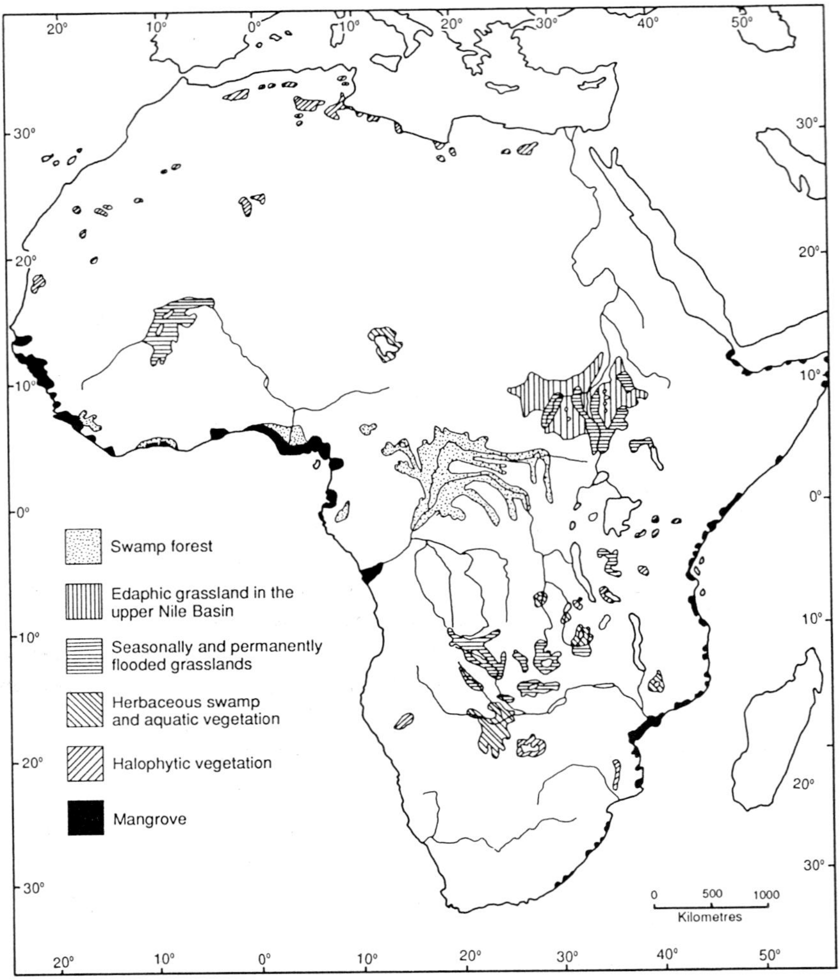

Figure 6. Map showing the major African wetlands (redrawn and modified from the maps in White 1983).

which are specifically adapted to, or are tolerant of flooding and waterlogged soils. The forests are diverse and often support a rich epiphytic flora (see photograph on p. 28). Because of their location and due to population pressures, forests and swamps are also often heavily exploited as shown in photographs on p. 29. Some reside in permanently-flooded areas whilst others are found in seasonally-flooded zones. Swamp forests are widespread in the Guineo-Congolian Region but extend into the adjacent transition zones, the Lake Victoria regional mosaic, and the wetter parts of the Zambezian region.

Table 1. Physiognomic classification of major wetland vegetation (White 1983).

Main vegetation type	Community	Sub-divisions
Forests	Swamp forest	Permanent swamp, Seasonal swamp
	Riparian forest	
Grasslands	Seasonally-flooded grasslands Permanently-flooded grasslands	
Herbaceous wetlands	Emergent plant communities	Bottom rooted swamp, Floating swamp
	Euhydrophyte communities	Floating-leaved, rooted vegetation, Submerged rooted and free-floating vegetation
	Surface-floating communities	
Mangroves		
Halophytic vegetation		

Permanent swamp forest. As the name implies, these occur on permanently flooded soils. They are relatively poor in species diversity, the most common genera being *Caraipa*, *Mitragyna*, *Nauclea*, *Pandanus*, *Phoenix*, *Raphia*, *Spondianthus*, *Symphonia*, *Uapaca*, and *Voacanga* with *Raphia* predominant in the wettest areas. They come into their own particularly in an equatorial climate such as is found in the Guineo-Congolian region.

Seasonal swamp forest. Although the species found in the seasonal swamp forest will depend upon the major phytochorion region, seasonality and flooding periodicity will regulate the zonation of species in any particular swamp. Most of the permanent swamp trees can be found together with genera of less flood-tolerance such as *Anthocleista*, *Croton*, *Diospyros*, *Ficus*, *Pseudospondias*, and *Rauvolfia*. Seasonal swamps may be found on the fringes of permanent swamps but largely occur in regions of distinct seasonality especially the Zambezian region.

Riparian forests

Riparian forests typically form gallery forests along watercourses in tropical and sub-tropical climatic zones where the watertable is high but standing water and waterlogging rarely occurs. In the Guineo-Congolian region riparian forests are not easily distinguishable from swamp forests but in the Zambezian and Sudanian regions, where climate is less humid and a more distinct seasonality exists, they are a feature of most rivers. They are easily distinguished along the banks of the Zambezi, for example. Genera include

Acacia, *Croton*, *Combretum*, *Diospyros*, *Ficus*, *Garcinia*, *Kigelia*, *Rauvolfia*, and *Zizyphus*. In the complex flora of the Lake Victoria regional mosaic, *Erythrina*, *Phoenix*, *Pseudospondias*, and *Spathodea* are characteristic.

Grasslands

Further reading on edaphic (flooded) grasslands include: Menaut (1983), Menaut and Cesear (1983), White (1983), Thompson (1985), Iltis and Lemoalle (1983), Vesey-Fitzgerald (1970, 1973), Ellenbrock (1987), Howell, P., Lock M., and Cobb, S. (1988) and Denny (1991).

Seasonally-flooded grasslands
Lacustrine floodplain grasslands are often restricted to a fringe vegetation between seasonal low- and high-water and are in intense competition with more aquatic vegetation. The riverine floodplains on the other hand, particularly those associated with the great rivers, include some of the vast areas of seasonally-flooded grasslands which are so important to the African economy and wildlife. They are formed from the regular spilling of the river water over the levees into the surrounding plains. This can be from once to several times a year. In a normal flooding regime the spilling phase is always much more rapid than the dissipation phase. Thompson (1985) quotes a six week flooding (spilling) and a six month dissipation phase as an example. The plant communities are a function more of water dissipation than spilling. The regularity of flooding, and its depth and duration, clearly affect the species distribution.

Although the major African floodplains are relatively well-known, there are huge areas of seasonally waterlogged land unconnected with major waterways and lakes, which are poorly documented. These are the black (cracking) clay soils (cotton soils) with a high (30–80%) montmorillonite content (Fig. 7) and the pan wetlands (Thompson 1985). The cracking clays expand in the rain season to become impervious and waterlogging ensues – sometimes for up to six months of the year. Pan wetlands are formed when there is an impermeable hardpan in the soil profile which causes water to collect over it during the rains. If the pan horizon comes to the surface, the resultant seasonal wetlands are commonly called dambos. Pan wetlands thus predominate in regions of high seasonal rainfall where land drainage is impeded.

The major floodplains are associated with rivers which have a seasonal rainfall catchment area particularly in the Zambezian, Sudanian, and Somalia-Masai centres of endemism and the adjacent transitions zones, together with the Lake Victoria regional mosaic, and the east coast where there is strongly seasonal rainfall. Where the dry season is very short or absent, such as in the Guineo-Congolian region, these grasslands are absent or constitute

Figure 7. A cracking-clay soil of the Upper Nile floodplain in the dry season. At the first rains the clay swells and forms an impervious layer which becomes inundated (photograph by P. Denny).

a very small part of the total wetland area. Rivers obviously include the Nile, Niger, Zaire, and Zambezi, but many of the lesser ones too.

The species composition and vegetation dynamics of the seasonally-flooded grasslands cannot easily be generalized as so many climatic and physiographic variables interplay to determine the actual community. In addition to these, heavy grazing, and anthropogenic activities, especially burning, further moderate the vegetation. Indeed, in many places the grassland is a fire-determined climax. In any particular area a continuous change in the vegetation with the soil water regime, a hydrosere, may be found. Boundaries between vegetation zones are diffuse and the communities range from that of open water, through swamp, to permanent and seasonal grasslands, to dry savanna. A hydrosere that can commonly be found is: *Vossia* → *Oryza* → *Echinochloa* → *Hyparrhenia rufa*, but there are an almost infinite number of variables to this sequence, and many more species interactions. A selection of figures redrawn from Thompson (1985) provides some concept of the variety (Fig. 8) but the reader is referred to White (1983) and Thompson (1985) for more detail.

Permanently-flooded grasslands

It can be seen from Fig. 8 that the hydrosere passes through seasonal- and permanently-flooded grasslands, the latter being composed of species more

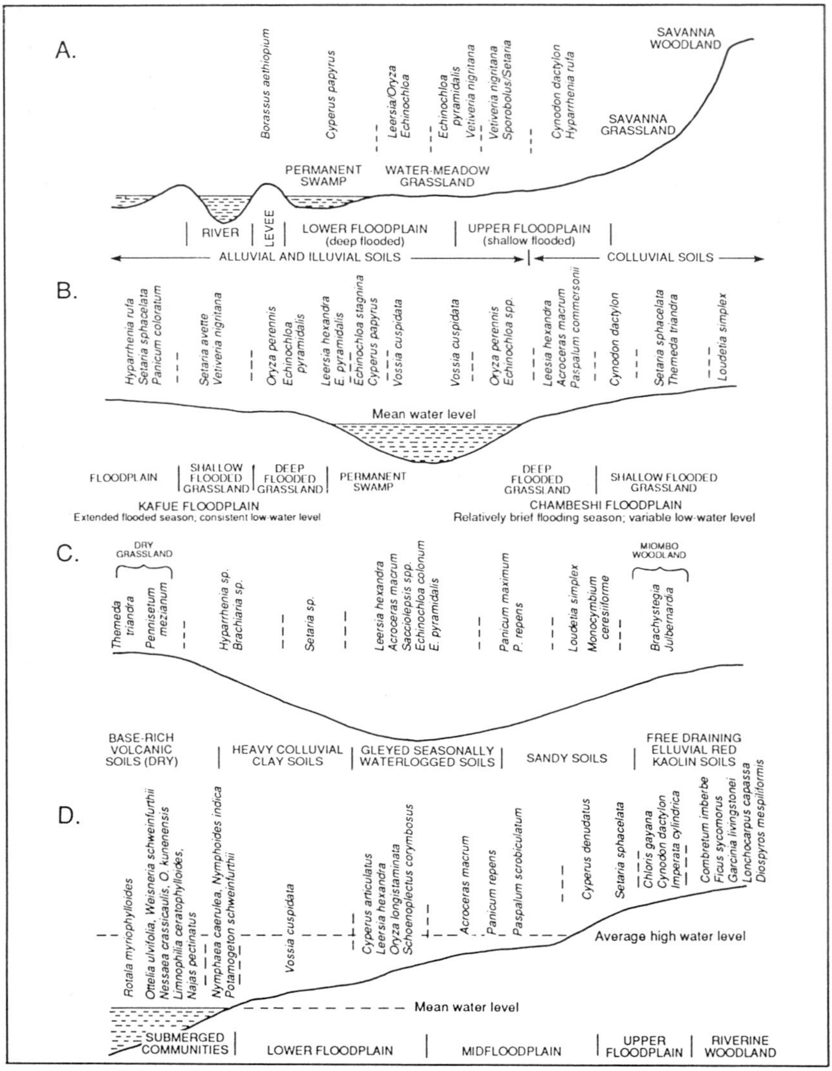

Figure 8. Diagram showing various types of hydrosere that can be found in Africa: (A) A floodplain in the upper Zaire River Basin, Shamba Province, Zaire, (B) Composite of two Zambian floodplains with different flooding regimes and soil types, (C) A composite illustrating some characteristic grasses of dambo hydroseres, (D) A section of the lower Boro River, Okavango Delta, Botswana (redrawn and compiled from Thompson 1985; Figs. 3.2, 3.7, 3.8 and 3.10).

adapted to the continuously waterlogged and shallow water zones. Again, the demarcation between the two edaphic grasslands is very flexible and the boundaries drift up and down the hydrosere depending upon water levels. In periods of drought, for example, the seasonally-flooded grasslands encroach upon the wetter grasslands as waterlevels recede. In the deeper water the grasses are in direct competition with vegetation of herbaceous wetlands. In fact, an arbitrary decision has to be made as to which category some of the grasses belong. *Vossia*, for example, is considered with swamp vegetation by Thompson (1985) but in this chapter is included in grasslands.

The number of taxa in the permanently-flooded grasslands is substantially less than in other grassland types. At the deep-water end of the hydrosere, in nutrient-poor areas, *Vossia cuspidata* may replace emergent swamp plants such as *Cyperus papyrus*. Along the Zaire river, for example, permanent swamps are dominated by *Vossia* with grasses such as *Brachiaria mutica*, *Panicum subalbidum*, *Echinochloa pyramidalis*, *E. scabrao*, *Leersia hexandra*, and *P. parvifolium* associated with it (White 1983). *Vossia* also occurs on the outside (lakeward) of swamp vegetation and river edges (e.g. along the Upper Nile, Denny 1984) forming a distinct fringe. *Miscanthidium*, likewise, is typical of low-nutrient sites which are too acid for other emergents. What actually determines the dominant deep-water grass is uncertain: *Oryza* predominates in the Kafue Flats; *E. scabra* in the Inland Delta of the Niger river, and *E. pyramidalis* in the edaphic grasslands between swamps and gallery forests of the wetter dambos of South Africa (Thompson 1985). *Echinochloa pyramidalis* and *E. stagnina* are important components of a number of other hydroseres including the Sudd vegetation of the Upper Nile (see later). *Leersia hexandra*, *Paspalum repens*, *Panicum repens*, and *Paspalidium geminatum* tend to occur in slightly shallower water, are very tolerant of water-level fluctuations, and can stand occasional drying out. They are generally lush, nutritious grasses much favored by herbivorous browsers.

Whilst a few of the grasses of the permanently-flooded grasslands are restricted to particular phytochoria (*Jardinea*, for example only occurs in the Guineo-Congolian region) most have a distribution largely independent of floristic and climatic boundaries. Permanent waterlogging and depth of water on the other hand, are quite critical variables.

Herbaceous wetlands

These wetlands (see Fig. 6 for generalized distribution pattern) include all the herbaceous vegetation found in the hydrosere beyond (deeper water) the flooded grasslands; particularly, swamp and open water communities. The following literature is recommended for further reading: Allanson 1979,

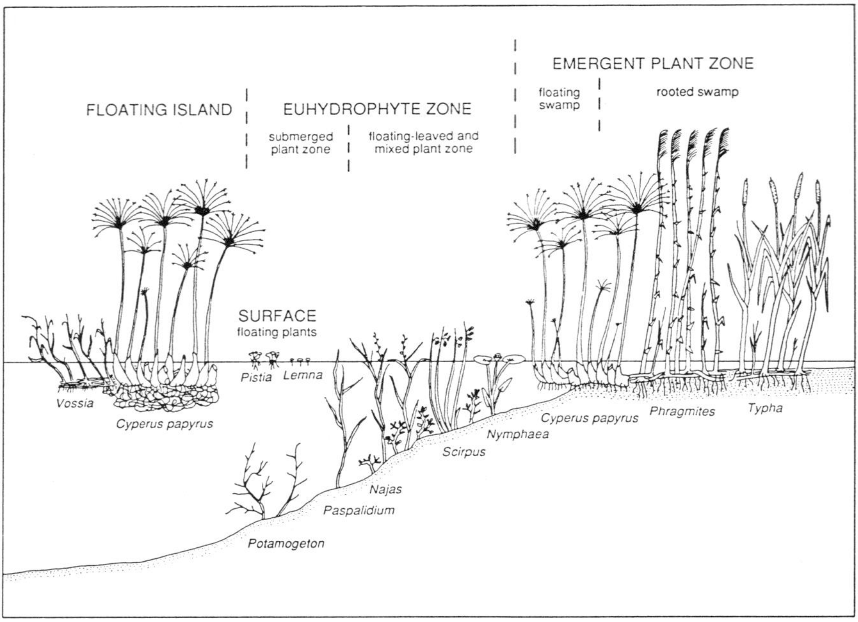

Figure 9. A schematic representation of a zonation of vegetation showing different plant life-forms (redrawn and modified from Denny 1985; Fig. 1.2).

Carmouze *et al.* 1983, Denny 1973, 1984, 1985a, 1991, Gaudet 1977a, 1977b, 1982, Howard-Williams 1979a, 1979b, 1979c, Lind and Morrison 1974, Symoens 1968, Symoens and Ohoto 1973, Thompson 1976, Thompson and Hamilton 1983, Van der Ben 1959, Van Meel 1952, and White 1983. There are many more and additional citations can be found in Denny (1985a).

A distinct zonation of vegetation is often apparent in herbaceous wetland hydroseres. This is described in detail by Denny (1985c) and need only be outlined here. A transect line taken from dry land or a flooded grassland through the vegetation to open water will normally pass through a range of plant life-forms associated with increasing depth of water. At first, there will be tall, emergent vegetation, and then rooted, floating-leaved plants, followed by submerged plants. On the surface of the water may be surface-floating plants. A schematic representation of a typical zonation is given in Fig. 9. On closer examination the vegetation is usually more complex and contains a wide range of life-forms. The main vegetation types and their associated life-forms are described below.

Emergent plant communities

This community is composed of robust, herbaceous plant which are normally anchored to the substratum (Fig. 10) but, in equatorial Africa particularly,

Figure 10. Emergent plants normally found at the water's edge. These beds were left on the exposed shore of Lake Kariba after the water level dropped during a recent drought. (a) *Cyperus articulatus*,

can form extensive rafts of vegetation. They are typically the components of swamp vegetation. Howard-Williams and Gaudet (1985) define herbaceous swamps thus: "... flat areas which are flooded to a shallow depth either permanently, or for most of the year, and which are densely covered with herbaceous vegetation whose shoots rise out of the water to a height of more than one metre." The term reed-swamp is used frequently for this type of vegetation (e.g. White 1983, Thompson 1985) but should be discouraged.

The most common emergents in African swamps are *Cyperus papyrus*, *Typha*, *Phragmites*, and *Cladium* with *Vossia* as a fringing plant. *Cyperus papyrus* (p. 30) is widely distributed and utilized (see p. 31) in central and eastern Africa and is a particular feature of swamps in the Lake Victoria regional mozaic. It is relatively rare in the Guino-Congolian region where it is replaced by *Cyrtosperma senegalense* (a giant Araceae) and *Vossia*. It used to be the dominant swamp plant of the Nile but anthropogenic pressures have eliminated it from all but a very isolated area in Egypt, and it is not normally found above about latitude 15° N. It is fairly widespread in southern Africa and occurs in Madagascar but it only reaches majestic heights (up to 9 m tall) and high biomass values in Ethiopia and East Africa: normally, it is from 4–5 m tall. The distribution and production of papyrus shows a climatic/altitude response (Thompson *et al.* 1979) and even in Uganda, does not occur above about 2,100 m (Denny 1973).

Typha taxa occur through Africa but their identification and taxonomy

Figure 10. (b) *Cyperus involucratus* with *Polygonum senegalense* behind. Shore of Lake Kariba.

have been somewhat confused until recently. There appears to be two distinct taxa in Africa: *T. domingensis* Pers. sensu lato, a taxon of world-wide distribution in tropical and warm-temperate climates and *T. capensis* Rohrb. found in North Africa and southern Africa. *T. capensis* is tetraploid and Smith believes it is actually the hybrid *T. angustifolia* L. × *latifolia* L. (T. × glauca Godr.), and/or *T. domingensis* × *latifolia*. Be that as it may, it is convenient for us to call it *T. capensis* and includes citations to *T. latifolia* and *T. latifolia* ssp. *capensis*. *T. domingensis* is a robust plant normally standing from 2–2.5 m high which predominates throughout warmer regions of Africa including, for example, great expanses in the Upper Nile region (Denny 1984, Mefit-Babtie 1983), Lake Chilwa (Howard-Williams and Walker 1974), and Lake Chad (Carmouze *et al.* 1983). *Typha capensis* is not quite so robust and occurs in the cooler regions of the continent including southern Africa, the Okavango swamps (Thompson 1985), and higher altitude regions of the tropical and equatorial zones (e.g. Lake Bunyonyi in Uganda at 1,950 m, Denny 1973).

Typha forms dense, monospecific stands and is often in competition with *Phragmites* and *Cyperus papyrus* but it is more drought tolerant and salt tolerant than the other plants. It has been suggested that it has a preference for nutrient-richer swamps. Beadle and Lind (1960) and Lock's (in Mefit-Babtie 1983 and Howell *et al.* 1988) studies in the swamps of the Upper Nile may support this (see later). Over recent years *Typha* has been making strong inroads into some swamps, for example, the swamps of the Upper Nile (Lock, in Mefit-Babtie 1983, Howell *et al.* 1988) and into shallow waterbodies (e.g. Lakes Jipe and Nyumba-ya-Mungu, Welsh and Denny 1978). One may speculate that the reason for this is the recent climatic changes which have brought drought to large areas of Africa. The greater tolerance of *Typha* to reduced soil waterlevels over extended periods may give it a competitive and survival advantage.

Phragmites is arguably the most abundant emergent swamp plant in Africa but this needs to be confirmed by extensive satellite vegetation studies. Three taxa occur in Africa but, like *Typha*, their taxonomy and identification is sometimes dubious. The cosmopolitan species *P. australis* (= *P. communis*) mainly occurs in the more temperate regions of Africa and in the Equatorial zones at higher altitudes (e.g. in Lake Bunyonyi). *Phragmites mauritianus* is more characteristic of riverbanks (Thompson 1985). The third taxon, *P. karka*, has only been recorded from the swamps of the Upper Nile where it forms very tall (5.5 m) dense stands of vegetation along the edges of channels (Denny 1984, Lock in Mefit-Babtie 1983, Howell *et al.* 1988). This will be examined in more detail later.

Cladium in Africa is represented by one taxon *C. mariscus* var. *jamaicense* (= *jamaicense*) and is not nearly so widespread or abundant as the other three emergent swamp plants. It has a restricted distribution in sites of higher nutrient status and tends to occur towards the limits of climatic tolerance of *Cyperus papyrus* where it can successfully compete. Thus it forms dense swards in Lake Bunyonyi (Denny 1974), the high altitude swamps in Rwanda

(Deuse 1966), and coastal fens in South Africa (Martin 1960). It also makes an occasional appearance in isolated patches in tropical and equatorial regions and, for example, has been recorded from a swamp at the north end of Lake Victoria (Lind and Visser 1962).

An important feature of African emergents which distinguish them from most others in the world is the ability of some to form floating rafts of vegetation. The two distinct swamp types thus formed are termed: bottom-rooted and floating swamps. The consequences of these two growth habits and their different effects on swamp and shallow-waterbody ecosystems are discussed in detail by Howard-Williams and Gaudet (1985) and Howard-Williams (1985). A diagrammatic representation of the two types of swamp is given in Fig. 11. A bottom-rooted swamp is the typical emergent swamp found throughout the world and a reasonable amount is known about its structure and functioning (e.g. Good *et al.* 1978, Greeson *et al.* 1979, Howard-Williams 1985). Floating swamps often develop from the floating fringes of rooted, permanent swamps and are particularly abundant in the Guineo-Congolian, Zambzian, and Sudanian regions and transitions zones including the Lake Victoria regional mozaic. *Cyperus papyrus* is the classic example of the emergent plant which forms rafts of floating rhizomes at the outer (lakeward) fringes of the swamp. During periods of rapid waterlevel

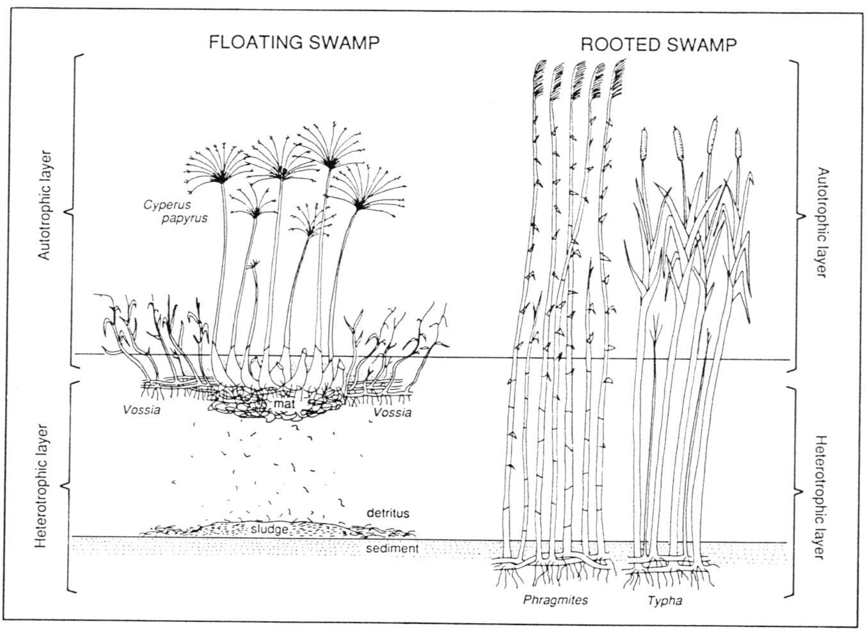

Figure 11. Schematic diagram showing the main difference between a floating and bottom-rooted swamp (after Howard-Williams and Gaudet 1985; Fig. 6.1).

fluctuation and stormy weather, these rafts will break away from the stable swamp together with fringe plants, such as *Vossia*, *Ludwigia*, *Polygonum*, etc., and form islands of floating vegetation. First encountered in the Nile by early explorers, they were termed Sudd (from the Arab word meaning 'blockage'). The islands are at the mercy of wind and water current and have the habit of blocking entire watercourses, harbours etc. where they can become relatively permanent fixtures. The water below a floating raft of vegetation is dark so the growth of submerged plants is suppressed and deoxygenation often excludes aquatic animals. The sudd is bound together by plant rhizomes (*Cyperus papyrus* and *Cladium*) or floating stems (*Vossia*, *Ludwigia*, *Echinochloa*) which form a suitable habitat for non-mat-forming plants. Studies in the Okavango Delta have also shown that organic detritus from the bottom of swamps can rise to the surface of the water on bubbles and pockets of gas (CO_2 and methane) and become colorized (Elling *et al.* 1990). Fuller accounts of floating and rooted swamps can be found in Denny (1985a), Thompson and Hamilton (1983), and Elling *et al.* (1990).

Euhydrophyte communities

The term *Euhydrophyte* is relatively new and therefore requires explanation and definition. It was proposed by Denny (1985c) to overcome the cumbersome phrase "... rooted, floating-leaved and submerged macrophytes" which is used for plants in the vegetation zones to the lakeward of emergents. Whilst emergents have a range of characters akin to terrestrial plants – look at the similarities between *Phragmites* and *Pennisetum*, for example – the floating-leaved and submerged taxa are more truly aquatic with major structural and physiological modifications for the aquatic environment. Hence the term *Euhydrophyte* (Gr. *Eu* = true or well; *hydrophyte* = waterplant).

Euhydrophytes are plants that are completely submerged (except for their inflorescences), or are anchored to the substratum with floating leaves, or have floating and submerged leaves (Fig. 12). Submerged, free-floating species are also included. They occupy the body of water in the euphotic zone not taken over by emergent plants or shaded out by the more permanent surface-floating vegetation. Examples are: *Chara*, *Fontinalis*, *Nymphaea*, *Ceratophyllum*, *Vallisneria*, *Potamogeton*, *Wolffiopsis*, and *Lemna trisulca*.

Euhydrophytes are found in waters throughout Africa where conditions are suitable but, whilst the biogeographical distribution of forest, grassland, and emergent wetland species may be roughly correlated with major phytochoria, euhydrophytes are not so easily placed. This is largely due to the water which provides a buffer to the more extremes of climate. Thus, their distribution is only controlled by climate insofar as it affects the water. For example, the water levels of rivers and lakes are often affected by rainfall distant from the immediate vicinity; and water temperature changes are less

Figure 12. A mixed bed of euhydrophytes in Lake Bunyonyi, S-W Uganda. The floating leaves belong to *Potamogeton bunyonyiensis* whilst submerged plants include *P. schweinfurthii* and *P. pectinatus* (photograph by P. Denny).

that in the surrounding environment. The amelioration of temperature may account for both the wide distribution of many euhydrophytes and their relatively low species diversity. Species such as *Ceratophyllum demersum*, *Hydrilla verticillata*, *Lemna trisulca*, *Vallisneria spiralis*, *Ottelia ulvifolia*, *Potamogeton panormitanus*, and *Potamogeton pectinatus* are found throughout Africa; *P. pectinatus* encroaching into estuarine and saline athalassic lakes.

Others have a restricted distribution. Two species of *Nymphaea* (*N. caerulea* and *N. lotus*) are common and have overlapping distributions but whereas *N. caerulea* tends to occur in the warm, tropical and subtropical areas, *N. lotus* prefers cooler waters. *Potamogeton schweinfurthii* is widely distributed, except in the cooler areas at higher altitudes and latitudes where it is replaced by *P. lucens* (Denny 1985b). *Potamogeton thunbergii* is confined to High Africa (Fig. 2), largely being replaced by *P. nodosus* in Low Africa. The distributions of *P. schweinfurthii* and *P. thunbergii* have wide areas of overlap and hybridization in common forming *P.* × *bunyonyiensis* (Denny and Lye 1973). *Potamogeton crispus* has a curious distribution. It is found mainly in the more temperate zones where it tends to grow in the winter, (e.g. Rogers and Breen 1980, 1982) and is absent from the Guineo-Congolian region and the Lake Victoria regional mozaic. However, it is a weed in Egyptian canals and has been reported as a serious weed in the Nile near Khartoum (Tag el Seed 1981). Clearly, the biology of the temperate and tropical types are somewhat different. *Potamogeton perfoliatus* is largely confined to the temperate regions of North Africa, higher altitudes in West Africa (Chad, and Air Mountains in Niger), and in the Nile down to lattitude 15° N.

There are always some taxa that have an extremely restricted distribution. In Africa, perhaps the best examples are provided by the family *Podostemaceae*. These are a difficult group as they are so small and occur in the most inaccessible places such as waterfalls. New species are still being found and it would appear that each waterfall has its own particular community with high species specificity.

Colonization of lakes and rivers by euhydrophytes depends upon a variety of factors as reviewed by Denny (1972, 1980, 1985b, 1985d). Where monsoonal and unimodal annual rainfalls prevail, waterlevels tend to fluctuate widely, and in man-made lakes drawdown is substantial. These conditions explain the dearth of euhydrophyte vegetation in man-made lakes such as Nyumba-ya-Mungu, Tanzania (Welsh and Denny 1978) and in many West African rivers. Large rivers such as the Zaire carry heavy burdens of particulates during seasonal spate and the euphotic zone is sometimes restricted to less than one meter. In warm waterbodies with a supply of nutrients algal blooms readily develop: thus, in lakes such as Lake George, Uganda, rapid light attenuation inhibits euhydrophyte growth. Even when water conditions appear suitable, competition from emergent and floating plants may be prohibitive and in the more exposed, wave-washed areas, instability of substrate deters colonization.

Euhydrophytes flourish in irrigation channels and are a particular problem in the Gezira irrigation scheme, a massive irrigation development between the confluence of the Blue and White Nile, just south of Kharkoum. They occur as a fringe vegetation to many swamps and provide an interface between swamp and open water, and in lake littorals can develop a distinct zonation of vegetation. In Lake Kivu for example, Van der Ben (1959) found the following zonation to 5 m depth of water: *Phragmites mauritianus* → *Nymphaea* → *Scirpus subulatus* → *Paspalidium geminatum* → *Potamogeton pectinatus*. Denny (1985b) attributes this species-poor community to high salinity. A much richer zonation and plant community (to a depth of 7 m) can be found in the high altitude and mesotrophic Lake Bunyonyi in Uganda. The typical zonation is *Potamogeton* × *bunyonyiensis* → *Nymphaea lotus* → *N. caerulea* → *Chara* sp. → *Lagarosiphon* sp. → *Potamogeton pectinatus* → *P. schweinfurthii* → *Hydrilla verticillata* → *Ceratophyllum demersum* (Denny 1973).

The interface zone of euhydrophytes between the fringes of swamps and open water is probably the most important zone for the success of inland freshwater fisheries and wildlife (Denny 1985d, 1991, Howard-Williams and Gaudet 1985, Howard-Williams and Thompson 1985). It acts as a nutrient

Table 2. The distribution of the main surface-floating plants (obligate acropleustophytes) in Africa (after Mitchell 1985a; Table 4.1).

Plant species	Distribution
Indigenous species	
Azolla pinnata var *Africana*	pan-African
Azolla nilotica	eastern tropical and sub-tropical
Pistia stratiotes	tropical and sub-tropical
Lemnaceae	pan-African
Introduced species	
Azolla filiculoides	South Africa
Salvinia molesta	central and east Africa, southern Africa
Eichhornia crassipes	Nile valley, Zaire, Tanzania, southern Africa

trap to suppress their transfer from the swamp to the open water and provides an invaluable habitat for invertebrate fauna, fish fry, and littoral fish. The rich aquatic fauna encourage a high diversity of bird life.

Surface-floating communities

Surface-floating communities are found in most bodies of water. In general, rivers are the least suitable for colonization because of their flushing effects (Mitchell 1985a) but, on the larger rivers such as the Nile and Zaire there is ample opportunity for establishment, and weed problems occur. The most suitable waterbodies are the man-made lakes, especially in their early years of existence when nutrient-loading is high and competition from other water-plants is low. As Africa has the most numerous and some of the largest man-made lakes in the world, and surface-floating plants grow in proliferation, it is faced with some of the greatest weed problems (Mitchell 1985b).

African surface-floating plants can be divided into those which are indigenous and those that have been introduced. The distribution of the most common species is given in Table 2. Of the indigenous species, *Pistia stratiotes* used to be widespread and create weed problems: indeed, it was a major component of the sudd. It was still common in the Nile in the early 1960's but, since then, it has been replaced largely by *Eichhornia crassipes* and is now found mainly in backwaters and inaccessible pools. On man-made lakes (e.g. Lake Volta), *Pistia* can sometimes become troublesome (Hall and Okali 1974). *Azolla pinnata* and *A. nilotica* are indigenous species and can form dense carpet of vegetation over standing waters but rarely become a pest. Members of the Lemnaceae are widespread and grow in association with emergents and other floating plants but are not mat-forming and are only occasionally troublesome.

The worst weed problems have evolved from two species introduced from South America (*Salvinia molesta* and *Eichhornia crassipes*) whose notoriety is worldwide. The literature on their impact in Africa is extensive and includes Little (1969), Gaudet (1979), Bond and Roberts (1978) and a list of citations in Denny (1985a). Optimal growth conditions for *Salvinia molesta* are 25 to 30°C, high light intensity and non-limiting supplies of nutrients (Mitchell 1985a). Under these conditions individual plants divide in about two days. Lake Kariba provided such conditions after impoundment and within three years of completion *Salvinia* covered over 1,000 km^2 – hence, Kariba Weed. No artificial methods of control were really effective but, as nutrient levels declined naturally, *Salvinia* growth decreased dramatically (Marshall and Junod 1981).

The beautiful Water Hyacinth (*Eichhornia crassipes*) was introduced to Egypt for aesthetic purposes between 1879 and 1892 and into South Africa at the turn of the present century (Wild 1961). A doubling time of 6.2 days in terms of plant numbers has been recorded under experimental conditions and in good conditions in tropical Africa it is normally around 10 days. It is a very successful mat-forming species and, as its growth is so rapid and its control so difficult and expensive, it is being considered as a possible source of energy.

Mangroves

Mangroves are separated from forests as a major physiognomic type by White (1983) and are dominated by trees or bushes occurring on coastal land periodically flooded by seawater. All true mangroves have either pneumatophores or are viviporous. A reading list about African mangroves can be found in White (1983) and some physiological aspects in Steinke *et al.* (1983), Steinke and Charles (1984), and Tomlinson (1987).

Mangroves occur around the shores of Africa particularly in the deltas of large rivers (Fig. 6). In West Africa, for example, they extend 190 km up the River Gambia and large stands can be found in the delta of the River Niger. The forcing factors that influence their occurrence and abundance are complicated and include: frequency and duration of flooding by seawater; soil type, and the ratio of freshwater to seawater. Temperature (cold tolerance) will control the limits of individual species. Although mangrove swamps are best developed in the equatorial regions, and are confined to the tropics in West Africa, on the east coast they extend up to the Gulf of Suez (28° N) and down as far as East London (33° S). Communities are different in the western and eastern coastal regions and may be considered separately.

In West Africa the major species are *Rhizophora mangle*, *R. harrisonii*, *R. racemosa*, *Avicennia germinans*, and *Laguncularia racemosa*. *Rhizophora* spp. account for 99% of the vegetation in the River Niger and attain heights of up to 45 m. A zonation of vegetation is apparent with *Rhizophora racemosa* (up to 40 m tall) as a pioneer species at the edge of the alluvial salt swamps; *R. harrisonii* (c. 6 m tall) in the middle region, and the small shrubby *R. mangle* to the drier limit of the *Rhizophora* zone. Behind the *Rhizophora* can be found *Avicennia germinans* followed by the aquatic grass, *Paspalum vaginatum*. Along the River Gambia a mozaic of vegetation is apparent as the flatness of the land creates large areas of swamp tidally inundated by sea-water (Giglioli and Thornton 1965). The fringes of the numerous creeks and inlets thus created are bordered by a gallery forest of *Rhizophora* behind which occurs *Avicennia* bush. The change-over of species is often very distinct and closely correlated with water depth.

The eastern coastal mangrove swamps have a different and greater diversity of species including *Rhizophora mucronata*, *Avicennia marina*, *Sonneratia alba*, *Ceriops tagal*, *Bruguiera gymnorrhiza*, *Xylocarpus granatum*, *X. moluccensis*, *Lumnitzera racemosa*, and *Heritiera littoralis* (White 1983). All nine species can be found in Kenya, Tanzania, and Mozambique but numbers thin out rapidly to the north and south with *Avicennia*, *Bruguiera*, and *Rhizophora* having the widest range. The zonation of species is not so distinct as in West Africa but White (1983) suggests a zonation from *Sonneratia* in the deepest water → *Rhizophora* → *Ceriops* → *Avicennia*, with wide variation from place to place. For example, at the mouths of rivers where the salinity of the water is less *Sonneratia* is replaced by *Rhizophora*. Further south where species diversity is restricted there is only a scattered distribution of mangroves and *Avicennia* tends to take over as a pioneer.

The mangrove forests of Africa are of substantial importance to the local economies the wood being used in construction work, for fuel, and as a source of tannin. Development pressures, however, are now seriously reducing the extent of swamps due to draining and conversion for agriculture, especially rice-growing.

Halophytic vegetation

A number of African wetland areas have a high salt content and support halophytic vegetation. Summaries of their types and distribution occur in White (1983) and Thompson (1985) whilst detailed accounts of vegetation in specific saline lakes can, for example, be found in Howard-Williams (1977, 1978) for Lake Swartvlei, Howard-Williams (1979b) for Lake Chilwa, and Howard-Williams (1980a) for the coastal lakes of Maputaland.

Coastal swamps will be under the influence of seawater but inland athalassic saline wetlands are created particularly in volcanic areas with high evaporation. The latter include the Makgadigadi in Botswana (the largest saltpan in the world) and numerous small and large bodies of water and edaphic grasslands in the Eastern Rift Valley.

Clearly, there will be a tolerance of particular species to different salinity values. In the standing waters of coastal lagoons, for example, the only

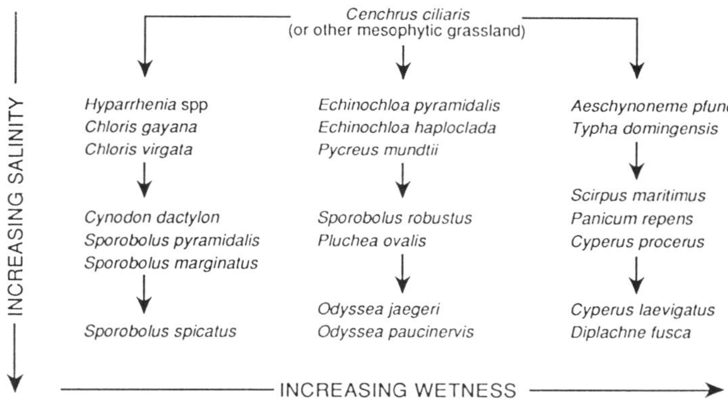

Figure 13. Tolerance of some edaphic wetland plants to increased salinity and wetness. The arrows represent possible directions for plant zonations (after Thompson 1985).

submerged "freshwater" macrophytes that can tolerate salinities up to 10,000 $\mu S\ cm^{-1}$ are *Potamogeton pectinatus* and *Najas marina*. Above this value specialist halophytic flora appear (see Table 2 in Denny 1985b). Plant communities of alkaline swamps and flats are given in Vesey-Fitzgerald (1963, 1970). The emergent plants *Cyperus laevigatus* and *Juncus maritimus* occur in standing waters whilst *Diplachne fusca* is more typical of semi-permanent, shallow water saline swamps. On the edaphic grasslands *Sporobolus spicatus* is very salt-tolerant, *S. robustus* is moderately tolerant, and *S. pyramidalis* mildly so. Thompson (1985, Fig. 3–13) constructed a flow diagram of tolerance of species in edaphic wetlands to increased salinity and wetness. This is reproduced in Fig. 13.

Distribution of major wetland types

Many of the large expanses of wetlands are associated with major drainage systems of the great rivers. Table 3 has been compiled to give some idea of their extent. However, the table cannot include the seasonal (temporary) wetlands unconnected with the major rivers as their extent is still largely unknown: nor can it include the enormous number of small wetlands. Thus, the Table vastly under-represents the total area of wetlands in Africa – the dambos of the Upper Nile alone may account for some 70,000 km^2 (Rzoska 1976b). With the seasonal and long-term fluctuations in water levels the distinction between areas of floodplain, swamp, and shallow water-body is very variable and somewhat academic so the Table should only be used as a guide to their respective areas. Having outlined the general characteristics and structure of African Wetlands, the following sections consider wetlands of three important geographical areas of Eastern, Western, and South Africa.

Table 3. An indication (km^2) of the extent of wetlands in Africa. Wetlands are divided into three main categories: floodplains, swamps and shallow water-bodies (less than 7 m deep). The divisions are very subjective as the vegetation is part of a hydrosere and waterlevels may vary dramatically from season to season. Figures have been rounded up to the nearest 100 km^2, and, at best, can only be taken as estimates of possible areas. Temporary floodplains such as pans and dambos have *not* been included as there is insufficient data. However, it must be remembered that their areas are very substantial: the dambos of the Upper Nile Valley alone account for an additional 70,000 km^2, or so. Data on the main drainage basins have been displayed separately and then areas amassed for the rest of Africa. This table has been compiled from Howard-Williams and Thompson (1985), Table 8.1. The reader is referred to that publication for a full account of wetland areas.

Geographical location	Floodplains	Swamps		Shallow water-bodies
Congo-Zaire drainage system	7,400	56,800		17,600
Niger-Benue drainage system	22,200	7,200		100
Nile drainage system	16,500	46,000		7,900
Zambezi drainage system	18,300	3,900		400
Chad drainage system	6,000		14,000	
Okavango drainage system	12,800	3,100		900
Mediterranean area	500	300		<100
western Africa	17,400	8,200		400
southern Africa	8,700	900		900
northern Rift valley (excluding the Nile)			2,400	

The trees in swamp forests sometimes support a rich epiphytic flora. Here, the epiphytic elephant's ear fern, *Platycerium elephantotis*, is growing in profusion (photograph by P. Denny).

Swamp forests are being cut and destroyed extensively in Africa. A pile of firewood from the very rare *Syzygium cordatum*, high altitude swamp forest tree in Uganda, lies by the roadside (photograph by P. Denny).

The edges of tropical swamps in Africa are often used for shifting, seasonal cultivation. This does little damage to the overall structure of the swamp and provides food for the people (photograph by P. Denny).

A swamp of *Cyperus papyrus*. The emergent stems can be up to nine metres tall (photograph by P. Denny).

Emergent swamp vegetation has many uses. Here *Cyperus papyrus* is being cut on a sustainable basis for thatching and matting (photograph by P. Denny).

The stems of *Cyperus papyrus* are used extensively for thatching in Uganda (photograph by P. Denny).

Eastern Africa

P. DENNY

Geographically, eastern Africa includes Uganda, Kenya, Tanzania, and the southern Sudan (Fig. 1). It lies within the equatorial or tropical climatic zones (Fig. 3) and embodies complex phytochoria centered around the Lake Victoria regional mozaic. To the north of Lake Victoria the regional mozaic grades into the Sudanian regional center, and to the east into the Somalia-Masai center of endemism. Under the influence of the Indian Ocean along the eastern coast, the flora is typical of the Zanzibar-Inhambane transition zone.

Forcing factors in wetlands

Climate

The climate and the geomorphological features of the area are clearly the determinants of, and forcing functions for, the wetlands. The climate-diagrams for the Lake Victoria Basin give a ready assessment of the main climatic features (Fig. 14). In the equatorial zone (Fig. 3) there are two rain seasons each year, each exceeding a 1,000 mm precipitation. The humid period is sustained throughout the year and the temperature fluctuates very little from month to month. This pattern can be seen clearly on the climate-diagram for Entebbe (Fig. 14). To the north and south of the equator, summer rainfall decreases with increasing latitude and drought seasons ensue. Thus, at Malakal, the rain season peaks from May to September and a severe arid period prevails for the rest of the year. Further exaggeration of these conditions has brought about the recent droughts of the Sahel (Sudanian) and Zambezian regions.

The predominant winds are the North-east and South-east Trade Winds so that to the east of Lake Victoria local climate is governed largely by the Indian Ocean and the land between the lake and sea. Moisture picked up from the Ocean is mainly precipitated on the hills and mountains on the eastern side of the Eastern Rift (e.g. Mt. Kenya, the Aberdare Mts., Mt.

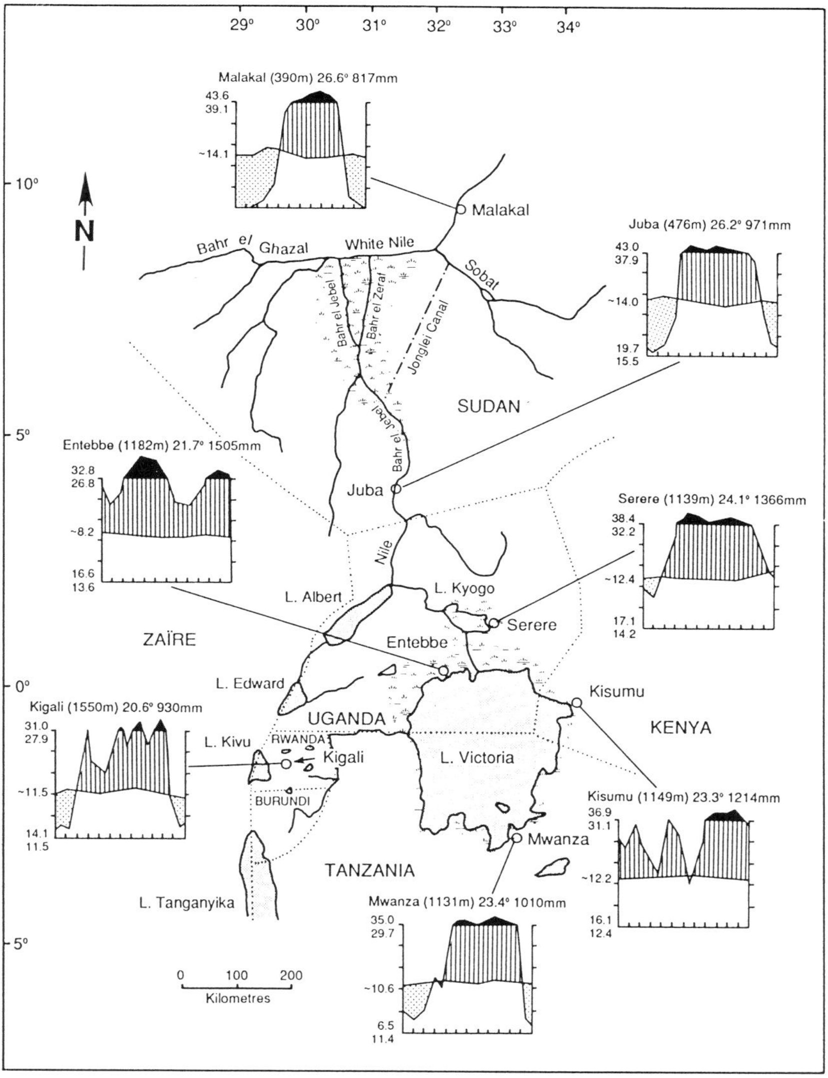

Figure 14. Climate-diagram map of the Lake Victoria and Upper Nile region (after Walter *et al.* 1975).

Kilimanjaro, and the Pare and Usumbara mountain ranges). The net result is that, even near the equator, the coastal region is fairly dry and the Eastern Rift Valley tends to have an arid period (see for example, the climate-diagrams for Serere and Mwanza) which becomes more pronounced to the north and south (e.g. Juba).

The air currents passing over Lake Victoria pick up moisture and the

Rwenzori Mountains and Virgunga volcanoes, which demarch the western edge of the Western Rift Valley, cause the moisture to be dropped rapidly. Thus in contrast to the eastern side, the west and north-west of the lake has continuous high humidity and heavy rainfall.

Geomorphology

Geomorphology determines the extent of the wetlands so a very simplified account of the major earth movements since the Miocene is appropriate. In Late Miocene vertical upwarping followed by some sinking of the land in central and eastern Africa created an enormous, raised shallow basin, the putative Lake Victoria Basin. In the Pleistocene, north-south rifting occurred to form the Western Rift Valley and tectonic activity, including folding and uplifting immediately to the west formed the Rwenzori mountain range. This was accompanied by intense volcanic activity. Rifting and volcanic activity was repeated on the eastern side of the basin to form the Eastern Rift Valley and some of the highest mountains (Mt. Kilimanjaro). Very deep lakes formed in the Western Rift and a range of shallow, often endorheic, mainly saline, lakes in the Eastern Rift (see Fig. 15 and Table 4). Further tectonic movement tilted the central basin and a large shallow endorheic lake was formed at an altitude of over 1,000 m. The northern rim of the basin was finally breached which provided an outlet to the north, the Nile, and established Lake Victoria.

Superimposed on these land movements numerous valleys were cut by drainage systems. Subsequent tectonic movement has raised, back-tilted, and drowned many of these. Thus in the hills of Kigezi there are numerous high altitude valleys in which water flow has been reduced by back-tilting: others have been plugged by volcanic activity to form valley lakes (e.g. Lake Bunyonyi). Around the northern shores of Lake Victoria numerous "drowned valleys" can be found. All these valleys make suitable habitats for the development of wetlands. The shallow, saucer-shape of the Lake Victoria Basin, the drowned valleys and the high rainfall, particularly around north-western Lake Victoria, ensures that Uganda has a large area of wetlands: some 6,500 km^2 of permanent swamps (Lind and Morrison 1974) and 8,000 km^2 of seasonal swamps.

The outflow from Lake Victoria, the Rippon Falls at Jinga, was dammed for hydro-electric power in the 1950's but this has had little effect on outflow rate. The Victoria Nile spreads out into Lake Kioga before it becomes constricted through Muchison Falls and joins the Albert Nile. From there, although initially maintaining its course within well-defined river banks, the gradient becomes less and less and the river spreads out to form the largest swamp in the world, the Sudd. It re-forms and is joined by the River Sobat

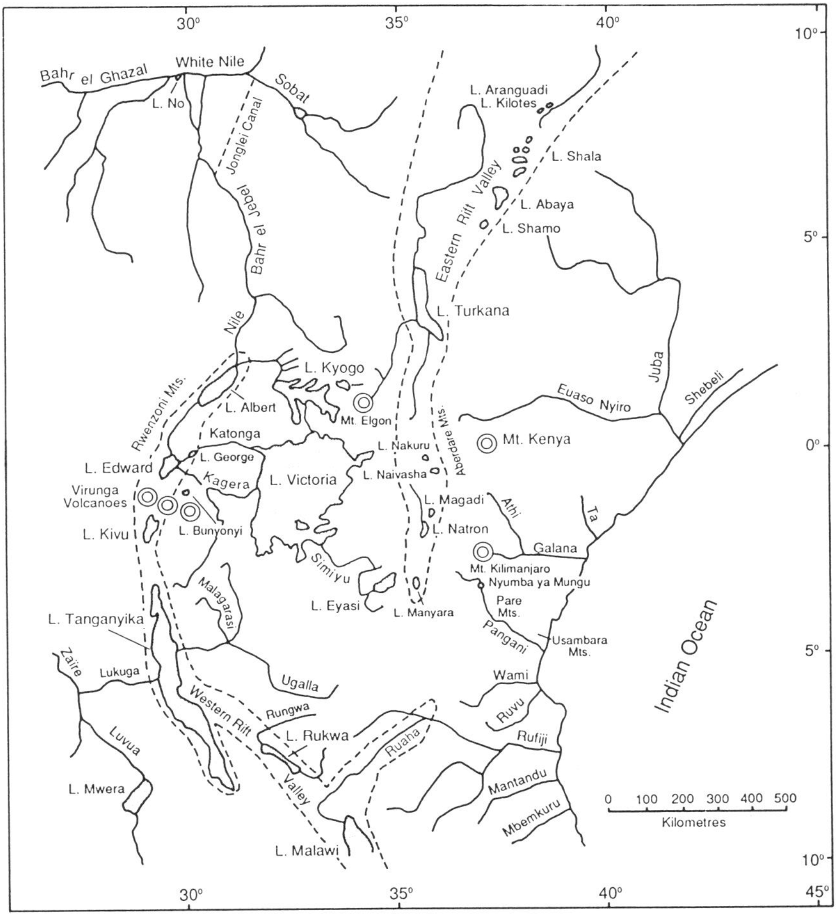

Figure 15. Map of eastern Africa showing the major waterbodies and physical features.

and Blue Nile, and thence travels north via Khartoum and Cairo to the Mediterranean Sea.

Nutrient supply

Although climate and geomorphology determine the extent of wetlands, other factors regulate the vegetation within a particular waterbody, the most obvious of which is nutrient supply. The ionic composition of the waters and underlying sediments, of course, will be indirectly governed by soil type, weathering, rainfall and, more recently, man's interference. So far in eastern Africa, man's effects in terms of pollution, fertilizer application, and sewage is small and localized compared with the much larger effects of natural phenomena. Much of the soil is iron-rich, nutrient-poor, well-weathered,

Table 4. Basic information on some of the larger waterbodies in eastern Africa. The data have been gathered from many sources and should be used only as a guide.

Water body	Latitude	Longitude	Altitude (m)	Area (km^2)	Maximum depth (m)
Albert	1° 40′N	30° 40′E	615	5,350	46
Bunyonyi	1° 18′S	29° 54′E	1,950	60	40
Edward	0° 25′S	29° 40′E	920	2,250	117
Eyasi	3° 30′S	35° 10′E	1,030	450*	
George	0° 00	30° 10′E	913	270	2.0
Kyoga	1° 30′N	33°E	1,036	4,500	4 (Avr.)
Kivu	1° 48′S	29°E	1,500	2,250*	
Magadi	1° 50′S	36° 20′E	600	50*	
Manyara	31° 35′S	35° 50′E	960	300*	
Mtera Reservoir	7° 40′S	36° 50′E	900	610	
Mutanda	1° 41′S	29° 39′E	1,791	25	56
Muchoya Swamp	1° 15′S	29° 50′E	2,256	7.5	–
Mulehe	1° 14′S	29° 43′E	1,803	2.8	6
Naivasha	0° 45′S	36° 20′E	1,890	180	18
Nakuru	0° 23′S	36° 05′E	1.758	40	2.8
Natron	2° 20′S	36°E	610	450*	
Nyumba-ya-Mungu Reservoir	3° 45′S	37° 25′E	670	180	41
Rukwa	8°S	32° 30′E	800	2,000*	
Tanganyika	3° 25′– 8° 45′S	30°E	773	34,000	1,470
Victoria	0° 20′N– 3° 0′S	31° 39′– 34° 53′E	1,240	68,800	79

*Rough estimates only.

lateritic soil. With the high annual precipitation, run-off waters are normally unbuffered and extremely oligotrophic. Indeed, Viner (1974) indicates that the observed differences in chemistry of run-off waters in Uganda is largely a function of climate, especially rainfall.

The aquatic vegetation itself moderates the nutrient status of the water. Thus, whilst rooted swamps tend to cycle nutrients from the sediment to the water, floating swamps have the reverse effect and remove nutrients from the water and lock them up in the plant biomass. The water below a floating mat of papyrus can become totally deoxygenated (Beadle 1981) whilst the swamp acts as nutrient filter, removing ions and lowering the pH value (Gaudet 1977b). Gaudet suggests that papyrus swamps could be viewed as large holding tanks. The implications of this in water purifications from sewage outflows has not escaped attention and Thompson (1976) has reported the efficiency of papyrus swamps near Kampala in removing enrichment from sewage.

During periods of high rainfall, when rivers and streams are in spate and the water-flow below floating swamps is increased, the de-oxygenated water together with any nutrients and suspended organic matter may be flushed out into the open lake (Howard-Williams and Gaudet 1985). It has been suggested that the euhydrophyte zone fringing the swamps absorb these nutrients and act as a valuable buffer against loss from the ecosystem (Denny 1985d, 1991). The nutrient dynamics of papyrus swamps have been discussed fully in Howard-Williams and Gaudet (1985).

Whilst the nutrient status of the water and muds *per se* affects the type of plants it can support (*Typha*, for example, requires higher levels of phosphate than *Cyperus papyrus*, and *Utricularia* tends to thrive in low-nitrogen, oligotrophic waters), total salt content regulates the distribution of plants. Thus, whilst *Potamogeton schweinfurthii* does not occur in waters of conductivities $>1000\,\mu S\,cm^{-1}$, *P. pectinatus* can tolerate salinities of up to $13{,}000\,\mu S\,cm^{-1}$. More particularly, it has been suggested that tolerance to water salinity is largely a function of alkalinity (*P. schweinfurthii*, for example, is intolerant of values >10.0 meq cm^{-1}) and the euhydrophyte communities of the lakes in the Rift Valley are thought to be regulated by it (Denny 1985b). Lake Kivu, with the highest alkalinity, has the poorest species diversity, and Lake Victoria, with the lowest, the greatest diversity. Many of the saline lakes in the Eastern Rift have no euhydrophytes and a very specialized emergent flora.

Water-level fluctuations

Finally, mention should be made of water-level fluctuations. Even in the equatorial zone where there is a bimodal rainfall pattern the water-levels of lakes and rivers change seasonally. Lake Victoria, for example, normally has an amplitude of about 1.5 m each year. This is mainly due to the inflow waters from its catchment area rather than direct rainfall. Water level changes in the Sudd are regulated by the rate of outflow from Lake Victoria whilst additional areas are affected by direct precipitation (see later).

Several of the more important aquatic grasses (e.g. *Vossia*, *Echinochloa stagnina*, *Leersia*) can tolerate fluctuations of a metre or more for they are rhizomatous and often stoloniferous, with their stems floating on the surface. *Miscanthidium*, on the other hand, does not have this facility and tends to be a shallow-flooded species. Bottom-rooted, emergent swamp plants such as *Phragmites* and *Typha*, can withstand a certain amount of flooding through physiological and biochemical adaptations. *Phragmites karka* occurs in the wetter part of the swamp whilst *Typha domingensis*, which does not tolerate deep-flooding but can withstand drying out, occurs in the drier areas. *Cyperus papyrus* is slightly different. In shallow water it is fixed to the bottom but its outer fringe often floats. The bed can sustain water-level changes of around a meter but if greater than this, the floating section will break away and form a floating island. This is encouraged further by the action of waves and wind, so typical of Lake Victoria, with the result that large, wind-driven islands cause navigational hazards. As papyrus has a very high growth rate (up to 12 kg m^{-2} y^{-1}, dry weight; Thompson *et al.* 1979), the fixed beds can re-generate at the new water-level if the changes are not too rapid or extreme.

Wetland types

It is not possible to cover all the wetlands of eastern Africa in this chapter and so some of the more important ones only are mentioned and special emphasis is given to the two largest: the Sudd and the northern swamps of the Lake Victoria basin. Broad accounts of the wetlands of Uganda can be found in Lind and Morrison (1974) and Beadle (1981), and the numerous peatlands and swamps of central and eastern Africa in Thompson and Hamilton (1983), and in Hamilton (1982).

The Western rift lakes
Eastern Africa contains extensive waterbodies mainly associated with the Rift Valley. The great lakes (Albert, Edward, Kivu, and Tanganyika) are deep fault lakes of the Western Rift with relatively steeply-sloping sides. This restricts the wetlands associated with them mainly to a littoral fringe. None of their vegetation has been much studied since the surveys by the Belgians in the 1940's and 1950's, i.e. for Lake Tanganyika (Van Meel 1952), Lake Albert (Robyns 1947–55, Hoier 1950) and Lakes Kivu, Edward, and Albert (Van der Ben 1959). The vegetation of Lake George, a small shallow lake connected to Lake Edward by the Kasinga Channel, has been described by Lock (1973) and the lake itself has been extensively studied as part of the International Biological Programme (IBP) (Greenwood and Lund 1973).

High altitude valley lakes
In the Kigezi Region of southeastern Uganda there are a series of high altitude valley lakes formed from volcanic activity and tectonic movement including Lake Bunyonyi and the nearby lakes Mutanda and Mullehe. Swamp vegetation is associated with them, particularly at their inflows. At these higher altitudes equatorial species are replaced by more temperate ones. Thus, *Cyperus papyrus* is only poorly developed and at the limit of its altitudinal distribution. It is largely replaced by *Cladium jamaicense*. *Phragmites australis* replaces *P. mauritianus* and *Typha capensis* (*T. latifolia*) replaces *T. domingensis*. Of the euhydrophytes, *Nymphaea alba* is found with *N. caerulea* and *Potamogeton lucens* with *P. schweinfurthii* (Denny 1973). Some of the upland valley lakes have become infilled or partially drained and may support dense swamp vegetation of papyrus at the lower altitudes and fen species such as *Pycreus nigricans* at altitudes above 2,000 m. At these altitudes decomposition is slow and it is only here that peat formation can occur (Hamilton 1982, Thompson and Hamilton 1983). Muchoya Swamp in

Kigezi, has been carefully studied by Morrison (1968) with special reference to peat formation over the last 12,000 years. Many of the valley swamps of Kigezi have been drained by man and are now important areas for intensive agriculture.

Athalassic saline lakes

Around the eastern side of Lake Victoria and to the south are saline lakes such as Nakuru, which is a famous reserve for the flamingo. Others include Magadi, Natron, Eyasi, and Manyara which is an important game reserve. Typical littoral plants of these lakes are *Cyperus laevigatus*, *Sporobolus spicatus* and *Dactyloctenium* sp. but there is very little information available on their vegetation. Details of their chemistry can be found in Talling and Talling (1965). Most of these lakes are shallow and liable to substantial seasonal water-level fluctuations and have extensive edaphic floodplains and pans associated with them. On occasions they dry out completely to form salt pans. Lake Rukwa, a large shallow saline lake in Tanzania, dominated by *Diplachne fusca*, is an example. There are expansive edaphic grasslands surrounding it, described fully by Vesey-Fitzgerald (1963), which become inundated in periods of high water. The littoral of the salt lakes and the surrounding floodplains provide very valuable grazing for animals and feeding places for birds.

Coastal swamps

Along the east coast of Africa there are well-developed mangrove swamps which have been mentioned earlier. Behind the mangroves are isolated patches of swamp forest dominated by *Barringtonia racemosa* with *Acrostichum aureum*, *Hibiscus tiliaceus*, *Pandanus* spp. and *Phoenix reclinata* (White 1983).

Man-made lakes

There is a number of man-made lakes (Msangi and Ellenbrock 1990), two of which are in Kenya on the Tana River (the Kamburu and Kindarumu Reservoirs) and seventeen in Tanzania. In African terms they are mainly small ($<10\,km^2$) and their total surface area accounts for only $863\,km^2$ (Bernacsek 1984). The Mtera Reservoir on the Great Ruaha River and Nyumba-ya-Mungu on the Pangani River are the two largest. Although many reservoirs have been plagues by infestations of surface-floating plants, those of East Africa have, so far, mainly avoided the problems. The regular draw-down of man-made lakes demands a rather special flora (see earlier in the chapter) but only that of Nyumba-ya-Munga has been studied in any depth (Welsh and Denny 1978). Nine years after impoundment of the Pangani River, *Typha domingensis* swamp was predominant and extensive, whilst

Cyperus alopecuroides and *Paspalidium geminatum* were common emergents. Submerged vegetation could not be found in the main area of the lake. Zonations of vegetation were described from different sites and it was concluded that drawdown and water depth were the main forcing functions for colonization. At the inflow region, beds of *Typha domingensis* were creating a serious weed problem.

Lake Naivasha

On the Mau Escarpment to the north-west of Nairobi is Lake Naivasha. From its position in the Eastern Rift it might have been expected to be saline but presumed subterranean inflow waters retains its freshness. It is one of the few lakes in East Africa in which the vegetation has been studied thoroughly and a number of publications has arisen (Gaudet 1976a and b, 1977a, 1977b, 1982, Gaudet and Muthuri 1981). Not only has Gaudet described the vegetation and studied the zonation, and the change in zonation with seasonal natural drawdown, but he has investigated the nutrient dynamics of the swamps and used the data to develop hypotheses on tropical swamp dynamics in general (Howard-Williams and Gaudet 1985). The lake fringe is dominated by *Cyperus papyrus*. During the lowering of the water-level a succession of vegetation with three distinct zones develops (i) a *Sphaeranthus suaveolens* dominated zone nearest the water's edge, (ii) a sedge zone dominated by *Cyperus papyrus*, *C. digitatus*, and *C. immensus*, and (iii) a composite zone dominated by *Conyza* spp. On re-flooding a sub-climax fringe community of papyrus becomes established once again (Gaudet 1977a).

The headwaters of the Nile: Lakes Victoria and Kioga

The swamps of Lakes Victoria and Kioga at the headwaters of the Nile have been studied more than most. Good accounts can be found in Eggeling (1935), Carter (1955), Lind and Visser (1962), and Gaudet (1975, 1976 a and b). The information from these, with some additional data, have been used subsequently to provide useful general accounts, and analysis of vegetation dynamics of papyrus swamps (Lind and Morrison 1974, Thompson 1976, Beadle 1981, Thompson and Hamilton 1983, and Denny 1985a).

The deeply dissected terrain which produce shallow-flooded valleys around Lake Victoria (particularly the northern shore) and Lake Kioga, make it ideal for swamp development. It is here that *Cyperus papyrus* swamps are most extensive. They are bottom-rooted in the shallow water (<3 m) and floating in the deeper valleys and around the outer (lakeward) fringes of vegetation.

A schematic representation of a hydrosere through a papyrus swamp has been constructed by Thompson (1976) from description by Eggeling (1935) and Lind and Visser (1962). Over deeper water the floating raft is dominated by *Cyperus papyrus* with the shade-tolerant fern, *Dryopteris striata* amongst the rhizomes. Various climbers may rise through the canopy. *Vossia cuspidata* often surrounds the raft and grows on the outer fringes of fixed swamps (Figs. 9 and 11). Euhydrophytes such as *Nymphaea caerulea*, *Trapa natans*, and *Potamogeton* spp. may colonise the lake bed if conditions are suitable (see earlier) whilst free-floating species, especially *Ceratophyllum demersum* can usually be found amongst the outside rhizomes.

In shallow water the rhizomatous mat of papyrus is fixed to, if not rooted in, the sediment. An organic layer, which can be several metres thick, builds up between the lake bottom and the underside of the mat (Beadle 1981). The papyrus zone may be many kilometres wide but with decreasing water depth it grades into a zone dominated by *Miscanthidium violaceum*, often with *Loudetia phragmitoides* (Lind and Morrison 1974, Fig 3.3). *Miscanthidium*, which cannot tolerate deep flooding, extends up to the shoreline. Lind and Morrison view this change-over of dominant species in terms of plant succession: the build-up of detritus, silt and organic matter by the papyrus mat reduces the depth of water and provides a suitable habitat for the colonisation and development of the *Miscanthidium*-dominated, mixed vegetation. Landward of the *Miscanthidium* zone progression to an edaphic grassland or swamp forest, or to both occurs. Narrow bands of swamp forest with *Phoenix reclinata*, *Raphia monbuttorum*, and *Mitragyna stipulosa* is a common site around Lake Victoria and in the inlets. In 1962 when the lake level rose and remained two metres higher, many of these trees were killed by the flooding and now stand as grotesque forms, with renewed colonisation behind.

The swamps of the Upper Nile: The Sudd

These are the massive swamps between approximately latitudes 6° to 9° 30′N where the Bahr el Jebel north of Bor bifurcates with the Bahr el Zeraf to make devious routes around Zeraf Island; joins the Bahr el Ghazal at Lake No and flows into the better-defined White Nile (Fig. 14). The gradient of land is very shallow and in places is only about 1 cm km^{-1}. The water spills laterally into lakes, side channels which in places form distinct river systems, and swamps. These account for some 16,000 km^2 of permanent swamp and 15,000 km^2 of seasonal swamp. In addition to this may be a further 70,000 km^2 of rain-induced floodplain. The hydrology of the Nile including the Upper Nile swamp region is well explained in Kashef (1981).

An excellent study by Migahid (1948) was the first real attempt to describe the vegetation of the central swamps. Since then the area has attracted periodic attention after renewed proposals to cut a canal, the Jonglei Canal, to the east of the main swamp (from Jonglei to the River Sobat). This precipitated a vegetation survey of the area by Sutcliffe (Equatorial Nile Project 1954; Vol. 1, pp. 140–142; 150–166) and a further account later (Sutcliffe 1974). The canal project did not proceed then, but was re-vitalized in the 1970's and further surveys were undertaken (Denny 1984). A full ecological survey was commissioned by the government of the Sudan in 1979 and this was carried out by Mefit-Babtie Srl. Their botanist, Dr Lock, conducted the rangeland and swamp investigations and had at his disposal satellite and aerial surveying facilities. Results are published in Mefit-Babtie (1983) and Howell *et al.* 1988. The cutting of the canal has started but has not been completed, owing to civil unrest.

The main water supply to the Sudd is derived from the large catchments of the Lake Victoria Basin. As a result of exceptionally heavy rains in 1961–1962, water-flow from the Victoria and Albert Niles have increased and been sustained, raising the water levels in the Sudd accordingly. This has practically drowned Zeraf Island and extended the area of permanent swamp to it's current high value. Thus, the early vegetation reports by Sutcliffe (1974) show marked differences from those of Denny (1984) and Howell *et al.* (1988).

The rains in the Upper Nile region commence in about May and have the double effect of (i) sealing the black cracking clays of the edaphic grasslands so that they become prone to flooding and (ii) raise the overall water level of the permanent swamp. At about the same time the rains in the Lake Victoria Basin increase the discharges from Lakes Victoria and Albert, and this causes peak flows through the swamps in about August-September. The net result is that from August to November there is massive overspill from the permanent waterways to the adjacent floodplains; followed by seasonal contraction.

This hydrological sequence produces seven main vegetation zones: (i) the euhydrophyte zone in permanent, open water, (ii) a *Vossia* zone fringing emergent swamp (iii) *Cyperus papyrus*, which is tolerant of deep flooding, bordering channels and open water (Fig. 16), (iv) *Typha domingensis* in shallow-flooded areas not prone to great water level fluctuations, (v) *Oryza longistaminata* and *Echinochloa* spp. which form seasonally river-flooded edaphic grasslands, and (vi) *Hyparrhenia* rufa which occurs in rain-flooded edaphic grasslands. These different zones are shown in Fig. 17 (redrawn from Mefit-Babtie 1983; Vol. 2, map 3). *Eichhornia crassipes* forms dense mats of floating vegetation in many areas.

The deep channels, where there is quite fast-flowing turbid water, are devoid of euhyrophytes, as are the deeper standing waters. However, the Lake No area gives an insight into typical vegetation of the permanent swamp (Denny 1984). In waters up to 2 m deep euhydrophytes are common, especially *Najas pectinata*. *Ceratophyllum demersum*, *Utricularia gibba* and

Figure 16. A floating swamp of papyrus in the Upper Nile. *Cyperus papyrus* was thought to cover most of the swamp but, in fact, it only occurs in the deeper water of the permanent swamp, boardering channels etc. (photograph by P. Denny).

members of the Lemnaceae tangle with rhizomes of floating vegetation. In the Bahr el Ghazal (a smaller river with lower turbidity) there is greater diversity of euhydrophyte species with extensive beds of *Potamogeton pectinatus*, *P. schweinfurthii*, and *P.* × *bunyonyiensis* with *Trapa natans* and *Nymphaea lotus* in the shallower water. The emergent vegetation surrounding the open waters is dominated by *Cyperus papyrus* with two or more metres wide bands of *Vossia* in front. *Eichhornia crassipes* fringes the outside of the *Vossia* zones and floating rafts of it move slowly downstream. *Pistia statiotes*, which used to be so common in the Nile is now mainly confined to backwaters where it is not in competition with *Eichhornia*. The creeping sedge, *Cyperus mundtii* spreads amongst the fringe vegetation. In the slightly drier areas of the channels *Phragmites karka* tends to replace the papyrus but it never becomes particularly extensive. Behind the *Cyperus papyrus* are extensive beds of *Typha domingensis*. Lock (in Mefit-Babtie 1983) has estimated that

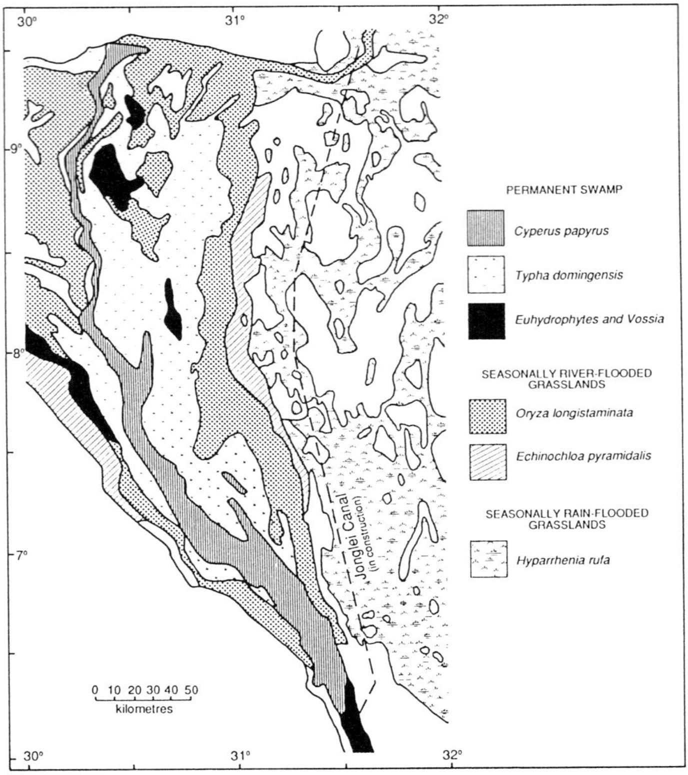

Figure 17. Map showing part of the Sudd to indicate the main vegetation zones. The area shows the bifurcation of the Bahr el Jebel and Bahr el Zeraf in the region of the flooded Zeraf Island. Areas shaded black indicate open water. The course of the river is not easily discernible but roughly follows the patches of euhydrophyte vegetation and the line of *Cyperus papyrus*. The map was compiled by Lock from mosaics of NASA satellite imagery and aerial transect surveys by Mefit-Babtie (redrawn and modified from Mefit-Babtie 1983; Vol. 2, Map 3.)

in the Sudd area, papyrus covers about 3,630 km^2 of swamp whilst *Typha* occupies some 12,500 km^2. Progressing along the hydrosere towards drier land are the floodplain grasslands. *Vossia* and *Echinochloa stagnina* occupy the wettest areas, whilst *Oryza longistaminata* predominates in the river-flooded edaphic grasslands (Mefit-Babtie 1983, Howell *et al.* 1988). It grows best when deeply flooded for long periods, possibly because of increased availability of nitrogen and phosphorus from flooded soils. *Echinochloa pyramidalis* is a common grass to the eastern side of the permanent swamp and can also withstand deep-flooding in the wet season. It tends to form a broad belt between the *Oryza* grasslands and the *Hyparrhenia* grasslands of the rain-flooded areas (Fig. 17). The river-flooded grasslands are estimated to extend for about 15,500 km^2. The rain-flooded grasslands, dominated by

Hyparrhenia rufa are extremely extensive to the east of the Sudd and are of paramount importance as grazing areas. They occur at the driest part of the hydrosere and are not strictly considered in wetlands.

Lock (in Mefit-Babtie 1983, Howell *et al.* 1988) attributes the main patterns of vegetation in the Sudd to depths of water, periods of flooding, and nutrient status. Generally, nutrient levels are higher at the inflow (southern) end of the swamp and at the water fringe vegetation rather than in the middle of beds.

Utilization and management

Eastern Africa has extensive and wide varieties of wetlands (approximately one sixth of Uganda is swamp) for which a management policy is now being developed. Decisions made which may affect them relate normally to an immediate practical problem, such as waterweed congestion, and the actions taken have limited consideration for the wetland ecosystem. Currently, the wetlands are under pressure from two major sources: (i) from encroachment by local communities and (ii) from large-scale schemes which affect the hydrology of a wide area. A source of concern in the future is the greater development of urban areas, industry and agriculture, with the consequent chemical pollution of the aquatic environment.

The Upland Valley Swamps in Kigezi, south-western Uganda, are clear examples of local utilization. Since the mid 1930's the Kigezi District, with its equitable climate, has been a region of intensive small farming. Inevitably, the fertile soils of the swamps have encouraged encroachment but they are prone to acidification on drainage and cannot yield their true potential without long-term planning, investment, and guidance. Often, careful management with rotation between swamp and farmland is preferable to mass clearance and drainage (which largely destroys soil quality). By rotation, the land remains fertile and productive and the habitat is conserved.

Large-scale regulation of water supply is a much greater problem. Already, dams and man-made lakes have irreversibly altered the environment. Wetlands have been drowned and the regulation of water flow has affected the natural cycle of flooding and retreat which is so necessary to maintain the ecological balance of the floodplains. Soil deterioration and erosion has become a serious problem in several areas. It must be assumed that these types of projects will increase and without the fullest consultation with wetland management experts, unnecessary destruction of the habitats will ensue. The Jonglei canal scheme typifies some of the problems. The decision to cut the canal was a political and engineering one made with full international consultation and support. The canal will affect the largest wetland in Africa, the Upper Nile Swamp. The swamp supports a number of nomadic tribes who rely particularly upon its floodplains for fisheries and cattle grazing; and it is a unique area for wildlife. The wetland is probably sufficiently extensive for the local climate to be affected by evapotranspiration from the vegetation. Yet sociologists, and swamp and rangeland ecologists were not an influential part of the initial discussions. How much better it would have been for them to be present from the beginning, and for them to be provided with sufficient resources to carry out thorough pre-construction surveys. To call in environmental consultants who spend, perhaps, a relatively short time in the field, and then are expected to write extensive reports, often pays only lip-service to the problem.

Eastern Africa wetlands are a priceless commodity which should be utilized for many different purposes. By good management they will also be conserved and their wildlife will be protected. Their floodplains provide valuable grazing land for cattle and game and recent studies have shown that their use for wildlife and cattle is not incompatible. The swamps support some of the most productive plants in the world and it is only a matter of time before these can be exploited properly in terms of, say, biogas, feedstuffs, and building material. Swamps act as a nutrient filter and are used already for domestic sewage treatment. With planning, this can be exploited to increase swamp plant production for harvest and be used in water purification. Constructed wetlands specifically designed for wastewater treatment is a cheap and very efficient way of treating effluent wastewater. In tropical environments they should be particularly effective and should be encouraged.

The interface zone between swamp edge and open water is often very rich in terms of secondary production. Many local fishermen set fishtraps and nets whilst predatory birds are attracted by the bountiful supply of benthic animals, amphibians, and fish. The inner swamp is often devoid of oxygen and therefore unproductive. The cutting of channels into the swamp will increase the interface zone and the fisheries will be improved significantly.

The extent of the wetlands in Eastern Africa, their relative inaccessibility, the civil unrest in some regions has, so far, protected them from excessive damage. If a wetland management board with a wide range of expertise were to be established, not only could the wetlands be exploited wisely for the benefit of all concerned, but it would be conserved for future generations. Management guidelines are discussed fully in Denny (1985a, 1989) and are outlined at the end of this chapter. Eastern Africa has one of the fastest-growing populations in the world. In order that the countries may prosper and the people may be fed adequately, pressures on the wetlands will increase proportionally. The absence of a management policy will lead inevitably to the decline and demise of the wetland habitat.

Western Africa

D.M. JOHN, C. LÉVÊQUE AND L.E. NEWTON

In this account the northernmost limit of West Africa is taken as 18° of northern latitude. Its eastern limit extends to about 22° E and so the region includes the main wetlands of the Chad basin that lie to the south and to the east of Lake Chad. To the south of this basin the boundary is the border of present-day Nigeria and the United Republic of Cameroon or Cameroun (see Fig. 1).

Forcing factors in wetlands

Climate

West African wetlands undergo a regular pattern of change in response to the pronounced seasonality of the region's weather. Each year there is a seasonal migration of two air masses and separating these is a roughly east-west zone of climatic instability variously referred to as the intertropical front, intertropical convergence, or monsoon front. This front is know as the surface discontinuity where the two air masses meet near ground level. Rainfall is generally greatest some 320 to about 480 km south of it. The northerly mass is of hot, dry, and stable continental or "Saharan" air, and the more southerly one is of moist, maritime or "Atlantic" air. Over the period December to March the discontinuity is south of about 8° to 9° N. This is the time when the northerly air mass dominates much of the region and consequently is the period of the major dry season in the forest zone. The dust-laden dry wind blowing from the north-east across the region is known as the Harmattan. Sometimes the Harmattan reaches as far south as the coastal region lying almost parallel to the equator (Ivory Coast to Nigeria). The surface discontinuity migrates northwards by at least 10° of latitude and by July or August reaches about 20° N. The predominant wind direction is then no longer north-easterly or easterly but south-westerly. In late August or September the discontinuity once more begins its southward migration.

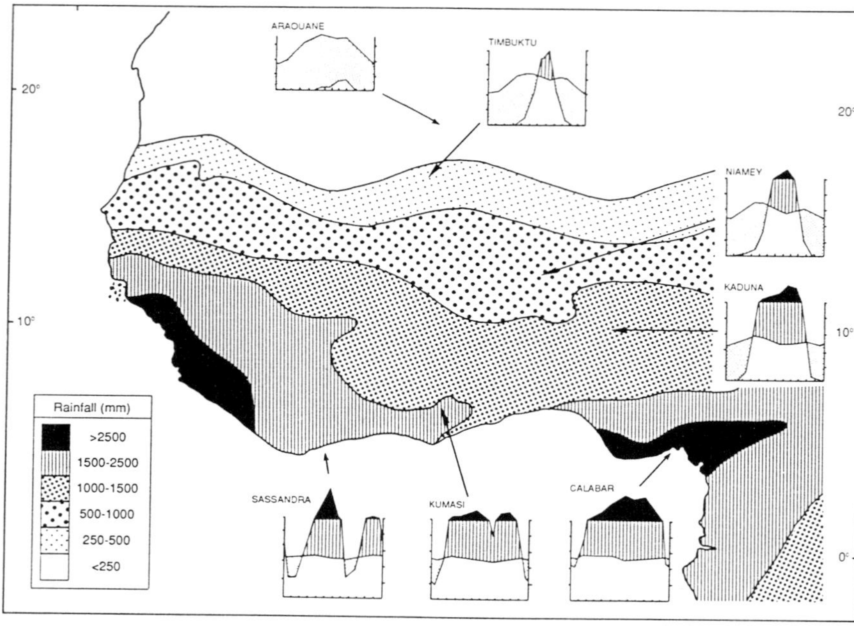

Figure 18. The approximate distribution of mean annual rainfall in tropical West Africa and the seasonal pattern of air temperature and rainfall at 7 stations using the climate diagrams of Walter *et al.* (1975) (after John 1986). For interpretation of such diagrams see Fig. 4.

The seasonal migration of the surface discontinuity separating the two opposing air masses governs the seasonal distribution of rainfall in West Africa. In the more southern parts of the region there is a distinct four-season climatic regime consisting of a major dry season (November/December – February/March), a minor dry season (August/September), a major rainy season (March/April – July/August), and a minor rainy season (September – November/December). The short dry season occurs when the main rains are far to the north. In latitudes higher than about 8° to 9° N the two rainy seasons merge so the year has a two-season climatic regime consisting of a short rainy season and a long dry season. The rainy season is not only shorter from south to north across the region but the annual rainfall also becomes progressively less. Yearly rainfall totals and the exact period of the rains do show some variation from this general pattern due to such factors as the direction of moisture-laden winds blowing from the coast, presence of coastal currents, coastal orientation, and interior relief. Rainfall data for West Africa (including temperature data) are summarized using climate diagrams (Fig. 18) of which some 392 were published for mainland Africa by Walter *et al.* (1975).

Differences in the rainfall pattern account for the changes in the vegetation with the discontinuous belt of forest giving way to different savanna types that get drier until finally the desert is reached. For a number of reasons, it is not possible to accommodate desert vegetation in a single physiognomic classification. The recent increase in desertification (from the early 1970's) in West Africa is probably mainly the result of the destruction of the vegetation by man and his domestic animals rather than to the deterioration of the climate (see White 1983, p. 23, for references). For further and more detailed consideration of the region's climate, see Griffiths (1972a) and Toupet (1968) and accounts in more general works on African climate (Thompson 1965, Walter *et al.* 1975).

Geomorphology

West Africa lies in what is termed Low Africa (Fig. 2) where sedimentary basins and upland plains range in altitude from about 150 to 600 m above mean sea level. In Low Africa lies the Sahara Desert and the catchments of various rivers including the Senegal, Niger, Shari, Logone, Volta, Zaire, and lower Nile. Land above 1,000 m forms the watershed areas and escarpments of the Niger basin (e.g. the Fouta Djalon, Upper Guinean Highlands, Jos Plateau, the Jebel Marra, and the saharan massifs of Ahaggar and Tibesti). Outcropping in these highlands are the ancient Precambrian rocks underlying the whole of the region. The main watersheds are shown in Fig. 19.

Changes in drainage patterns that have been brought about in the last million or so years have been caused almost wholly by variations in climate. These climatic variations have often been extreme with differences in rainfall and evaporation causing the expansion, contraction, or even complete disappearance of river systems and areas of standing water within basins along with any connections between them. Some of the lakes present in the late Pleistocene between 14° and 22° N were enormous, but the only large natural lake remaining in this region is Lake Chad. Lacustrine deposits around this lake indicate that it has undergone considerable expansion and contraction at different times in the past in response to climatic change and was probably once connected with the basins of the Niger, Nile, and Zaire (Servant and Servant 1983). Further testimony to recent climatic change is the presence of water courses that rarely, if ever, now flow (e.g. Tilemsi and Azaouak tributaries of the River Niger). Indeed, the present course of the River Niger dates back no more than 5,000 to 6,000 years.

Wetland types

West African rivers and floodplains

West Africa is divisible into two very broad physiographic regions whose boundary runs roughly in an east-west direction at about 12° N. To the north

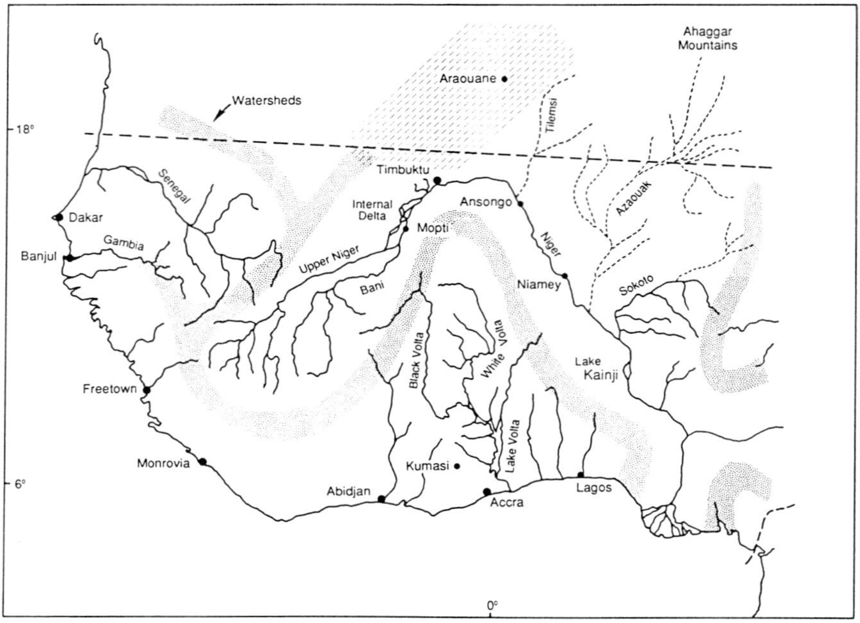

Figure 19. The principal rivers and lakes in tropical West Africa as defined here (see Fig. 1). Also shown are other important features mentioned in the text such as principal centres, watersheds, and the position of Pleistocene lakes (modified after John 1986). See Fig. 23 for details of rivers, floodplains, and more important lakes in the Chad basin.

lies a sedimentary plain and to the south a series of highlands distinct from each other and from the coast. The highlands separate those rivers flowing directly into the Atlantic Ocean from those flowing northwards for at least part of their course. These rivers are torrential in nature where the relief is greatest, flowing through narrow gorge-like valleys, with the water course punctuated by rapids and falls wherever the bedrock is exposed. In areas of low relief the main channel of many rivers commonly subdivides into many anastomosing watercourses resulting in the isolation of narrow sand and sometimes vegetated islands. Commonly rivers overspill their channels during the rainy season and flood low lying areas. These floodplains support profitable fisheries and are agriculturally very productive (see Welcomme 1979).

Most rivers are of the flood type in which pulses of increased flow are transmitted throughout the drainage network. The damming of several major rivers within the past 30 years has resulted in a reduced and more stable flow downriver of the lake or series of lakes. Thus the lower course of the Volta, Niger, and Bandama rivers has become transformed from that of a flood type into what has been termed a reservoir type of river. Certain physio-chemical changes take place in the water below the dams. These have been studied immediately downriver of the Akosombo Dam on the River Volta (Obeng-Asamoa 1979) and the Kainji Dam on the River Niger (Sagua 1979). The only investigation of the effect of river damming on the aquatic vegetation has been undertaken in the lower reaches of the River Volta by Hall and Pople (1968). In contrast, much more attention has been given to the often deleterious effects on downriver agriculture and fisheries in the River Volta (Lawson 1970, Grove 1985, Hilton and Kowu-Tsri 1970) and in the River Niger (Adeniji 1973, Grove 1985, Sagua 1979).

The flood or flow regimes of rivers are different between the north and south of the region, with the interpretation of the flow regime patterns complicated by river size and differences in surface relief. In the north the run-off tends to be concentrated into a few floods in August and September, whereas further south it is more likely to be more evenly spread throughout the year. Often above a latitude of about 12° N small streams remain completely dry for at least six months of the year (Ledger 1964). Such differences clearly reflect the seasonal distribution of rainfall. Rodier (1961) considers West African rivers as falling into two major flood regime types based on their seasonal flow patterns: the equatorial regime types of the extreme south with two periods of high water each year, and the tropical regime types of the rest of the region where there is a single high water season. He divided still further the tropical regime rivers based mainly on the exact time of high and low water and the relative length of these two periods. Iltis and Lévêque (1982) have used the number as well as the time and size of the flood or flood peak(s) to further subdivided rivers in the Ivory Coast having an equatorial flood regime. They also recognize a so-called montane type with a single flood period (March–October) which seems to closely correspond with what Rodier (1964) has termed the classical tropical type. Most of the longer rivers in West Africa (e.g. River Niger, River Volta, River Comoe) have mixed flood regimes. The situation is especially complicated in the lower course of the River Niger whose drainage areas are widely separated. Much attention has been focused on this river along with its associated floodplains and so it is singled out for special treatment in the following section.

The Niger. The River Niger is the third longest river in Africa (4,200 km long) and has a watershed covering about 1,250,000 km^2. It issues from a deep ravine on the landward side of the Fouta Djalon highlands (Guinea–Sierra Leone border) at about 1,000 m above mean sea level. From its source some 240 km from the Atlantic the river flows northeastward to a vast lowland area of swamps lying at the edge of the Sahara Desert. This area of some 20,000 to 30,000 km^2 is known as the internal delta and is formed by the Niger and its second largest tributary, the River Bani (1,110 km long).

The many channels of the Niger come together below Dire and from Kabara (the port of Timbuktu) to the rock sill at Tosaya the river flows east-northeast. After leaving the internal delta at Tosaya it flows south-east through successively more humid zones where its discharge is supplemented mainly by left bank tributaries. Downriver of Niamey begins the second main drainage area of the river with northward flowing tributaries rising in the highlands of Benin (formerly Dahomey). Other rivers contribute to its flow through Nigeria including its third major tributary, the River Sokoto. During the rainy season the lower valley of this tributary becomes inundated and forms an extensive floodplain. Some information is provided by Holden and Green (1960) on the hydroclimate and biology of this tributary. In Nigeria the main river has been dammed at Kainji and further downriver at Jebba. Downriver of Jebba is the confluence of the Niger with its main tributary, the River Benue. This tributary delivers annually a volume of water about equal to that of the main river. Its own floodplain is vast, covering an area of 3,100 km^2 at the peak of the flood. About 250 km from the Gulf of Guinea begins the coastal delta which is most densely forested. This is the largest delta in Africa (ca. 36,260 km^2) and has a coastal fringe covering 7,500 km^2 of brackish-water swamps (mostly mangrove-dominated).

The flood regime of the internal delta is dependent on the seasonality and magnitude of rainfall in the headwaters of the Niger and Bani rivers as local rainfall is negligible. During the six month flood season (May – September) in the delta the rivers overspill their channels resulting in the formation of many temporary lakes and ponds as well as the filling of river arms and creeks (Fig. 20). The flood peak is flattened as the passage of the floodwaters is hindered by the vast area of swamp vegetation that lies between Lake Debo and the port of Timbuktu (Kabara). It thus takes the flood peak about 4 months to travel from Kiafarabe to Gao. The single annual flood peak of the upper drainage area deposits most of its silt in the extensive swamps of the delta. Salts are also removed from the water and are probably absorbed by the aquatic vegetation, involved in clay synthesis, or accumulate as bottom deposits on the floors of the many temporary lakes and ponds. The water leaving the delta is thus comparatively clear, silt-free, and low in dissolved salts. Downriver the conductivity of the water increases possibly due to evaporation, solutes derived from dust carried on the Harmattan wind, and the influx of more solute-rich waters from the tributaries in the lower drainage area. This flood water is often referred to as the 'black flood' when it arrives in the Kainji area (see Fig. 19) of Nigeria in November (peak about January/February) after taking 6 to 7 months to travel a distance of about 2,000 km. In Nigeria there is a second flood season which begins in August and reaches a peak in the Kainji area in September. This is the major of the biannual floods and represents the run-off of the local rains principally en-

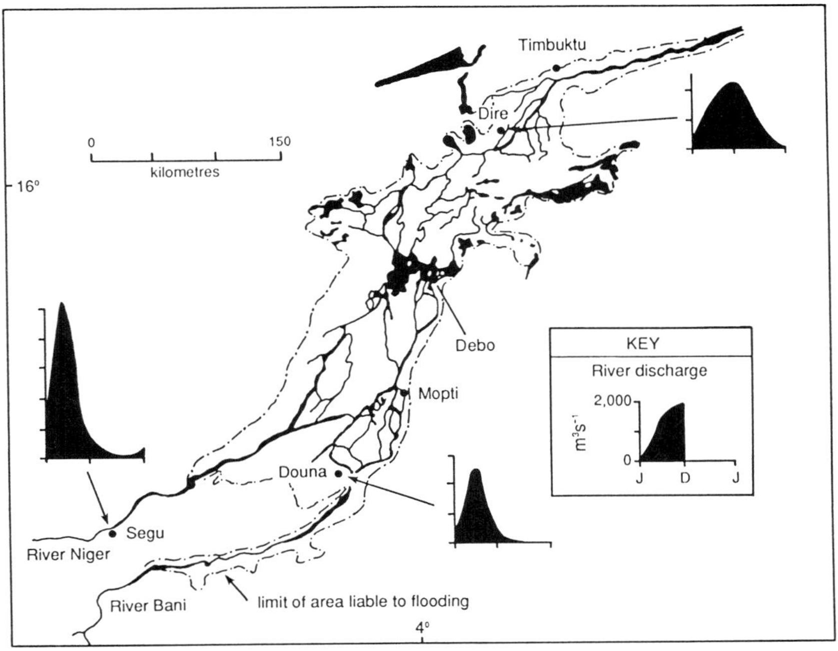

Figure 20. The internal delta of the River Niger showing its many lakes and channels. Indicated is the seasonal discharge at 3 stations within the delta. Modified after Welcomme (1979) and Rzóska (1985) (by kind permission of E. Schweizerbartsche/Gebrüder Borntraeger, publishers).

tering downriver of Niamey (mainly from the rivers Malendo and Sokoto). It is heavily silt-laden (mostly kaolinitic colloids) and this imparts a milky coloration to the water and hence is commonly referred to as the 'white flood'. These two floods significantly affect the ecology of the lower river and its adjacent floodplains as well as the two large Nigerian lakes known as Lake Kainji (see separate entry) and Lake Jebba.

Some of the earliest investigations of the Niger were undertaken in the 1930's principally by French scientists working in the internal delta, first at Diafarabe and later at Mopti. Several publications provide information on the water chemistry of the internal delta (e.g. Daget 1954, 1957, Blanc *et al.* 1955) and there also exist early accounts (Chevalier 1932, Duong-Huu-Thoi 1950a,b) of the associated vegetation. A number of studies (Cook 1968, Imevbore 1970, White 1965, Imevbore and Bakare 1974, Imevbore and Visser 1969) were undertaken in the area of the Niger Valley to be flooded following the closure of the Kainji Dam. The only longitudinal study of the river was carried out along a 2,700 km stretch (Koulikoron to confluence of the River Benue) during the 1969 to 1970 British Hovercraft Expedition (Grove 1972). For a review of much of the information on the water quality, hydrobiology, and other aspects of the Niger, see chapters (e.g. Rzoska) in Grove (1985).

The River Niger has a considerable variety of vegetation types associated with it. Within the river itself there are truly aquatic plants, or euhydrophytes, that form essentially similar associations throughout the length of the river. Marginal and floodplain vegetation, however, varies along the length of the river because of the influence of river flood cycles and local edaphic and climatic factors. Only Cook (1965, 1968) has attempted to define aquatic plant associations using the Braun-Blanquet system, though other authors have described associations of species without using formal phytosociological nomenclature (for a survey see Newton 1986).

In his account of the internal delta, Chevalier (1932) described three types of plant community which he called associations (not *sensu* Braun-Blanquet). All submerged and floating plants formed one association, occurring in shallow water with a sandy bed and weak or non-existent currents. Partly submerged and marginal vegetation dominated by grasses was included in the second association, called by Chevalier the aquatic prairie. Extensive stands of *Echinochloa stagnina* are the most conspicuous element of the aquatic prairie. The third association was the vegetation on banks of seasonally exposed sand. Plants in this association ranged from semi-aquatic species to terrestrial species tolerant of short periods of flooding, according to the length of immersion/exposure in the annual flood cycle. Duong-Huu-Thoi (1950a,b) included the internal delta in a general study of the vegetation of what was then the French Sudan, but he did not give a detailed account of the submerged or floating plants. He distinguished two kinds of community associated with the River Niger: marginal vegetation lining permanent water courses, and the vegetation of the floodplain. In studying zonation and succession he related differences between northern and southern ends of the delta to differences in soil, as well as to the different climatic conditions that would influence the later stages of succession towards terrestrial vegetation.

The only other section of the River Niger that has been studied in any detail is in the area now flooded behind the Kainji Dam, Nigeria, where there have been studies of vegetation before and after impoundment. This section differs from the internal delta in two main ways. One is the biannual flood regime ("black flood" and "white flood"), in contrast to the single annual flood cycle of the internal delta. The other is the arrival of the flood water in a different climatic season since it takes several months for the main flood wave (the "black flood") to travel along this lengthy river. Cook (1965, 1968) described a number of plant associations, ranging from highly specialized aquatic communities to groups of terrestrial and semi-aquatic species occurring on seasonally flooded low-lying land near the river. In all cases the occurrence of a particular association in any one location was related to the flood cycle of the river, and Cook's classification of life forms is based on the degree of submergence in relation to normal completion of the generative cycle. Imevbore and Bakare (1974) recognized two types of wet season vegetation in the Kainji area, both dominated by grasses and sedges: fringing swamps along indentations in the river shoreline, containing different vegetation types depending mainly on the soil, and swamps in old river beds that obtain their water from rainfall and the general water table.

Finally, away from lagoons and channels, the coastal delta of the river is covered by very extensive swamps, with the freshwater swamp forest giving way to a coastal fringe of brackish-water swamps dominated by mangroves (see Adejuwon 1973, p. 134). The ground in such forest is often very irregular, with frequent patches of open water present even in the dry season; the whole area is flooded in the rainy season. There is no detailed account of freshwater swamps in the Niger delta so they cannot be compared with those in other West African deltas, e.g. account by Adejuwon (1973) of the delta of the River Ogun in western Nigeria. Swamp forests have generally fewer species than forest occurring on well drained ground away from flood zones, and there are fewer very large trees. The main canopy is often rather open and in the gaps are dense tangles of shrubs and lianes whilst a number of the trees possess stilt roots. Particularly characteristic are the *Raphia* palm (*Raphia vinifera*) and various climbing palms such as *Calamus deerratus*. Much of the vegetation in the Niger delta has been disturbed by human activity. Ahn (1958) describes the regrowth stages in similar disturbed forest in western Ghana, where he distinguished between swamp forb regrowth, swamp thicket, and secondary swamp forest.

Swamp forest sometimes gives way to swampy savanna and areas of grassy swamp vegetation (Adejuwon 1973). The dominance of grasslands in some parts of West Africa (e.g. Ivory Coast, Benin, Ghana, Zaire) is accounted for by the alternate waterlogging and drying out of very shallow soils (see White 1983, p. 84, 178). Such grasslands are believed to represent an edaphic climax. The principal grasses include species of *Anadelphia* and *Jardinea*, Cyperaceae are well represented, the moss *Sphagnum* is often present, and also occur herbs such as *Lycopodium* spp., *Mesamthemum radicans*, and species of *Burmannia*, *Drosera*, and *Xyris*.

The Lower Senegal Valley. The Lower Senegal Valley of the Senegal river offers a large variety of wetlands: floodplains, fresh and brackish-waters, temporary marshes, paddy fields, lake lagoons near the coast, and mangrove swamps (Fig. 21). Development projects since 1963 have profoundly affected the ecosystem and have prevented large areas from being flooded.

1. *The Delta.* The so-called "Delta of the Senegal River" includes the Djouj

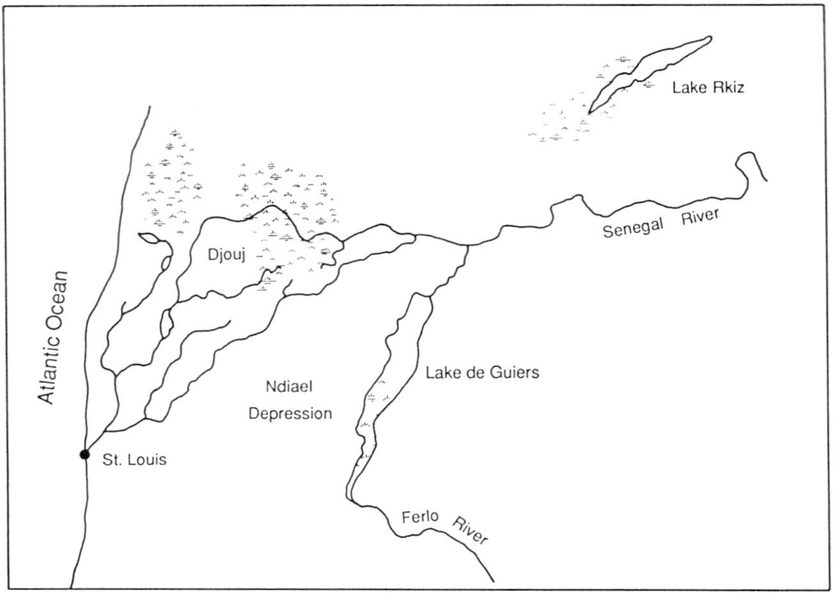

Figure 21. An outline map showing the main wetlands of the Lower Senegal Valley.

National Park. It covers 145 km^2 of marshes, temporary lakes, permanent watercourses, and floodplains, and may be considered as the last of the great settling basins of the delta remaining in a natural state. Floodwater from the river to the delta is controlled by dams at the river outflow points. The site has been protected totally since 1971 and is exceptionally important as a feeding and resting area for many aquatic birds.

The Ndiael depression (150 km^2) is normally filled by a combination of rainwater and flood water from the Senegal River. Since 1969 however, development has changed the course of the river water and the depression now receives only summer rainwater.

2. *Lake de Guiers*. The main wetland of the region is Lake de Guiers. It was originally part of the seasonally flooded Ferlo valley which flooded from August to November (maximum in October), and had low water for the rest of the year (Rochette 1974). At the end of the last century, saline invasions from the sea occurred up the Senegal estuary and into the lake during low water. In 1947 a flood gate was built near Richard Toll which allows flood waters to fill the lake. After maximum flood the gate is closed. This prevents the waters returning to the Senegal valley and stops brackish-water incursions. Finally, in 1957 a dam was built at the south end of the lake (Keur

Momar dam) to stop water flowing into the Ferlo valley where much used to be lost through evaporation.

Clearly, nowadays Lake de Guiers is artificially maintained. It occupies a flat depression with a bottom some 2 m below sea level. It is about 50 to 60 km long and 7 km wide. The surface area is 300 km^2 during flood and 120 km^2 at low water. The water volume at maximum capacity is about 800×10^6 m^3. In terms of water budget, 80% of the input is from the flood water of the Senegal River, 11% from rain, and 9% from drainage of sugar cane plantations. Eighty-two percent of the total water loss is through evaporation (Cogels and Gac 1982) and the remainder is used for irrigation of sugar cane plantations to the north of the lake.

The lake is sub-divided into two regions. The north lake, 30 km long and from 5 to 7 km wide, has a large open water zone. Salinity varies from 100 mg dm^{-3} in October to 500 mg dm^{-3} in July. The south lake is narrow with many small islands, and salinity ranges from 1,000 to 4,000 mg dm^{-3}. Water temperatures range from 22.5°C in January to 31°C in June and follows a similar cycle to Lake Chad.

Information on vegetation can be found in Adam (1964), Trochain, (1940), and Reizer (1974). After building the flood gate in 1947, *Typha australis* developed enormously and was considered a nuisance. The *Typha* beds have partly receded since 1972 due to the sahelian droughts and have been further reduced by burning. Detailed information on the management and economics of the region can be found in Ba *et al.* (1983).

Coastal lagoons of the Ivory Coast. Some limited amount of information exists for many West African lagoons (see John and Lawson 1990, for review), but by far the most intensively investigated are those in the Ivory Coast. These are situated along the north coast of the Gulf of Guinea between 2° 50′ and 5° 25′ W and occupy an area of approximately 1200 km^2 (see Dufour 1987). The climate is equatorial with 2,000 mm rainfall a year distributed in two rainy seasons: long rains from April to July and short rains from October to November. Three main lagoons are distinguishable (Fig. 22):

(1) The Grand Lahou Lagoon (200 km^2) is very shallow and receives the Bandama, the largest river in the Ivory Coast;

(2) The Ébrié Lagoon (560 km^2) which has the city of Abidjan on its shores, is 130 km long, less than 7 km wide, and 4.6 m keep. Since 1950 the water regime has been completely altered by the construction of the Vridi canal (Varlet 1978) to allow larger boats to enter Abidjan harbour. The lagoon has been intensively studied (Durand and Chantraine 1982, Dufour *et al.* 1985).

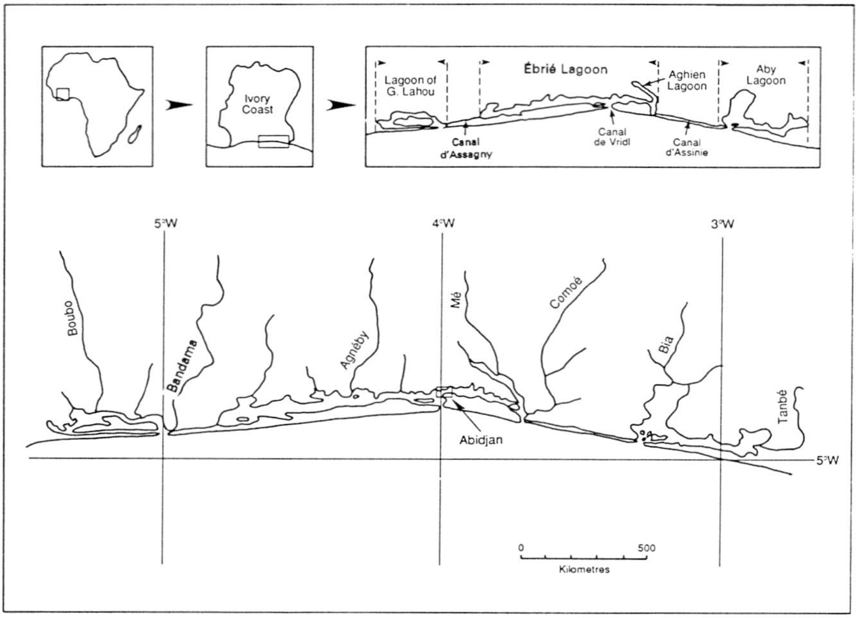

Figure 22. The lagoon areas of the Ivory Coast (redrawn from Durand and Chantraine 1982).

(3) The Aby Lagoon (424 km^2), which starts about 30 km inland, differs from the other two in having a reduced exchange with seawater.

Originally, the three lagoons were separated but they have now been connected by canals (Canal d'Assagny and Canal d'Assinie) to allow boat movement. Each lagoon has a different hydrological regime depending upon its morphology, freshwater inflows, and the amount of exchange with seawater.

1. *General characteristics*. The waters of the Ébrié lagoon are renewed regularly. The average annual inflow of freshwater and seawater is respectively 4 and 14 times the total volume of the lagoon but the freshwater inflow varies substantially from year to year (Durand and Chantraine 1982). As a result of the complex hydrodynamics the water salinity varies spatially and seasonally. Near the Vridi channel salinity measurements are around 20% near to Abidjan they are lower than 10 and at the extremities of the lagoon, less than 5‰. In the Grand Lahou lagoon salinity varies from 3 to 10‰ in the western area and between 0 and 25‰ in the east. The salinities of Aby lagoon are low (2 to 5‰) and relatively stable (Durand and Chantraine 1982, Durand and Skubich 1982). Water temperatures range from 27 to 31°C with an average of 29°C. Dissolved oxygen at the water surface is generally from 4 to 7 mg dm^{-3} but the vertical distribution varies greatly according to season.

Anoxic conditions are sometimes noted, mainly around the polluted bays near Abidjan, and in the central basin of the Aby lagoon. Generally, the less saline waters are characterized by a poor nutrient content and a rapid cycling of the nutrients (Dufour 1984, Lemasson and Pages 1980, Lemasson *et al.* 1980, 1981, 1982, Pagès and Lemasson 1981a,b). Bacteria and their role in mineralization of organic matter is discussed by Guiral (1984).

2. *Biota.* The main primary producers in the Ébrié lagoon are phytoplankton with a mean annual biomass of around 16.5 mg Chl. a m^{-3} (Dufour 1984, Dufour and Durand 1982). The gross primary production was estimated to be 1,400 g O_2 m^{-3} y^{-1} (Dufour 1982a,b,c, 1984, Pagès and Lemasson 1981a,b, Pagès *et al.* 1981a,b).

The zooplankton is largely represented by the copepod *Acartia clausi* (Pagano and Saint Jean 1983, Saint Jean and Pagano 1983, 1984). In the dry season when the freshwater is low, the influence of the sea encourages marine plankton. However, in the seasonal river spate, freshwater Cladocera occur. Molluscs are fairly abundant in the benthic fauna together with some crustaceans, crabs, and shrimps.

The productivity of the lagoons is generally high and supports a large community of fish (Daget and Durand 1968). Many species, freshwater and seawater species, spend part of their life-cycle in the brackish-waters of the lagoons where they are fished. Fish production for the three lagoons is between 15,000 and 20,000 tonnes · y^{-1}, the catches being dominated by *Ethmalosa* (Charles-Dominique 1982).

3. *Use and management.* The lagoons of the Ivory Coast are becoming increasingly polluted. The environment has been irreversibly modified by the cutting of channels, extraction of gravel, building of dykes, etc. Domestic and industrial waste waters from Abidjan and its suburbs enter the Ébrié lagoon, and land surrounding all the lagoons transfer insecticides and chemical fertilizers in their run-off (Arfi *et al.* 1981, Guiral 1984). If the situation is not considered catastrophic now, it could be so in the near future.

Natural lakes and inundated dunes in the Chad basin. The Chad basin (2,300,000 km^2) is divided into different endorheic basins, the larger being the Lake Chad basin (700,000 km^2) situated at an altitude of 281 m a.s.l. between latitudes 6° and 15° N and longitudes 7° and 25° E. This basin includes Lake Chad itself, the rivers Shari (Chari) and Logone, and their associated floodplains (Fig. 23).

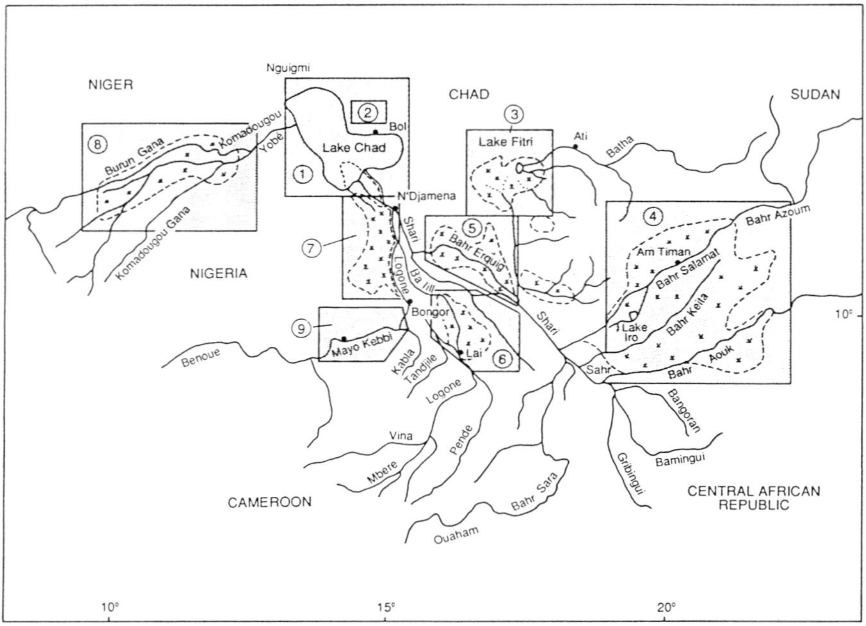

Figure 23. The Chad Basin showing the location of the main wetlands: 1. Lake Chad, 2. Kanem lakes, 3. Lake Fitri, 4. Salamat floodplains, 5. Massenya floodplains, 6. Ba Illi floodplain, 7. Yaere floodplain, 8. Komadougou-Yobe floodplain, 8. Mayo Kebi lakes.

Lake Chad 1. *General characteristics.* There are few natural lakes in West Africa of any size with the exception of Lake Chad (Table 5). Between 1964 and 1978 multidisciplinary research on Lake Chad provided a diversity of data which are almost unique for tropical Africa. A synthesis of the limnological studies has been published (Carmouze *et al.* 1983) and can be consulted for more detail. It contains a large bibliography of the lake including hydrology, geology, climatology, limnochemistry, flora, and fauna, etc. Only a brief account is given below.

Lake Chad is situated in the Sahelian zone. The climate is tropical with a dry, hot season from March to June, a rainy season from June to October and a dry, cool season from November to February. The mean annual rainfall on the lake is 320 mm. Insolation is high and a monthly average of 275 to 310 h, and a mean daily radiation of around 550 cal m^{-2} d^{-1}. The water temperature follows a seasonal cycle in accordance with the climate, with a minimum in January (18 to 19°C) and a maximum in June (31°C).

Lake Chad has an estimated mean depth of only 3 m and thus the total volume is relatively small. As a result, the total surface area and water level is controlled by the water budget which shows seasonal and year to year change. The water budget has been estimated for the period 1954 to 1972

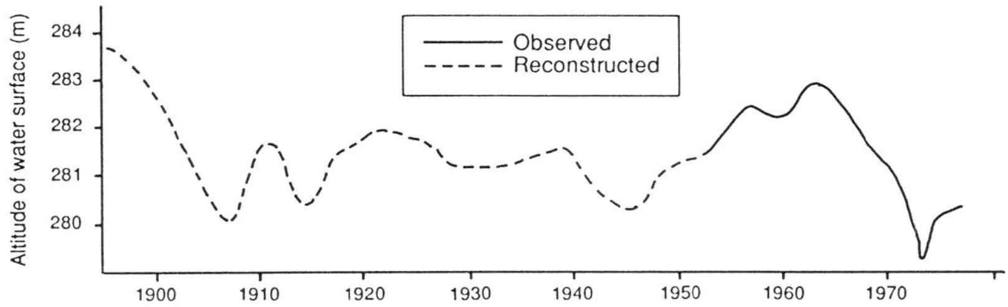

Figure 24. Changes in mean annual water level of Lake Chad from 1895 to 1975.

(Carmouze 1976a,b). Most of the input (87%) comes from the river tributaries ($45.1 \times 10^9\ m^3\ y^{-1}$) and, to a lesser extent (13%), from the rains ($6.35 \times 10^9\ m^3\ y^{-1}$). Evaporation is high and accounts for 92% of the output ($44 \times 10^9\ m^3\ y^{-1}$) with the remainder (8%) lost to seepage. As the lake is endorheic, there is no outflow. The spate in the tributaries starts in June and reaches a maximum around mid-November. Eighty five percent of the annual discharge occurs between July and December and the lake level rises to a maximum in January. The level then falls by up to a metre to a minimum in July.

The water budget is subordinate to the long-term climatic changes and, more specifically, to the rains on the drainage basin of the tributaries. As a result, the lake expands and contracts. Fig. 24 shows the water level fluctuations over the last 100 years. Three main states of the lake may be distinguished according to water level (Fig. 25):

(1) Great Chad, with a surface water level altitude of 283 m and a surface area estimate of 25,000 km^2. This state occurred between 1963 and 1965.
(2) Normal Chad, with a surface water level altitude of 281–282 m and a surface area of around 20,000 km^2. This occurred between 1965 and 1971.
(3) Lesser Chad, with a surface water level altitude of 280 m. The north and south basins became separated and the north basin dried up entirely. These extreme conditions have prevailed since 1975.

During the Normal Chad period three main types of landscapes occurred: (1) open water areas devoid of vegetation; (2) reed islands formed from fixed vegetation (*Cyperus papyrus*, *Phragmites australis*) and (3) archipelagoes consisting of about a thousand sandy islands which are the dune crests of a settled, partly submerged erg (sand desert, usually in the form of dunes). Because of differing environmental conditions several natural regions could be distinguished (Fig. 26). In spite of its endorheism, the water of Lake Chad is fresh. The salinity of the Shari River entering the lake is about 40 to 50 mg dm^{-3}. It increases from the Shari delta: the open waters of the

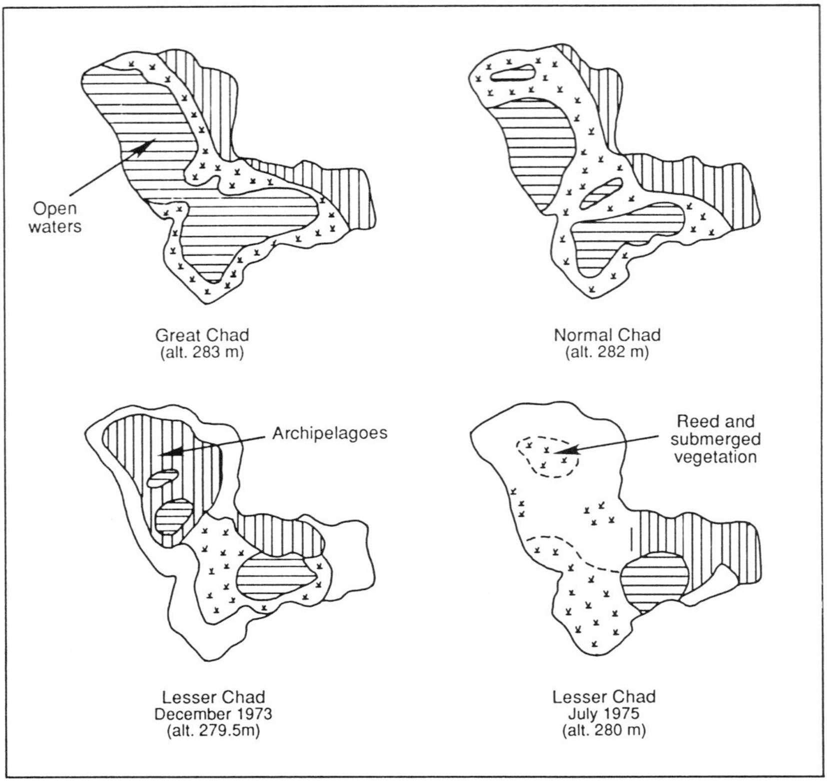

Figure 25. The three main states of Lake Chad: Great, Normal and lesser Chad, in relation to water levels.

south basin being 1.2 to 1.5 times more concentrated than the river while the water of the north archipelago is from 10 to 20 times higher (Carmouze 1976a, Roche 1980). On average, waters from the north basin (625 mg dm^{-3}) are 4 times more saline than those of the south basin. The pH ranges from 7 to 8 in the Shari River and open waters of the south basin, reaches 8.5 in the eastern archipelago and rises to 9 in the extreme north.

These figures show that, contrary to expectations, Lake Chad is not a basin of high salt content and the salinity of the water changes only slightly from year to year. This peculiar situation results from several factors which combine to maintain the dissolved salt stocks proportionally close to those of the water volume:

(1) The salinity of the river inflow is low (60 mg dm^{-3}).
(2) The climato-geographical regulation of the salinity results in a concentration of the river water by a factor of about 10.8, which is not very high for a closed lake in an arid zone. This is because seepage losses are

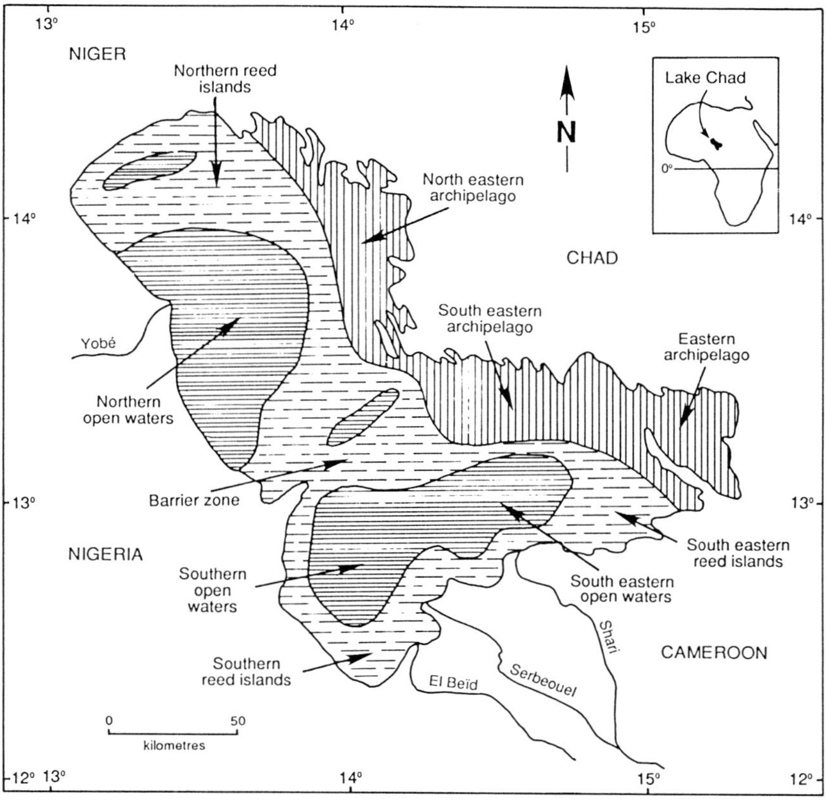

Figure 26. Lake Chad: the main types of landscape. The mean water level altitude at this stage is 282 m above sea level.

relatively important, mainly in the north basin where the waters are more salty.

(3) Considerable biogeochemical sedimentation occurs of SiO_4H_4, Ca, Mg, HCO_3/CO_3 and, to a lesser extent, K. The geochemical sedimentation is due to neoformation of smectites (Carmouze *et al.* 1977, Carmouze 1976a) and precipitation of calcite. The clayey neoformations are favoured by relatively high concentrations of SiO_4H_4 in the Shari River (20 mg dm^{-3}) and by supplies of solids rich in iron and aluminium hydroxides. The salinity of the water where calcite precipitation occurs need not be very high because the predominant anions are HCO_3 and CO_3. Molluscs and aquatic macrophytes contributions to the biochemical sedimentation, involving mainly Ca, Mg, K, and HCO_3, are important because the biomasses of these groups are substantial. The processes of

chemical sedimentation lead to a 45% decrease from the value that the salinity would reach by climato-geographical regulation alone.

2. *Biota.* The flora and fauna of Lake Chad were relatively well studied during the Normal Chad period of 1965 to 1971. It resulted in a synthesis paper (Carmouze *et al.* 1972) describing communities, biomasses, and the main ecological zones observed in the lake (see Carmouze *et al.* 1983). More than 1,000 species of algae were identified. The algal flora was dominated qualitatively by desmids and diatoms whilst the Cyanophyceae were important quantitatively (Iltis 1977). A model was established from primary production measurements made over several years in the eastern archipelago (Lemoalle 1973, 1979a) and remote sensing was used to evaluate the production in the south basin (Lemoalle 1979b).

Among the aquatic macrophytes, *Vossia cuspidata* was very abundant in the Shari delta and *Cyperus papyrus* in the south basin. They were both progressively replaced by *Typha australis* towards the north as a function of increased salinity. *Phragmites australis*, and a diversity of euhydrophyte genera included *Potamogeton*, *Ceratophyllum*, *Vallisneria*, *Utricularia* and *Nymphaea* were widely distributed.

The zooplankton diversity was fairly rich with 8 species of *Cladocera* and 4 species of copepods. There were several rotifer species but they only represented a small percentage of the total biomass. The zooplankton biomass for 1971 was estimated to be 12,200 g dry weight with an annual production of 860,000 tonnes (Gras and Sain-Jean 1983).

The benthic fauna was essentially composed of oligochaetes, molluscs (representing a small number of species), and a large number of larval insect species. The benthic biomass was estimated to be 71,000 tonnes for the entire lake in 1970. Molluscs accounted for over 90% of the biomass and their annual production was estimated to be 280,000 tonnes of organic matter (Lévêque 1972, 1973).

One hundred and twenty fish species were recorded from Lake Chad, most of which occurred also in the Nile and the Niger rivers. There is no endemism and some species seasonally migrate to spawn in the Shari river system. The fisheries (Figs. 27 and 28) were well developed with annual catches of over 100,000 tonnes (Durand 1980).

3. *The drying period.* The drying period of the lake, which started with the drought of 1972, had a major effect on the biotic environment. The water level dropped progressively by more than 2 m and the volume decreased by nearly half. As a result, the north basin dried up in 1975 and now began what is termed the Lesser Chad period with only the south basin containing water. The evolution of the macrophytic vegetation was characterized by

Figure 27. A fisherman's camp on a reed island in Lake Chad. The *Phragmites* is cut to form a floating mat upon which the fishermen live temporarily (photograph by C. Lévêque).

Figure 28. "Kadei" boat made out of papyrus in use on Lake Chad (photograph by C. Lévêque).

a general increase in the south basin and a spectacular development of *Aeschynomene elaphroxylon*, and prairies of *Vossia cuspidata*. In the phytoplankton a considerable development of euglenoids occurred whilst the benthic mollusc population dropped dramatically. The general reduction in benthic organisms was probably the result of the effect of turbulence on the much shallower water leading to the appearance of a layer of very soft mud. Severe mass mortalities amongst the fish population occurred owing to oxygen depletion (Benech *et al.* 1976). Fish communities changed from a lacustrine type to a palustrine type and the number of species was reduced. This Lesser Chad period was characterized by relative stability but the new lake, however, was not the homologue of the previous one as seasonal changes were much more marked and the influence of the fluvial system was increased.

Kanem lakes. Kanem, in the north-eastern region of the Lake Chad basin, has hundreds of small lakes and temporary pools. Their surface areas vary from a few hundred square metres to two kilometres square, and their depths do not normally exceed 2 m. They occupy depressions of an old settled erg and are fed by rainwater and groundwater. As they are in the same climatic zone as Lake Chad, evaporation is high. Although some are fresh, most are salty with salinity values reaching 200 mg dm^{-3}, and pH ranges from 8.7 to 10.4 (Maglione 1969, 1976). Salt deposits called "natron" are removed by the local people and used for domestic purposes or are exported to bordering countries (Fig. 29).

The salt lakes are surrounded by *Cyperus laevigatus*. The phytoplankton biomass is dominated by Cyanophyta with a high abundance of *Oscillatoria* (= *Spirulina*, = *Arthrospira*) *platensis* (Iltis 1973b, 1974, 1975). Indeed, the cyanophytes are cropped by the local people as food. There is a positive correlation between algal density and water salinity. Copepoda and Cladocera occur in the freshwater lakes but disappear with increased salinity. In the saline lakes, three species of Rotifera occur but fish are absent. Most of the lakes dried completely during the drought in 1973.

Inundated zones of the Shari River system. Along the Shari River and its tributaries, large floodplains develop during the rainy season. A permanent but poorly known lake in the floodplain is Lake Iro (100 km^2). The hydrological regime of the Shari is of the tropical type, flooding from July to December with a maximum in October/November (Billon *et al.* 1974). Two main floodplains are recognized: (1) the Salamat floodplain along the tributaries of the east bank, and (2) the Massenya floodplains along the Shari itself.

The Salamat floodplain is poorly studied although the surface area could extend for many thousands of square kilometers. The inundation starts in

Figure 29. Salt deposits called "natron" around one of the small Kanem lakes in the northeastern region of the Lake Chad basin. Local people collect the salt for domestic use and for trade (photograph by C. Lévêque).

June with the first rains and is then expanded by the arrival of the fluvial water (September/October). As usual in such a climate, there is large year to year variation in rainfall.

The Massenya floodplain (15,000 km^2) is smaller than the Salamat. Its yearly water budget has been calculated by Gac (1980) and is as follows: river input, 1.7×10^9 m^3; precipitation, 11.5×10^9 m^3; evaporation, 12.4×10^9 m^3; outflow, 0.8×10^9 m^3. It has been estimated that 240,000 tons of suspended solids from upstream erosion settles out in the floodplain each year.

The inundated zones are covered by *Hyparrhenia rufa* with *Cymbopogon* sp., *Echinochloa pyramidalis*, *E. stagnina*, and *Oryza barthii* in the more marshy areas.

Inundation zones of the Logone River system. The Logone river is boarded on the east bank by the Ba Illi floodplain, and on the west by the Yaéré floodplain of North Cameroun (see Fig. 23). The Ba Illi has a complex hydrological regime (Gac 1980) where erosion has been estimated to be 80,000 tons a year and deposition of some 500,000 tons of fine sediment a year. The Yaéré floodplain is probably the most studied floodplain in the Chad basin. It covers 81,000 km^2 and the water budget is, in some ways,

Figure 30. Traditional fisheries in the Logone River floodplain during the fall (photograph by C. Lévêque).

similar to the Massenya floodplain (Gac 1980). Annual river input is 3.2×10^9 m^3, precipitation averages 8.5×10^9 m^3, evaporation is 10.6×10^9 m^3, and the outflow averages 1.1×10^9 m^3. Inundation has been studied by remote sensing (Benech *et al.* 1982) and lasts from July to December.

The Yaéré floodplain is a grassy savanna without trees. The most common grasses are *Hyparrhenia rufa*, *Echinochloa colona*, *Panicum anabaptistum*, and *Eragrostis atrovirens*. Many fishes including some migratory species (e.g. *Alestes baremoze*) spawn in, or close to, the Yaéré floodplain (Fig. 30) which provides shelter and food for the juveniles. As the floods recede, juveniles migrate through the El Beïd River to Lake Chad (Benech and Ouensiere 1982, 1983a, b) where they add to the fish stock and are fished by the local population.

Lake Fitri. Lake Fitri, a small lake (800 km^2) situated to the east of Lake Chad and has many features similar to it. Conductivity ranges from 100 to 140 μS cm^{-1} and its water chemistry is similar to Lake Chad and, from the few data available, the flora and fauna have many species in common.

The lakes of the Mayo Kebi River. The Mayo Kebi is a southern tributary of the Benoue River and the only link between the Chad and Niger basins (see Fig. 23). It originates as an overflow from the Logone River and, during flood, water flows into a depression to form the Toubouris lakes. The outflow forms the Mayo Kebi River which progresses as a series of rapids and falls through another series of lakes (Trene and Léré) and then into the Benoue River.

The Toubouris lakes (Fianga, N'Gara, and Tikem) are very shallow (4–5 m depth) and range in surface area from 40 to 140 km^2. There is little information on these lakes but Lake Léré has been more closely examined (Lévêque 1971, Gras and Sain-Jean 1971, Dejoux *et al.* 1971). The invertebrate fauna seems similar to Lake Chad but the fish fauna is more closely related to the Niger. There is a report of a manatee (*Trichechus senegalensis*) from Lake Léré.

The Komadougou Yobe basin. The Yobe is the only tributary of Lake Chad north of the basin. During the rainy season a large floodplain (10,000 km^2) expands along the river and a marshy zone, similar to that of the inner delta, develops.

Man-made lakes. Since the early 1960's, the region's landscape has become dramatically transformed by the creation of vast man-made lakes formed by the damming of several major rivers. These damming schemes were undertaken primarily for the generation of electrical power and as reservoirs, allied to which were opportunities for flood control, irrigation, improvement of transportation and communication, and the development of inland fisheries and recreational activities. The first of these schemes involved the damming in May 1964 of the River Volta in Ghana. Over the next 15 years there followed other ambitious schemes principally in Nigeria (e.g. Lake Kainji, Lake Tiga) and the Ivory Coast (Lake Kossou).

Some of the dams most recently built in Nigeria are within the lower drainage area of the River Niger (e.g. Bakori Dam on the River Sokoto, Kiri Dam on a tributary of the River Benue, Jebba Dam directly across the course of the River Niger) whereas others are in the headwaters of the Niger (e.g. Guinea, Mali). Such damming schemes and the control of water flow through the internal delta of the Niger by means of dykes and canals have implications that extend far downriver. No doubt the low levels of lakes such as Lake Volta and Lake Kainji since about the mid-1970's is to some extent related to upriver schemes for water conservation as well as to the drought affecting a large part of Sub-Saharan Africa. These and other problems (e.g. the initial explosive development of aquatic weeds, increased incidence of waterborne diseases) were not fully anticipated in the planning of ambitious projects involving the impoundment of large rivers.

Much attention has been focused on the larger man-made lakes with often the emphasis on ecological changes accompanying the transformation of a river into a large body of standing water. This process of change is sometimes referred to as "lacustrinization" and is well documented for Lake Volta (see Entz 1969, Obeng 1973, 1981, Lawson *et al.* 1969) and some Nigerian lakes including Lake Kainji (El-Zarka 1973, Imevbore and Abegoke 1975) and Lake Asejire (Egborge 1974, 1979). Though Lake Kossou in the Ivory Coast is the second largest man-made lake in West Africa, it is not considered here in detail as there is very little information available on its vegetation (Mulligan 1972, Troare 1980). Some brief mention only is made by Whyte (1975) of the fish and vegetation along the shores of an unusual crater lake in Ghana, Lake Bosumtwi. For a fuller account of the vegetation associated with man-made lakes and other West African water bodies, see chapters by John and Newton in John (1986).

1. *Lake Volta.* The closure of the dam across the River Volta at Akosombo on 8 May 1964 led to the formation in Ghana of the largest man-made lake in Africa. Five years elapsed before it assumed its full size (maximum surface area 8,845 km) which was reached shortly after the 1969 rainy season. The lake when filled to capacity extends about 400 km northwards of the dam. Mean depth is about 18.6 m, the maximum depth is about 75 m, and the shoreline is very long (5,271 km) due to its complex dendritic shape (Fig. 31). It is divisible into the following sectors or regions: the main north-south axis, the major "arms" marking what was formerly the lower reaches of the River Volta and its tributaries, the shallow littoral areas originally cleared in some places of trees, and the 24 km long gorge area where the lake narrows at its southernmost end. Much of the lake is surrounded by savanna (Guinea savanna) which is relatively dry at the northern end. Only in the Afram and Pawmpawm arms in the extreme south is the lake bordered by dry semi-deciduous forest.

The lake level fluctuates in years of average rainfall by 3 to 4 m and probably reflects the inflow of floodwater draining the more northerly parts of the Volta basin rather than further south where the lake lies just within the forest zone. In June and July it is at its lowest yearly level and a rapid rise occurs from August through to about October. For a period of about a month the lake level stabilizes before a gradual drop begins over the dry season. Each year the lake is drawn down so as to compensate for the incoming floodwater associated with the next rainy season and to feed the turbines. There is thus a drawdown area which amounts to 850 km^2, averages 100 m in width, and is more extensive in the shallower northern sector of the lake.

The lake was very extensively studied in the decade following the closure of the Akosombo Dam by national institutions, a number of United Nations bodies as well as the cooperative multidisciplinary programme known as *The Volta Basin Research Project* of the University of Ghana, Legon. Some pre-

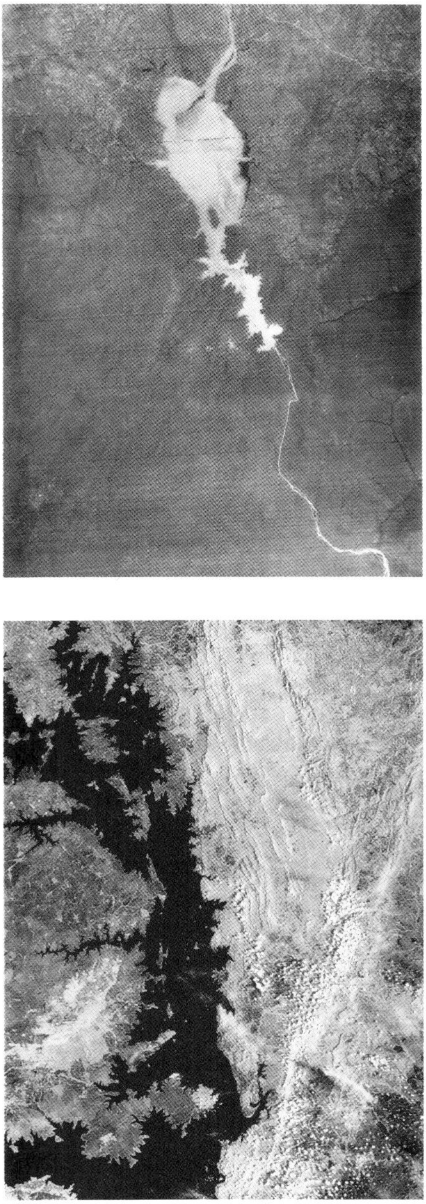

Figure 31. Top: LANDSAT image (Infra-red) of the north-south axis of Lake Volta showing also its very irregular outline; this was taken during the dry season (February 1976) and hence the virtual absence of any cloud cover. Bottom: LANDSAT image (red) of Lake Kainji taken about the time (November 1975) when it was at its highest level. Note the turbidity patterns in the lake water which are evident in this photograph taken with film sensitive especially to the red wavelengths of light (by courtesy of the National Aeronautical and Space Administration, U.S. Government).

impoundment surveys of the aquatic plants were carried out in the Volta basin (Lawson 1964, Hall *et al.* 1969). Summaries of the research findings covering the early period of the lakes existence are provided by Entz (1969), Obeng (1973, 1981), and Lawson *et al.* (1969). Many of the early, and often not widely circulated, research reports are mentioned in a comprehensive bibliography prepared by Brooks (1970).

The annual fluctuation in water level has not allowed the establishment of rooted euhydrophyte communities in the lake. Submerged colonies of the free-floating *Ceratophyllum demersum* and surface-floating *Pistia stratiotes* and *Lemna paucicostata* appeared shortly after impoundment. There was some sudd formation with *Vossia cuspidata* as a pioneer (Ewer 1966, Lawson 1967), but this declined later (Paperna 1969). Recession in amounts of *Pistia* has been accounted for by changes in water chemistry as the lake matured (Lawson *et al.* 1969). Fairly extensive mats of *Pistia* are now confined to the Afram and Pawmpawm arms of the lake. Following the report of an aphid-transmitted virus disease affecting *Pistia* in western Nigeria (Pettet and Pettet 1970), Hall and Okali (1974) studied seasonal development of colonies in the Pawmpawm arm of Lake Volta. They recognized four growth phases in the annual cycle, including a drastic die back in the dry season and related these to nutrient conditions in the water; they found no sign of the virus disease.

Marginal vegetation in the drawdown area is unable to achieve any great stability because of the alternation in flooding and exposure, and zonation results from this annual cycle (Hall 1970). The majority of the plants are flood-tolerant terrestrial species growing at higher levels. The few species that have successfully colonized the lower levels are those semi-aquatic species capable of sufficiently vigorous growth to keep pace with the seasonal rise in water level, notably *Polygonum senegalense* and *Vossia cuspidata*. Details of zonation vary in different locations around the lake but Hall *et al.* (1971) described three broad zones that can usually be recognized: annual forb zone (exposed for 35–45 weeks each year), perennial grass/*Polygonum* zone (exposed for 10–35 weeks each year), and sedge zone (exposed for about 10 weeks each year).

2. *Lake Kainji.* This lake was formed following the closure on 2 August 1968 of the Kainji Dam in Nigeria which is about 1,000 km from the coastal delta of the River Niger. The lake reached its maximum size in just 2.5 months (maximum surface area: 1,280 km). It extends some 137 km northwards of the dam, its mean depth is 12.3 m, the maximum depth is about 50 m, and the lake shore is about 3,720 km in length. The lake is divisible into three sectors or basins (Fig. 31): (1) a fairly deep-water southern basin (25 m deep) that is close to the dam and most affected by the release of

water, (2) a wide central basin accounting for about 70% of the lake surface, this portion is very open and is influenced by wind action; and (3) a northern basin where conditions most closely resemble those of the river. This lake lies in the savanna zone and about one third of the lake basin (ca. 433 km^2) was cleared of vegetation before impoundment.

Water level in the lake varies by about 9 to 10 m each year which leaves a very large drawdown area that has been estimated to be more than 653 km^2 in August. It is most extensive on the eastern side of the lake where the slope is most gentle. Also when the water level of the lake is low, a former floodplain is exposed (Foge Island which is the largest island in Fig. 31). After August, the water level rises rapidly and the highest level is usually reached during the first few months of the dry season (November – February/March). This seasonal pattern is dependent on management operations at the dam site and time of entry into the lake of the two annual floods of the River Niger (see entry for River Niger). These seasonal floods account for the rapid through-flow with mean water retention time of the lake only 76 days. Rapid turnover and large annual fluctuations in water level have an important influence on the ecology of the lake as they bring about significant changes in water quality properties (especially turbidity).

Some of the pre-impoundment, multidisciplinary research is reported in *The First Scientific Report of the Kainji Biological Research Team* (White 1965), the two-volumed *Kainji: a Nigerian Man-made Lake* (Vol. 1: Visser 1970; Vol. 2: Mabogunjie 1973), and a symposial volume entitled *The Ecology of Lake Kainji: the Transition from River to Lake* (Imevbore and Adegoke 1975). For a review of some of the early research findings on this lake, see El-Zarka (1973). The results of much of the research carried out in the 1970's were presented at an international conference entitled *Kainji Lake and River Basin Development in Africa* held at Ibadan (Nigeria) in 1977 and the proceedings published in two volumes in 1979 (Anon. 1979). Visser (1970a) has produced a useful bibliography on the River Niger which makes special mention of the many obscure reports dealing with Lake Kainji.

As was the case in Lake Volta, and as predicted by Cook (1965, 1968), annual flucutuation in water level preclude the establishment of extensive populations of rooted euhydrophytes. Following closure of the Kainji Dam, there was an initial development of floating vegetation, including *Pistia stratiotes* and some sudd communities. The sudd often accummulated in the north-eastern corner of the central basin, where it was carried by the prevailing south-westerly wind. This development was short-lived, and after the first year only 0.5% of the lake surface was covered by vegetation, mostly *Pistia* and a marginal fringe of *Echinochloa* (probably *E. pyramidalis*) spreading from the banks (Imevbore 1971, 1975). Although the euhydrophytes have practically disappeared, the *Echinochloa* fringes have continued to spread, as revealed by surveys of areas to the north-east and south-east of Foge Island in 1976 and 1977 using Land Satellite imagery. In a survey at high water in April 1977, Chachu (1979) found that an increasing area of the lake and its drawdown (28 and 40% of the lake surface from 1972 to 1983) was covered by *Echinochloa* (which they refer to the species *E. stagnina*).

Vegetation on the drawdown area is mainly influenced by the annual cycle of rise and fall in lake level (Hall 1975, Imevbore and Bakare 1974). Much of the drawdown area is colonized by essentially terrestrial species as the water level falls, successful colonization being enhanced by the onset of the rainy season at this time. This terrestrial vegetation is later killed with the coming of the white flood, though Morton and Obot (1984) show that *Echinochloa* seedlings also become established during the exposed phase. Chaudhry and Chachu (1979) showed that some species are more abundant on one type of geological formation than another, though these formations were only loosely associated with different soil types.

Dynamics of wetland vegetation

Four main kinds of change can be recognized in wetland vegetation: seasonal or short-term, long-term successional, changes resulting from major long-term climatic shifts, and those following large-scale human interference. There is very little information on changes in West African aquatic vegetation arising from long-term climatic changes, apart from some observations on Lake Chad relating to the differences between the 'Normal Chad' stage and 'Little Chad' stage. The large scale human interference referred to here include the damming of rivers to form impoundment lakes, the effects of which are mentioned earlier.

Seasonal changes are seen on all land that is subject to regular annual flooding, including the drawdown areas around man-made lakes. The alternating inundation and exposure is an important factor additional to the effects of the climatic seasons. Duong-Huu-Thoi (1950b) commented that one cannot talk of one plant community in such situations, but rather of two communities, one in the wet season and one in the dry season. As the water level rises in the flood season, terrestrial vegetation characterising the exposed phase of the floodplain is inundated and killed. Aquatic species rapidly cover the newly flooded areas, possibly stimulated by the release of soluble nutrients from the inundated soil. Some of these flood season plants are rooted (e.g. *Echinochloa stagnina*) whilst others are floating species (e.g. *Pistia stratiotes*) responding to an increase in available water surface. When the water level falls the aquatic vegetation is stranded and killed by desiccation. The newly exposed area is rapidly invaded by terrestrial species, especially grasses. An important factor determining the nature of the communities in the floodplain areas is the correlation between the cycle of water level changes and cycle of general climatic conditions in the region. For example, the water level of Lake Kainji rises during the dry season and falls in the rainy season (Chaudhry and Chachu 1979). This allows a rapid invasion by terrestrial plants in the low water phase and offers considerable agricultural potential.

The zonation observed in aquatic and semi-aquatic vegetation associated with rivers and closed inland water bodies has led some authors to regard the zones as stages in succession (e.g. Ake Assi 1977, Berghen 1982, Duong-Huu-Thoi 1950b). However, there have been no long-term studies of succession in West African wetland communities, and nobody has demonstrated that one association really does give way to another over a period of time in a sequence representing the classic hydrosere. One situation which undoubtedly does represent succession is sudd formation, though even this process has not been studied on a long-term basis in West Africa. Various stages can be found, ranging from pure colonies of pioneer species, such as *Cyperus papyrus*, *Pistia stratiotes* and *Vossia cuspidata*, to dense floating mats of vegetation containing semi-aquatic and even terrestrial species (Hall *et al.* 1969, Lawson *et al.* 1969, Okali and Hall 1974).

Economic aspects of wetland plants

Up until the mid-1980's, West African countries were fortunate to have very few aquatic weed problems in natural water bodies, and so far man-made lakes have remained more or less free of major weed infestations. Problems arising from physical effects of water weeds, such as blocking navigation channels and interfering with fishing, are mainly restricted to Lake Chad, the coastal delta of the River Niger, and lakes and lagoons in some coastal areas. Extensive sudd formation occurs especially in Lake Chad with *Cyperus papyrus* as the chief pioneer species. The species that has caused great problems on man-made lakes elsewhere in the tropics, namely *Eichhornia crassipes*, was until recently still rare in West Africa. Unfortunately *Eichhornia crassipes* is present and spreading in coastal lagoons, small impoundments and lakes in Ghana and Nigeria as well as intervening countries. It was apparently introduced into the region as an ornamental plant and has recently escaped (see Lowe 1987). Hall *et al.* (1969) indicate that the indigenous aquatics are of a kind that do not respond to impoundment by explosive growth. Imevbore (1975) suggests that the hydrology of Lake Kainji may be the most important environmental factor responsible for preventing the development of free-floating aquatic weeds within it. The lake has a short water retention time, with all the water changed about 4 times a year, resulting in a large and rapid drawdown during which time aquatic vegetation is stranded and killed on the extensive, seasonally exposed shoreline. Lake Kainji has, however, a potential problem with the spread of the marginal *Echinochloa* communities which, it is thought, could reduce the life expectancy of the lake by bringing about water displacement and an increase in silting.

Although man-made lakes in West Africa have not suffered the physical disadvantages of being covered with impenetrable floating vegetation, a more insidious problem has arisen and is associated with the relatively modest development of aquatic vegetation. This is the increased incidence of several human diseases in lakeside communities. There are many reports that the invertebrate vectors of some diseases, such as bilharzia and malaria, have increased in numbers due to the increase in aquatic vegetation (Betterton 1984, El-Zarka 1973, Grove 1985, Klumpp and Chu 1977, Obeng 1969b, Obei 1973, Paperna 1969, Petr 1968). Various methods of aquatic weed control have been suggested, but there has been no large scale investigation of suitable methods for use under West African conditions (Thomas and Tait 1984).

On the positive side, wetland vegetation can play an important role in agriculture and the fishing industry. Fish production is regarded as an important secondary function of man-made lakes (Obeng 1969a), and there have been various studies on the role of aquatic vegetation in increasing it (Dejoux 1983, Frempong and Nijjhar 1973, Obeng 1969b, Petr 1968). Aquatic vegetation provides shelter and breeding grounds for fish, as well as oxygenating the water and absorbing compounds from it. Some fish species feed directly on the macrophytes, whilst others feed on the epiphytic algae and rich invertebrate fauna to be found associated with them.

Marginal flood zones along major rivers are widely used for grazing cattle at low water, and some semi-aquatic grasses are gathered for use as fodder. Drawdown areas of newly created lakes now offer new grazing land and, possibly, new land for cultivation (Chaudhry and Chachu 1979, Kaul 1975). Such is the potential grazing and fodder value of the fringing *Echinochloa* populations around Lake Kainji that Chachu (1979) suggested control rather than eradication for what is seen as a troublesome weed. *Echinochloa pyramidalis* and *E. stagnina* both have a high nutrient content, and Rose Innes (1977) suggested that they could be exploited to greater advantage with good management. Morton and Obot (1984) estimate that a cropping rate of about 75% should allow *Echinochloa* populations around Lake Kainji to maintain their annual production and give a sustained yield. They have estimated the biomass of the above ground shoots of this annual grass as 10.83 t ha^{-1} in shallow water and 54.16 t ha^{-1} in water 8 m deep.

Flooded areas bordering West African rivers are widely used for rice production (Cook 1968, Dalziel 1937). The West African Rice Development

Association (WARDA) has investigated the possibility of increasing rice yields by using species of the floating fern *Azolla* as green manure (Hove *et al.* 1983). The nitrogen-fixing activity of the symbiotic cyanophyte *Anabaena azollae*, that lives in cavities within *Azolla* leaves, could raise crop yields and reduce fertilizer costs.

Future research

The major wetlands of West Africa are associated with the endorheic system of Lake Chad, upper floodplains of larger rivers, and the coastal riverine floodplain of the River Niger. A review of published information on the inland waters of West Africa (John 1986) draws attention to enormous gaps in our knowledge of these vast wetland areas. Many smaller wetland areas, such as those around lakes and along smaller rivers, are yet to be investigated (e.g., many listed in Nigeria by Ita *et al.* 1985). Information is especially lacking for the more inaccessible parts of the region (e.g. the Toubouris lakes). There is still much scope for descriptive information and collections of plants from wetland areas. Many of the descriptive accounts of wetland vegetation in West Africa are confined to a single period of the year, and thus little is known of seasonal changes. There are still few long-term studies on the vegetation in which dynamic aspects are considered.

The lack of basic information on dynamic and functional aspects of the region's wetlands means that it is difficult, if not impossible, to make meaningful proposals for their conservation and management. In spite of the dearth of basic floristic and ecological knowledge the region has seen the implementation of projects designed to exploit them. The projects are often based largely on principles established by studies on similar environments elsewhere in Africa or in other tropical areas. Dam construction for hydroelectric schemes on large rivers has had profound effects upon associated wetland habitats both upstream and downstream of the dam sites. Control of water movement and drainage by the construction of dykes and canals in 'reclamation' programmes aimed at the development of intensive agriculture has led, in the long-term, to loss of wetland habitat. Such reclamation inevitably leads to almost complete loss of productive fisheries and has far reaching effects on the wildlife of an area.

Wherever development projects have been completed there is a great need for continuing studies to monitor the effects on the environment. It is important to realize that any project can influence the fauna and flora over a very extensive area. For instance, more information is required on the floristic changes occurring downstream of dam sites due to the controlled flow of the river. Downstream wetlands inevitably disappear unless some management policy is adopted to cause flooding, perhaps by providing channels or deepening existing ones, in order to maintain the natural complex structure and zonation. The most productive wetlands, for fisheries and wildlife, are those with plentiful feeding areas and refugia for the fauna (see Howard-Williams and Thompson, 1985, p. 222).

A fruitful field of research might be to consider ways of making use of the seemingly prolific natural productivity of aquatic plants. It has been suggested, for example, that if wetlands used for grazing cattle are converted to cultivation, then deep water aquatics could be harvested to provide fodder and so compensate for the lost grazing pasture. Aquatic grases such as *Echinochloa* species are already grazed by cattle at low water as well as being used as fodder. These grasses are also used in the region for thatching and in dyeing and soap making. Many aquatic and semi-aquatic macrophytes have a variety of uses; these are summarized by Newton (1986, Table 29). Much remains to be discovered concerning the ecology, productivity, chemical properties, etc. of aquatic plants.

South Africa

C.M. BREEN, J. HEEG AND M. SEAMAN

The nature and distribution of wetlands in South Africa reflect the semi-arid climate and the disparate seasonal and spatial distribution of precipitation (Figs. 32 and 33). Sixty-five percent of the country, the central and western sector, receives less than 500 mm of rain annually and twenty-one percent receives less than 200 mm. Only a comparatively narrow region along the eastern and southern coastline is moderately well watered (Department of Water Affairs 1986).

At the broadest level, wetlands can be separated on the basis of their association with river systems (river source sponges, marshes, swamps, and floodplains) and endorheic shallow depressions (pans) in the landscape which are usually not associated with rivers or streams of notable size (Noble and Hemens 1978).

River source sponges or mires are common in the mountainous region in the east and south of the country. They are seepage areas on slopes which are seasonally or perennially waterlogged and they play an important role in regulating runoff from catchments (Jacot-Guillarmod 1962). Vegetation is dominated by sedges, other hydrophilous angiosperms, and occasionally mosses. Peat of up to 10,000 years old has been found in some sponges although the water is of neutral pH. Marshes, locally termed vleis, develop in flat reaches of rivers where waterlogging occurs seasonally. They are widely distributed over the country except in the arid western region (Fig. 34). The emergent vegetation is dominated by *Phragmites australis*, *Typha latifolia*, *Scirpus* sp., *Cyperus* sp., and *Oryza longistaminata*. Perennially waterlogged wetlands (swamps) are uncommon and are confined to the high rainfall north-eastern coastal zone where they are dominated by *Cyperus papyrus* and swamp forest species such as *Ficus trichopoda*. Floodplains have restricted distribution because there are few large rivers. The best examples are found on the Limpopo, Pongolo, Mkuze, Gamtoos, Olifants, and Orange rivers (Fig. 34).

Endorheic pans occur extensively in the drier parts of the country as well

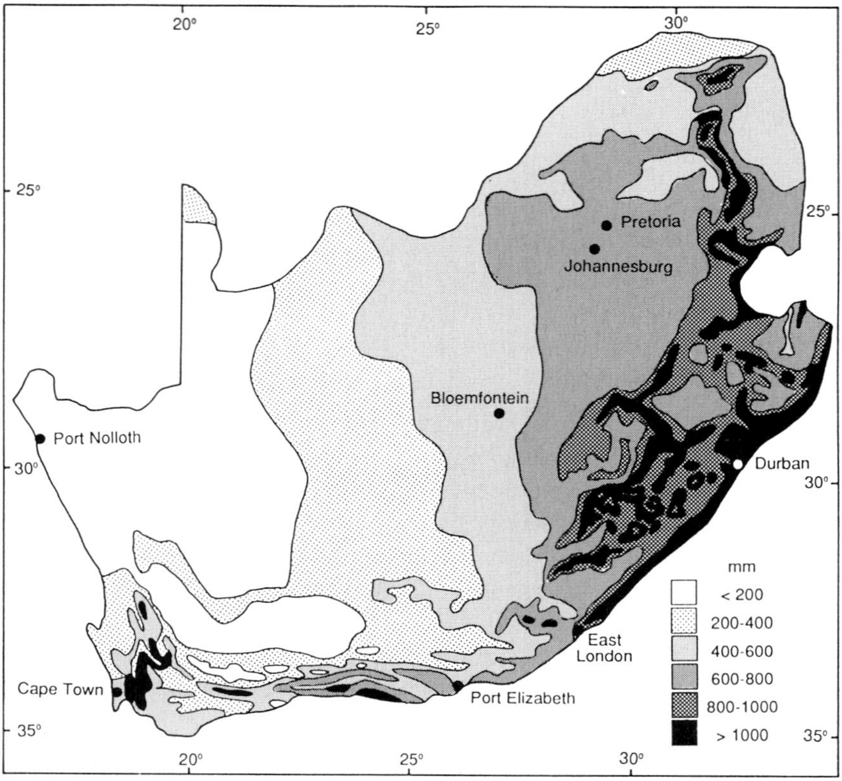

Figure 32. Mean annual precipitation over South Africa (adapted from Department of Water Affairs 1986).

as a small area in the wetter south-eastern Transvaal (Fig. 35). The origin of these oval depressions is uncertain but trampling by large herds of game and erosion during dry periods were probably important contributory factors. They are characteristic of the Kalahari, the western and north-central Orange Free State, and western and south-eastern Transvaal, forming the so called pan belt of southern Africa (Noble and Hemens 1978).

The Commission of Enquiry into water matters in 1970 recognized for the first time in South Africa, that water was required for environmental management (Department of Water Affairs 1986). The Commission considered that Lake St. Lucia, a wetland of international significance, and the Kruger National Park were the only two cases where water was required for management and it was estimated that 220,000,000 cubic metres per year would be required. Since then, however, there has been growing acceptance of the view that, in the utilization of the water resources of South Africa, provision must be made for the reasonable needs of nature conservation.

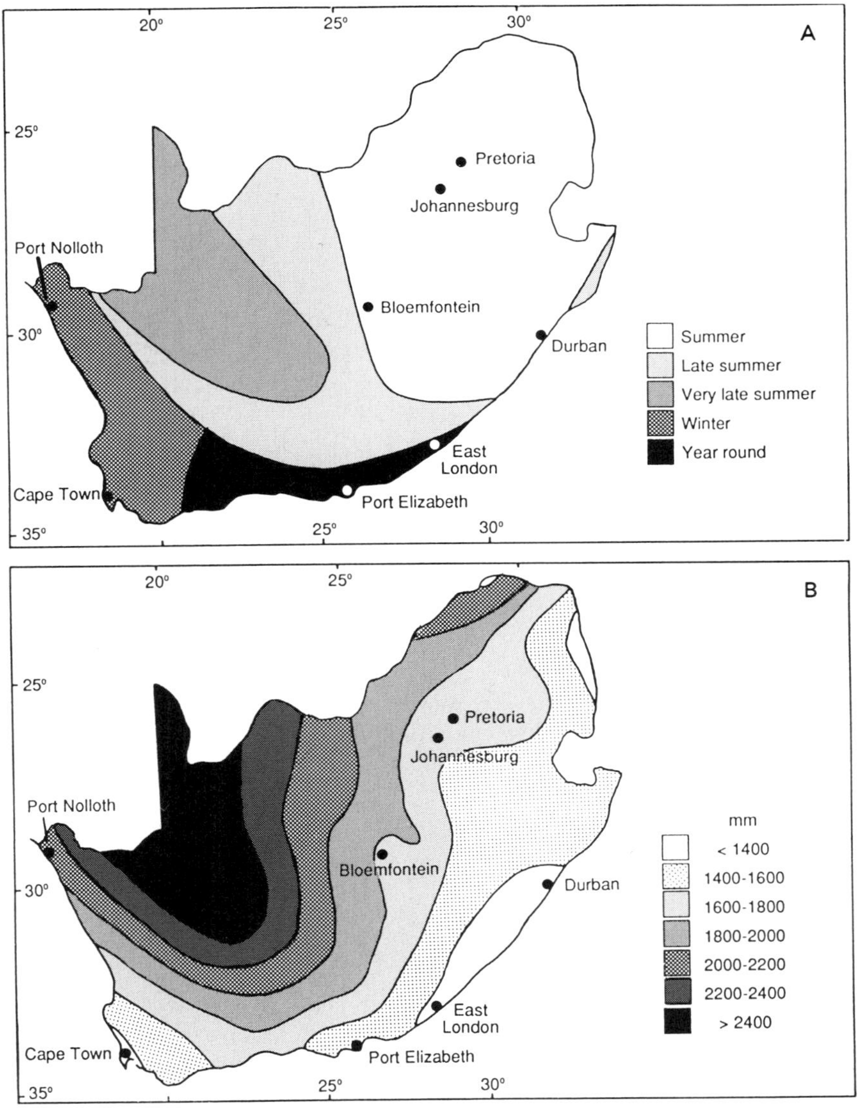

Figure 33. Seasonal rainfall regions (A) and mean annual evaporation (B) over South Africa (adapted from Department of Water Affairs 1986).

Although the total freshwater requirements of wetlands, lakes, and estuaries have yet to be defined it has been estimated that it amounts to 5% of the virgin mean annual runoff of rivers selected for study and could be as high as 15% of the utilizable resources. The water requirements for environmental

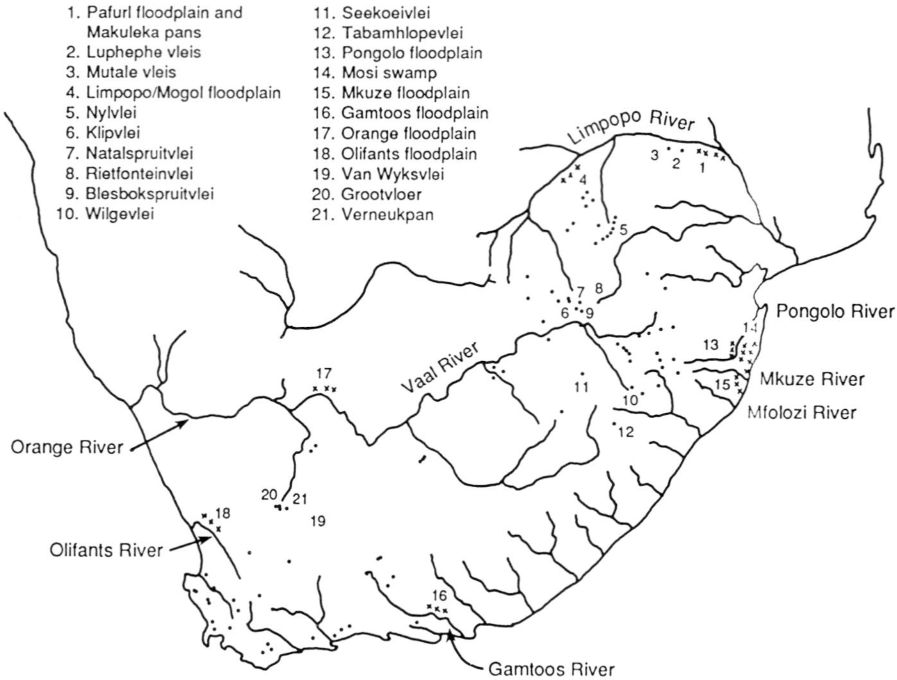

Figure 34. Major marshes, locally termed vleis (dots) and floodplains (crosses) in South Africa (adapted from Noble and Hemens 1978).

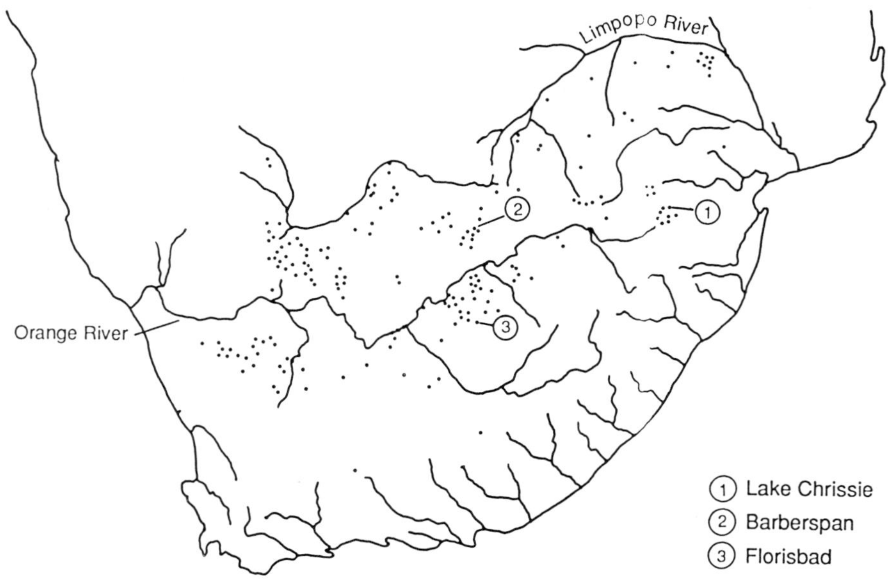

Figure 35. Endorheic pans and lakes in the interior of South Africa (adapted from Noble and Hemens 1978).

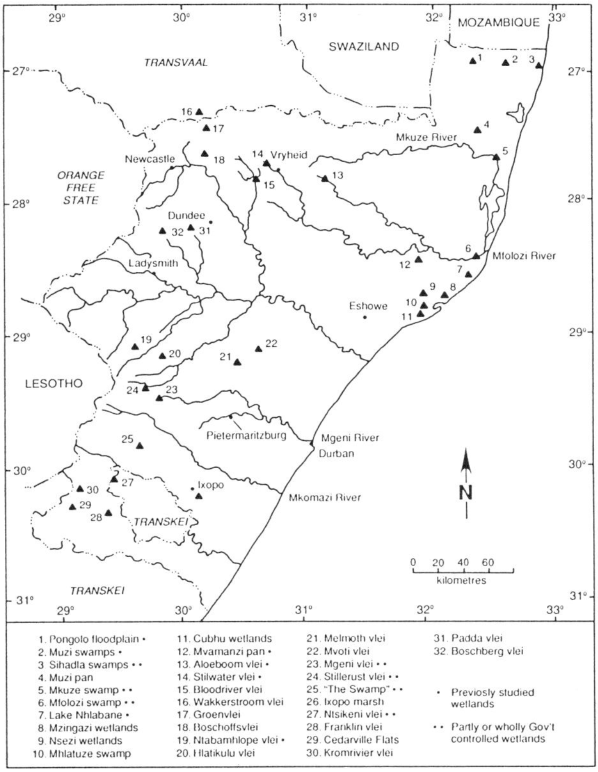

Figure 36. Major wetlands of the Natal region of South Africa (adapted from Begg 1989).

management have been estimated for each of the 22 drainage regions in South Africa (Table 5).

Wetlands in South Africa have been extensively modified by agriculture. By referring to soil maps of the Tugela Basin compiled by van der Eyk *et al.* (1969), and noting the location of hydromorphic soils, it was shown that 16% of this catchment was covered by wetlands prior to human influence. However, at the time of the survey in 1965, 34% had been destroyed by overgrazing and sheet erosion, or drained by gulley erosion (Scotney 1978). These losses may have doubled by now. In a comprehensive survey of the 10,000 km^2 Mfolozi catchment, Begg (1988) estimated that 58% of the original wetland has been lost, and that only 2% of the catchment is presently occupied by wetland.

Begg (1989) has described the location, status and functions of the priority wetlands of Natal (Fig. 36). The twenty-four priority wetlands cover 1,114 km^2 and include vleis in the headwater regions of rivers and swamps in the lower reaches. Approximately 65% of the priority wetlands are privately

Table 5. Estimated water requirement for environmental management (total of estuaries, lakes and nature conservation). Values are million cubic metres per year (modified from Department of Water Affairs 1986).

Primary drainage region	Major rivers	1980	1900	2000	2010
A	Crocodile	7	8	9	10
B	Olifants	46	47	48	49
C	Vaal	40	40	41	41
D	Orange	550	552	554	556
E	Olifants	77	77	77	77
F	Groen	1	1	1	1
G	Great Berg	143	143	143	143
H	Breede	149	149	149	149
J	Gouritz	40	40	40	40
K	Krom	70	70	70	70
L	Gamtoos	37	37	37	37
M	Swartkops	19	19	19	19
N	Sundays	17	17	17	17
Q	Great Fish	38	38	38	38
R	Buffalo	51	51	51	51
S	Great Kei	76	76	76	76
T	Mzimkulu	742	742	742	742
U	Mgeni	134	134	134	131
V	Tugela	230	230	230	230
W	Mfolozi	411	411	411	411
X	Komati	47	47	47	47
	TOTAL	2,946	2,949	2,954	2,958

owned, for example Blood river vlei, which is 6,640 ha in extent, has 54 different farmers owning part of the wetland. None of the privately owned wetlands have generally accepted management plans and all show degradation consequent upon the individualistic action of one or more landowners. Restoration procedures are currently being implemented by the Department of Agriculture (Directorate of Resource Conservation). The principal characteristics of these systems and a qualitative assessment of their functions and values are presented in Tables 6 and 7.

Policy proposals for the wetlands of the north-eastern sector of South Africa (Natal and KwaZulu) have been drafted and accepted by state departments, conservation agencies and other interested parties (Begg 1990). Issues addressed in the proposals are wetland definition, policy options, incentives for private protection of wetlands, improving government protection, sustainable use, restoration and creation of wetlands and policy implementation. These proposals will, in due course, form the basis of a national policy for wetlands.

Summary accounts of wetlands in Natal, the Cape, the Transvaal and the Orange Free State are given below. Publications of broad interest include *African Wetlands and Shallow Water Bodies: Directory* (Burgis and Symoens

1987), *Lake Sibaya* (Allanson 1979), *Studies on the Ecology of Maputaland* (Bruton and Cooper 1980), *Biogeography and Ecology of Southern Africa* (Werger 1988), *Perspectives in Southern Hemisphere Limnology* (Davies and Walmsley 1985), and *Wetlands of Natal* (Begg 1986) *Ecology and Conservation of Wetlands in South Africa* (Walmsley and Botten 1987). Those of a more specific nature by a number of authors including: B. R. Allanson, C. M. Breen, H. D. Furness, J. Heeg, and K. H. Rogers can be found in the list of references. A recent review by Whitlow (1985) provides a valuable insight into the Dambos in Zimbabwe.

Wetland types

Floodplain wetlands

Extensive floodplains are uncommon in South Africa because most of the rivers are short with low mean annual run off. A few rivers have short floodplains developed in their middle or lower reaches where the river attains grade and floods overtop the banks inundating the area on either side (Fig. 34). Noble and Hemens (1978) recognized three types of floodplain: Karoo salt flats, Floodplain vleis, and Storage floodplains. The former are very similar to another type of wetland locally referred to as pans and have not been studied in any detail. Summary accounts of the Pongolo and Mkuze river floodplains (storage floodplains) and the Nyl river floodplain (floodplain vlei) are presented.

The Pongolo river floodplain. One of the better studied wetlands in southern Africa is the Pongolo floodplain which lies within the Maputaland Plain. The floodplain is situated between latitudes 25°50′S and 27°26′S and longitude 32°04′E and 32°18′E (Figs. 34 and 37). It has a slope of 1: 3,000 and occupies an area of about 3,000 ha. Over its length of about 50 km it varies in width between 0.8 and 4.8 km. The characteristics of the area are fully described in Heeg and Breen (1982).

1. *Geology and geomorphology*. The Maputaland Plain lies to the east of the Lebombo Mountains between longitudes 32°E and 33°E and latitudes 26°50′S and 27°50′S. During the Pleistocene, sandy material which forms the Port Durnford beds was deposited over the nearly flat surface of Miocene rocks. This was followed by general lowerings of sea levels causing the coastline to shift progressively to the east. Where the shoreline was located for any length of time through the sea level being static, a system of longshore dunes developed. It is these dunes with their typical north-south orientation, subsequently much modified by wind action to produce the sands of recent ages, which characterize the landscape of the Maputaland Plain. The sea level change also affected flow velocities of the rivers of the area causing

Table 6. An overview of selected characteristics of the priority wetlands of the Natal region.

Name of wetland	Lat. ° ′S	Long. ° ′E	Area (ha)	River system	Catchment name	Catchment size (km^2)	Mean annual run-off ($m^3 \times 10^6$)
Pongolo flood plain	27 13	32 14	13,000	Pongolo	Pongolo	7,831	1,082
Muzi swamp	27 05	32 35	15,000	Muzi	Maputo	N/R	N/R
Mkuze swamp system	27 41	32 30	42,000	Mkuze	Mkuze	4,800	236
Mfolozi swamp	28 29	32 18	9,059#	Mfolozi	Mfolozi	10,075	887
Aloeboom vlei	27 50	31 06	142	Black Mfolozi	Mfolozi	48	6
Mvamanzi pan	28 25	32 01	390	Mvamanzi	Mfolozi	134	4
Stilwater vlei	27 47	30 44	1,828	White Mfolozi	Mfolozi	117	13
Mhlatuze swamp system	28 48	31 49	5,557#	Mhlatuze	Mhlatuze	4,170	620
Blood river vlei	27 49	30 34	6,540	Buffalo	Tugela	557	54
Padda vlei	27 09	30 02	912	Wasbank	Tugela	57	5
Boschoffs vlei	27 40	30 14	1,850	Dorpspruit	Tugela	526	78
Groen vlei	27 27	30 11	762	Slang	Tugela	269	38
Wakkerstroom vlei	27 21	30 08	1,000	Thaka	Tugela	207	30
Melmoth vlei (1)	28 18	30 16	104	Myamvubu	Tugela	4	N/R
Hlatikulu vlei	29 15	29 41	733	Nsonga	Tugela	150	41
Boschberg vlei	28 15	29 49	1,400	Sundays	Tugela	196	18
Ntabamhlope vlei	29 03	29 39	295	Little Bushmans	Tugela	34	3
Stillersust vlei	29 03	29 44	225	Mooi	Tugela	116	33
Mvoti vlei	29 09	30 35	2,800	Mvoti	Mvoti	316	40
Mgeni vlei	29 29	29 49	270	Mgeni	Mgeni	11	N/R
Franklin vlei	30 17	29 27	5,244	Mzintlava	Mzimvubu	377	38
Kromrivier vlei	30 15	29 13	1,087	Tswilika	Mzimvubu	288	41
Ntsikeni vlei	30 08	29 38	1,114	Lubhukwini	Mzimkulu	75	22
"The Swamp"	29 47	29 36	115#	Pholela	Mzimkulu	230	74
		Total	111,427				

(1) = Selected as the most important of the Myamvubu vlei systems.
= Extant portion only.
N/R = No record.
(+) = Only a small portion of the system is under this form of ownership.
(K) = Tribal authority (KwaZulu).
(I) = Transfer Government.
(WM) = Wakkerstroom Municipality (Transvaal).

them to deposit alluvial material at successively different levels. The alluvium now forms river terraces whilst the infilling forms the present day Pongolo floodplain. Marine cretaceous deposits underlie the floodplain and the groundwater is saline. Lakes which receive seepage can become quite saline (<500 to $>5{,}000$ $\mu S\ cm^{-1}$) during the dry winter season but summer floods flush them out and replenish them with low-conductivity, turbid water (Heeg *et al.* 1978). Lakes which are incompletely flushed clear rapidly because of flocculation brought about by residual ionic concentration (Akhurst and Breen 1988).

Table 6. Continued. Functions and values are given in *Table* 7 (from Begg 1989).

Altitude at outlet (m a.s.l.)	Approx. perimeter (km)	Average width (m)	Approx. length (km)	Landownership			Most characteristic *genera* or family of vegetation
				Private	State	Communal	
20	216	1,344	54		*	* (K)	*Cyperus*, *Echinochloa*
30	586	1,123	45			* (K)	*Cyperus*, *Digitaria*
5	364	3,542	45	(+)	*	* (K)	Papyrus, *Phragmites*
1	312	6,800	28	*	*		Papyrus, *Ficus*
1,135	19	183	6	*			Juncaceae, Poaceae
31	32	312	7			* (K)	*Potamogeton*, *Cynodon*
1,150	84	866	11	*			Gramineae
2	471	N/R	21	*	(+)	* (K)	Papyrus, *Barringtonia*
1,183	334	1,330	18	*			*Phragmites*, Poaceae
1,330	48	430	9	*			Poaceae
1,167	80	875	12	*			Poaceae
1,740	50	430	6	*			Cyperaceae
1,737	56	726	8	(+)		* (WM)	*Phragmites*, *Typha*
1,595	8	280	2	*			Poaceae
1,561	56	275	7	*			Cyperaceae/Poaceae
1,250	88	297	8	*			Poaceae
1,440	42	260	7	*	*		*Phragmites*, *Typha*
1,631	40	724	2		*		Poaceae
6954	192	720	19	*	*		*Phragmites*
1,828	13	755	4	(+)	*		*Carex*
1,498	340	652	32	*			*Phragmites/Poaceae*
1,627	88	394	11	*			*Poaceae*
1,752	81	690	11		* (*T*)		*Carex*, Poaceae
1,460	5	495	2	*	*		Poaceae

2. *Climate*. The climate of the Plain is described as warm to hot, humid subtropical. It receives some rain throughout the year but the winters are distinctly drier than the summers. The mean annual rainfall for seven weather stations on the Pongolo floodplain varies from 485 mm to 642 mm with an overall average of 574 mm (Heeg and Breen 1982). The plain is frost free and has high summer temperatures which can rise to over 40°C. The whole area is subjected to considerable wind, particularly from September to December when the average daily run is 230 to 240 km day^{-1}. The high temperatures and wind runs contribute to a high rate of evaporation.

3. *Vegetation*. The floodplain includes some riparian forest, rapidly draining hygrophilous grasslands, marshes, and a series of depressions (Fig. 38) which capture and retain water when the river overflows its banks (Furness and Breen 1980). There are about 90 small lakes with a collective area of about

Table 7. A qualitative assessment of the functions and values of the priority wetlands of the Natal region according to the benefits that accrue from existing land uses (from Begg 1989).

	Water storage	Streamflow regulation	Flood attenuation	Sediment trapping	Waste assimilation	Wildlife protection	Recreation	Agriculture	Silviculture	Most rational management options	
										Total protection	Multi-use objectives
Pongolo floodplain	3	2	3	3	2	3	2	3	0	(*)	*
Muzi swamp	3	2	3	3	2	2	1	2	0	(*)	*
Mkuze swamp system	3	2	3	3	2	3	1	1	0	(*)	*
Mfolosi swamp	2	1	3	3	1	3	1	3	0	(*)	*
Aloeboom vlei	2	2	2	2	3	7	0	2	0		*
Mvamanzi pan	2	1	2	2	1	3	0	2	0		*
Stilwater vlei	3	2	3	2	1	2	0	2	0		*
Mhlatuze swamp system	3	3	3	2	3	2	2	1	0	(*)	*
Blood River vlei	3	3	2	2	1	3	2	2	0		*
Paddavlei	2	2	2	1	1	1	0	2	0		*
Boschoffsvlei	1	1	2	—	1	1	0	2	0		*
Groenvlei	1	1	2	2	2	1	0	2	0		*
Wakkerstroom vlei	3	2	2	1	1	2	0	2	0		*
Myamvubu vlei systems	2	2	1	1	1	3	1	1	0	(*)	*
Hlatikulu vlei	2	2	2	1	1	2	0	2	0		*
Boschberg vlei	2	2	2	2	2	2	0	2	0		*
Ntabamhlope vlei	3	3	2	2	1	2	0	1	0	*	
Stillerust vlei	1	1	1	0	0	2	1	0	0	*	
Mvoti vlei	3	3	2	2	1	2	1	2	2	(*)	*
Mgeni vlei	2	3	1	2	2	3	1	0	0	*	
Franklin vlei	2	2	2	2	2	3	1	3	0		*
Kromrivier vlei	2	2	3	2	2	1	1	2	0		*
Ntsikeni vlei	3	3	2	2	2	3	1	1	0	*	
"The Swamp"	0	0	I	0	0	2	1	0	0	*	

*Provisional rating:
0 = unimportant,
1 = low value,
2 = moderate value,
3 = high value,
(*) = certain parts only,
(*) = preferred option.

2,200 ha when the floodwaters subside. Most are shallow and support dense growths of euhydrophytes during the winter months (Musil *et al.* 1973, Rogers and Breen 1980).

Six communities have been recognized on the floodplain and are grouped according to their relative periods of exposure and inundation (Fig. 38). The contribution of each community to the vegetation cover on the floodplain is

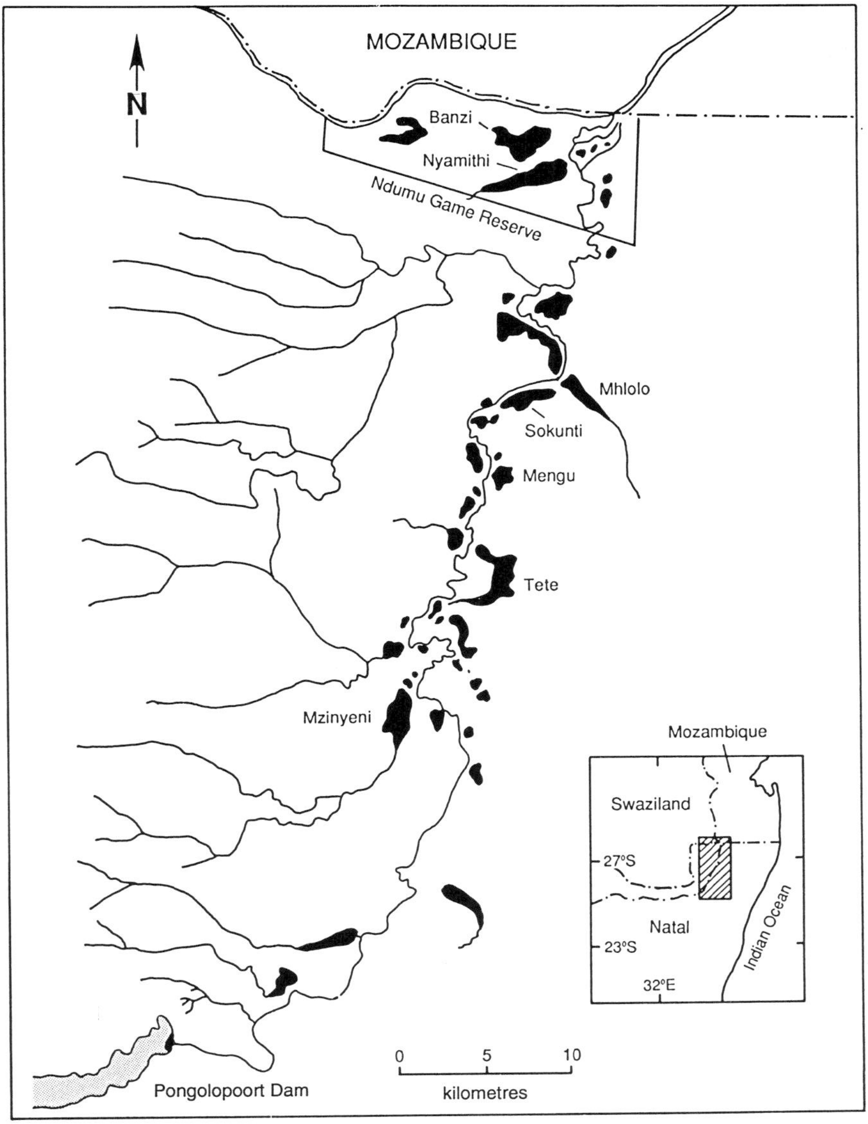

Figure 37. The Pongolo river floodplain showing the major floodplain lakes and the Pongolapoort dam.

illustrated in Table 8. Approximately 42% of the floodplain is covered by undisturbed vegetation while the remaining 58% is made up of areas disturbed principally by cultivation. Disturbance has increased notably in recent times (Wolansky and Roberts 1989).

1. The *Acacia xanthophloea – Dyschoriste depressa* community occurs near

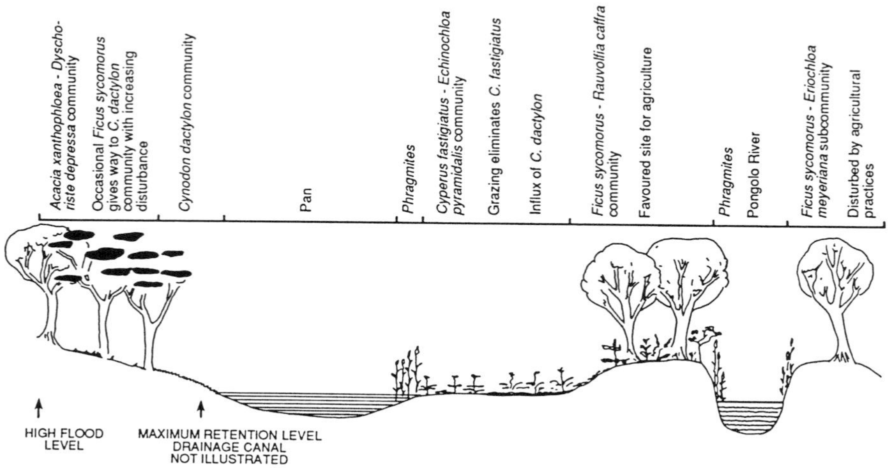

Figure 38. Diagrammatic section across the Pongolo river floodplain showing the distribution of the different plant communities (A) and the relationship of the communities to periods of exposure and inundation (B) (adapted from Furness and Breen 1980).

the outer edge of the floodplain under drier conditions and occupies a total area of about 128 ha.

2. The *Ficus sycomorus – Rauvolfia caffra* forest community grows along the levees of the Pongolo and Usutu rivers. The *Ficus* may reach heights of 18 m. Other large trees in the community include *Syzygium guineense* and *Trichilia emetica*. In the Ndumu Game Reserve, the forest occupies about 246 ha but outside the Reserve it has suffered from cutting and burning and only about 160 ha remain.
3. The *Cynodon dactylon* community occurs on areas which are alternately regularly inundated and exposed, and is especially well developed around shallow pans subjected to periodic flooding. Such 'lawns' or 'meadows' account for about 171 ha of floodplain.
4. *Cyperus fastigiatus – Echinochloa pyramidalis* community tends to occupy marshy areas rather than pan margins and tolerates longer periods of inundation. It is a very widespread community and covers some 2,471 ha.
5. There are two *Phragmites* communities each with a distinct species. Both occur in the wettest areas with *Phragmites australis* having a preference for flat, swampy areas and *P. mauritianus* favoring river banks, inlet/outlet channels and pan margins where there is a fluctuation in water level. The two communities together account for approximately 234 ha of wetland.
6. The euhydrophyte communities which may either be permanent or seasonal, occur within the zone of colonization of open waters. Permanent communities consist mainly of *Trapa bispinosa* and various *Nymphaea*

Table 8. Areal cover of land forms and pan communities on the Pongolo river floodplain (modified from Furness and Breen 1980).

Landform and plant community	Area (ha)	% of total area
Total area	13,000	100
Seasonally flooded	10,800	83
Lake area	2,200	17
Plant communities		
Disturbed – cultivated	7,540	5
Acacia xanthophloea-Dyschoriste depressa	128	1.0
Ficus sycomorus-Rauvolfia caffra	246	1.9
Cynodon dactylon	171	1.3
Cyperus fastigiatus-Echinochloa pyramidalis	2,471	19.0
Phragmites australis-Phragmites mauritianus	234	1.8

species. They are best developed in those pans where the water level is not subject to extensive seasonal fluctuations. The seasonal communities consist largely of *Potamogeton crispus* and *Najas pectinata* and normally occur where a reasonable depth of water is still retained in the dry season. The turions of *P. crispus* are extremely sensitive to desiccation and large plant standing crops can only develop where the pans do not dry out. Large turions are selectively grazed by waterfowl which stimulates production of small turions which are less easily grazed. The result is a stable grazing system (Rogers 1984, Rogers and Breen 1990a,b). The longevity of plant lifespan is enhanced by removal of epiphyton by snail grazing, but once senescence sets in snail grazing is the principal process leading to detritus production (Rogers and Breen 1983).

The sequence of exposure and submergence and of production and decomposition of the floodplain plant communities provides a continuous source of detritus for aquatic organisms (Furness and Breen 1982, Rogers and Breen 1982, Buchan and Breen 1988).

4. *Avifauna*. The floodplain supports a variety of birdlife which utilizes it as a feeding or breeding habitat. Ducks and pelicans feed on the pans during winter and spring, but their degree of dependence on the system cannot readily be assessed. A total of 30 endangered bird species included in the South African Red Data Book (Siegfried *et al.* 1976) are known to occur on the floodplain. The following water birds are known to occur on the floodplain: white pelican (*Pelecanus onocrotalus*), Goliath heron (*Ardea goliath*),

rufous-bellied heron (*Butorides rufiventris*), white-backed night heron (*Gorsachius leuconotus*), open-bill stork (*Anastomus lamelligerus*), yellowbilled stork (*Mycteria ibis*), woolly-necked stork (*Ciconia episcopus*), greater flamingo (*Phoenicopterus ruber*), lesser flamingo (*Phoenicopterus minor*), white stork (*Ciconia ciconia*), African fish eagle (*Haliaeetus vocifer*), lesser Jacana (*Microparra capensis*), fishing owl (*Scotopelia peli*), white-winged plover (*Vanellus crassirostris*), Caspian tern (*Hydroprogne caspia*).

5. *Use and management*. The population resident around the floodplain is increasing at 4.5% per annum and is expected to double every 15 years. Increasing environmental degradation by over-exploitation of resources on the floodplain and adjoining areas is clearly evident. Use of floodplain resources centers around cultivation, grazing, fishing and collection of water and building materials.

Cultivation is the dominant activity and most families work fields on and off the floodplain to minimize the risk of crop failure from floods and droughts. The area of land cultivated on the floodplain almost doubled since 1970 and has increased by 10% in the areas surrounding the floodplain. Produce (maize, sugar cane, mangoes, tomatoes, and pumpkins) are marketed locally and some is resold further afield. Accurate estimates of the economic returns from agrarian practices are not available.

The numbers of domestic stock (principally cattle) have increased markedly since eradication of Ngana and the improvement of veterinary services during the first half of this century. Currently, about 18,500 cattle use the floodplain for grazing, yielding an annual return to the local economy worth about R 1,755,000. Only 4% is realized through sale of stock while most (86%) reflects the value of milk produced and consumed by the families owning stock. The remainder (10%) reflects the provision of draft power for which cash exchange may take place (Buchan 1988).

Only 16% of the population own cattle and with both cultivation and livestock increasing there is growing evidence of conflict. Because most of the floodplain fields are either not fenced or fenced inadequately, cattle have to be herded continuously and are usually removed from the floodplain after fields have been planted, notwithstanding that better grazing is available on the floodplain. Favoured grazing areas are the stands of *Echinochloa*, *Cynodon*, and *Phragmites* (Furness and Breen 1985, Buchan 1988).

Conflict also arises in the demands for release of water from the dam. Agriculturalists desire floods before the advent of summer rains (September) at a time when access to grazing on the floodplain is most important.

The lakes support a wide diversity of fish and it is estimated that stocks could yield about 500 tons of fish per annum. The principal fishing techniques are mono trapping in channels during floods, gill netting, and thrust-basket fishing when lake depths are shallow. Several fish species are flood-dependent spawners and are sexually immature in September. They require summer floods (October – February) for successful breeding. During the severe drought (1981–1984) those lakes which retained water developed an icthyofauna dominated by cichlids. Diversity of the fauna was reinstated after the floods by immigration from lakes in the Ndumu Game Reserve and in Mozambique.

Aquatic macrophyte communities have recovered slowly after the drought. Stands of *Nymphaea* redeveloped very quickly from tubers buried in the mud and *Najas pectinata* developed from seeds. *Potamogeton crispus* has, however, been extremely slow to recolonize.

Construction of the Pongolopoort Dam (Fig. 37) for irrigation upstream of the floodplain now permits regulation of downstream floods. From the agricultural viewpoint it is considered that floods before the summer rains (September) would be most beneficial, but it would restrict grazing at a critical time, adversely affect submerged plant production (and hence waterfowl), and would reduce fish spawning. Therefore, it is proposed to continue with a pattern of summer floods and attention is being directed to analysis of optimal flood height and duration and the development of models which will facilitate prediction of stage heights and duration and the biotic responses (Drewes 1988, Slinger 1988). A system of locally elected water committees facilitates reaching agreement on the ration of water from the upstream impoundment and to inform local people of decisions reached.

The Mkuze wetland system. The Mkuze river cuts across part of the Maputaland coastal plain before turning south and discharging into Lake St Lucia (Fig. 34 and 39). Sediment carried by the Mkuze river is deposited when the river attains grade on the coastal plain. In its passage across this plain this river has cut through the north-south orientated dune ridges and superimposed the fluvial landforms characteristic of a floodplain, thereby enhancing the diversity of wetland systems. Alluviation and hard pan development have created impervious layers in the dune slacks so that they retain water and form a mosaic of wetlands of varying degrees of depth and duration of period of water retention.

The coastal plain has a high rainfall and water drains from an east-west divide southwards between the dune ridges, forming a number of streams with varied hydrology. In the east, where the rainfall is highest (Maud 1980), the Mbazwane stream flowing south is dammed by levees formed by the Mkuze river, adding further to the complexity of the system locally referred to as the 'Mkuze Swamps'. The wetland system includes a mosaic of open water, swamp forest, marshes, floodplain, and hygrophilous grasslands covering an area of about 40,600 ha (Fig. 40).

1. *Hydrology.* The hydrology of the system is incompletely known. Many

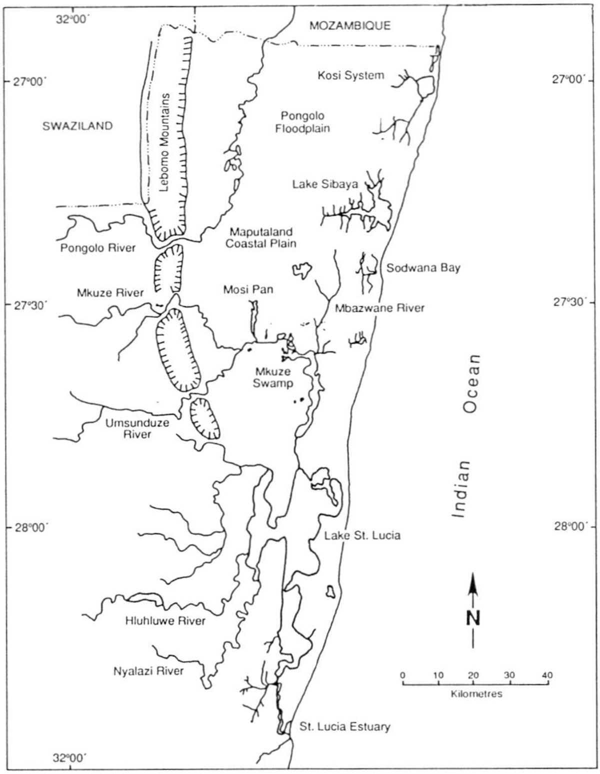

Figure 39. The southern sector of the Maputaland coastal plain showing the location of the Pongolo River Floodplain, the Mkuze Swamp System and Lake St Lucia.

of the small wetlands in the dune slacks are replenished by local rainfall but the major part of the system receives its water from the Mkuze river (Hutchinson and Pitman 1973), which carries a high silt load, and the much smaller Mbazwane river, which is not gauged and which drains the leached sandy coastal plain.

2. *Vegetation.* Fourteen plant formations dominated by emergent species have been recognized (Fig. 41) and their distribution is determined by processes such as subsidence, channel switching, siltation, and salinization. Monospecific stands of *Sporobolus virginicus* occur in the southern part of the system where saline water from Lake St Lucia pushes up into the wetlands. *Hyphaene coriacea* occurs in the grasslands fringing the wetlands where the water table is high. The *Cyperus natalensis*, *Scirpus nodosus*, and *Cyperus corymbosus – Ischaemum arcuatum* communities occur on sandy substrata in the dune slacks that are not under the influence of inorganic sediment from the Mkuze river.

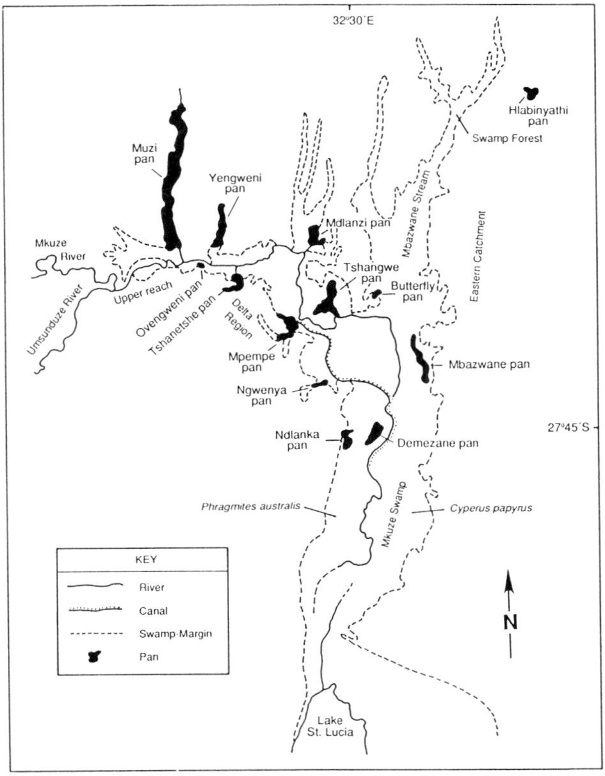

Figure 40. Mkuze Swamp System showing the extent and the canal from Mpempe pan.

Swamp communities dominated by floating mats of *Cyperus papyrus* (Breen and Stormanns 1991) and *Ficus trichopoda* are best developed in drainage lines receiving sediment-free water from the sandy coastal plain. In the west, where fluvial landforms characteristic of floodplains dominate, communities similar to those on the Pongolo river floodplain occur. These include the *Phragmites australis*, *Phragmites mauritianus*, *Typha latifolia*, *Acacia xanthophloea*, *Cynodon dactylon*, *Echinochloa pyramidalis*, *Cyperus immensus*, and the *Ficus sycomorus – Rauvolfia caffra* communites (Figs. 38 and 41).

3. *Vertebrate fauna*. The avifauna of the Mkuze wetland system, and its immediate surrounds is remarkable for the number of rare species it contains and the fact that in most cases these populations represent the largest or only concentration of the species in Natal and South Africa (Johnson 1986). Species which are associated with wetlands are: white pelican (*Pelecanus onocrotalus*), pinkbacked pelican (*Pelecanus rufescens*), rufous-bellied heron (*Butorides rufiventris*), woolly-necked stork (*Ciconia episcopus*), open-bill

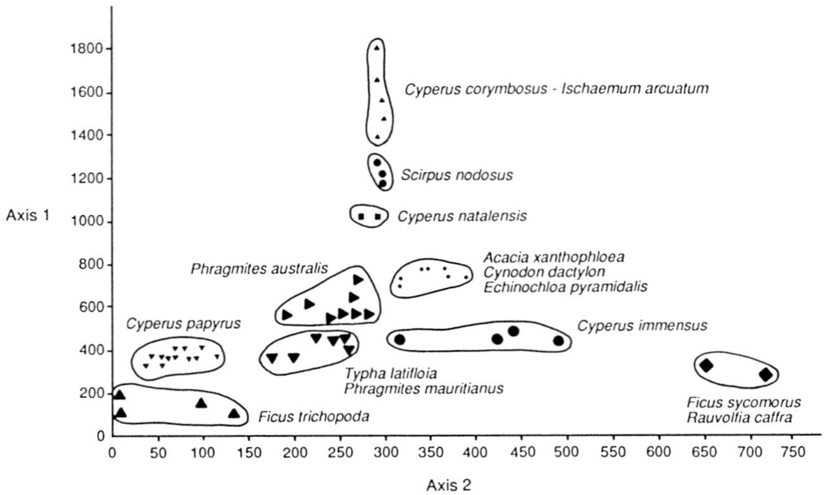

Figure 41. Two dimensional ordination using detrended correspondence analysis (DECORANA) showing the principal plant communities of the Mkuze Wetland system. Refer to text for interpretation (from Stormanns and Breen 1987).

stork (*Anastomus lamelligerus*), yellowbilled stork (*Mycteria ibis*), saddlebilled stork (*Ephippiorhynchus senegalensis*), greater flamingo (*Phoenicopterus ruber*), lesser flamingo (*Phoenicopterus minor*), African finfoot (*Podica senegalensis*), pygmy goose (*Nettapus auritus*), lesser Jacana (*Microparra capensis*).

Also noteworthy is that it is the only recorded breeding site in Natal for the whiskered tern (*Chlidonias hybridus*). The reptiles and amphibians of the system are incompletely known but Bourquin (1986) suggests that it is the southern-most major wetland habitat for a great variety of amphibians (78% of the known Natal species), all the known Natal terrapins, 40% of the snake species known in Natal, and a small percentage of lizards. Among these are included the following 'red-data' species: *Varanus niloticus*, *V. exanthematicus*, *Bradypodion setaroi*, *Python sebae*, *Lycophidion semiannule*, *Natriciteres variegata*, and *Dasypeltis medici*. It is a potentially important habitat for crocodiles (*Crocodylus niloticus*) as it provides freshwater habitats when salinities rise in Lake St Lucia.

4. *Use and management.* The largest portion of the swamp system has recently come under the control of the provincial conservation agency, the Natal Parks Board. Most of the wetland to the west and south of the Mkuze river is under the jurisdiction of the KwaZulu administation and is not conserved. Local inhabitants fish in the lakes and cultivate crops for local consumption on the levees and around the periphery of the system. Cattle graze extensively on the floodplain and in the wetlands to the north. Plant

resources such as reeds and palm sap are harvested on a limited scale.

Hypersaline conditions in Lake St Lucia in the 1960's led to the dredging of a 13.5 km canal southwards from the Mpempe pan. The purpose was to allow water from the Mkuze river to bypass the swamps so that more would reach the lake. The canal has eroded the Mpempe pan and drained a large wetland area to the west. It has also reduced flooding and recharging of swamps and the pan areas (Goodman 1987).

Since 1980, the local commercial farmers have seriously modified the functioning of the system by illegal activities including clearing of riverine forest, diversion of river flow, excavation of canals, and raising the overflow level of pans. The consequences of these actions are documented by Goodman (1987). Remedial measures are being evaluated at present.

The Nyl river floodplain. The Nyl River floodplain (Fig. 42) lies at approximately 20°30′S and 25°30′E and is an example of a floodplain vlei. These systems lack the levees and pans which characterize storage floodplains such as those of the Pongolo and Mkuze rivers and the floodwaters simply inundate areas bordering the main channel. The Nyl river floodplain is probably the largest floodplain vlei in South Africa and during exceptional years, 16,500 ha becomes inundated, although more usually it is between 8,000 and 9,000 ha. The system has been described by Tarboton (1979, 1987).

1. *Geology and geomorphology.* The floodplain is underlain over much of its course by Stormberg basalts of the Karoo Sequence whereas the catchment areas in the adjacent Waterberg hills are underlain by felsites of the Rooiberg Group and coarse-grained sediments of the Waterberg Group. The formation of the floodplain in which the subsiding Karoo beds on the downthrust side produced a basin-type structure in the Nyl river's path, led to a reduced flow rate and the deposition of alluvium (Wagner 1927).

2. *Hydrology.* Run-off is strongly seasonal with most (85%) occurring in summer. Maximum run-offs occur between November and May, but the peak run-off month can vary annually. The Groot Nyl has the most sustained winter run-off and Badseloop the least.

The amount of flooding has been subjectively assessed according to the area flooded (limited, moderate, or extensive), the duration of the flood (brief = dried out in 3–4 months, sustained = water persisted through to next rainy season), and the timing of the flood (early = November–December, mid-season = January–February, late = March–April). In the period between 1960–61 and 1986–87, extensive and sustained flooding occurred in about two out of three years but sustained flooding only in about one year in three. A breakdown of the 27 year period by incidence of flooding is: extensive, sustained, early = 3; extensive, sustained, mid-season = 3; moder-

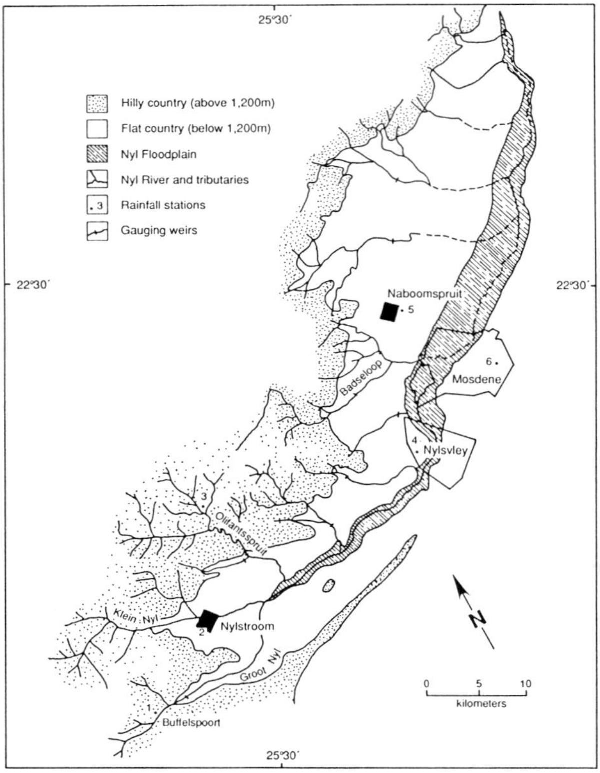

Figure 42. The location and major features of the Nyl river floodplain.

ate, sustained, early = 1; moderate, sustained, mid-season = 2; moderate, sustained, late = 1; moderate, brief, early = 4; moderate, brief, mid-season = 1; limited, brief, early = 2; no flood = 10.

3. *Vegetation*. The vegetation of the system has been incompletely described. Extensive stands of *Oryza longistaminata* occur throughout the floodplain. In deeper water habitats species of *Nymphaea*, *Polygonum*, *Potamogeton*, and *Ludwigia* are common. The emergent *Phragmites* also forms extensive stands.

4. *Fauna*. The birds are the only animals which have been studied. One hundred waterbird species were recorded on the floodplain during the period 1941–1987. Eleven are considered resident, and 34 are vagrants. Fifty-five species are visitors and fall into the following categories: regular visitor, breeds (common 22, fairly common 12, scarce 6); regular visitor, probably breeds (scarce 1); regular visitor, non-breeding (common 2, fairly comon 5, scarce 2); regular visitor, probably breeds (scarce 1).

Fifty-seven species breed on the floodplain and 23 of the species recorded are listed in the current South African Red Data Book – Birds (Brooke 1984). One species has been listed as extinct, one as vulnerable, 13 as rare, and eight as indeterminate. Fourteen of the listed species occur as non-breeding vagrants on the floodplain, but the floodplain has particular conservation significance for six species as it supports the only known population of rufous-bellied heron, the largest known number of South African breeding populations of the dwarf bittern, little bittern, and bittern, and it is one of relatively few known localities in South Africa where the pygmy goose and black stork breed.

There are several others, though not listed Red Data species, which have restricted breeding ranges in South Africa and for which the Nyl floodplain supports the largest known breeding populations. They include great white egret (up to 250 pairs), black egret (up to 40 pairs), squacco heron (up to 550 pairs), lesser gallinule (up to 1,000 pairs) and lesser moorhen (up to 8,000 pairs).

Three families are best represented amongst the waterbirds. Among the ducks and geese (Anatidae), 17 of the 20 southern African species occur along the Nyl and 15 breed there. Of the herons (Ardeidae), 17 of the 19 southern African species occur and 11 breed on the Nyl. Twelve of the 19 southern African species of crakes and rails (Rallidae) occur there and 9 breed. Of the 91 waterbird species known to breed in southern Africa (Maclean 1985), 84 (92%) have been recorded on the Nyl floodplain and 53 (63%) breed or have bred in the system.

Crude attempts to count the numbers of waterbirds present in various flood years have provided the following estimated maximum numbers: ducks 19,000, herons and egrets 12,000, crakes and rails 43,000, total water bird species 80,000. The occurrence of these large populations and of the species of particular conservation significance is dependent on the extent, duration and timing of flooding. In general, extensive sustained flooding supports the highest numbers and greatest species diversity. Timing of flooding is important for migratory species (e.g. dwarf bittern, lesser gallinule, lesser moorhen). In years of early or mid-season flood, they arrive on the floodplain in December–January and breed before departing in April, even when suitable habitat and conditions persist. They are entirely absent in years of no flood and in years of late flood. Other endangered vertebrate species include the only known South African population of Johnson's topminnow (*Aplocheilichthyes johnstoni*), which occur in the Groot Nyl catchment, and a large population of the African python (*Python sebae*).

5. *Use and management.* Virtually all of the Nyl floodplain is on privately owned farmland and only about 300 of the 16,000 ha is located in the Provincial Nylsvley Nature Reserve. All the farms along the floodplain were zoned

as an Underground Water Control Area in 1971 (restricting removal and use of Groundwater) and in 1973 many of the farms were proclaimed as Private Nature Reserves. These Acts do not, however, prevent the floodplain from being modified by the construction of dykes, impoundments and dams, or the extraction of sand. The floodplain is used mainly for grazing and is regarded by landowners as an important winter resource in the respect. Many dams and dykes have been built on the floodplain to retain flood water for drinking purposes for stock, These not only retain flood water but change water depth and profoundly affect plant communities.

Aerial spraying of redbilled quelea roosts with toxic chemicals has potentially serious consequences for the floodplain. These birds are a major pest for grain-farmers and the flocks, which may number millions, often roost in the *Phragmites* reedbeds on the floodplain. There have been several instances in the past decade in which the spraying of quelea roosts on the floodplain has also destroyed whole breeding colonies of herons, egrets, and cormorants which occur alongside the Quelea. Similarly, massive die-offs of fish and associated foraging birds have occurred in waters that were contaminated by the spraying. The greatest threat to the system, however, is the impoundment of water and the consequent reduction of the frequency, depth, and duration of flooding. There are currently 37 dams in the catchment of the Klein Nyl river and 28 in the Groot Nyl river catchment.

Endorheic pans. Noble and Hemens (1978) summarised the general features of forty-three endorheic pans and lakes in the interior of South Africa. They ranged from small (3 ha) and temporary to quite large (2,000 ha) permanent waterbodies that vary from brackish to very saline. There are, however, many thousands of small pans in the Orange Free State (Geldenhuys 1982, Seaman and Kok 1987) and Allan (1987) identified about 360 in an area of 740 square km around Lake Chrissie (Fig. 35).

Pans are characteristic of the western, drier, part of southern Africa where the greatest concentrations are in the southern Kalahari region of Botswana and the northern Cape Province and western Orange Free State of South Africa. Most have inundation periods of weeks or months but a small group of 313 pans in the Lake Chrissie area of the wetter south-eastern Transvaal has longer inundation periods. Depending on permanence of water, vegetation varies from reeds through grasses to no vegetation. Sparse scrub occurs on the more saline pans. Invertebrates vary from an ubiquitous littoral fauna to specific temporary-water crustaceans. Birds are the most important vertebrates at pans. Many waterbirds utilize the pans for feeding, breeding, and roosting and partly as a resource on the north-south migratory fly-way. Although pans have an obvious importance to wildlife and are the most productive parts of an otherwise arid landscape, none are specifically protected. The vast number of pans makes them a common commodity and

Table 9. The size distribution of endorheic pans in the Lake Chrissie area (modified from Allan 1987).

Area (ha)	Number of pans	% of total
1–9	224	72
10–19	36	12
20–49	26	8
50–99	12	4
100–200	11	3
>200	4	1
TOTAL	313	100

there is no control of their modification or destruction. However, the clayey nature of their soil base limits their value for cultivation. Very little is known of these pans and summary accounts of the pans in the Lake Chrissie (Allan 1987, Hutchinson *et al.* 1932) and western Orange Free State areas are presented.

Pans in the Lake Chrissie area. The Lake Chrissie 'panveld' lies between 26°10′ and 26°30′ south and 30°05′ and 30°25′ east (Fig. 35). The topography is flat to rolling and altitude varies between 1,600 and 1,800 m. The area is poorly drained and is located above the headwaters of several major rivers. Annual rainfall is relatively high (750 mm) and evaporation relatively low (1,250 mm).

1. *Geomorphology.* The size distribution of 313 pans depicted on 1:50,000 maps of the area is presented in Table 9. Although the greatest proportion (72%) are smaller than 10 ha, 5% exceed 10 ha, and the largest, Lake Chrissie, is 1,045 ha in extent. Because of the density of pans, individual catchments are small and the average ratio of pan area to catchment area is 1:7.4. Large pans (750 ha) have a ratio of 1:7.0 and small pans (10–20 ha) a ratio of 1:9.8. Characteristics of selected pans are presented in Table 10.

2. *Vegetation.* Pans were classified using a combination of vegetation physiognomy, species composition, and salinity. In reed pans, over 90% of the pan basin was covered by *Phragmites australis*. A narrow zone of open water with dense growths of *Lagarosiphon muscoides*, *Potamogeton thunbergii* and *P. pectinatus* surrounded the central reedbed.

The vegetation of sedge pans was dominated by *Eleocharis palustris* and tall, emergent stands of *Schoenoplectus corymbosus*. Areas of open water were colonized by the submergent species *L. muscoides*, *P. thunbergii* and *P. pectinatus*. Other abundant species included: *Echinochloa jubata*, *Odontelytrum abyssinicum*, a *Panicum* sp. and *Leersia hexandra*. The aquatic grass, *Odontelytrum abyssinicum*, which was found in six flooded sedge pans, has only been collected once before in South Africa.

Table 10. Characteristics of endorheic pans in the Lake Chrissie area (adapted from Allan 1987).

Characteristic	Pan type Reed pans	Sedge pans	Open pans	Salt pans
Number of pans studied	10	19	40	1
Range of sizes (ha)				
10–19	–	12	7	–
20–49	7	5	14	–
50–99	1	1	81	1
100–200	1	1	8	–
>200	1	–	3	–
Pan area:catchment area	1:3.9	1:10	1:7.2	
Substratum	Organic with clay and silt	Organic	Clay, silt, sand, rock	Rock, sand
Water depth (m)	>2	1	<1 small	Shallow
Water level stability	Relatively stable	Unstable	Unstable	–
Permanence	Permanent	Semipermanent	Impermanent	–
Conductivity (μ cm^1)				
mean	2,437	835	3,076	>10,000
range	1,125–3,495	215–2,450	213–8,140	
pH				
mean	7.8	7.5	7.9	10
range	7–8.5	7–9.5	7–9.5	–

The basins of open pans are devoid of vegetation except for the grass *Diplachne fusca* which colonizes very shallow and waterlogged areas. Salt pans are also devoid of vegetation except for the salt tolerant sedge *Schoenoplectus triqueter* which dominates the shoreline vegetation.

3. *Vertebrate fauna.* Birds are the most evident vertebrates as pans do not provide important habitat for fish or reptiles. Ephemeral sedge pans are breeding grounds for bullfrogs but no amphibians in the Transvaal are entirely restricted to highveld pan habitats. Otters were found in several of the larger more permanent pans but no mammals appear to be restricted to pans. The pre-eminent value of pans is for wildlife breeding, feeding, and roosting sites for vast flocks of waterbirds.

A total of 73 avian species use pans for feeding, 57 are regular visitors, and 15 of those have been shown to breed in pans. Twelve other species are non-breeding visitors and 4 are rare vagrants. Four rare and endangered avian species have also been recorded. Among these were a single individual of Baillon's crake and sixty-five individuals of the chestnutbanded plover out of a total estimated South Africa population of 500. Flocks of several thousand greater and lesser flamingoes also have been observed in pans.

In addition to the aquatic species, several mainly terrestrial species use extensive *Phragmites* beds for roosting and breeding. Species that only roost on the pans are the European swallow, pied starling, cattle egret, eastern redfooted kestrel and longtailed widow. Species that also breed in pans are the cape weaver, masked weaver, and red bishop.

4. *Invertebrate fauna*. In their early study, Hutchinson *et al.* (1932), found a variety of rotifers, ostracods, cladocerans, and copepods. The fauna was characterised by *Daphnia pulex*, *Daphnia gibba*, a number of chydorids, *Metadiaptomus transvaalensis*, and *Lovenula excellens*, of which only the latter is restricted to the pan environment and is endemic to the Lake Chrissie pans.

5. *Use and management*. The area is wholly in private ownership and the major form of land use is pastoral farming. Consequently, grazing and trampling of the pan margins is common and fences traverse pans. The worst affected were reed pans. Water is abstracted from some pans and alien fish species (carp and bass) have been introduced into deeper permanent pans. The majority are, however, relatively unmodified by human influences.

Pans of the western Orange Free State. Although endorheic pans are widespread in southern Africa, the greatest density occurs in a broad band in the western Orange Free State between Welkom, Bloemfontein, and Kimberley between 27° and 30°S and 24° and 27°E (Fig. 43). Annual rainfall is generally less than 600 mm (Fig. 32) and annual evaporation exceeds 1,800 mm and most pans seasonally dry. Many of the larger pans (e.g. Florisbad; Fig. 35) are also saline and the brines contain high concentrations of sodium chloride and sulphates of sodium, calcium, and magnesium (Seaman 1987). This group of pans are generally referred to as 'Salt pans.' This summary account that follows was taken from Geldenhuys (1982), Seaman (1987), and Seaman and Kok (1987).

1. *Geology and gemomorphology*. The terrain of the western Orange Free State is flat. Little water drains from the area and pans occur mainly on the Ecca Group of the Karoo Sequence. A combination of factors is thought to contribute to the origin of the pans. These include the presence of bowl-shaped dolerite corridors, lithological variation, poor drainage, low rainfall with high variation in daily and seasonal temperature, and strong winds in the dry season. Large scale erosion of rock in the area, probably due to thermic distention because NaCl in the underlying rocks, erodes faster than the other rock components. The large amount of material produced by erosion is then removed by wind during the dry season and by animal trampling in the wet season (Geldenhuys 1982).

The density of pans is high with 8,803 having been recorded in an area of 41,819 km^2. That pattern of distribution is summarized in Figure 43. Since most pans are small (<2 ha), the total area covered by them is only 2.9%

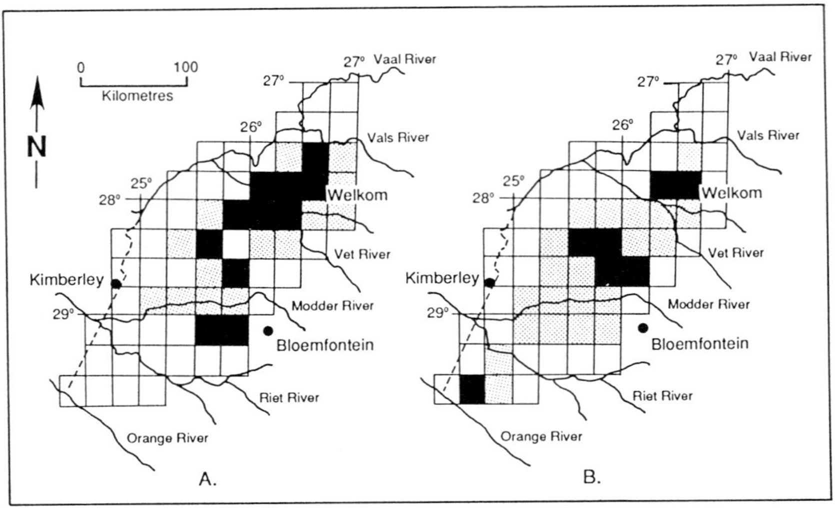

Figure 43. Maps of the western Orange Free State showing spatial variation in pan density (A) and area occupied by pans (B) in sixteenth degree squares. Unshaded areas represent squares where pan density and area occupied by pans are respectively lower than the mean of the 62 squares ($x = 19.23$ pans per 100 km^2; $x = 292$ ha pan area per 100 km^2). Shaded and black squares indicate area where the numbers and pan areas respectively exceed twice the respective means (from Geldenhuys 1982).

of the landscape. In some areas, the density of pans is extremely high. Most pans (56%) are <2 ha, 38% are between 2 and 25 ha, and 6% are >25 ha. Two pans exceed 3,000 ha in size. It is estimated that only 10% of the pans retain water for sufficient time for waterfowl to breed successfully in years where rainfall equals the long term average. Since rainfall is extremely variable the success of breeding varies markedly from year to year. Breen (1991) has suggested that because of the wide distribution of these types of wetlands, suitable habitats are likely to be available for nomadic species at shorter intervals than suggested by local rainfall patterns.

2. *Vegetation.* Geldenhuys (1976, 1982) has recognised six pan types on the basis of the presence and nature of emergent vegetation after two to three months of inundation.

1. Bare pans are those which have a distinct high water line and less than one percent of the flooded area is covered with emergent vegetation. Submerged plants such as Characeae and *Zannichellia* may occur on the pan bottoms. The littoral zone may be devoid of vegetation, consist of similar vegetaion to the adjacent veld, or may include hydrophytic components such as clumps of Cyperaceae. They represent about 18% of the pans.
2. Sedge pans have emergent vegetation, predominantly Cyperaceae, that

covers the entire water surface and intermingles with the adjacent terrestrial vegetation. The high water lines of sedge pans are frequently difficult to identify. Sedge pans are relatively small (0.2 ha), and the water is at most moderately brackish. They are uncommon and represent only 4% of the total.

3. Scrub pans usually have indistinct high water lines. They are rarely flooded and are covered, even when dry, with halophytic dwarf shrubs or scrub such as *Salsola aphylla* and *Suaeda fruticosa*. Hydrophytes are absent around the perimeter. When flooded the water becomes hypersaline and crystallized salt may accumulate to the point where it is extracted commercially. Scrub pans, more common in the drier southern area, are uncommon and represent only 3% of the total.
4. Mixed grass pans have indistinct high water lines. Emergents, consisting of various moisture-loving grasses (mainly *Eragrostis* spp.) cover the entire pan and the littoral zone and intermingle with adjacent terrestrial vegetation. Depending on water depth the grasses may be sparsely distributed or form a dense cover. These pans are fairly common, comprising 13% of the total and salinity is generally low.
5. Closed *Diplachne* pans have high water lines that are clearly defined and emergent grasses, particularly *Diplachne fusca*, cover more than 90% of the water surface. This grass species forms homogenous stands of sparsely distributed to fairly dense cover. Unlike the tuft structure of the grasses of mixed grass pans, *D. fusca* is a creeping grass. This is the most common type of pan representing 34% of the total but they are uncommon in the northeast. Salinity may be high but salt mining has not been recorded.
6. Open *Diplachne* pans are similar to Closed *Diplachne* pans, except that emergent grasses cover only a portion of the water surface. They represent 27% of the pans and together with the Closed *Diplachne* pans represent the most common type of pans.

3. *Vertebrate fauna*. As in the Lake Chrissie area, birds are the most abundant vertebrates. Fish and aquatic reptiles are absent and mammals only utilize the pans for drinking-water and as salt licks.

The bullfrog *Pyxicephalus adspersus* is a widespread inhabitant of Grass and Sedge pans where young post-metamorphic individuals are found in large numbers from about a month after inundation.

Only the avifauna has been studied in any detail and value of the pans as wetland habitat for birds (Table 11) lies primarily in providing wintering grounds for palaearctic waders as well as breeding habitats for certain duck species. The pans are widely used by the greater and lesser flamingo (numbers exceeding 7,000 at a single pan) although they do not breed at the pans. Forty three waterbird species have been recorded. Two groups, 12 species of ducks and 14 species of small waders are predominant and, apart from

Table 11. Percentage waterfowl composition at different pan types in the western Orange Free State (from Geldenhuys 1982).

Waterfowl	Bare	Sedge	Scrub	Mixed grass	Closed *Diplachne*	Open *Diplachne*
Whitefaced duck	0	0	0	0	1	0
Fulvous whistling duck	0	0	0	0	1	0
Egyptian goose	4	0	12	52	28	22
South African shelduck	80	2	79	0	15	36
Yellow-billed duck	0	71	0	21	10	9
Cape teal	8	0	4	0	6	6
Redbilled teal	2	18	3	13	17	9
Cape shoveller	2	9	2	0	9	7
Southern pochard	0	0	0	0	0	3
Knob-billed duck	0	0	0	0	1	0
Spurwinged goose	3	0	0	14	14	7
Maccoa duck	1	0	0	0	0	0

birds such as the white stork, sacred ibis, spoonbill, two flamingo species, and the redknobbed coot, other birds were relatively uncommon. Breeding was recorded in six duck species and six other waterbirds. The ducks appear to require specific breeding habitats. Fourteen breeding species were recorded in Sedge and Scrub pans, 17 in Mixed grass pans, 21 in Bare pans, 28 in Closed *Diplachne* pans, and 32 in Open *Diplachne* pans. No breeding was recorded in Scrub Pans. The pans are favoured by shelduck, redbilled teal, cape shoveller and cape teal and are important breeding grounds for the latter three species.

4. *Invertebrate fauna.* Endemic and common species of Anostracans, Notostracans and Conchostracans occur where the mean period of inundation is less than one month. *Paradiaptomus schulzei* and *Ceriodaphnia rigaudi* were common species in the pans but uncommon in other local water bodies. At the end of an inundation, when salinity rises dramatically, there appears to be insufficient time for the establishment of salt-tolerant species and the existing community merely dies out.

5. *Use and management.* Virtually all pans are in private ownership and no area has been set aside for protection. Neither are there any restictions on the use to which they may be put. Most pans are on mixed pastoral/crop farms, where the unsuitability of their clayey soils generally precludes cultivation. They serve as a periodic source of drinking water for stock and numerous pans have been modified to extend this useful period. Others are altered by roads, fences, garbage dumps and the addition of effluent water from mines. Grass pans are used for livestock grazing and where soil mineralization is not too high some are cultivated. Many of the larger pans are

Table 12. The principal physical features of South African coastal lakes (data from Begg 1978, Hill 1969, Mepham 1987, Noble and Hemens 1978, and Whitfield *et al.* 1983).

System	Location	Length (km)	Width (km)	Shorelength (km)	Area (km^2)	Depth (in m) (mean/max)	Volume ($10\ m^3$)
Kosi Lakes	32°50′S 27°00′E						
Sifungwe		2.0	2.0	6	3.0	8.0/18.0	0.1
Nhlange		7.3	6.1	23	31.0	7.0/31.0	0.2
Amanzimnyama		–	–	–	1.5	2.0 –	–
Lake Sibaya	27°25′S 32°40′E	18.7	18.3	89	34.0	12.6/43.0	0.8
Lake St Lucia	28°50′S 32°30′E	60.0	6.0	350	350.0	1.0/2.0	330.0
Nhlabane	28°38′S 32°16′E	–	1.0	14	1.2	–/–	10.0
Mzingazi Lake	28°45′S 32°05′E	5.0	3.0	20	12.6	6.0/14.0	47.5
Lake Nsezi	28°45′S 32°00′E	5.0	2.0	–	10.0	– 5.0	–
Lake Cubhu	28°50′S 32°00′E	–	–	–	5.0	– –	–
Wilderness Lakes	34°30′S 32°40′E						
Groenvlei		3.7	0.9	–	2.5	3.7/5.5	–
Swartvlei		9.0	3.0		8.8	5.5/16.7	48.0
Rondevlei					1.3	– 5.0	–
Langvlei					2.2	– 4.0	–
Eilandvlei					1.4	– 6.5	–
DeHoopvlei	34°31′S 20°23′E	18.0	1.0	–	6.2	1.1/7.7	–
Soetendalsvlei		7.5	2.5	–	20.0	3.0	–
Zeekoeivlei	34°06′S 18°30′E	2.0	1.0	–	2.2	3.6 –	–
Sandvlei	34°05′S 18°28′E	3.4	2.1	–	4.0	– 5.0	–
Verlorenvlei	32°19′S 18°21′E	13.5	1.4	–	10.0	2.5/5.0	–

Table 13. Selected physico-chemical and biotic properties of the coastal lakes of South Africa (data from Begg 1978, Boshoff and Palmer 1981, Mepham 1987 and Noble and Hemens 1978).

System	Secchi depth (m)	pH (m S/m)	Conductivity (‰)	Salinity macrophytes	Dominant aquatic	Dominant fish species	Other vertebrates
Kosi Lakes					Submerged macrophytes uncommon	Varied marine and freshwater – 39 resident estuarine and 9 freshwater species	Generally rather poor
Sifungwe	3.7	–	80	Stratified	*Potamogeton pectinatus*		
Nhlange	1.2	–	40	Max 4.2	*Phragmites australis* *Scirpus littoralis*		
Amanzimyama	1.0	–	9	–	–	–	
Lake Sibaya	3.0	8.2	6	–	*Potamogeton* spp. *Myriophyllum spicatum*	*Oreochromis mossambicus* *Clarias gariepinus*	Generally poor
Lake St Lucia	0.5	–	10	0–120	*Zostera capensis* *Ruppia spp.*	82 species *Pomedasys*, *Argyrosomus*, *Mugil*	Registered as a wetland under Ramsar convention Rich avifauna, *Crocodyllus niloticus*
Mzingazi Lake	1.3	7.4	30	–	*P. pectinatus*, *P. schweinfurthii*	*Clarias gariepinus*, *O. mossambicus*	
Lake Nsezi	–	–	–	–	–	–	–
Lake Cubhu	–	–	–	–	–	–	–
Wilderness Lakes							
Groenvlei	–	8.7	–	2–3	Floating leaved macrophytes absent.	Predominantly marine or estuarine	An important wetland area for birds
Swartvlei	1.1	8.0	–	1–20	Dense fringed *Phragmites*, *Scirpus*	(12 species with the freshwater	
Rondevlei	3.0–1.4	8.6	–	12–16	*littoralis*, and *Juncus kraussii.*	*Oreochromis mossambicus*	

Langvlei	5.0–2.5	7.8	–	8–13	Submerged *P. pectinatus*, *Ruppia* sp., and *Chara* sp.		
Eilandvlei	0.9–3.0	7.8	–	4–10			
DeHoopvlei	–	9.0	–	5–11	*P. pectinatus*	*Sandelia capensis*, *O. mossambicus*	Registered as a wetland under Ramsar convention
Soetendalsvlei	–	–	–	–	–	–	–
Zeekoeivlei	–	–	–	3–33	*P. pectinatus*	Introduction *Cyprinus carpio* and *O. mossambicus*	Rich avifauna particularly associated Rondevlei
Sandvlei	0.2–1.8	8.4	3–27	2–19	*P. pectinatus*, *P. australis*	Estuarine species with recent fresh-water introductions	150 bird sp. Important refuge for waders
Verlorenvlei	–	7.4	–	–	*M. spicatum*, *Phragmites sp.*	*Cyprinus carpio* and *O. mossambicus*	Rich avifauna
Nhlabane	–	–	2–4	Formerly slightly saline now fresh	*P. pectinatus*, *Utricularia* sp., and emergents *P. australis*, *Cyperus*	Formerly *Mugil cephalus*, *Megalops cyprinoides* now freshwater spp.	Duck and geese common

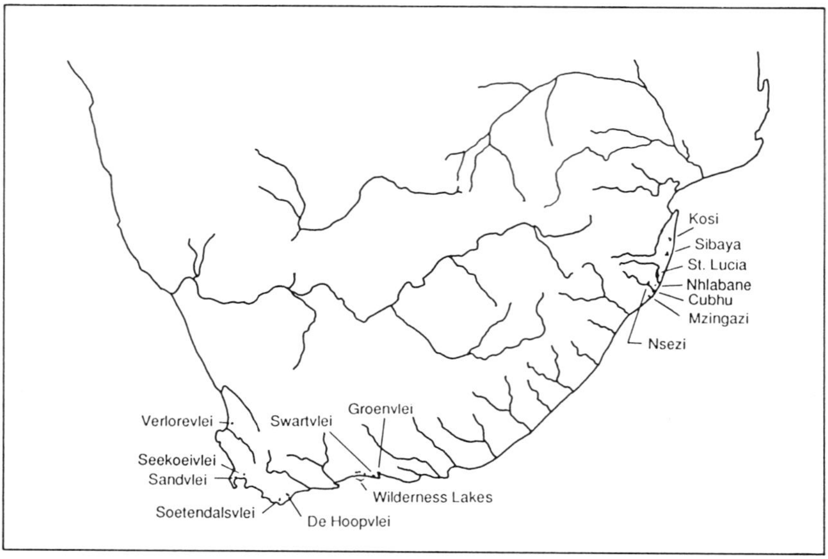

Figure 44. Coastal and estuarine lakes in South Africa (from Noble and Hemens 1978).

saline and the brine is pumped into crystallisation basins. Florisbad pans, for example, has about 100 wind pumps pumping brine.

Coastal lakes. With the exception of the numerous endorheic pans of the interior there are few natural lakes in South Africa, most are near the coast (Fig. 44), and they occur in two groups. On the north-east and south western coasts, lakes originated where rivers have been dammed by coastal dunes (Hill 1969 and 1975, Orme 1973). Since the valleys were incised when sea level was lower than at present, some of the lakes are quite deep (e.g. Lake Sibaya, Table 12), but most are quite shallow (e.g. Lake St Lucia; mean depth 1 m, Table 12) as a result of considerable infilling. Orme (1973) suggests that lakes on the Maputaland coastal plain may have decreased by as much as 60% in areal extent since being isolated in the Holocene.

Although Noble and Hemens (1978) separated the lakes along the coast into two types, coastal and estuarine lakes, there is a continuum of variation between those with a very strongly developed marine influence (e.g. Lake St Lucia in which salinity may rise to 120% during droughts) to those, such as Lake Sibaya, which are completely isolated from the sea.

The principal physico-chemical and biotic properties of the coastal and estuarine lakes are summarized in Table 13. It is evident that they are extremely varied in shape, size, and depth. All are essentially open water systems with emergent plants around the edges and varying degrees of development of submerged vegetation. As such, they are generally regarded as lakes (deep water habitats) rather than wetlands.

Wetland use and conservation

P. DENNY

Wetlands of Africa are a valuable commodity open to use, abuse, and exploitation. They remained practically untouched until about forty years ago when a number of developments has put them under pressure. The increasing demand for water and electricity, for example, has resulted in a large number of dams being built. Off-shoots from these developments are varied and include expansion of inland fisheries, irrigation schemes, and (sometimes) improved transport along waterways. However, large man-made lakes not only flood valleys and their concomitant wetlands, but regulate the water flow in catchment and drainage areas. The natural cycle of flooding and drainage which provides the rich and varied floodplains may thus be disrupted, and the floodplain ecosystem destroyed. Water-borne diseases such as schistosomiasis, onchocerciasis and malaria, become more widespread and remedial action, often involving obnoxious chemicals, deteriorates the environment further.

As the human population of Africa increases, drainage and irrigation of the land and encroachment onto wetlands become priorities for large-scale agricultural programs and subsistence farmers alike. The pattern is all too common and is the price of "progress". Specific examples are mentioned in the preceding text and a number of cases is discussed in detail in Denny (1985a).

Parallel problems arose in the developed world a century or so earlier so that their wetlands are now a shadow of their former expanse. The sadness is that in the developed countries the few remaining wetlands are under greater threat than ever before: politics and economics, apparently, know no boundaries. In Britain, for example one of the most valued wetlands – Havergate Marshes in East Anglia – was once threatened with drainage to provide additional land for agriculture. Yet, gross excesses of cereals and dairy products in the European Community (EC) are an embarrassment.

Conservationists tell us that the African wetlands are one of the few remaining "wild and remote places" where wildlife lives in harmony with the natural environment; a refuge from man, where rare plants and animals have a chance of survival; where large numbers of animals and tens of millions of migratory birds can feed and breed undisturbed. We all have a responsibility to be concerned for, and protect such habitats.

But what right have developed countries (who cannot manage their own wetlands satisfactorily) to demand the protection of wetlands in Africa? At least in Africa there is a clear and obvious need to improve the standard of living of the people and exploitation of the wetlands is one way to achieve this. The truth is, that many of the wetlands of Europe and North American have already been degraded or destroyed. This is not so in Africa and there is still the opportunity to manage wisely and bequeath to future generations a unique and valuable asset.

One of the main reasons given for the preservation of African wetlands is, indeed, the protection of its wildlife. But this is unlikely to impress local inhabitants whose main concern may be day-to-day survival. Nor is it likely to impress politicians who must put the welfare of their people before more esoteric aspirations. On the positive side there are organizations such as the World Conservation Union (IUCN), the Scientific Committee on Problems of the Environment (SCOPE), and the United Nations Environmental Program (UNEP) to attract funds and draw global attention to the plight of the wetlands. However, their energies, surveys, and research activities until quite recently were directed towards the wildlife (which frequently does not seem to include plants) rather than the whole ecosystem; and they rarely considered possible multipurpose development. Their approach now is more enlightened and programmes of study into the functioning, values, and uses of wetlands are progressing. The United Nations conference on the environment and development (popularly known as the "Earth Conference") held in Rio de Janeiro, Brazil (June 1992), should give additional momentum to these types of study.

Projects, particularly those relating to swamp and floodplain fisheries and to grazing by ungulates clearly have a more direct relevance to local requirements. But it is a little surprising that detailed studies on secondary producers such as fish are not paralleled by equivalent studies on the primary producers upon which the ecosystem depends. To take one example: the floodplain vegetation and the euhydrophyte interface between swamp and open water together provide over 36% of the total African freshwater fish catch (Welcomme 1972). In the Bangweulu swamps, approximately 50% of the catch (12,000 to 15,000 tons y^{-1}) are derived from the swamp interface (Bazigos *et al.* 1976) and in Lake Chilwa 30% of the fish catch comes from the surrounding swamps (Furse, *et al.* 1979). Yet, by glancing through the text the reader must be only too aware of the limitations of our knowledge on wetland vegetation. The importance of the interface zone to fisheries and wildlife cannot be over emphasized (Denny 1985a). Indeed, the richness of the swamp fisheries can often be judged by the number of predatory birds.

The euhydrophytes provide breeding sites for fish, a substrate and nutrient supply for the Aufwuchs community, which in its turn is food for the invertebrate grazer upon which many fish feed. The vegetation bed acts as a refuge for the fish against predators. However, the species composition, production, and dynamics of the interface zone has barely been touched upon. Logically, in studies of secondary production, particularly when a regular harvesting of biomass takes place (or is anticipated), a complementary study of the primary producers should be undertaken.

What is the answer? The African wetlands are a highly valued asset which have many demands made upon them. It is idealistic folly to proclaim that they must be preserved. Preservation implies ossification and the preservationists will inevitably be at loggerheads with the developers. On the other hand, the management of wetlands as multipurpose commodities for sustainable development is practical.

The varied interests and demands on wetlands include: conservation, wildlife and tourism, sites for leisure activities, agriculture, fisheries, water resources (including storage and regulation of waterflow), and nutrient filters for tertiary sewage treatments. Industrial and agrochemical pollution, of course, is a permanent threat. Whilst some activities may be antagonistic, others are complementary. Development and conservation are not necessarily mutually exclusive and the aspirations of both need to be appreciated more fully.

Before a wetland is irreversibly changed there is a number of clearly identifiable procedural steps which can be taken (Mitchell *et al.* 1985).

First of all, the broad but relevant guiding principles which explain the need for change must be established. Secondly, the various criteria to be used in selecting options must be determined; specific objections must be considered and the possible combination of options assessed: then options can be arranged in an order of priority. Thirdly, the most suitable option (or options in an integrated system) can be implemented and progress monitored.

Different stages concern different sections of the community. The initiation is usually the prerogative of politicians and industrialists. A multidisciplinary group of experts drawn from industrial, ecological, and socio-economic backgrounds should then be brought in. This is followed by maximum consultation and collaboration with the local community. Thus, from an early stage the project can be based as far as possible, on community participation and local direction. In the past, large-scale projects have been devised with regretable lack of communication at the local level which has led, all too frequently, to discontent and suffering.

Whilst the setting of management goals should involve the wishes of the local population, the carrying out of research to fulfil these goals must be by experts. However, in order that the correct decisions may be made, a firm data base is required. Unfortunately, biological and environmental data for most of Africa's wetlands are totally inadequate. Studies on community structure and energy flow with the primary producers (aquatic macrophytes, Aufwuchs and phytoplankton) as the starting point are urgently required. Until these data are obtained and collected it is not possible to make valid predictions. It is not too late. Hopefully the wetlands of Africa, which include some of the most beautiful and remote areas of the continent, can thus be managed wisely to the benefit of all.

References

Adam, J. -G. (1964) Contribution à l'étude de la végétation du lac de Guiers (Sénégal). Bulletin de l'Institut Français d'Afrique Noire, sér. A, 26: 1–72.

Adejuwon, J. O. (1973) The ecological status of freshwater swamp savannas in the forest zone of Nigeria. Journal of the West African Science Association 16: 133–154.

Adeniji, H. A. (1973) Preliminary investigation into the composition and seasonal variation of the plankton of Kainji Lake, Nigeria. Geophysical Monographs (American Geophysical Union) 17: 617–619.

Ahn, P. (1958) Regrowth and swamp vegetation in the western forest areas of Ghana. Journal of the West African Science Association 4: 163–173.

Aké Assi, L. (1977) *Salvinia nymphellula* Desv. (Salviniacées) fleau en extension vers l'ouest de l'Afrique intertropicale. Bulletin de l'Institut Fondamental d'Afrique Noire, sér. A, 39: 555–562.

Akhurst, E. G. J. and Breen, C. M. (1988) Ionic content as a factor influencing turbidity in two floodplain lakes after a flood. Hydrobiologia 160: 19–31.

Allan, D. (1987) The conservation status of the pans in the Lake Chrissie area. p. 91–100. *In*: R. D. Walmsley and M. L. Botton (eds.). Ecology and Conservation of Wetlands in South Africa. Ecosystems Occasional Report Series No. 28, Council for Scientific and Industrial Research, Pretoria, South Africa.

Allanson, B. R. (ed.) (1979) Lake Sibaya. Dr. W. Junk Publishers, The Hague, The Netherlands. 364 pp.

Allanson, B. R., Hart, R. C., O'Keeffe, J. H. and Robarts, R. D. (1990) Inland Waters of Southern Africa: An Ecological Perspective. Kluwer Academic Publishers, Dordrecht, The Netherlands. 458 pp.

Anon (1979) Proceedings of the International Conference on Kainji Lake and River Basins Development in Africa. Vol I, p. viii + 1–249, Vol. II, p. viii + 251–621. Lake Kainji Research Institute, New Bussa, Nigeria.

Arfi, R., Dufour, P. and Maurer, D. (1981) Phytoplancton et pollution: premières études en baie de Biétri (Côte d'Ivoire). Traitement mathématique des données. Oceanologica Acta 4: 319–330.

Ba, A. T., Cogels, F. X., Doyen, A., Niang, C. I., Singleton, M., Tibesar, A., Vincke, P. P. and Wane, O. (1983) Le lac de Guiers. Problématique d'environnement et de développement. Actes du Colloque de l'Institut des Sciences de l'Environnement, 9–11 mai 1983, Dakar. Administration générale de la Coopération au Développement. 506 pp.

Bazigos, G. P., Grant, C. D. and Williams, C. (1976) Revised fisheries time series of the natural waters of Zambia (1966–1974). UN/FAO report ZAM/73/009, Rome, Italy.

Beadle, L. C. (1981) The Inland Waters of Tropical Africa: An Introduction to Tropical Limnology. (2nd edition). Longmans, London and New York. 475 pp.

Beadle, L. C. and Lind, E. M. (1960) Research on the swamps of Uganda. Uganda Journal 24: 84–87.

Begg, G. (1978) The Estuaries of Natal. Natal Town and Regional Planning Commission Report, Vol. 41. Pietermaritzburg, South Africa. 657 pp.

Begg, G. W. (1986) The Wetlands of Natal (Part 1). An overview of their extent, role and present status. Natal Town and Regional Planning Commission Report No. 68. Pietermaritzburg, South Africa. 114 pp.

Begg, G. W. (1988) The Wetlands of Natal (Part 2). The distribution extent, and status of wetlands in the Mfolozi catchment. Natal Town and Regional Planning Commission Report No. 71. Pietermaritzburg, South Africa. 278 pp.

Begg, G. W. (1989) The Wetlands of Natal (Part 3). The location, status and function of the priority wetlands of Natal. Natal Town and Regional Planning commission Repot No. 73. Pietermaritzburg, South Africa. 256 pp.

Begg, G. W. (1990) The Wetlands of Natal (Part 4). Policy Proposals for the Wetlands of Natal and KwaZulu. Natal Town and Regional Planning Commission Report No. 75. Pietermaritzburg, South Africa. 86 pp.

Benech, V. and Quensiere, J. (1982) Migrations de poissons vers le lac Tchad à la décrue de la plaine inondée du Nord-Cameroun. 1. – Méthodologie d'echantillonnage et résultats généraux. Revue d'Hydrobiologie Tropicale 15: 253–270.

Benech, V. and Quensiere, J. (1983a) Migrations de poissons vers le lac à la décrue de la plaine inondée du Nord-Cameroun. 2. – Comportement et rythme d'activité des principales espèces. Revue d'Hydrobiologie Tropicale 16: 79–101.

Benech, V. and Quensiere, J. (1983b) Migrations de poissons vers le lac Tchad à la décrue de la plaine inondée du Nord-Cameroun. 3. – Variations annuelles en fonction de l'hydrologie. Revue d'Hydrobiologie Tropicale 16: 287–316.

Benech, V., Lemoalle, J. and Quensiere, J. (1976) Mortalités de poisson et conditions de milieu dans le lac Tchad au cours d'une période de sécheresse. Cahiers de l'Office de la Recherche Scientifique et Technique Outre-mer, sér. Hydrobiologie 10: 119–130.

Benech, V., Quensiere, J. and Vidy, G. (1982) Hydrologie et physicochimie des eaux de la plaine d'inondation du Nord-Cameroun. Cahiers de l'Office de la Recherche Scientifique et Technique Outre-mer, sér. Hydrobiologie 19: 19–35.

Berghen, C. Van Den (1982) Premier aperçu sur la végétation aquatique de la Basse Casamance (Sénégal meridional). p. 266–271. *In*: J. J. Symoens, S. Hooper and P. Compere (eds.), Studies on Aquatic Vascular Plants. Royal Botanical Society of Belgium, Brussels, Belgium.

Bernacsek, G. M. (1984) Guidelines for dam design and operation to optimize fish production in impounded river basins (based on a review of the ecological effects of large dams in Africa). CIFA Technical Paper II: 98 pp.

Betterton, C. (1984) Ecological studies on the snail hosts of schistosomiasis in the South Chad Irrigation Project area, Borno state, Northern Nigeria. Journal of Arid Environments 7: 43–57.

Billon, B., Guiscafre, J., Herband, J. and Oberlin, G. (1974) Le bassin du fleuve Chari. Monographies hydrologiques, l'Office de la Recherche Scientifique et Technique Outre-mer No. 2. 450 pp.

Blanc, M., Daget, J. and Aubenton, F. D. (1955) L'exploration des eaux douces dans le basin du Moyen Niger. Bulletin de l'Institut Français d'Afrique Noire, sér. A, 17: 1157–1173.

Bond, W. J. and Roberts, M. G. (1978) The colonization of Cabora Bassa, Mocambique a new man-made lake, by floating African macrophytes. Hydrobiologia 60: 243–259.

Boughley, A. S. (1957) The physiognomic delimitation of West African vegetation types. Journal of the West African Science Association 3: 148–165.

Bourquin, O. (1986) Reptiles and amphibians. p. 45–50. *In*: C. H. Stormanns and C. M. Breen (compilers). Proceedings of the Greater Mkuze Swamp System Symposium and Workshop. Institute of Natural Resources Investigational Report No. 22, University of Natal, Pietermaritzburg, South Africa.

Breen, C. M. (1991) Are intermittently flooded wetlands of arid environments important conservation sites? Madoqua 17: 61–65.

Breen, C. M. and Beggs, A. W. (1989) The Conservation status of Southern African wetlands. p. 245–263. *In*: B. J. Huntley (ed.). Biotic Diversity in Southern Africa: Concepts and Conservation. Oxford University Press, Cape Town, South Africa.

Breen, C. M. and Stormanns, C. -H. (1991) Observations on the growth and production of *Cyperus papyrus* in a subtropical swamp. Internationale Vereinigung für Theoretische und Angewandte Limnologie 24: 2722–2725.

Breen, C. M., Rogers, K. H. and Ashton, P. J. (1988) Vegetation processes in swamps and flooded plains. p. 223–247. *In*: J. J. Symoens (ed.). Vegetation of Inland Waters. Kluwer Academic Publishers, Dordrecht, The Netherlands.

Brooke, R. K. (1984) South African Red Data Book – Birds. South African National Programme Report No. 97. Council for Scientific and Industrial Research, Pretoria, South Africa. 213 pp.

Brooks, M. (1970) A bibliography of hydrological work in Ghana including references to onchocerciasis, bilharziasis and dracontiasis (including yellow fever, malaria and trypanosomiasis). Ghana Journal of Science 10: 49–67.

Bruton, M. N. and Cooper, K. H. (eds.) (1980) Studies on the Ecology of Maputaland. Rhodes University, Grahamstown, South Africa. 560 pp.

Buchan, A. J. (1988) Grazing resources of the Makitini and Pongolo floodplain: production, use and significance in the local socio-economy. Institute of Natural Resources Investigational Report No. 32. University of Natal, Pietermaritzburg, South Africa. 30 pp.

Buchan, A. J. and Breen, C. M. (1988) Leaf material production in *Nymphaea lotus* L.: implications for sustained detritus production. Internationale Vereinigang für Theoretische und Angewante Limnologie 23: 1940–1944.

Burgis, M. J. and Symoens, J. J. (1987) African Wetlands and Shallow Water Bodies. Éditions de l'ORSTOM. Institut Français de Recherche Scientifique pour le Dévelopment en Coopération No. 211. Paris, France. 650 pp.

Carmouze, J. P. (1976a) Grands traits de l'hydrologie et de l'hydrochimie du lac Tchad. Cahiers de l'Office de la Recherche Scientifique et Technique Outre-mer, sér. Hydrobiologie 10: 33–56.

Carmouze, J. P. (1976b) La régulation hydrogéochimique du lac Tchad. Travaux et Documents. L'Office de la Recherche Scientifique et Technique Outre-mer, No. 58. 418 pp.

Carmouze, J. P., Dejoux, C., Durand, J.-R., Gras, R., Iltis, A., Lauzanne, L., Lemoalle, J., Lévéque, C., Loubens, G. and Saintjean, L. (1972) Grandes zones écologiques du lac Tchad. Cahiers de l'Office de la Recherche Scientifique et Technique Outre-mer, sér. Hydrobiologie 6: 103–169.

Carmouze, J. P., Durand, J.-R. and Lévêque, C. (eds.) (1983) Lake Chad. Dr. W. Junk Monographiae Biologicae 53, Dordrecht, The Netherlands. 575 pp.

Carmouze, J. P., Pedro, G. and Berrier, Y. (1977) Sur la nature des smectites de néoformation du lac Tchad et leur distribution spatiale en fonction des conditions hydrogéochimiques. Compte Rendu Hebdomadaire des Séances de l'Académie des Sciences, Paris, sér. D 284: 615–618.

Carter, G. S. (1955) The Papyrus Swamps of Uganda. W. Heffer and Sons, Cambridge, England. 25 pp.

Caumette, P., Pagano, M. and Saint-Jean, L. (1983) Repartition verticale du phytoplancton, des bacterias et du zooplankton dans un milieu stratifié en Baie de Biétri (Lagune Ébrié, Côte d'Ivoire). Relations trophiques. Hydrobiologia 106: 135–148.

Chachu, R. E. O. (1979) The vascular flora of Lake Kainji. p. 479–487. *In*: Proceedings of the International Conference on Kainji Lake and River Basins Development in Africa, Vol. II. Lake Kainji Research Institute, New Bussa, Nigeria.

Charles-Dominique, E. (1982) Exposé synoptique des données biologiques sur l'ethmalose, *Ethmalosa fimbriata* S. Bowdich, 1825. Revue d'Hydrobiologie Tropicale 15: 373–397.

Chaudhry, A. B. and Chachu, R. E. O. (1979) Preliminary study of some indigenous vascular flora in the drawdown area of Kainji Lake and its possible significance to lakeshore farming. p. 529–539. *In*: Proceedings of the International Conference on Kainji Lake and River Basins Development in Africa. Vol. II. Lake Kainji Research Institute, New Bussa, Nigeria.

Chevalier, A. (1932) Les associations végétales du lit du Moyen-Niger. Compte Rendu Sommaire des Séances. Société de Biogéographie. Paris 9: 73–77.

Cogels, F. X. and Gac, J. -Y. (1982) Le lac de Guiers: fonctionnement, bilans hydriques. Evaporation d'une nappe d'eau libre en zone sahélienne (Sénégal). Cahiers de l'Office de la Recherche Scientifique et Technique Outre-mer, sér. Geologie 12: 41–60.

Cook, C. D. K. (1965) The aquatic and marsh plant communities of the reservoir site. p. 21–42. *In*: E. White (ed.). The First Scientific Report of the Kainji Biological Research Team. University of Liverpool, Liverpool, United Kingdom.

Cook, C. D. K. (1968) The vegetation of the Kainji reservoir site in northern Nigeria. Vegetation 15: 225–243.

Daget, J. (1954) Les poissons de Niger supérieur. Mémoire de l'Institut Français d'Afrique Noire 36: 1–391.

Daget, J. (1957) Données recentes sur la biologie des poissons dans le delta central du Niger. Hydrobiologia 9: 321–347.

Daget, J. and Durand, J. R. (1968) Étude du peuplement de poissons d'un milieu saumâtre poikilohalin, la baie de Cocody en Côte d'Ivoire. Cahiers de l'Office de la Recherche Scientifique et Technique Outre-mer, sér. 2: 91–112.

Dalziel, J. M. (1937) The Useful Plants of West Tropical Africa. Crown Agents, London, United Kingdom.

Davies, B. R. and Gass F. (1988) African Wetlands and Shallow Water Bodies Bibliography. Éditions de l'ORSTOM. Institut Français de Recherche Scientifique pour le Dévelopment en Coopération No. 211. Paris, France. 502 pp.

Davies, B. R. and Walker, K. F. (eds.) (1986) The Ecology of River Systems. Dr. W. Junk Publishers, The Hague, The Netherlands. 793 pp.

Davies, B. R. and Walmsley, R. D. (eds.) (1985) Perspectives in Southern Hemisphere Limnology. Proceedings of a Symposium held in Wilderness, South Africa, July 3–13, 1984. Developments in Hydrobiology, Series 28. D. W. Junk, Dordrecht, The Netherlands. 263 pp.

Dejoux, C. (1983) The fauna associated with the aquatic vegetation. p. 273–292. *In*: J. P. Carmouze, J. R. Durand and C. Lévéque (eds.). Lake Chad. Dr. W. Junk, Dordrecht, The Netherlands.

Dejoux, C., Lauzanne, L and Lévêque, C. (1971) Prospection hydrobiologique du lac de Léré. IV – Faune benthique. Cahiers de l'Office de la Recherche Scientifique et Technique Outre-mer, sér. Hydrobiologie 5: 179–185.

Denny, P. (1972) Sites of nutrient absorption in aquatic macrophytes. Journal of Ecology 60: 819–829.

Denny, P. (1973) Lakes of south-western Uganda. II: Vegetation studies on Lake Bunyonyi. Freshwater Biology 3: 123–135.

Denny P. (1980) Solute movement in submerged angiosperms. Biological Review of the Cambridge Philosophical Society 55: 65–92.

Denny, P. (1984) Permanent swamp vegetation of the Upper Nile. Hydrobiologia 110: 79–90.

Denny, P. (ed.) (1985a) The Ecology and Management of African Wetland Vegetation. Dr. W. Junk, Dordrecht, The Netherlands. 344 pp.

Denny, P. (1985b) Submerged and floating-leaved aquatic macrophytes (euhydrophytes). p. 19–42. *In*: P. Denny (ed.). The Ecology and Management of African Wetland Vegetation. Dr. W. Junk, Dordrecht, The Netherlands.

Denny, P. (1985c) Wetland plants and associated plant life-forms. p. 1–18. *In*: P. Denny (ed.). The Ecology and Management of African Wetland Vegetation. Dr. W. Junk, Dordrecht, The Netherlands.

Denny, P. (1985d) The structure and functioning of African euhydrophyte communities. p. 125–151. *In*: P. Denny (ed.). The Ecology and Management of African Wetland Vegetation. Dr. W. Junk, Dordrecht, The Netherlands.

Denny, P. (1989) Wetlands. *In*: Strategic Resources Planning in Uganda. UNEP Report. Nairobi, Kenya. Vol. IX. 103 pp.

Denny, P. (1991) African wetlands. p. 115–148. *In*: M. Finlayson and M. Moser (eds.). Wetlands. Facts on File Ltd., Oxford, United Kingdom.

Denny, P. and Lye, K. A. (1973) The *Potamogeton schweinfurthii* complex in Uganda. Kew Bulletin 28: 117–120.

Department of Water Affairs. (1986) Management of the Water Resources of the Republic of South Africa. Department of Water Affairs, Pretoria, South Africa.

Deuse, P. (1966) Contribution à l'étude des tourbières du Rwanda et du Burundi. p. 53–115. *In*: Treizième Rapport d'Institut pour la Recherche Scientifique en Afrique Centrale (I.R.S.A.C.). 1960–1964.

Drewes, A. K. (1988) TWISK 53 8: Analysis of water management options for a seasonally inundated wetland – a system dynamics approach. Technical Report, Centre for Advanced Computing and Decision Support, Council for Scientific and Industrial Research, Pretoria, South Africa. 252 pp.

Dufour, P. (1982a) Influence des conditions de milieu sur la biodégradation des matières organiques dans une lagune tropicale. Oceanologica Acta 5: 355–363.

Dufour, P. (1982b) Les frontières naturelles et humaines du système lagunaire Ébrié. Incidences sur l'hydroclimat. Hydrobiologia 94: 105–120.

Dufour, P. (1982c) Modèles semi-empiriques de la production phytoplanctonique en milieu lagunaire tropical (Côte d'Ivoire). Acta Oecologica, Oecologica Genéralis 3: 223–239.

Dufour, P. (1984) La biomasse végétale des lagunes ivoiriennes. Exemple de la lagune Ébrié (Côte d'Ivoire). Revue d'Hydrobiologie Tropicale 17: 207–233.

Dufour, P. and Durand, J.-R. (1982) Production végétale des lagunes ivoiriennes. Revue d' Hydrobiologie Tropicale 15: 209–230.

Dufour, P., Chantraine, J. M. and Durand, J. R. (1985) Impact of man on Ébrié lagoon ecosystem (Ivory Coast). p. 467–484. *In*: L. N. Chao and W. W. Kirby-Smith (eds.). Utilization of Coastal Ecosystems: Planning, Pollution and Productivity. Vol. 1. University of Rio Grande de Sul, Brazil.

Dufour, P., Cremoux, J. L. and Slepoukha, M. (1981) Contrôle nutritif de la biomasse du seston dans une lagune tropicale de Côte d'Ivoire. 1. Étude méthodologique et premiers résultats. Journal of Experimental Marine Biology and Ecology 51: 247–267.

Dufour, P., Lemasson, L. and Cremoux, J. L. (1981) Contrôle nutritif de la biomasse du seston dans une lagune tropicale de Côte d'Ivoire. II. Variations géographiques et saisonnières. Journal of Experimental Marine Biology and Ecology 51: 269–284.

Dugan, P. J. (ed.) (1990) Wetland Conservation. A Review of Current Issues and Required Action. IUCN, Glands, Switzerland. 96 pp.

Duong-Huu-Thoï. (1950a) Introduction à l'étude de la végétation du Soudan Français. p. 49–54. *In*: Conferência Internacional dos Africanistas ocidentais 2A. Vol. II, Trabalhos apresentados à 2A. Secçao (Meio Biologico), la. Parte, Lisboa, Portugal.

Duong-Huu-Thoï. (1950b) Étude préliminaire de la végétation du Delta Central Nigerien. p. 55–156. *In*: Conferência Internacional dos Africanistas ocidentais 2A. Vol. II, Trabalhos apresentados à 2A. Secçao (Meio Biologico), la. Parte, Lisboa, Portugal.

Durand, J.-R. (1980) Evolution des captures totales (1962–1977) et devenir des pêcheries de la région du lac Tchad. Cahiers de l'Office de la Recherche Scientifique et Technique, sér. Hydrobiologie 13: 93–111.

Durand, J.-R. and Chantraine, J. M. (1982) L'environnement climatique des lagunes ivoiriennes. Revue d'Hydrobiologie Tropicale 15: 85–113.

Durand, J.-R. and Lévêque, C. (1980) Flore et Faune Aquatiques de l'Afrique Sahelo-Soudanienne. I. Office de la Recherche Scientifique et Technique Outer-mer Collection Initiations – Documentation Techniques No. 45. Paris, France. 389 pp.

Durand, J.-R. and Skubich, M. (1982) Les lagunes ivoiriennes. Aquaculture 27: 211–250.

Egborge, A. B. M. (1974) The seasonal variation and distribution of phytoplankton in the River Oshun, Nigeria. Freshwater Biology 4: 177–191.

Egborge, A. B. M. (1979) Observations on the diurnal changes in some physico-chemical variables of Lake Asejire – a new impoundment in Nigeria. Polskie Archwum Hydrobiologii 26: 301–311.

Eggeling, W. J. (1935) The vegetation of Namanve Swamp, Uganda. Journal of Ecology 23: 422–435.

Ellenbroek, G. A. (1987) Ecology and Productivity of an African Wetland System. The Kafue Flats, Zambia. Dr. W. Junk Publishers, The Hague, The Netherlands. 267 pp.

Ellery, K. Ellery, W. N., Rogers, K. H. and Walker, B. H. (1990) Formation, colonization and fate of floating sudds in the Maunachira river system of the Okavango Delta, Botswana. Aquatic Botany 38: 315–330.
El-Zarka, S. E. 1973. Kainji Lake, Nigeria. Geophysical Monographs (American Geophysical Union) 17: 197–219.
Entz, B. (1969) Limnological conditions in Volta Lake, the greatest man-made lake of Africa. UNESCO Nature and Resources 5: 9–16.
Equatorial Nile Project (1954) Being the report of the Jonglei Investigation Team, Sudan Government, London, United Kingdom.
Ewer, D. W. (1966) Biological investigations on the Volta Lake. May 1964 to May 1965. p. 21–32. *In*: R. H. Lowe-McConnell (ed.). Man-Made Lakes. Institute of Biology No. 15. Academic Press, London, United Kingdom.
Finlayson, M. and Moser, M. (eds.) (1991) Wetlands. Facts on File Ltd., Oxford, United Kingdom. 224 pp.
Frempong, E. and Nijjhar, B. (1973) Some preliminary observations on the fauna and flora of Barekese Lake, Ghana. Bulletin de l'Institut Fondamental d'Afrique Noire, sér. A, 35: 67–78.
Furness, H. D. and Breen, C. M. (1980). The vegetation of seasonally flooded areas of the Pongolo river floodplain. Bothalia 13: 217–231.
Furness, H. D. and Breen, C. M. (1982) Decomposition of *Cynodon dactylon* (L.) Pers. in seasonally flooded areas of the Pongolo River floodplain: pattern and significance of dry matter and nutrient loss. Hydrobiologia 97: 119–126.
Furness, H. D. and Breen, C. M. (1985) Interaction between period of exposure, grazing and crop growth rate of *Cynodon dactylon* (L.) Pers. on seasonally flooded areas of the Pongolo River floodplain. Hydrobiologia 126: 65–73.
Gac, J. Y. (1980) Géochimie comparée du bassin du lac Tchad. Bilan de l'altération de l'érosion et de la sédimentation. Travaux et Documents. L'Office de la Recherche Scientifique et Technique Outre-mer, No. 123. 251 pp.
Gaudet, J. J. (1975) Mineral concentration in papyrus in various African swamps. Journal of Ecology 63: 483–491.
Gaudet, J. J. (1976a) Nutrient relationships in the detritus of a tropical swamp. Archiv für Hydrobiologie 78: 213–239.
Gaudet, J. J. (1976b) *Salvinia* infestation on Lake Naivasha in East Africa. p. 193–209. *In*: C. K. Varshney and J. Rzóska (eds.). Aquatic Weeds in South East Asia. Dr. W. Junk, The Hague, The Netherlands.
Gaudet, J. J. (1977a) Natural drawdown on Lake Naivasha, Kenya, and the formation of papyrus swamps. Aquatic Botany 3: 1–47.
Gaudet, J. J. (1977b) Uptake accumulation and loss of nutrients by papyrus in tropical swamps. Ecology 58: 415–422.
Gaudet, J. J. (1979) Aquatic weeds in African man-made lakes. PANS 25: 279–286.
Gaudet, J. J. (1982) Nutrient dynamics of pappyrus swamps. p. 305–319. *In*: B. Gopal, R. E. Turner, R. G. Wetzel and D. F. Whigham (eds.). Wetlands: Ecology and Management. Proceedings of the 1st International Wetlands Conference, New Delhi, 10–17 September 1980. National Institute of Ecology, New Delhi, India.
Gaudet, J. J. and Muthuri, F. M. (1981) Nutrient relationships in shallow water in African lake, Lake Naivasha. Oecologia 49: 109–118.
Geldenhuys, J. N. (1976) Relative abundance of waterfowl in the Orange Free State. Ostrich 47: 27–54.
Geldenhuys, J. N. (1982) Classification of the pans of the western Orange Free State according to vegetation structure, with reference to avifaunal communities. South African Journal of Wildlife Research 12: 55–62.
Giglioli, M. E. C. and Thornton, I. (1965) The mangrove swamps of Keneba, Lower Gambia River Basin. I. Descriptive notes on the climate, the mangrove swamps and the physical communities.

Good, R. E., Whigham, D. F. and Simpson, R. L. (eds.) (1978) Freshwater Wetlands. Academic Press, New York, USA. 378 pp.

Goodman, P. S. (1987) Mkuze floodplain and swamps – wetland management case study. African Wildlife 41: 269–270.

Gras, R. and Saint-Jean, L. (1971) Prospection hydrobiologique du lac de Lérec et mares avoisinantes. III. Cladocéres et Copépodes. Cahiers de l'Office de la Recherche Scientifique et Technique Outre-mer, sér. Hydrobiologie 5: 175–178.

Gras, R. and Saint-Jean, L. (1983) Production du zooplancton du Lac Tchad. Revue d'Hydrobiologie Tropicale 16: 57–77.

Greenwood, P. H. and Lund, J. W. G. (1973) Introductory Remarks. Proceedings of the Royal Society of London, ser. B. 1984: 229–233 *et seq*.

Greeson, P. E., Clark, J. R. and Clark, J. E. (eds.) (1979) Wetland Functions and Values: The State of Our Understanding. American Water Resources Association, Minneapolis, Minnesota, USA. 674 pp.

Griffiths, J. F. (1972a) Nigeria. p. 167–192. *In*: J. F. Griffiths (ed.). World Survey of Climatology. Vol. 10. Climates of Africa. Elsevier, Amsterdam, The Netherlands.

Griffiths, J. F. (1972b) General introduction. p. 1–35. *In*: J. F. Griffiths (ed.). World Survey of Climatology. Vol. 10. Climates of Africa. Elsevier, Amsterdam, The Netherlands.

Grove, A. T. (1972) The dissolved and solid load carried by some West African rivers: Senegal, Niger, Benue and Shari. Journal of Hydrology 16: 277–300.

Grove, A. T. (ed.) (1985) The Niger and Its Neighbours: Environmental History and Hydrobiology, Human Use and Health Hazards of the Major West African Rivers. A. A. Balkema, Rotterdam, The Netherlands. 331 pp.

Guiral, D. (1984) Devenir de la matière organique particulaire dans un milieu eutrophe tropical (baie de Biétri, lagune Ébrié, Côte d'Ivoire). Revue d'Hydrobiologie Tropicale 17: 191–206.

Hall, J. B. (1970) Observations on the vegetation of the Lake Volta drawdown area. University of Ghana Volta Basin Research Project. Technical Report x35. (mimeograph). Legon, Ghana. 21 pp.

Hall, J. B. (1975) The vascular flora of Lake Kainji and its shores. p. 149–156. *In*: A. M. A. Imevbore and D. S. Adegoke (eds.). The Ecology of Lake Kainji. University of Ife Press. Ife, Nigeria.

Hall, J. B. and Okali, D. U. U. (1974) Phenology and productivity of *Pistia stratiotes* L. on the Volta Lake, Ghana. Journal of Applied Ecology 11: 709–726.

Hall, J. B. and Pople, W. (1968) Recent vegetation changes in the lower Volta River. Ghana Journal of Science 8: 24–29.

Hall, J. B. and Swaine, M. D. (1981) Distribution and ecology of vascular plants in a tropical rain forest vegetation in Ghana. Dr. W. Junk, Dordrecht, The Netherlands. 383 pp.

Hall, J. B., Laing, E., Hossain, M. and Lawson, G. W. (1969) Observations on aquatic weeds in the Volta Basin. p. 331–336. *In*: L. E. Obeng (ed.). Man-made Lakes: The Accra Symposium. Ghana Universities Press, Accra, Ghana.

Hall, J. B., Pierce, P. C. and Lawson, G. W. (1971) Common plants of the Volta Lake. University of Ghana, Legon, Ghana. 123 pp.

Hamilton, A. C. (1982) Environmental History of East Africa. Academic Press, London. 328 pp.

Heeg, J. and Breen, C. M. (1982) Man and the Pongolo floodplain. South African National Programme Report No. 56. Council for Scientific and Industrial Research, Pretoria, South Africa. 117 pp.

Heeg, J., Breen, C. M., Colvin, P. M., Furness, H. D. and Musil, C. F. (1978) On the dissolved solids of the Pongolo floodplain pans. Journal Limnological Society of South Africa 4: 59–64.

Hill, B. J. (1969) The bathymetry and possible origin of Lakes Sibaya, Nhlange and Sifungwe in Zululand (Natal). Transactions of the Royal Society of South Africa 38: 205–216.

Hill, B. J. (1975) The origin of southern African coastal lakes. Transactions of the Royal Society of South Africa 41: 225–240.

Hilton, T. E. and Kowu-Tsri, J. Y. (1970) The impact of the Volta scheme on the lower Volta floodplains. Journal of Tropical Geography 30: 29–37.

Hoier, R. (1950) Exploration du Parc National Albert. Institut Parcs Nationaux Congo Belge, Bruxelles, Belgium. 172 pp.

Holden, M. J. and Green, J. (1960) The hydrology and plankton of the River Sokoto. Journal of Animal Ecology 29: 65–84.

Hove, C. van, Diara, H. F. and Godard, P. (1983) *Azolla* in West Africa. Azolla Project, West Africa Development Association. Court-St-Etienne, Belgium. 53 pp.

Howard-Williams, C. (1977) The distribution of nutrients in Swartvlei, a southern Cape coastal lake. Water SA, 3: 213–217.

Howard-Williams, C. (1979a) Distribution, biomass and role of aquatic macrophytes in Lake Sibaya. p. 88–107. *In*: B. R. Allanson (ed.). Lake Sibaya. Dr. W. Junk Biological Monographs, The Hague, The Netherlands.

Howard-Williams, C. (1979b) Interactions between swamp and lake. p. 231–245. *In*: M. Kalk, A. J. McLachlan and C. Howard-Williams (eds.). Lake Chilwa. Dr. W. Junk, The Hague, The Netherlands.

Howard-Williams, C. (1979c) The distribution of aquatic macrophytes in Lake Chilwa: annual and long term environmental fuctuations. p. 105–122. *In*: M. Kalk, A. J. McLachlan and C. Howard-Williams (eds.). Lake Chilwa. Dr. W. Junk, The Hague, The Netherlands.

Howard-Williams, C. (1980a) Aquatic macrohytes of the coastal wetlands of Maputaland. p. 42–51. *In*: M. N. Bruton and K. H. Cooper (eds.). Studies on the Ecology of Maputaland. Rhodes University Press, Grahamstown, South Africa.

Howard-Williams, C. (1985) Cycling and retention of nitrogen and phosphorus in wetlands: a theoretical and applied perspective. Freshwater Biology 15: 391–431.

Howard-Williams, C. and Gaudet, J. J. (1985) The structure and functioning of African swamps. p. 153–175. *In*: P. Denny (ed.). The Ecology and Management of African Wetland Vegetation. Dr. W. Junk, The Hague, The Netherlands.

Howard-Williams, C. and Thompson, K. (1985) The conservation and management of African wetlands. p. 203–230. *In*: P. Denny (ed.). The Ecology and Management of African Wetland Vegetation. Dr. W. Junk, The Hague, The Netherlands.

Howard-Williams, C. and Walker, B. H. (1974) The vegetation of a tropical African lake: classification and ordination of the vegetation of Lake Chilwa (Malawi). Journal of Ecology 62: 831–854.

Howell, P., Lock, M. and Cobb, S. (1988) The Jonglei Canal. Cambridge University Press, Cambridge, United Kingdom. 537 pp.

Hutchinson, G. E., Pickford, G. E. and Schuurman, J. F. M. (1932) A contribution to the hydrobiology of pans and other inland waters of South Africa. Archiv für Hydrobiologie 24: 1–154.

Hutchinson, I. G. P. and Pitman, W. V. (1973) Climatology and hydrology of the St Lucia lake system. St Lucia Lake Research Report, Vol. 1, Natal Provincial Administration, Hydrological Research Unit, Pietermaritzburg, South Africa.

Ita, E. O., Sado, E. K., Balogun, J. K., Pandogari, A. and Ibitoye, B. (1985) Inventory survey of Nigeria inland waters and their fishery resources. I. A preliminary checklist of inland water bodies in Nigeria with special reference to ponds, lakes, reservoirs and major rivers. Kainji Lake Research Institute, Technical Set. 14: iv-51.

Iltis, A. (1973a) Algues des eaux natronées du Kanem (Tchad). Premiéres parties. Cahiers de l'Office de la Recherche Scientifique et Technique Outre-mer, sér. Hydrobiologie 6: 173–246.

Iltis, A. (1973b) Phytoplancton des eaux natronées du Kanem (Tchad). VI. Estimation des biomasses et groupes d'algues dominants. Cahiers de l'Office de la Recherche Scientifique et Technique Outre-mer, sér. Hydrobiologie 7: 164–194.

Iltis, A. (1974) Phytoplancton des eaux natronées du Kanem (Tchad). VIII. Classification des milieux étudiés et espèces caractéristiques. Cahiers de l'Office de la Recherche Scientifique et Technique Outre-mer, sér. Hydrobiologie 8: 81–91.

Iltis, A. (1975) Phytoplancton des eaux natronées du Kanem (Tchad). X. Conclusions. Cahiers de l'Office de la Recherche Scientifique et Technique Outre-mer, sér. Hydrobiologie 9: 13–17.
Iltis, A. (1977) Peuplements phytoplanctoniques du lac Tchad. I. Stade Tchad Normal (Février 1971 et Janvier 1972): II. Stade Petit Tchad (Avril 1974, Novembre 1974 et Février 1975): III. Remarques générales. Cahiers de l'Office de la Recherche Scientifique et Technique Outre-mer, sér. Hydrobiologie 11: 33–52, 53–72, 189–199.
Iltis, A. and Lemoalle, J. (1983) The aquatic vegetation of Lake Chad. p. 125–144. *In*: J. -P. Carmouze, J. -R. Durand and C. Lévêque (eds.). Lake Chad: Ecology and Productivity of a Shallow Tropical Ecosystem. Dr. W. Junk Monographiae Biologicae 53, Dordrecht, The Netherlands.
Iltis, A. and Lévêque, C. (1982) Caractéristiques physico-chimiques des rivières de Côte d'Ivoire. Revue d'Hydrobiologie Tropicale 15: 115–130.
Imevbore, A. M. A. (1970) The chemistry of the River Niger in the Kainji reservoir area. Archiv Hydrobiologie 67: 412–431.
Imevbore, A. M. A. (1971) Floating vegetation of Lake Kainji, Nigeria. Nature, London 230: 599–600. (Reprinted in Imevbore and Adegoke, 1975).
Imevbore, A. M. A. (1975) Floating vegetation of Lake Kainji. p. 146–148. *In*: A. M. A. Imevbore and O. S. Adegoke (eds.). The Ecology of Lake Kainji: The Transition from River to Lake. University of Ife Press, Ife, Nigeria.
Imevbore, A. M. A. and Adegoke, S. O. (eds.) (1975) The Ecology of Lake Kainji. The Transition from River to Lake. University of Ife Press, Ife, Nigeria.
Imevbore, A. M. A. and Bakare, O. (1974) A pre-impoundment study of swamps in the Kainjii Lake basin. African Journal of Tropical Hydrobiology and Fisheries 3: 79–93.
Imevbore, A. M. A. and Visser, S. A. (1969) A study of microbiological and chemical stratification of the Niger River within the future Kainji Lake area. p. 94–102. *In*: L. E. Obeng (ed.). Man-made Lakes: The Accra Symposium. Ghana Universities Press, Accra, Ghana.
Jacot-Guillarmod, A. (1962) The bogs and sponges of the Basutoland mountains. South African Journal of Science 58: 179–182.
John, D. M. (1986) The Inland Waters of Tropical West Africa. Ergebnisse der Limnologie. Archiv für Hydrobiologie Beihefte 23. E. Schweizerbart'che, Stuttgart, Germany. 244 pp.
Johnson, D. N. (1986) The ornithological importance of the Sodwana Bay State Forest. p. 33–44. *In*: C. -H. Stormanns and C. M. Breen (compilers). Proceedings of the Greater Mkuze Swamp System Syposium and Workshop. Institute of Natural Resources Investigational Report No. 22. University of Natal, Pietermaritzburg, South Africa.
Kalk, M., McLachlan, A. J. and Howard-Williams, C. (eds.) (1979) Lake Chilwa, Studies of Change in a Tropical Ecosystem. Dr. W. Junk Monographiae Biologicae 35, The Hague, The Netherlands. 462 pp.
Kashef, A. -A. I. (1981) The Nile – one river and nine countries. Journal of Hydrology 53: 53–71.
Kaul, S. N. (1975) Agricultural Potential. FAO, Rome, Italy.
Kingdon, J. (1990) Island Africa, The Evolution of Africa's Rare Animals and Plants. Collins, London, United Kingdom. 287 pp.
Klumpp, R. K. and Chu, K. Y. (1977) Ecological studies of *Bulinus rohlfsi*, the intermediate host of *Schistosoma haematobium* in the Volta Lake. Bulletin World Health Organization 55: 715–730.
Lawson, C. W. (1964) A preliminary report on the aquatic weeds in the Volta Basin. Volta Basin Research Project Technical Report No. X6 (mimeograph). Legon, Ghana.
Lawson, G. W. (1967) 'Sudd' formation on the Volta Lake. Bulletin de l'Institut Fondamental d'Afrique Noire, sér. A, 29: 1–4.
Lawson, G. W. (1970) Lessons of the Volta – a new man-made lake in tropical Africa. Biological Conservation 2: 90–96
Lawson, G. W. (ed.) (1986) Plant Ecology in West Africa: Systems and Processes. John Wiley and Sons, New York, USA. 357 pp.

Lawson, G. W., Petr, T., Biswas, S., Biswas, E. R. I. and Reynolds, J. D. (1969) Hydrobiological work of the Volta Basin Research Project 1963–1968. Bulletin de l'Institut Fondamental d'Afrique Noire, sér. A, 31: 965–1003.

Ledger, D. C. (1964) Some hydrological characteristics of West African rivers. Transactions Institute British Geographers 35: 73–90.

Lemasson, L. and Pagès, J. (1980) Note on the excretion of dissolved organic phosphorus in tropical brackish water. Estuarine and Coastal Marine Science 11: 511–523.

Lemasson, L., Pagès, J. and Cremoux, J. L. (1980) Inorganic phosphate uptake in a brackish tropical lagoon. Estuarine and Coastal Marine Science 11: 547–561.

Lemasson, L., Pagès, J. and Cremoux, J. L. (1982) Exchanges d'éléments nutritifs dissous entre l'eau et le sédiment dans une lagune tropicale saumâtre. Revue d'Océanographie Tropicale 17: 45–58.

Lemasson, L., Pagès, J., Dufour, P. and Cremoux, J. L. (1981) Matière organique particulaire et biomasse dans une lagune tropicale. Revue d'Hydrobiologie Tropicale 14: 191–212.

Lemoalle, J. (1973) L'énergie lumineuse et l'activité photosynthétique du phytoplancton dans le lac Tchad. Cahiers de l'Office de la Recherche Scientifique et Technique Outre-mer, sér. Hydrobiologie 7: 95–116.

Lemoalle, J. (1979a) Biomasse et production phytoplanctoniques du lac Tchad (1968–1976). Relations avec les conditions de milieu. L'Office de la Recherche Scientifique et Technique Outre-mer, Paris, France. 311 pp.

Lemoalle, J. (1979b) Application des données Landsat à l'estimation de la production du phytoplancton dans le lac Tchad. Cahiers de l'Office de la Recherche Scientifique et Technique Outre-mer, sér. Hydrobiologie 13: 35–46.

Lévêque, C. (1971) Prospection hydrobiologique du lac le Léré et des mares avoisinantes. I. Milieu physique. Cahiers de l'Office de la Recherche Scientifique et Technique Outre-mer, sér. Hydrobiologie 5: 161–169.

Lévêque, C. (1972) Mollusques benthiques du lac Tchad: écologie, étude des peuplements et estimation des biomasses. Cahiers de l'Office de la Recherche Scientifique et Technique Outre-mer, sér. Hydrobiologie 6: 3–45.

Lévêque, C. (1973) Dynamique des peuplements, biologie, et estimation de la production des mollusques benthiques du lac Tchad. Cahiers de l'Office de la Recherche Scientifique et Technique Outre-mer, sér. Hydrobiologie 7: 117–147.

Lind, E. M. and Morrison, M. E. S. (1974) East African Vegetation. Longmans, London, United Kingdom. 257 pp.

Lind, E. M. and Visser, S. A. (1962) A study of a swamp at the north end of Lake Victoria. Journal of Ecology 50: 599–613.

Little, E. C. S. (1969) Weeds and man-made lakes. p. 284–291. *In*: L. E. Obeng (ed.). Man-Made Lakes: The Accra Symposium. Ghana University Press, Accra, Ghana.

Lock, J. M. (1973) The aquatic vegetation of Lake George, Uganda. Phytocoenologia 1: 250–262.

Lowe, K. (1987) Editorial. Nigerian Field 52, 65.

Luther, H. and Rzóska, J. (1971) *Project Aqua.* IBP Handbook No. 21. Blackwell Scientific Publications, Oxford, United Kingdom. 239 pp.

Mabogunjie, A. L. (ed.) (1973) Kainji: A Nigerian Man-made Lake, Socio-Economic Conditions. Kainji Lake Studies Vol. 2. Ibadan Universities Press, Ibadan, Nigeria.

Maclean, G. L. (1985) Roberts' Birds of Southern Africa. The Trustees of the John Voelcker Bird Book Fund, Cape Town, South Africa. 848 pp.

Maglione, G. (1969) Première données sur le régime hydro-géochimique des lacs permanents du Kanem (Tchad). Cahiers de l'Office de la Recherche Scientifique et Technique Outre-mer, sér. Hydrobiologie 3: 121–141.

Maglione, G. (1976) Géochimie des évaporites et silicates néoformés en milieu continental confiné. Les dépressions interdunaires du Tchad. Travaux et Documents. L'Office de la Recherche Scientifique et Technique Outre-mer, No. 50. 335 pp.

Marshall, B. E. and Junod, F. J. R. (1981) The decline of *Salvinia molesta* on Lake Kariba. Hydrobiologia 83: 477–484.

Martin, A. R. H. (1960) The ecology of Groenvlei, a South African fen. I: The primary communities. Journal of Ecology 48: 55–71.

Maud, R. R. (1968) Quaternary geomorphology and soil formation along the Natal coast. Zeitschrift für Geomorphologie, suppl. 7: 155–199.

Maud, R. R. (1980) The climate and geology of Maputaland. p. 1–7. *In*: N. M. Bruton and K. H. Cooper (eds.). Studies on the Ecology of Maputaland. Rhodes University, Grahamstown, and the Natal Branch of the Wildlife Society of Southern Africa, Grahamstown, South Africa.

Mefit-Babtie (1983) Development Studies in the Jonglei Canal Area. Swamp Ecology Survey. A report to the Jonglei Executive Organ, Government of the Sudan. Vols. 1–4.

Menaut, J. -C. (1983) The vegetation of Africa savannas. p. 109–149. *In*: F. Bouliere (ed.). Tropical Savannas. Elsevier, Amsterdam, The Netherlands.

Menaut, J. -C. (1983) The structure and dynamics of a West African Savanna. p. 80–100. *In*: B. J. Huntley and B. H. Walker (eds.). Ecology of Tropical Savannas. Springer-Verlag, Berlin, West Germany.

Mepham, J. S. (1987) Southern Africa/Afrique Australe, p. 457–593. *In*: M. J. Burgis and J. J. Symoens (eds.). African Wetlands and Shallow Water Bodies. Éditions de L'ORSTOM, Institut Français de Recherche Scientifique pour le Dévelopment en Coopération No. 211, Paris, France.

Mepham, R. H. and Mepham, J. S. *In press*. A Directory of African Wetlands.

Migahid, A. M. (1947) An ecological study of the 'sudd' swamps of the Upper Nile. Proceedings of the Egyptian Academy of Science 3: 57–86.

Mitchell, D. S. (1978) Freshwater Plants. p. 1113–1138. *In*: M. J. A. Werger (ed.). Biogeography and Ecology of Southern Africa. Dr. W. Junk, The Hague, The Netherlands.

Mitchell, D. S. (1985a) Surface-floating aquatic macrophytes. p. 109–124. *In*: P. Denny (ed.). The Ecology and Management of African Wetland Vegetation. Dr. W. Junk, Dordrecht, The Netherlands.

Mitchell, D. S. (1985b) African aquatic weeds and their management. p. 177–202. *In*: P. Denny (ed.). The Ecology and Management of African Wetland Vegetation. Dr. W. Junk, Dordrecht, The Netherlands.

Mitchell, D. S., Denny, P. and Howard-Williams, C. (1985) African Wetland vegetation – concluding perspective. p. 231–238. *In*: P. Denny (ed.). The Ecology and Management of African Wetland Vegetation. Dr. W. Junk, Dordrecht, The Netherlands.

Morrison, M. E. S. (1968) Vegetation and climate in the uplands of south-western Uganda during the later pleistocene period. I. Muchoya Swamp, Kigezi District. Journal of Ecology 56: 363–384.

Morton, A. J. and Obot, E. A. (1984) The control of *Echinochloa stagnina* (Ratz.) P. Beauv. by harvesting for dry season fodder in Lake Kainji, Nigeria – a modelling approach. Journal of Applied Ecology 21: 687–694.

Mulligan, H. F. (1972) Report on Growth of *Pistia stratiotes* in Kossou Lake, Ivory Coast. FAO Lake Kusson Fishery Development Project. FI:DP/IVC/2671/526/1. Rome, Italy. 7 pp (also issued in French).

Musil, C. F., Grunow, J. O. and Bornman, C. H. (1973) Classification and ordination of aquatic macrophytes in the Pongolo river pans, Natal. Bothalia 11: 181–190.

Newton, L. E. (1986) General features of West African waters: the aquatic macrophytes. p. 160–181. *In*: D. M. John (ed.). The Inland Waters of Tropical West Africa. An Introduction and Botanical Review. Ergebnisse der Limnologie. Archiv für Hydrobiologie Beihefte 23. E. Schweizerbatche, Stuttgart, Germany.

Noble, R. G. and Hemens, J. (1978) Inland water ecosystems in South Africa – a review of research needs. South African National Scientific Programmes Report No. 34. Council for Scientific and Industrial Research, Pretoria, South Africa. 150 pp.

Obeng, L. E. (ed.) (1969a) Man-made Lakes: The Accra Symposium. Ghana Universities Press, Accra, Ghana. 398 pp.

Obeng, L. E. (1969b) The invertebrate fauna of aquatic plants of the Volta Lake in relation to the spread of helminth parasites. p. 320–325. *In*: L. E. Obeng (ed.). Man-made Lakes: The Accra Symposium. Ghana Universities Press, Accra, Ghana.

Obeng, L. E. (1973) Volta Lake: physical and biological aspects. Geophysical Monographs (American Geophysical Union) 17: 87–98.

Obeng, L. E. (1981) Man's impact on tropical rivers. p. 265–288. *In*: M. A. Lock and D. D. Williams (eds.). Perspectives in Running Water Ecology. Plenum Press, New York, USA.

Obeng-Asamoa, E. K. (1979) The effect of the Akosombo Dam on the oxygen regime of the Volta waters below the dam. p. 170–177. *In*: Proceedings of the International Conference on Kainji Lake and River Basins Development in Africa, Vol. I. Kainji Lake Research Institute, New Bussa, Nigeria.

Odei, M. A. (1973) Observations on some weeds of malacological importance in the Volta Lake. Bulletin de l'Institut Fondamental de l'Afrique Noire, sér. A, 35: 57–66.

Okali, D. U. U. and Hall, J. B. (1974) Colonization of *Pistia stratiotes* L. mats by *Scirpus cubensis* Poeppig & Kunth on the Volta Lake. Ghana Journal of Agricultural Science 7: 31–36.

Orme, A. R. (1973) Barrier and lagoon systems along the Zululand Coast, South Africa. p. 181–217. *In*: Coates, D. R. (ed.). Coastal Geomorphology. State University, New York, USA.

Pagano, M. and Saint Jean, L. (1983) Croissance en poids d'*Acartia clausi* en lagune Ébrié (Côte d'Ivoire). Revue d'Hydrobiologie Tropicale 16: 151–163.

Pagès, J. and Lemasson, L. (1981a) Production et utilisation du carbone organique dissous dans une lagune tropicale. Revue d'Hydrobiologie Tropicale 14: 83–101.

Pagès, J. and Lemasson, L. (1981b) Mesure de la production primaire dans une lagune tropicale. II. Bilan de la production par la méthode au ^{14}C. Revue d'Hydrobiologie Tropicale 14: 213–222.

Pagès, J., Lemasson, L. and Dufour, P. (1981a) Primary production in a brackish tropical lagoon. II. Effect of light, as studied at some stations by the ^{14}C method. Revue d'Hydrobiologie Tropicale 14: 3–15.

Pagès, J., Lemasson, L. and Dufour, P. (1981b) Variabilité des évaluations du seston dans une lagune tropicale. Revue d'Hydrobiologie Tropicale 14: 17–30.

Paperna, I. (1969) Aquatic weeds, snails and transmission of bilharzia in the new man-made Volta Lake in Ghana. Bulletin de l'Institut Fondamental de l'Afrique Noire, sér. A, 31: 487–499.

Petr, T. (1968) Population changes in aquatic invertebrates living in two water plants in a tropical man-made lake. Hydrobiologia 32: 449–485.

Pettet, A. and Pettet, S. J. (1970) Biological control of *Pistia stratiotes* L. in Western State, Nigeria. Nature, London 226: 282.

Reizer, C. (1974) Définition d'une politique d'aménagement d'un écosysteme aquatique complexe par l'étude de son environement abiotique, biotique et anthropique: le fleuve Sénégal moyen et inferieur. Thesis, Arlon, Belgium.

Robyns, W. (1947–1955) Flore des spermatophytes du Parc National Albert, vols. 1–3. Institut des Parcs Nationaux du Congo Belge, Bruxelles, Belgium.

Roche, M. A. (1980) Traçage naturel salin et isotopique des eaux du système hydrologique du lac Tchad. Travaux et Documents. L'Office de la Recherche Scientifique et Technique Outremer, No 117. 389 pp.

Rochette, C. (1974) Le bassin du fleuve Sénégal. Monographie hydrologique No. 1. L'Office de la Recherche Scientifique et Technique Outre-mer, Paris, France. 325 pp.

Rodier, J. (1964) Régimes hydrologiques de l'Afrique-Noire à l'Ouest du Congo. Édition provisoire 17. Paris, France. 302 pp.

Rogers, K. H. (1984) The role of *Potamogeton crispus* in the Pongolo river floodplain ecosystem. Unpublished Ph.D. dissertation, University of Natal, Pietermaritzburg, South Africa. 198 pp.

Rogers, K. H. and Breen, C. M. (1980) Growth and reproduction of *Potamogeton crispus* in a South African lake. Journal of Ecology 68: 561–571.

Rogers, K. H. and Breen, C. M. (1982) Decomposition of *Potamogeton crispus* L.: the effects of drying on the pattern of mass and nutrient loss. Aquatic Botany 12: 1–12.

Rogers, K. H. and Breen, C. M. (1983) An investigation of macrophyte, epiphyte and grazer interactions. p. 217–226. *In*: R. C. Wetzel, (ed.). Periphyton of Freshwater Ecosystems. Dr. W. Junk, The Hague, The Netherlands.

Rogers, K. H. and Breen, C. M. (1990a) Waterfowl of a sub-tropical African floodplain. 1. Seasonality of community composition and food resources. Wetlands Ecology and Management 1: 85–97.

Rogers, K. H. and Breen, C. M. (1990b) Waterfowl of a sub-tropical African floodplain. 2. Stability of the grazing system. Wetlands Ecology and Management 1: 99–109.

Rose Innes, R. (1977) A Manual of Ghana Grasses. Ministry of Overseas Development, Surbiton, United Kingdom.

Rzóska, J. (ed.) (1976a) The Nile, Biology of an Ancient River. Dr. W. Junk Monographiae Biologicae 29, The Hague, The Netherlands. 417 pp.

Rzóska, J. (1976b) Descent to the Sudan Plains. p. 197–214. *In*: J. Rzóska (ed.). The Nile, Biology of an Ancient River. Dr. W. Junk, The Hague, The Netherlands.

Rzóska, J. (1985) The water quality and hydrobiology of the Niger. p. 77–99. *In*: A. T. Grove (ed.). The Niger and Its Neighbours: Evolution, Ecology and Utilization. A. A. Balkema, Rotterdam, The Netherlands.

Sagua, V. O. (1979) Some uses and effects of multipurpose large dams in tropical Africa – the Kainji experience. p. 44–54. *In*: Proceedings of the International Conference on Kainji Lake and River Basins Development in Africa. Vol. 1. Kainji Lake Research Institute, New Bussa, Nigeria.

Saint-Jean, L. and Pagano, M. (1983) Rythme journalier de ponte chez *Acartia clausi* en lagune Ébrié (Côte d'Ivoire). Revue d'Hydrobiologie Tropicale 16: 145–150.

Sain-Jean, L. and Pagano, M. (1984) Influence de la salinité, de la température et de la quantité de particules en suspension sur la croissance et la production d'oeufs d'*Acartia clausi* en lagune Ébrié (Côte d'Ivoire). Revue d'Hydrobiologie Tropicale 17: 235–244.

Seaman, M. T. (1987) Pans of the Orange Free State. African Wildlife 41: 237–238.

Seaman, M. T. and Kok, D. J. (1987) Ecological diversity in Orange Free State Pans. p. 260–273. *In*: R. D. Walmsley and M. L. Botten (compilers). Ecology and Conservation of Wetlands in South Africa. Ecosystems Occasional Report Series No. 28, Council for Scientific and Industrial Research, Pretoria, South Africa.

Servant, M. and Servant, S. (1983) Paleolimnology of an upper quaternary endorheic lake in Chad Basin. p. 11–26. *In*: J. -P. Carmouze, J. -R. Durand and C. Lévêque (eds.). Lake Chad: Ecology and Productivity of a Shallow Tropical Ecosystem. Dr. W. Junk Monographiae Biologiceae 53, Dordrecht, The Netherlands.

Scotney, D. M. (1978) The present situation in Natal. p. 16–34. *In*: Symposium on the relationship between agriculture and environmental conservation in Natal and KwaZulu, Durban, 19–20 October 1978. Durban, South Africa.

Showers, V. (1979) World Facts and Figures. J. Wiley, New York, USA.

Siegfried, W. R., Frost, P. G. H., Cooper, J. and Kemp, A. C. (1976) South African red data book – Aves. South African National Programme Report No. 7. Council for Scientific and Industrial Research, Pretoria, South Africa. 108 pp.

Slinger, J. H. (1988) A mathematical modelling approach to effective flood management of a freshwater wetland and associated pastoral system. M.Sc. Thesis. University of Natal, Pietermaritzburg, South Africa. 181 pp.

Steinke, T. D. and Charles, L. M. (1984) Productivity and phenology of *Avicennia marina* (Forsk.) Vierh. and *Bruguiera gymnorrhiza* (L.) Lam. in Mgeni Estuary, South Africa. p. 25–36. *In*: H. J. Teas (ed.). Physiology and Management of Mangroves. Dr. W. Junk, The Hague, The Netherlands.

Steinke, T. D., Naidoo, G. and Charles, L. M. (1983) Degradation of mangrove leaf and stem tissues *in situ* in Mgeni Estuary, South Africa. *p*. 141–150. *In*: H. J. Teas (ed.). Biology and Ecology of Mangroves. Dr. W. Junk, The Hague, The Netherlands.

Stormanns, C-H. and Breen, C. M. (1987) The use of two multivariate techniques in the analysis of wetland vegetation for conservation evaluation. p. 274–282. *In*: R. D. Walmsley and M. L. Botten (compilers). Ecology and Conservation of Wetlands in South Africa. Ecosystems Occasional Report Series No. 28, Council for Scientific and Industrial Research, Pretoria, South Africa.

Sutcliffe, J. V. (1974) A hydrological study of the southern Sudd region of the Upper Nile. Hydrological Science Bulletin 19: 237–255.

Symoens, J. J. (1968) Exploration hydrobiologique du Lac Bangweolo et du Laupula. Cercle Hydrobiologie 11 (1). 191 pp.

Symoens, J. J. and Ohoto, E. (1973) Les éléments phytogeographiques da la flore macrophytique aquatique et semi-aquatique du Haut-Katanga. Verh. Internat. Verein. Limnol. 18: 1385–1394.

Symoens, J. J., Burgis, M. and Gaudet, J. J. (eds.) (1981) The Ecology and Utilization of African Inland Waters. UNEP Reports and Proceedings, Series 1. Nairobi, Kenya. 191 pp.

Tag el Seed, M. (ed.) (1981) The Blue Nile Invaded. Sudan Environment 1: 13.

Talling, J. F. and Talling, I. B. (1965) The chemical composition of African lake waters. Internationale Revue der gesamten Hydrobiologie und Hydrographie 50: 421–463.

Tarboton, W. R. (1979) The Nyl floodplains and its birds. Bokmakierie 31: 2–31.

Tarboton, W. R. (1987) The Nyl floodplain: its significance, phenology and conservation status. p. 101–114. *In*: R. D. Walmsley and M. L. Botten (compilers). Ecology and Conservation of Wetlands in South Africa. Ecosystems Occasional Report Series No. 28, Council for Scientific and Industrial Research, Pretoria, South Africa.

Thomas, J. D. and Tait, A. I. (1984) Control of the snail hosts of schistosomiasis by environmental manipulation: a field and laboratory appraisal in the Ibadan area, Nigeria. Philosophical Transactions of the Royal Society, London, ser. B 305: 201–253.

Thompson, B. W. (1965) The Climate of Africa. Nairobi, Kenya. 132 maps.

Thompson, K. (1976) Swamp development in the headwaters of the White Nile. p. 77–196. *In*: J. Rzósak (ed.). The Nile, Biology of an Ancient River. Dr. W. Junk Monographiae Biologicae, 29, The Hague, The Netherlands.

Thompson, K. (1985) Emergent plants of permanent and seasonally-flooded wetlands. p. 43–107. *In*: P. Denny (ed.). The Ecology and Management of African Wetland Vegetation. Dr. W. Junk, The Hague, The Netherlands.

Thompson, K. and Hamilton, A. C. (1983) Peatlands and swamps of the African continent. p. 331–373. *In*: A. J. P. Gore (ed.). Mires: Swamp, Bog, Fen and Moor. Elsevier, Ecosystems of the World Series, Vol. 4B (Regional Studies). Amsterdam, The Netherlands.

Thompson, K., Howard-Williams, C. and Mitchell, D. (1985) Cross-indexed bibliography of African wetland plants and vegetation. p. 237–316. *In*: P. Denny (ed.). The Ecology and Management of African Wetland Vegetation. Dr. W. Junk, The Hague, The Netherlands.

Thompson, K., Shewry, P. R. and Woolhouse, H. W. (1979) Papyrus swamp development in the Upemba Basin, Zaire: studies of population structure in *Cyperus papyrus* stands. Botanical Journal of the Linnean Society 78: 299–316.

Thornton, J. A. (ed.) (1982) Lake McIlwaine. Dr. W. Junk, The Hague, The Netherlands. 251 pp.

Tomlinson, P. B. (1986) The Botany of Mangroves. Cambridge University Press, Cambridge, United Kingdom. 413 pp.

Toupet, C. (1968) Major climatic elements. p. 1–3, pls 10–13. *In*: International Atlas of West Africa. OAU Scientific, Technical and Research Commission, Paris, France.

Traore, K. (1980) Caractéristiques limnologiques du lac de Kossou (Côte d'Ivoire). I. Hydroclimat et aperçu hydrochimique du lac de Kossou. Annales de l'Université d' Abidjan, sér. E, 12: 29–69.

Trochain, J. (1940) Contribution à l'étude de la végétation du Sénégal. Mémoires de l'Institut Français d'Afrique Noire 2: 1–433.

Van Der Ben, D. (1959) La végétation des rives des lacs Kivu, Edouard et Albert. *In*: Résultats Scientifiques. Exploration Hydrobiologique des Lacs Kivu, Edouard et Albert (1952–1954). Institut Royal des Sciences Naturelles de Belgique, Bruxelles, Belgium.

Van Der Eyk, J. J., MacVicar, C. N. and De Villiers, J. M. (1969) Soils of the Tugela Basin – a study in sub-tropical Africa. Natal Town and Regional Planning Report 15: 1–263.

Van Meel, L. (1952) Le milieu végétal. p. 51–68. *In*: Exploration Hydrobiologique du Lac Tanganyika (1946–1947). Résultats Scientifiques. I. Institut Royal des Sciences Naturalles de Belgique, Bruxelles, Belgium.

Varlet, F. (1978) Le régime de la lagune Ébrié, Côte d'Ivoire. Trait physiques essentiels. Travaux et Documents. L'Office de la Recherche Scientifique et Technique Outre-mer 83: 1–162.

Vesey-Fitzgerald, D. F. (1963) Central African grasslands. Journal of Ecology 51: 243–274.

Vesey-Fitzgerald, D. F. (1970) The origin and distribution of valley grasslands in East Africa. Journal of Ecology 58: 51–75.

Vesey-Fitzgerald, D. F. (1973) East African Grasslands. East African Publishing House, Nairobi, Kenya. 95 pp.

Viner, A. B. (1974) The supply of minerals to tropical rivers and lakes. p. 227–261. *In*: G. Olson (ed.). An Introduction to Land–Water Relationships. Academic Press, New York, New York, USA.

Visser, S. A. (ed.) (1970a) Bibliography on the River Niger with special reference to the Kainji dam. p. 113–126. *In*: S. A. Visser (ed.). Kainji: A Nigerian Man-made Lake. Kainji Lake Studies Vol. 1, Ecology. Ibadan Universities Press, Ibadan, Nigeria.

Visser, S. A. (1970b) Kainji: A Nigerian Man-made Lake. Kainji Lake Studies, Vol. 1, Ecology. Ibadan Universities Press, Ibadan, Nigeria.

Wagner, P. A. (1927) The geology of the northeastern part of the Springbok Flats and surrounding country. The Government Printer, Pretoria, South Africa.

Walmsley, R. D. and Botten, M. L. (compilers) (1987) Ecology and Conservation of Wetlands in South Africa. Ecosystems Occasional Report Series No. 28, Council for Scientific and Industrial Research, Pretoria, South Africa. 313 pp.

Walmsley, R. D. and Roberts, C. P. R. (1989) Changing patterns of resource use on the Pongolo River Floodplain. Occasional Report Series No. 36, Ecosystem Programmes, Foundation for Research Development, Council for Scientific and Industrial Research, Pretoria. 181 pp.

Walter, H. (1971) Ecology of Tropical and Subtropical Vegetation. Oliver and Boyd, Edinburgh, United Kingdom. 539 pp.

Walter, H. and Lieth, H. (1960–67) Klimadiagramm–Weltatlas. Jena, Fischer, Germany.

Walter, H., Harnickell, E. and Mueller-Dombois, D. (1975) Climate-diagram maps of the individual continents and the ecological climatic regions of the earth. Springer-Verlag, Berlin, Germany. 36 pp.

Welcomme, R. L. (1972) A brief review of the floodplain fisheries of Africa. FAO Committee for Inland Fisheries of Africa. Report CIFA/72/S.17.

Welcomme, R. L. (1979) Fisheries Ecology of Floodplain Rivers. Longman, London, United Kingdom. 317 pp.

Welsh, R. and Denny, P. (1978) The vegetation of Nyumbaya Mungu reservoir, Tanzania. Biological Journal of the Linnean Society 10: 67–92.

Werger, M. J. A. (1978) Biogeography and Ecology of Southern Africa. (2 Volumes) Dr. W. Junk, The Hague, The Netherlands. 1439 pp.

White, E. (ed.) (1965) The First Scientific Report of the Kainji Biological Research Team. University of Liverpool, Liverpool, United Kingdom. 356 pp.

White, F. (1983) The Vegetation of Africa: a Descriptive Memoir to Accompany the Unesco/AETFAT/UNSO Vegetation Map of Africa. UNESCO, Paris, France. 356 pp.

Whitfield, A. K., Allanson, B. R. and Heinecken, T. J. E. (1983) Estuaries of the Cape Part II: Synopsis of available information on individual systems. Report No. 22; Swartvlei (CMS 11). 62 pp. *In*: A. E. F. Heydorn and J. R. Grindley (eds.). Council for Scientific and Industrial Research, Research Report 421, Stellenbosch, South Africa.

Whitlow, J. R. (1985) Dambos in Zimbabwe: a review. Zeitschrift für Geomorphologie, N. F., 52: 115–146.

Whyte, S. A. (1975) Distribution, trophic relationships and breeding habits of the fish population in a tropical lake basin (Lake Bosumtwi – Ghana). Journal of Zoology, London 177: 25–56.

Wickens, G. E. (1976) Jebel Marra. Her Majesty's Stationary Office, London, United Kingdom. 368 pp.

Wild, H. (1961) Harmful aquatic plants in Africa and Madagascar. CCTA/CSA Project 14; reprinted from Kirkia, 2: 1–66.

Wetlands of southern Europe and North Africa: Mediterranean wetlands

R.H. BRITTON AND A.J. CRIVELLI

Abstract

Wetlands of the south of Europe and north Africa extend from Bulgaria to the Iberian peninsula, and from Tunisia to Morocco. The three main environmental factors explaining the distribution of the wetlands are: climate, topography and geology, and tides. The principal geomorphological formations containing wetlands are described in detail and numerous examples are given. They are: river deltas, coastal lagoons, riverine floodplains, inland freshwater lakes, man-made reservoirs, athalassic salt basins, intertidal systems, permanent river channels, and seasonally-flooded river channels. Many inventories and classifications of wetlands of the region have been carried out, but few of them are comparable. A new classification is proposed for the whole study area. The factors determining the ecological characteristics of the emergent and submerged vegetation of Mediterranean wetlands are identified and their impacts on the composition of the vegetation are analyzed in detail. Six vegetation categories have been determined: halophytic vegetation, emergent reedswamp communities, riverine forests, dwarf rush communities, and submerged and floating vegetation of freshwater habitats. Invertebrates, fish, birds, and mammals occurring in Mediterranean wetlands are listed and the factors responsible for their presence or absence are analyzed. As a general rule, birds are very well known, but information on invertebrates, fish, and mammals is still poor. The usage of and conservation problems of Mediterranean wetlands show that most of them are overexploited and degraded, and few are protected. Many organisms depending on these wetlands are nowadays threatened with extinction. Although the distribution and size of the Mediterranean wetlands is well known, the functioning of the ecosystems in the Mediterranean region remains poorly understood. This is a major handicap at the moment, when ways of better managing or even restoring certain Mediterranean wetlands are being talked of.

D.F. Whigham et al. (eds.), Wetlands of the World I, 129–194.

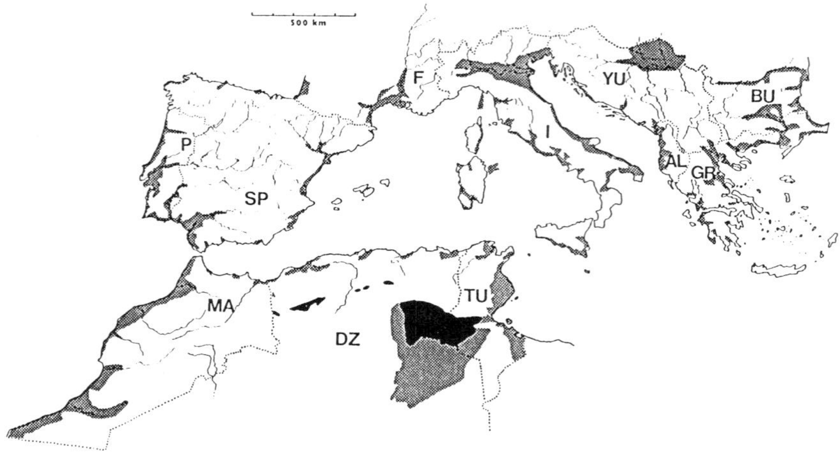

Figure 1. Map of study area. Land less than 200 m above sea level is shaded, and endorheic basins are in black. P: Portugal; SP: Spain; F: France; I: Italy; YU: Yugoslavia; AL: Albania; GR: Greece; BU: Bulgaria; TU: Tunisia; DZ: Algeria and MA: Morocco.

Introduction

The area covered by this chapter (Fig. 1) comprises those parts of southern Europe and North Africa which lie within the basin of the Mediterranean Sea or which have a basically Mediterranean type of climate. On the north side of the Mediterranean, the whole of the Iberian peninsula is included since, although much of it drains towards the Atlantic, the climate is basically Mediterranean, except in the north and west where continental wetlands are few. In Italy and France the northern limit of the region is formed by the mountain ranges of the Massif Central and the Alps, so that the upper parts of the valleys of the Rhône and Po which drain to the Mediterranean are excluded. In Yugoslavia and Albania only the coastal areas forming part of the Mediterranean catchment are included; the extensive wetlands of the Pannonian Basin in Yugoslavia will be dealt with in the Central European chapter in Volume 2. The whole of Greece and its islands and the Black Sea coastal fringe of Bulgaria complete the European part of the area covered in this chapter.

In North Africa the criterion for inclusion is the existence of a Mediterranean climate with annual rainfall occurring within the winter months. Egypt, which has a desert climate, and Libya which has only a very limited area of Mediterranean climate are excluded. In the remaining countries, which constitute the Maghreb (Tunisia, Algeria, and Morocco), the southern limit of the area is formed by the Sahara, where rainfall is no longer a regular annual event, and where wetlands are virtually absent. The study area covers a rather narrow range of latitudes from about 30°N in Morocco to about 46°N in northern Italy.

We will start this chapter by describing the main factors responsible for the distribution of wetlands in the Mediterranean region, describing the geomorphological formations in which they occur. We will then continue with inventories and classifications which have been carried out in the study area and comment on their characteristics. A detailed description of the plant and animal communities of Mediterranean wetlands precedes a review of wetland uses and conservation problems. Suggestions for improving our knowledge of Mediterranean wetlands ends the chapter.

Wetland types

Within the study area, three main environmental factors determine the distribution of wetlands and restrict the number of types that occur to a limited range of regionally characteristic wetlands. These factors are: climate, topography and geology, and marine tides.

Climate

Many attempts have been made to define the Mediterranean climate (e.g. Emberger 1954, Aschmann 1973, Nahel 1981). The fundamental characteristics are that it is "a non-tropical climate with seasonal variation in photoperiod, with the rainfall concentrated in the cold or cooler part of the year; the summer, the hottest season, being dry" (translation from Emberger 1954).

Evaporation greatly exceeds rainfall during the summer months, and generally on an annual basis as well. Typically the winters are mild, prolonged freezing is rare, and plant growth can continue throughout the year.

Although most of the area covered in this Chapter can be considered to have a Mediterranean type climate, there are strong regional variations. In areas bordering the Atlantic, rainfall is more evenly spread over the year and the annual range of temperature is small. In northern coastal regions of Spain and Portugal these tendencies are particularly marked and the climate and vegetation are more closely allied to those of north-west Europe than to the Mediterranean region. There is also a marked west to east and north to south decrease in winter temperature across the Mediterranean basin. The lowest mean January temperatures within the area occur in northern Greece, where wetlands freeze regularly every winter. In North Africa, temperatures rarely fall below zero at sea level but lakes in the High Atlas mountains at altitudes of over 1,000 m regularly freeze for short periods in winter. Freezing also regularly occurs in inland areas of Spain, which have a steppe-like climate, and at high altitudes in the Pyrenees and Apennines.

Total rainfall also varies mainly on a north to south gradient, but there are strong local variations due to topography. Highest rainfall (>1,000 mm) occurs along the Atlantic seaboard and where mountain ranges approach the coast (e.g. Gulf of Genoa and southwestern Greece). Lowest rainfall occurs in areas bordering the Sahara in North Africa and in inland Spain.

As well as being highly seasonal, the Mediterranean climate is characterized by great year to year variation, particularly in the timing and quantity of rainfall. In the north and west there is a tendency for some years to have greater than average summer precipitation. The climate in such years resembles that of northern temperate Europe. Year to year variation in the quantity of precipitation increases as the mean annual rainfall decreases. As one progresses towards the driest region (Sahara), rainfall ceases to be an annual event, and only intermittent heavy rains occur at intervals of several years. In mountainous areas mean annual rainfall is higher and there is less year to year variation than in lowland areas.

A final feature of the Mediterranean climate is that locally strong winds occur. These greatly increase evaporation from open water, cause wind seiches on the open sea and on inland waters, and have direct effects on some wetland plants and animals.

Since annual evaporation greatly exceeds annual rainfall except in humid mountainous areas, wetlands only occur in topographic depressions which collect and retain water from a surrounding catchment. Moreover, since rainfall and evaporation show great seasonal variation, wetlands which do not have a connection to a permanent source of water (e.g. the sea or a large perennial river), show pronounced draw-down in water level in the summer months. In many shallow palustrine and riverine wetlands the water table regularly falls below the soil surface for periods of a month or more each year. Because of the large year to year variability in rainfall, however, the timing and duration of the draw-down within any wetland system is not fixed. Some wetlands remain permanently flooded for several years and a period of prolonged summer draw-down occurs during which time there is only brief winter submergence. In North Africa and drier areas of inland Spain, many wetlands are normally dry and only flood after exceptionally wet winters, which may only occur at intervals of several years.

The excess of evaporation over rainfall allows the build up of dissolved salts even in inland wetlands, and in the more arid parts of the region most wetlands are saline. The seasonality of the rainfall/evaporation deficit also causes variation in salinity in non-freshwater wetlands. Many isolated coastal lagoons regularly evaporate to dryness and to saturation with sodium chloride. As with water level, salinity may show both seasonal and long-term year to year variation.

Summer drought and high summer salinities make this season unfavorable for many wetland organisms. Fortunately, winter temperatures are sufficiently high in most parts of the region to allow for plant and animal growth. Some wetland organisms, which grow and reproduce in the summer in northern Europe, perform these functions in winter and early spring in the Mediterranean region, and pass the summer months in quiescent stages (Nourrisson and Aguesse 1961, Champeau 1971). Since most Mediterranean wetlands remain ice-free, they are particularly important as wintering grounds for waterfowl. In the summer the breeding birds using Mediterranean wetlands include species which are adapted to exploiting conditions of draw-down or high salinity (e.g. certain Ardeidae and flamingos). Wetlands which remain flooded in summer assume added importance as breeding sites.

Topography and geology

The Mediterranean region, being at the point of contact between the African and Eurasian tectonic plates, is very mountainous. Around most of the basin, mountains descend to the coast and a coastal plain, if it exists at all, is very narrow and discontinuous (Fig. 1). Extensive low-lying lands are mainly restricted to alluvial flood plains of the major rivers (Guadalquivir and Ebro in Spain, Rhône in France, Po in Italy, and Axios and Evros in Greece). Smaller floodplains not associated with major river systems occur in southern Italy and along the Languedoc-Roussillon coast of France. In North Africa an extensive, but discontinuous, coastal plain exists along the Atlantic coast of Morocco, but the Mediterranean coasts of Morocco, Algeria, and northern Tunisia are mountainous. The eastern coast of Tunisia, facing the Gulf of Gabes, has a continuous coastal plain of varying width. In the north it is the flood plain of the River Medjerda, while in the south the coastal plain extends inland to include the huge interior drainage basin of the Chott Djerid and Chott Melrir which lie below present-day sea level (Fig. 1). Endorheic drainage basins also occur further west into Algeria and at high altitudes on the Plain of Chotts between the two main ranges of the Atlas mountains. A few such basins, of small size, also occur in interior Spain. Most non-riverine wetlands occur on these coastal plains or within interior drainage basins, and wetlands consequently have a patchy and discontinuous distribution.

The uplands of the Mediterranean are mainly composed of porous limestone in which the water table lies many meters below the surface. Rivers are generally short and steep and many cease flowing in the summer months.

Lakes are rather rare and mainly confined to karstic, volcanic or tectonic basins. Only very limited areas of the uplands were subject to glaciation during the Pleistocene, and this process has not been important in the formation of wetlands within the region. Glacial outwash was, however, responsible for the creation of some of the coastal plains on which wetlands are now situated (e.g. the plain of the Crau in southern France).

Marine tides

The effect of Atlantic tides diminishes rapidly as one enters the Mediterranean through the Straits of Gibraltar. Throughout most of the Mediterranean the vertical tidal amplitude is less than 50 cm per cycle. Variations in sea level tidal amplitude are caused by wind surges, changes in barometric pressure, or seasonal variation in evaporation and rainfall are generally greater than this. Intertidal wetlands are therefore mostly absent from the coasts of the Mediterranean.

Two areas within the western basin of the Mediterranean have a greater than average tidal amplitude and show limited development of intertidal flats and salt marsh. These are at the northern end of the Adriatic, especially near Trieste, and along the Tunisian coast between Sfax and the island of Djerba. Although there is a moderate tidal range (ca. 2 m) along the Atlantic coast of Spain, Portugal, and Morocco, intertidal wetlands are restricted in occurrence by the generally steep coastlines. Tidal wetlands which occur along these coastlines are mostly associated with existing river estuaries (e.g. River Tejo, Portugal), or with relict estuaries (e.g. Merja Zerga, Morocco), and are mostly of rather limited extent. For the region as a whole, therefore, intertidal wetlands are of limited importance as compared to other coastal regions of the world.

Geomorphological setting of the major wetland types

Wetlands occur within the Mediterranean in a limited number of geomorphological formations, but within each of these formations a range of wetland types may be found. For example, river deltas may contain areas of salt marsh, freshwater marsh, freshwater and salt water lagoons, and forested wetlands.

The major geomorphological formations containing wetlands are: River deltas, Coastal lagoons, Riverine flood plains, Inland freshwater lakes, Man-made reservoirs, Athalassic salt basins, Intertidal systems, Permanent river channels, and Seasonally flooded river channels.

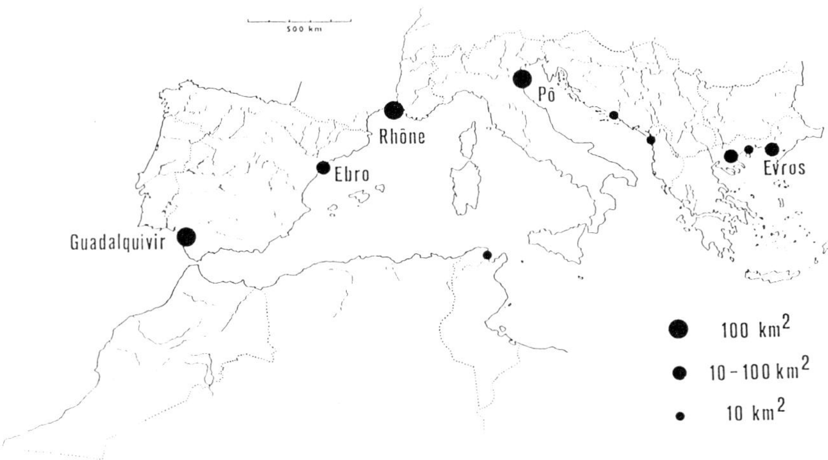

Figure 2. Distribution of major deltaic wetlands within the Mediterranean region.

River deltas

Deltaic wetlands now constitute some of the most extensive and varied wetland areas remaining around the north of the Mediterranean basin (Bethemont and Villain-Gandossi 1987). They are widely distributed from Spain to Greece (Fig. 2), but no large examples occur in the Maghreb, where most rivers are short, highly seasonal, and do not provide sufficient sediment for delta formation. The rivers forming extensive deltas rise in mountainous areas of high rainfall (e.g. Guadalquivir from the Sierra Nevada of Spain), or in humid areas outside the region of Mediterranean climate (e.g. Rhône and Po which rise in the Alps). These rivers are perennial and do not show pronounced seasonal variation in discharge, as do rivers originating entirely within the Mediterranean.

The formation of extensive deltaic sediments appears to be favored by the small amplitude of marine tides within the Mediterranean. The resulting low current velocities allow the formation of structures such as offshore sand banks, and the deposition of alluvial sediments in shallow water. Delta formation was also favored by considerable glacial outwash from rivers draining the Alps, and later by increased erosion following human clearance of the Mediterranean forest. Suspended sediment loads in the rivers are among the highest recorded (Bradbury 1981), and alluvial deposits up to 60 m thick have been deposited in the Rhône delta in quaternary times (DDA 1970).

All the large delta regions of Europe are now protected from flooding by artificial river levees, and in nearly all cases the sediment load in the river has been intercepted by the construction of reservoirs upstream. As a consequence, the rate of delta formation has been considerably reduced and cases

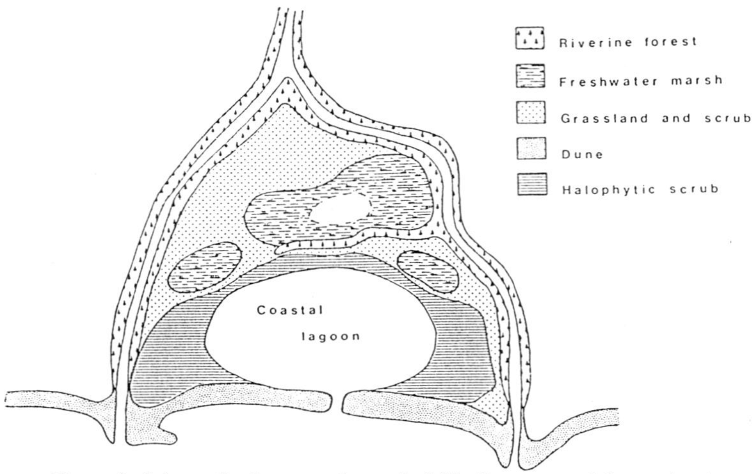

Figure 3. Schematic diagram of a typical Mediterranean delta wetland.

of coastal erosion are now reported from delta areas, where once the coastline was advancing (DDA 1970).

The rivers along the Atlantic coasts of Spain and Portugal, where the tidal range is of the order of 2 m, form estuaries, some of them drowned (e.g. Rio Aveiro, Portugal), rather than deltas. The Rio Guadalquivir in southern Spain is, however, sufficiently similar to Mediterranean deltas to be included here. This river system enters the sea with several other smaller rivers, the entrance to the sea being partially blocked by a coastal dune ridge, behind which riverine wetlands have developed.

All the delta areas show similar structural zonation of wetland types (Fig. 3). In reality most deltas have been extensively modified by man and much of the former wetlands have been converted to agriculture. Inland from the coast the following zones are found:

1. Coastal dunes – These vary in importance from a narrow spit with scarce Ammophiletum to extensive wooded dunes with *Juniperus phoenicea* or *Pinus* spp. Relict coastal dunes well inland attest to the progression of the coast line. In some deltas, inland dunes have also formed from alluvial sands. Inland dunes are usually covered with dry deciduous woodland.

2. Coastal lagoons – These may still connect to the sea via gaps in the dune ridge or may be isolated. They are typically brackish with a salinity between 5 g l^{-1} to 40 g l^{-1} and the vegetation is usually dominated by *Ruppia* spp. (Verhoeven 1979, 1980a, 1980b) and various algae (*Ulva* spp. and *Chaetomorpha linum*). When the coastal lagoon is cut off from the sea and river water predominates, they may have typical freshwater vegetation (e.g. *Potamogeton pectinatus*). Maximum depths are usually less than 2 m.

3. Salt marshes – These occur immediately behind the dune ridge and surrounding the coastal lagoons. They can extend far inland since a saline water table underlies all these deltas and salt is brought to the surface in the summer months by evaporation. Halophytic wetlands show a gradation from wet to dry vegetation types and from communities tolerant of highly saline conditions to *Phragmites* communities that grade into freshwater marshes. Vegetation of more saline areas is sparse and dominated by Chenopodiaceae, while salt tolerant rushes (e.g. *Scirpus maritimus* and *Juncus maritimus*) are typical of the less saline end of the spectrum. All deltaic salt marshes normally dry out in the summer. Because of their limited agricultural potential, extensive areas of halophytic wetlands remain although many have been modified by drainage attempts.

4. Freshwater marshes – Freshwater marshes form along old river courses or where flood water is impounded behind levees. Depending on the depth and permanence of water, they may vary from shallow lakes with open water and only submerged vegetation, to dense reed swamp, or grazed wet meadows. These have mostly been modified by man and input of water is now mainly artificial as the rivers have invariably been diked and no longer flood naturally.

5. Riverine forest – Deciduous forests with *Populus*, *Salix*, *Fraxinus*, and *Ulmus* formerly occurred on the flood plains of the lower reaches of all Mediterranean rivers. They must have been particularly extensive in delta areas, where the flat topography meant that the influence of river water was felt far from the river course itself. Construction of dikes, conversion of agriculture, grazing and felling for timber have now reduced these forests to a few isolated relict stands.

Coastal lagoons

These occur all around the Mediterranean and along the Atlantic coast of Morocco wherever there are extensive coastal plains (Fig. 4). The largest areas are along the southern coast of France where they stretch for some 200 km between the Rhône delta and the Pyrenees, and to the north and south of the Po delta at the head of the Adriatic.

Lagoons were formed by the processes of coastal sedimentation and dune formation and some occur on former marine terraces now lying above sea level. Coastal accretion has isolated many lagoons from the sea and these are now freshwater lakes (e.g. Lac Biguglia in Corsica), while others still have open connections to the sea and have a salinity similar to that of seawater. Brackish water lagoons occur where there is only limited communication with the open sea, and where there is some input of freshwater. The salinity of these lagoons varies seasonally according to the relative inputs of

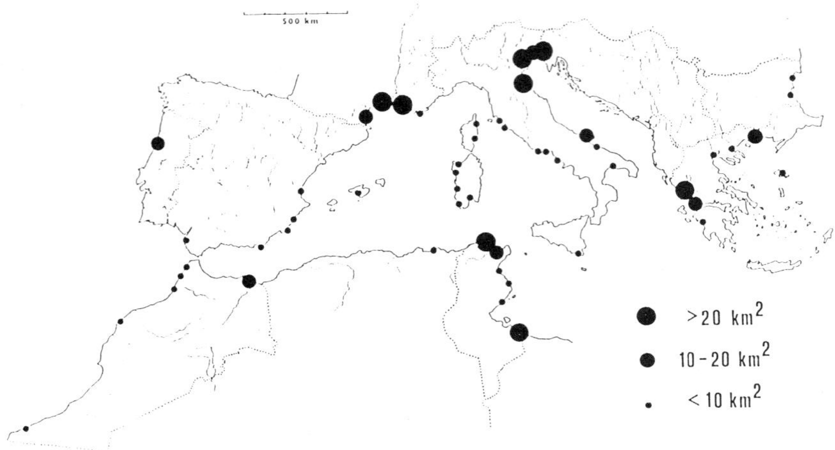

Figure 4. Distribution of coastal lagoons within the Mediterranean region.

sea water and freshwater. Natural hypersaline lagoons are now rare as most have been converted to salinas, but seasonally flooded lagoons (Sebkhets), which become highly saline before drying out are frequent along the coasts of Tunisia and Morocco.

The flora and fauna of coastal lagoons is highly variable and depends on salinity and water depth. In lagoons with a wide connection to sea, and particularly on tidal coasts, the fauna is essentially marine and the vegetation is composed mainly of *Zostera noltii*, with *Ruppia cirrhosa* in more sheltered inshore areas and *Cymodocea nodosa* in deeper permanently flooded areas. Examples are the lagoons of the Messolonghion region of western Greece and the Merga Zerga in Morocco.

In brackish water lagoons, the submerged vegetation is composed of *Ruppia* spp. or *Potamogeton pectinatus* and there may be peripheral wetlands dominated by *Salicornia* spp. or *Scirpus maritimus*.

Freshwater lagoons are isolated from the sea and generally have extensive marginal reed swamp and beds of submerged or floating aquatic macrophytes typical of permanently flooded shallow freshwaters in the Mediterranean (Grillas 1990). A few lagoons occur on sands which have become leached, or which derive from non-calcareous rocks, and these lagoons are of mesotrophic status. Examples are found along the Ligurian coast of Italy. Shallow coastal lagoons that have become isolated from the sea frequently progress to the state of coastal freshwater marshes with little or no open water. Because of their potential fertility, many such sites have been partially or completely drained and transformed into agricultural land (e.g. the Agoulinitsa lagoon in southern Greece, the Drana lagoon in the Evros delta, northeastern Greece, and the Albufera deValencia in Spain). The project in the Agoulinitsa lagoon never worked, however, because of the saline soils and there are proposals to re-convert it to lagoon (N. Morgan, personal communication). Examples remaining relatively intact occur in Mallorca (Albufera de Alcudia) and in Algeria (Lac Oubeira wetland complex). Commercial salinas are a characteristic and widespread form of man-modified coastal lagoon which is described in the section on wetland use.

Riverine floodplains

The most extensive wetlands within the Mediterranean region undoubtedly occurred on lowland flood plains before the advent of man. Two thousand years of hydraulic engineering have, however, reduced them to a few tiny isolated remnants. The remainder has been converted to farmland or urban development. Three types of flood plain wetlands can be distinguished:

1. Oxbows – Former river courses dating back several centuries occur in some delta areas (e.g. Rhône and Messolonghion), but in the lower and middle reaches of major river systems most oxbows have long since disappeared. Where they occur in deltas, oxbows form shallow eutrophic lakes, or have progressed by siltation to the stage of seasonally flooded freshwater marshes. Oxbows of recent age in the coastal region resemble brackish water coastal lagoons. The riverine vegetation of oxbows is composed of *Populus alba*, *P. nigra*, *Salix alba*, and *Salix* spp.. The frequently abundant submerged vegetation is mainly represented by *Nymphea alba*, *Nuphar luteum*, *Ceratophyllum* spp., and *Potamogeton pectinatus*.

2. Floodplain freshwater marshes – These too are now very rare as most have long since been drained for agriculture. Large lowland rivers are now not only diked, but also regulated by reservoirs to control flooding. Freshwater marshes once occupied those parts of the flood plain that were flooded too frequently to allow tree growth, or where trees had been removed by felling or grazing. All these marshes were flooded seasonally in winter and spring, but dried out to varying extents in the summer. The wetter areas were colonized by *Phragmites*, *Typha*, and *Scirpus* spp., which gave way in less frequently flooded parts to *Carex* and *Cladium* and finally to wet herb-rich meadows (*Lythrum* sp., *Salicornia* sp., and *Epilobium* sp.). The latter graded into forested wetlands and were probably largely maintained by grazing by domestic stock. These marshes must once have occupied vast areas in the flood plains of all the major river systems in southern Europe and the wetter parts of North Africa, but it is difficult to judge their former extent since old documents do not distinguish between forested and non-forested wetlands. Remaining examples are found in a fragmentary and highly modified state on the river Tejo in Portugal, in the Languedoc and

Crau in southern France, in the Po valley in northern Italy, and in northern Algeria. Probably the most intact system now constitutes the Parque Nacional de las Tablas de Daimiel on the River Guadiana in central Spain.

3. Forested bottom lands – Riverine gallery forests are the driest of the freshwater wetland types. They are dependent on the presence of groundwater close to the soil surface throughout the growing season. They also flood in the winter. These forests are clearly distinguished from the evergreen sclerophyll upland forests by the dominance of deciduous species of which *Populus alba* is the most abundant and widespread.

Riverine forests have all but disappeared in southern Europe. The most extensive remnants being on the Moraca River (Lake Skadar, Yugoslavia) and on the river Strymon (Lake Kerkini, northern Greece). Small fragments occur along all the main river courses, but all have been modified by flood control schemes or by felling. In many cases native deciduous woodland has been replaced by plantations of non-indigenous *Populus* sp. and *Eucalyptus* spp. There is a virgin bottom land floodplain present in Thrace, dominated by *Fraxinus holotricha*, the last in Greece.

Because of the highly seasonal nature of North African rivers it seems doubtful that extensive riverine forest occurred there except for a few tiny fragments found on some of the inflows to the Garaet Ichkeul wetland in northern Tunisia.

Freshwater lakes

Glacial erosion and deposition were the main lake forming processes in north temperate regions. In the Mediterranean region, however, glaciation was limited to the highest parts of the Sierra Nevada, Pyrenees, Apennines, Dinarique Alps, and the Atlas. Within each of those areas there are numerous small cirque lakes (see Ferrari *et al.* 1975 for the Apennines, Laville 1975 for the Pyrenees, Dumont *et al.* 1973, and Morgan 1982b for the Atlas). Cirque lakes are typically nutrient-poor, steep-sided, deep and have little or no emergent vegetation. Some of the larger mountain lakes of Morocco are, however, eutrophic, shallow, and have abundant emergent (*Phragmites*) and submerged vegetation such as *Myriophyllum spicatum*, *Potamogeton pectinatus*, and *Ranunculus* spp. (Morgan 1982b). Some of these are in limestone areas and may be partially of karstic origin.

Outside of the glaciated areas, natural freshwater lakes are uncommon. A group of large lakes in central Italy are mainly of volcanic origin and occupy calderas (e.g. Lago di Bolsena, Lago di Bracciano, Lago de Nemi). Lago di Trasimeno is said to be a solution lake (Hutchinson 1957). These lakes are more lowland in character and are shallow and eutrophic. Lago di Trasimeno (Taticchi 1968) has a wide marginal band of *Phragmites* and extensive submerged vegetation (*Ceratophyllum demersum*, *Potamogeton lucens*, *P. perfoliatus*, *P. pectinatus*, *Nymphaea alba*, *Salvinia natans* etc.).

Karstic lakes dating from the tertiary era occur in Yugoslavia, Albania, and north-western Greece. These include Lake Ohrid, Lake Skadar, the Lakes Megali and Mikri Prespa, and Lake Vegoritis, which are the largest and deepest lakes within the region (Stankovic 1931). Some, such as Lake Ohrid, partially occupy tectonic basins and are geologically very old (Stankovic 1960). Although calcareous, they are typically nutrient-poor and have a very limited zone of littoral vegetation. Mikri Prespa and Skadar Lakes are exceptions, in being rather shallow mesotrophic and having an extensive littoral *Phragmites* zone (Koussouris *et al.* 1987, 1989, Petrovic 1981). The emergent vegetation is composed of *P. australis*, *Typha latifolia*, *Scirpus lacustris*, and *Iris pseudacorus*. Submerged plants (*Nymphaea alba*, *Ranunculus* sp., *Nymphoides peltata*, and *Myriophyllum verticillatum*) are abundant in the littoral zone (Pavlides 1985). Some of the largest shallow eutrophic lakes in Greece have been completely drained for agricultural development (e.g. Lake Voivis and the Agoulinitsa Lagoon).

In southern France there are a few natural freshwater inland lakes but none are very large. They are usually shallow and eutrophic and occupy karstic or structural depressions. In lowland Spain, there are no large freshwater lakes but inland drainage basins with numerous small shallow lagoons are found in three areas (Andalucia, Castilla and south-west of Zaragossa). Those lagoons are saline and seasonal and are therefore classed as inland salt lakes, but some, particularly in Andalucia are permanent and freshwater (Amat 1984). The emergent vegetation is represented by *P. australis*, *T. angustifolia*, *T. latifolia*, *Scirpus maritimus*, and *Eleocharis palustris*, and the submerged plants by *Ranunculus baudotii*, *Zannichellia peltata*, and *Chara* spp. (Rivas-Martinez *et al.* 1980).

In North Africa, permanent freshwater lakes have always been scarce outside of the sub-humid mountainous areas, and the largest examples have now been drained (e.g. Lake Fetzara, Algeria). The only large lowland freshwater lakes that remain, either dry out periodically or are in coastal areas and receive some inflow of saline water in periods of drought. Garaet Ichkeul in Tunisia receives water from the marine Gulf of Bizerta when freshwater inflow ceases in the summer, and it then becomes oligo- to mesohaline. A similar situation was found in the complex of Lakes Oubeira, Melah, and Tonga in Algeria (Skinner and Smart 1984, Stevenson *et al.* 1988, Guelorget *et al.* 1989), but return flow of saline water is now prevented by sluices. The submerged vegetation is mainly composed of *Ceratophyllum demersum* and *N. alba* and the emergent plants by *P. australis* and *S. lacustris* (Kadid 1989). Lac Kelbia in Tunisia is a freshwater lake which remains flooded for periods of several years at a time, since it is supplied by a large catchment, but periodically it dries out completely and in the draw down phase it becomes saline (Zaouali 1976).

Man-made reservoirs

The insatiable demand for freshwater in the semi-arid climate of the Mediterranean countries has led to the building of so many water storage reservoirs that their number and area far exceeds all the remaining natural freshwater wetlands. Moreover, construction of new reservoirs is continuing, particularly in the Maghreb.

Reservoirs, in general, have deleterious effects on downstream wetlands, as a result of flow regulation with reduced flooding and sediment deposition which is required for the development of coastal wetland systems. On the other hand, some reservoirs have themselves developed into wetland systems of considerable wildlife value. Reservoir construction has introduced permanent standing water into arid landscapes where none existed before and some such man-made lakes are much used by migratory waterfowl. To some extent, reservoirs have replaced natural freshwater wetlands, as feeding and roosting sites for these birds. Examples are the huge reservoir complexes on the River Guadiana and Tejo in western Spain which now hold internationally important concentrations of ducks and coots, and the Esla Reservoir in central Spain which has a wintering population of between 5,000 and 10,000 greylag geese (*Anser anser*) (Carp 1980). In North Africa, Boughzoud lake in Algeria is of international importance for waterfowl, thus compensating in part for the loss of most of natural permanent freshwater lakes in the Maghreb (Chalibi 1990). In general, however, most reservoirs in the Mediterranean have been constructed in upland areas and have steep shorelines, are subject to severe drawdown on a seasonal or even diurnal basis, and of very little wildlife value.

Athalassic salt lakes

The term athalassic is used in the sense of Williams (1981) to describe saline waters which are isolated from the sea, or which were once connected to the sea, but which have dried out before being re-flooded by water of non-marine origin. As a wetland class, they are restricted to more arid areas, particularly to the Maghreb and interior Spain, where the annual rainfall is less than 400 mm (Fig. 5). No large inland salt lakes are known from the other countries of southern Europe, although some small examples are found. Salinas resemble natural athalassic salt lakes in their fauna and flora, but the periodicity and duration of flooding is much more regular (Fig. 6).

Natural salt lakes mostly occupy endorheic drainage basins of tectonic

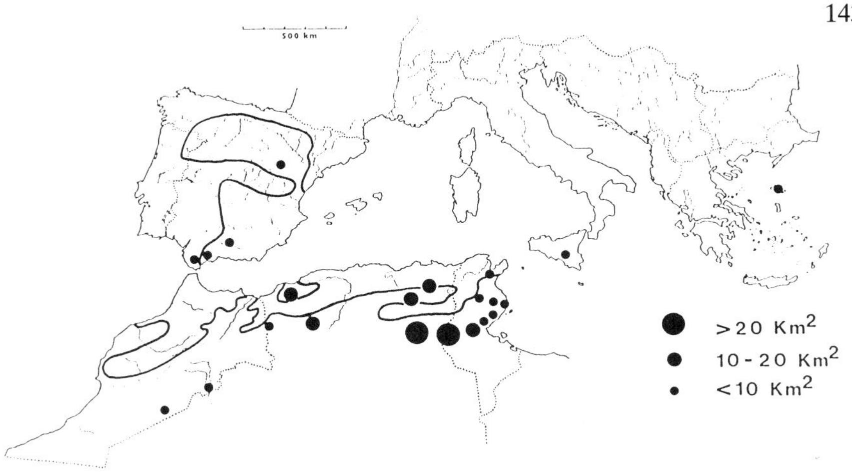

Figure 5. Distribution of athallassic salt lakes within the Mediterranean region, in relation to the 400 mm precipitation isohyet.

origin but some in North Africa have periodic outflow at times of extreme high water level. They vary from sites lying below present sea level (e.g. Chott Melrir, Algeria at −30 m) to upland basins at over 1,000 m (e.g. Plain of Chotts, Algeria and Laguna de Gallocanta, Spain). There is evidence that many of the basins now occupied by salt lakes in North Africa were once extensive freshwater lakes when the climate was much more humid than at present (Hutchinson 1957).

All of the salt lakes are shallow and at least occasionally dry out completely. The periodicity of flooding depends both on the climate and on the ratio of the area of the catchment to the area of the lake. In general, there is a north to south gradient in flooding regime. The most northerly large

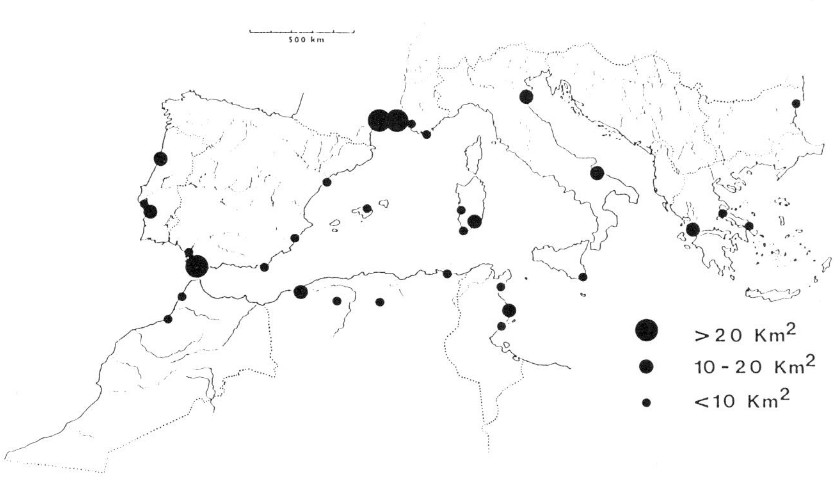

Figure 6. Distribution of salinas within the Mediterranean region.

example, the Laguna de Gallocanta has only dried out three times this century (Comin *et al.* 1983, 1990). Those in Andalucia and the extreme north of Tunisia and Algeria are typically seasonal and dry out in summer and reflood in most winters. Further south in North Africa, salt lakes are mostly irregularly flooded and may remain dry for several years. Following rainfalls, they may hold water for weeks, months or even years (Amat 1982, Morgan and Boy 1982).

Chemically, all those lakes which have been studied are typical sodium chloride waters (Hutchinson 1957). Ionic proportions are in the following order: $Na > Mg > Ca$ and $Cl > SO_4 > CO_3$. Salts are derived from weathering of ancient marine sediments, or in the case of some the basins lying below sea level, from relict salt deposits left from Pleistocene marine transgression. Salinity varies not only from basin to basin but also according to the state of evaporation. When full, most have salinities $<5\ g\ l^{-1}$, but many evaporate to saturation with $NaCl > 300\ g\ l^{-1}$ before drying out.

The most irregularly flooded lakes, which also tend to be the most saline, are devoid of aquatic vegetation, and have a crust of halite or anhydrite covering the lake floor. The most saline of the vegetated salt lakes have a marginal band of Chenopodiaceae (*Salicornia* spp. and *Arthrocnemum* spp.), which develops particularly after drawdown, and submerged *Ruppia* spp. and *Characeae*. In less saline conditions the marginal vegetation is composed of *S. maritimus* or *Juncus maritimus* and the submerged flora is more species-rich (*R. baudotii*, *Zanichellia pedunculata*, *Chara* spp.). The aquatic fauna of salt lakes within the Mediterranean region is composed of a limited range of salt and drought resistant groups; *Artemia*, Copepoda, Ostracoda, Cladocera, and Diptera larvae are characteristic (Baltanas *et al.* 1990, Alonso 1990).

Tidal wetlands

These wetlands have a very restricted distribution because of the low tidal range. They are rather localized along the Atlantic coasts and occur in river estuaries, or in sheltered bays (Fig. 7). Tidal systems occurring within the region can be divided into five classes:

1. Permanently flooded estuaries of variable salinity.
2. Unvegetated sand and mud flats, exposed at low tide.
3. Vegetated flats dominated by *Zostera noltii*, *Z. nana*, or *Ruppia maritima*.
4. Salt marshes which flooded at most high tides; and are dominated by *Spartina maritima* and *Salicornia herbacea*.
5. Salt marshes that flood only at spring tides. *Arthrocnemum* sp. or

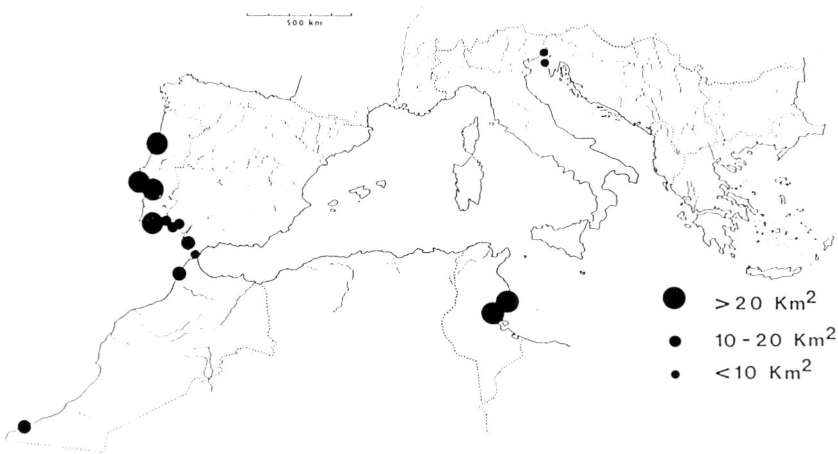

Figure 7. Distribution of main tidal wetlands (salt marsh and inter-tidal flats) within the Mediterranean region.

grazed grass swards (*Halimione portulacoides* and *Puccinellia* spp.) are characteristic.

Tidal sand and mud flats form the largest remaining area of wetland. In most sites, the upper marsh has been more or less modified, either by drainage and diking, or transformation into salinas (particularly in Portugal). The form of the tidal wetlands is quite variable and depends primarily on local topography. Many of the estuaries are oversized in relation to present day river flows, and they then take the form of tidal lagoons with a narrow exit to the sea, which may become partially blocked by a sand bar. This is the case of the Aveiro estuary in Portugal, where the exit is so narrow that water is retained in the lagoon at all stages of the tide. In Morocco, freshwater input is seasonal, the rivers dry in summer, and there is only a limited development of salt marsh vegetation (*Salicornia arabica*, *Tamarix riffensis*). In extreme cases, (e.g. Puerto Cansado, in the extreme south of the Morocco), the estuary is a relict from a period of more humid climate, and the river now flows only irregularly. At this site the upper marsh is replaced by an unvegetated salt pan (sebkhet). Because of the seasonality of rivers in Morocco there is a tendency for tidal inlets to become blocked by a sand bar at times of low discharge. Such systems then develop into seasonally flooded saline coastal lagoons until another river flood once more makes a breach in the sand bar. Two areas of tidal wetlands (the Northern Adriatic and the Gulf of Gabes) occur at the head of shallow, gently shelving bays. In both cases the extent of salt marsh is rather limited, and in Tunisia, it appears that *Spartina* dominated salt marsh is absent.

Permanent river channels

Although they are more widespread than lakes and marshes, there are few data on riverine wetlands and for most countries there has been no inventory. Phytosociologists have documented the vegetation of these wetlands, but have not generally carried out wide ranging inventories of sites. There are probably therefore botanically interesting riverine wetlands which have not been surveyed, particularly in the more remote parts of the region.

Throughout much of the region the terrain is mountainous, and rivers are short and precipitous. The smaller 1st and 2nd order streams only flow during periods of rainfall and there is little development of wetland vegetation. On gentler slopes, rivers within the region are characterized by well-developed flood plains. Those flood plains formed during periods when the river flow and sedimentation were natural. Most rivers now occupy down-cut channels with lateral erosion during flood periods (Paskoff 1973). Down-cutting of the river channel is particularly marked in the more arid regions of central Spain and North Africa, where variation in rainfall is enormous and severe erosion accompanies occasional flash floods. In these areas flood plain wetlands are restricted to coastal areas of very low gradient. The largest permanent river systems are nearly all now highly modified by embankment, canalization, reservoir construction, water and gravel extraction, and in some cases, by domestic and industrial pollution.

In fast flowing rivers, and particularly those with great seasonal variation in discharge, the bed is frequently scoured by floods and even in the lower sections there is little accumulation of fine-grained sediments. The bottom of such rivers is composed of large pebbles with some accumulations of gravel and coarse sand. Shallow rivers with fluctuating discharge often form braided channels with extensive shingle islands (e.g. Durance, France; Upper Po, Italy). Depending on the frequency and severity of flood waters these shingle banks may be unvegetated or develop a covering of annual or perennial vegetation. Typical perennial woody communities include *Salix* spp., *Nerium oleander* and the introduced *Amorpha fruticosa*. Annual vegetation forming on shingle banks is mainly composed of ruderals.

In deep and especially in turbid rivers, aquatic vegetation is restricted to a narrow marginal band of *Phalaris* sp. and isolated beds of resistant submerged species such as *Nuphar* sp. and *Potamogeton pectinatus*.

The best development of submerged and floating vegetation occurs in spring-fed rivers with clear water and little seasonal variation in river flow. Here fine-grained sediments can accumulate and a community of plants with a trailing habit is found (e.g. *Sparganium* spp., *Potamogeton coloratus*, and *Ranunculus* spp.). There is also usually a marginal fringe of a herb-rich helophyte community (*Typha* spp. and *Phragmites australis*). Such rivers occur particularly in limestone areas of low gradient (e.g. River Sorgue, France), but must be rather rare in the Mediterranean as a whole.

Seasonally flooded river channels

A study of a small sample of river networks in Mediterranean France, on topographic maps at a scale of 1:50000 (Institut Géographique National), showed that about 75% of 1st order streams were depicted as being seasonally or irregularly flowing. For 2nd order streams the proportion of permanent to seasonally flowing was roughly equal, while 75% of 3rd order streams were permanent. All streams of 4th order or more were shown as permanent.

Comparable maps of North Africa (Morocco and Tunisia) revealed that, under the much more arid climatic conditions, the proportion of seasonally or irregularly flowing streams was much higher. Nearly all 1st and 2nd order streams flowed temporarily (97%), and only about 20% of the 3rd and 4th flowed continuously. Only largest order streams had continuous flow. The total length of temporarily flowing streams therefore greatly exceeds that of permanent rivers, and in North Africa the latter are even rare.

Small temporary streams, particularly in upland areas, generally have little aquatic vegetation. Aquatic epilithic bryophytes, Characeae in pools and small species of *Juncus* on wet gravel are typical. However, an aquatic invertebrate fauna rapidly develops after flooding, from resistant stages remaining in the stream bed or by immigration (Legier and Talin 1973).

Larger temporary streams, especially on low gradients, dry out to form a series of isolated standing water pools, which may be perennial. These pools frequently become saline in the more arid parts of the region, and then develop marginal wetland communities of Chenopodiaceae and submerged beds of *Ruppia* sp. and *Characeae. Tamarix* sp. scrub is very characteristic of slightly saline temporary stream beds, particularly in North Africa. Freshwater pools in stream beds develop communities of ruderals, and annual helophytes and hydrophytes (aquatic species include: *Plantago aquatica*, *Eleocharis palustris*, *Callitriche* spp., and *Isoetes* spp.).

Geographical distribution of wetland types

In the absence of comprehensive published inventories for each country, it is not possible to give accurate figures for the areas of numbers of different wetland types in the region. A minimal estimate has, however,been calculated for some countries, based on published and unpublished inventories, supplemented where possible by additional map search (Table 1). It has not been possible to estimate the lengths or areas of riverine wetlands, and these have been omitted. Small sites, especially freshwater marshes, ponds and forested wetlands, are certainly under-represented as complete inventories are not available.

The sources used for compiling Table 1 were Carp (1980) and Scott(1980) for the whole region, Morgan and Boy (1982) and Morgan (1982a, b) for North Africa. MOPU (1984) was used for Spain and phytosociological maps at a scale of 1:50,000 for France. The figures for Italy are derived from an unpublished inventory produced by the Ministry of Agriculture in 1972, which was reported to be incomplete. Similarly, the Greek totals are derived in part from the inventories of Dorikos (1981) and Heliotis (1988). Nowaks (1980) was used as source of information on Albanian wetlands. No recent comprehensive information was available for Portugal or Bulgaria.

The largest areas of remaining wetlands are athalassic salt lakes in North Africa. This category is almost confined to the arid parts of North Africa, and is hardly represented in Europe (Fig. 5). Coastal lagoons are the next most abundant type and occur in all countries. The largest concentrations of lagoons occur along the Mediterranean coast of France, on the Adriatic coast of Italy, in north east Greece and in Tunisia (Fig. 4). Most lagoons are connected to the sea, and are brackish or hypersaline, freshwater lagoons that have survived drainage are rather rare. Seasonally flooded coastal lagoons are with few exceptions restricted to the North African countries. Salinas are present in all countries, the largest areas occurring in the more industrialised countries of southern Europe (Fig. 6).

The majority of intertidal wetlands occur along the Atlantic coasts of Spain and Portugal, particularly around the Gulf of Cadiz (Fig. 7). A surprisingly large area of intertidal flats and marshes exists in the Gulf of Gabes in Tunisia, and this exceeds the area along the Atlantic coast of Morocco.

Inland freshwater lakes are most abundant in Italy and the Balkans, but are scarce in France, Spain, and lowland North Africa. The figure for Italy is only for that part of the lakes with a depth of less than 6 m. The total area of lakes would be much higher, since most of the lakes are deep.

The palustrine systems (non-tidal salt marsh, freshwater marsh, and forested wetland) are also certainly under estimated. It has not usually been possible to distinguish non-tidal salt marsh from the larger wetland units within which it occurs and its area is included in that of saline coastal lagoons and athalassic salt lakes. The figures for freshwater marshes show that they are now reduced and most occur in a few large sites. For forested wetlands, the remaining examples are now so fragmented that it is impossible to give a realistic estimate of their distribution and extent.

Table 1. Minimal estimated areas (km^2) of wetlands. 0 = absent, ? = present but area unknown.

	Portugal	Spain	France	Italy	Albania	Yugoslavia	Greece	Bulgaria	Tunisia	Algeria	Morocco
Estuaries			0	2	0	?	0	0	0	0	>17
Intertidal flats	655	204	0	?	0	?	0	0	281	0	>31
Intertidal saltmarsh			0?	?	0	?	0	0	59		34
Coastal Lagoons	140		938	115	?		292		659	37	216
Freshwater	?		37	?	?	?	?		?	21	0
Saltwater	100		664	?	?		?		536	9	103
Seasonal	0		0	?	0	0	?	0	85	0	108
Salinas	40	150	237	84	8	<1	47	3	38	8	5
Non-tidal saltmarsh	?		208	*	?	?	94		*	*	*
Freshwater lakes	?		5	30	<350		1641		112	>20	14
Reservoirs	?		36	?	?	?	125	?	?	33	>75
Athalassic salt lakes	?	>55	0	<1	0	?	0	0	7525	3589	416
Freshwater marshes	?	>65	203	15	?	?	53	?	51	290	2
Forested wetlands	?	?	<10	>3	?	?	3	?	<5	<1	<1

* Included in category of athalassic salt lakes and coastal lagoons.

Wetland inventory and classification

Until recently there have been few attempts to inventory the wetlands in any Mediterranean countries and most surveys were done to identify sites of importance for breeding and wintering waterfowl (e.g. Olney 1965, Carp 1980, Scott 1980, Yesou and Trolliet 1983). All extensive wetland areas were identified but small sites holding few birds, and certain wetland types of limited ornithological interest (e.g. peatlands, streams, riverine forest) were excluded.

A list of wetlands of international limnological value was prepared for Project Aqua (Luther and Rzoska 1971). They included 28 sites within the countries covered by this chapter and some were the same as those listed in the inventories of wetlands of international waterfowl importance. Others (deep lakes, underground waters and springs) were of no value to aquatic birds and were not included.

National wetland inventories are now being undertaken at the national or regional level in many countries. Inventories which have appeared so far have covered limited area (e.g. Britton and Podlejski (1981) for the Rhône Delta, MOPU (1984) for the Andalucia region of Spain), or are obviously incomplete (e.g. Dorikos (1981) for Greece, Morgan and Boy (1982) for North Africa).

Phytosociological maps of all natural and semi-natural vegetation have been prepared for the entire Mediterranean region of France at a scale of 1:50,000, and for certain regions of great floristic interest (e.g. Camargue) at a scale of 1:10,000 (Lavagne and Moutte 1980). These allow the delimitation of all major areas of wetland vegetation and are being used in a national survey of sites of wildlife value (ZNIEFF, Zones Nationales d'Intérêt Florisique et Faunistique).

In Spain, it would appear that an attempt at complete inventory of sites, including running waters, is being undertaken, but maps (1:50,000) are being used as the data base. Consequently, information on each site will be limited to a classification into broad wetland type, area, altitude, location etc. Map-based inventory is likely to overlook many palustrine and especially forested wetlands.

All wetland inventories have used some sort of classification of wetlands and have ascribed individual sites to particular wetland classes. These classifications have generally been very simple, and are based on topographic characteristics which could be ascertained from maps rather than on vegetational or hydrochemical attributes. The classifications used in these inventories are mostly linear and non-hierarchical, so that large wetland systems are ascribed to several classes. Morgan and Boy (1982), however, describe hierarchical classification, based on multivariate analysis, for North African

Table 2. Classification of wetlands used in Italian wetland inventory (1972, unpublished).

1. Natural Wetlands
    1. Large inland lakes and littorals
    2. Small inland lakes
    3. Mountain lakes
    4. Coastal lakes
    5. "Valli"
    6. Lagoons
    7. Marshes
    8. Ponds
    9. Swamps
    10. Peat-bogs
    11. River banks and river beds
    12. Estuaries
    13. Deltas
2. Artificial Wetlands
    1. Expansion tanks
    2. Reservoirs
    3. Sedimentation basins
    4. Salinas

wetlands. The range of wetland types that are covered is limited and the system is not applicable to the Mediterranean region as a whole.

The simplest inventories, based largely on information obtained from maps, (e.g. Dorikos 1981) have divided wetlands into a few broad classes: lakes (including reservoirs and lake littorals), coastal lagoons (including salinas and partially enclosed bays), marshes (both fresh and salt), river deltas (complex wetlands which may include all the above categories), and riverine wetlands.

In the inventory of Andalucian wetlands (MOPU 1984), lakes and lagoons were lumped together, but salinas and artificial reservoirs were distinguished, and rivers were divided into two classes: main rivers and tributaries.

An unpublished inventory of Italian wetlands produced by the Department of Land Reclamation of the Ministry of Agriculture in 1972 divided wetlands into artificial and natural areas, thirteen categories being recognized in the latter and four in the former (Table 2). Large lakes were those with an area of greater than 3 km^2, but only those parts with a depth of less than 6 m were included in the inventory. Mountain lakes were defined as lying above 750 m altitude, whereas coastal lakes were within 1 km of the sea, either with or without a connection to the sea. "Valli" are coastal lagoons which have been transformed into extensive fish farms, usually by the construction of surrounding dikes, and may be fresh or salt water. The term lagoon was used in a restricted sense to describe partially enclosed shallow coastal areas, bounded by off shore islands, and having a high salt content (e.g. Laguna di Venezia). The terms marshes and ponds were both used to describe

Table 3. Wetlands types used by Yesou and Trolliet (1983).

1.	Open sea
2.	Unvegetated intertidal flats
3.	Vegetated intertidal flats
4.	Salt marsh
5.	Coastal lagoons (and adjoining salt marsh)
6.	Lowland rivers
7.	Lakes (>100 ha, >2 m deep)
8.	Ponds (<100 ha, <2 m deep)
9.	Marshes, swamps and bogs
10.	Wet meadow

shallow areas with emergent vegetation, the former being seasonally flooded, and the latter permanent, while the term swamps was reserved especially for forested wetlands. The term estuaries was used for the lower reaches of rivers with saline influence, whether there was significant tidal movement or not. Among the artificial wetland classes, expansion tanks was a term used to describe artificial flood plains used to regulate stream flow.

The system used by Yesou and Trolliet (1983) included 10 wetland classes (Table 3) which were of importance for wintering waterfowl in France. Peat bogs were included with marshes, since they are of little importance for waterfowl. Open water bodies <100 ha were classed as lakes if the shoreline was mainly unvegetated. Wet meadows are important feeding grounds for wintering waterfowl, but many of the sites included in this category are not true wetlands, as flooding is of short duration in the winter, and vegetation is composed of xerophytes or even agricultural crops.

The classifications used by Carp (1980) and Scott (1980) are similar and both adapted from that originally proposed by Isakov (1966) and Eber (1969). The system used by Scott is compared in Table 4 to that used in this paper and to the higher order categories used by Cowardin *et al.* (1979) for the United States. The main differences between the classification used in this study and that of Scott, is that we have deliberately excluded marine wetlands, and we have subdivided artificial wetlands (reservoirs and salinas) from their natural counterparts. Scott included a category of wet meadows, which have been lumped with freshwater marshes in this study, but did not recognize riverine forests as wetlands, although he includes a number of forested sites, which are used by colonial nesting herons.

The wetland categories can be readily fitted into the Systems and Subsystems of Cowadin *et al.* (1979), with a few exceptions. Although the estuarine system was retained to include mixohaline or hyperhaline coastal wetlands a separate "non-tidal" subsystem was required to accommodate the wetlands of the Mediterranean coast. In view of the importance of salinity as a factor determining the flora and fauna of Mediterranean wetlands, it was felt that

Table 4. Wetland classification system used in this study, compared to that of Scott (1980), and the systems and subsystems of Cowardin *et al.* (1979).

	Cowardin *et al.*	Scott	This study
System	Subsystem		
		Open sea, offshore	–
	Subtidal	Open sea, offshore	Not included
		Bays, straights	–
Marine			
		Rocky coasts	–
	Intertidal	Beaches (unconsolidated)	Not included
		Small islands	–
	Subtidal	Estuaries, deltas	Estuaries
		Coastal mud flats	Intertidal flats
Estuarine	Intertidal	Intertidal saltmarsh	Intertidal salt-marsh
		Coastal lagoons, saltwater	Coastal lagoons, saltwater
	Non-tidal	Coastal lagoons, freshwater	Coastal lagoons, freshwater
		Raised salt-marsh	Non-tidal salt-marsh
		–	Salinas
	Tidal	Estuaries	Estuaries
Riverine	Lower perennial	Rivers, slow flowing	Rivers
	Upper perennial	Rivers, fast flowing	
	Intermittent	–	Intermittent
Lacustrine	Limnetic	Freshwater lakes and reservoirs	Freshwater lakes
		–	Reservoirs
	Littoral	Salt lakes	Athalassic salt lakes
Palustrine		Freshwater marshes	Freshwater marshes
		Wet meadows	
		Forest	Forested wetland

freshwater and saline wetlands should be distinguished for all wetland types. On the other hand, since most inventories of wetlands in the region have been carried out without the aid of aerial photographs, it was not possible to distinguish between the limnetic and littoral subsystems of lakes, and these were lumped together. This constraint also prevents the use of the lower order categories used in the U.S. system and leads to the adoption of a rather simplified classification for wetlands of the Mediterranean region.

Data on mean salinities and annual range of salinities are rarely available for Mediterranean wetlands, and in any case, salinity is highly variable from year to year depending on rainfall and evaporation. It has not, therefore, been possible to incorporate the various classifications for salinity classes of wetlands into the classification. The most widely used classification is the Venice System (Caspers 1959) which divides waters into eight major classes

Table 5. The Venice System of classification of saline waters.

Class	Salinity range $g\,l^{-1}$
Hyperhaline	>40
Euhaline	30–40
Mixohaline	0.5–40
Mixo-Euhaline	>30 (But > adjacent sea)
Mixo-Polyhaline	18–30
Mixo-Mesohaline	5–18
Mixo-Oligohaline	0.5–5
Freshwater	<0.5

(Table 5). In this paper, many of the sites classed as freshwater marshes, and some freshwater coastal lagoons would fall within the oligohaline salinity range of the Venice System, at least in the summer months. Most glycophytes of Mediterranean wetlands can withstand salinities in excess of $1\,g\,l^{-1}$, and the major change in the composition of both the flora and fauna occurs between the oligohaline and mesohaline classes.

Aguesse (1957) proposed a salinity classification for Mediterranean wetlands in which both mean salinity and the annual range of variation were included. The classes for mean salinity were essentially the same as those in the Venice System, except that he used an upper limit of $16\,g\,l^{-1}$ for the division between the meso and polyhaline classes. Three further classes were proposed to describe the annual range of salinity. Oligopoikilohaline wetlands were those in which the maximum salinity was in the same class as the mean, mesopoikilohaline sites have a maximum salinity in the class superior to the mean, and polypoikilohaline sites have a difference of two or more classes between mean and maximum salinity. Thus an oligopoikilohaline oligohaline wetland would be one whose salinity remained within the oligohaline range throughout the year, where as apolypoikilohaline mesohaline site would have mean salinity in the mesohaline range, but would have a maximum salinity in the hyperhaline class. Year to year variation in wetlands of varying salinity is so great that this system is of limited application in wetland inventory. Where the data exist it is preferable, perhaps, to give the mean annual salinity and the maximum and minimum salinity when describing a particular wetland.

Few peatlands, if any, exist in the region but there has been a report (N. C. Morgan, personal communication) of the existence of a Sphagnum bog in Greek Macedonia.

Ecological characteristics of wetland vegetation

Factors determining the composition of vegetation

The main factors which determine the species of plant which occur within wetlands in the Mediterranean region are: depth of flooding, periodicity of flooding, water or soil salinity, grazing intensity, strength of water currents, base-status of water, soil or sediment type, and annual temperature range.

Species occurrence within any one wetland is determined by the sensitivity or tolerance of individual species to these factors and to competitive interactions between the plants themselves. Plant species which tend to occur together have been classified by phytosociologists into plant communities. Within Mediterranean wetlands, however, many of the above factors can vary enormously during the course of a year, from year to year, or over a longer time period. Wetland vegetation is, therefore, highly dynamic and shows long-term non-successional changes in species composition and abundance. The species assemblage present at any one time is often as much a reflection of past events as it is of present environmental conditions. In such circumstances, the value of the concept of the plant community as a stable entity is of less value than when describing upland vegetation such as forest. This is particularly the case with submerged hydrophytes of freshwaters, which are largely short-lived annual species.

Salinity and period of flooding are factors which influence the composition of both emergent and submerged plant communities, where as grazing by domestic and wild animals is a major determinant of the composition of reedswamp communities. Water depth is critical for certain submerged plants and emergent species of short stature, but some perennial emergents such as *Phragmites* will tolerate a wide range of water depths. Most wetlands of the region are highly calcareous, and there are few species of plant which are restricted to base-poor wetlands.

The vegetation of Mediterranean wetlands can be conveniently classed into the following categories: halophytic vegetation, including 3 categories of submerged and 6 categories of emergent (6 subcategories) types; emergent reedswamp communities (4 categories); wet meadow communities; riverine forests; dwarf rush communities; submerged and floating vegetation of freshwater habitats (2 categories).

Halophytic vegetation

The main environmental factors controlling halophytic vegetation are water salinity in the growing season and the depth and period of flooding. Halophyte communities are generally species poor, particularly with high salinities, and the upper limit of tolerance for angiosperms is about 100 g l^{-1} NaCl.

Sea-grass communities – These occur in the lowest parts of saline wetlands where fluctuations in water level and salinity are least. They intergrade into marine communities dominated by algae and *Posidonia oceanica*.

In intertidal wetlands of the Atlantic coast, *Zostera marina* communities occur on mud flats where they are exposed briefly. In the Mediterranean, where *Z. marina* is rare, this vegetation is dominated by *Cymodocea nodosa* and *Z. noltii*. These communities are typical of sheltered marine bays and permanently flooded coastal lagoons, where the salinity does not vary much from that of sea water. Where the two species of *Zostera* occur, *Z. marina* occupies the lower more permanently flooded areas.

Ruppia cirrhosa community – In shallow lagoons and bays where the salinity and temperature are higher, *Zostera* communities are replaced by a vegetation dominated by *Ruppia cirrhosa* and algae such as *Chaetomorpha linum* and *Cladophora*. The charophyte, *Laprothamnium papulosum* is also a frequent associate and, in the lagoons of the northern Adriatic coast, it forms dense monospecific stands with *R. cirrhosa* being restricted to the shallower margins (Ferrari *et al.* 1972). This vegetation type is perennial but plant biomass is much reduced in winter by grazing and storm damage. The community occurs in salinities from about 15 g l^{-1} to 80 g l^{-1} and in depths from less than 10 cm to about 2 m. This vegetation will not withstand prolonged or frequent periods of drying out (Verhoeven 1979, 1980). In waters of lower salinity it is replaced by *Potamogeton pectinatus* dominated vegetation. In shallow areas which dries out in early summer, it is replaced by the submerged annual halophytes described below. Both *Ruppia* and *Zostera* dominated vegetation types are important for herbivorous waterfowl such as widgeon (*Anas penelope*) and are also important spawning and nursery grounds for coastal fish species.

Submerged annual halophyte community – This vegetation type occurs in shallow saline pans and ditches which flood in winter but dry up before midsummer, and on the margins of larger bodies of salt water. Small, annual species such as *Ruppia maritima* (particularly var. *brevirostris*), *Althenia filiformis*, *Tolypella glomerata*, and *Chara* spp., are dominant. The maximum salinity range recorded by Verhoeven (1980) for *Ruppia maritima* in the Camargue was 1–42 g l^{-1}. At lower salinities this vegetation is replaced by one dominated by *Ranunculus baudotii* and *Zannichellia* spp. At higher elevations on the shore, or later in the year in the same place, this shallow vegetation type is often replaced by the one described next.

Emergent annual halophyte community – This vegetation type is an emergent counterpart of the one just described. Most emergent species are annuals which germinate when the water level is below the soil surface. It is domin-

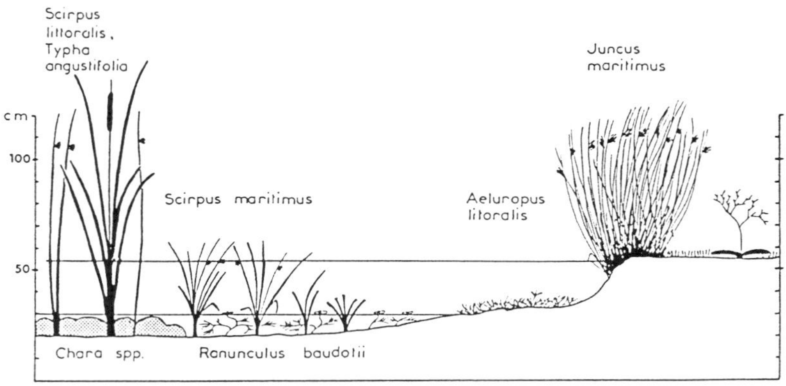

Figure 8. Typical zonation of plant communities in a seasonally flooded oligohaline marsh.

ated by annual Chenopodiaceae species of which *Salicornia europaea*, *Bassia hirsuta*, *Salsola* spp., and *Suaeda* spp. are the most important. In habit and species composition, it is similar to the annual *Salicornia* community of the intertidal wetlands of the Atlantic coast. In intertidal wetlands it occurs between the *Spartina* marsh at higher elevations and the mud flats, and which are exposed to almost daily submergence. In non-tidal Mediterranean wetlands, this vegetation occurs on exposed flats between the perennial *Arthrocnemum* communities and the submerged *Ruppia* zone, where there is infrequent flooding during the summer. *Suaeda* and *Salsola* primarily occur on strand lines of organic debris and on soils of low organic content. *Salsola* communities are also characteristic of irregularly flooded North African salt lakes where they may cover extensive areas following draw down. These communities are highly ephemeral and their occurrence and composition at any one site varies from year to year according to the hydraulic regime and the composition of the seed bank.

Although not an annual, the grass *Aeluropus littoralis* is important because it often forms a monospecific carpet within the annual community in areas where soil salinities are between 320–420 mg 100 g^{-1} soil (Basset 1978). Its presence represent a successional stage which occurs under more stable water level conditions and in areas less exposed to wave action. It typically occurs in the central parts of small seasonally flooded saline basins or on the margins of larger oligohaline seasonal marshes, where it occupies a zone between the *Scirpus maritimus* or *Ranunculetum baudotii* and the *Juncetum maritimus* zone (Fig. 8). During winter flooding, *Aeluropus* survives as dormant surface rhizomes. In oligohaline sites, it is probably maintained by grazing which eliminates taller species such as *Scirpus maritimus* (Fig. 8). Under heavy grazing, *Aeluropus* is reduced in density but it is not eliminated (Basset 1978).

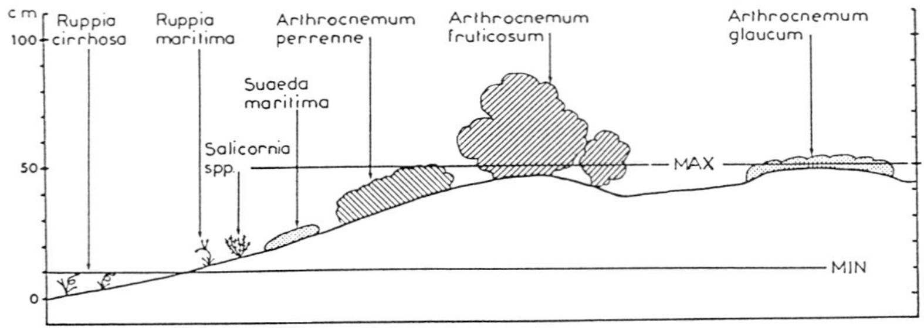

Figure 9. Typical zonation of plant communities in a mesohaline non-tidal salt marsh. Max and Min represent average annual water levels.

Spartina communities – *Spartina* dominated areas are almost restricted to tidal wetlands and extend into the northern Mediterranean and in tidal wetlands in the Adriatic (Beeftink 1968). The vegetation is restricted to a rather narrow zone around the high water mark. In Southern Europe *Spartina anglica* is replaced by *Arthrocnemum fruticosum* communities at higher elevation on the shore. In Morocco, *Arthrocnemum* extends right down to the limit of perennial vegetation and the *Spartina* zone is very restricted. *Halimione portulacoides* is generally present within the *Spartina* community and it often forms more or less pure stands along creek margins.

Perennial *Arthrocnemum* salt marsh – This community is composed mainly of plant species whose center of distribution lies within the Mediterranean basin. In some locations, it also extends along the Atlantic coasts of Europe and North Africa. In tidal wetlands, communities dominated by *Arthrocnemum fruticosum* and *Limonietum monopetalum* replace the *Spartina* community at higher elevations and they occur up to the extreme high water level. In non-tidal wetlands where the *Spartina* community is scarce or absent, the intermediate position between annual Chenopodiaceae and the *A. fruticosum* zone is occupied by a community composed almost entirely of *A. perenne* (Fig. 9), a prostrate species that is more resistant to flooding than *A. fruticosum*. *Arthrocnemum perenne* typically remains flooded for 2–6 months in the winter, while *A. fruticosum* is generally flooded for less than 2 months to a depth of less than 50 cm. *Arthrocnemum fruticosum* is a shrub 20–80 cm high, which forms a community of generally high vegetation cover, but with low species richness. *Arthrocnemum fruticosum* and *A. perenne* communities occur in salinities from about $10\,g\,l^{-1}$ up to about $40\,g\,l^{-1}$. At higher salinities they are replaced by *A. glaucum* or *Halocnemum* communities, while at lower salinities *Juncus maritimus* communities occupy zones of similar flooding depths and duration.

Dwarf salt scrub – Two species (*A. glaucum* and *Halocnemum strobilaceum*) form structurally similar communities on highly saline flats which are flooded for a period of a few weeks to a depth of 5–30 cm, either seasonally or at irregular intervals. Both plants form low (5–20 cm high) tussocks interspersed with extensive areas of bare soil on which a salt efflorescence develops during the dry season. Aeolian and water-borne sediment becomes trapped within the plant tussocks and leads to the formation of a hummocky terrain. Other species are infrequent and are mainly small shallow-rooted annuals (e.g. *Sphenopus divaricatus* and *Hordeum marinum*) which exploit the superficial soil layer from which salts have been leached by rainfall. Soil salinity ranges from 500–600 ms 100 g^{-1} in the summer months (Basset 1978).

During flooding, blue-green algal mats form on the bare ground and an ephemeral zooplankton community hatches from dormant eggs lying in the sediment. If flooding is prolonged, *Chara* spp. and *Ruppia maritima* may also become established. An erect bushy form of *A. glaucum* occurs in saline pans in the Camargue which flood to a depth of over 50 cm (Molinier and Tallon 1970). This may be an adaptation to long term rise in water level.

Arthrocnemum and *Halocnemum* scrub dominated wetlands occupy vast areas of saline soils in deltaic regions of the Mediterranean. They also occur on the margins of coastal lagoons, particularly in depressions isolated from the main basin (Fig. 9), and around inland salt lakes in North Africa. *Arthrocneum glaucum* occurs throughout the region, but *H. strobilaceum* is restricted to the Balkans and North Africa. Where the two species occur together, *H. strobilaceum* is considered to be the pioneer and is replaced by *A. glaucum* once build up of hummocks has occurred (Ayyal and El Ghareeb 1982).

Halophytic rush communities – One of the most characteristic features of the oligohaline and mesohaline wetlands is a narrow marginal band of vegetation dominated by *Juncus maritimus*. This zone marks an abrupt transition between wetland and upland halophytic vegetation. This band is normally only a few meters wide and often occurs on steep slopes formed by wave action during winter flooding. On the landward side, the vegetation is replaced by terrestrial communities but *J. maritimus* may be present as a minor species. On the wetland side there is an abrupt transition to a variety of aquatic communities whose composition varies according to water salinity and grazing pressure. *Tamarix* spp., *Triglochin maritima*, and *Juncus gerardii* are frequent associates in the *J. maritimus* belt, and under conditions of low salinity and heavy grazing *J. gerardii* may form extensive stands that grade into wet-meadows. *Juncus maritimus* is a Mediterranean species that also occurs at the upper limit of tidal salt marshes along the Atlantic coast in areas where salinity ranges from 5–40 g l^{-1} (Braun-Blanquet 1951). Because of its unpalatable nature, *J. maritimus* even occurs in wetlands where all other emergent species have been eliminated by grazing pressure.

Juncus subulatus and *A. fruticosum* occur at higher salinities and in areas subject to longer periods of flooding than *J. maritimus* dominated sites. Like

J. maritimus, however, it tends to form dense monospecific stands. In terms of zonation patterns, it would occur between the *A. fruticosum* and *A. perenne* zones.

Tamarix scrub – Scrub woodland with a shrub layer composed entirely of *Tamarix* spp. occupies large areas of saline wetlands, particularly in delta areas and along intermittent saline streams. Different species of *Tamarix* occur through the Mediterranean, but all have similar structure and occur in the same types of biotope. *Tamarix africana* and *T. canariensis* occur especially in Spain and Portugal, *T. gallica* in France, and *T. tetrandra* in the Balkans. The presence of *Tamarix* appears to depend on the existence of suitable conditions for seedling germination and establishment for, once established, they thrive under a wide range of salinities and water levels (Marks 1950). *Tamarix* seeds are widely dispersed but are short-lived. A suitable germination bed consists of bare wet mud which must be present at the time the seeds are shed (Waisel 1972). The seedlings are tolerant of a wide range of salinities (Ungar 1974) but require a prolonged period of soil moisture to survive. Establishment occurs particularly on the margins of water bodies having a very slow summer draw down, or after a period of summer rainfall.

Understorey vegetation associated with *Tamarix* wetlands is very variable. In the driest or most saline sites there may not be any understorey. Surface soil salinity is also affected by the excretion of salt from *Tamarix* foliage and some halophytes cannot successfully establish. At the other extreme, *Tamarix* can survive flooding for up to 6 months or more to a depth of up to 1 m. Under these conditions, the understorey is generally composed of submerged associations of *Ranunculus baudotii* and *Zannichellia* spp. with sparse *Scirpus maritimus* developing in the summer. *Scirpus maritimus* is the most frequent understorey species in seasonally flooded *Tamarix* scrub but is replaced by *Juncus maritimus* where flooding is of brief duration. *Tamarix* bushes are often associated with the narrow marginal fringe of *J. maritimus* that encircles oligo- and mesohaline seasonally flooded wetlands.

Emergent reedswamp communities

Macrophytes of freshwater and oligohaline wetlands, also called marshes, are mainly tall herbaceous reed-like species that tend to occur in more or less monospecific stands which form distinct zones. Depth is the main factor determining zonation and species composition but grazing pressure from domestic animals (cattle, *Bos taurus*; horses, *Equus caballus*; sheep, *Ovis aries*), wild mammals (deer, *Cervus elaphus*; wild pigs, *Sus scrofa*; muskrat, *Ondatra zibethicus*; coypu, *Myocastor coypus*; beaver, *Castor fiber*), and birds (coots, *Fulica atra*; Anatidae, *Anas* spp., *Aythya* spp. and *Anser* spp.) are also important. Grazing pressure is also related to the water level regime which also controls the re-establishment of emergent species from the seed bank. The interactions between plants and herbivores, however, are complex and only partly understood (Duncan and D'Herbes 1982).

Most freshwater wetlands are grazed by domestic stock in the summer months when the production of upland vegetation is limited by drought. Only those wetlands which are permanently flooded and too deep, or which are reserved for other purposes (e.g. hunting preserves or exploited reedstands), are exempt. The impacts of rodents are, in contrast, greater in deep permanent marshes. Heavy grazing pressure opens up the vegetation and allows the establishment of less competitive plants including shallow water submerged species. Some marshes are managed artificially by reed cutting or other mechanical treatment to maintain early successional stages which are favored by waterfowl (Duncan *et al.* 1982).

Four types of reedswamp communities can be distinguished. *Scirpus maritimus* dominated wetlands are the most salt tolerant and occur in oligohaline (lagoons) and freshwater wetlands, especially those managed for waterfowl where competition from other species is reduced by grazing. *Scirpus maritimus* may be 70 cm high and it tolerates salinities of at least $20\,g\,l^{-1}$ and in freshwater areas it occurs in waters with alkalinity of 4 mequ. l^{-1} as Ca CO_3 (Corre 1961, Podlejski 1981). *Scirpus maritimus* occurs primarily on the margins of the larger and deeper wetlands where the maximum winter flooding depth is about 70 cm. In deeper water *Scirpus maritimus* is replaced by taller emergent species in freshwater sites, and by the submerged *Ruppia* communities in water bodies where the salinity exceeds $10–15\,g\,l^{-1}$.

In freshwater and oligohaline conditions the presence of *S. maritimus* is favored by light grazing which eliminates taller growing and more palatable species, especially *Phragmites*. In the absence of grazing, the taller species often replace *S. maritimus* even the shallow water areas. *S. maritimus* is more resistant to grazing but under heavy grazing both density and stem length are reduced and it can be eliminated and replaced by submerged communities or short grass swards of *Aeluropus* sp. or *Paspalum* sp. (Basset 1980, Duncan and D'Herbes 1982). When it is grazed, an understorey of submerged species, particularly *Ranunculus baudotii*, *Zannichellia* spp., and *Chara* spp. can develop. The rhizomes of *S. maritimus* are favored as winter food by wild pigs and by greylag geese (*Anser anser*), which can cause local extinctions. *S. maritimus* occurs in vegetation dominated by taller species and it forms low density understorey with little biomass.

Tall *Scirpus/Typha* marshes are characteristic of the deeper parts of many permanently or seasonally flooded base-rich wetlands where there is freshwater input. They occur particularly in areas where *Phragmites australis*, a superior competitor, has been eliminated by grazing or where the water is too deep for *Phragmites* to grow.

A mixed community of *Scirpus littoralis*, *Scirpus tabernaemontani*, *Typha angustifolia*, and *T. domingensis* is highly characteristic of grazed oligohaline seasonally flooded wetlands in water to about 1 m deep. *Phragmites australis* is often present as stunted individuals, but it may replace the taller species if grazing is excluded. This is rather an open community with plants as high as 2.0 m in summer. The species all die back almost entirely in winter. There is frequently an understorey of *Chara* spp. *Typha* is favored by seasonal draw down which allows establishment of seedlings; but is gradually eliminated by coypu and muskrat damage under permanent flooding (Kohli 1981).

A pioneer community of *Typha* (especially *T. latifolia*) and Lemnaceae is characteristic of disturbed wetlands (e.g. dredged drainage canals). This pioneer vegetation is replaced by *P. australis* under a non-grazing regime.

Scirpus lacustris is typical of rather deep (1–2 m) and exposed lake shores and the deeper parts of marshes. It generally occurs as dense monospecific stands that have little associated submerged vegetation, *Chara aspera* is the typical species. In many lakes, the emergent species in shallow water are often eliminated by grazing. *Scirpus lacustris* in these situations forms an offshore fringe that is separated from the shore by a zone with only submerged species. This situation is found in several North African lakes.

Phragmites reedswamps are widespread because *P. australis* is the most competitive reedswamp species and it occurs in a wide range of salinity and flooding conditions. It often forms dense monospecific stands which shade out almost all other plant species. *Phragmites* is, however, extremely sensitive to grazing by large mammalian herbivores, because its meristem is apical, rather than basal. It is rapidly eliminated and replaced by less palatable but more resistant species of the *Scirpus*/*Typha* community, even under moderate grazing pressure. Of the commonly occurring reedswamp plants of the Mediterranean region, *Phragmites* is the species preferred by cattle and horses (Duncan and D'Herbes 1980). Grazing causes branching of the shoots and weakening of subsequent year's growth. Trampling also damages the rhizomes (Haslam 1971a, b).

Phragmites australis tolerates salinities up to about $10\,g\,l^{-1}$ above which it is replaced by *Scirpus maritimus*. In nutrient-poor wetlands, particularly those with a slight water current, communities of *Cladium mariscus* in frequently flooded marshes, or *Molinia caerulea* in drier situations, out-compete *P. australis*.

Phragmites australis is tolerant of a wide range of water level regimes, and is restricted by water table conditions which must be above, or near the soil surface for at least part of the year. It grows in permanently flooded sites to a summer depth of about 1.5 m and in south-eastern Europe it also occurs as floating rafts (Plaur) in even deeper water. In tidal regions, it will withstand daily submergence and draw down provided the salinity is low. Optimal development occurs in wetlands that are submerged to a depth of 1 m or more in winter, which dry out partially in summer, and which have a water table that does not descend more than about 1 m below the soil surface. Under such conditions there are few associated species. The seasonal cycle of flooding and drying out allows for rapid decomposition and prevents the build-up of a deep anaerobic litter layer. If other environmental conditions (e.g. water regime, salinity, grazing) remain stable, such stands can persist, perhaps for centuries, with no successional changes either towards shrub invasion or to open lake conditions. *Phragmites* clones as old as 1000 years have been reported (Haslam 1971a).

In wetlands that flood only briefly, and where the water table descends well below the soil surface in summer, *Phragmites* stands are invaded by other species and it may eventually be replaced by more competitive associations. In freshwater and in the absence of grazing, succession under drier conditions is towards vegetation dominated by *Salix* spp. and *Alnus* spp. scrub and finally towards riverine forest. Under lighter grazing *Phragmites* is replaced in drier conditions by one of the wet-meadow communities described below.

Phragmites reedswamps occur throughout the region but are scarce in North Africa where they are limited by salinity, grazing, and drought. Wetland drainage, salinization, and increasing exploitation of wetlands for grazing, pisciculture, and hunting have greatly reducing the extent of this vegetation type. Avifauna which depends on this biotope has also decreased alarmingly in some countries (Duhautois 1984).

Saw-sedge marshes dominated by *Cladium mariscus* can form dense, almost monospecific, stands in permanently wet areas. In the Mediterranean region, however, such vegetation is extremely localized, and it is not entirely clear what environmental factors favor its dominance over the more frequent *Phragmites*, and *Scirpus*/*Typha* wetlands.

The largest known Mediterranean *Cladium* marshes are in the Tablas de Daimiel in central Spain, the Marais de la Crau in southern France, around some lakes on the Tyrrhenian coast of Italy, and on Crete. In North Africa, out of 67 wetlands surveyed by Morgan (1982 a, b), *Cladium* was recorded only from the Bas Loukkos in Morocco. All these localities are freshwater (salinity $<0.2\ g\ l^{-1}$), and probably base-rich, but there is some indication of a tendency towards mesotrophy compared to neighboring *P. australis* marshes. The soil pH of *Cladium* wetlands in France was 6.2–7.8 compared to >7.8 for adjacent *Phragmites* and *Typha* stands (Moubayed 1977). Low nutrient status is thought to be a factor favoring *Cladium* over *Phragmites* in northern Europe (Wheeler 1980).

At least two of the Mediterranean *Cladium* marshes (Tablas de Daimiel and Marais de la Crau) are fed by underground springs. Molinier and Tallon (1950) considered the Cladietum of the Marais de la Crau to be a glacial relict community that was able to survive under a Mediterranean climate due to the influx of cool spring water. Many of the associated plant species (e.g. *Gentiana pneumonanthe*, *Lathyrus palustris*, and *Thalictrum flavum*) are rare in the Mediterranean but common in northern Europe. The inflow of spring water also maintains a more constant water level in these marshes as compared to wetlands dependent on precipitation and runoff. *Cladium mariscus* wetlands may remain in permanently flooded areas or where the water table may descend below the soil surface for several months in the year. The soil surface, however, normally remains saturated. As a result, this is the only vegetation type in the Mediterranean in which there are appreciable accumulations of peat up to 1.2 m depth in the Marais de la Crau.

Nearly all *C. mariscus* wetlands are at least periodically grazed, mown or burnt. Pure stands tend to occur in the deeper and more permanently flooded parts of the marshes. In drier areas, a Molinietum vegetation will replace it and isolated tussocks of *Carex elata* are also characteristic of drier *Cladium* marshes. This species occasionally forms a narrow band between the Cladietum and the Molinietum. In certain marshes, *C. mariscus* is replaced by *Scirpus lacustris* or *Phragmites australis* in deeper water but at Tablas de Daimiel this zonation is reversed apparently due to local variation in nutrient loading or grazing pressure (Coronado-Castillo *et al.* 1974).

Wet meadow communities

This vegetation type is a heterogeneous assemblage of plant species which occurs in areas which are shallow flooded freshwater in winter and which dry up but remain moist in summer. They are maintained by at least occasional burning, mowing, or grazing which prevents shrub invasion and discourages the growth of *Phragmites australis*. This vegetation type is widespread in northern Europe but restricted in the Mediterranean, where summer drought limits the growth of many of the species. This vegetation type often occupies a marginal position between reedswamps, described in a previous section, and upland vegetation. It is, therefore, analogous to the *Juncus maritimus* vegetation in oligohaline wetlands. This vegetation may also occur in partly drained wetlands where the water table remains near the soil surface and where grazing occurs.

Wet meadows are dominated by Gramineae, Juncaceae, Cyperaceae, and a variety of tall forbs. When standing water remains until at least May, an association (Butometum umbellati) dominated by *Butomus umbellatus*, *Eleocharis palustris*, *Iris pseudacorus*, and *Juncus gerardii* occurs. *Alisma* spp. and *Althaea officinalis* are common associated forbs. Introduced species of *Paspalum*, which are graminoid weeds of rice fields, have invaded this association in many parts of southern Europe.

Stands dominated by *Molinia caerulea* form another wet meadow association on the margins of, particularly *Cladium*, grazed reedswamps from coastal Spain and Algeria to northern Greece. The *Molinia* stands are flooded to about 20 cm from October to March or April but dry in the summer. *Molinia caerulea* is intolerant of any salinity (Moubayed 1977). The association forms on base-rich peat, unlike most other meadow associations which form on mineral soils. Molinier and Tallon (1950) considered the Molinietum of the Marais de la Crau in southern France to be a glacial relict community because of the presence of a number of associated species typical of northern Europe. In drier situations the Molinietum grades into vegetation dominated by *Carex* spp. and *Scirpus holoschoenus*.

The most widespread of the *Carex* dominated meadow communities in the Mediterranean is the Leucoio Caricetum which mostly replaces the Caricetum elatae of northern Europe. Characteristic species include *Carex riparia*, *Leucoium aestivum*, *Althaea officinalis*, *Lysimachia vulgaris*, *Lythrum salicaria*, and *Oenanthe lachenalii*. Under drier conditions one finds a Juncetum subnodulosi meadow association in which *Juncus subnodulosus* occurs together with amphibious grasses such as *Agrostis stolonifera* and *Alopecurus geniculatus*. These meadows contain many completely terrestrial plant species and form a transition between wetland and upland sites.

Riverine forests

Riverine forests occur along flowing water courses, either on adjoining floodplains, or on sand and gravel islands of braided river channels. These areas are flooded by winter rainfall or spring snow melt and the water table is below the soil surface for much of the summer. The characteristic association of floodplains of slow-flowing lowland rivers (up to 150 m altitude in France) is the Populetum albae, a community restricted to the Mediterranean region. This forest dominated by *Populus alba*, *Ulmus* spp., and *Alnus glutinosa*. In wetter areas *Salix alba* grows to a height of 15–20 m. There is normally a tall shrub layer (e.g. *Cornus sanguineum*, *Laurus nobilis*, *Amorpha fructicosa*) and vines are well represented (e.g. *Smilax aspera*, *Vitis vinifera*, *Hedera helix*, *Clematis* spp.). The cover of forbs and grasses depends on the intensity of grazing and flooding. In extreme cases the forest floor is bare but swards dominated by grasses such as *Brachypodium sylvaticum* are common. Soils in this type of wetland are typically a gleyed base-rich silt.

In the Balkans, the Populetum albae is replaced by the Platanetum orientalis association which is dominated by *Platanus orientalis*. Structurally it is very similar to Populetum albae, and shares many of the associated plant species. In Crete, *Castanea sativa* dominates the more acidic soils.

In faster flowing rivers on gravel soils, the Populetum albae is replaced by an association in which *Salix purpurea* is the main tree species (Saponarieto-Salicetum purpureae). This is more of a shrub woodland (3–5 m high) with several other species including *Vitex anguscastus*, *Nerium oleander*, and *Salix eleagnos*. The forb layer is mainly composed of ruderal annuals. This association is frequently subject to flash floods, and is very much a pioneer community. In arid regions, *Salix purpurea* is replaced along intermittent streams by *Tamarix* spp.

Riverine forests are now highly fragmented in the Mediterranean region and nowhere are there large intact stands. The largest remnants occur in the middle Po valley and in the lower Rhone, but small vestiges are scattered along all river courses in southern Europe. Riverine forests have probably always been scarce in North Africa because of climatic factors, and are now almost non-existent.

Dwarf rush communities

These are rare and localized communities, which are always of very limited extent. They occur throughout the Mediterranean region, in situations as diverse as dune slacks and small rock basins in upland areas. They are of interest because they contain a number of plant species endemic to Mediterranean wetlands. They occur in small, seasonally flooded freshwater basins, normally on base-poor rocks, where heavy grazing and trampling during summer draw down maintain a short sparse vegetation cover with much bare soil. The species composition is highly variable and numerous phytosociological associations have been named and all are in the Alliance Isoetion (Braun-Blanquet 1951).

Characteristic plant forms are bryophytes, annual dwarf-rush species, dwarf helophytes, and dwarf aquatic pteridophytes which develop during winter flooding and die back in summer. At least seven species of *Isoetes* endemic to the Mediterranean region are described from these communities. Other typical plants are *Juncus bufonius*, *J. pygmaeus*, *Damasonium* spp., *Elatine* spp., and *Lythrum thymifolium*, but most of these are restricted to particular associations.

Submerged and floating vegetation of freshwater habitats

Wetlands of shallow water habitats are highly dynamic and often transient. As a result, it has been difficult to classify the vegetation into acceptable phytosociological units. They often rely on disturbance in the form of grazing, drawdown, dredging or reed-cutting for their continued existence. A high proportion of the species are annuals or short-lived perennials, which tend to disappear from one year to another to be replaced by other species. Some communities also show seasonal succession, with pioneer species of low stature and a rapid reproductive cycle in winter and early spring being replaced by slower growing, but more competitive species later in the year. In contrast, the hydrophyte communities of deep-water habitats, beyond the depth limit for emergent vegetation, are more stable and persist for many years although the species composition may change from one year to another. Most vegetation in these two types of conditions belong to the order Potametalia (Braun Blanquet 1951). The class Littorelletea (Den Hartog and Segal 1964) is restricted to high altitude oligotrophic lakes within the Mediterranean region, but the dwarf-rush communities (previous section) are their counterpart in seasonally flooded wetlands. Three associations will be described.

Submerged vegetation also occurs in deep water which is here defined as water beyond the depth tolerance of emergent vegetation, usually from 1 to 2 m. In permanent freshwater lakes, submerged vegetation occurs between 1 and 2 m and belongs mainly to the order Magnopotametalia (Den Hartog and Segal 1964). Typical are broad-leaved *Potamogeton* species such as *P. perfoliatus* and *P. lucens*, and *Myriophyllum verticillatum*. This association is rather restricted because inland freshwater lakes tend to be deep and steep sided. In some lakes the association is further restricted by large seasonal variation in water level.

In freshwater to slightly saline coastal lagoons, which are more typical of the Mediterranean region, the submerged vegetation is species poor and stands tend to be monospecific. The most widespread species are *Potamogeton pectinatus* and *Myriophyllum spicatum*. These two species tend to exclude each other either spatially or temporally, and form dense beds with 100% cover, and few associated species. *Potamogeton pectinatus* has a higher salinity tolerance than *M. spicatum*, and occurs in salinities up to 10–15 g l^{-1}, above which it is replaced by *Ruppia* dominated communities. Both species are characteristic of permanently flooded wetlands, and will not withstand prolonged periods of drying out. If drawdown persists for more than 1–2 months in the summer, and the soil dries out completely the perennating organs (tubers in the case of *P. pectinatus* and terrestrial plantlets for *M.*

spicatum), do not survive, and these species are replaced by pioneer species of *Chara*, or by shallow water communities.

Almost pure stands of *Chara* spp. are characteristic of open water freshwater habitats which suffer prolonged summer drawdown. Once established this vegetation may resist invasion by *P. pectinatus* and *M. spicatum* for many seasons without draw down. Typical deep water species are *C. hispida* and *C. aculeolata*, and in sites with a deep organic mud substrate, *Nitellopsis obtusa*. These are robust species, which have been assigned to the association Magnocharetum (Corillion 1975).

Stands of vegetation dominated by *Ceratophyllum* spp. are typical of disturbed and particularly nutrient-enriched sites (e.g. drainage ditches). It also occurs in sheltered parts of freshwater lagoons beyond the reed fringe. In both situations they are often associated with floating vegetation, either free-floating Lemnetea or rooted Nymphaeion.

Submerged communities in shallow water must compete with taller emergent species. Vigorous stands, therefore, only occur in areas where the emergent vegetation has been controlled either by human intervention, or by grazing. The vegetation type is most typical in areas which are dry from May to July and reflood in autumn or winter. The plant species are essentially "r" strategists with short generation times, high reproductive investment, good powers of dispersal, or ability to survive as propagules for prolonged periods of conditions unfavorable for plant growth. All are of low competitive ability.

Pioneer species such as Characeae appear first following flooding, and are gradually replaced during a seasonal succession by slower growing plants such as *Potamogeton* spp. If water remains for prolonged periods into summer, species characteristic of deeper water (e.g. *P. pectinatus*) will appear but do not usually have time to become dominant before drying out. The composition and relative abundance of the different species within the associations depends largely on chance events of colonisation. In calcareous waters characteristic species are *Chara vulgaris* and *C. aspera*, *Tolypella glomerata*, *Nitella flexilis*, *Potamogeton pusillus*, *Callitriche* spp., and *Zannichellia* spp. Communities with *Callitriche* and *Zannichellia* are more typical of waters with some slight salinity (Van Vierssen 1982), and grade into an association with *Ranunculus baudotii* (Callitricho Ranunculetum baudotii). *Ranunculus baudotii* occurs in a range of salinities from freshwater to at least 15 g l^{-1}, and forms a transition between the freshwater communities and the *Ruppia* dominated vegetation of brackish water described earlier. *Utricularia* spp. are characteristic of shallow humic waters (e.g. clearings within reedswamp and overgrown ditches).

Floating-leaved associations include two distinct types: Nymphaeion, which consists of rooted plants with floating leaves, and Lemnetea, which consists of free-floating micro-hydrophytes (pleustrophytes). Rooted floating-leaved associations are rather uncommon in the Mediterranean region, due to large annual fluctuations in water level and wave action caused by strong winds. No floating leaved species are tolerant of any salinity which further restricts their occurrence. Associations of Nymphaceae and *Potamogeton natans* are restricted to stable, sheltered conditions, such as found in canals and slow flowing perennial rivers. *Nymphaea* communities (*Nuphar luteum*, *Nymphaea alba*, *Nymphoides peltata*) are also reported from many of the freshwater lagoons in the lower Po valley, where water levels are maintained artificially (Carp 1980).

An association of *Trapa natans* and *Salvinia natans* is found particularly in the Balkans where it covers large areas of the open water of some freshwater lagoons with rather stable water level (e.g. Lake Mitrikou in northern Greece).

The Lemnetea consists of pioneer communities of disturbed environments. They are particularly prevalent in drainage ditches which have been cleared of emergent vegetation, or in clearings within *Typha* and *Phragmites* reedswamp. The species composition is very variable, but always includes one or more species of Lemnaceae (*Lemna minor*, *L. triscula*, and *Spirodela polyrrhiza*), with *L. triscula* being partially or wholly submerged). Other plants include *Riccia* spp., *Hydrocharis morsus-ranae*, and *Azolla* spp., the latter growing mostly in the winter months. *Ceratophyllum demersum* frequently occurs as a submerged understorey.

Wetland animals

Factors affecting fauna

Factors which determine the composition of wetland vegetation also affect the fauna. The periodic drying out of many Mediterranean wetlands has produced many adaptations to drought including: production of stages resistant to desiccation, small scale movements into permanent water in summer, migration out of the Mediterranean region, nomadism and opportunistic breeding cycles, and breeding in winter and early spring. Different groups of animals clearly use different strategies to escape the effects of drought. Aquatic birds are mostly migrants and only a few Anatidae and shorebirds breed in the Mediterranean. Birds which do remain to breed are either erratic in their occurrence and exploit ephemeral water supplies (*Anas angustirostris*, *Tadorna casarca*, *Himantopus himantopus* and some of the Ardeidae), or depend on permanent water such as the sea or large lakes (e.g. pelicans, cormorants, terns, and gulls). By regulating water levels in many wetlands, particularly salinas, man has made breeding conditions more stable for some of the aquatic birds in the Mediterranean (e.g. flamingos; Johnson 1989).

The fish fauna of wetlands depends primarily on the availability of permanent water during dry periods. The fish fauna of many wetlands is thus composed of migratory euryhaline species. Typical freshwater fauna of cyprinids or salmonids are restricted to wetlands connected to permanent rivers or lakes.

Invertebrates mostly survive drought by producing stages which are resistant to desiccation. These may be eggs, which remain in diapause until reflooding, or winged adult stages which fly off in search of permanent water. A few dragonflies undergo long distance migrations between the more arid parts of the region and southern Europe. Many species breed in the winter or early spring, as do a few fish species (Crivelli 1981a, 1981b, Crivelli and Britton 1987).

Fauna of saline wetlands

At high salinities, osmotic pressure and direct toxicity reduces the fauna to a few highly adapted species (Britton and Johnson 1987). Fish are absent from waters with a salinity in excess of about 80 g l^{-1}, the most resistant being *Atherina* spp. The absence of fish allows the development of the brine shrimp, *Artemia* spp., which are eliminated from waters of low salinity by fish and invertebrate predation. *Artemia* occurs over a salinity range up to 300 g l^{-1} (MacDonald and Browne 1989). Where salinity is greater than about 120 g l^{-1}, the only other invertebrate species are Diptera larvae (*Ephydra* spp. and Dolichopodidae), a copepod (*Cletocamptus retrogresses*), Microturbellaria, Nematoda, and Protozoa (particularly *Fabrea salina*). At lower salinities Chironomidae larvae, water beetles and a number of other Copepoda species become abundant. All these animals are of freshwater or terrestrial origin and have stages resistant to desiccation. Animals of marine origin (e.g. Mollusca, Polychaeta) do not occur in salinities in excess of 80 g l^{-1} probably because of the change in ionic composition due to differential precipitation of ions such as Ca, BO_3, CO_3, and SO_4 (Nixon 1969).

This fauna just described is found particularly in salinas, where the rather stable and predictable salinity and water level favor its survival. The invertebrates are consumed by a limited range of bird species, of which flamingos (*Phoenicopterus ruber*), shelduck (*Tadorna tadorna*), avocet (*Avocetta recurvirostra*), kentish plover (*Charadrius alexandrinus*), and in certain areas, black-headed gulls (*Larus ridibundus*), are the most abundant. A wide variety of shorebirds use the lower salinity lagoons in salinas on migration in winter and on migration (Blondel and Isenmann 1981). Wind seiches in these shallow lagoons exposes large areas of the bottom and its invertebrates to bird predation, thus mimicking the tidal regime of Atlantic coastal mud flats.

In athalassic salt lakes, isolation from permanent water, a long period of drying conditions, and the great seasonal variation in salinity all play a role in determining the composition of the fauna. After filling with rain water, salinity conditions are often oligohaline, and the fauna comprises a few freshwater species resistant to desiccation or species with long distance powers of dispersal; Cladocera (*Daphnia* and *Moina*), Coleoptera, and Corixidae are characteristic. As the salinity increases by evaporation, this fauna is replaced by that found in high salinity lagoons in salinas, in particular *Artemia* (Beadle 1943, Morgan 1982a, b).

The intertidal flats of the Atlantic coast have a similar fauna to mixohaline coastal lagoons. The fauna survives the low tide periods by burrowing into the substrate or by migrating with the water mass. The most abundant species of intertidal invertebrates are gastropods *Hydrobia* and *Cerithium* (Britton 1985), polychaetes, and amphipod crustaceans. Because of the regular exposure of the substrate and its fauna, these wetlands are much more important to shore birds than the coastal lagoons of the Mediterranean coast. Consequently the Atlantic coast of the Iberian peninsula and Morocco constitutes a major flyway for migratory shorebirds (Cramp and Simmons 1983). Salt marshes are flooded too briefly at each tide to develop much of an aquatic fauna. Amphipods and fish invade at each submergence, but the bulk of the fauna is composed of terrestrial arthropods and molluscs. The same is true of non-tidal salt marsh in the summer months (Bigot 1963), but during winter submergence these develop a temporary aquatic fauna. The water is generally only oligohaline at this season and the fauna comprises mainly freshwater species of zooplankton and Corixidae that are similar to those in temporary freshwaters.

Fauna of freshwater wetlands

The major factors influencing the composition of the fauna of freshwater habitats are the periodicity and depth of flooding, the degree of isolation (which determines whether fish and drought sensitive organisms can colonise), and the structure of the vegetation and substrate.

Isolated temporarily flooded pools and marshes with open water develop characteristic winter and spring fauna of drought tolerant forms that include giant planktonic species which are eliminated by fish predation in non-isolated sites. The planktonic species include *Anostraca* similar to *Artemia* of saline habitats (e.g. *Tanymastix*, *Branchipus*, *Cheirocephalus*), Conchostraca (e.g. *Imnadia*), Nosostraca (e.g. *Triops*), and giant Copepoda (e.g. *Mixodiaptomus*, *Hemidiaptomus*) (Champeau 1966, 1971, Nourrisson and Aguesse

1961, Aguesse 1958). Large species of Cladocera (e.g. *D. magna*) and the bright red Copepoda, *Arctodiaptomus wierzesjkii* are also characteristic and often very abundant. Free-swimming invertebrate predators (e.g. Odonata, Dytiscidae, Notonectidae) are frequently present in high numbers. Some of these have drought resistant forms while others arrive as winged adults. A few other drought resistant groups are also characteristic, including a few species of gastropod (Marazanof 1970), Culicidae (Rioux and Arnold 1955), Trichoptera, and Chironomidae (Tourenq 1975). The latter invade as flying adults, and there would appear to be few truly drought resistant species. The absence of fish also favors the use of such wetlands as breeding sites by Amphibia. Characteristic species of temporary pools are *Bufo calamita* and *Pelobates* spp., which occur particularly in coastal sites. All species of Anura are more abundant in shallow wetlands with few or no fish than in lakes.

When fish can enter a seasonally flooded wetland from adjoining permanent water, much of the characteristic invertebrate fauna disappears through predation. Fish frequently enter such wetlands in spring, and fish predation becomes more intense as the water temperature increases rapidly. As a result, a change in fauna from one dominated by large zooplankton species in the winter, to a benthos dominated summer fauna is characteristic (Pont *et al.* in press). The fish which enter seasonally flooded freshwaters include carp (*Cyprinus carpio*), sticklebacks (*Gasterosteus aculeatus*), rudd (*Scardinius erythropthalmus*), tench (*Tinca tinca*), pike (*Esox lucius*), salt-smelt (*Atherina boyeri*), and the introduced mosquito-fish (*Gambusia affinis*) and pumpkinseed (*Lepomis gibbosus*). All these species breed in seasonally flooded marshes but it is not known what factors influence their migrations into such wetlands nor what role seasonal marshes play in their population dynamics. Endemic species of Cyprinodont (*Aphanius* spp., *Valencia* spp.) were once widespread in shallow wetlands of the region but these have declined considerably and many may be extinct in some areas due to drainage and competition from *Gambusia* (Van Vierssen *et al.* 1984, Bianco 1987, Fernandez Delgado *et al.* 1988).

Dense zooplankton communities of seasonally flooded marshes are exploited by shoveler (*Anas clypeata*) and other duck species in the winter (Pirot and Pont 1987). In summer, the fauna becomes concentrated by falling water levels and such concentrations are exploited by a variety of bird species, but particularly by herons such as little egret (*Egretta garzetta*) (Hafner *et al.* 1982). Falling water levels make such marshes precarious for breeding waterfowl and breeding densities and success are generally low.

Seasonally flooded streams have a more reduced fauna as the current prevents the development of zooplankton. Benthos, however, rapidly invades from permanent pools or from resistant stages in the substrate (Legier and Talin 1973, Bouzidi *et al.* 1984). In seasonally flooded mountain streams with a substrate which is frequently swept away in flash floods, the fauna must be very impoverished or non-existent.

Unlike the invertebrate fauna of seasonally flooded marshes, there are few endemic species in permanent freshwater areas and most are wide spread palaearctic species. In permanent waters, fish are the keystone predators and the invertebrate fauna is composed mainly of small zooplankton species and cryptic or distasteful benthos. Detailed studies of the composition and production of the benthos and zooplankton have been carried out both in high altitude oligotrophic lakes (e.g. Laville 1975) and in shallow lowland eutrophic lakes (e.g. Morgan 1980) but quantitative studies of the fauna of emergent reed swamps are lacking.

In lowland lakes, Cyprinidae make up the bulk of the fish fauna (Crivelli 1990), while salmonids, both natural and introduced, are restricted to high altitude or deep oligotrophic lakes. A large number of freshwater fish are endemic to large permanent rivers, or ancient tectonic lakes, especially in the Balkans (e.g. *Pachylon pictum*, *Barbus prespensis*, *Rutilus rubilio*). Lake Ohrid in Yugoslavia is one example and is of interest for the large number of endemic aquatic invertebrates, including sponges, amphipods, ostracods, and especially gastropods (Stankovic 1960).

Permanent water bodies are more favorable as breeding sites for aquatic birds than seasonal marshes. Major colonies of pelicans (*Pelecanus crispus*, *P. onocrotalus*), herons (e.g. *Egretta garzetta*, *Ardeola ralloides*), spoonbills (*Platalea leucorodia*), cormorants (*Phalacrocorax pygmeus*), and ibis (*Plegadis falcinellus*) occur around permanently flooded marshes, lakes, and lagoon systems. Lowland lakes and freshwater marshes are one of the major wintering areas for the western Palaearctic populations of waterfowl. The principal wintering species in the region are *Anser anser*, *Anas crecca*, *A. platyrhynchos*, *A. clypeata*, *A. strepera*, *Athya ferina*, *Netta rufina*, and *Fulica atra* (Atkinson-Willes 1976, Carp 1980, Scott 1980). Larger bodies of water are used mainly as day roosts and birds spend nights feeding in shallow seasonal marshes or farmland (Tamisier 1966). These water bodies also serve as staging posts between Africa and their northern breeding grounds for certain shorebirds (e.g. *Limosa limosa*, *Philomachus pugnax*, and *Tringa ochropus*) which prefer freshwaters.

The ecology of the rivers of the Mediterranean region has not been extensively studied. Recent accounts for two river systems are given Chiaudani *et al.* (1984) for the Po, and Prat *et al.* (1984) for the Llobregat in northern Spain. Both rivers have been highly modified in their lower reaches and have an impoverished invertebrate fauna. Ephemeroptera, Plecoptera, and Simuliidae have all disappeared and have been replaced by Chironomidae because of pollution in the lower sections of the Llobregat. The Po has a rich fish fauna, with 27 native and 7 introduced species. Ten species are migratory or euryhaline, but construction of dams and alteration of river flow have reduced the runs of some, especially shad (*Alosa fallax nilotica*) and sturgeons (*Acipenser naccari*, *A. sturio*). Recently 38 fish species in 21 families have been identified in the drainage basin of the Ebro River (Spain). Of these 27 are native and 11 have been introduced (De Sostoa and Lobon-Cervia 1989).

Wetland use and conservation

In the summer, vegetated wetlands stand out as oases of greenery in the parched upland landscape of the Mediterranean region. These areas have been exploited by man since the beginnings of civilization. The major forms of wetland exploitations are: drainage and conversion to agriculture, grazing, water storage, fisheries and pisciculture, mineral exploitation, hunting, harvest of wetland vegetation, tourism and water sports, and nature conservation.

Drainage

Many wetlands of southern Europe and North Africa have been drained and no longer exist and their former existence can be readily determined on maps by place names and by the relict network of drainage ditches. In many places of former wetlands, there are now fertile agricultural lands. Wetlands have also been drained for purposes of providing land for industrial or residential development, to prevent flooding of upstream areas, and to control malaria.

Major wetland drainage began in Italy at the time of the Etruscans (before 5th century B.C.) who drained a series of marshlands along the Tyrrhenian coast by the construction of a canal system which still functions today. Wetland drainage accelerated in the time of the Romans, who succeeded in creating agricultural land from riverine wetlands in all the major valleys of Italy and in their provinces in France, Spain, and North Africa. With the fall of the Roman Empire, most of the drainage works fell into disuse and wetlands developed again. Drainage was renewed again following the establishment of powerful monasteries in the Middle Ages. Wetland drainage increased after the Renaissance, often with the aid of engineers brought from Holland. Most of this early drainage was undertaken to provide irrigated agricultural land and to prevent flooding by the construction of levees, and the straightening of river courses.

A further acceleration took place about 1850 with the introduction of steam-powered machinery. This enabled large scale engineering works to be undertaken at reasonable cost and many of the major dikes and canals in

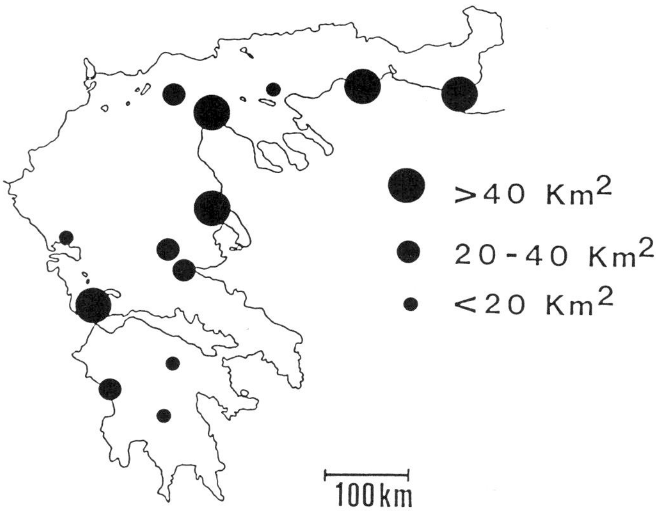

Figure 10. Map showing major wetland sites in mainland Greece which have disappeared, or have been severely modified since 1950.

existence today date from that period. The impetus for drainage in the latter part of the 19th and beginning of the 20th centuries was largely to eradicate malaria, which was then endemic in all the Mediterranean countries. This aim was finally achieved by the 1940s, but many wetlands had been drained.

In Greece, Albania, Bulgaria, and North Africa the most intensive period of wetland drainage began in the 1950s. In Albania, of an estimated total of 2000 km^2 of marshes existing in the 1940s, 600 km^2 had been drained by 1980 (Nowak 1980), including three of the most important wetland sites for waterfowl. Most of the major wetlands in Greece have suffered some degree of drainage since the 1950s (Fig. 10), and many large sites have disappeared completely.

The area of land drained annually within the countries of the European Economic Community is still increasing (Baldock 1984) but it is now certainly decreasing in the Mediterranean region as most areas had already been under cultivation from earlier drainage projects. Few of the wetlands that remain are being drained because the economic benefits from drainage are often not realized.

Soil salinisation is the major obstacle facing attempts to convert wetlands to agricultural land in the region, particularly in coastal areas where the underlying saline water comes to the surface by capillary action once the superficial freshwater has been drained away. Agriculture is only possible with intensive irrigation, especially the cultivation of rice, which recreates freshwater wetland conditions. Rice culture is extensive in the Ebro and

Guadalquivir deltas in Spain, in the Rhone Delta (France), in the Axios delta (Greece), in the valley of the Po (Italy), and of the Strymon (Greece), but in many areas, particularly in North Africa, irrigation water is not available in sufficient quantities in summer to allow such practices.

Drainage for industrial and residential purposes currently poses a greater threat to the remaining wetlands than does reclamation for agriculture. About 46 km^2 of mostly saline wetlands have been destroyed in France since the 1960s to make way for the new port of Fos and adjoining heavy industry near Marseille. Smaller industrial development projects on coastal wetland sites are planned, or have taken place in all countries. Many wetlands have been drained for hotel development along the coasts.

Drainage of wetlands has most affected shallow seasonally flooded freshwater marshes. Brackish wetlands and deep lakes and lagoons have been modified rather than drained and highly saline ephemeral wetlands have mostly been left intact. Riverine forest has undoubtedly been the wetland type most affected by drainage, so that only tiny fragments now remain. In most cases, however, it is likely that the forest was felled and converted to seasonal grazing marshes before being drained. Most other wetlands which have been drained were either seasonally flooded wet meadows, reed swamps, or shallow freshwater lakes. In some river deltas, extensive areas of seasonally flooded halophytic scrub have been drained by the construction of polders to prevent flooding from the sea or the river. These areas have, however, proved difficult to convert to agricultural land.

Grazing

Most remaining wetlands with emergent vegetation are grazed by domestic stock. This represents one of the major economic benefits derived from wetland exploitation in the region. Figures are not available for the total number or economic value of animals using wetlands as the wetlands are generally only used for summer grazing. In winter, when wetland vegetation dies back or is covered by water, animals are normally moved to upland pastures or are fed supplementary food.

Stock densities in the Camargue are of the order of 0.2 to 0.5 horses ha^{-1}, or 0.3 to 3 cattle ha^{-1} of wetland. In winter these same animals require 2 to 5 ha per individual, of poor upland range (0.2 to 0.5 animals ha^{-1}) if they are not to receive supplementary feed (Duncan and D'Herbes 1982, Gordon *et al.* 1990). The effect of grazing on the composition of the emergent vegetation has already been described.

Cattle are undoubtedly the most widespread and abundant domestic grazing animal. Wetlands in Spain, Portugal, and southern France are used to rear bulls for bullfights but elsewhere the emphasis is on meat production.

Rustic local breeds, adapted to wetland conditions, are the norm. Following cattle, mixed sheep and goat herds are the most abundant grazers, especially in North Africa. Unlike other domestic stock these will not enter water to graze, and are therefore restricted to feeding in seasonally flooded wetlands after drawdown. They will, however, consume halophytic plants (e.g. *Tamarix* and *Suaeda*) which are rejected by cattle and horses. A few horses, donkeys and mules are found in wetlands almost everywhere where grazing is practiced but horses are only numerically important in the Guadalquivir and Camargue where distinct breeds of white horses feed extensively in the freshwater and brackish marshes (Vlassis 1978). Buffalo (*Bubalus bubalis*) have been introduced into wetlands in Greece, Tunisia, and Italy and are capable of feeding in deeper water and on softer bottoms than other large grazing animals.

Muskrat (*Ondatra zibethicus*) and Coypu (*Myocastor coypus*) have been introduced in the past for fur breeding in several parts of Europe and have escaped into many natural wetlands. The Muskrat is spreading rapidly and has only just reached the Mediterranean region (Niethammer and Krap 1982). Once it becomes established, it is likely to have considerable impact on the wetland vegetation, as Coypu already has in the marshes where it occurs.

Water storage

The capacity of existing wetlands for water storage has not been greatly exploited in the region. Water is pumped from some natural lake sand marshes in the summer to supply local irrigation requirements, but most water storage schemes have taken the form of reservoir construction in upland valleys. These provide water for industrial and domestic consumption and for irrigation of lowland agricultural areas which often occupy former wetlands areas. Some existing natural lakes have also been regulated by dam construction. In addition to water supply schemes, barrages constructed for hydro-electric purposes are also frequent within the region. These include both upland reservoirs and low-head barrages on large lowland rivers.

Water storage is the key to economic development in the region. Rainfall occurs in the winter months when the demand for water for irrigation is low but in the dry summer most agriculture is impossible without irrigation. There is a similar disparity in the seasonal supply for water destined for domestic consumption. The maximum demand occurs at the height of the tourist season, in July and August which are the driest months of the year. It is not surprising, therefore, to find that water storage and river regulation schemes have been practiced on an immense scale in all the countries.

Some idea of the scale of these works is given by the total areas of reservoirs in a few different countries (Table 1). Since reservoirs are mostly in upland areas they are subject to enormous seasonal draw down or short-term changes in water level. Because of the dramatic changes in water levels, few wetlands have developed in reservoirs.

The major effects of water storage schemes on wetlands have been to divert water away from sites downstream and to decrease siltation in deltas and estuaries. Some formerly perennial river systems now cease to flow for most of the summer (e.g. Durance in France, Nestos in Greece). In the delta of the Guadalquivir, the combined effects of prolonged drought and upstream reservoir construction lead to a catastrophic drying out the marshes and only emergency action saved the aquatic flora and fauna from extinction (WWF 1984). A similar alarming drop in water level is reported from the Tablas de Daimiel, also in Spain. In Tunisia, upstream reservoir construction has caused increased periods of drought in the intermittently flooded Lake Kelbia while the brackish water lagoon, Garaet Ichkeul, is threatened with increasing salinity because of proposed diversion of part of the freshwater inflow for irrigation (Hollis 1983, 1986).

Fisheries and pisciculture

Nearly every permanent water in the region is exploited by fishermen. The only exceptions are hypersaline lagoons in salines, and a few nature or hunting reserves. Coastal lagoons, because of the importance of their extent and their high productivity, are quantitatively the most important type of wetlands for fisheries (Crivelli 1991).

A sophisticated type of traditional fishery is practiced in certain lagoons in Italy, particularly along the Adriatic coast. The lagoons, known as valli, are enclosed by artificial dikes and equipped with artificially controllable inlets for freshwater and sea water. The fishery depends on migration of catadromous fish species from the sea in the spring and their subsequent return in the autumn. The passages connecting the lagoons to the sea are equipped with permanent fish traps which allow the migration of the young fish into the lagoon but retain fish of commercial size returning to the sea. The principal fish species caught in Mediterranean coastal lagoon fisheries are eels (*Anguilla anguilla*), grey mullet (*Mugil* spp.), sea bass (*Dicentrarchus labrax*), gilt-head (*Sparus auratus*), sole (*Solea vulgaris*), and sand smelt (*Atherina* spp.). In the Italian valli, eels dominate both numerically and economically.

Yields of eels in Italian valli average about 30–40 kg ha^{-1} yr^{-1} (Huet 1970, Gatto *et al.* 1982, Ardizonne *et al.* 1988). In French coastal lagoons, where fishing takes place mainly in the lagoon rather than at the exits, the total yield of all species has been estimated betwen 24 and 250 kg ha^{-1} yr^{-1}

(Quignard and Zaouali, 1980). The productivity of the lagoon fishery has been estimated to be 8–10 times that of the unproductive Mediterranean sea. The economic value of the lagoon fishery along the French Mediterranean coast exceeds that of the Mediterranean trawling fleet (Amanieu 1972).

In recent years, extensive lagoon fisheries have been supplemented by intensive aquaculture. Perhaps the most successful has been the development of oyster and mussel culture on artificial supports in certain lagoons on the French coast where the total annual production is of 1,500 t yr^{-1}, or 6.5 t ha^{-1} yr^{-1} (Amanieu 1972). Intensive rearing of eels is now being practiced in Italian lagoons alongside traditional valli culture, while in France cage culture of *Salmo* spp., *Dicentrarchus labrax*, and *Sparus auratus* are developing, but are as yet on a small scale.

A major problem facing lagoon fisheries in some regions is excessive algal growth, aggravated by run-off of agricultural fertilizers or domestic sewage. Algal blooms can produce extensive deoxygenation in the high summer temperatures and massive fish kills have been reported (e.g. Boutiere *et al.* 1982). Serious algal blooms have become almost an annual event in some lagoons.

Fisheries in freshwater lakes are of less economic value than coastal lagoons, but assume regional importance in the Balkans. In shallow eutrophic lakes, carp and eels are the main commercial species but other species of Cyprinidae may form the bulk of the catch (Crivelli 1990). In the deeper lakes such as Ohrid, salmonids may form a substantial part of the catch (Stankovic 1960). Lake fisheries use both fyke nets and, increasingly, nylon gill nets which have led to over-fishing in some lakes (Crivelli 1990). Extensive pond culture of carp has been attempted in some freshwater wetlands in southern France but has not been very successful because of the low esteem for this fish.

Rivers of the region support mainly sport fisheries or small scale net fisheries. A number of fish species have been introduced, including North American centrarchids (*Lepomis gibbosus*), salmonids (*Salmo gairdneri*), and catfish (*Ictalurus melas*). Carp have also been widely introduced, for example in North Africa (Kraiem 1983). Such introduction can cause the decline or possible disappearance of the indigenous fauna.

Mineral exploitation

From time immemorial, man has evaporated sea water to produce salt for direct consumption and for food preservation (Baas-Becking 1931). In semi-arid climates, such as that found in the Mediterranean, this has traditionally been accomplished by solar evaporation in small shallow ponds during the summer months. In recent times the demand for sodium chloride and other minerals that can be extracted from sea-water has increased dramatically as a raw material for use in the chemical industry (detergents, plastics, organochlorides, road-salt etc.). As a result, what was once a scattered craft industry has been transformed into a major industry which is still expanding. The transformation dates from the end of the 19th century when the Solvay process for converting salt to soda was invented. The new salt works (salinas) cover areas of hundreds or thousands of hectares, and are highly mechanized. At the same time, small traditionally operated salinas are being abandoned. All of the salinas constructed around the Mediterranean coast occupy the sites of former natural wetlands. For the most part, they have replaced permanently or seasonally flooded brackish water lagoons but in some cases they have been built on freshwater marshes or by enclosing areas of shallow sea.

The distribution of the major existing salinas is shown in Fig. 7 and the approximate total areas of salinas in each country is given in Table 1. The yield of salt per hectare of salina depends on the operating conditions and the climate, so that production greatly from year to year. For the Salin de Giraud in France, the yield varies from 23.3 t ha^{-1} to 91.3 t ha^{-1} depending on the year (Gleize 1978).

Artificial wetlands associated with salinas are of great value as wildlife habitat, particularly for aquatic birds (Hoffmann 1964). At any one time, about half of the 60,000 to 70,000 greater flamingos (*Phoenicopterus ruber*) in western Mediterranean are found in salines. The salinas of the Camargue are the only regular breeding site of this population (Johnson 1984). The transformation of a wetland into a salina does not constitute a wetland loss, but merely a change of wetland type and it may even lead to an enhancement of wildlife value.

The operation of the salina results in the creation of a series of ponds with nearly constant water level and salinity which ranges from about 40 g l^{-1} up to 300 g l^{-1}, in the summer. A highly distinctive flora and fauna develops at each salinity. In salinities >100 g l^{-1} the fauna consists mainly of the brine shrimp (*Artemia*) and the bottom is composed of evaporites. At lower salinities the bottom is frequently covered by laminated mats of Cyanophyceae, and Chironomidae and Copepoda compose most of the fauna. In the lowest salinity lagoons a flora and an invertebrate fauna typical of natural coastal lagoons is found (Britton 1985).

Hunting

The largely ice-free wetlands of the region are the wintering grounds for a major portion of the western Palaearctic populations of several species of

Table 6. Numbers of hunters recorded for each country for the years 1980-1981 (from Lampio 1983).

Country	Total hunters	Waterfowl hunters
Portugal	250,000	?
Spain	1,050,000	30,000
France*	1,850,000	500,000
Italy	1,600,000	250,000
Yugoslavia	198,000	?
Albania	?	?
Greece	260,000	52,000
Bulgaria	40,000	
Malta	12,000	?
Morocco	26,000	?
Algeria	?	?
Tunisia	7,900	500

*All of France, including areas outside of Mediterranean.

waterfowl, particularly ducks and geese (Smart 1976, Scott 1980, Joensen *et al.* 1987, Monval and Pirot 1989). These areas are also important staging areas for migratory water and land birds moving between Europe and tropical Africa in the spring and autumn (Cramp and Simmons 1977, 1980, 1983).

These populations of birds are heavily exploited by hunters in all the countries around the Mediterranean, but particularly in southern Europe. The numbers of hunters in southern Europe (Table 6) is probably higher than in any other comparably sized area in the world and hunting evidently has a considerable influence on the distribution and abundance of waterfowl using the wetlands.

Based both on the numbers of hunters per km (Baledent 1973) and the estimated annual bag (Office Nationale de la Chasse 1976), hunting pressure is considered excessive along the Mediterranean coast of France. The annual harvest has been estimated to be of the order of 1 million to 3.5 million ducks (Office Nationale de la Chasse 1976). This compares to a January wintering total of 0.46 to 0.93 million ducks over the same area (Yesou *et al.* 1983). The authors, however, consider the harvest figures to be unreliable and probably too low! The annual value of the carcasses alone is estimated at 70 million French francs. The value of hunting rights can also be extremely high (1,000 to 40,000 francs ha^{-1} $year^{-1}$) so that it is evident that waterfowl hunting is a major economic activity in France and also in other southern European countries. No attempt has been made, however, to assess the overall economic value of waterfowl hunting and there are no reliable estimates of annual hunting bags from any country.

Hunting is regulated in all countries by the issue of licenses, by close seasons, and by the protection of certain species. In some countries there are also bag limits, prohibition of the sale of game, and limitations on the methods of hunting (Lampio 1983). Enforcement of hunting regulations is, however, poor compared to northern Europe and North America and infringement of hunting regulations is widespread in some countries and largely unregulated.

The species which can be legally shot varies from country to country, although there has been a move towards harmonization of laws concerning bird protection within member states of the EEC (Journal Officeal des Communautés Europeennes 1979). Most duck species and coot can be shot in all countries. Swans are protected everywhere and geese are protected only in some countries. Similarly with shorebirds, there is great variation from country to country. In general, smaller species are protected as are avocets and stilts.

As well as having direct effects on the fauna through the killing of birds and disturbance, hunters control and manage vast areas of wetlands and thus exert indirect influences on the entire flora and fauna. Many of the largest and most famous of the wetlands of southern Europe owe their continued existence to the economic benefits derived from hunting. Without the revenue obtained from this source many would have long since been drained. For example, the Coto Donana (part of the Guadalquivir delta) was once a royal hunting reserve but is now managed as a National Park. Most of the freshwater marshes in the deltas of the Rhône and Po are still managed as hunting estates and many are in private ownership.

Such marshes are managed in an empirical manner by hunting organizations to attract waterfowl and to facilitate shooting. Such management is particularly prevalent in France and Italy but almost non-existent in North Africa. Management of hunting marshes involves control of vegetation and water level. Tall-growing macrophytes are controlled to produce open water and the development of submerged plants which form the basis of the ducks' diet. This is normally done by cutting or crushing the vegetation by tractor or by grazing. Most managers aim for about a 50:50 mix of open water and reed swamp. The pattern of circular holes in the reed swamp, each being about 100 m in diameter, with a shooting butt in the center, is one particularly favored design. This greatly fragments the reed swamp, and is detrimental to breeding birds requiring dense reed cover for nesting (e.g. *Ardea purpurea*, *Botaurus stellatus*).

Where water depth and other interests (e.g. fisheries) permit, hunting marshes are dried out at intervals of at least every 3 to 5 years. This is done to arrest the build up of anoxic sediments and to favor the growth of submerged plants. Draw down obviously takes place in the close season, in spring or early summer, but water is always put back in by mid or late summer in advance of the hunting season which starts as early as July 14 in parts of France. Such a water regime is very unfavorable for many aquatic breeding birds which build nests at or near the water surface, (e.g. *Podiceps cristatus*, *Himantopus himantopus*, and *Chlidonias* spp.) and which are faced either with a rapid drying out of their habitat or subsequent flooding later in the season.

Other management practices include building of shooting butts, control of predators, and provision of nesting boxes. Re-stocking with hand-reared birds, usually mallards, is practiced on some marshes and live and artificial decoys are very widely used. Artificial feeding is less prevalent, and the cultivation of food crops specifically for waterfowl would appear to be unknown. However, stubble fields, particularly rice fields, are frequently flooded after harvest to attract ducks. Most hunting is done in the daytime, or on the morning and evening flights to the hunting marshes from nearby reserves. If a non-hunting zone, where birds can rest during the day without disturbance, is absent from a wetland complex, the pressure of hunting may cause the abandonment by waterfowl. This has happened to some of the lagoons on the Languedoc coast of France (Tamisier and Saint-Gerard 1981).

Harvest of wetland vegetation

Compared to removal by grazing animals, plant harvest is a very minor activity. Firewood, and some commercial timber, is harvested from forested wetlands and these activities have, in the past, lead to the disappearance of most riverine forests. Nowadays the remaining areas of natural forest are so small and of only minor economic importance. Native riverine forest has, however, been replaced in many areas by plantations of hybrid poplar (*Populus* sp.) which are harvested commercially for timber. *Tamarix*, and even *Arthrocnemum*, is harvested for fuel in arid regions where there is little other wood available. In the past, *Salsola* spp. were harvested in many coastal and inland saline areas and were burnt to produce soda ash. Soda is now, however, produced from salt and other sources, and this activity has died out.

Reed cutting on a commercial scale is practiced in southern France, and probably elsewhere in the Mediterranean. The main use is for the fabrication of screens and wind breaks but *Phragmites* is also used for the manufacture of cellulose. *Cladium* and *Scirpus* were formerly harvested as litter for animals kept indoors in the winter but this has been largely replaced by the use of cereal straw. Some hay is cut from wet meadows (e.g. *Molinia*), this is of only marginal agricultural value.

Algae are harvested from coastal lagoons in Portugal (de Sousa 1976) and in Tunisia, where they are used as an organic fertilizer on agricultural fields. The main genera are *Enteromorpha*, *Ulva*, and *Chaetomorpha*. Harvest of algae has been proposed for lagoons in southern France, with the aim of reducing the incidence of summer deoxygenation. The algae were to be used as animal food (Barnabe 1980).

Tourism and water sports

The Mediterranean coastal fringe is the most important tourist region in the world and accounts for some 30% of international world tourism (UNEP 1977). Most tourists stay close to the sea itself, but inland waters such as lakes and rivers also attract large numbers of recreational visitors. Even marshlands are now attracting tourists in search of wilderness and contacts with nature which are increasingly difficult to find in densely populated northern Europe. The Camargue annually attracts some 1 to 1.5 million tourists in the three summer months (Richez 1981), and over 40,000 visited the headquarters of the Parc Regional Naturel, one of the conservation bodies in the delta (Anon 1980). With the exception of some of the inland salt lakes, which lie in inhospitable areas, all Mediterranean wetlands are, therefore, receiving increasing pressure from tourism.

The effects of tourism on wetlands are manifold. At the extreme, wetlands are filled in for hotel development (many examples along the Spanish coast) or coastal lagoons are transformed into marinas (e.g. La Grande Motte complex in southern France). Even when wetlands are left more or less intact, their proximity to a tourist complex brings about almost inevitable degradation of the environment. Piecemeal development, often un-authorised, of camping sites, summer residences, and al fresco catering establishments, is a familiar feature of the Mediterranean coastline. Wetlands, because they are frequently in public ownership, and are thought of as worthless, are particularly prone to such degradation. The simple passage of people is sufficient to cause deterioration of wetland values. For example, there are few sandy beaches left in the western Mediterranean where shorebirds can breed without disturbance. Faced with increasing tourist demand, managers of wetland nature reserves are now having to restrict visitors to areas where they are less likely to cause damage or disturbance to the environment. This is done by providing of hides, pathways, and information centers.

Wetlands, such as coastal lagoons, are increasingly being used for water-based sports (e.g. water-skiing and surf-boarding), even when in close proximity to the sea. Such activities not only require a certain infrastructure (e.g. car parks, ramps, towing systems) which often destroy limited areas of wetland but also lead to the abandonment of the site by aquatic birds.

A less obvious but potentially more damaging effect of tourism on wetlands is the increased demand for water caused by the annual influx of visitors. This can lead to depletion of aquifers, draw down of rivers, and the drying out of wetlands as described earlier. This huge increase in the population that takes place in the Mediterranean coastal fringe for a few months each summer poses enormous problems for sewage disposal. For many campers there are no facilities whatsoever and lake side vegetation forms a suitable screen until excess trampling destroys it. The simplest and most widespread solution to sewage disposal is to pour it into the sea but beach pollution has forced some authorities to seek other solutions. Wetlands are increasingly being used for sewage treatment, either by modification into custom built lagoon systems (e.g. Stes Maries de la Mer, France), or by simple discharge into an existing wetland (e.g. Lake Sedjoumi, Tunis). Such wetlands show an impoverishment of the flora and fauna, typical of waters with heavy organic pollution, but their high productivity may continue to attract waterfowl. Tourist development on the Languedoc coast of France led to filling in, drainage and spraying of numerous small wetlands in an attempt to control the mosquito problem, a problem of nuisance rather than public health.

Nature conservation

All of the countries of the region have made some effort towards conserving representative examples of their wetlands but the degree of commitment varies widely from country to country.

The impetus for wetland conservation can come from many quarters. Private initiative is responsible for the creation of a vast area of wetland hunting estates in southern Europe. While these have no legal status, the economic and cultural benefits derived from hunting ensure the continued existence of semi-natural biotopes where otherwise there could be agricultural or industrial development. Some such estates are of large size, and include wetlands judged to be of international value for waterfowl conservation.

Regional or local nature reserves or hunting reserves, administered by locally elected bodies or by local hunting organizations, are found particularly in the European countries of the region. These generally have some legal status, which in theory protects the biotope from degradation or development, as well as laying down restrictions on hunting and on public access. In some countries (e.g. France) national hunting organizations have established networks of wildlife refuges, including wetlands.

At the national level, most of the countries of the region have at least one wetland site with the status of national nature reserve or national park. Examples are the Coto Donana in Spain, the Camargue in France, Garaet Ichkeul in Tunisia, and Mikri Prespa in Greece. Despite the protection of national park or equivalent status, several are threatened by human activities

Table 7. Numbers and areas of wetlands nominated for inclusion in the Ramsar Convention by signatory nations (RAMSAR Bureau, pers. comm.).

	Year of ratification	N	Approx. area (km)
Portugal	1980	2	306
Spain	1980	17	1301
France	1986	1	850
Italy	1976	65	566
Yugoslavia	1977	2	181
Greece	1975	11	1076
Bulgaria	1975	4	21
Tunisia	1980	1	126
Algeria	1983	2	84
Morocco	1980	4	106
Malta	1988	1	0.1

operating outside the park boundaries, especially water removal. The degree to which human activities are allowed in some reserves (e.g. Ichkeul and Prespa) but are prohibited in others (e.g. Donana).

A few countries have adopted laws for the overall protection and conservation of wetlands. The Italian senate passed a law in 1972 recommending that the remaining wetlands should be preserved and protected. In 1980, the Spanish council of ministers passed an accord for the protection of the countries' estuaries, rias, and coastal wetlands, and called for the drawing up of a national wetland inventory. In 1975 the French government, alarmed by the rate at which the coastline was being developed and industrialized, passed a law aimed at public acquisition of remaining intact areas of coastline and lake shore, with the intent of preventing further indiscriminate urbanization. The Conservatoire du Littoral thus formed now owns and manages a considerable area of wetlands, including several sites on the Mediterranean coast and in Corsica.

Certainly the most important event for the conservation of wetlands within the Mediterranean region, however, is the almost unanimous ratification by all states of the Ramsar Convention on Wetlands of International Importance, especially as waterfowl habitat (Carp 1972). Only Albania has so far failed to sign the convention.

Signatory states are pledged to adopt a policy for the conservation and wise use of their wetland resources and must nominate at least one site to the list of internationally important wetlands. Nominated sites must be protected against significant habitat destruction, but if this proves impossible states reserve the right to substitute an alternative site of equal value. So far, no wetland has been removed from the list.

Up to the present, a total of 89 wetland sites covering nearly 4,600 km^2 have been nominated for inclusion within the convention (Table 7). These wetlands cover almost the entire range of types occurring within the region, from riverine forest and high altitude lakes, to tidal salt marsh and coastal lagoons. Inland salt lakes and river systems are, however, under-represented; the former are probably the least threatened of the wetland types in the region, and river systems are difficult to protect without control over the entire catchment, which is seldom possible.

In addition to this recent international and governmental action on wetland conservation, public interest and involvement in nature conservation has increased greatly in Mediterranean countries in recent years, particularly among the younger generation. Despite these optimistic trends, however, the future for Mediterranean wetlands and their flora and fauna is still uncertain. The major problems facing wetland conservation are water extraction and diversion, water pollution, continued drainage, excessive hunting, tourist pressure and the pressure from rapidly increasing human population in the less developed countries of region. Even international conventions may be powerless against such threats.

Recommendations

Existing national and international inventories provide a good base for making decisions on conservation of wetlands as habitat for waterfowl. Objective criteria exist for evaluating individual sites as wintering or breeding sites (Smart 1976, Scott 1980), and sufficient data on bird populations are available for most sites to enable these criteria to be implemented. It is unlikely that any wetland of major importance for waterfowl within the region is not adequately documented in the various inventories.

For other components of the aquatic fauna and flora, the existing data are less comprehensive and the information required for the establishment of an adequate network of representative sites for the conservation of these components is incomplete. For certain types of wetland (e.g. coastal lagoons, large salt or freshwater marsh systems), it can be reasonably assumed that a network of protected sites chosen largely on ornithological grounds is likely to adequately represent the range of variation of wetland vegetation and fauna. For wetland types which have little value for waterfowl, such as riverine systems, forested wetlands and the small isolated peat lands of the region, however, the existing inventories certainly do not cover the range of variation. Further survey and inventory is required for these wetland types. This is particularly the case for forested wetlands, which as previously stated, now only exist as isolated relicts. Remote sensing would be required for the initial identification of sites. Aerial photography has been under-utilized in wetland inventory work in the Mediterranean region. Remote sensing would certainly have to be followed up by ground survey, in order to identify distinct regional vegetation types, or sites of value for particular species.

Another urgent need within the region is for additional information on the present status of the many endemic fish species. Many are probably threatened with extinction by wetland degradation and by competition from introduced fish species. Existing wetland reserves do not adequately cover fish habitats and additional sites may be required to ensure the survival of some species. Captive breeding programmes might be appropriate short term measures, especially for those species which inhabit degraded wetlands. Further information is required on the role of wetlands, particularly seasonally flooded types, in the dynamics of fish populations. Finally, the most urgent requirement is a better understanding of the functional processes specific to Mediterranean wetlands, in order to be able to manage them correctly as renewable resources.

References

Aguesse, P. (1957) La classification des eaux poikilohalines, sa difficulté en Camargue, nouvelle tentative de classification. Vie et Milieu 8: 341–365.

Aguesse, P. (1958) La diversité des milieux aquatiques de Camargue et son influence sur le peuplement en Copépodes, Cladocères et Coléoptères hydrocanthares. Comptes rendus du congres – Sociétés savantes, Paris 1958: 487–493.

Alonso, M. (1990) Anostraca, Cladocera and Copepoda of Spanish saline lakes. Hydrobiologia 197: 221:232.

Amanieu, M. (1972) Ecologie et exploitation des étangs et lagunes saumâtres du littoral françcais. Annales-Société royale zoologique de Belgique 103: 79–94.

Amat, J. A. (1984) Ecologia de las lagunas Andaluzas. pp. 101–108. *In*: Las zonas humedas en Andalucia. Ministerio de Obras Publicas y Urbanismo, Sevilla, Spain.

Anon. (1980) Courrier du Parc Naturel Régional de Camargue 1981. 20–21:30.

Ardizzone, G. D., Cataudella, S. and Rossi, R., 1988. Management of coastal lagoon fisheries and aquaculture in Italy. FAO Fisheries Technical Paper, 293. Rome, Italy, 103 pp.

Aschmann, H. (1973) Distribution and peculiarity of Mediterranean ecosystems. pp. 11–20. *In*: F. D. Di Castri and H. A. Mooney (eds.). Mediterranean Type Ecosystems. Origin and Structure. Springer-Verlag, Berlin, West Germany.

Atkinson-Willes, G. L. (1976) Criteria for selecting wetlands of international importance. pp. 199–254. *In*: M. Smart (ed.). Proceedings International Conference Conservation Wetlands and Waterfowl. Heiligenhafen, West Germany, 1974.

Ayyal, M. A. and El Ghareeb, R. (1982) Salt marsh vegetation of the Western Mediterranean desert of Egypt. Vegetatio 49: 3–19.

Baas-Becking, L. S. M. (1931) Historical notes on salt and salt manufacture. Scientific Monographs – Research Institute for Land Reclamation and Improvement 32: 434–446.

Baldock, D. (1984) Wetland Drainage in Europe. The effects of agricultural policy in four EEC countries. Institute for European Environmental Policy. International Institute for Environment and Development. London, United Kingdon. 166 pp.

Baledent, M. H. (1973) Chasse et tourisme en France, situations et perspectives. Dissertation, Université d'Aix-Marseille, France.

Baltanas, A., Montes, C. and Martino, P. (1990) Distribution patterns of ostracods in iberian saline lakes. Influence of ecological factors. Hydrobiologia 197: 207–210.
Barnabe, G. (1980) Les lagunes méditerranéenes et aquaculture: perspectives et problèmes. Bulletin mensuel – Office National de la Chasse, numéro spécial scientifique et technique, Novembre 1980: 117–126.
Basset, P. A. (1978) Vegetation of a Camargue pasture. Journal of Ecology 66: 803–827.
Basset, P. A. (1980) Some effects of grazing on vegetation dynamics in the Camargue, France. Vegetatio 43: 173–184.
Beadle, J. C. (1943) An ecological survey of some inland waters in Algeria. Journal Linnaean Society Zoology 41: 218–242.
Beeftink, W. G. (1968) Die Systematik der Europaischen Salzpflanzengesellschaften. pp. 239–263. *In*: R. Tuxen (ed.). Pflanzensoziologie Systematik, Junk, The Hague, The Netherlands.
Bethemont, J. and Villain-Gaudossi, C. (1987) Les deltas mediterranéens. Vienne, France. 387 pp.
Bianco, P. G. (1987) Précision sur la distribution de *Cobitis teania* Linnaeus (Cobitidae) et de *Valencia hispanica* (Valenciennes, 1846) (Cyprinontidae) dans les eaux douces d'Italie. Cybium 11: 207–212.
Bigot, L. (1963) Observations sur les variations de biomasses des principaux groupes d'invertébrés de la "sansouire" camarguaise. Terre Vie 17: 319–334.
Blondel, J. and Isenmann, P. (1981) Guide des Oiseaux de Camarque. Delachaux and Niestlé, Neuchâtel, Switzerland. 344 pp.
Boutiere, H., de Bovee, F., Delille, D., Fiala, M., Gros, C., Jacques, G., Knoepffler, M., Labat, J. P., Panouse, M. and Soyer, J. (1982) Effet d'une crise dystrophique dans l'Etang de Salse-Leucate. pp. 231–242. *In*: P. Lasserre and H. Postma (eds.). Les Lagunes Côtières. Oceanologica Acta 5 Supplément 1982. Proceedings International Symposium on Coastal Lagoons. Bordeaux, France 1981.
Bouzidi, A., Legier, P. and Cazaubon, A. (1984) Les cours d'eau temporaires du massif des Maures: hydrologie, hydrochimie, communautés algales et animales. Ecologia Mediterranea 10: 93–119.
Bradbury, D. E. (1981) Physical geography of the Mediterranean lands. pp. 53–62. *In*: F. Di Castri, D. W. Goodall, and R. L. Specht (eds.). Ecosystems of the World. 11. Mediterranean-Type Shrublands. Elsevier, Amsterdam, The Netherlands.
Braun-Blanquet, J. (1951) Pflanzensoziologie. Grundzüge der Vegetationskunde. Springer-Verlag, Wien, Austria. 631 pp.
Britton, R. H. (1985) Life cycle and production of *Hydrobia acuta* Drap. in a Mediterranean hypersaline lagoon. Hydrobiologia 122: 219–230.
Britton, R. H. and Johnson, A. R. (1987) An ecological account of a Mediterranean Salina: the Salin de Giraud, Camargue (S. France). Biological Conservation 42: 185–230.
Britton, R. H. and Podlejski, V. D. (1981) Inventory and classification of the wetlands of the Camargue. Aquatic Botany 10: 195–228.
Carp, E. (1972) Proceedings of the International Conference on the Conservation of Wetlands and Waterfowl, Ramsar, Iran, 30 January–3 February 1971. International Waterfowl Research Bureau, Slimbridge, England. 303 pp.
Carp, E. (1980) A Directory of Wetlands of International Importance in the Western Palaearctic. United National Environment Programme, and International Union for Conservation of Nature and Natural Resources. Gland, Switzerland. 506 pp.
Caspers, H. (1959) Estratta Dall' Archivo di Oceanografia e Limnologia 9 Supplement: 153–169.
Chalabi, B. (1990) Contribution a l'étude de l'importance des zones humides algériennes pour la protection de l'avifaune. Dissertation, Institut national Agronomique, Alger. 133 pp.
Champeau, A. (1966) Contribution a l'étude écologique de la faune des eaux temporaires de la Haute Camargue. Archivio di Oceanografia e Limnologia 14: 309–357.
Champeau, A. (1971) Recherches sur l'adaptation a la vie latente des copépodes cyclopides et harpaticides des eaux temporaires provençales. Bulletin de la Société d'Ecologie 2: 151–167.
Chiaudani, G. and Marchetti, R. (1984) Po. pp. 401–436. *In*: B. A. Whitton (ed.). Ecology of European Rivers. Blackwell, Oxford, England.

Comin, F. A., Alonso, M., Lopez, P. and Comelles, M. (1983) Limnology of Gallocanta lake, Aragon, northeastern Spain. Hydrobiologia 105: 207–221.
Comin, F. A., Julia, R., Comin, M. P. and Plana, F. (1990) Hydrogeochemistry of lake Gallocanta (Aragon, NE Spain). Hydrobiologia 197: 51–66.
Corillion, R. (1975) Flore des charophytes (Characées) du massif Armoricain et des contrees voisines d'Europe occidentale. Jouve, Paris, France. 216 pp.
Coronada Castillo, R., Leon Jiminez, F. and Morillo Fernandez, C. (1974) Guia del Parque Nacional de las Tablas de Daimiel. ICONA, Madrid, Spain. 168 pp.
Corre, J. J. (1961) Une zone de terrains sales en bordure de l'étang de Mauguio. Bulletin des Séries de la Carte Phytogéographique. Sér. B6: 105–151.
Cowardin, L. M., Carter, V., Golet, F. C. and LaRoe, E. T. (1979) Classification of wetlands and deep water habitats of the United States. United States Department of Interior, Washington, U.S.A. 103 pp.
Cramp, S. and Simmons, K. E. L. (1977, 1980, 1983) Handbook of the Birds of Europe, the Middle East and North Africa: The Birds of the Western Palaearctic. Vols I–III. Oxford University Press, Oxford, England. 2305 pp.
Crivelli, A. J. (1981a) Les peuplements de poissons de la Camargue. Revue d'Ecologie 35: 617–671.
Crivelli, A. J. (1981b) The biology of the common carp, *Cyprinus carpio* L. in the Camargue, southern France. Journal of Fish Biology 18: 271–290.
Crivelli, A. J. (1990) Fisheries decline in the freshwater lakes of northern Greece with special attention for Lake Mikri Prespa. pp. 230–247. *In*: W. L. T. van Densen, B. Steinmetz and R. H. Hughes (eds.). Management of Freshwater Fisheries. Proceedings of a Symposium organized by the European Inland Advisory Committee, Goteborg, Sweden, 31 May–3rd June, 1988. PUDOC, Wageningen, The Netherlands.
Crivelli, A. J.(1992) Fisheries of the Mediterranean wetlands. Will they survive beyond the year 2000? pp. 237–253. *In*: Proceedings of "Fisheries in the year 2000" Conference, organized by the Institute of Fisheries Management, 10–14th September 1990. London, England.
Crivelli, A. J. and Britton, R. H. (1987) Life history adaptations of *Gasterosteus aculeatus* in a Mediterranean wetland. Environmental Biology of Fishes 18: 109–125.
Den Hartog, C. and Segal, S. (1964) A new classification of the waterplant communities. Acta Botanica Neerlandica 13: 367–393.
De Sostoa, A. and Lobon-Cervia, J. (1989) Fish and fisheries of the River Ebro: actual state and recent history. pp. 233–247. *In*: G. E. Petts (ed.). Historical Change of Large Alluvial Rivers: Western Europe. John Wiley and Sons, Ltd., London, England.
De Sousa, T. T. (1976) Os Molicos. Quantidade de molicos; seu valor licencas de apantha et transporte. Arquivo do Distrito de Aveiro 2: 57–64.
D.D.A. (1970) Camargue. Etude hydrogéologique, pédologique et de salinité. Direction départmentale de l'Agriculture des Bouches du Rhône, Marseille, France. Maps and unnumbered reports.
Dorikos, S. (1981) The Basic Wetlands of the Country. Department Water and Natural Resources, Athens, Greece. 499 pp. (In Greek)
Duhautois, L. (1984) Heron pourpre butors: le declin. Courrier de la Nature 92: 21–29.
Dumont, H. J., Miron, I., Dall'Asta, U., Decraemer, W., Claus, C. and Somers, D. (1973) Limnological aspects of some Moroccan Atlas lakes, with reference to some physical and chemical variables, the nature and distribution of the phyto- and zooplankton, including a note for the development of an inland fishery. Internationale Revue der gesamten Hydrobiologie 58: 33–60.
Duncan, P. and D'Herbes. (1982) The use of domestic herbivores in the management of wetlands for water birds in the Camargue, France. pp. 51–67. *In*: D. A. Scott (ed.). Managing Wetlands and Their Birds. International Waterfowl Research Bureau, Slimbridge, England.
Duncan, P., Hoffmannn, L., Lambert, R. and Walmsley, J. G. (1982) Management of a day roost for wintering waterfowl, especially teal *Anas crecca* and gadwall *Anas strepera*, in the Camargue, France. pp. 73–82. *In*: D. A. Scott (ed.). Managing Wetlands and Their Birds. International Waterfowl Research Bureau, Slimbridge, England.

Eber, G. (1969) Zum vorläufigen Schema der Typologie und Klassification von Wasservogelbiotopen. Ornithologische Mitteilungen 21: 69–78.

Emberger, L. (1955) Une classification biogéographique des climats. Recueil des Travaux – Laboratoire de Botanique, Géologie et Zoologie, Université de Montpellier, Série Botanique 7: 3–43.

Fernandez-Delgado, C., Hernando, J. A., Herrera, M. and Bellido, M. (1988) Age, growth and reproduction of *Aphanius iberus* (Cuv. & Val., 1846) in the lower reaches of the Guadalquivir river (Southwest Spain). Freshwater Biology 20: 227–234.

Ferrari, C., Pivola, A. and Piccoli, F. (1972) Saggio cartografio della vegetazione delle Valli di Comacchio. Atti – Accademia della Scienze di Ferrara 48: 1–18.

Gatto, M., Laniado, E. and Rossi, R. (1982) The management of eels in the Valli di Comacchio Lagoon. pp. 303–308. *In*: P. Laserre and H. Postma (eds.). Les Lagunes Côtières. Oceanologia Acta 5 Suppl. 1982. Proceedings International Symposium on Coastal Lagoons, Bordeaux, France 1981.

Gleize, L. (1978) Exploitation de Salin-de-Giraud par la Compagnie des Salins du Midi et de l'Est. Courrier du Parc Naturel Régional de Camargue. 13: 6–14.

Gordon, I. J., Duncan, P., Grillas, P. and Lecomte, T. (1990) The use of domestic herbivores in the conservation of the biological richness of European wetlands. Bulletin d'Ecologie, 21: 49–60.

Grillas, P. (1990). Distribution of submerged macrophytes in the Camargue in relation to environmental factors. Journal of Vegetation Science 1: 393–400.

Guélorget, O., Frisoni, G-F., Ximenes, M-C. and Perthuisot, J-P. (1989) Expressions biogéologiques du confinement dans une lagune méditerranéenne: le lac Melah (Algerie). Revue d'Hydrobiologie Tropicale 22: 87–99.

Hafner, H., Boy, V. and Gory, G. (1982) Feeding methods, flock size and feeding success in the little egret *Egretta garzetta* and squacco heron *Ardeola ralloides* in Camargue, southern France. Ardea 70: 45–54.

Haslam, S. M. (1971a) Community regulation in *Phragmites communis* Trin. I Monodominant stands. Journal of Ecology 59: 65–74.

Haslam, S. M. (1971b) Community regulation in *Phragmites communis* Trin. II Monodominant stands. Journal of Ecology 59: 75–88.

Heliotis, F. D. (1988) An inventory and review of the wetland resources of Greece. Wetlands 8: 15–31.

Hoffmann, L. (1964) La valeur des salins comme milieux biologiques. International Union for the Conservation of Nature and Natural Resources. Publication New Series 3: 410–413.

Hollis, G. E., Ed.. (1983) A feasibility study for a sluice on the Oued Tindja in the Ichkeul Wetland National Park, Tunisia. A report on aspects of the hydrology, sedimentology and ecology of the project. Parts I–III. University College London. Contract Env 676 UK(H) Environmental Research Programme, London, England. 55 pp.

Hollis, G. E. (1986) The modelling and management of the internationally important wetland at Garaet El Ichkeul, Tunisia. International Waterfowl and Wetland Research Bureau, Special Publication No. 4, Slimbridge, England. 121 pp.

Hutchinson, G. E. (1957) A Treatise on Limnology. Vol. I. Geography, Physics and Chemistry. John Wiley and Sons, New York. 1015 pp.

Isakov, Y. A. (1967) Problems concerning typology and evaluation survey of waterfowl habitats. Proceedings of the Meetings of International Cooperation in Waterfowl Research, Jablona, Poland 1966. International Waterfowl Research Bureau, Tour du Valat, France. 225 pp.

Joensen, A. H., Preus, N. O. and Hansen, P. (1987) Waterfowl and raptors wintering in Western Greece and Peloponnese, 1986–1987. Natura Jutlandica 22: 89–100.

Johnson, A. (1984) Etho-écologie du flamant rose (*Phoenicopterus ruber roseus* Pallas) en Camargue, et dans l'ouest paléarctique. Dissertation, Université de Toulouse, France.

Johnson, A. R. (1989) Population studies and conservation of Greater flamingos in the Camargue. pp 49–63. *In*: A. L. Spaans (ed.) Wetlands en watervogels PUDOC, Wageningen, The Netherlands.

Journal Officiel des Communautés Européennes, 1979. Directive of the Council of the European Economic Community – Conservation of Wild Birds. Luxembourg 2nd. April 1979. No. L 103/1, 25/4/79. 5 pp.

Kadid, Y. (1989) Contribution à l'étude de la végétation aquatique du lac Tonga, Park National d'El Kala. Dissertation, Institut National Agronomique, Alger. 100 pp.

Kohli, E. (1981) Untersuchung zum Einfluss der Nutrias (*Myocastor coypus*, Molina) auf die natürliche Vegetation der Camargue. Dissertation, Université de Bern, Switzerland. 69 pp.

Koussouris, T. S., Diapoulis, A. C. and Balopoulos, E. T. (1987) Limnological situations in two shallow greek lakes (Kastoria and Mikri Prespa lakes). GeoJournal 14: 377–379.

Koussouris, T. S., Diapoulis, A. C. and Balopoulos, E. T. (1989) Assessing the trophic status of Lake Mikri Prespa, Greece. Annales de Limnologie 25: 17–24.

Kraiem, M. M. (1983) Les poissons d'eau douce de Tunisie: inventaire commenté et répartition géographique. Bulletin de l'Institut National Scientifique et Technique d'Oceanographie et de Pêche de Salammbo 10: 107–124.

Lampio, T. (1983) Waterfowl hunting in Europe, North America and some African and Asian countries in 1980–81. International Waterfowl Research Bureau Special Publication No. 3, Slimbridge, England. 35 pp.

Laville, H. (1975) Production d'un Chironomus semivoltine (*Chironmus commutatus* Str.) dans le lac de Port de Bielh (Pyrénées Centrales). Annales de limnologie 11: 13–48.

Legier, D. and Talin, J., 1973. Comparaison de ruisseaux permanents et temporaires de la Provence calcaire. Annales de limnologie 9: 273–292.

Luther, H. and Rzóska, J. (1971) Project Aqua, a source book of inland waters proposed for conservation. International Biological Programme Handbook No. 21. IUCN Occasional Paper No. 2. Blackwell, London, England. 239 pp.

MacDonald, G. H. and Browne, R. A. (1989) Population dynamics of an asexual brine shrimp Artemia population. Journal of Experimental Marine Biology and Ecology 133: 169–188.

Marazanof, F. (1970) Contribution à l'étude de l'écologie des Mollusques des eaux douces et saumâtres de Camargue. I. Milieux – Espèces. Annales de limnologie 5: 201–323.

Marks, J. B. (1950) Vegetation and soil relations in the lower Colorado Desert. Ecology 31: 176–193.

Molinier, R. and Devaux, J. P. (1978) Carte phytosociologique de la Camargue au 1/50,000 e. Biologies Ecologie Méditerranéene 5: 159–195.

Molinier, R. and Tallon, G. (1950) La végétation de la Crau. Revue Générale de Botanique 56, 57: 1–111.

Molinier, R. and Tallon, G. (1970) Prodrome des unites phytosociologiques observées en Camargue. Buletin de la Muséum d'Histoire Naturelle de Marseille 30: 5–110.

Monval, J-Y. and Pirot, J-Y. (1989) Results of the IWRB International Waterfowl Census 1967–1986. International Waterfowl and Wetland Research Bureau, Special Publication No. 8, Slimbridge, England. 145 pp.

MOPU. (1982) Las zonas humedas en Andalucia. Ministerio de Obras Publicas y Urbanismo, Sevilla, Spain. 236 pp.

Morgan, N. C. (1980) Secondary production. pp. 247–340. *In*: E. D. LeCren and R. H. Lowe-McConnell (eds.). The Functioning of Freshwater Ecosystems. International Biological Programme Publication No. 22. Cambridge University Press, Cambridge, England. 588 pp.

Morgan, N. C. (1982a) An ecological survey of standing waters in North West Africa. II. Site descriptions for Tunisia and Algeria. Biological Conservation 24: 81–111.

Morgan, N. C. (1982b) An ecological survey of standing waters in North West Africa. III. Site descriptions for Morocco. Biological Conservation 24: 161–182.

Morgan, N. C. and Boy, V. (1982) An ecological survey of standing waters in North West Africa. I. Rapid survey and classification. Biological Conservation 24: 5–44.

Moubayed, Z. (1977) Etude écologique des marias du sud de la Crau (Bouches du Rhône). Analyse des peuplements d'invertébrés dulcaquicoles et de leurs relations avec l'hydrologie, la végétation et les influences humaines. Dissertation, Université Aix-Marseille, France. 222 pp.

Nahel, I. (1981) The Mediterranean climate from a biological viewpoint. pp. 63–86. *In*: F. DiCastri, D. W. Goodall and R. L. Specht (eds.). Ecosystems of the World. 11. Mediterranean-Type Shrublands. Elsevier, Amsterdam, The Netherlands.

Niethammer, J. and Krapp, F. (1982) Handbuch der Säugetiere Europas. Bd II. Rodentia. Akademische Verlaggesellschaft, Wiesbaden, West Germany. 649 pp.

Nixon, S. W. (1969) Characteristics of some hypersaline ecosystems. Thesis, Univerity of North Carolina, Chapel Hill, North Carolina, USA. 401 pp.

Nourrisson, M. and Aguesse, P. (1961) Cycle annuel des Phyllopodes d'une mare temporaire de Camargue. Bulletin de la Société Zoologique de France 86: 754–762.

Nowak, E. (1980) Wasservögel und Feuchtgebiete Albaniens: status, Veränderungen, Nützung und Schutz. Beiträge zur Vogelkunde, Jena (DDR) 1980: 65–103.

Office National de la Chasse. (1976) Enquête statistique sur les tableaux de chasse à tir pour la saison 1974–75: premier résultats. Bulletin mensuel – office National de la Chasse. Numéro spécial scientifque et technique 5. 83 pp.

Olney, P. J. S. (1965) Project Mar. List of European and North African Wetlands of international importance. International Union for the Conservation of Nature and Natural Resources. Publication New Series No. 5, Morges, Switzerland. 89 pp.

Paskoff, R. P. (1973) Geomorophological processes and characteristic land forms. pp. 53–60. *In*: F. DiCastri and H. A. Mooney (eds.). Mediterranean Type Ecosystems. Origin and Structure. Springer-Verlag, Berlin, West Germany.

Pavlides, G. (1985) Geobotanical study of the National Park of lakes Prespa (N.W. Greece). Institute of Systematic Botany and Phytogeography, Aristotelian University, Thessaloniki (In Greek).

Petrović, G. (1981) Chemical investigations of water and sediments of Lake Skadar. pp. 68–96. *In*: G. S. Karaman and A. M. Beeton (eds.). The Biota and Limnology of Lake Skadar. Titograd, Yugoslavia.

Pirot, J-Y. and Pont, D. (1987) Le canard souchet (*Anas clypeata* L.) hivernant en Camargue: alimentation, comportement et dispersion nocturne. Revue d'Ecologie 42: 59–79.

Podlejski, V. D., 1981. Observations sur *Scirpus maritimus* L. en Camargue. Ecologia Mediterranea 7: 63–78.

Pons, A., Toni, C. and Triat, H. (1979) Edification de la Camargue et histoire holocène de sa végétation. Revue d'Ecologie (Terre Vie) Supplément 2: 13–30.

Pont, D., Crivelli, A. J. and Guillot, F. (1991) The impact of three-spined sticklebacks on the zooplankton of a fish-free pool. Freshwater Biology 26: 149–163.

Prat, N. F., Puig, M. A., Gonzalez, G., Tort, M. F. and Estrada, M. (1984) Llobregat. pp. 527–552. *In*: B. A. Whitton (ed.). Ecology of European Rivers. Blackwell, Oxford, England.

Quignard, J. P. and Zaouali, J. (1980) Les lagunes periméditerranéenes. Bibliographie annotée. Première partie: les étangs français de Canet à Thau. Bulletin de l'Institut National scientifique et technique d'oéanographie et de pêche de Salammbo 10: 107–124.

Richez, G. (1981) Aspects du tourisme en basse Camargue. pp. 56–58. *In*: La Camargue, ses Problèmes son Avenir. Comité d'Expansion économique Bouches du Rhône, Duplicated report.

Rioux, J. A. and Arnold, M. (1955) Les Culicidés de Camargue (étude systématique et écologique). Terre Vie 9: 244–286.

Rivas-Martinez, S., Costa, M., Castroviejo, S. and Valdez, E. (1980) Vegetacion de Donana (Huelva, Espana). Lazaroa 2: 51–90.

Rodriguez, L. G., Noguiera, L. C., Miralles, J. M. and Nogueira, H. C. (1984) Cabo de Gata. Guia de la Naturaleze Perfil Ecologica de una Zona Arida. Editorial Everest, Madrid, Spain. 174 pp.

Scott, D. A. (1980) A preliminary inventory of wetlands of international importance for waterfowl in West Europe and Northwest Africa. International Waterfowl Research Bureau Special Publication No. 2, Slimbridge, England. 127 pp.

Senra, A. (1982) Evolucion y situacion actual de las zonas humedas Andaluzas. pp. 23–42. *In*:

Las zonas humedas en Andalucia. Ministeriode Obras Publicas y Urbanismo. MOPU, Sevilla, Spain.

Skinner, J. and Smart, M. (1984) The El Kala wetlands of Algeria and their use by waterfowl. Wildfowl 35: 106–118.

Smart, M. (1976) Proceedings of the International Conference on Conservation of Wetlands and Waterfowl, Heiligenhafen, Federal Republic Germany 2–6 December 1974. International Waterfowl Research Bureau, Slimbridge, England. 492 pp.

Stankovic, S. (1931) Les particularités limnologiques des lacs égéens. Verhandlungend. Internationalen Vereinigung Limnologie 5: 158–196.

Stankovic, S. (1960) The Balkan Lake Ohrid and its Living World. Monographiae Biologicae IX. Junk Publishers Den Haag, The Netherlands. 357 pp.

Stevenson, A. C., Skinner, J., Hollis, G. H. and Smart, M. (1988) The El Kala National Park and environs, Algeria: an ecological evaluation. Environmental Conservation 15: 335–348.

Tachicchi, M. I. (1968) Vicendestagionali delle communita littoranee del Lago Trasimeno. Rivista di Idrobiologia 7: 195–302.

Tamisier, A. (1966) Dispersion crépusculaire des Sarcelles d'hiver (Anas crecca) en recherche de nourriture. Terre Vie 20: 316–337.

Tamisier, A. and Saint-Gerard, T.(1981) Stationnements d'oiseaux d'eau et chasse de nuit dans les départments côtiers de France. Alauda 49: 81–93.

Tourenq, J. N. (1975) Recherches écologiques sur les Chironomidés (Diptères) de Camargue. Dissertation, Université de Toulouse, France. 414 pp.

UNEP (1977) Le Bassin Mediterraneen. Cadre géographique et socio-économique du Plan Bleu. Programme des Nations Unies pour L'Environnement. Rapport Préliminaire. Geneva, Switzerland. 135 pp.

Ungar, I. A. (1974) Inland halophytes of the United States. pp. 233–305. *In*: R. J. Reimold and W. H. Queen (eds.). Ecology of Halophytes. Academic Press, New York, USA.

Van Vierssen, W. (1982) The ecology of communities dominated by *Zannichellia* taxa in western Europe. II. Distribution, synecology and productivity aspects in relation to environmental factors. Aquatic Botany 12: 63–198.

Van Vierssen, W., Van der Zee, J. R. and Van Kessel, C. M. (1984) The creation of sanctuaries for the endangered fish species *Aphanius apodus*, *Aphanius iberus* and *Valencia hispanica* (Cyprinodontidae) in the western Mediterranean. Report Nature conservation Department, Agricultural University Wageningen, The Netherlands. 6 pp.

Verhoeven, J. T. A. (1979) The ecology of *Ruppia*-dominated communities in western Europe. I. Distribution of *Ruppia* representatives in relation to their autoecology. Aquatic Botany 6: 197–268.

Verhoeven, J. T. A. (1980a) The ecology of *Ruppia*-dominated communities in western Europe. II. Synecological classification. Structure and dynamics of the macroflora and macrofauna communities. Aquatic Botany 8: 1–85.

Verhoeven, J. T. A. (1980b) The ecology of *Ruppia* dominated communities in western Europe. III – Aspects of production, consumption and decomposition. Aquatic Botany 8: 209–253.

Vlassis, G. (1978) Le cheval Camargue à travers le temps. Courrier du Parc Naturel Régional de Camargue 15: 8–10.

Waisel, Y. (1972) Biology of Halophytes. Academic Press, New York, USA. 395 pp.

Wheeler, B. D., 1980. Plant communities of rich-fen systems in England and Wales. II Communities of calcareous mires. Journal of Ecology 68: 365–395.

Williams, W. D. (1981) 1. Inland salt lakes: An introduction. Hydrobiologica 81: 1–14.

WWF (1984) Spain's wetlands drying up. WWF News 32 Nov–Dec 1984: 4–5.

Yesou, P., Trolliet, B. and South, M., 1983. Anatidés et zones humides de France métropolitaine. Bulletin mensuel de l'office nationale de la Chasse. Décembre 1983: 1–315.

Zaouali, J. (1976) Contribution à l'étude écologique de la sebkha Kelbia (Tunisie centrale). Archives de l'Institut Pasteur de Tunis 3: 271–380.

Wetlands of Australia: Northern (tropical) Australia

C. MAX FINLAYSON AND ISABELL VON OERTZEN

Abstract

Ecological information for Australian wetlands is reviewed and placed in context with the climate and geography, and a summary of wetland classifications. Temperate and tropical Australian wetlands are dealt with in two sections. In both sections wetlands are classified on the basis of physiography. For temperate Australia, information on the flora and fauna is reviewed on the basis of wetland classification within each of the major Australian Drainage Divisions, whereas for northern Australia, ecological information is summarized according to the major wetland types. We commented on biological aspects of conservation issues, although we recognize it is only one aspect of conservation. Finally, recommendations are made on the broad directions that conservation of Australian wetlands should take. These include: (i) the need to examine wetlands from a total catchment and drainage basin perspective, (ii) to consider conservation as one of several uses for a wetland; (iii) to unify the status of reserve systems between the Australian States; (iv) to give equal attention to all developments hazardous to wetlands rather than focussing on those with a high media profile; (v) in temperate Australia, the need for more information on arid zone wetlands to assess their conservation status and value, and an immediate expansion of current research activities and/or control of threats posed by feral animals and alien weeds.

Introduction

This section covers the monsoonal coastal region and a large part of the arid interior of the continent. Except for the north-eastern coast, the region is neither heavily populated nor extensively developed. This has resulted in an uneven collection of ecological information from wetlands. In general, biologists and conservation authorities have given more attention to wetlands

D.F. Whigham et al. (eds.), Wetlands of the World I, 195–304.

located near major population centres or to those in scenically attractive and relatively accessible places. As an example, the Arafura Swamp in Arnhem Land has received very little attention, despite it being one of the few large, permanent swamps in the north of the Northern Territory. An account of the distribution of wetlands and the main geographic features of northern Australia is presented.

We then provide an historical overview of systems used to classify wetlands, describe the ecological characteristics of the major wetland types, provide details of wetland inventories, discuss conservation issues and major threats to wetlands, and present steps that need consideration if undesirable impacts are to be avoided, or remedied. The wider implications of these steps in the realms of socio-economic and political factors that have a direct bearing, if not controlling influence, over conservation issues have not been considered. Whilst outside the scope of this review it is recognised that these factors have to be considered by conservation authorities and, if required, adjustments made to reflect them in management strategies. Despite this recognition the review has been restricted to the biological or ecological aspects of wetland conservation and all comments and recommendations are made within this restriction.

Study area

Northern or tropical Australia contains all areas north of the Exmouth–Pilbara region in Western Australia, large parts of the Northern Territory, and all areas north of Rockhampton in Queensland (Fig. 1). The area is relatively sparsely populated with major centres in Townsville, Rockhampton, Cairns, and Mt Isa in Queensland, Darwin in the Northern Territory, and Port Hedland and Karratha in Western Australia. Agricultural development occurs along the Queensland coast with sugar cane being the main crop, while much of the inland region is used for open-range cattle grazing. Mining developments (e.g. copper, lead and zinc at Mount Isa, iron ore in the Pilbara) have been responsible for inducing sizable populations to settle in otherwise remote areas.

Climate

The climate of tropical Australia has been described by Ramage (1971), and Lee and Neal (1984). The monsoon-influenced coastal regions have only two seasons, known locally as the Wet and the Dry. The Wet season commences late in the year (Nov–Dec) and lasts for 3–4 months; both onset and duration vary from year to year. Locations nearer the equator generally have the

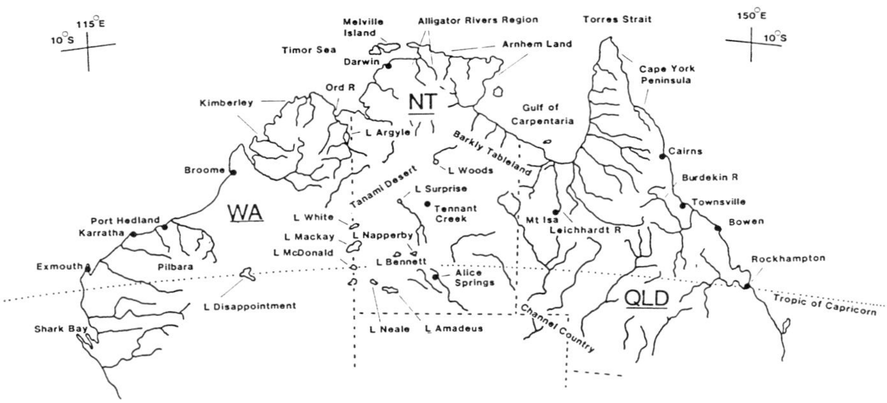

Figure 1. Map of northern (tropical) Australia.

smaller range between annual dates of onset of the Wet season (Nicholls 1984).

The mean monthly rainfall values for Darwin are given in Table 1, but these figures disguise the considerable variation in timing and duration of the monsoonal rains. Very little rain falls during the Dry seasonal, although the amount that does fall is more variable than during the Wet season (Taylor and Tulloch 1985).

The mean monthly rainfall recorded at several centres (Table 1) illustrates the marked seasonal and spatial differences that occur in northern Australia. Alice Springs, in the interior, has a similar seasonal pattern of rainfall to Darwin, but the total amount is considerably less. There is a general increase in rainfall from south to north over Western Australia and the Northern Territory, and from south-east to north-east over Queensland.

The monsoonal northern region is warm to hot all year round while in the south, it is mild during the Dry season (Table 1). Overnight frost can occur at these latter sites which have a continental desert type climate. In the Wet season high temperatures in Darwin are accompanied by high relative humidities of about 80% compared to about 30% in Alice Springs (Table 1). Near the coast, cloud cover is greatest during the warm Wet season-decreasing over the dry interior and allowing overnight radiative cooling.

The arid zone of Central Australia has low annual rainfall, high evaporation, and a large annual variation. The regional pattern of climate has been described by Slayter (1962) who identified the high pressure belt of the southeastern trade-winds as the dominating influence. During the winter months (Apr–Oct) this belt lies over southern Australia and brings clear days to the centre, but every 7–10 days a low pressure trough passes across the region. If this coincides with an inflow of moist air, light and usually

Table 1. Mean monthly rainfall (mm), mean maximum (Max) and minimum (Min) temperatures (°C), mean monthly evaporation (Evap.) in mm, and mean relative humidity (Humid) in % at 9 AM and 3 PM for selected centres (Lee and Neal 1984, Bureau of Meteorology 1975, 1986, Hall *et al.* 1981).

		J	F	M	A	M	J	J	A	S	0	N	D
Darwin	Rainfall	399	337	279	96	16	3	1	3	14	59	130	241
12°26′S	Max Temp	32	32	32	33	32	31	30	31	33	33	33	33
130°52′E	Min Temp	25	25	24	24	22	20	19	20	23	25	25	25
	Evap.	186	168	186	210	217	210	217	217	240	248	240	217
	Humid 9AM	79	81	81	75	66	62	60	65	68	68	70	74
	Humid 3PM	68	69	65	51	42	39	35	40	45	50	55	63
Broome	Rainfall	165	153	97	28	25	23	5	2	2	1	12	65
17°57′S	Max Temp	33	33	34	34	31	29	29	30	32	33	34	34
122°13′E	Min Temp	26	26	26	23	19	16	14	15	18	22	25	27
	Evap.	279	224	217	240	217	180	217	217	270	279	300	310
	Humid 9AM	68	73	67	54	47	48	46	43	46	51	56	62
	Humid 3PM	62	65	57	42	37	35	32	31	40	51	55	59
Townsville	Rainfall	288	297	199	74	32	30	16	13	16	30	47	131
19°15′S	Max Temp	31	31	30	29	27	26	25	26	28	29	31	31
146°46′E	Min Temp	24	24	32	20	17	15	13	15	17	21	23	24
	Evap.	279	196	217	210	186	150	186	217	240	279	300	279
	Humid 9AM	68	74	71	66	65	65	63	62	57	58	59	62
	Humid 3PM	62	66	63	57	54	52	47	51	51	52	56	58
Alice	Rainfall	43	41	30	16	17	14	11	11	9	20	25	35
Springs	Max Temp	37	36	33	29	23	20	19	22	26	31	34	35
23°36′S	Min Temp	22	21	18	14	9	6	5	7	10	15	18	20
133°36′E	Evap.	403	336	310	240	155	120	124	155	240	310	330	372
	Humid 9AM	31	32	36	41	55	63	60	46	33	24	24	24
	Humid 3PM	19	19	21	23	31	34	30	25	19	15	15	16
Brisbane	Rainfall	158	164	148	89	69	66	55	46	49	69	93	129
27°28′S	Max Temp	30	30	29	27	25	22	22	23	25	27	28	29
153°2′E	Min Temp	21	21	20	18	14	12	10	11	14	17	19	21
	Evap.	176	142	140	114	81	64	70	98	128	152	168	193
	Humid 9AM	66	69	71	70	71	72	70	66	62	60	60	62
Perth	Rainfall	8	11	20	46	126	185	175	142	81	55	21	15
31°57′S	Max Temp	30	30	28	24	21	19	18	18	19	22	25	28
151°12′E	Min Temp	19	19	17	14	12	11	9	9	10	12	14	17
	Evap.	285	242	213	132	94	69	75	87	118	173	211	275
	Humid 9AM	50	53	56	65	72	78	79	74	68	60	53	51
Sydney	Rainfall	96	110	125	134	129	130	111	79	70	81	75	130
33°52′S	Max Temp	26	26	25	23	20	18	17	18	20	22	24	25
151°12′E	Min Temp	19	19	18	15	12	10	8	9	11	14	16	18
	Evap.	217	177	157	126	94	85	93	116	141	168	193	252
	Humid 9AM	65	70	71	70	71	74	68	66	62	63	62	63
	Humid 3PM	60	61	61	57	57	59	51	51	47	54	54	58
Adelaide	Rainfall	20	19	24	44	69	67	66	62	51	45	31	26
34°56′S	Max Temp	29	28	26	23	18	16	15	16	18	21	24	25
138°35′E	Min Temp	17	17	16	14	11	9	8	10	12	13	15	
	Evap.	254	216	180	120	79	56	60	78	110	164	196	242
	Humid 9AM	42	45	40	58	69	76	77	71	62	53	46	43

Table 1. Continued.

		J	F	M	A	M	J	J	A	S	0	N	D
Charlottes	Rainfall	143	132	154	183	201	223	215	251	213	260	210	158
Pass	Max Temp	18	17	15	11	6	3	2	3	5	9	11	15
36°26′S	Min Temp	6	6	4	1	−2	−5	−6	−4	−3	−1	1	3
148°20′E	Humid 9AM	66	68	67	64	75	84	77	83	77	67	69	58
	Humid 3PM	46	42	57	67	77	90	76	90	93	68	55	49
Melbourne	Rainfall	48	48	54	58	56	52	49	49	59	69	59	58
37°49′S	Max Temp	27	26	24	21	17	15	14	15	17	20	22	24
144°58′E	Min Temp	15	15	14	11	9	7	6	7	8	10	11	13
	Evap.	206	181	140	91	57	37	44	62	86	127	152	189
	Humid 9AM	59	63	66	72	78	82	81	75	68	63	60	61
Hobart	Rainfall	50	41	47	55	49	61	52	48	52	64	55	57
42°53′S	Max Temp	22	22	20	18	14	12	12	13	15	17	18	20
147°20′E	Min Temp	12	12	11	9	7	5	5	5	6	8	9	11
	Evap.	142	123	92	59	36	20	24	43	59	90	121	142
	Humid 9AM	59	63	66	69	75	78	78	73	66	62	59	58

localised rain can fall. During summer the south-east trade-winds are interrupted by intrusions of moist air from the equatorial (tropical low-pressure) trough to the north. Rainfall from violent convectional thunderstorms occurs – this comprises most of the annual rainfall of Central Australia (Table 1). Cyclonic depressions that move into Central Australia from the tropical coast are an occasional, but significant source of summer rainfall.

Drainage pattern

About two-thirds of total runoff from Australia occurs in northern Australia, far from existing major population centres and development demands (Australian Water Resources Council 1976). Highest runoff, 130 ML y^{-1}, comes from the Gulf of Carpentaria region (Fig. 2). The north-east coast with 90 ML and the Timor Sea region with 80 ML runoff, however, have a much greater exploitable yield of surface water (Australian Water Resources Council 1976). The extensive Western Plateau region has no significant runoff as rainfall is generally low and evaporation can exceed 4 m y^{-1}.

Compared with rivers in most other continents, Australian rivers carry relatively small volumes of water and are ephemeral or only flow seasonally. The Burdekin River, Queensland, has a catchment of 130,000 km^2. Its maximum instantaneous peak discharge is 40,000 $m^3 s^{-1}$, the highest recorded for any river in Australia (Fleming 1981a). However, about 90% of the total annual flow occurs between January and April (Fleming 1981b). The interior lowlands have an endorheic drainage pattern with runoff restricted to the warm Wet season. During periods of drought these rivers are little more than a chain of elongated waterholes. They typically have irregular flows,

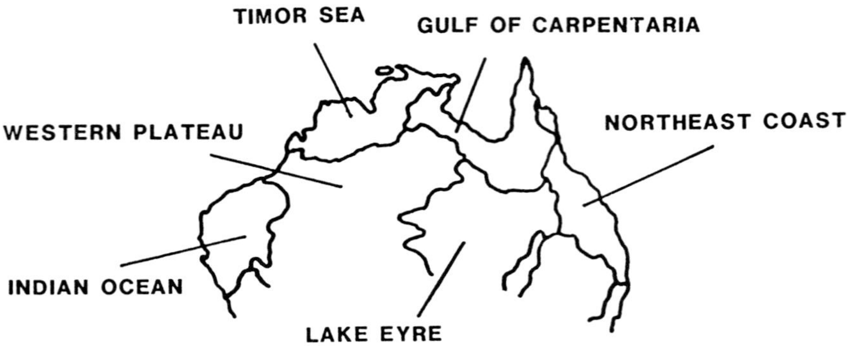

Figure 2. Drainage regions of northern Australia (adapted from Australian Water Resources Council 1976).

low gradients, and spread out over vast areas of flat country and into normally dry inland lakes (playas) (Paijmans *et al.* 1985).

A number of dams and reservoirs have been constructed to conserve surface water (Table 2). The largest with a storage capacity of 5,720 Mm^3 is Lake Argyle on the Ord River, Western Australia – constructed in 1971 for a proposed irrigation development of 70,000 ha (Australian Water Resources Council 1976).

Wetland distribution

Paijmans *et al.* (1985) defined wetlands as "land permanently or temporarily under water or waterlogged. Temporary wetlands must have surface water or waterlogging of sufficient frequency and/or duration to affect the biota. Thus, the occurrence, at least sometimes, of hydrophytic vegetation or use by waterbirds are necessary attributes". This definition does not contain a

Table 2. Dams and reservoirs in northern Australia (Australian Bureau of Statistics 1985). Fairbairn is located about 30 km south of the Tropic of Capricorn.

Dam or reservoir	Location	Capacity (10^6 m^3)
Lake Argyle	Ord River, WA	5720
Darwin River	Darwin River, NT	259
Ross River	Ross River, Qld	417
Tinarroo Falls	Barron River, Qld	407
Koombooloomba	Tully River, Qld	201
Eungella	Broken River, Qld	131
Julius	Leichhardt River, Qld	127
Moondarra	Leichhardt River, Qld	107
Fairbairn	Nogoa River, Qld	1440
Burdekin Falls	Burdekin River, Qld	1860

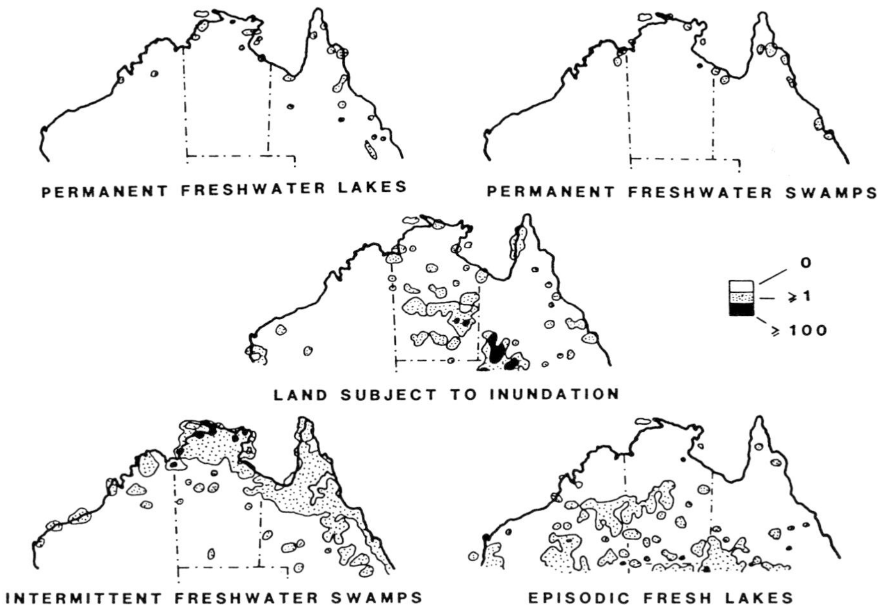

Figure 3. Distribution and frequency of occurrence of wetlands in Australia (adapted from Paijmans *et al.* 1985).

depth criterion but is otherwise similar to the so-called "Ramsar definition" (Lyster 1985) of wetlands as "areas of marsh, fen, peatland or water, whether natural or artificial, permanent or temporary, with water that is static or flowing, fresh, brackish or salt, including areas of marine water the depth of which at low tide does not exceed 6 metres".

In an attempt to compile an overview of wetland distribution, Paijmans *et al.* (1985) summarized information derived from 1:250,000 topographical maps on to a 1:2,500,000 map of the continent. While giving a broad indication of wetland distribution the map contains many uncertainties and is described by the authors themselves as "too large, too detailed and too inaccurate". Thus, they also considered wetland distribution in terms of the broad geographical settings of Jennings and Mabbutt (1977) (Fig. 3).

Along the Queensland coast the upland areas contain permanent and seasonal wetlands – floodplain lakes, billabongs (oxbow lakes), swamps, waterholes, and river flats liable to flooding occur in alluvial valley tracts. There are also extensive tidal flats and mangroves, some backing on to seasonal swamps. Floodplain lakes, billabongs, and waterholes occur in major deltas, notably the Burdekin near Townsville. On Cape York Peninsula seasonal swamps in shallow depressions are common while some volcanic craters contain permanent lakes and swamps.

The lowlands along the Gulf of Carpentaria contain intermittent or seasonal swamps in shallow pans, permanent waterholes in channels, seasonal billabongs, lakes, and swamps on the upper portion of rivers that drain to the Gulf of Carpentaria. Supratidal lakes, up to 30 km inland and with very little vegetation, and narrow intertidal flats with a fringe of mangroves, occur along the coast.

The lowlands surrounding the Arnhem Land Plateau have numerous floodplain lakes, waterholes, and swamps along the major rivers and permanent or seasonal swamps on alluvial plains near the sea. The coastal plains to the east of Darwin contain extensive floodplains along rivers (e.g., the Adelaide and the Alligator Rivers) originating from the Arnhem Land Plateau. Coastal wetlands include intertidal flats with mangroves and supratidal flats, either bare or with sparse vegetation. The Kimberley coast is very rugged and coastal wetlands are confined to supratidal flats near large rivers.

In the north-west Pilbara area of Western Australia, the drainage system contains waterholes along river channels, and extensive lakes or areas liable to flooding along the river valleys. Intermittently flooded lakes are found between linear dunes and along the main river channels. Supratidal flats are usually bare of vegetation, though some of the saline intertidal flats contain mangroves and algal mats. The large inland arid zone, occupying nearly half the continent, has low rainfall with high evaporation and generally low hydrologic gradients and porous soils. Drainage is thus uncoordinated and characterised by saline intermittently flooded and episodic wetlands. Some of these lakes (e.g. Lake Disappointment, 1,400 km^2 and Lake Mackay, 2,500 km^2) are very large. Further to the east the lowlands contain extensive inland drainage basins with many rivers flowing into the Channel Country which contains areas liable to seasonal flooding with many waterholes and billabongs.

Wetland classification

An overall classification of Australian wetlands has been proposed by Paijmans *et al.* (1985). The system was deliberately loosely defined and lacking in detail compared to the system introduced by the United States Fish and Wildlife Service (Cowardin *et al.* 1979). Paijmans *et al.* (1985) did not adopt this system, regarding it as too detailed for application to Australian wetlands which lack detailed description. Problems of gradation between wetland types, lack of seasonal information, and inconsistent and indefinite terms have not been fully resolved in the system they proposed. In the context of northern Australia, the degree of permanence and salinity of many inland lakes is an aspect that needs further consideration.

In addition to the hierarchical system proposed by Paijmans *et al.* (1985)

other classification attempts that cover northern Australian wetlands include a vegetation structure and floristic scheme by Briggs (1981), a general and broad-scale classification of wetlands in Queensland by Stanton (1975), and a preliminary classification of wetlands in the Pilbara region of Western Australia using physical characteristics by Masini (1986). The main details of these schemes are given below.

Hierarchical classification of northern Australian wetlands

Paijmans *et al.* (1985) have adopted a simple and loosely defined hierarchical classification of categories, classes and subclasses (Table 3), that are controlled by hydrologic and vegetation characteristics.

Lakes
Lakes are areas of open water generally more than 1 m deep when full, and with little or no persistent emergent vegetation. They are divided into four classes (Table 3) based on their degree of permanence, although the distinction between intermittent and episodic lakes is not clear-cut.

Swamps
Swamps differ from lakes in having persistent emergent vegetation and being generally less than 1 m deep. Permanent, seasonal, intermittent, and episodic classes are recognised (Table 3).

Land subject to inundation
These areas are either seasonally or intermittently flooded and differ from swamps in not containing water long enough to allow hydrophytic vegetation to develop.

River and creek channels
River and creek channels are divided into the same four classes but there is a lot of hydrologic integration between them.

Tidal flats
Tidal flats are divided into three classes according to the frequency and nature of flooding. Intertidal flats, inundated by most high tides are biologically rich and are dominated by mangroves. Supratidal flats are covered only at spring tides, or even less frequently, and comprise bare surfaces or limited plant cover. The third class has spring tidal and less frequent flooding combined with seasonal freshwater flooding.

Table 3. Categories, classes, and subclasses of the hierarchical classification system devised by Paijmans *et al.* (1985).

I. Lakes
- i) Permanent and near-permanent
  - a) Floodplain lakes including billabongs and waterholes in channels
  - b) Lakes of coastal dunes and beach ridge plains
  - c) Lakes in terminal drainage basins
  - d) Lakes associated with lava flows
  - e) Crater lakes
  - f) Karst lakes
  - g) Glacial lakes
  - h) Man-made lakes
- ii) Seasonal
  - a) Floodplain lakes
  - b) Terminal drainage basin lakes
- iii) Intermittent
  - a) Floodplain lakes
  - b) Coastal dune lakes
  - c) Lakes in terminal drainage depressions
  - d) Man-made lakes
- iv) Episodic
  - a) Lakes in terminal drainage depressions
  - b) Lakes on present or former floodplains

II. Swamps
- i) Permanent
  - a) Floodplain swamps
  - b) Swamps of coastal dunes and beach ridge plains
  - c) Swamps in terminal drainage depressions
  - d) Swamps associated with lava flows
  - e) Crater swamps
  - f) High-mountain swamps
  - h) Swamps fed by springs
- ii) Seasonal
  - a) Floodplain swamps (Other classes rare)
- iii) Intermittent
  - a) Floodplain swamps
  - b) Swamps in terminal drainage depressions
- iv) Episodic

III. Land subject to inundation
- i) Seasonally
  - a) Floodplains
  - b) River and creek banks
- ii) Intermittently
  - a) Floodplains
  - b) River and creek banks

IV. River and creek channels
- i) Permanent and near-permanent
  - a) Rocky
  - b) Sandy
  - c) Silty/clayey

Table 3. Continued.

ii) Seasonal
a) Rocky
b) Sandy
c) Silty/clayey
iii) Intermittent
a) Rocky
b) Sandy
c) Silty/clayey
iv) Episodic
a) Rocky
b) Sandy
c) Silty/clayey
V. Tidal flats
i) Daily tidal flooding
a) Intertidal flats of open coasts
b) Intertidal estuarine flats
c) Intertidal stream banks
ii) Spring tidal and less frequent flooding
a) Supratidal surfaces
b) Supratidal stream banks
c) Saline pools
iii) Spring tidal and less frequent flooding and seasonal freshwater flooding
a) Supratidal flats
b) Brackish pools and billabongs
VI. Coastal water bodies
i) Permanently open to the sea
ii) Intermittently open to the sea
iii) Rarely open to the sea

Coastal water bodies

This category includes three classes (permanently, intermittently, or rarely open to the sea) and encompasses estuaries, inlets, lagoons, and some lakes. The vegetation and hydrology of lakes and swamps are considered in more detail by Paijmans *et al.* (1985). They also present a digest of published information on vegetation of other wetland types.

Structural and floristic classification of wetland vegetation

Briggs (1981) has classified the Australian freshwater wetland flora by structural (Specht 1970) and then by floristic characteristics (Table 4). Wetland vegetation is defined as "vegetation in shallow, non-tidal water and on land subject to inundation. Terrestrial plants in dry swamps, lacustrine plants confined to water more than two metres deep and phytoplankton are excluded . . .". Saline and coastal wetland communities were not considered. Whilst not strictly a classification of wetlands this system does differentiate the major

Table 4. Structural and floristic classification of wetland vegetation (Briggs 1981).

Swamp forests
mesophyll palm vine-forests
paperbark swamp forests
swamp sclerophyll forests
Swamp woodlands
paperbark swamp woodlands
swamp sclerophyll woodlands
Swamp scrubs and heaths
swamp scrubs
swamp heaths
Swamp shrublands
lignum shrublands
chenopod shrublands
samphire shrublands
Sedgelands
Eleocharis sedgelands
Baumea sedgelands
button-grass sedgelands
Carex sedgelands
Swamp glasslands
wet grasslands
Phragmites grasslands
Typha grasslands
tussock grasslands
canegrass grasslands
Swamp herblands
floating and floating-leaved herblands
submerged and emergent herblands

wetland types through their vegetation assemblages. The word "swamp" has been used in a general manner to designate wetland communities.

Classification of Queensland wetlands

A generalised classification scheme for Queensland has been proposed by Stanton (1975). Due to the lack of information on many of the wetlands, the classification was broad-scaled (Table 5) with two major classes (tidal and inland waters) with the latter being divided into saline and fresh sub-classes. The terms permanent, seasonal, semi-permanent, and intermittent are used but not quantitatively defined. Permanent and seasonal wetlands imply surface water is present for most of every year; water levels in seasonal wetlands can decrease to below the ground surface. Semi-permanent implies irregular inundation from either heavy local rain or general flooding, while intermittent refers to less frequent inundation.

Despite the vagueness of the terms in this scheme, it nevertheless results

Table 5. A preliminary classification of wetlands in Queensland (from Stanton 1975).

- A. Tidal wetlands
  - I) Mangroves
  - II) Salt marshes
  - III) Salt mudflats
  - IV) Saltwater meadow
- B. Inland waters
  - I) Fresh
    - a. Permanent and seasonal
      - lakes, swamps, and marshes of coastal sand
      - dunes
      - fluviatile lacustrine plains
      - coastal and subcoastal swamps, shallow lakes, marshes, and meadows
      - sand-dune, higher rainfall swamp forests
      - argillaceous sediment, higher rainfall swamp forests
      - marine plain freshwater to brackish swamps and lagoons
      - basalt lakes, swamps, and springs
      - higher rainfall, lateral lakes, and back swamps
      - higher rainfall, infertile sediment heath, and sedge swamps
      - elongated lakes in braided stream channels
    - b. semi-permanent
      - floodplains and deltas of larger rivers
      - lateral lakes and back swamps
      - fresh/saline large lakes with limited outflow
      - internal drainage basins
    - c. intermittent
      - inland clay-pan lakes and small vegetated swamps
      - distributary swamps and back swamps of large inland rivers
      - terminal floodplains of inland rivers
  - II) Saline
    - a. permanent or semi-permanent
    - b. intermittent
      - terminal floodplains and lakes
      - large or small terminal lakes

in a simple framework around which to examine the extent, distribution and type of wetlands in Queensland.

Classification of inland waters of the Pilbara, Western Australia

Masini (1986) surveyed, during 1983, the relatively poorly described inland waters of the Fortescue and De Gray River systems of the Pilbara. The objectives of the survey were to produce an inventory of permanent and ephemeral inland surface waters, to classify these according to significant physical and biological characteristics, and to establish conservation and management priorities. Nine subjective characteristics of inland water are used in a classification based on physical characteristics (Table 6). The framework established by this classification was used to describe the water quality, flora, and fauna of the wetlands.

Table 6. Classification of inland waters of the Pilbara area of Western Australia (from Masini 1986).

Category	Physical characteristics
Spring systems	Interconnected pools with water flowing down medium to steep gradients, generally permanent, fed by aquifers.
Permanent/semi-permanent	Occur on outer edges of meanders or river pools on narrow river sections, initially filled by river flows, sustained by localised drainage or direct link to water table.
Headwater streams	Drain small catchments, headwater streams, and drainage channels, generally narrow, shallow, and fast flowing.
Primary river channels	Spasmodic influxes of large volumes of water, usually wide, dry most of year, contain shallow ephemeral pools.
Adjoining pools	Ephemeral to intermittent pools in depressions adjacent to rivers, include cut off meanders and wind deflation hollows between vegetation hummocks.
Ephemeral claypans	Areas of sheet runoff collecting on fine-grained soils, shallow but could be very large and expansive.
Semi-permanent claypans	Shallow, low-permeabilitity, and flow-through, deeper and more persistent than ephemeral claypans.
Tidal areas	Region of river with large diurnal fluctuations in depth and area, active erosion and deposition.
Man-made surface waters	Dams, sewage ponds, etc.

Ecological characteristics of wetlands

The main characteristics and ecological variables associated with seagrass meadows, mangrove swamps, salt-marshes and flats, seasonally inundated floodplains and billabongs, freshwater swamps, and lakes are discussed in this section. The terms adopted are comparable, though not the same, to those of Paijmans *et al.* (1985) as theirs were considered to be either too detailed or not appropriate for our purposes. The choice of categories (Table 3) is based on the extent of available information and does not necessarily reflect their ecological importance. Very little detailed information is available on the characteristics of the large, ephemeral lakes or land subject to intermittent inundation (i.e., wetland categories that occur in the arid interior part of northern Australia). The crater lakes of Cape York Peninsula (Timms 1976), waterholes, pools and channels of the Pilbara (Masini 1986), and river channels of the Kimberley (Miles and Burbidge 1975, Kabay and Burbidge 1977) are not included due to their relatively unknown biological characteristics.

Seagrass meadows

Seagrasses are marine flowering plants that are able to grow completely submerged and have an anchoring system that withstands wave and tidal movements. Of the 12 genera of angiosperms regarded as seagrasses 11 are

Table 7. Seagrasses of northern Australia (Den Hartog 1970, Walker and Prince 1987, and S. Jacobs, pers. comm.).

Amphibolis antarctica, *Cymodocea angustata*, *C. rotundata*, *C. serrulata*, *Enhalus acoroides*, *Halodule pinifolia*, *H. uninervis*, *Halophila decipiens*, *H. ovalis*, *H. ovata*, *H. spinulosa*, *H. tricostata*, *Posidonia australis*, *Syringodium isoetifolium*, *Thalassodendron ciliatum*, *Thalassia hemprichii*, *Zostera capricorni*

found in Australia; 8 of these occur along the Queensland coast (Den Hartog 1970) and 9 in northern Western Australia (Walker and Prince 1987). A list of species found in northern Australia is given in Table 7. The species in northern Western Australia do not fit clearly into precise biogeographic provinces, but there are similarities to the flora of New Guinea, Torres Strait, and Gulf of Carpentaria (Walker and Prince 1987). The high seagrass diversity along this coast has been attributed by these authors to the general suitability of the coast for seagrass growth, a high degree of habitat partitioning, and a range of species from both the north and south available for colonisation.

Seagrass meadows are found in northern Australia in shallow water that is susceptible to disruption by high winds, heavy seas, and low salinities from freshwater runoff associated with cyclones (Spain and Heinsohn 1973, Heinsohn and Spain 1974). They occur in reef, inter-reef, and offshore situations and in habitats extending from intertidal to subtidal (Lanyon 1986). Their structure and species composition vary considerably and Wake (1975) reported them to be of relatively low biomass (1–280 g dry weight m^{-2}). Seagrass meadows, however, support a rich and diverse fauna and flora and constitute a major food source in coastal waters (Lanyon 1986). They are generally considered to be major nursery grounds for commercial prawn species (Penaeidae) and provide a direct food source for two large vertebrates, the dugong (*Dugong dugon*) and the green turtle (*Chelonia mydas*).

A conceptual successional model developed by Birch and Birch (1984) ranked the seagrasses into four groups. The first group contained the pioneer Halophila, the second the stenohaline *Cymodocea serrulata* and *Syringodium isoetifolium*, the third the euryhaline *Halodule uninvervis* and probably *Thalassia hemiprichii*, and the fourth *Enhalus acoroides*. These groups were noted to be generally consistent with field observations of resistance to disturbance, tolerance of emersion and low salinity. The field observations were combined with an objective classification of strategies evolved by seagrasses to cope with disturbance and stress to develop a preliminary model of intertidal tropical seagrass distribution.

Birch and Birch (1984) have presented a comprehensive account of the course of recolonisation by seagrass of an intertidal area near Townsville (Queensland) following devastation by a cyclone in December 1971. The successional pattern outlined illustrated several clear trends. The Halophila, after being dominant early in the successional sequence, had an exponential decrease and after 9 years reached a steady state of abundance of about 5% frequency. In contrast, *Cymodocea serrulata* increased to about 3% frequency until replaced by the coralline alga *Halimeda opuntia*. At this stage (1980–82) mean dry matter of seagrass decreased from 186 to 99 g m^{-2}. The frequency of *Halodule uninervis* increased linearly to about 40% frequency and showed no sign of stabilisation at this level. From the results it was suggested that a density-dependent competitive system was operating. It was also considered that a major disturbance could lead to an increase in yield followed by a decrease, as demonstrated by the fall in dry matter weight between 1980–82 back to the pre-cyclone level. *Halophila ovata* and *H. ovalis* were the main pioneer species and eventually were restricted to the two ends of the intertidal zone, indicating a tolerance to substrate variability.

In the Gulf of Carpentaria, Poiner *et al.* (1987) found 906 km^2 of intertidal and shallow subtidal areas fringing 670 km of coastline supporting seagrass communities. Open-coastline communities dominated by monospecific stands of *Halophila ovalis* and *H. uninervis* intertidally, and *C. serrulata* and *Syringodium isoetifolium* subtidally, occurred along 75% of the coastline. Other communities recognised were: the *Enhalus acoroides* dominated shallow embayments, thin-leaved *H. uninervis* dominated intertidal community, and the *H. ovalis* reef-flat communites.

Coles *et al.* (1987) reported 13 seagrass species along the north-eastern coast of Queensland. *Halodule uninervis* and *H. ovalis*, relatively small and shallow-rooted pioneering species, were the most common in the coastal waters. *Enhalus acoroides* and *Thalassia hemprichii*, common in Torres Strait, Bridges *et al.* 1982, were rare, being replaced by dense stands (200–300 shoots m^{-2} averaging 30 g m^{-2}) of *Cymodocea serrulata*. The species composition and zonation suggested that the seasonal reduction in salinity and high turbidity associated with the summer rainfall pattern were responsible for determining seagrass species distribution in coastal waters.

Bridges *et al.* (1982) described the occurrence of 12 species of seagrasses found in Torres Strait. The most common species throughout this region were *Halophila ovalis* and *Halodule* spp., which are relatively small, shallow-rooted species that can invade and grow in newly deposited sand, and *Enhalus acoroides* which is thought to be able to withstand periodic and partial covering by shifting substrates. Environmental stress, associated with length of tidal exposure, swift currents, water turbidity, and substrate suitability in addition to grazing by dugongs and turtles was regarded as the major mechanism underlying the distribution of seagrasses in this area.

The dugong (*Dugong dugon*) is strictly marine and is considered to occupy an important position in the shallow water ecosystems of tropical oceans

(Heinsohn *et al.* 1977). The animals frequent shallow bays and channels that are protected against strong winds and heavy seas, and that contain extensive seagrass meadows. They feed predominantly on seagrasses, though algae are eaten when seagrasses are not available, for example following destruction of seagrass beds by cyclonic storms (Marsh *et al.* 1982). Heinsohn and Birch (1972) found at least six seagrass species in the stomachs of 15 dugongs, while Marsh *et al.* (1982) found *Halophila*, *Halodule*, and *Cymodocea* species to be the most common in the stomachs of 95 specimens. All genera, however, occurred in widely ranging proportions. Rhizomatous material was present in all stomachs. These authors concluded that while dugongs select particular habitats within the available plant communities they also eat a wide range of seagrasses. Wake (1975) concluded that they do not graze selectively, but rather in accordance with species availability.

Grazing by dugongs is considered as being partly responsible for the low biomass densities of seagrass meadows as they dig the plants from the substrate, leaving a distinctive feeding trail (Anderson and Birtles 1978). In Shark Bay (Western Australia) where the tall species *Amphibolis antarctica*, *Cymodocea angustata* and *Posidonia australis* predominate dugongs actually crop the plants (Logan and Cebulski 1970).

The nutritional value of seagrass is not known, though Birch (1975) found a relatively low calorific content, 1,400–3,250 cal. g^{-1}, and nitrogen concentrations, 0.3–1.3% (dry weight), in four *Halophila* species. Phosphorus concentrations ranged from 0.07–0.26%, and like nitrogen, higher levels were found in the rhizomes than in the leaves. In Shark Bay, Anderson (1986) recorded phosphorus concentrations of 0.54–1.73% in *Halodule uninervis* and 1.18–2.65% in *Amphibolus antarctica*.

The green turtle (*Chelonia mydas*) is common in Queensland waters and, like other marine turtles, it utilises a wide range of habitats, including seagrass meadows. Green turtles in Torres Strait are capable of digesting a wide range of soft algae and seagrasses, the actual intake being determined by the food available at their site of residence (Garnett *et al.* 1985). Unlike dugongs the turtles only crop the leaves of seagrasses and do not dig up entire plants. Heinsohn *et al.* (1977) have briefly summarised the information available on turtles in Queensland.

Nearshore, intertidal, and estuarine seagrass communities are important habitats for commercial penaeid prawn species including *Penaeus esculentus*, *P. semisulcatus*, *Metapenaeus endeavouri* and *M. ensis* (Coles *et al.* 1987, Poiner *et al.* 1987). Postlarval and juvenile stages of these prawns are found almost exclusively in seagrass communities. Pointer *et al.* (1987) found the lowest abundance of juvenile prawns in river mouths and the highest abundance in *Enhalus acoroides*-dominated sheltered embayments, though Coles *et al.* (1987) point out that the relationship between juvenile prawns and seagrass is probably more complex than a simple function of seagrass species composition or density.

Mangrove swamps

Vegetation. In this section we review the distribution of mangroves in northern Australia, address environmental factors known to affect their distribution and the zonation of species, and review information about their productivity. Details of associated fauna are presented in the next section. The term mangrove is used loosely, especially as it is difficult to define precisely what constitutes a mangrove or the mangal. In a following discussion citations are given wherever there is a reference to a species being a mangrove or resident in a mangrove swamp.

About 6,000 km of the mainland and another 1,000 km of island coast lines are fringed by mangrove swamps (Galloway 1982). They range from a narrow coastal fringe to extensive forests and extend up to 40 km inland along rivers. The most extensive communities occur along the northeast coast of Queensland, along the Arnhem Land coast, and around Melville Island (Stanton 1975, Bunt *et al.* 1982a, Galloway 1982). Mangroves cover 4,540 km^2 in Queensland, 2,520 km^2 in Western Australia, and 4,120 km^2 in the Northern Territory (Galloway 1982).

Beadle (1981) mapped 27 species of mangroves and Wells (1983) presents distribution maps of 33 species. The number of species decreases from east to west and becomes progressively fewer from north to south. Bunt and Williams (1980) have listed over 30 species and at least 30 different vegetation associations (Table 8). They also demonstrated that variation within zones was at least as great as that between zones and that, with appropriate freshwater influences landward, species can penetrate to the waters' edge. In a further study, Bunt *et al.* (1982b) listed 45 mangrove plant species on the criteria of including any species that is frequent in areas subject to tidal inundation and that grows in association with species indisputably recognised as mangroves. They have included species of *Barringtonia* and *Diospyros* as mangroves.

All mangroves are subject to flooding by seawater and in many cases by freshwater from rain and stormwater run-off. The most extensive areas are associated with freshwater influences (Macnae 1966, Stanton 1975, Galloway 1982, Bunt 1984). High-energy wave action prevents the deposition of silt that is largely responsible for the development of tall and floristically rich communities, such as those in north eastern Queensland (Macnae 1966, Stanton 1975, Saenger *et al.* 1977). Shelter from waves also prevents erosion of silt previously deposited. In well sheltered conditions, such as those where

Table 8. Generalised mangrove zonation scheme (Beadle 1981). Comments about the species are given in parentheses.

Zone	Species-groups
Outer zone-flooded at all tides	1. *Sonneratia caseolaris* (deep soft mud) 2. *Avicennia marina* or *A. eucalyptifolia* (firm substrate) 3. *Rhizophora stylosa*, *Acanthus ilicifolius* (coral reefs and understory), *Aegialitis annulata* (understory)
Middle zone-flooded by medium high and spring tides	4. *Bruguiera gymnorhiza* (dominant), *B. parviflora* (locally dominant in waterlogged areas), *Xylocarpus granatum* (occasional to rare), *Aegiceras corniculatum* (subsidiary to locally common), *Avicennia marina* (subsidiary to locally common), *Acanthus ilicifolius* (understory), *Aegialitis annulata* (understory)
Flooded by spring tides	5. *Ceriops tagal* (dominant), *Bruguiera exaristata* (sometimes subdominant), *Avicennia marina* (occasional to rare or absent), *Lumnitzera spp.* (higher ground, understory)
Inner zone-flooded	6. This zone is variable in composition and is sometimes absent. The most common species are: *Camptostemon schultzii*, *Cynometra ramiflora*, *Excoecaria agallocha*, *Heritiera littoratis*, *Osbornia octodonta*, *Scyphiphora hydrophyllacea*, *Xylocarpus australasicus*

a wave-built barrier separates areas from the open sea, mangrove forests can extend out into open water (Galloway 1982).

Adaptations to inundation and to waterlogged soil are shown by the plant root systems; particularly in the development of aerial roots. Examples are pneumatophores in *Sonneratia* and *Avicennia*, stilt roots in *Rhizophora*, or the mass of roots above the mud surface that is common in *Aegialitis*, *Aegiceras*, and *Excoecaria* (Beadle 1981). The aerial roots (pneumatophores) are composed of aerenchymous tissue that allows the passage of oxygen into the root system which is generally located in an anaerobic environment.

Beadle (1981) considers mangroves to be obligate halophytes whereas Galloway (1982) regards them as facultative halophytes favored by salinity excluding competing species. They exhibit different degrees of salt-tolerance with *Avicennia marina* having a tolerance range from the outer seaward margins to the inner landward fringe (Macnae 1968). In hypersaline patches, stunted or shrubby types can occur (Bunt 1984).

Zonation in mangrove communities (Fig. 4) is regarded as being primarily determined by tidal levels and inundation periods (Chapman 1977, Bunt *et al.* 1982a), though not all communities are zoned (Buckley 1982). Factors like salinity, exposure, and substrate can be important influences (Chapman 1977, Bunt *et al.* 1982a).

Relatively simple zonation patterns have been proposed by a number of

Figure 4. Mangrove zonation along the coastline and a tidal creek in the Alligator River Region in the Northern Territory. A sparsely vegetated salt flat is on the landward side of the mangroves.

authors (e.g. Macnae 1966, Saenger *et al.* 1977, Beadle 1981). The generalized scheme presented by the latter is shown in Table 8.

Three zones are recognised – outer or pioneer, middle, and landward. The outer or pioneer zone is commonly monospecific with *Sonneratia caseolaris*, 8–10 m high, growing on soft, deep mud, tolerating the salinity range from seawater to freshwater. On firmer substrates *Avicennia marina* var. *resinifera* in forests to 10 m high, flooded to depths of 2 m at high tide, is found near coral reefs and generally with no associated species except *Aegiceras corniculatum* where there are strong freshwater influences. The inner limit of the outer zone is dominated by *Rhizophora* spp. *Rhizophora stylosa* is the most common and forms forests 6–12 m high. It can form the outer fringe of the mangrove swamps when the two outermost zones are absent due to unsuitable changes in slope. At the inner fringe the shrubs *Aegialitis annulata* and *Acanthus ilicifolius* can form a discontinuous understory. The middle zone occurs above the level of medium high tides and is dominated by pure stands of *Bruguiera gymnorhiza* in forests to 30 m high. *Xylocarpus granatum* may occur as an associate species. The upper limit of this zone is dominated by *Ceriops tagal* in well drained soils that are only flooded by spring tides. *C. tagal* may be co-dominant with *Avicennia marina* and *Lumnitzera racemosa*.

The landward zone is generally well developed with a variety of species in forests to 12 m high. Under open canopies an understorey of saline mudflat or dune species may develop. *Hibiscus tiliaceus*, *Pemphis acidula*, and *Thespesia populnea*, sometimes regarded as mangroves, may occur. *Nypa fruticans*, a palm, occurs along tidal reaches of some rivers.

Bunt and Williams (1980) demonstrated that north Queensland tidal forests did not consist of individual species confined to particular zones. Rather, individual species each consistently occupy their own section of the overall tidal range and variation within and between zones can be very large. Most species can be found in pure stands, even if restricted in area, but there is a complex of associations determined by various combinations of eleven main indicator species (Table 9).

Faced with the complexity of interactions, Bunt *et al.* (1928b) used numerical methods of classification and ordination to simplify their results. As a result, they were able to group a number of species and sites. Furthermore, they showed that the mangrove flora was richer than previously indicated, that it had a strong floristic affinity to the mangroves of Papua New Guinea, and species distribution was strongly influenced by the extent of freshwater influence, either from rainfall or from rivers. The resultant distribution and zonation patterns suggest that mangroves are "opportunistic colonisers" of available habitats, a characteristic already suggested by Thom (1975) and Stoddart (1980).

Further details on the mangrove communities can be found in Dowling and McDonald (1982) for Queensland, Kenneally (1982) for Western Australia, Woodroffe *et al.* (1985) for the Northern Territory, and Wells (1982) for the northern coasts. The marine environments, including the mangroves, of the northern part of Western Australia have been investigated (Thom *et al.* 1975, Semeniuk 1980 and 1981, Semeniuk *et al.* 1982) with an emphasis on geomorphic processes and vegetational dynamics.

Detailed information on litter fall from mangroves in north-eastern Queensland has been presented by Duke *et al.* (1981). For all species, leaves accounted for the greatest proportion of litter, varying from 40% of the total for *Bruguiera parviflora* to 75% for *Avicennia* sp. The mean yields of total litter fall showed little interspecific variation and were comparable with values reported elsewhere. The maximum yields were at sites occupied by *Rhizophora apiculata*, with overall variation from 1.04–6.36 g dry weight $m^{-2} d^{-1}$, or 380–1960 $g\,m^{-2}\,y^{-1}$. In Darwin harbour, Woodroffe (1985) re-

Table 9. Associations of mangroves determined by combination of 11 indicator species (Bunt and Williams 1980).

Group No.	Defining species
1.	*Bruguiera gymnorhiza*, *Acrostichum* sp., *Rhizophora apiculata*
2.	*B. gymnorhiza*, *Acrostichum* sp., *Ceriops decandra*
3.	*B. gymnorhiza*, *Acrostichum* sp.
4.	*B. gymnorhiza*, *Xylocarpus granatum*, *Lumnitzera littorea*
5.	*B. gymnorhiza*, *X. granatum*, *Ceriops tagal*
6.	*B. gymnorhiza*, *X. granatum*, *C. decandra*
7.	*B. gymnorhiza*, *X. granatum*, *Rhizophora apiculata*
8.	*B. gymnorhiza*, *X. granatum*
9.	*B. gymnorhiza*, *R. apiculata*, *Rhizophora stylosa*
10.	*B. gymnorhiza*, *R. apiculata*
11.	*B. gymnorhiza*, *R. stylosa*
12.	*B. gymnorhiza*, *C. tagal*
13.	*B. gymnorhiza*
14.	*R. stylosa*, *C. tagal*, *Lumnitzera littorea*
15.	*C. tagal*, *L. littorrea*, *Excoecaria agallocha*
16.	*C. tagal*, *L. littorea*
17.	*C. tagal*, *Rhizophora lamarckii*
18.	*C. tagal*, *R. stylosa*
19.	*C. tagal*, *Acrostichum* sp.
20.	*C. tagal*, *Avicennia marina*
21.	*C. tagal*
22.	*R. stylosa*, *R. apiculata*
23.	*R. stylosa*
24.	*R. lamarckii*
25.	*Excoecaria agallocha*, *A. marina*
26.	*E. agallocha*
27.	*R. apiculata*
28.	*X. granatum*
29.	(residual group – no defining species)
30.	*R. apiculata*, *C. tagal*

corded litter fall values of around 10 g m^{-2} d^{-1} in the creek bank *Avicennia marina* community. Over nine months of study, more than 1,000 g m^{-2} litter fall was recorded; a reasonably high value and comparable to productive mangrove systems elsewhere.

The productivity of mangroves, as expressed by litter fall rates, is reasonably well known, but the fate of this material is less well known. Whether or not it is exported or recycled in-situ depends, to a large extent, on the tidal regime. In northern Queensland it is unusual to see litter accumulate amongst the mangroves (Boto and Bunt 1981).

In a mangrove estuary in northern Queensland, Boto and Bunt (1981) estimated that litter export was about 3,560 t dry weight or 19.5 kg ha^{-1} d^{-1}. This represented 9.9 kg C ha^{-1} d^{-1} and 0.06 kg N ha^{-1} d^{-1}. The low level of nitrogen loss was not unexpected in view of low nitrogen levels in senescing leaves that contribute most to the total litter fall. Nitrogen values in all tissues varied from 0.1–0.5% dry weight and averaged 0.3%. Annually, 1,800 t C and 11 t N were exported in the litter. The amount of carbon exported as particulate matter was generally small, 1.1 t particulate organic carbon per tidal cycle, representing about 300 t C y^{-1}; there was, however, considerable error attached to this value. The total export amounted to 2,100 t y^{-1} or 11.5 kg C ha^{-1} d^{-1}. The particulate nitrogen export was estimated as 8 t y^{-1} (also with considerable error). The export of dissolved organic matter was ignored, but as creek water can contain up to 30 mg l^{-1} dissolved carbon it may be a significant amount. Boto and Bunt (1982) estimated that the total nitrogen, phosphorus and carbon export of 0.1, 0.007 and 11.5 kg ha^{-1} d^{-1}, respectively represented 14, 12, and 46% of the production of these forests.

Fauna. Information on vertebrate and invertebrate fauna, in the mangrove zone is summarised by Saenger *et al.* (1977), Milward (1982), and Hutchings and Recher (1982, 1983). The following description draws heavily on these reports.

Over 200 species of birds have been recorded from mangrove habitats with 14 virtually restricted to the mangal and 12 using it as a primary habitat (Schodde *et al.* 1982). On Cape York Peninsula, 10 species are considered mangrove "specialists" and seven are endemic to mangrove habitats, though none have major morphological adaptations to the environment. The composition, structure, and origin of the avifauna of mangroves is reviewed by Schodde *et al.* (1982) and a species list is given by Saenger *et al.* (1977).

There are no mammals endemic to mangrove forests, though many obtain part of their food there. They include a number of rodents (*Rattus colletti*, *Mus musculus*, *Melomys* spp., *Mesembriomys* spp. and *Conilurus* spp.), bandicoots (*Perameles* and *Isoodon* spp.), flying foxes (*Pteropis policephalus*, *P. alecto*, and sometimes *P. conspicillatus*). Only one mammal, the rat *Xeromys myoides*, utilises these forests as a primary habitat (Magnusson *et al.* 1976). The introduced Asian water buffalo (*Bubalus bubalus*), pig (*Sus scrofa*), and cattle (*Bos taurus*) are also found in mangrove areas.

Reptiles are common in mangrove swamps, though many only use them as a secondary habitat. The pythons (*Liasis fuscus* and *L. divaceus*) are attracted by colonies of flying foxes, while the mangrove monitor (*Varanus indicus*), the file snake (*Acrochordus* sp.), the bockadam (*Cerbeurs rhynchops*), the white-bellied mangrove snake (*Fordonia leucobalia*), and the mangrove snake (*Myron richardsonii*) occur regularly. The best known reptile, however, is the estuarine or saltwater crocodile, *Crocodylus porosus*. It is an opportunistic feeder with the younger ones eating mainly invertebrates, whereas the older, larger adults eat more vertebrates. Since commercial hunting ceased in 1971 there is evidence that numbers, sizes, and total biomass have increased (Webb *et al.* 1983). During the Wet season they may venture into the freshwater swamps and may even remain there (Jenkins and Forbes 1985). Extensive surveys have been conducted across northern Australia to determine population sizes (see summary by Messel and Vorlicek 1986). The extent of population recovery to 30,000–40,000 individuals was regarded as sufficient, with adequate controls, to justify the transfer of this species from Appendix I to Appendix II of C.I.T.E.S to enable crocodile farming and export of skins (Webb *et al.* 1984). Some reservations, however, have been expressed by Messel and Vorlicek (1986) about the long-term future of this species outside reserves and national parks.

Fishes are a conspicuous feature of mangrove swamps, though few species are restricted to this habitat. Large numbers invade the mangrove forests at high tide and retreat to deeper water at low tide. The major group belongs to the gobiid sub-family Oxcidercinae and includes mud skippers (*Periophthalmus* and *Periophthalmodon* spp.), and members of the *Boleophthalmus* and *Scartelaos* genera. The mud skippers have bulging eyes, can crawl on the mud, and can survive for long periods out of water. A diverse fish fauna occurs in creeks and lagoons in the mangroves, and in the adjacent estuaries (e.g. the commercially important silver barramundi, *Lates calcarifer*). Estuarine species that frequent mangroves are toados (*Torquigener hamiltonii*), mullet (*Myxus elongatus*), and the fortesques (*Centrapogon australis*). A list of fish species in mangroves is given by Saenger *et al.* (1977).

The invertebrate fauna is richer than the vertebrate fauna. Saenger *et al.* (1977) and Hutchings and Recher (1982) provide extensive species lists and descriptions of the invertebrates. Insects, particularly biting midges and mosquitoes, are common. Most studies of these are, however, either purely taxonomic or related to the transmission of disease and very little general ecological information has been collected. Spiders are similarly rich in species. The dominant groups are the orb web weavers (*Teragnatha* and *Eriophora* spp.), the wolf spiders (*Geolycosa* spp.), and the allied hunting spiders of the Pisauridae.

Within the intertidal zone the fauna is dominated by polychaetes, crustaceans, and molluscs. Encrusting animals occur at the edge of the mangrove forest, are inundated by each tide, provide shelter for a rich and mobile fauna of polychaetes, crustaceans, and gastropods, and are dominated by the oyster (*Saccostrea commercialis*) and barnacles.

Molluscs, particularly gastropods, live on the surface of the forest floor and are zoned according to the pattern of tidal inundation. There is, however, very little information on the abundances of these animals. A further group of animals, both sedentary and mobile species, live in the sediment of the forest floor. A diverse fauna of polychaetes occurs in the less consolidated

Table 10. Plant species recorded in salt-marsh communities of northern Australia (adapted from Saenger *et al.* 1977). Species with an * are found only on the north-west coast of Western Australia.

Family	Common species
Aizoaceae	*Sesuvium portulacastrum*, *Trianthema turgidifolia*
Batidaceae	*Batis argillicola*
Caryophyllaceae	*Spergularia rubra*
Chenopodiaceae	*Halosarcia arbusculum**, *H. halocnemoides var. pergranulatum*, *H leiostachyum*, *Scleroleana astrocarpa**, *Enchylaena tomentosa*, *Hemichroa diandra**, *Rhagodia baccata**, *Sarcocornia quinqueflora*, *Salsola kali*, *Suaeda australis*, *Tecticornia australasica*
Convolvulaceae	*Wilsonia backhousei**
Frankeniaceae	*Frankenia pauciflora**
Plumbaginaceae	*Limonium salicorneacea*
Poaceae	*Sporobolus virginicus*, *Xerochloa barbata*

sediments at the seaward margin of the forests. Molluscs and crustaceans, commonly bivalves and crabs, are also well represented in this habitat. The burrows of the mud lobster (*Thalassina anomola*) and the mud crab (*Scylla serrata*) are conspicuous. A further group of invertebrates, dominated by wood-borer teredinid molluscs is also common, but not necessarily restricted to mangrove areas.

Coastal salt-marshes and flats

Salt-marshes are not a predominant feature of coastal tropical regions of Australia (Stanton 1975, Saenger *et al.* 1977). Salt-marshes are usually located in the upper-tidal zone behind a fringe of mangroves and adjacent to salt-flats. In general though, areas containing extensive mangrove swamps do not have well-developed salt-marshes (Saenger *et al.* 1977). The salt-flats are usually devoid of vegetation and are encrusted with salt deposits (Macnae 1966, Bunt 1984). Seasonal vegetation (e.g. *Tecticornia australasica*) can occur in some salt-flats (Saenger *et al.* 1977).

Saenger *et al.* (1977) lists 20 salt-marsh plant species occurring in northern Australia (Table 10). The arid northwest of Western Australia contains eight species not found elsewhere in the north, but some are found in temperate Australia. Specht (1981) lists nine species in tropical Australia and 13 in the subtropical region.

The most extensive salt-marshes and salt-flats occur along the Arnhem Land-Gulf of Carpentaria coast (Love 1981), and parts of the eastern Queensland coast (Stanton 1975). The following general description of the vegetation draws very heavily on that of Specht (1981). *Halosarcia leiostachyum* and *H. halocnemoides* occur on mudflats that are only infrequently inundated by seawater. Associated herbaceous species include *Trianthema turgidiflora*, *Sesuvium portulacastrum*, *Tecticornia australasica*, *Frankenia pauciflora*, and the grasses *Sporobolus virginicus* and *Xerochloa barbata* in less saline areas. Very rare species include *Scleroleana astrocarpa* and *Hemichroa diandra* in Western Australia and *Cressa australie*, *Epaltes australis*, and *Limonium australie* in Queensland. *Sarcocornia quinqueflora* occurs in highly saline areas. Under less saline conditions, *Halosarcia* spp. are not abundant and less salt-tolerant species invade the mudflats (e.g. *Sporobolus virginicus* and *Xerochloa barbata*).

A generalised zonation pattern for northern salt-marsh plants, adapted from Saenger *et al.* (1977), characterises the zonation of species as being related to their tolerance of salt and of waterlogged substrates. A ranking of species from the mangrove zone to the salt-flat, based on decreasing salinity, includes *Halosarcia leiostachyum*, *Tecticornia australasica*, *Batis argillicola*, *Suaeda australis*, and *Sporobolus virginicus*.

In Princess Charlotte Bay, Queensland, Elsol and Saenger (1983) describe the vegetation on supra-tidal and high-tidal flats. The former have cracking, strongly alkaline soils that support a large number of herbaceous species, grasses, and occasional shrubs. The most common grasses include *Rottboellia exaltata*, *Themeda quadrivalvis*, and *Xerochloa barbata* and the introduced *Sorghum laxiflorum*. The lowest limit of this unit corresponds to the upper limit of the high-tidal flats and contains a fringe of chenopods dominated by *Halosarcia* spp. The high-tidal flats have saline-alkaline cracking clays and uniform fine-textured soils and are characteristically devoid of vegetation.

Spencley (1976) described supra-tidal flats as being unvegetated, hypersaline, infrequently submerged by spring or storm tides, and experiencing a dry climate. The impetus for the development of these flats was considered to be a break in the vegetation canopy, probably caused by cyclones, and the establishment of hypersaline conditions. The lack of bare saline flats in the wetter, cyclone-prone areas is attributed to flushing of surface salts by rainfall run-off and tidal inundation. In dryer areas, salt accumulation prevents vegetation from being established.

Spenceley (1976) recognised two main types of salt-flats: relict and contemporary. The relict flats, such as those in the Burdekin delta (Queensland) have been caused by a decrease in local sea level and contemporary flats are considered to be the result of hypersaline conditions. Examples of the latter process in the Townsville region are discussed by Spenceley (1976).

Seasonally inundated floodplain lakes and billabongs

Floodplain lakes, as described by Paijmans *et al.* (1985), can be either seasonally flooded or permanent. To be permanent they must be deep enough to retain water between floods; as most are relatively shallow this is not common, although subsurface flow can sustain some through the Dry Season. The lakes receive their water through channels connecting with the main stream, from overbank flow, from local rainfall, and from underground sources. If flushing does not occur regularly they may become saline by evaporative concentration of solutes.

The seasonally-covered floodplain lakes in the Northern Territory undergo a pronounced wet-dry cycle. They are filled directly by overbank flow or indirectly by backflow from large streams. During the Dry they generally lack surface water, except for a few permanent billabongs (or waterholes) and swamps. The seasonal wet-dry hydrological cycle on the Magela Creek floodplain in the Alligator Rivers Region, east of Darwin has been schematically described by Sanderson *et al.* (1983), and slightly adapted by Finlayson *et al.* (1988a, 1990). The schematic presentation (Fig. 5) was based on four years of observations leading to the identification of five hydrological phases (intermittent storms and initial surface wetting, prolonged rain and creek flow causing wide-scale flooding, cessation of rain and drying out, cessation of all flow and water recession, and the dry phase). These hydrological changes and their timing have very significant effects on the biota of the floodplain system (Finlayson *et al.* 1988a, 1990).

The classification of permanent water bodies in the tropics is somewhat confused. Strictly speaking, billabongs occur in anabranches and waterholes occur in main river channels (Paijmans *et al.* 1985), but these terms are commonly used interchangeably. The billabongs of the Magela Creek system have been classified by Walker *et al.* (1984) as channel (depressions in flow channels), backflow (located on small tributaries and initially filled by water from the main channel), or floodplain billabongs (generally residual features of infilled deep channels on the floodplain). The three categories have different water quality characteristics (Walker and Tyler 1984). Whilst billabongs are discrete wetland entities they are associated with the seasonal floodplains and are reviewed with the floodplain lakes in the following discussions.

The herbaceous, woodland, and forest swamp categories of Paijmans *et al.* (1985) are commonly found on the floodplains of northern Australia and are considered with them rather than as separate entities, especially as swamps usually occur as the terminal phase of seasonal floodplain lakes.

Finlayson *et al.* (1988b) described the ecological characteristics of the floodplains on the lowlands between the Kimberley and Arnhem Land Plateaus. In a further report, Finlayson *et al.* (1990) summarised the extensive biological investigations on the Magela Creek floodplain (Fig. 6). The following description of vegetation and fauna is largely based on those reports. The vegetation section reviews environmental factors and threats that affect the occurrence and distribution of plant species, presents details of the occurrence and productivity of the Magela Creek floodplain species, and reviews problems of alien plant invasions. The fauna section reviews the diversity and abundance of the vertebrates, their ability to overcome seasonal

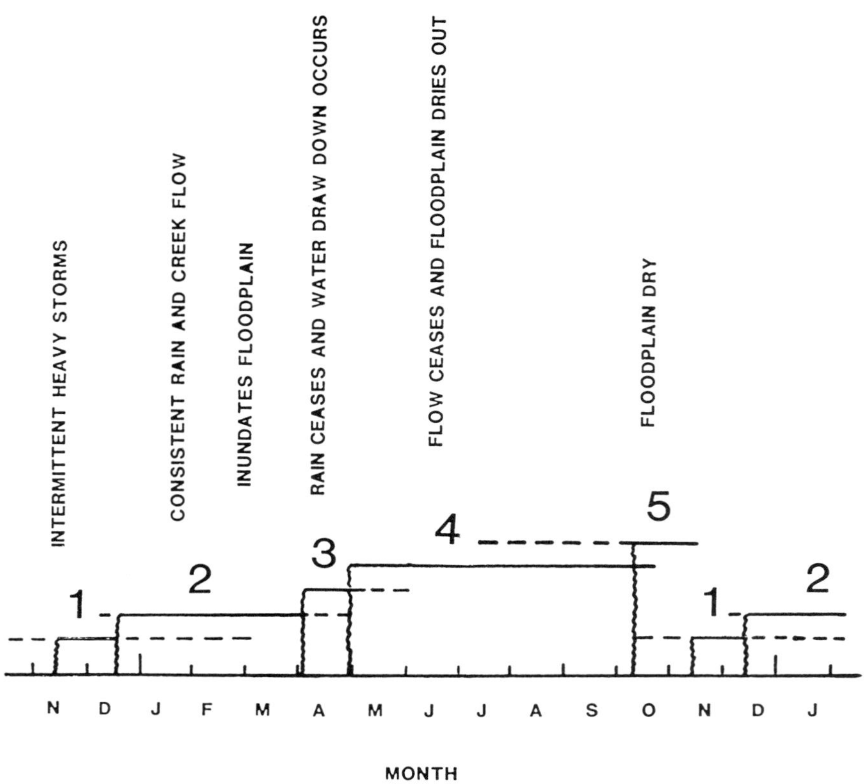

Figure 5. A schematic presentation of the seasonal cycle on the Magela Creek floodplain based on observations from 1975–85 (adapted from Sanderson *et al.* 1983).

changes in food availability, and then discusses the invertebrates and the problems associated with the occurrence of alien vertebrates.

Vegetation. As assessment of the distribution of major species across the floodplains enables several broad generalisations to be made. *Oryza meridionalis* grasslands, *Melaleuca* spp. woodlands, and open forests are extensive and occur on most, if not all floodplains. *Oryza* is an annual species that grows in seasonally inundated areas and survives the harsh dry season drought as seed in the soil. There are about five *Melaleuca* species on the floodplains and some grow in areas that flood to a depth of 2–3 m during the Wet season. Other common genera include sedges (*Eleocharis* and *Fimbristylis*), water lilies (*Nymphaea* spp.), the lotus lily (*Nelumbo nucifera*),

Figure 6. A seasonally inundated floodplain in the Alligator Rivers Region, Northern Territory during the Wet season. The vegetation communities are mainly mixed grasses and sedges and *Melaleuca* forest.

and aquatic grasses (*Hymenachne acutigluma*, *Pseudoraphis spinescens*, and *Hygrochloa aquatica*).

The large emergent species *Phragmites karka* is common on floodplains that have not been subjected to intensive grazing and trampling pressure by the feral buffalo, *Bubalus bubalus*. It has been estimated that 3,000 to 4,000 ha of reed "swamp" in Kakadu National Park (in the Alligator Rivers Region) have been destroyed by these animals (Hill and Webb 1982). *Phragmites karka* is now restricted to a fringe of plants along riverbanks in the Park. Initial evidence suggests that removal of buffaloes from the floodplains can be followed by rapid colonisation by *Urochloa mutica*, a grass introduced from north Africa, and subsequent suppression of native species. In some places, however, the native grass *H. acutigluma* has re-established. *Typha domingensis* is not found on all floodplains, but it is not known whether its distribution is influenced by the extent of seasonal water fluctuations or if it has suffered from the presence of buffaloes. The floating *Pistia stratiotes* has a disjunct distribution pattern, being common to the west and south of Darwin, less common in the Alligator Rivers Region, and abundant in the permanent Arafura Swamps further to the east.

Of 125 plant species from the Magela Creek system considered by Finlayson *et al.* (1990), six are restricted to permanent billabongs and swamps, 50 occur in seasonally inundated areas (43 of which are water plants as defined by Cook *et al.* 1974), 27 are common to both permanent waterbodies and seasonally inundated areas (24 are water plants), and 42 are terrestrial herb or sedge species that occur on areas exposed during the Dry season. These species can be divided into 4 groups on the basis of life histories that enable them to survive on the floodplain:

1. Phanerophytic perennials (35 species, includes the emergent grasses *Pseudoraphis spinescens* and *Hymenachne acutigluma*, and *Melaleuca* spp.);
2. Geophytic perennials (10 species, includes *Eleocharis* and *Nymphaea* species);
3. True annuals (57 species, 15 are water plants and 42 terrestrial herbs, grasses or sedges);
4. Facultative annuals (24 species, includes the submerged species that occur in seasonally inundated areas and rely on seeds to survive).

The introduced plants *Salvinia molesta* (from South America) and *Mimosa pigra* (from South and Central America) also occur on the floodplains, and undoubtedly pose a major threat to the present structure and character of these areas (Cowie *et al.* 1988). *Salvinia molesta* has been found at a number of places since it was first recorded in 1976. The most recent record is from the Magela Creek floodplain in September 1983 (Finlayson 1984a). A combination of manual and chemical control methods have been successfully used to control other infestations in the Northern Territory but it was decided that the only practicable option for control in this case was to use the biological-control agent *Cyrtobagous salviniae*. While the control agent has established on plants in the Magela system it has not yet been as successful as it has been elsewhere in Australia (e.g. Room *et al.* 1984).

A greater threat to the floodplains, however, is *Mimosa pigra*, a vigorous, prickly shrub that occurs in unispecific stands, shading out native vegetation, and reducing the number of potential breeding sites for animals such as the magpie goose (*Anseranas semipalmata*) and the saltwater crocodile (*Crocodylus porosus*). It is mainly concentrated along the Adelaide River where it covers an estimated 8,000 ha (Considine 1985). It has a large output (12,000 m^{-1} y^{-1}) of small hard-coated seeds that are dispersed by water within and between river systems in mud adhering to vehicles and animals (Lonsdale *et al.* 1988). Total wetland coverage is in excess of 30,000 ha with infestations occurring on all of the major floodplain systems including those contained in Kakadu National Park, a World Heritage area.

Manual and herbicide control measures have been implemented, but with only limited success. Research into biological control is being undertaken with, so far, two species of seed-eating beetle, *Acanthoscelides* spp. and a foliage feeder, *Chlamisus* sp. being released (M. Lonsdale personal communication).

A detailed analysis of plant distribution is available for the Magela Creek system. Williams (1979) sought to establish a relationship between vegetation and patterns of water flow, and distinguished and described six vegetation types (Table 11). However, as the survey times did not correspond with the peak Wet season vegetation he did not distinguish the boundaries of communities dominated by annual species such as *Oryza meridionalis*. Morley (1981) recognised 36 communities on one part of the floodplain, but the classification was not reproducible in subsequent Wet seasons (Sanderson *et al.* 1983). In an attempt to present a broad and generalised vegetation classification, Finlayson *et al.* (1988a, 1990) used peak Wet season data from several years to describe and map 10 communities (Table 11) on the floodplain. The seasonal wet-dry cycle and associated species changes were described for the main communities. Throughout these descriptions it is recognised that species dominance can change markedly in some localities from one year to the next, and there is a succession in dominance after flooding and during the drying out phase.

A characteristic vegetation association that is under threat from the grazing habitats of buffaloes is the floating grass mats (Hill and Webb 1982). The mats have pronounced vegetation zones and are usually formed on a base of *Pistia stratiotes*. They include *Leersia hexandra*, *Phragmites karka*, *Typha domingensis*, *Hymenochaeta grossa*, and *Cyperus platystylis*.

Changes in standing crop of dominant grass species on the Magela Creek floodplain appear related to changes in water levels (Finlayson 1988). *Pseudoraphis spinescens* grows rapidly as water levels increase during the Wet season, reaches the water surface, flowers, sets seed, and senesces. The peak standing crop at the end of the Wet season was 1.7 kg dry weight m^{-2}. As the water level decreases it reverts to a turf-like form that persists across the floodplain surface during the Dry season. Minimum standing crop was 212 g m^{-2}. Standing crop of *Hymenachne acutigluma*, growing as a perennial, decreased markedly from 1.29 to 0.23 kg m^{-2} following the first heavy rains of the Wet when it changed from a creeping semi-erect form to a fully erect form. Following stabilisation and decreases in water levels, there was an increase in standing crop to 1.41 kg m^{-2} with a decrease later in the Dry to

Table 11. Vegetation communities and major plant species recognised by Finlayson *et al.* (1986b) and Williams (1979) on the Magela Creek floodplain. Dominant species and genera are given in parentheses.

Finlayson *et al.*	Williams
1. *Melaleuca* open forest and woodland (*M. viridiflora, M. cajaputi, M. leucodendra*	1. Mixed herbfield (*Eleocharis* sp., *Fimbristylis* sp., *Oryza meridionalis*
2. *Melaleuca* open woodland (*M. cajaputi, M. leucodendra*)	2. Grassland (*Pseudoraphis spinescens*)
3. Mixed swamp (*Nelumbo nucifera, Hymenachne acutigluma Hymenochaeta grossa, Ludwigia adscendens*)	3. Undulating annual swamp and grassland (*P. spinescens, Panicum* sp., *L. adscendens, Polygonum attenuatum, Nymphoides indica*
4. *Oryza* grassland (*Oryza meridionalis, Digitaria* sp., *Melaleuca* sp., *Hygrochloa aquatica*)	4. Forest (*Melaleuca nervosa, M., viridiflora M. cajaputi, M. leucodendra*)
5. *Hymenachne* grassland (*H. acutigluma*)	5. Annual swamp (*Eleocharis* sp., *N. indica*)
6. *Pseudoraphis* grassland (*P. spinescens, Nymphaea* spp.)	6. Perennial swamp (*Chara* sp., *Nelumbo nucifera, H. acutigluma*)
7. *Hymenachne – Eleocharis* swamp (*H. acutigluma. Eleocharis* spp.)	
8. Mixed grass/sedgeland (*H. acutigluma, Eleocharis* spp, *P. spinescens, O. meridionalis*)	
9. *Eleocharis* sedgeland (*Eleocharis* spp., *Glinus oppositifolius*, (*Coldenia procumbens*)	
10. Open water communiity (*Hydrilla verticillata, Najas tenuifolia, Ceratophyllum demersum, Utricularia* spp.)	

0.55 kg m^{-2}. This species also occurs as an annual on sites that dry out completely. *Oryza meridionalis* is an annual that germinates following the first storms, grows rapidly and reaches a peak standing crop, 0.51 kg m^{-2}, at the end of the Wet.

Litter fall from the extensive *Melaleuca* woodlands and forests (Fig. 7) contributes a large amount of material to the detrital pool on the Magela floodplain (Finlayson 1988). Litter fall is dominated by leaves, representing 70% of the total annual fall of 8.1 t ha^{-1} at one site. Peaks of litter fall occur during the Wet season and mid-Dry season and are probably under the influence of high winds. The actual amount of litter that accumulates on the ground over an annual cycle is not very large with physical removal occurring during the early stages of the Wet season. The amount of litter that falls varies spatially with annual values ranging from 8 to >15 t ha^{-1}.

Figure 7. Billabong surrounded by *Melaleuca* forest during the Dry season. The sand stream (to the right) connects to the main creek and can channel water into and out of the billabong, depending on the flow conditions during the Wet season.

Fauna. The seasonal floodplains contain high standing biomasses of vertebrates including freshwater and saltwater crocodiles, other reptiles (e.g. file snakes and turtles), freshwater fishes, and many species of waterbirds. A summary of available data (Finlayson *et al.* 1988b) conveys an impression of high standing biomass, but unfortunately the data are expressed in numerical units that are not easily relatable (Table 12). The large, long-lived animals exploit these wetlands by being highly mobile and/or having mechanisms that allow them to withstand the seasonal changes in food and nutrient availability. Large animal body size is usually related to long generation time and the ability to tolerate short term environmental disturbance (Pianka 1983). Animals that do not have the mobility to successfully exploit sequences of highly productive patches have to withstand low food availabilities, at least on a seasonal basis. They overcome this problem by physiological adaptations to periodic or constant low metabolism and slow growth, or by food habits that reduce their dependence on food from the aquatic environment.

The freshwater crocodile, *Crocodylus johnstoni*, displays reduced feeding activity during the Dry season, apparently independently of temperature changes (Webb *et al.* 1982). Somewhat surprisingly, they obtain about 40% of their food from the terrestrial environment (Webb *et al.* 1982). Similarly, freshwater turtles depend heavily on vegetable foods of terrestrial origin.

Table 12. Estimates of the abundances of a variety of wetlands animals in the wetlands of the Top End of the Northern Territory (from Finlayson *et al.* 1988a). Estimates of crocodile abundances are from Webb *et al.* (1983) and Bayliss *et al.* (1986). Turtle abundances are the maximum values recorded from a variety of habitats (Legler personal communication), *Acrochordus* data are from Shine (1986), the waterbird data from Bayliss (personal communication) and the fish data are mid-wet season standard net samples taken by Bishop *et al.* (1991). The buffalo data are maximum abundance reported by Graham *et al.* (1982) for the South Alligator Rivers floodplain.

Species	Maximum size	Abundance
1. Crocodiles		
Crocodylus johnstoni	3 m long	mean = 1.57/pool (max. = 61)
C. porosus	7 m long	Tidal Adelaide River:
		a. downstream: mean = 135 per 10 km
		b. upstream: mean = 54 per 10 km
		c. side creeks: mean = 44 per 20 km
2. Turtles		
Carettochelys insculpta	0.7 m	17 per 100 m of stream
Chelodina rugosa	0.4 m (shell)	9 per 100 m of stream
Elseya dentata	0.35 m (shell)	38 per 100 m of stream
E. latisternum	0.2 m (shell)	8 per 100 m of stream
Emydura australis	0.3 m (shell)	10 per 100 m of stream
3. File snake		
Acrochordus arafurae	2.5 m long	100 per ha^{-1}, >50 kg ha^{-1}
4. Waterfowl		
Anseranas semipalmata	2.5 kg	279 km^2 (dry season)
Dendrocygna arcuata		88 km^2 (dry season)
D. eytoni		ibid
5. Fish (all species)		
Lowland muddy lagoon	mean = 4392 g	
Corridor lagoon	mean = 1179 g	
Floodplain lagoon	mean = 9912 g	
6. Buffalo (South Alligator River)		15 km^2

Emydura australis and *Elseya latisternum* are omnivorous, whereas *Elseya dentata* and *Carettochelys insculpta* subsist in the Dry season on a diet of leaves, flowers, and fruits. Turtles such as *Chelodina rugosa* that live in seasonally wet-dry habitats aestivate over the Dry season. Surface feeding fish such as *Melanotaenia splendida inornata* and *M. nigrans* depend on food entering the aquatic environment from terrestrial sources, while *Hephaestus fuliginosus* and *Syncomistes butleri* scavenge material of terrestrial origin (Finlayson *et al.* 1988b). Larger species such as silver barrumundi (*Lates calcarifer*) are highly mobile and move between the marine estuarine areas and the freshwater floodplains during the Wet season.

Many birds species (e.g. *Anseranas semipalmata*, *Tadorna radjah*, *Dendrocygna arcuata*, and *D. eytoni*) move between the swamps and floodplains, presumably as a result of changes in the availability of food and water. Changes in abundance of *A. semipalmata* have been well documented (Frith and Davies 1961). Bayliss (unpublished) has surveyed the seasonal movement of waterbirds between the floodplain systems whilst Morton *et al.* (1984) have studied movements between floodplains of the Alligator Rivers area.

Details of the fauna of the Magela Creek system (floodplains, billabongs, and creeks) are summarised by Finlayson *et al.* (1990). Marchant (1982) carried out the first scientific collection of invertebrates and identified 90 taxa from five billabongs. High numbers of taxa and individuals were present at the end of the Wet season when food and shelter was plentiful among the well-developed water plant communities. During the Dry season the invertebrates adopt different survival strategies (e.g., aestivating in the mud, or having resistant eggs or pupae). The microcrustacea of the billabongs are regarded as a rich assemblage of cladoceran and copepod species, many of them occurring among the aquatic vegetation (Julli 1986). Tait *et al.* (1984) presented a checklist of 227 rotifers, 14 copepods, 35 cladocerans, and five ostracods.

Shine (1986) investigated the food habits, habitats, and general biology of file snakes (*Acrochordus arafurae*), sand goannas (*Varanus panoptes*, and *V. gouldii*), and water goannas (*V. mertensi* and *V. mitchelli*). Most attention was directed to file snakes that are restricted to billabongs during the Dry season where they congregate around the fringing trees and grass mats (Shine and Lambeck 1985). In the Wet season they move out of billabongs to shallow inundated grasslands. Females were more commonly caught than males in these surveys with more than 1,000 in each of the billabongs. Both freshwater (*C. johnstoni*) and saltwater crocodiles (*C. porosus*) occur in the Magela system, but few nests have been found (Messel *et al.* 1979, Grigg and Taylor 1980, Jenkins and Forbes 1985).

Tyler *et al.* (1983) identified nine genera and 24 species of frogs in the Magela Creek system. In general, they are totally inactive during the Dry season and most remain beneath the ground or seek shelter beneath logs and leaf litter to avoid dehydration. The greatest densities occur on poorly drained sandy soils and the least on well drained gravelly soils (Tyler and Cook 1987). Substantial rainfall is required to soften the soil before they surface to disperse and spawn.

Information on fish collected by Bishop *et al.* (1986) points to a relationship between habitat and age/size of species. Smaller juveniles were most abundant in the muddy billabongs on the lowlands and floodplain, larger juveniles and smaller adults in areas that connect the floodplain and lowland billabongs during the Wet, and larger adults in floodplain billabongs.

Of the larger species, *L. calcarifer* exhibits great mobility, breeding in sea water at the mouth of the river system and then either staying in the salt water or swimming upstream to fresh water. Juveniles spend their early weeks in brackish coastal swamps with some migrating upstream to freshwater areas on floodplains.

The diversity and abundance of fish populations of the Alligator Rivers Region have been described by Bishop *et al.* (1986) and Bishop and Forbes (1989). The latter contains a comparison with other northern areas. Bishop and Forbes (1991) recognised three groups of species according to their association with fresh water:

1. Marine or estuarine vagrants that move into fresh water for short times and distances.
2. Species that migrate to estuarine regions for considerable times and distance.
3. Species that live in fresh water for their entire lives.

Fish migration occurs during the Wet season. Recolonisation of the lowland sandy creeks and backflow billabongs in the early-Wet results in the most obvious seasonal changes in fish community structure. Movement occurs in both an upstream and downstream direction from Dry season refuge areas on the floodplains and upper escarpment area.

Based on diversity-drainage area relationships, the diversity of freshwater fish in tropical floodplains is higher than in temperate areas of Australia and in floodplains elsewhere. About 50 species occur in the Alligator Rivers Region, though only a few (e.g. the blackanal-finned grunter (*Pingalla* nov. sp.) and the Arnhem Land blue-eye (*Pseudomugil tennellus*) are endemic.

The floodplain systems have been subject to disturbance by feral animals, particularly water buffalo which were released during attempts to establish settlements in the Northern Territory from 1827 to 1949. The number of animals is now about 280,000 (Graham *et al.* 1982) and substantial circumstantial evidence has been collected on the threat they pose to floodplains and billabongs (Fogarty 1982). Taylor and Friend (1984), and Friend and Taylor (1984) have related ground surface features attributable to buffaloes to vegetation structure and plant life, and to the abundance of small animal species.

Fogarty (1982) has assessed the extent of buffalo impact on floodplains of the Northern Territory. He reported open plains heavily grazed and pugged (Fig. 8), networks of tracks and swim channels, reduction in floating grass mats and vegetation cover, and saltwater intrusion. Specific details of each floodplain are presented and highlight the loss of natural vegetation and invasion of alien plants, increased water salinity and turbidity, and destruction of crocodile breeding areas. Quantitative evidence of the effect of buffaloes on the floodplain should come from the recently conducted experiment on the South Alligator floodplain where buffaloes were excluded from one area (Taylor, unpublished). In National Parks or reserves it has been ac-

Figure 8. A dry floodplain with a deep buffalo wallow in the foreground. Such ground features are still found on these plains despite attempts to reduce the numbers of animals over the past decade.

cepted that buffaloes are incompatible with wetland conservation objectives and steps are being taken to remove them, usually by shooting from helicopters.

In addition to feral buffalo, feral pigs and horses are also present on floodplains. Pigs are relatively common and caused obvious but not quantified damage around billabongs and amongst the *Melaleuca* forests. No effective means of control for this animal has been devised.

The introduced cane toad, *Bufo marinus*, is spreading from the Queensland eastern coast through the river systems of the Gulf of Carpentaria and will, with or without assistance from man, reach the northern floodplains (Freeland and Martin 1985). Its likely effects on the native fauna and the wetland environments are discussed by Freeland (1985).

Freshwater swamps

Freshwater swamps, as described by Paijmans *et al.* (1985), are found along the east coast of Queensland and include herbaceous, woodland, and forest categories. As the differentiation between floodplain lakes and swamps is not precise, the floodplain lakes along the Queensland coast are included with the swamps. The seasonal floodplain lakes of the Northern Territory are not, however, included (see previous section). The extensive Arafura

Swamps on the Glyde and Goyder Rivers in Arnhem Land are also not reviewed because of the paucity of information. Information in this section concentrates on the occurrence and diversity of aquatic plants, presence of alien plant species, and the use of the habitats by waterbirds.

Herbaceous swamps are characterised by plants common on many of the floodplain lakes (e.g. *Eleocharis* spp. and *Lersia hexandra*) and by the introduced *Eichhornia crassipes* and *Urochlea mutica*. The grasses *Pseudoraphis spinescens*, *Hymenachne acutigluma*, and *Oryza* sp. also occur. The woodland and forest swamps are generally shallower than the herbaceous swamps and are dominated by *Melaleuca* spp., particularly *M. quinquenervia*. Where present, the ground layer vegetation of these swamps consists of tall sedges, or else is dominated by *E. crassipes*, *L. hexandra*, *Phragmites australis*, *Philydrum lanuginosum*, or *Isachne globosa*. The composition of the ground layer vegetation is influenced by conditions of waterlogging, fire frequency, and tree density.

The vegetation of the lagoons and freshwater swamps inland of the Townsville–Bowen region has been described by Perry (1953) and reproduced by Nicholls (1981) in a description of the vegetation of the lower Burdekin valley. The lagoons contain the water lilies *Nymphaea gigantea* (?) and *Nelumbo nucifera*, *E. crassipes*, *Ottelia ovalifolia*, and *P. spinescens*. In some places, *Cyperus* spp. and *Eleocharis* spp. are common along with *L. hexandra*. The swamps are dominated by *Melaleuca* spp. and to a lesser extent by *Nauclea orientalis* and *Livistona* sp. A list of the aquatic plants, with an emphasis on weed species, found in the same region was prepared by Finlayson and Mitchell (1981). They regarded the introduced *Salvinia molesta* and *E. crassipes* as the species most liable to be major weeds in the planned expansion to the Burdekin irrigation system. Potential problems associated with other species (e.g. *Urochloa mutica*, *Hydrilla verticillata* – a native species –, and *Echinochloa crus-galli*) were also discussed.

The numerous small waterholes, lakes, and swamps inland from Townsville were surveyed by Paijmans (1978). Semi-permanent swamps had the richest flora, although recurrent associations or communities were not observed. The vegetation was greatly influenced by the depth and duration of inundation. One of the least disturbed wetlands was Minnamoolka Swamp covering about 1,000 ha and supporting a rich and varied birdlife (Stanton 1975). The dominant plant was *Eleocharis dulcis*, though that could be replaced by the grass *P. spinescens* when the swamp dried out. Mention has already been made of the presence of the noxious weeds *S. molesta* and *E. crassipes* in the swamps. In several localities they are a dominant part of the aquatic flora and are regarded as serious threats (Finlayson and Gillies 1982, Finlayson and Mitchell 1982).

Blackman and Locke (1985) have determined the persistence of components within a swamp at Mt St John, Townsville. This swamp has been changed from deep to shallow water by silt deposition behind an artificial retaining wall (Paijmans *et al.* 1985). Over the period 1964–75 there was a decline in the area covered by *Nymphoides indica* and *Urochloa mutica* and an increase in *Nelumbo nucifera*, *Marsilea* spp., and *E. crassipes*. *Salvinia molesta* and *Polygonum* spp. emerged and disappeared again over this period. Over the same period the numbers of waterfowl fell from about 5,000 to around 500 due to changes in water depth and aquatic plants. The distribution, seasonal occurrence, and persistence of the vegetation components are being examined in relation to seasonal changes in numbers, distribution, and habitat utilisation by the brolga (*Grus rubicundus*), a large crane.

The coastal swamps of the Burdekin–Townsville region are important habitats for large numbers of waterbirds (Blackman and Locke 1985). They provide a refuge for many species during the Dry season and during droughts, and also support resident populations of some species (e.g. *G. rubicundus*). The swamps undergo a seasonal cycle that is not unlike that of the Northern Territory seasonal floodplains. With the first storms of the Wet season the swamps begin to fill and vegetation development depends on the life-history characteristics of individual species. Waterbirds disperse from the permanent lagoons to the shallow seasonally inundated areas. Storms and heavy rainfall cause widespread flooding and further development of the vegetation. Many species of waterbird nest when the swamps have reached their peak water level and the vegetation is lush and abundant. The mechanisms that trigger the reproductive cycle are still not explained, but it is evident that the young are hatched into an environment suitable for their survival. Other species that nest in the swamps include magpie goose (*Anseranas semipalmata*), black duck (*Anas superciliosa*), water and grass-whistling ducks (*Dendrocygna arcuata* and *D. eytoni*), and occasionally black swan (*Cygnus atratus*).

The invertebrate fauna of these swamps has not been intensively investigated. A list of molluscs found at Mt St John was presented by Blair and Finlayson (1981) as part of a study on avian schistosomes, though this was not comprehensive.

Lakes

Permanent man-made lakes are important features of tropical Australia. They vary in size from small stock-watering dams (or tanks) to the 5720×10^6 m^3 Lake Argyle on the Ord River, Western Australia (Table 2). Their purpose is to ensure continuity of water supply in areas of highly seasonal and/or unreliable rainfall. Prominent examples are the Ross River Dam (Townsville) and Lake Moondarra (Mount Isa) (Fig. 9) built to supply water for urban, industrial, and mining purposes, and Tinaroo Dam and Lake Argyle built for irrigation. The ecological characteristics of these water

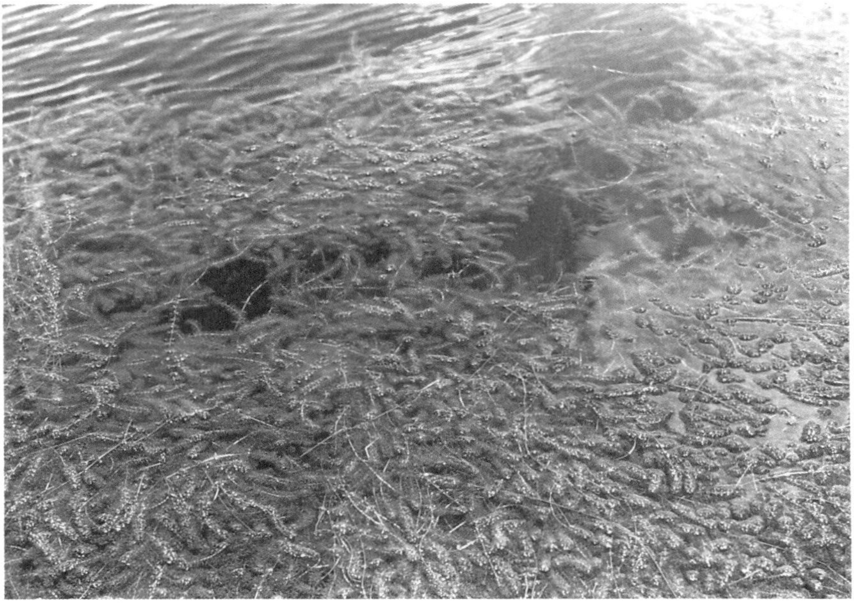

Figure 9. Hydrilla verticillata growing in Lake Moondarra, Queensland. This species along with the floating weed *Salvinia molesta* covered around 1,200 ha of the lake.

bodies have not received a great deal of attention, with the exceptions of Ross River Dam (Finlayson and Gillies 1982), and Lake Moondarra and other dams near Mount Isa (Farrell *et al.* 1979, Finlayson 1980, Finlayson *et al.* 1984a,b). Details of the plants include occurrence and productivity of native and alien species, and restrictions on their growth in artificial lakes. The importance of intermittent natural lakes in the arid region is also highlighted. Specific examples of both artificial and natural lakes are given and their vegetation and vertebrate fauna briefly reviewed.

The Ross River Dam was built in 1973 to supply water to the city of Townsville and the surrounding area. At the augmented stage I level of construction in 1978 it was relatively shallow with a mean water depth of 3.2 m. Consequently it contained a large number of aquatic plants (Finlayson and Gillies 1982) dominated by the submerged species *Hydrilla verticillata*, *Potamogeton javanicus*, and *Ceratophyllum demersum*. Floating-leaved species and *Typha* spp. were present around the periphery. Eleven species of fish were found with *Glossamia gilli* and *Craterocephalus stercusmuscarum* the most plentiful. The lake has also been stocked with silver perch (*Bidyanus bidyanus*) and sleepy cod (*Oxyeleotris lineolatus*).

The water quality, dissolved oxygen, and temperature, and aquatic vegetation of Lake Moondarra were described by Farrell *et al.* (1979) and

Finlayson *et al.* (1980, 1984a). This lake, located downstream of the base metal mining complex at Mount Isa, received a continuous inflow of "secondary-treated" sewage containing 27 mg N l^{-1} and 10 mg P l^{-1} (Finlayson *et al.* 1982). For much of the year this waste was the only inflow and averaged 6 ML day^{-1}. The vegetation was dominated by the introduced floating weed *S. molesta* and the native submerged *H. verticillata*. Over a four year period the former spread to cover 330 ha of the total lake area of about 2,500 ha, with a maximum biomass of 167 t fresh weight ha^{-1} or 0.8 kg dry weight m^{-2} (Finlayson *et al.* 1984a), despite the implementation of a number of chemical control programs (Farrell 1978, Finlayson and Farrell 1983). It was finally brought under control in 1981 by a combination of decreasing water levels, stranding many plants, and by the activity of a released biological-control agent, *Cyrtobagous salviniae* (Room *et al.* 1981, Finlayson and Mitchell 1982).

Hydrilla verticillata formed extensive stands in water to 6–7 m deep and covered about 1,000 ha. Dry weight standing crop values varied from 0.06–2.97 kg m^{-2} (Finlayson *et al.* 1984a). The maximum values were reasonably high for this species. Other common submerged species included *P. crispus*, *Najas tenuifolia*, and *Myriophyllum verrucosum*. The former, although not as common as *H. verticillata*, had a maximum standing crop of 6.41 kg m^{-2}. Submerged plants in Lake Moondarra were able to photosynthesise at depths allowing irradiances of only 30 μE m^{-2} s^{-1} (Finlayson *et al.* 1984a). The maximum photosynthetic rate recorded for *H. verticillata* was ca. 0.7 mg Chl a h^{-1} and for *M. verrucosum* the maximum was almost double that value and well above the range previously reported by Westlake (1975) for submerged plants.

Except for the fringing grasses *P. spinescens* and *Cynodon dactylon*, emergent species were not well developed, unlike Lake Kununurra (the diversion dam near Lake Argyle on the Ord River) where *T. domingensis* is a major weed. The treated sewage inflow was assumed to be partly responsible for the development of the extensive plant populations in Lake Moondarra. Nutrient values of the submerged *H. verticillata* reached 4.4% N and 0.4% P (dry weight) though mean values were around 1.9–2.9% N and 0.2–0.25% P whilst for *S. molesta* the range was 1.3–2.5% N and 0.2–0.3% P (Finlayson *et al.* 1984a).

Water temperatures near the surface among *S. molesta* plants varied from about 12–30°C (Finlayson 1984b). The highest recorded growth rates were between 17–26% day^{-1} and resultant doubling times of about 3–4 days. However, only 5% of the variance in growth rates was explained by an Arrhenius equation relating water temperature to growth rate, suggesting limitation by other factors (Toerien *et al.* 1983). The nature of these other factors is not known, especially as the nutrient content of the water was considerably higher (Farrell *et al.* 1979) than the half saturation constants (i.e. levels required to maintain half the maximum growth rates) of 15.6 g N L^{-1} and 1.0 g P L^{-1} calculated by Toerien *et al.* (1983) from glasshouse studies. It was worth noting that at a nearby sewage lagoon with much higher nutrient concentrations than in the lake, much faster growth rates were recorded, 31–43 hours doubling times (Finlayson 1984b), and that temperature explained 76% of the variance (Toerien *et al.* 1983).

The aquatic vegetation of other man-made lakes in the vicinity of Mount Isa was similar to that of Lake Moondarra (Table 13), with the exception of the extensive mats of *S. molesta* (Finlayson 1980, Finlayson *et al.* 1984b). Despite the inflow of treated sewage to Lake Moondarra it did not have higher nitrogen and phosphorus levels in the water than the other lakes which did not receive a similar inflow of sewage. This situation may reflect the dynamic nature of the lake ecosystem that resulted in rapid assimilation of sewage nutrients by the vegetation, including the algae, and sediments.

A characteristic feature of these lakes is the large draw-down of water between seasons and between years. In Lake Moondarra this can exceed 4.5 m. Needless to say, this has a major influence on the aquatic plants. Mention has already been made of *S. molesta* plants being stranded by falling water levels (Finlayson and Mitchell 1982). Following relatively high water levels between 1976–79 the water level fell more than 5 m, stranding and dessicating the large biomass of submerged plants. Similarly, the riparian vegetation, dominated by the river red gum, *Eucalyptus camaldulensis*, that typically develops around these lakes (Finlayson 1980) can suffer from excessively high or low water levels.

The potential for removing nutrients and heavy metals from Lake Moondarra was assessed by Finlayson *et al.* (1984a). Using standing crop values and tissue concentrations of nitrogen, phosphorus, copper, lead, and zinc the biomass and area of *S. molesta* and *H. verticillata* that would need to be harvested to balance the input of nutrients and heavy metals from the sewage inflow was calculated. Harvesting of *Hydrilla* was considered to be more efficient than harvesting *Salvinia* for removal of nitrogen, phosphorus and zinc whereas for copper and lead the reverse was true. The amount of vegetation (dry weight) that was required to be harvested, however, was high (ca. 2,700 t of *Hydrilla* for removal of 52 t nitrogen and 7,968 t for removal of 20 t phosphorus). As these values represented a large proportion of, and possible exceeded, the total standing crop in the lake, the capacity of the plants to recover to enable further harvesting had to be considered. Removal of too great a proportion of the biomass would reduce productivity and cause a decrease in the total amount of nutrient or metal absorbed from the water. It was concluded that the removal of nutrients and heavy metals

Table 13. Aquatic and wetland plants recorded in six man-made lakes in north Queensland (adapted from Finlayson and Gillies 1982 and Finlayson et al. 1984).

	Ross River Dam	Rifle Creek	Lake Moondarra	Lake Corella	Lake Mary Katherine
1. Vegetation with floating leaves					
Eichhornia crassipes		X			
Marsilea sp.	X		X		
Nymphaea capensis	X				
N. gigantea	X		X		X
Nymphoides indica	X		X	X	X
Potamogeton javanicus	X				
P. tricarinatus	X				
Salvinia molesta			X		
Spirodela oligorrhiza		X		X	X
2. Submerged vegetation					
Ceratophyllum demersum	X				
Hydrilla verticillata	X	X	X	X	X
Myriophyllum verrucosum	X	X	X	X	X
Najas tenuifolia	X	X	X	X	X
Potamogeton crispus	X	X	X	X	X
Utricularia exoleta	*X*				
Vallisneria sp.	X	X	X		
3. Emergent vegetation					
Cynodon dactylon	X	X	X	X	X
Cyperus spp.	X	X	X	X	X
Bulbostylis barbata			X		
Fimbristylis spp.		X	X		
Ipomoea spp.					X
Juncus aridicola					X
Ludwigia spp.	X		X	X	X
Polygonum spp.	X	X	X	X	X
Pseudoraphis spinescens	X	X	X	X	X
Typha spp.	X	X	X	X	X
Scirpus squarrosus			X		

by harvesting plants could not be wholly successful if the sewage inflow continued.

Intermittent natural lakes occur in the arid region of uncoordinated drainage. In the Northern Territory two significant examples are Lake Surprise and Lake Woods, both located in the Tanami Desert (see Finlayson *et al.* 1988b). Both require substantial rainfall events before they fill, but once full will hold water for up to three years. The flooding of the lakes causes fish populations to increase rapidly and attracts thousands of waterbirds (e.g. magpie geese (*A. semipalmata*), cormorants (*Phalacrocorax* spp.), herons (*Ardea* spp.), and pelicans (*P. conspicillatus*)).

Lake Woods, when dry, supports a plant community dominated by *Eleocharis pallens* and *Psoralea cinerea* and is fringed by *Eucalyptus microtheca* and scattered remnants of *Muehlenbeckia cunninghamii*. The vegetation has been greatly modified by 100 years of intensive cattle grazing. At Lake Surprise the vegetation has not been grazed and consists of an open *Eucalyptus microtheca* woodland over a sedgeland dominated by *Cyperus vaginatus*.

Salt lakes such as Lake Neale and Lake Amadeus in the Amadeus Basin, Lake Bennett and Napperby Lake in the Burt Plain, and the extensive playas (salt-pans) of Lake Mackay, Lake White, and Lake McDonald also occur in the arid region. They are relics of larger lakes and have undergone alternate wetting and drying over long periods. They are filled by local rainfall events that enable the development of large populations of brine shrimps (*Parartemia* sp.) that are exploited by birds such as black-winged stilts (*Himantopus himantopus*) and red-necked avocets (*Recurvirostra novaehollandiae*). Fringing the lakes is a band of *Halosarcia* spp. (samphire) that gives way to a sparse and low tree community of *Melaleuca glomerata*.

The 200 or so dune lakes on Cape York Peninsula have not been intensively investigated, partly as a result of their isolation, but Timms (1986a) has reported on the limnological characteristics of nine of them. The water was acidic and humic with sodium and chloride the dominant ions. The aquatic plants, microcrustacea, macroinvertebrate, and vertebrate species recorded were generally also common elsewhere in the tropics.

Wetland conservation

Wetland inventory

No consolidated inventory of the whole of northern Australian wetlands exists (see section on *Wetland classification*). With the large areas involved and responsibility for conservation of wetlands divided between several governments and their departments it may not be possible to obtain a standardised inventory of northern Australian wetlands. This should not, however, preclude inventories being done within the political boundaries by the States, with common objectives and methods. Preferably though, a national survey by one organisation (e.g. CSIRO) should be done.

Wetland conservation and threats

The conservation status of wetlands in northern Australia has recently been reviewed by Lane and McComb (1988) for Western Australia, Finlayson *et al.* (1988b) for the Northern Territory, and Arthington and Hergerl (1988) for Queensland. These authors identified a number of threats to wetlands that basically fall into two categories – threats derived from the invasion and spread of alien plants and feral animals, and threats directly associated with land use (e.g. agriculture, mining, urban developments, or the tourist industry). The former threats are often enhanced by, if not closely linked to those associated with land use patterns.

With the exception of the mining and exporting of iron ore from the Pilbara, the northern part of Western Australia does not support extensive industrial development or a large population. Hence, threats to wetlands from these activities are not great and their overall conservation status is generally good. The much greater population and development in northern Queensland, in contrast, places these wetlands under a much greater threat, particularly those near major urban centres. Cattle grazing and the introduction of alien plants and animals such as the buffalo (Fig. 8) could be responsible for bringing about undesirable change to wetlands, but in many areas (e.g. northern Western Australia), very little comprehensive information is available. It is not unreasonable to expect, however, that wetlands in northern Australia have all been affected in some way by about 100 years of unrestricted grazing.

Grazing is of particular concern in the arid zone where cattle are forced to rely largely on wetlands for their Dry season food and water supplies. Very little, however, is known about plant successional changes and other ecological processes associated with cattle grazing on wetlands. In the Northern Territory, buffalo grazing on wetlands has been regulated in an attempt to reduce environmental disturbance and possible destruction of important breeding areas or habitats of native animals (e.g. crocodiles and waterfowl).

In both Queensland and the Northern Territory the nature of wetlands is compromised by alien weeds such as *Mimosa pigra*, *Salvinia molesta*, and *Eichhornia crassipes*. In the Northern Territory, *Mimosa* has completely changed the nature of some floodplains from grassland to shrubland.

The floating plants *Pistia stratiotes*, *E. crassipes*, and *S. molesta* are prevalent weed species in coastal freshwater wetlands in Queensland (Mitchell 1978, Finlayson 1979, Finlayson and Gillies 1982, Finlayson and Mitchell 1981, 1982), although only the latter is currently a major problem in the Northern Territory (I. Miller, personal communication). They all have the capability of completely covering areas of open water and leading to deoxygenation and changes in the chemical status of the water and in the life-cycle and habitats of native biota. As described earlier, biological control has been used successfully on some infestations of *Salvinia* (Room *et al.* 1981). A similar program has been mounted against *Eichhornia*, but it has not achieved the successes of the *Salvinia* program.

Other important, but poorly studied, weeds include *Hyptis suaveolens*, *Cassia* spp., *Sida* spp., and *Urochloa mutica*. The latter is rampant on the floodplains of the Northern Territory and freshwater swamps of Queensland. Like many weed species it prevents the establishment of native plants and can disrupt the breeding and feeding patterns of native animals although it was originally introduced to improve stock food. It is less of a problem when grazed by cattle, but this activity is not usually compatible with conservation objectives. Attempts to find a biological control agent for *Hyptis* are proceeding, although there are no agents of any promise currently available (I. Cowie, personal communication).

Feral animals, particularly the Asian water buffalo in the Northern Territory, are regarded by conservation authorities as a major threat. A survey of the impact of buffaloes on wetlands reported open floodplains heavily grazed and pug-marked, networks of tracks and swim-channels, reduction in floating grass mats and vegetation cover, especially around billabongs, bank erosion and slumping, and premature drainage of freshwater (Graham *et al.* 1982). The latter may be partly a natural process, but it is exacerbated by buffaloes breaking down the banks or levees that separate saline and freshwater wetlands. In the parks and reserves of the Northern Territory, buffaloes are being removed, though the response of both native and alien plant species to this action is not yet known.

The introduced cane toad (*Bufo marinus*) is regarded as a potential threat in the Northern Territory (Freeland 1985, Freeland and Martin 1985). It has been present in Queensland coastal wetlands since 1935, but its impact on the wetland fauna is still not understood. Recent introductions of aquarium fishes in Queensland are also viewed with concern, though again, the implications are not understood (Arthington *et al.* 1984).

Agricultural development (e.g. the sugar cane industry) and activities associated with mining developments can lead to nutrient enrichment and/or pollution of wetlands from either surface runoff and/or discharge of waste water. Wetlands have often been simply regarded as wastelands and therefore suitable sites for waste disposal. Urban sewage, treated or untreated, has often been discharged into saline and freshwater wetlands. Problems of eutrophication and pollution commonly occur near large towns or cities, but may also be associated with specific and isolated activities such as the Rum Jungle uranium/copper mine, near Darwin, in the Northern Territory. This operation caused pollution that since 1983 has cost the Australian Government at least A $16.2 million to remedy (Allen 1985). Considerable effort has been expended in the Northern Territory to ensure that two uranium mining operations (Fig. 10) in the Alligator Rivers Region do not leave similar legacies.

Interference with the hydrological regimes of both coastal and inland wetlands can result in the degradation of the vegetation, changes in sediment transport and deposition, erosion, nutrient enrichment and pollution, and disruption of animal migration patterns. Such interference can occur as a result of constructing dams, weirs and barrages for irrigation, urban and industrial water supply, and for flood mitigation. Large-scale drainage of wetlands for urban and agricultural development (e.g. sugar cane cultivation in Queensland) can have similar effects. Mangrove swamps near major urban centres are also under threat due to pressure to "reclaim" land and to eradicate "pest" problems such as mosquitoes.

Figure 10. Ranger Uranium Mine at Jabiru in the Alligator Rivers Region, Northern Territory. This mine is located adjacent to Magela Creek and is upstream of the area generally referred to as the Magela floodplain. A great deal of controversy has surrounded plans to release stored runoff water from the mine site to the creek.

Increasing tourism and recreational pressure pose an ever increasing threat to wetlands in both heavily and sparsely populated areas. Tourism facilities, particularly holiday resorts and marinas, are developed at the expense of wetland habitats and wetland resources can be over-utilised (e.g. excessive fishing pressure on the popular angling species such as *Lates calcarifer* (silver barramundi), as a result of increased recreational activity.

Recommendations for wetland conservations

The geographical area covered by this review is both large and sparsely populated. Nevertheless, many wetlands have been disturbed, or are threatened with disturbance that could alter the pathways and levels of nutrient flow through the system, or cause an influx of toxicants. These changes could cause a drastic reduction in, or total elimination of, one or more of the major biotic components, or a reduction in the diversity of wetland types. Whatever the type of disturbance, management for conservation purposes should be designed to minimise unacceptable impact on native species or "natural" habitats. Determining what is an unacceptable impact is obviously a difficult task and must involve the myriad of societal considerations in addition to conservation objectives.

With this view in mind the following general recommendations are presented for consideration when assessing the nature of threats and the conservation status of wetlands in Northern Australia.

1. Extension of the park and reserve system is one way of initiating the processes that are required to enable wetland species and habitats to be conserved. By itself, however, the proclamation of reserves may not achieve a great deal. It is also necessary to develop and implement management practices that take into account clearly defined conservation objectives. This could involve preliminary work to compile species inventories, including rare or endangered species, and population studies to determine if a particular habitat or species is under- or over-represented in the conservation area. If this were to be done a list of priority areas or objectives could be established and used as a guide when determining conservation policies.

Instead of opportunistic proclamation of available land as reserves, an assessment of the need to preserve habitats or conserve species is required. The possible conflict between conservation and recreational usage also needs to be considered, a particularly relevant fact when dealing with activities such as barramundi fishing. A singular obsession with one or the other aspect could result in non-profitable conflict and ineffectual management of the reserve system. Thus, the question of using reserves as havens for threatened or endangered species, conservation of habitats, or as recreational assets needs constant appraising and re-appraising.

2. Feral animals are present in many wetlands, and in some instances have caused considerable disturbance to the "natural" system. The most prominent example, the Asian water buffalo on the coastal floodplains of the Northern Territory, has received a large amount of attention and it is now generally accepted that buffalo will be eradicated from conservation reserves. As with any management strategy, however, the success and effects of this action need to be monitored and, if required, adjustments made. The impact of other feral animals on wetlands (e.g. cane toads, pigs, horses, aquarium fishes) has generally not received a great deal of attention. The extent of disturbance caused by these animals needs to be assessed before conservation strategies can be implemented to prevent or reduce further undesirable changes.

3. Weeds, particularly alien species, pose a major threat to the conservation status of wetlands. The potential of species such as *Mimosa pigra* and *Salvinia molesta* to cause problems is well established and it is generally accepted that they should be controlled, if not eradicated. The status of other species (e.g. *Urochloa mutica*) is not as clear and should be assessed on both a local and a regional basis. An immediate expansion of current control and research activities should be coupled with this assessment, especially if wildlife breeding and feeding areas are under threat. Unless the problem of weed invasion is addressed the nature and conservation status of wetlands could be drastically altered.

4. Agricultural development often results in diffuse sources of pollution that can have a significant impact on wetlands. Whilst it is difficult to control diffuse source pollution (e.g. from sugar cane farms along the Queensland coast), attempts should be made to limit the impact of nutrient and pesticide runoff onto wetlands, especially those that are classified or function as reserves or conservation areas. To be fully effective this should involve management of the entire catchment and perhaps the application of rigid controls such as those that are currently used to regulate uranium mining and processing in the Northern Territory. If the nature of the problem is assessed prior to development and adequate controls devised, the need for future remedial action could be avoided. Point sources of pollution can be readily identified and are often, at least locally, extremely detrimental to the integrity of wetlands. Whilst discharge of relatively dilute effluent to wetland can be an acceptable disposal technique, the more concentrated effluent (e.g. untreated sewage or sugar mill waste), or those that contain hazardous materials (e.g. mineral processing waste water) should be treated to reduce, if not eliminate, any detriment to the wetland. It is important that equal attention is given to all developments that may threaten wetlands, rather than focussing solely on ventures (e.g. uranium mining) that have a high media profile.

5. A further effort is required to assess the conservation value and status of wetlands (e.g. Lake Surprise and Lake Wood) in the arid zone. Information that can be used to develop and implement management strategies is urgently required to enable decisions to be made on the problems of economic use of the land (e.g. cattle grazing) and conservation objectives.

6. Once wetland habitats have been described and species behavioural patterns examined, investigations are required to determine the interrelationships between species and their habitats. This should be directed specifically towards determining the effect of potential changes (e.g. in nutrient loadings or alteration of the hydrological balance) that could come about as a result of economic developments such as tourist facilities near coastal wetlands or irrigation schemes upstream of wetlands. To be fully beneficial the programme should include experimental testing of hypotheses, generated after assessing available descriptive information, and the development of multi-faceted management capabilities.

7. The continued expansion of recreational activities into wetlands is likely to be a major problem for conservation authorities. The main areas of concern that need careful consideration are the extent of commercial and recreational fishing and the provision of tourist facilities in wetlands. These problems can only be properly assessed after the development of conservation strategies that are based on carefully designed and compiled data bases and that consider the potential impact of all threats to particular wetland types and the need to conserve wetland species or even preserve wetland habitats.

Wetlands of Australia: Southern (temperate) Australia

S. W. L. JACOBS AND MARGARET A. BROCK

Introduction

This second section covers the area of Australia south of the Tropic of Capricorn (23°26.5′S) and encompasses the greater part of the arid interior, an extensive coastline with coastal plains, and upland areas. The most intensive agricultural areas are covered as are regions with both the highest and lowest densities of population and industry. Information availability has been influenced by conservation issues and pressures for exploitation. Exploitation pressures, in particular, have been responsible for the gathering of information on particular wetlands including: (i) those near areas of high population density (e.g. Goodrick 1970, Pressey 1981, 1987a,b), (ii) fragile habitats such as the mound springs of the Lake Eyre Basin (Greenslade *et al.* (1985), and (iii) areas of political significance (e.g. Knights (1980) for the Macquarie Marshes, Thompson (1986), Pressey (1986), and the Murray-Darling Basin Ministerial Council (1987) for the Murray-Darling system).

Study area

About 60% of Australia is south of the Tropic of Capricorn (Fig. 11), including all of New South Wales (N.S.W.), Victoria (Vic.), Tasmania (Tas.), and South Australia (S.A.), and the southern 60% of Western Australia (W.A.), 25% of Queensland (Qld.), and 15% of the Northern Territory (N.T.). Offshore islands total much less than 1% of the land area and have not generally been included in large scale studies of wetlands. Temperate Australia is a flat peneplain with a range of low mountains on the east coast extending south into Tasmania. Some of the Lake Eyre basin is below sea level while most of the remainder is of low relief with occasional ranges usually less than 1,000 m. The south east portion is the most densely populated and the two major cities, Sydney and Melbourne, contain 33% of the

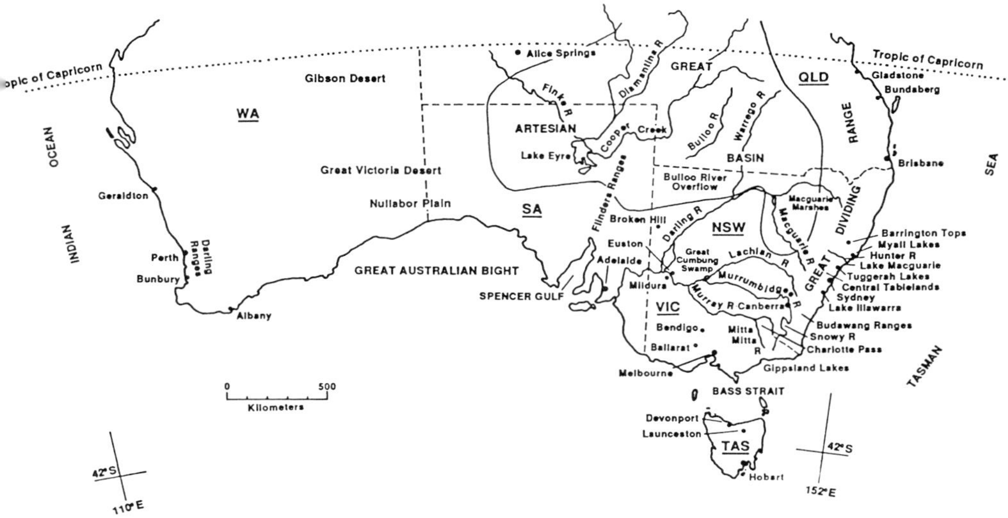

Figure 11. Map of southern (temperate) Australia south of the tropic of Capricorn showing major political boundaries, the larger river systems, and place names mentioned in the text.

total population of Australia while New South Wales and Victoria account for 60% (Castles 1986). The Queensland coast has several urban areas (e.g. Brisbane, Bundaberg, and Gladstone) but is primarily an agricultural area where tropical cash crops are grown. The remainder of the coastal region and most of the inland region is used for beef production. Wool production is common in parts of the inland.

Sugarcane, cash crops, and dairying are major activities on north coastal N.S.W. while higher parts of the Great Dividing Range (Fig. 11) are important sheep and cattle areas. To the west of the Great Dividing Range are large areas for wheat and similar dryland crops, sheep, and cattle. Irrigation is important in the southern areas west of the ranges.

Dairying and associated enterprises are the commonest agricultural pursuits in Victoria, followed by sheep, wheat, and cattle in the drier areas, and extensive irrigation areas in the north. Melbourne is the largest city but there are several large rural centres (e.g. Ballarat, Bendigo, and Mildura).

Wool is the main agricultural product of S.A. with wheat and cattle important towards the south and cattle towards the north. Nearly all the population in concentrated in and around Adelaide.

Sheep, cattle, and field crops are the major agricultural enterprises of W.A. Most people live in and near Perth and the larger rural towns are small even by Australian standards (e.g. Geraldton, Bunbury, and Albany).

Tasmania is wetter and colder than the mainland and dairying and beef production are the most significant agricultural activities. The population is concentrated on the northern and eastern coasts near Hobart, Launceston, and Devonport.

Climate

Southern Australia spans several climatic areas: wet tropical climate in coastal Qld., alpine areas in Tas., northern Vic., and southern N.S.W.; arid areas in S.A., western N.S.W., and northern and eastern W.A.; temperate areas in coastal N.S.W., Tas., and Vic.; and mediterranean areas in coastal S.A. and southern W.A. On the east coast, precipitation falls predominantly in the summer north of about 31°S, and in the winter south of 34°S. Areas in between generally have a bimodal or even distribution of rainfall. On the west coast winter rainfall dominates south of about 30°S and rainfall is erratic rather than strictly seasonal north of 30°S. Inland arid areas also have erratic rainfall patterns. The southern coastline receives predominantly winter rainfall.

The combination of long coastline and few mountain barriers means that oceans influence the climate of much of Australia. The currents off Australian shores are neither as distinctly cold, nor as warm, nor as persistent as those off other continents and consequently there are less extremes in the range of climates. The southern part of the continent lies in the path of high pressure systems that move from west to east. The centres of these pressure systems move from an average latitude of about 29°S in late summer to about 37°S in late winter (Linacre and Hobbs 1977).

The climate is notable for its high temperature and drought resulting from relatively cloudless skies. Alice Springs (Fig. 11) has an annual average of 9.8 hours bright sunshine per day, Perth 7.8, Sydney 6.6, and both Melbourne and Hobart 5.7. Temperature, rainfall, evaporation, and humidity data are summarised for several areas in Table 1. The critical factor for most biological activity in Australia is the variability of the rainfall.

Tasmania is far enough south to be strongly influenced by the continuous westerly winds and has less seasonality than the southern coast of the mainland. There is a substantial alpine and sub-alpine area in Tasmania where there is snow for most of the winter and rain can occur at almost any time of the year.

Drainage patterns

Australia is a very dry continent and all rivers carry relatively small volumes of water and many are strongly seasonal and variable. The total run-off from all Australian catchments is smaller than the annual average flow of at least 15 of the world's individual rivers (Brown 1983). Australia has been divided into twelve drainage regions (Fig. 12, Australian Water Resources Council

Figure 12. Drainage Divisions of Australia (redrawn after Australian Water Resources Council 1976).

1976), six of these falling entirely within the area south of the Tropic of Capricorn and four partly so. Table 14 has been adapted from Walker (1985) and is a summary of area, mean annual run off, potential exploitable use (without considering conservation values), and actual water usage for each drainage region.

Table 14. Hydrological data for Australian Drainage Divisions (Fig. 2) from Walker (1975). Flow data are million megalitres per year. Figures less than 0.1 ML $\times 10^6$ year^{-1} and 1% are regarded as not significant. Area is in thousands of km^2. Runoff, Yield, and Mean use values are ML $\times 10^6$ per year. Percent is the percentage of the exploitable yield that is used.

Drainage division	Area	Runoff	Yield	Mean use	Percent
North-east Coast	450	91.5	26.6	1.7	6
South-east Coast	274	45.5	15.1	2.5	17
Tasmania	68	53.4	35.5	0.3	<1
Murray-Darling	1,062	22.6	13.4	11.8	88
South Australian Gulf	82	1.0	0.3	0.3	100
South-west Coast	314	6.7	1.8	0.6	33
Indian Ocean	519	4.0	0.3	<0.1	–
Timor Sea	547	81.2	16.0	0.1	<1
Gulf of Carpentaria	638	130.5	28.7	<0.1	<1
Lake Eyre	1,170	3.3	0.1	<0.1	–
Bulloo-Bancannia	101	0.6	<0.1	<0.1	–
Western Plateau	2,455	<0.1	<0.1	<0.1	–
Totals	7,680	440.3	137.8	17.3	13

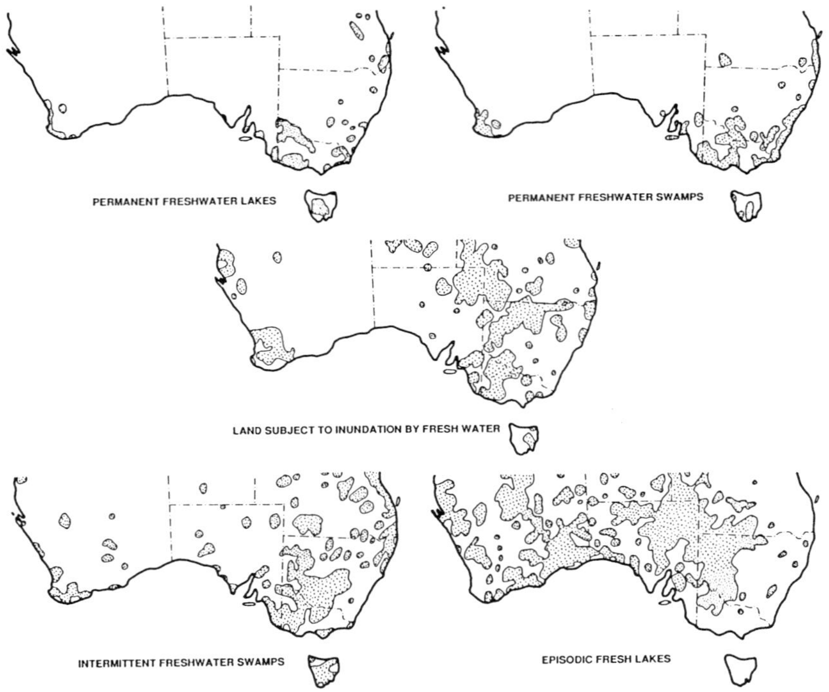

Figure 13. Distribution of wetlands in Australia south of the Tropic of Capricon (modified after Paijmans *et al.* 1985): a. permanent freshwater lakes, b. permanent freshwater swamps, c. Iand subject to inundation, d. intermittent freshwater swamps, e. episodic fresh lakes.

Rivers can be classified by their flow characteristics. Some have a strong discharge throughout the year with one or two distinct maxima; others have a sustained discharge and seasonal peaks following rainfall or snow-melt; still others approach either extreme but their flows tend to be less predictable. All Australian rivers fit into this last category (Walker 1985).

Wetland definition and distribution

In both sections of this chapter, wetlands are defined as: "land permanently or temporarily under water or waterlogged . . . ." (after Paijmans *et al.* 1985). This could be interpreted to include shallow marine areas which we include in this review. Paijmans *et al.* (1985) mapped the wetlands of Australia at 1:2,500,000, compiling the information from 1:250,000 maps. These maps are valuable reference points for further research and planning. We have redrawn these maps (Fig. 13) to present a rough idea of the distribution of permanent freshwater lakes, permanent freshwater swamps, land subject to inundation, intermittent freshwater swamps, and episodic fresh lakes. The distribution of permanent and near-permanent wetlands, and "generally dry wetlands" in Paijmans *et al.*'s maps tend to over-estimate the perennity of wetlands, as well as including several man-made storages. Accordingly our modifications have resulted in reduced areas in each category with only a few additions. Despite the severe limitations of scale, the maps indicate the general distribution of wetlands and their lack of perennity.

Wetland classification

Classification systems

Australian wetlands have been classified on the basis of geographical and/or physical features (e.g. Jacobs 1983, Riley *et al.* 1984, Paijmans *et al.* 1985), the structure of vegetation (e.g. Briggs 1981, Kirkpatrick and Harwood 1983a, Blackhall 1986), or a mixture of both (e.g. Riggert 1966, Beadle 1981, Mills 1983, Thompson 1986, Norman and Corrick 1988). In the first section of this chapter the classification system used by Paijmans *et al.* (1985) and others that deal specifically with the northern area of Australia are described. In this section we concentrate on schemes applied in southern Australia.

Most classifications have arisen from detailed localised studies, often with political rather than geographical boundaries, and hence they tend to rely heavily on characteristics important to that particular study. For example, many studies of birds concentrate on vegetation structure as the basis for habitat classification (e.g. Mills 1983 and Blackhall 1986), whereas De Deckker (1982) uses the relative suitability of wetlands as sites for paleolimnological investigations.

General classification systems

Attempts to classify wetland vegetation in the whole of Australia are those by Beadle (1981), and Briggs (1981). Briggs' system has already been discussed (Table 4). Beadle's (1981) impressive work on Australian vegetation uses an hierarchical classification based initially on geographical and then subsequently on a mixture of geographical, floristic, and structural criteria (Table 15). Beadle has three major divisions, (i) inland watercourse, floodplain and discharge areas, (ii) communities in fresh or brackish water, mainly on coastal lowlands; including lagoons, lakes, rivers, swamps and flooded areas, and (iii) communities on mudflats. The first has five "communities" and 23 "alliances", the second four communities and 25 alliances, and the third five communities and 12 alliances. Although one of the more complete classifications, it is difficult to use out of its original context.

Classifications of wetlands relying on vegetation structure and floristics are often difficult to adapt to larger scale studies. The degree of precision involved when genera or species are involved in the definitions means lack

Table 15. Wetland classification extracted from Beadle (1981).

1. Inland watercourse, flood-plain, and discharge areas
   - 1.1 Submerged communities
   - 1.2 Swamp communities
     - 1.2.1 *Phragmites australis* Alliance
     - 1.2.2 *Typha domingensis* Alliance
     - 1.2.3 *Marsilea drummondii* Alliance
     - 1.2.4 *Muehlenbeckia cunninghamii* Alliance
     - 1.2.5 *Eleocharis pallens* Alliance
     - 1.2.6 *Chenopodium auricomum* Alliance
     - 1.2.7 *Eremophila maculata* Alliance
     - 1.2.8 *Eragrostis australasica* Alliance
   - 1.3 Communities on clay in flooded areas, mainly channel country and playas
     - 1.3.1 ephemeral communities in the Channel Country
     - 1.3.2 communities on and around playas
       - 1.3.2.1 *Sarcocornia-Sclerostegia* Alliance
       - 1.3.2.2 *Frankenia* spp. Alliance
       - 1.3.2.3 communities of subsaline zone
       - 1.3.2.4 communities on upper beaches
       - 1.3.2.5 communities on sandy areas around playas
   - 1.4 *Eucalyptus* communities fringing watercourses and on floodplains
     - 1.4.1 *Eucalyptus camaldulensis* Alliance
     - 1.4.2 *Eucalyptus rudis* Alliance
     - 1.4.3 *Eucalyptus microtheca* Alliance
     - 1.4.4 *Eucalyptus largiflorens* Alliance
   - 1.5 Communities of minor watercourses and other small irrigated areas.
     - 1.5.1 minor watercourses
     - 1.5.2 semi-permanent rock holes and deep gorges
     - 1.5.3 rock-crevice and boulder communities
2. Communities In fresh or brackish water, mainly on coastal lowlands; Including lagoons, lakes, rivers, swamps, and flooded areas.
   - 2.1 Tropics
     - 2.1.1 *Nymphaea gigantea - Nelumbo nucifera* Alliance
     - 2.1.2 grassland – sedgeland communities
     - 2.1.3 *Pandanus* spp. Alliance
     - 2.1.4 *Livistona humilis* Alliance
     - 2.1.5 *Lophostemon lactiflua – Grevillea pteridifolia – Banksia dentata* Alliance
   - 2.2 communities dominated by *Melaleuca*
     - 2.2.1 *Melaleuca leucadendra* Alliance
     - 2.2.2 *Melaleuca viridiflora* Alliance
     - 2.2.3 *Melaleuca minutiflora* Alliance
   - 2.3 Communities in the east from south-eastern Queensland to South Australia and Tasmania
     - 2.3.1 communities of coastal brackish lakes and estuaries
     - 2.3.2 communities of fresh water lakes and lagoons
     - 2.3.3 the sedgelands
       - 2.3.3.1 *Baumea juncea* Alliance
       - 2.3.3.2 *Caloraphus minor – Leptocarpus tenax* Alliance
       - 2.3.3.3 *Gymnoschoenus sphaerocephalus* Alliance
       - 2.3.3.4 *Gahnia trifida – G. filum* Alliance
       - 2.3.3.5 *Lomandra dura – L. effusa* Alliance
     - 2.3.4 Communities associated with rivers

Table 15. Continued.

2.4 Communities in south-western Western Australia
2.4.1 Communities in permanent and semi-permanent fresh water
2.4.2 sedgelands
2.4.2.1 *Leptocarpus aristatus* Alliance
2.4.2.2 *Evandra – Anarthria – Lyginia* spp. Alliance
2.4.3 communities dominated by Melaleuca
2.4.3.1 *Melaleuca raphiophylla* Alliance
2.4.3.2 *Melaleuca preissiana* Alliance
2.4.3.3 other species of *Melaleuca*
2.4.4 communities dominated by *Banksia*
3. Communities on mudflats
3.1 The seagrasses and marine meadows
3.1.1 communities in the north-east
3.1.2 communities in the south
3.2 Mangroves and mangrove communities (mangals)
3.2.1 tropics
3.2.1.1 *Sonneratia caseolaris* Alliance
3.2.1.2 *Avicennia marina* var. *resinifera* Alliance
3.2.1.3 *Rhizophora* spp. Alliance
3.2.1.4 *Bruguiera* spp. Alliance
3.2.1.5 *Ceriops tagal* Alliance
3.2.1.6 the inner zone (landward fringe)
3.2.1.7 *Nypa fruticans* Alliance
3.2.2 south of the tropic
3.2.2.1 *Avicennia marina* var. *australasica* Alliance
3.3 Mangrove islands
3.4 Communities adjoining mangroves on the landward side
3.5 Samphire, sedgeland, and grassland communities
3.5.1 tropics
3.5.2 south of the tropics

of flexibility which often requires change in the classification of a particular wetland after flood, drought, or fire (all common in the Australian climate). As a result there have been several classification systems developed for States or regions that use geographical and physical criteria and are adaptable. Examples are Stanton (1975) for Queensland (Table 5), Jacobs (1983), and Pressey and Harris (1988) for New South Wales, Semeniuk (1987) for the Darling region of Western Australia, Lane and McComb (1988) for Western Australia, and Lothian and Williams (1988) for South Australia. These systems are all similar, differing mainly in the degree of subdivision of the categories with Stanton (1975) being the most divided and Lane and McComb (1988), and Lothian and Williams (1988) having equally brief systems. The system of Jacobs (1983) is intermediate between the extremes and is appropriate for Australian temperate wetlands. The classification used here is:

1. Coastal wetlands
   (a) upland swamps
   (b) rivers and tributaries
   (c) floodplain swamps and billabongs

   (d) coastal lagoons and lakes
   (e) estuaries
      (i) mangroves
      (ii) seagrass meadows
      (iii) salt marsh
2. Mountain lakes and swamps
   (a) perennial lakes
   (b) perennial swamps
   (c) ephemeral lakes/swamps
3. Inland rivers
   (a) rivers
      (i) perennial, including anabranches
      (ii) ephemeral
   (b) billabongs – floodplains
   (c) swamps – overflow or terminating
4. Inland lakes
5. Mound springs
6. Man-made storages, canal systems, dams, channels, drains, bores, bore-drains, farm storages, rice fields, storage swamps.

Regional classification systems

There are several excellent classification systems developed for detailed localised studies. The best known is that of Goodrick (1970) for wetlands of coastal New South Wales. It is based on a mixture of geographic, lithologic, physical, structural, and floristic characteristics. The same author (Goodrick 1984, 1985) used a land systems approach for the classification of wetlands of north western New South Wales. This used related land units defined as " areas of wetlands with similar geomorphology and hydrology and recurring patterns of landforms, soils and vegetation. Thus, wetland system boundaries delineate complexes of related land units". This land systems approach lends itself well to mapping and could prove more useful in the future.

Campbell (1983) used geographic criteria and vegetation structure in classifying wetlands in a study of bogs and mires of Australasia. The system is biased towards bogs and mires in New Zealand and is inappropriate for much of the mainland areas of Australia, especially the arid, semi-arid, and tropics.

Mills (1983), in a classification modified from Riggert (1966) and the Wetlands Classification Committee of the United States Fish and Wildlife Service, uses salinity and situation (coastal or inland) as initial categories followed by characters such as water depth and permanence, and vegetation structure. This is effective for classifying wetlands as to suitability for waterbirds but has not yet been used in Australia for other purposes.

Riley *et al.* (1984) produced a precise geomorphological hierarchical classification for the wetlands of New South Wales. With seven levels in the hierarchy and more than 1,500 possible final categories it is far too finely divided to use for biological data though only the top five levels of the hierarchy could be used but even this results in about 150 categories.

Norman and Corrick (1988) have produced a pseudo-hierarchical classification for wetlands in Victoria based initially on depth and permanence of water, followed by a mixture of structural and floristic-based categories. It is too specific to adapt for general purposes. For example, *Muehlenbeckia cunninghamii* (Lignum) communities, although essentially very similar to each other, appear in three of the six different "categories" yet alpine meadows, and *Marsilea* spp. (Nardoo) dominated communities of the semi-arid north west, would both fall into the same "subcategory".

Kirkpatrick and Tyler (1988) use a modification of the classification in Kirkpatrick and Harwood (1983a) for Tasmania. This Tasmanian scheme avoids some problems encountered by Norman and Corrick (1988) for Victoria by the use of 14 categories and by avoiding any hierarchical structure.

The landsystem approach to wetland classification was used by Semeniuk (1987) classifying waterbird habitats of the "Darling System" in Western Australia. Although for a comparatively small area, this is probably indicative of the way mapping and survey techniques will develop.

Ecological characteristics of wetlands

The distribution and ecological characteristics of temperate wetlands are given by drainage divisions (shown in Fig. 12) within each classification category outlined in the section on Wetland Classification. There is at least one review for each State; for Queensland: Stanton (1975) and Arthington and Hergerl (1988); New South Wales: Miles (1975), and Pressey and Harris (1988); Victoria: Smith (1975a), and Norman and Corrick (1988); Tasmania: Smith (1975d), and Kirkpatrick and Tyler (1988); Northern Territory: Finlayson *et al.* (1988b); South Australia: Smith (1975b), Warcup (1982), and Lothian and Williams (1988); Western Australia: Smith (1975c), and Lane and McComb (1988). Many of these reviews are gathered in a single volume (McComb and Lake 1988).

Most waterbird species are mobile and move in response to wetland availability. Hence to avoid confusion we discuss waterbird usage under more general groupings than the rest of the information. Many birds use the wide variety of wetlands in temperate Australia. These include waterfowl (family Anatidae), grebe (Podicipedidae), pelican, darter, and cormorants (order

Pelecaniformes), herons, egrets, ibis, and spoonbills (Ciconiiformes), and rails, crakes, coot, and swamphen (Rallidae).

Wetland descriptions

Coastal wetlands

These occur in all coastal drainage divisions: southern area of the North-east Coast, South-east Coast, Tasmania, South Australian Gulf, South West Coast, and Indian Ocean. Because of the urban pressure on coastal wetlands, there is a larger literature and a greater number of surveys than for any of the other types. Despite this, many types are little studied, especially the upland swamps.

Birds of coastal areas move freely from habitat to habitat and between drainage areas. Coastal wetlands have some specialist fish raptors (e.g. sea eagle and harriers) also recorded from inland wetlands. Many coastal and some inland wetland communities are visited by seabirds (e.g. gulls and terns). The Australian shelduck (*Tadorna tadornoides*) is a coastal specialist occurring on dunes, salt flats, and freshwater habitats. Shelduck also are widespread on some inland waterbodies, particularly salt lakes in the south-west of Western Australia. Cape Barren geese (*Cereopsis novaehollandiae*) occur also in coastal salt and fresh waters. They inhabit areas along the south of the South-west Coast and Tasmania, migrating annually across Bass Strait. Magpie geese (*Anseranas semipalmata*) were common in the South-west Coast Division in the last century but are now restricted to the tropics except for a small area around Bool Lagoon in South Australia where they have been re-introduced.

Inundated floodplains and claypans, although they may lack aquatic macrophytes, are often rich in invertebrates and support large numbers of waterfowl; musk (*Biziura lobata*) and blue-billed duck (*Oxyura australis*) both favour deeper permanent swamps, black duck (*Anas superciliosa*) and grey teal (*A. gibberifrons*) favour inundated coastal floodplains, while chestnut teal (*A. castanea*) prefer mangroves and salt marshes (Briggs 1983).

Mangroves and estuarine saltmarshes do not have waterbird species that are characteristic of these communities alone. Cosmopolitan highly mobile species of vertebrate and invertebrate feeding ducks (e.g. black duck, teal, cormorants, and darters) are common but vegetation feeders are absent (Rallidae).

Upland swamps

Upland swamps are most common in the North-east Coast, South-east Coast and Tasmanian Drainage Divisions (Fig. 12). They are situated on the edge of the highlands at the heads of the coastal rivers. Many often do not have free-standing water but may absorb large volumes of water, in the peat-like accumulations of organic matter, that is released slowly, helping to maintain flow in tributaries. They are frequently difficult of access, especially from the lowland wet areas. They differ from the rest of the wetlands, often being described simply as wet heaths and having more in common with the surrounding dryland vegetation and fauna than with that of the other wetlands. Consequently, the information available is in treatments of the surrounding vegetation. Publications dealing with the vegetation include Fraser and Vickery (1939), and Dodson *et al.* (1986) for the Barrington Tops area (South-east Coast Division); Davis (1936, 1941), Pidgeon (1938), Porter (1984), Young (1986), and Keith (1984, 1985) for the Sydney sandstone areas; and Pickard and Jacobs (1983) for the Budawang Ranges (South-east Coast Division). The vegetation of the swamps includes a large proportion of monocotyledonous genera from the families Cyperaceae and Restionaceae and common genera include *Baumea*, *Gahnia*, *Lepidosperma* and *Gymnoschoenus* (Cyperaceae), *Leptocarpus*, and *Lepyrodia* (Restionaceae). Other species belong to *Leptospermum* (Myrtaceae), *Hakea* and *Banksia* (Proteaceae), and *Epacris* and *Leucopogon* (Epacridaceae).

Mostly the aquatic fauna, like the flora, reflects that of the neighbouring vegetation. The exceptions are the upland swamps of Tasmania. These have been reviewed by Kirkpatrick and Tyler (1988), and Williams (1974), the latter containing reviews of the ecology and biogeography of the fish, crustaceans, phytoplankton, and amphibians. More recent studies have concentrated on particular groups or habitats (e.g. the macrobenthic fauna by Fulton (1983a,b) and eels by Sloane (1984)). In other respects the aquatic fauna is similar to that described later for mountain lakes and swamps.

Rivers and tributaries

Rivers occur all along the coast except in the Western Plateau Drainage Division (Fig. 12). Although water from the Murray-Darling drainage division reaches the coast, the rivers, and tributaries are classified as inland rivers here. Where the flow rate is high there is little characteristic vegetation in or around the channels. The vegetation of the banks reflects the soil type and the rainfall of the country through which the rivers flow. As the channels broaden, slopes lessen, and flows decrease, more alluvium is deposited and the nutrient content of the water increases.

Typical submerged plant species in these habitats include *Vallisneria gigantea*, *Hydrilla verticillata* and various species of *Potamogeton* (Fig. 14). Free-floating species or species with floating leaves are not common. River banks are frequently lined with emergent species such as *Phragmites australis*, *Bolboschoenus fluviatilis*, *Schoenoplectus mucronatus*, *S. validus*, *Triglochin procera*, and *Persicaria* spp.

Although rivers and their tributaries are most closely associated with settlement, comparatively little has been published about them. Most infor-

Figure 14. The Goulburn River, a tributary of the Hunter River of New South Wales. The trees on the bank are *Casuarina cunninghamiana* with a lone introduced *Salix babylonica* on the lower bank. Emergents include species of *Bolboschoenus*, *Schoenoplectus*, and *Phragmites*. Submerged plants include species of *Potamogeton*, *Vallisneria*, and *Hydrilla*. Although flowing through land partly cleared for agriculture, these habitats are frequented by the platypus and many species of waterbirds.

mation exists in the "grey" literature as reports, environmental impact statements, and theses all of which can be difficult to trace. Walker (1985) reviews some relevant publications, mainly on limnology and submerged fauna, but notes the relative paucity of studies and that the studies that do exist are concentrated near centers of high population. The lack of information on vegetation is related to the difficulty in obtaining reliable maps of river beds. Even where adjacent wetlands are well documented, the channels have received scant attention.

It is difficult to generalise about the fauna of rivers draining such a wide latitudinal range. Modification by man has resulted in an altered fauna. Walker (1985) considers the effects of river regulation on the coastal rivers in all drainage areas and he discusses the effects of altered flow regimes on invertebrates, and native and introduced fish populations. Walker gives an entry to the more specific literature on faunal changes in river systems. Williams (1983) suggested that macro-invertebrate diversity is often high but may decrease in the lower reaches of rivers. One anomaly is the presence of atyid shrimps (family Atyidae) which are unusual river inhabitants because of their planktonic larvae; these may breed in areas and times of low flow. The invertebrate communities of temperate Australian streams are compared with those in the northern hemisphere by Lake *et al.* (1985).

Fish diversity also varies with type of habitat and geographical location. In general, diversity decreases in coastal rivers from north to south with the lowest diversity in the southwest of Western Australia. Here the low diversity may be attributed to an increase in water salinity as a result of the clearing of native vegetation (Froend *et al.* 1987). In many coastal rivers and creeks migratory species of fish (e.g. grayling (*Prototroctes maraena*), bass (*Macquaia colonorum*), and some galaxids), eels, and lampreys move between marine and freshwater environments. Many of the 19 introduced species of fish with self-maintaining populations occur in coastal rivers and creeks. The effects of all of these are not known but must include competition, predation, hybridization with native species, and altering of the invertebrate communities (Fletcher 1986). Brown trout (*Salmo trutta*), carp (*Cyprinus carpio*), and mosquitofish (*Gambusia affinis*) are the best known of the species thought to have adverse effects on native communities.

For Queensland (North-east Coast Division) there is a paucity of published information about the channels themselves. Neither Stanton (1975) nor Arthington and Hergerl (1988) report studies or information on the vegetation of these habitats. Similarly in New South Wales (South-east Coast Division), there is little information on the biota of the bed or channel of the river itself (Fig. 15). In many cases the river channel is actually mapped (e.g. Pressey 1987a,b) but no further mention is made of it. Neither Miles (1975) nor Pressey and Harris (1988) report any studies on river channels.

Likewise in Victoria (South-east Coast Division) wetland surveys by Corrick and Norman (1980), Corrick (1981, 1982), and Norman and Corrick (1988) specifically exclude the rivers. Tasmanian surveys are much the same though at least some of these list submerged species (e.g. Kirkpatrick and Harwood 1983b) even if they do not include data from river channels. The South Australian Gulf Division has no studies of the river beds (Smith 1875b, Lothian and Williams 1988) nor do the South-west Coast or Indian Ocean drainage divisions (Smith 1975c, Lane and McComb 1988).

Floodplain swamps and billabongs

These wetlands, the most diverse in both the structure and composition of vegetation, occur in all drainage areas except for the Western Plateau Division. The word billabong (Fig. 16) is of aboriginal origin and is used in Australia to describe permanent or semipermanent areas of open water on riverine floodplains. On inland rivers they often are oxbows. They have been better studied than other wetland groups because many occur near urban

Figure 15. The Lachlan River in western New South Wales, part of the Murray-Darling system. The Lachlan only rarely flows through to the Murrumbidgee (and thence into the Murray) and is here photographed where the channels have started to divided just prior to the terminating Great Cumbung Swamp. The trees are River Red Gum (*Eucalyptus camaldulensis*) and there are a few scattered emergent species of *Persicaria*, *Cyperus* and *Phragmites*. Submerged species are no longer common here, possibly because of the introduced European Carp (*Cyprinus carpio*).

and/or prime agricultural areas, and they are often rich in waterbirds. Many floodplains and billabongs have been modified, or destroyed by alteration of water level, grazing, draining, and reclamation. They receive increased nutrient loads from urban and rural developments in their catchments and support a large number of introduced species.

Low-lying areas of the North-east Coast and the northern part of the South-east Coast Divisions have been well studied and documented. Coaldrake (1961) provided maps of the land systems and information on geology, physiography, climate, soils, and vegetation but only commented briefly on man and other animals. The wetter areas of these lowlands are dominated by trees of *Melaleuca quinquenervia* with an understorey dominated by sedges (Cyperaceae) and species of Restionaceae. Common understorey genera include *Caustis*, *Fimbristylis*, *Schoenus*, *Baumea*, *Callistemon*, and *Banksia*.

The floodplains of the South-east Coast Division have been reasonably well documented. Goodrick (1970) produced a classical vegetation survey of the region, and Paijmans (1978a) provides a brief survey. The most significant

Figure 16. A billabong on the floodplain of the Dawson River in central coastal Queensland. The aquatics include *Ottelia ovalifolia*, *Aponogeton elongatus*, and *Hydrilla verticillata*. The trees include the smooth-barked River Red Gum (*Eucalyptus camaldulensis*), the rough-barked Coolabah (*E. microtheca*) and a yet to be described palm species of the genus *Livistona*.

work on the vegetation is that of Pressey (1981, 1987a,b) and Pressey and Griffith (1987), which includes information on size, diversity, and interspersion of vegetation/habitat types, cover, condition of marginal vegetation, water supply, and types of alteration and land use in catchments. Pressey's reports concentrate on the marginal and emergent vegetation and do not treat submerged vegetation. Myerscough and Carolin (1986) document the vegetation of the Eurunderie Sand Mass (Myall Lakes area) which includes a dune-swale complex with the swales varying from almost permanently wet to almost permanently dry. The little information available on limnology is provided by Timms (1970) and Johnson (1985). Timms treats the zooplankton but information is scant for other elements of the fauna as most temperate region studies have concentrated on inland billabongs of the River Murray system (e.g. Hillman 1986) rather than on coastal billabongs and swamps.

Some of the Victorian coastal swamps and other wetlands have been surveyed by Corrick (1981, 1982) and Corrick and Norman (1980). Waterbirds are emphasised and the vegetation is treated in less detail than in the reports by Pressey. Coastal swamps of Tasmania have been mapped and described by Kirkpatrick and Harwood (1983a) and Kirkpatrick and Tyler (1988). Tasmanian coastal wetlands include some of the least disturbed but several Tasmanian rivers are heavily polluted by effluent and runoff from abandoned and current mine workings, with major effects on the biota (Kirkpatrick and Tyler 1988). The Murray-Darling Drainage Division has a brief frontage on the southern mainland coast. The wetlands have been mapped by Pressey (1986) who discusses their geomorphology and significance to waterbirds but does not deal with the vegetation. Thompson (1986) includes the vegetation in his survey of the River Murray wetlands. The South Australian Gulf Division has very few floodplain swamps and billabongs, and Lothian and Williams (1988) do not have a category for these in their classification. The Darling System of the South-west Coast Division has been studied by Riggert (1966), Majer (1979), and Semeniuk (1987). Lane and Burbidge (1978) surveyed waterbirds, and Brock and Pen (1984) provide a good description and inventory of the Canning River wetlands. The rivers of the Indian Ocean Division are all ephemeral with irregular and highly variable flows. There have been no studies of floodplain swamps and billabongs.

Coastal lagoons and lakes

The nature and biota of coastal dune lakes (Fig. 17) are summarised by Timms (1986b) who classifies lakes according to their relationship to the surrounding dunes, watertable, and the sea. Six lake-types are identified: (i) perched dune lakes (formed in a natural hollow with the water retained by a podsolised layer above the level of groundwater in surrounding country), (ii) lowland dune lakes (in swales or gutters at or close to sea level), (iii) watertable windows (drowned valley or interdune space), (iv) dune-contact lakes (between dunes and adjacent rock), (v) marine-contact lakes (connected to sea), and (vi) frontal dune ponds (wind-created hollows in frontal dunes).

The biology of these lakes has been studied by Bayly *et al.* (1975), Timms (1982), and Arthington *et al.* (1986). Water can vary from brackish to more saline than sea water and the lakes may be permanently or temporarily open to the sea or adjoining estuary. In some areas lagoons are opened mechanically at intervals, or dredged channels are constructed. In the northern part of the east coast, lagoons, or lakes that are more frequently open have a submerged vegetation of *Zostera capricorni* which, in the south, is replaced by *Z. muelleri*. Other lagoons in all of the southern drainage divisions usually have *Ruppia* spp., charophytes (e.g. *Chara* spp. *Nitella* spp., *Lamprothamnium* spp.) and, along the southern coasts, species of *Lepilaena*. The edge communities, when present, usually consist of emergents such as *Phragmites australis*, *Juncus kraussii*, *Bolboschoenus* spp., and *Schoenoplectus* spp.

The southern areas of the North-east Coast Division have been mapped and described by Coaldrake (1961), Stanton (1975), Arthington *et al.* (1986),

Figure 17. A small lagoon behind coastal dunes on the southern coast of Victoria. Emergents include species of *Bolboschoenus*, *Typha*, and *Phragmites*. Submerged species include species of algae and *Ruppia*. These habitats are important for many species of waterbird for both feeding and breeding.

and Arthington and Hergerl (1988). Arthington *et al.* (1986) studied the limnology of many of the dunal wetlands and areas of open water that are scientifically interesting for their invertebrate, fish, and frog fauna. They are mostly unproductive in terms of biomass and are not important areas for waterbirds (Arthington and Hergerl 1988). Timms (1986b) suggested that the biota is distinctive and includes a sparse phytoplankton dominated by desmids, a characteristic copepod (*Calomoecia tasmanica*) and a few microcrustaceans, odonates, trichopterans, chironomids, and fish. The general absence of groups such as planarians, rotifers, ostracods, amphipods, and molluscs is also characteristic (Timms 1986). The South-east Coast division has a few studies, many of them of marine contact lakes. Lake formation and chemistry are discussed by Timms (1970, 1986b), and Myerscough and Carolin (1986) report on the Myall Lakes although they do not include their unpublished information on the submerged vegetation. Atkinson *et al.* (1981) consider the vegetation and limnology of the lakes in some detail. Lake Macquarie vegetation has been mapped and well described by Wood (1959b) whose study is one of the few to include algae (including phytoplankton).

Benson (1986) omits the submerged lake flora in his vegetation map of the Lake Macquarie region but includes the submerged aquatics of the small surrounding swamps. The submerged vegetation of the Tuggerah Lakes is mapped and described by Higginson (1966, 1970) who also describes the changes since settlement. Lake Illawarra is close to a major industrial area and a University and has received a large amount of attention, including its own bibliography (Mills 1985). The submerged vegetation has been described by Harris *et al.* (1980), the emergent by Mills (1983), and Clarke and Yassini (1985). Ducker *et al.* (1977) surveyed the submerged vegetation of the Gippsland Lakes system providing another of the few algal studies for such wetlands.

Corrick (1981, 1982) provides some information on waterbird habitats further south in this division. Bayly and Williams (1966a,b) supply chemical data for some coastal lakes in western Victoria. The Southeastern Wetlands Committee (1984) describe the wetlands of South Australia.

Coastal lagoons and lakes in the Tasmanian division are treated, again by Kirkpatrick and Harwood (1983a), and by Kirkpatrick and Tyler (1988). The Murray-Darling division has only one large coastal lagoon/lake complex described by Thompson (1986), and Pressey (1986) though again the submerged vegetation has been virtually ignored. The South Australian Gulf division has few coastal lakes, and these are briefly mentioned by Laut *et al.* (1977). There are no coastal lagoons or swamps in the Western Plateau division. The South-west Coast Division is little studied but the studies by Brock and Shiel (1983), and Brock and Lane (1983) include some coastal lagoons. Bunn and Brock (1984), and Lane and McComb (1988) provide an entry into other literature. The Indian Ocean Division has few coastal lakes and neither Smith (1975c) nor Lane and McComb (1988) list any studies of them.

Estuaries

Estuaries in southern Australia have been reasonably well studied and the vegetation can be divided into: (i) mangroves, (ii) seagrass meadows, and (iii) salt marshes. Although the three are quite different in structure and species composition, they often occur together. Some of the more important estuary studies include Warren (1975), Hodgkin *et al.* (1981), Brock and Pen (1984), and John (1987). Hegerl and Timmins (1973), and Shine *et al.* (1973) have surveyed some small Queensland estuaries.

The fauna of estuaries is not easily described in terms of the more permanent vegetation because of the tidal mobility of most of the fauna. Bayly (1975, 1980) reviews the work on estuaries and emphasises the role of detritus and bacteria as sources of food for filter-feeding animals. Phytoplankton are often at low densities because light penetration for photosynthesis is often limited. Zooplankton are represented by a number of estuarine calanoid copepods such as *Gladioferens* spp., *Gippslandia estuarina*, and *Sulcanus conflictus* which are often abundant. Some marine zooplankton (copepods and cladocerans) also penetrate estuarine waters (Bayly 1980). The few studies on benthic animals and plants are reviewed by Bayly (1975); foraminiferans, crabs, and mussels have been the subject of more detailed studies (see references in Bayly 1975, 1980).

Australian estuaries serve as nurseries for many species of fish and some crustaceans. Bayly (1980) reviews the Australian work on fish and larger crustaceans. Potter *et al.* (1983) and Lenanton (1984) report studies on fish species in Western Australian estuaries. Most fish in estuaries are essentially marine species that can tolerate some degree of salinity change; they inhabit estuaries for feeding, breeding, or growth of young. The Australian black bream, *Acanthocarpus butcheri* is one of the few species that is permanently resident in estuaries. Estuaries in southern Australia are also important for migratory species of eels, lampreys, and fish.

Mangroves. Of the 25 species of mangroves that occur in the North-east Coast Division (Beadle 1981) only the following six reach the northern limit of the South-east Coast Division: *Bruguiera gymnorhiza*, *Rhizophora stylosa*, *Excoecaria agallocha*, *Aegiceras corniculatum*, *Avicennia marina*, and *Hibiscus tiliaceus*. South of this Division only *A. marina* occurs. Love (1981), Clough (1982), and Hutchings and Saenger (1987) provide critical sources of information and an entry into the literature on Australian mangroves (Fig. 18).

Seagrass meadows. Prior to den Hartog (1970), Wood (1959a,b) was almost the sole reference on seagrasses. There are 30–36 species of seagrasses in Australia (Kuo and McComb 1989). Larkum *et al.* (1989) provides the most up to date assessment of Australian seagrasses. Surveys of the seagrasses are provided by Young and Kirkman (1975), Cambridge (1975), and West *et al.* (1985). Species of seagrasses are found in all drainage divisions with the largest number and areas of seagrass in the Indian Ocean and South-west Coast. The South Australian Gulf also has large seagrass beds. Species of *Halodule*, *Halophila*, *Posidonia*, *Zostera*, *Heterozostera*, and *Amphibolis* occur in the South-east Coast. Of these *Halodule* and *Posidonia* are not found in Tasmania. In the Indian Ocean and South-west Coast Divisions species of *Syringodium* and *Thalassodendron* are added to the list. These areas are also centres of diversity for *Posidonia*.

Saltmarshes. Saltmarshes (Fig. 19) are frequent on low-lying areas adjacent to saline water or mangroves. Common saltmarsh species include *Sarcocornia*

Figure 18. A stand of Grey Mangrove (*Avicennia marina*) on Botany Bay, an estuary immediately to the south of Sydney, New South Wales. A second mangrove species (*Aegiceras corniculatum*) occurs immediately to the landward of this stand. Deeper water off the edge contains extensive stands of the seagrasses *Posidonia australis* and *Zostera capricorni*.

quinqueflora, *Halosarcia* spp., *Sporobolus virginicus*, *Zoysia macrantha*, *Samolus repens*, and *Triglochin striata*. *Casuarina glauca* (in the east) or *C. obesa* (in the west) may grow in and/or on the landward margins of the marsh. The best general references are Beadle (1981), Saenger *et al.* (1977), Adam (1981), and Hutchings and Saenger (1987). Durrington (1977) and Batianoff and McDonald (1980) briefly describe and map saltmarshes in the North-east Coast.

Clarke and Hannon (1967, 1969, 1970, 1971) and Kratochvil *et al.* (1973) are the most detailed series of papers dealing extensively with saltmarsh around Sydney. They record data on soils, climate, interaction between species, and growth in relation to salinity and waterlogging. Other studies from the South-east Coast include those of Goodrick (1970), Pressey (1981), Pressey and Griffith (1987), and Adam *et al.* (1988). Bridgewater (1975) provides one of the few phytosociological treatments of Australian wetlands.

Tasmanian saltmarshes have been treated by Kirkpatrick and Glasby (1981), and Kirkpatrick and Harwood (1983a,b). Studies of the Murray-Darling saltmarshes are mentioned in Thompson (1986), and Pressey (1986). The South Australian Gulf division has very large areas of saltmarsh that

Figure 19. Saltmarsh on Botany Bay, an estuary immediately to the south of Sydney, New South Wales. The shrubs are landward remnants of the mangrove *Aegiceras corniculatum* with the trees in the background being *Casuarina glauca*. The ground cover is largely composed of samphire (*Sarcocornia quinqueflora*) with smaller amounts of *Sporobolus virginicus*, *Samolus repens*, and *Suaeda australis*.

may merge into more inland saline communities dominated by species of Salicorneae and other chenopods. Laut *et al.* (1977) map these communities but little else is published on these saltmarshes. In contrast, the South-west Coast saltmarshes are well studied (Congdon and McComb 1980, Hodgkin *et al.* 1981, Brock and Pen 1984, Penn 1987). There are no published accounts of saltmarshes in the Indian Ocean Division.

Mountain lakes and swamps

Australia has comparatively few upland perennial lakes and these rarely develop a characteristic flora. The lakes are mostly shallow and subject to large fluctuations in area and depth. Consequently, most have few submerged species and those that do occur are adapted to seasonal water-level fluctuations. The most common species in the lakes are *Ruppia megacarpa*, *Vallisneria gigantea*, *Lepilaena* spp., and *Myriophyllum* spp.. Marginal emergent vegetation is not well developed and is usually restricted to the hardiest species, such as *P. australis*, *Typha* spp., and members of the Cyperaceae. Algal blooms are common, especially where lakes are surrounded by agricultural land. Most lakes are found in the southern areas of the South-east Coast or Murray-Darling Divisions and in Tasmania. A good general reference is Williams (1983). The alpine lakes are treated by Costin (1954), Browne *et al.* (1944), Dulhunty (1945), Timms (1980b), and Raine (1982). Timms states that the benthic fauna of some glacial lakes near Mt Kosciusko is of biogeographical significance. The lower altitude lakes are few in number and not extensively studied, though the problems with algal blooms have been studied (May 1970, 1972).

The scarcity of information on the fauna of mountain lakes may reflect a paucity of wetlands on the very limited areas of mountains. Lakes that have low nutrient levels are unproductive and support few waterfowl. Tasmanian lakes, however, have been treated more thoroughly by Tyler (1974), Timms (1980a), Williams (1964, 1974, 1980), and Kirkpatrick and Tyler (1988).

Of the perennial lakes, the Kosciusko glacial lakes have been best studied. Bayly (1970b) and Benzie (1984) discuss the zooplankton and Timms (1980b) the benthic fauna. An assemblage of species typical of these lakes includes the oligochaete worm, *Antipodrilus davidis*, the phreatoicid isopod, *Metaphreatoicus australis*, the chironomid, *Chironomus oppositus*, the pea-shelled mussel, *Pisidium tasmanicum*, and an unidentified gammarid amphipod (Timms 1980b). Several endemic species of worms and molluscs have been described from these lakes. Timms also indicates that predation by fish (especially introduced species) may affect species numbers.

Perennial swamps (Fig. 20) in the higher country are similar to coastal swamps in species composition but generally support more species of the families Cyperaceae and Juncaceae as well as species of *Ranunculus*, *Villarsia*, *Glossostigma*, *Limosella*, *Lythrum*, and *Montia*. These swamps occur in the South-east Coast Division, in adjoining high altitude areas in the Murray-Darling and Tasmania Divisions. The alpine swamps have been studied by Costin (1954, 1957), Costin *et al.* (1979), and Beadle (1981). Other swamps are less well studied but information, mainly on limnology and bird usage, is available from Timms (1970), Briggs (1980), and White (1986). The Tasmanian swamps are dealt with by Kirkpatrick and Harwood (1983a), and Kirkpatrick and Tyler (1988).

The fauna of upland swamps is also sparsely described. Timms (1970) and White (1986) are the best sources of data. Timms surveyed a number of wetlands on the Northern Tablelands of New South Wales and examined the distribution of aquatic invertebrates in relation to altitude, water chemistry, turbidity, and the age of the locality. White focused on waterbirds but included some detail of invertebrate and plant communities.

Ephemeral lakes/swamps (Figs. 21, 22) can be difficult to distinguish from their more perennial counterparts. Many are ephemeral on a long term cycle (e.g. 10–15 years) and the dry period is probably important in their com-

Figure 20. A permanent swamp at Barrington Tops on the Great Dividing Range, NW of Sydney, New South Wales. The swamp is largely vegetated with species of Cyperaceae, Juncaceae, and Restionaceae with low-growing species of *Ranunculus*, *Sphagnum*, and *Montia*. The swamp is at a comparatively high altitude (c. 1,300 m) and receives substantial rainfall; water is released continuously into the streams draining into the coastal tributaries.

munity changes: such lakes are superficially similar to the perennial equivalents. The more ephemeral wetlands sometimes support few plant species and the area may be covered with one species of *Amphibromus* and *Eleocharis* or species of the family Cyperaceae. Although some aquatic species are present only when habitats are wet the composition of the aquatic flora may vary with different periods of inundation or different grazing regimes (Brock 1988, Brock and Casonova in press, Casonova and Brock in press). There seems to be little specialized literature on these habitats and the best sources of information are small sections of many of the references listed in the previous section. With the exception of parts of the alpine region (Costin 1954, Costin *et al.* 1979) most of these habitats have been heavily modified by grazing.

Inland rivers

There are three types of inland riverine systems: (1) rivers (Fig. 23), (2) billabongs, and (3) swamps. Only three drainage divisions (Murray-Darling, Bulloo-Bancannia, and Lake Eyre) have rivers whose origins are in areas of comparatively higher rainfall and then flow inland through areas of lower

Figure 21. Lake Callabonna, a salt lake in NE South Australia, is one of a series of low-lying (some below sea level) endorheic lakes. The lakes rarely support macrophytes even on the rare occasions they hold water but, when full, support a specialised fauna with many endemics and provide good feeding and breeding conditions for many waterbirds. The lakes are surrounded by a flora dominated by species from the families Chenopodiaceae (including many species of samphire), Aizoaceae, and Gramineae.

rainfall. Only from the well-studied Murray-Darling (Fig. 15) does some water actually reach the sea. The Western Plateau Division has some dry water courses that even occasionally hold water: these are similar to the ephemeral rivers described below but so rarely hold water that they are of little significance to aquatic animals or plants.

Twelve of the 19 Australian species of waterfowl occur in the Murray-Darling Division. Different wetland types are used at times for feeding, breeding, moulting, or as refuges (Frith 1977, Braithwaite 1975, 1980). Large numbers of waterfowl occur in inland Lignum (*Muehlenbeckia cunninghamii*) swamps when flooded and some of the less common species, such as the freckled duck (*Stictonetta naevosa*) breed there (Briggs 1983). Australian shelduck, pink-eared duck (*Malacorhynchus membranaceus*), grey teal, Pacific black duck, and maned duck (*Chenonetta jubata*) all inhabit the *Eucalyptus camaldulensis* forests that line most river channels. Swans (*Cygnus atratus*) prefer the more permanent parts of the river systems where submerged and emergent plants are well established. Breeding rookeries for ibis, heron, egret, and spoonbill are limited and are concentrated in areas of inland

Figure 22. Much of the inland receives little rainfall and there are numerous endorheic drainage areas with swamps or small lakes at their centre. This one is in the Strzelecki Desert in NE South Australia, south of the Simpson Desert. The plant with the strap-like leaves is the Darling Lily (*Crinum flaccidum*), and the glaucous Canegrass is *Eragrostis australasica*. Ground cover includes species of *Marsilea* (nardoo), *Diplachne*, and *Uranthoecium*. Animals include shrimps and frogs adapted to the erratic rainfall.

wetlands. Many of their mixed-species rookeries are vulnerable to land drainage and clearing.

Rivers

1. *Perennial rivers (including anabranches).* Only the Murray-Darling Drainage Division includes true perennial rivers although the Lake Eyre Division has some that have perennial headwaters. Towards their source, perennial rivers are lined with *Casuarina cunninghamiana* and support submerged aquatics such as *Vallisneria gigantea* or *Potamogeton* spp. Emergents including *Typha* spp., *Phragmites australis*, *Triglochin procera*, or *Eleocharis* spp. grow in protected sites. After the rivers meet the plains, the channels are characteristically lined with *Eucalyptus camaldulensis*, with a scattering of *Acacia stenophylla*, *E. microtheca*, or *E. largiflorens* and the water becomes turbid from suspended soil particles and dissolved organic matter. The reduction in light penetration, coupled with fluctuating levels and flows, means that few submerged aquatic species grow in the river channels. European

Figure 23. Towards the headwaters of the Cudgegong River in central New South Wales, a tributary of the Macquarie River. Trees of River Red Gum (*Eucalyptus camaldulensis*) line one bank and the introduced *Salix babylonica* the other. Submerged plants include species of *Vallisneria* and *Potamogeton*. Even though the valleys are cropped and grazed, these rivers still support native fish, platypus, and waterbirds.

carp is variously claimed to have greatly increased the turbidity of inland rivers and to be partly responsible for the paucity of submerged aquatic vegetation; they are also claimed to have caused little damage (Fletcher 1986).

Anabranches, with their less rapidly fluctuating flows, may support sparse submerged aquatics like *V. gigantea*, *Myriophyllum* spp., or *Potamogeton* spp., as well as patches of emergents such as *Typha* spp. and *Phragmites australis*.

The Murray-Darling system, which has been studied quite intensively, has erratic flows, largely as a result of overcommitment to agricultural and urban usages (Walker 1985). General studies include those of McLennan and Moore (1976) and Brown (1983). There is a voluminous grey literature (e.g. Fleming 1982, New South Wales Department of Water Resources 1987, references in Pressey 1986). The two most comprehensive reports are those of Thompson (1986) and Pressey (1986). Ecological research on the Murray is reviewed by Walker (1986b).

The diverse phytoplanktonic flora of the Murray River contrasts with the

Darling where this community is limited by the suspended sediments it carries (Walker and Hillman 1981, Shiel *et al.* 1982, Shiel 1981, Shiel and Walker 1985, 1986a,b). In the Murray the assemblage of phytoplankton may alter with changes in the flow regime and with influences from tributaries and billabongs. Early in a flooding cycle algae flushed from impoundments and surrounding wetlands predominate. The high flows of winter and spring show a dominance of diatoms with several species of *Melosira*. Low flows of summer and autumn produce blooms of cyanobacteria (e.g. *Anabaena* and *Anacystis*), green algae (e.g. *Scenedesmus* and *Volvox*), and chrysophytes (e.g. *Cryptomonas* and *Synura*) (Walker 1986a).

Like the phytoplankton, the zooplankton of the Murray-Darling system reflects flow regimes. The Murray's zooplankton groups are typically lacustrine with calanoid copepods and cladocerans dominant. In contrast the zooplankton of the Darling are riverine with rotifers dominating. The lower reaches of the Murray and Lake Alexandrina have a mixed assemblage of zooplankters (Shiel 1986, Walker 1986a). The differences between the Murray and the Darling are attributed to the regulation of the former (Shiel 1986). Of the rotifers, Brachionidae is the most widespread family and has most endemic species and the widest range of morphological variants (e.g. the genus *Brachionus* has 44 species and subspecies). Most of the cladocerans are herbivores or detritivores with predatory groups notably absent. A large proportion (60%) of the cladocerans are restricted in distribution, occurring attached to plants, to substrate in billabongs, or flushed into rivers. The most abundant copepods are calanoid copepods from the Centropagidae. Of these, *Boeckella triarticulata* is ubiquitous in the Murray River and it is the most widespread Australasian copepod species. Cyclopoid copepods are not well represented in rivers and lakes, but are reasonably common in littoral and benthic habitats in billabongs. Ostracods have been infrequently recorded from the river (Shiel 1986).

Only the common benthic invertebrate species are well known taxonomically and for these ecological information is scarce. Little taxonomic or ecological information is available for less common groups but a general guide can be found in Williams (1980, 1983) and Walker (1986c). Benthic habitats are mainly associated with submerged plants and wood rather than the sediments. A decrease in the proportion of insects with distance down the Murray and less obvious increases in annelids and crustaceans was recorded by Mackay *et al.* (1983). Coleoptera, Diptera, Ephemeroptera, Hemiptera, Lepidoptera, Odonata, Plecoptera, and Tricoptera are all well represented.

Molluscs have been surveyed by Smith (1978) and Smith and Kershaw (1979). Some gastropod species may have been eliminated by flow regulation (Walker 1985). Bivalves include the species of the common genera *Sphaerium* and *Pisidium* and the basket shell *Corbiculina australis* which may cause pipeline blockages (Sainty and Jacobs 1981, Walker 1986c). Walker (1981, 1985) has studied the bivalves *Velesunio ambiguus*, *Alathyria jacksonia* and *A. ondola*. Of the freshwater mussels, *V. ambiguus* is a floodplain species and weir construction appears to have favored it over the riverine *Alathyria* spp.

Crustaceans are represented by the yabbie (*Cherax destructor*) and the Murray crayfish (*Euastacus armatus*). The yabbie occurs in all aquatic environments except fast-flowing rivers. Commercial fishing for the yabbie declined in the 1970's, possibly because of the increase in European carp. The Murray crayfish is a species of fast-flowing deep waters and their numbers have declined since weir construction (1920–1940) which decreased suitable habitats, favouring the yabbie. Annelids are commonly represented by the oligochaetes (*Tubifex*, *Branchuria*) and glosiphonid leeches. Coelenterates, sponges, bryozoans, turbellarians, nematodes, and mites are groups commonly represented in this river system (Williams 1980, 1983).

Fish of the Murray-Darling are discussed by Lake (1975), Pollard *et al.* (1980), Glover (1982), and Cadwallader (1986). Of the 50 species recorded, nine are introduced from outside Australia and 13 are marine or estuarine species found only in the lower reaches of the river. Twenty two native species from 12 families spend their whole life-cycle in this system. Environmental change, fishing, and introduced species have caused most native species to decline.

Information on frogs and reptiles is available from Cogger (1983), Tyler (1976), Cheesman (1978), and Thompson (1983). Frogs from two families occur along the river; the tree-dwelling family Hylidae is represented by eight species of *Litoria* and the ground-dwelling family Leptodactylidae by 21 species in seven genera. Although several species are restricted to corridors along the river, most species have flexible modes of development appropriate to a changeable environment (Tyler 1976). Reptiles include tiger (*Notechis scutatus*), copperhead (*Austrelaps superbus*), and red-bellied black (*Pseudechis porphyriacus*) snakes, the golden water skink (*Sphenomorphus quoyii*), and the riverine Murray short-necked (*Emydura macquarii*), broad-shelled (*Chelodina expansa*), and the floodplain and swampland long-necked (*C. longicollis*) turtles.

The platypus (*Ornithorhynchus anatinus*) and the water rat (*Hydromys chrysogaster*) are the two aquatic native mammals found in temperate inland rivers. The platypus, one of the surviving monotremes, is carnivorous and feeds on insect larvae, crustaceans, and molluscs in river sediments. It is locally common in the upper Murray but was once more widespread. The platypus also occurs in rivers and streams in the Great Dividing Range of eastern Australia and Tasmania. The water rat, a rodent, is locally common

Figure 24. Yandama Creek flows from NW New South Wales into South Australia towards Lake Callabonna. Flows are uncommon, highly variable, and rarely reach the Lake. The trees are mostly Coolabah (*Eucalyptus microtheca*) with some River Red Gum (*E. camaldulensis*). The occasional very heavy flow may kill trees long-established after decades of low flows.

particularly in the irrigation areas. Fish and benthic invertebrates are the preferred prey.

2. *Ephemeral rivers*. Many ephemeral rivers (Fig. 24) only carry water after rain, with a few shallow pools persisting longer. Others, with catchment heads in higher rainfall areas, have almost permanent headwaters, but are ephemeral for the majority of their length with occasionally more perennial waterholes. The channels are usually lined with *Eucalyptus camaldulensis*, *E. microtheca*, or *E. largiflorens*. Aquatics are confined to the more permanent pools. The more permanent the pool the greater the range of aquatic species. As in perennial rivers, *Potamogeton* spp., *Myriophyllum* spp., and *Vallisneria gigantea* are the more common submerged aquatics but there is usually a greater range of emergents with species of *Cyperus*, *Eleocharis*, *Persicaria*, and *Juncus* in addition to those found in perennial rivers. Algal blooms are common when nutrient levels rise as the water levels fall.

There have been few studies on this type of wetland though Goodrick (1984, 1985) provides an excellent basis for further studies of wetlands in the Murray-Darling and Bulloo-Bancannia Divisions. Ephemeral rivers are also mentioned in references on perennial rivers and on swamps. Bayly and Williams (1973) provide some useful information and Boulton and Suter (1986) review what little has been done on Australian ephemeral streams. They conclude that Australian ephemeral streams have a greater macroinvertebrate species richness than expected from studies elsewhere in the world.

Billabongs – floodplains

Billabongs of inland areas are oxbows or river bends that are often quite deep, fill during floods, and dry slowly. Plant species are similar to the main rivers but emergents are more common. *Muehlenbeckia cunninghamii* often occurs on the banks. If domestic stocking rates are not too high, *Ottelia ovalifolia* or *Potamogeton tricarinatus* may grow in the shallow water. Edge communities of species such as *Damasonium minus*, *Marsilea* spp., *Crinum flaccidum*, *Sporobolus mitchellii*, and strand species such as *Heliotropium curassavicum* or *Glinus lotoides* may also establish. Billabongs are distinct limnologically from the main channels (Hillman 1986) due to the more sluggish flows of shorter duration and slower drying periods. A diverse community of benthic invertebrates occurs in billabong and floodplain habitats (Shiel 1980, Walker 1986a). Hillman (1986), Thompson (1986), and Pressey (1986) provide further information.

Swamps – overflow or terminating

The overflow swamps, especially in the Murray-Darling Division, are well studied because of their significance to both irrigated agriculture and to conservation issues. Some inland rivers, even the perennial ones, only join up with the main channel during major floods; most of the time they terminate in large swamps or playas. Others may overflow in large floods, inundating huge areas (sometimes in excess of 100 km^2). Normally the Lachlan River (New South Wales) terminates in the Great Cumbung Swamp, the Macquarie River (New South Wales) in the Macquarie Marshes, and the Bulloo River (Queensland) before the Bulloo Overflow. The Bulloo Overflow fills only occasionally and the water may take ten years or more to disappear. When full, many swamps have large expanses of turbid, shallow water and wave action, and turbidity usually limit submerged vegetation. As the water levels drop, a ground cover of species of *Marsilea*, *Cyperus*, and *Eleocharis* develops with *Atriplex* spp. and *Eragrostis australasica* in the marginal depressions. Stands of *Muehlenbeckia cunninghamii*, *Eucalyptus microtheca* or *E. largiflorens*, or *E. camaldulensis* develop along the deeper channels.

The large variation in structure, appearance, and vegetation of these swamps is described by Beadle (1981) with other useful contributions from

Goodrick (1984, 1985), Knights (1980), Paijmans (1978a), and Pressey *et al.* (1984). Briggs *et al.* (1985) and Briggs and Maher (1985) provide information on water chemistry and macrophyte productivity for waterfowl habitats. Linacre *et al.* (1970) showed that water loss from these swamps may be reduced by vegetation. Management plans are available for some of these wetlands (e.g. New South Wales Department of Water Resources and National Parks and Wildlife Service 1986).

Inland lakes

Inland lakes are well studied, thanks largely to W. D. Williams and his associates. Good general references to this complex include Bayly and Williams (1966a,b, 1973), Bayly (1970a), Timms (1980a), Williams (1967, 1980, 1981a, 1983, and other references in McComb and Lake (1988). Inland lakes occur in all southern mainland drainage divisions. They are mostly ephemeral but some seasonally ephemeral. Others flood irregularly and some may remain full for many years before drying. Some perennial lakes occur in the wetter regions of the South-east Coast, Murray-Darling, South Australian Gulf, and South-west Coast Drainage Divisions.

Many of the lakes are endorheic, some are exorheic and range from fresh to quite saline. Wave action and fluctuating water levels limit the vegetation in larger lakes but species of *Ruppia*, *Lepilaena*, and the charophyte *Lamprothamnium papulosum* are reasonably widespread, especially in slightly saline lakes. The immediate areas draining into these lakes usually support swamp species such as *Eucalyptus camaldulensis*, *Muehlenbeckia cunninghamii*, or *Eragrostis australasica* if the salinity levels are not too high, and species of Chenopodiaceae (especially Salicorneae) and sometimes *Muehlenbeckia coccoloboides* for lakes of higher salinity.

Waterbird species on inland lakes reflect the other types of wetlands in the region as most species are mobile and use a combination of wetland types. The shelduck, for example, is common on inland salt lakes in Western Australia but requires a freshwater source as well. The birds of inland lakes are thus similar to those described for inland rivers.

The fauna of Australian salt lakes has been studied with increasing intensity over the last 20 years. De Deckker (1983) and Williams (1984) summarise the major characteristics and generalities from a wide range of studies. The fauna differs from that of salt lakes elsewhere in the world in that it has many endemic elements (e.g. two crustacean genera, the brine shrimp *Parartemia*, the isopod *Haloniscus*, and the gastropod mollusc genus *Coxiella*). Within Australia the fauna shows some marked regional differences from east to west. Many of the salt lake species are closely related to freshwater species (Williams 1983). Benthic microbial communities, common in saline lakes, are described by Bauld (1986). These communities are dominated by photosynthetic prokaryotes (cyanobacteria) and, sometimes, by eukaryotic microalgae. Other communities of micro-organisms of particular note are those of some meromictic lakes in Tasmania. There is a finely stratified community of micro-organisms around the abrupt boundary between stagnant bottom waters and mixed upper layers; this community includes pigmented and non-pigmented sulphur bacteria, some cyanobacteria, and eukaryotic algae (Croome 1986).

The fauna of the relatively few freshwater inland lakes is also reasonably well studied. The studies on zooplankton communities in freshwater lakes and ponds are well summarised by Mitchell (1987), and Geddes (1986) deals with these communities in farm dams. There are over 600 species of rotifers known from Australian inland waters, with 15% endemicity (Shiel and Koste 1986). The diversity of species in freshwater lakes is much higher than in salt lakes. However, the species diversity of the benthic fauna is low compared with freshwater lakes outside Australia. Reasons suggested include geographical isolation, the paucity and relative youthfulness of the lakes, and the lack of well defined seasons. Zooplankton species are also limited compared with northern hemisphere counterparts; in general an Australian inland lake will contain less than three calanoid copepods, two cyclopoid copepods, three cladoceran species, and five rotifers. Calanoid copepods of the Centropagidae are diverse and widespread in southern Australia and many of the species are endemic (Williams 1983, Mitchell 1987).

The variety of inland waters in which fish are distributed include freshwater lakes and rivers, saline lakes, cave-pools, artesian wells, and large temporary lakes (Williams 1983). Some species are remarkably tolerant of salinity, these include the Lake Eyre hardyhead (*Craterocephalus eyresii*), the mosquito fish and the common galaxid (*Galaxias maculatus*), all of which occur in salinities above that of seawater as well as in freshwater.

Inland lakes are not very common in the South-east Coast Division. In the south west of the region there are several lakes described by Bayly and Williams (1973), Williams (1981b), Southeastern Wetlands Committee (1984), and Lothian and Williams (1988). Detailed information is available on the macrophytes and their growth (Brock 1981, 1982a,b). These lakes are more varied than in any other region of the country as they include crater lakes, spring-fed lakes, and lakes fed from local run-off. The latter group may start as essentially fresh, becoming increasingly saline as the season progresses; others are saline and salinity increases as the water level falls. Information on the limnology (Bayly and Williams 1964, 1966b, Tamuly 1970, Timms 1974) and the prehistory (Dodson 1974, 1975) is available for the volcanic lakes.

The Murray-Darling Drainage Division has inland lakes that fall roughly into two variants. The southern lakes (in Victoria and South Australia) are either nearly perennial or regularly ephemeral whereas in the north lakes are ephemeral and intermittently filled. The southern lakes have been a topical issue because of increased salinity. There have been numerous studies, mostly unpublished (some listed in Norman and Corrick 1988) and the best treatment is by Williams (1967).

Intermittent lakes further north have received less attention. Goodrick (1984, 1985) provides some information on vegetation and Williams *et al.* (1970) provide limnological information for these lakes from the Murray-Darling and Bulloo-Bancannia Divisions. The lakes of the South Australian Gulf Division have mostly been included in the studies of the south west of the South-east Coast Division.

The Lake Eyre Drainage Division has the largest, most spectacular, and most intermittent of the inland saline lakes. On the rare occasions when these lakes fill they support abundant life until the salinity levels increase. When partially full the salinity levels are often too high for life but the salinity can be quite variable (Dulhunty 1974). The filling of the lakes in 1949–50 and 1971–74 resulted in blooms of papers on the subject (see bibliographies of Smith 1975b, Warcup 1982, Lothian and Williams 1988).

Lakes are uncommon in the porous limestone of the southern half of the Western Plateau. The intermittent lakes occurring closer to the Tropic of Capricorn are relatively inaccessible areas and have been little studied.

The inland lakes of the South-West Coast Division have been well documented. The formation of the lakes further from the coast have been studied by Killigrew and Gilkes (1974); vegetation and limnological studies include those by Congdon and McComb (1976), Gordon *et al.* (1981), Brock and Lane (1983) and Brock and Shiel (1983). An index of the research on these inland waters has been produced by Bunn and Brock (1984).

Mound springs

Mound springs form a minute percentage of wetlands yet they represent (or represented) the only permanent and reliable water source in much of the drier regions of the Murray-Darling, Bulloo-Bancannia, and Lake Eyre Drainage Divisions. A comparatively large number of surveys has been done on mound springs, the amount of information increasing as the number of active springs decrease.

Mound springs are natural outlets of the Great Artesian Basin, one of the largest artesian basins in the world with an area of about 1.8 million square kilometres. They are usually associated with fractures and fault lines and often associated with mounds of various sizes. The mounds are built up from the minerals precipitated from the springs, the weaker the spring the larger the mound tends to be, up to several square kilometres. The salinity of the water varies, as does the rate of flow. In many instances, the rates of flow have decreased and many springs have ceased flowing altogether, due to the increasing number of bores that tap this underground resource. Most of the springs are used by stock, feral and native animals. Well-developed vegetation and larger surrounding wetlands were common before artesian bores were drilled and stock introduced. At that time these springs would have made a more significant contribution to waterbird habitats and refuges. The survey of South Australia's mound springs (Greenslade *et al.* 1985) provides information on many aspects of mound springs, including inventories of the flora and fauna. Other treatments include those of Bayly and Williams (1973), Ponder (1986), and Lothian and Williams (1988). The southern mound springs characteristically have species such as *Cyperus gymnocaulos*, *C. laevigatus*, *Schoenoplectus pungens*, *Typha domingensis*, and *Phragmites australis*.

The aquatic microflora of mound springs is virtually undocumented. Ponder (1986) suggests that it may differ from nearby waterholes because the carbonate-rich waters are likely to influence species composition. The zooplankton and benthic invertebrates have been reported by Mitchell (1985), Ponder (1985, 1986), and Ponder and Hershler (1984). The aquatic microfauna has widely dispersed elements (in particular insect larvae of the Odonata, Hemiptera, Coleoptera, and Diptera) together with less well dispersed elements that, in general, do not have resistant stages. Mitchell (1985) suggests that this latter group forms a unique mound spring assemblage with species of limited distribution and some endemic species, including hydroiid gastropods, the phreatoicid isopod *Phreatomerus latipes*, and amphipod and ostracod species. Many of the springs also have microcrustaceans and insect species that also occur in temporary pools but Mitchell did not find this mound spring assemblage in temporary pools. Ponder and Hersler (1984) and Ponder (1985) have studied the small snails (Hydrobiidae), finding endemic genera and species, and postulating the evolution within the group.

Crustaceans are represented both by common species such as the yabbie (*Cherax* spp.) and by endemics. The endemic isopod *Phreatomerus latipes* belongs to a monotypic genus but has relatives common in Tasmania and the southeast and southwest of Australia. Atyid prawns occur in some springs. Of the smaller crustaceans, amphipods favour the less disturbed springs and steady water flow, and several endemic ostracod species occur in the Lake Eyre group of springs (Ponder 1986). With the exception of one chydorid cladoceran the microcrustaceans of the Lake Eyre springs species are widespread. Aquatic oligochaetes occur in many springs. A flatworm from the Lake Eyre springs is the first Australian record in the order Macrostomida (Ponder 1986). Fish from the South Australian springs are well studied (Ivanstoff and Glover 1974, Glover and Ingliss 1971, Glover 1973, 1979, 1982, Glover and Sim 1978a,b) but there is a paucity of information on fish of Queensland springs. There are eight species of fish recorded from South

Australian mound springs, most of which occur in other waterbodies in the region. Dispersal is most likely by floodwaters and by streams and pools associated with man-made bores (Mitchell 1985). The desert goby, *Chlamydogobius eremius*, is common in springs of the Lake Eyre group. Only two species are endemic, the Dalhousie catfish (*Neosilurus* sp.) and the Dalhousie hardyhead (*Craterocephalus dalhousiensis*). The introduced mosquito fish is common in many Queensland springs and may threaten native species (Ponder 1986).

Tadpoles and frogs have been observed around some springs (Ponder 1986). Although 22 reptile species are recorded, Thompson (1985) suggests that water from the springs is not important in their distribution but the limestone may provide refuge sites. Records of the birds of mound springs are sparse, although Badman (1979, 1985), and Badman and May (1983) provide some records. Few birds were recorded on the springs themselves and Badman (1985) suggests that artesian bores and creek beds are more important bird habitats. Only springs with large areas of shallow open water supported many waterbirds. Fenced, and therefore well-vegetated, springs had a greater diversity of species and were the only ones where nesting was recorded.

Man-made storages, canal systems, channels, drains, bores, bore-drains, farm storages, rice fields, storage swamps

Construction alters pre-existing systems. For example, the construction of the Dartmouth Dam on the Mitta Mitta river has had a catastrophic effect on the benthic fauna (Blyth *et al.* 1984, Doeg 1984). Effects include the increase in numbers of large kangaroos since european settlement, largely due to the installation of a much more reliable network of watering points across the country. Farm dams are useful habitats for waterbirds. Species that are hunted are aided by an increase in number and dispersion of suitable habitats. The waterbird populations on artificial waterbodies depend largely on the age, size, and depth of the wetland. The types and extent of particular habitats are important, those with an array of depths and plant communities have rich invertebrate communities and more waterbird species. Larger artificial waterbodies often are favored by diving ducks, swan, coot (*Fulica atra*), terns, and gulls (Broome and Jarman 1983).

Man-made wetlands are too diverse to generalise. Each has a faunal composition that reflects not only its limnological characteristics, but also its age and the neighbouring wetlands that act as sources for colonization. Often man-made wetlands provide habitats for organisms in areas that would naturally support only an ephemeral aquatic community (Fig. 25). Although this may be of some advantage to the extended permanent wetland communities in the short term it may alter water tables and natural drainage regimes.

Figure 25. A recently constructed lake at Penrith, west of Sydney, New South Wales. Part of a scheme to convert sand and gravel pits into a series of lakes for recreation and conservation. Already the lakes have good submerged growth of native species and marginal stands of emergents are starting to develop. The waterbodies already attract large numbers of water birds.

The salinisation of a large number of inland wetlands, particularly in the southwest of Western Australia and in irrigation areas in the eastern States exemplifies this. The ecological effects of river regulation in the various drainage divisions are well reviewed by Walker (1985). Man-made lakes are discussed by several authors in Williams (1980). Castles (1986) lists 64 dams and reservoirs with a capacity of more than 100 million cubic meters of water. The largest are Lakes Gordon and Pedder in Tasmania with a combined capacity of 11,728 million cubic meters of water. Other specialized man-made wetlands have been treated to varying degrees. Waste stabilisation ponds have been examined by Mitchell and Williams (1982a,b) and farm dams by Geddes (1986) and Timms (1980c). Less common plant species rarely seem to benefit from man-made storages. For example, *Eriocaulon carsonii* and *Utricularia* sp., endemic to mound springs, have not yet been recorded from any artesian bore or bore drain. Most aspects of man-made storages and structures have been covered by Bayly and Williams (1973), Williams (1974, 1980, 1983), Sainty and Jacobs (1981), De Deckker and Williams (1986), and McComb and Lake (1988).

Wetland use and conservation

Table 14 summarises the potential and actual water usage of run-off from the drainage divisions. The Murray-Darling and South Australian Gulf div-

Table 16. Agricultural use of water by State. Figures are not available for individual drainage basins but some idea of the importance of each can be obtained from Table 14. Figures for Queensland and Western Australia include figures north of the Tropic of Capricorn as it is difficult to separate the data; approximately half of Western Australia's irrigation areas are north of the tropic and possibly more in Queensland (Castle 1986). The high figure for "other" crops in Queensland is mostly made up by Sugarcane (c. 104,000 ha).

Use	Area (ha)					
	N.S.W.	Vic.	Qld.	S.A.	Tas.	W.A.
Pastures(+hay)	289,243	469,373	32,953	46,578	18,506	13,932
Cereals	244,755	30,868	34,242	2,501	1,668	1,455
Vegetables	13,217	19,580	20,167	6,412	13,147	4,132
Fruit	24,218	27,915	9,548	29,319	2,166	4,382
Other	90,102	7,353	157,473	2,143	4,418	1,797
Total	661,535	555,089	254,383	86,953	39,905	25,698

isions are over-committed when the variability of rainfall is considered (Brown 1983). Water is already diverted from the South-east Coast Division to the Murray-Darling to provide water for irrigation and power for hydro-electricity schemes. Table 16 summarises the use of irrigation water State by State.

The only attempt at an inventory of all Australian wetlands suffered, as the authors admit, from the size of the task and the scale available (Paijmans *et al.* 1985). Most inventories are on a State (or smaller) basis as conservation and management of wetlands is controlled at the State level. Stanton (1975) has compiled a useful wetland inventory for Queensland (part of each of North-east Coast, Murray-Darling, Bulloo-Bancannia, and Lake Eyre Divisions). Large areas of New South Wales (part of each of South-East Coast, Murray-Darling, and Bulloo-Bancannia Divisions) have been covered by Pressey (1981) and co-workers, Goodrick (1984, 1985), Shorthouse (1984), and West *et al.* (1985). Corrick (1981, 1982) and co-workers are producing similar inventories for Victoria (part of South-east Coast and Murray-Darling divisions). Tasmania has been well served by Kirkpatrick and co-workers (Kirkpatrick and Glasby 1981, Kirkpatrick and Harwood 1983a, 1983b, Kirkpatrick and Tyler 1988). South Australia (South Australian Gulf, part of Lake Eyre and Western Plateau Divisions) has no equivalent survey though Laut *et al.* (1977) goes part of the way toward supplying this information. Western Australia (South-west Coast, part of Indian Ocean and Western Plateau Divisions) has similar studies completed by several different people (see references in Bunn and Brock 1984 and Lane and McComb 1988), particularly for the more critical areas of the South-west Coast, especially the estuaries and coastal plains.

Major impacts

Major impacts of river impoundment and pollution have been reviewed by Walker (1985). Williams (1980) reviews the problem associated with general management of water resources. In Australia the three main ecological problems facing wetlands are: (i) pollution, (ii) alteration in flow, (iii) salinity.

Pollution. For pollution, we use a definition modified from Bayly and Williams (1973): "a significant and deleterious change in the natural character of water resulting from the addition, directly or indirectly, of material or heat by man". Pollution can be urban or agricultural and both have two main types of action: (a) via chemical toxins, and (b) reduction in the amount of dissolved oxygen. Bayly and Williams (1973) discuss the various limnological aspects of pollution in Australia. In Australia urban pollution areas are mainly coastal, and in areas of reasonably high density of population in rural Victoria (Murray-Darling Division). Some effort is being made to minimise the effects of urban pollution. Agricultural pollution is more widespread and less obvious. Efforts to minimise the effects of agricultural pollution, are only successful when economic returns are influenced (e.g. reduction in use of toxic long-life pesticides).

Bowmer (1981) states that eutrophication is widespread in Australian waters but the natural condition of many of our wetlands would be classified as eutrophic by world standards. However, more research is needed to understand the situation before we become too complacent about increasing nutrient loads. Problems of one alpine wetland receiving extra nutrient load from a resort area have been assessed by Finlayson *et al.* (1986), who measured the uptake of phosphorus and nitrogen throughout the year and commented on introduced weedy species.

Alterations in flow. Walker (1985) summarises the effects of flow alterations. The effects on breeding cycles of water birds and fish as well as other aquatic animals are most obvious. Where possible, flow allocations are being made for conservation purposes.

Salinity. Salinity is a problem that has hit the Murray-Darling and Southwest Coast Divisions hardest. The problem is the establishment of large agricultural areas on a large prehistoric saline "lake" though there still seems to be some argument as to the actual pathway(s) by which salt enters the system. Baldwin *et al.* (1939), Storrier and Stannard (1980), and Grieve (1987) all supply useful accounts of the salinity problem. Baldwin *et al.* (1939) state that salinity first became noticeable as a problem in about 1909, about 20 years after the commencement of limited irrigation and about 14 years before the enlargement of the scheme in that area.

Up to 80% of South Australia's water supply comes from the lower River Murray which makes salinity in the Murray-Darling of particular concern. At present irrigation water is not returned to the river but is directed to

Figure 26. Water Hyacinth (*Eichhornia crassipes*) blocking a floodplain billabong on the-Hawkesbury River west of Sydney, New South Wales. Although biological control programs have been established for this and other troublesome species, they work much better in some habitats than others. Populations such as this are usually sprayed with herbicide with resulting continued disturbance to the few remaining native species in the habitats.

low-lying areas and allowed to evaporate. This practise increases both the permanence and salinity of many of the wetlands of semi-arid areas. In areas where salinity is exacerbated by raising the water table, the attempt to reduce river salinity may remove even more land from agricultural and conservation use. Salinity is a problem in other low-lying areas wherever extensive clearing has taken place and subsequent agricultural practices have resulted in a raising of the water table. Froend *et al.* (1987) document one example from W.A.

Minor impacts

Aquatic weeds (Fig. 26), feral animals, and urban spread are minor only in the sense that they occur on a smaller scale and they have comparatively easier solutions. The solutions still cost money and, at best, are usually only partially implemented.

Aquatic weeds are both an agricultural and a conservation problem. Some species, notably *Eichhornia crassipes*, *Salvinia molesta*, *Alternanthera philoxeroides*, and *Ludwigia peruviana* have the ability to invade natural wetlands (Jacobs and Sainty 1987). The first three species have all been the target for biological control programmes (Harley 1981) that have been at least partially effective in preventing widespread monospecific stands. *Ludwigia peruviana* is starting to spread into wetlands on the South-East Coast Division and control methods are being investigated. Other management options are treated in Sainty and Jacobs (1981).

In the southern wetlands the feral pig (*Sus scrofa*) is of major concern as a potential reservoir of disease, especially exotic disease. Pigs also extensively damage vegetation in intermittently wet areas. Several control options are being investigated. The cane toad (*Bufo marinus*) is naturalized in Queensland and has the potential to move much further south and west; it competes with native amphibians for food and habitat. There are 19 species of exotic fish established in Australia and some of these have deleterious effects on the environment (Fletcher 1986). As well as aquarium escapes, stocking of sport fish such as trout (*Salmo* spp.) alters the habitat for native fish. The mallard (*Anas platyrhynchos*) and domestic ducks (*A. domestica*) are potentially a problem for conservation of the native pacific black duck (*A. superciliosa*); these three species interbreed and threaten the genetic stock of the native black duck (P. Jarman personal communication). The use of waterholes by domestic stock alters them and changes the composition of the aquatic communities.

Urban development, although unavoidable, can be planned and controlled. Like other areas of the world, urban spread in Australia ranges from haphazard to integrated planned development. As a result of increasing public awareness of conservation issues, wetlands are given more protection than a decade ago.

Conservation status

Conservation status of Australian wetlands in each State has been reviewed by McComb and Lake (1988). Relevant legislation in Australia is enacted on a State basis. There is an overall (commonwealth or national) approach for wetlands of international significance; Michaelis and O'Brien (1988) list 25 such wetlands from southern Australia. While all these wetlands are protected in some way the degree of control and restriction varies; for example, some are still logged or grazed.

Arthington and Hergerl (1988) regard the south west of Queensland as the area most deficient in conserved wetlands; present land use and tenure in many inland areas makes reservation difficult. Although reserving wetlands under pressure in New South Wales is in progress, Pressey and Harris (1988) feel that more survey work and an integrated approach is needed to replace the current piecemeal procedures. Norman and Corrick (1988) also recommend further survey and inventory work for Victorian wetlands before conservation policies are implemented. Victoria, with the highest population density and smallest area could be considered to be lagging in conserving wetlands. In Tasmania, 36 wetlands of known high conservation significance are identified by Kirkpatrick and Tyler (1988). The problem here is the fragility of the reserves system and susceptibility to the "tyranny of small decisions". The situation in South Australia is stronger as 64 important wetlands have been identified and 40% reserved and some progress is still being made (Lothian and Williams 1988). The critical areas of Western Australia have been studied and some action taken although inventories, research, management and an integrated approach to wetland problems are still required (Lane and McComb 1988).

Recommendations for wetland conservation

The problems confronting Australian wetlands are complex and need more than simple identification and reservation of individual wetlands. Some recommendations are to:

1. Examine wetlands from a drainage basin and whole catchment perspective rather than from political boundaries. Legislation would still require State action but an integrated approach is essential. The establishment of a commission responsible for the Murray-Darling system shows this is possible. This new initiative was established because of salinity problems and overcommitment of the waters; the results are yet to be seen.

2. Establish national standards, principles, or guidelines. The status and stability of reserve systems varies between States and, with time, between governments within States. All reserves can be revoked or altered by legislation but the ease with which this occurs in some States is ludicrous.

3. Lease some wetlands areas for enterprises that depend on maintenance of the wetland for their success as an alternative to government control. State, rather than private, management of conservation areas may reduce flexibility. Examples include private management of reserves for hunting, fishing, or birdwatching. Some private leasing schemes have been tried and some are directly operated by State Governments in important wetland reserves (Yugovic 1985). A similar approach to the use of rangelands has been in operation for some years in western New South Wales. Problems with such systems have largely been due to political climates when the conditions of the lease are not enforced, or ideological climates where conservation can be suddenly switched to exploitation. Experience has shown that the latter switch happens much more readily and easily than the switch from exploitation to conservation.

4. Preserve an adequate sample of unaltered systems while they are available. Relatively unaltered systems occur in most dryland areas.

5. Accept that altered ecosystems are also suitable for conservation. Rejection of altered ecosystems as suitable for conservation is common in

Australia. Water and wetlands are scarce resources and compromises will be necessary. Where systems have been modified or managed, subsequent management needs to ensure flexibility so that environmental damage can be minimised and conservation values (retained or created) maximised. Possibilities include the creation of new nesting sites for waterbirds, controlling introduced predators, and more careful disposal of used or contaminated water from irrigation systems. Of course there will be a price but it may prove far cheaper than the alternatives.

Acknowledgment

Some of the figures in this text are reproduced with the permission of the Supervising Scientist for the Alligator Rivers Region.

References

Adam, P. (1981) Australian saltmarshes. Wetlands (Australia) 1: 8–10.

Adam, P. (1990) Saltmarsh Ecology. Cambridge University Press, Cambridge, England. 461 pp.

Adam, P., Wilson, N. C. and Huntley, B. (1988) The phytosociology of saltmarsh vegetation in New South Wales. Wetlands (Australia) 7: 35–85.

Allen, C. C. (1985) The rehabilitation of Rum Jungle. p. 197–218. *In*: J. W. Lawrie (ed.). Proceedings North Australia Mining Rehabilitation Workshop No. 9 Weipa, Queensland, Comalco Aluminium Ltd., Queensland, Australia.

Anderson, P. K. (1986) Dugongs of Shark Bay, Australia – seasonal migration, water temperature, and forage. National Geographic Research 2: 473–490.

Anderson, P. K. and Birtles, A. (1978) Behaviour and ecology of the dugong, *Dugong dugon* (Sirenia): Observations in the Shoalwater and Cleveland Bays, Queensland. Australian Wildlife Research 5: 1–23.

Arthington, A. H., Burton, H. B, Williams, R. W. and Outridge, P. M. (1986) Ecology of humic and non-humic dune lakes, Fraser Island, with emphasis on the effects of sand infilling in Lake Wabby. Australian Journal of Marine and Freshwater Research 37: 743–764.

Arthington, A. H. and Hergerl, E. J. (1988) The distribution, conservation status and management problems of Queensland's athalassic and tidal wetlands. p. 59–101. *In*: A. J. McComb and P. S. Lake (eds.). The Conservation of Australian Wetlands. Surrey Beatty & Sons, Chipping Norton, Australia.

Arthington, A. H., McKay, R.J., Russel, D. J. and Milton, D. A. (1984) Occurrence of the introduced cichlid *Oreochromis mossambicus* (Peters) in Queensland. Australian Journal of Marine and Freshwater Research 35: 267–272.

Atkinson, G., Hutchings, P., Johnson, M., Johnson, W. D. and Melville, M. D. (1981) An ecological investigation of the Myall Lakes region. Australian Journal of Ecology 6: 299–327.

Australian Bureau of Statistics (1985) Year Book Australia 1985. Australian Government Publishing Service, Canberra, Australia. 705 pp.

Australian Water Resources Council (1976) Review of Australia's water resources (1975). Australian Government Publishing Service, Canberra, Australia. 170 pp.

Badman, F. J. (1979) Birds of the southern and western Lake Eyre drainage. South Australian Ornitholigist 28: 29–81.

Badman, F. J. (1985) Birds of the mound springs south and west of Lake Eyre with special reference to the Coward Springs area. p. 84–92. *In*: P. Greenslade, L. Joseph and A. Reeves (eds.). South Australian Mound Springs. Nature Conservation Society of South Australia, Adelaide, Australia.

Badman, F. J. and May, I. A. (1983) Waders in northern South Australia. South Australian Ornithologist 29: 29–39.

Baldwin, J. G., Burrill, G. H. and Freedman, J. R. (1939) A soil survey of part of the Kerang District, Victoria. CSIRO Bulletin No. 125, Melbourne, Australia. 43 pp.

Batianoff, G. N. and McDonald, T. J. (1980) Capricorn Coast sand dune and headland vegetation. Botany Branch, Department of Primary Industries, Brisbane, Technical Bulletin No. 6, Brisbane, Australia. 71 pp.

Bauld, J. (1986) Benthic microbial communities of Australian saline lakes. p. 95–112. *In*: P. De Deckker and W. D. Williams (eds.). Limnology in Australia. Dr. W. Junk, Dordrecht, The Netherlands.

Bayliss, P., Webb, A. J. W., Whitehead, P. J., Dempsey, K. and Smith, A. (1986) Estimating the abundance of saltwater crocodiles, *Crocodylus porosus* Schneider, in tidal wetlands of the Northern Territory: A mark-recapture experiment to correct spotlight counts to absolute numbers and the calibration of helicopter and spotlight counts. Australian Wildlife Research 13: 309–320.

Bayly, I. A. E. (1970a) Further studies on some saline lakes of south-east Australia. Australian Journal of Marine and Freshwater Research 21: 117–129.

Bayly, I. A. E. (1970b) A note on the zooplankton of the Kosciusko region. Bulletin of the Australian Society of Limnology 3: 25–28.

Bayly, I. A. E. (1975) Australian estuaries. Proceedings of the Ecological Society of Australia 8: 41–67.

Bayly, I. A. E. (1980) Estuaries and coastal lakes. p. 391–396. *In*: W. D. Williams (ed.). An Ecological Basis for Water Resources Management. Australian National University Press, Canberra, Australia.

Bayly, I. A. E., Ebsworth, E. P. and Hang Fong Wan (1975) Studies on the lakes of Fraser Island, Queensland. Australian Journal of Marine and Freshwater Research 26: 1–13.

Bayly, I. A. E. and Williams, W. D. (1964) Chemical and biological observations on some volcanic lakes in the south-east of South Australia. Australian Journal of Marine and Freshwater Research 15: 123–132.

Bayly, I. A. E. and Williams, W. D. (1966a) Chemical and biological studies on some saline lakes of south-east Australia. Australian Journal of Marine and Freshwater Research 17: 177–228.

Bayly, I. A. E. and Williams, W. D. (1966b) Further chemical observations on some saline lakes of south-east Australia. Australian Journal of Marine and Freshwater Research 17: 229–237.

Bayly, I. A. E. and Williams, W. D. (1973) Inland Waters and Their Ecology. Longman Australia, Hawthorn, Australia. 316 pp.

Beadle, N. C. W. (1981) The Vegetation of Australia. Cambridge University Press, London, England. 690 pp.

Benson, D. H. (1986) The vegetation of the Gosford and Lake Macquarie 1:100 000 vegetation map sheet. Cunninghamia 1: 467–489.

Benzie, J. H.A. (1984) Zooplankton of an Australian high alpine lake, Lake Cootapatamba, Kosciusko Range. Australian Journal of Marine and Freshwater Research 35: 691–702.

Birch, W. R. (1975) Some chemical and calorific properties of tropical marine angiosperms compared with those of other plants. Journal of Applied Ecology 12: 201–212.

Birch, W. R. and Birch, M. (1984) Succession and pattern of tropical intertidal seagrasses in Cockle Bay, Queensland, Australia: A decade of observations. Aquatic Botany 19: 343–367.

Bishop, K. A. and Forbes, M. A. (1991) The freshwater fishes of northern Australia. pp. 79–107. *In*: C. D. Haynes, M. A. Ridpath, and M. A. J. Williams (eds.). Monsoonal Australia: Landscape, Ecology and Man in the Northern Lowlands. A. A. Balkema, Amsterdam, The Netherlands.

Bishop, K. A., Allen, S. A., Pollard, D. A. and Cook, M. J. (1986) Ecological studies on the fishes of the Alligator Rivers Region, Northern Territory. Supervising Scientist for the Alligator Rivers Region Research Report 4, Australian Government Publishing Service, Canberra, Australia. 55 pp.

Blackhall, S. A. (1986) A survey to determine waterbird usage and conservation significance of selected Tasmanian wetlands. National Parks and Wildlife Service, Tasmania, Occasional Paper no. 14, Hobart, Australia.

Blackman, J. G. and Locke, D. K. (1985) Quantitative analysis of seasonal wetlands in the Burdekin-Townsville region with special reference to waterbird habitat. Proceedings of the Ecological Society of Australia 13: 139–152.

Blair, D. and Finlayson, C. M. (1981) Observation on the habitat and biology of a Lymnaeid snail, *Austropeplea vinosa* (Gastropoda: Pulmonata), an intermediate host for avian schistosomes in tropical Australia. Australian Journal of Marine and Freshwater Research 32: 757–767.

Blyth, J. D., Doeg, T. J. and St Clair, R. M. (1984) Responses of the macro-invertebrate fauna of the Mitta Mitta River, Victoria, to the construction and operation of the Dartmouth Dam. 1. Construction and initial filling period. Occasional Papers from the Museum of Victoria 1: 83–100.

Boto, K. G. and Bunt, J. S. (1981) Tidal export of particulate organic matter from a northern Australian mangrove system. Estuarine, Coastal and Shelf Science 13: 247–255.

Boto, K. G. and Bunt, J. S. (1982) Carbon export from mangroves. p. 105–110. *In*: I. E. Galbally and J. R. Freney (eds.), The Cycling of Carbon, Nitrogen, Sulphur and Phosphorus in Terrestrial and Aquatic Ecosystems. Australian Academy of Science, Canberra, Australia.

Boulton, A. J. and Suter, P. J. (1986) Ecology of temporary streams – an Australian perspective. p. 313–327. *In*: P. De Deckker and W. D. Williams (eds.). Limnology in Australia. Dr. W. Junk, Dordrecht, The Netherlands.

Bowmer, K. H. (1981) Nutrient enrichment, eutrophication. p. 491–500. *In*: G. R. Sainty and S. W. L. Jacobs. Waterplants of New South Wales. Water Resources Commission of New South Wales.

Braithwaite, L. W. (1975) Managing waterfowl in Australia. Proceedings of the Ecological Society of Australia 8: 107–128.

Braithwaite, L. W. (1980) Waterfowl resources and their management. p. 287–294. *In*: W. D. Williams (ed.). An Ecological Basis for Water Resource Management. Australian National University Press, Canberra, Australia.

Bridges, K. W., Phillips, R. C. and Young, P.C. (1982) Patterns of some seagrass distributions in the Torres Strait, Queensland. Australian Journal of Marine and Freshwater Research 33: 273–283.

Bridgewater, P. B. (1975) Peripheral vegetation of Westernport Bay. Proceedings of the Royal Society of Victoria 87: 69–78.

Briggs, S. V. (1980) Chemical studies on four wetlands on the Northern Tablelands of New South Wales. Australian Journal of Marine and Freshwater Research 31: 729–736.

Briggs, S. V. (1981) Freshwater wetlands. p. 335–360. *In*: R. H. Groves (ed.). Australian Vegetation. Cambridge University Press, London, England. 449 pp.

Briggs, S. V. (1983) Ducks, geese and swan. p. 45–51. *In*: C. Haigh (ed.). Wetlands. New South Wales National Parks and Wildlife Service, Sydney, Australia.

Briggs, S. V. and Maher, M. T. (1985) Limnological studies of waterfowl habitat in south-western New South Wales. II. Aquatic macrophyte productivity. Australian Journal of Marine and Freshwater Research 36: 707–715.

Briggs, S. V., Maher, M. T. and Carpenter, S. M. (1985) Limnological studies of waterfowl habitat in south-western New South Wales. I. Water chemistry. Australian Journal of Marine and Freshwater Research 36: 56–67.

Brock, M. A. (1981) The ecology of halophytes in the south-east of South Australia. p. 23–32. *In*: W. D. Williams (ed.). Salt Lakes. Dr. W. Junk, The Hague, The Netherlands.

Brock, M. A. (1982a) Biology of the salinity tolerant genus *Ruppia* L. in saline lakes in South

Australia. I. Morphological variation within and between species and ecophysiology. Aquatic Botany 13: 219–248.
Brock, M. A. (1982b) Biology of the salinity tolerant genus *Ruppia* L. in saline lakes in South Australia. II. Population ecology and reproductive biology. Aquatic Botany 13: 249–268.
Brock, M. A. (1988) Flexibility of life cycle pattern as a mechanism for tolerance of fluctuations of environmental conditions by aquatic plants. Verhandlungen Internationale Vereinigung für Theoretische und Angewandte Limnologie 23: 1949–1953.
Brock, M. A. and Casanova, M. T. (*In press*) Plant survival in temporary waters: A comparison of charophytes and angiosperms. Verhandlungen Internationale Vereinigung für Theoretische und Angewandte Limnologie 24.
Brock, M. A. and Lane, J. A. K. (1983) The aquatic macrophyte flora of saline wetlands in Western Australia in relation to salinity and permanence. Hydrobiologia 105: 63–76.
Brock, M. A. and Pen, L. J. (1984) Ecological studies of the Canning River wetland. Report to the City of Canning, City of Canning, Perth, Australia. 165 pp.
Brock, M. A. and Shiel, R. J. (1983) The composition of aquatic communities in saline wetlands in Western Australia. Hydrobiologia 105: 77–84.
Broome, L. S. and Jarman, P. J. (1983) Waterbirds on the natural and artificial waterbodies in the Namoi Valley, New South Wales. The Emu 83: 92–104.
Brown, J. A. H. (1983) Australia's surface water resources. Water 2000: Consultant report No. 1. Australian Government Publishing Service, Canberra, Australia. 177 pp.
Browne, W. R., Dulhunty, J. A. and Maze, W. H. (1944) Notes on the geology, physiography and glaciology of the Kosciusko area and the country north of it. Proceedings of the Linnean Society of New South Wales 69: 238–252.
Buckley, R. C. (1982) Patterns in north Queensland mangrove vegetation. Australian Journal of Ecology 7: 103–106.
Bunn, S. E. and Brock, M. A. (eds.) (1984) An Inventory of Inland Aquatic Research in Western Australia. Department of Conservation and Environment, Perth, Australia. 102 pp.
Bunt, J. S. (1984) Australian tropical marine ecosystems. p. 125–145. *In*: D. Parkes (ed.). Northern Australia: The Arenas of Life and Ecosystems on Half a Continent. Academic Press, Sydney, Australia.
Bunt, J. S. and Williams, W. T. (1980) Studies in the analysis of data from Australian tidal forests (mangroves). I. Vegetational sequences and their graphic representation. Australian Journal of Ecology 5: 385–390.
Bunt, J. S., Williams, W. T. and Clay, H. J. (1982a) River water salinity and the distribution of mangrove species along several rivers in north Queensland. Australian Journal of Botany 30: 401–412.
Bunt, J. S., Williams, W. T. and Duke, N. C. (1982b) Mangrove distribution in north-east Australia. Journal of Biogeography 9: 111–120.
Bureau of Meteorology (1975) Climatic averages, New South Wales. Australian Government Publishing Service, Canberra, Australia. 136 pp.
Bureau of Meteorology (1986) What's the Weather. Department of Science, Canberra, Australia. 24 pp.
Cadwallader, P. L. (1986) Fish of the Murray-Darling System. p. 679–694. *In*: B. R. Davies and K. F. Walker (eds.). The Ecology of River Systems. Dr. W. Junk, Dordrecht, The Netherlands.
Cambridge, M. L. (1975) Seagrasses of southwestern Australia with special reference to the ecology of *Posidonia australis* Hook. f. in a polluted environment. Aquatic Botany 1: 149–161.
Campbell, E. A. (1983) Mires of Australasia. p. 153–180. *In*: A. J. P. Gore (ed.). Ecosystems of the World 4B. Mires: Swamp, Bog, Fen and Moor, Regional Studies. Elsevier, Amsterdam, The Netherlands.
Casanova, M. T. and Brock, M. A. (1990) Charophyte germination and establishment from the seedbank of an Australian temporary lake. Aquatic Botany 36: 247–254.

Castles, I. (1986) Year Book Australia 1986. Australian Bureau of Statistics, No 70, Canberra, Australia. 771 pp.

Chapman, V. J., (ed.) (1977) Ecosystems of the World. I. Wet Coastal Systems. Elsevier, Amsterdam, The Netherlands. 428 pp.

Cheesman, B. C. (1978) Ecological studies of freshwater turtles in south-eastern Australia. PhD. thesis, Monash University, Melbourne, Australia. 428 pp.

Clarke, A. and Yassini, I. (1985) Foreshore vegetation of Lake Illawarra. Lake Illawarra Management Committee, Wollongong, Australia. 26 pp.

Clarke, L. D. and Hannon, N. J. (1967) The mangrove swamps and salt marsh communities of the Sydney District. I. Vegetation, soils and climate. Journal of Ecology 55: 753–771.

Clarke, L. D. and Hannon, N. J. (1969) The mangrove swamps and salt marsh communities of the Sydney district. II. The holocoenotic complex with particular reference to physiography. Journal of Ecology 57: 213–234.

Clarke, L. D. and Hannon, N. J. (1970) The mangrove swamps and salt marsh communities of the Sydney district. III. Plant growth in relation to salinity and waterlogging. Journal of Ecology 58: 351–369.

Clarke, L. D. and Hannon, N. J. (1971) The mangrove swamps and salt marsh communities of the Sydney district. IV. The significance of species interaction. Journal of Ecology 59: 535–553.

Clough, B. F. (ed.) (1982) Mangrove Ecosystems in Australia. Australian National University Press, Canberra, Australia. 302 pp.

Coaldrake, J. E. (1961) The ecosystem of the coastal lowlands ("Wallum") of southern Queensland. CSIRO Australia Bulletin No. 283, Melbourne, Australia. 138 pp.

Cogger, H. G. (1983) Reptiles and Amphibians of Australia. 2nd ed. Reed, Sydney, Australia. 84 pp.

Coles, R. G., Lee Long, W. J., Squire, L. C. and Bibby, J. M. (1987) Distribution of seagrasses and associated juvenile commercial penaeid prawns in north-eastern Queensland waters. Australian Journal of Marine and Freshwater Research 38: 103–119.

Congdon, R. A. and McComb, A. J. (1976) The nutrients and plants of Lake Joondalup, a mildly eutrophic lake experiencing large seasonal changes in volume. Journal of the Royal Society of Western Australia 59: 14–23.

Congdon, R. A. and McComb, A. J. (1980) Productivity and nutrient content of *Juncus kraussii* in an estuarine marsh in south-western Australia. Australian Journal of Ecology 5: 221–234.

Considine, M. L. (1985) Sensitive plant invades the north. Ecos 45: 10–11.

Cook, C. D. K., Gut, B. J., Rix, E. M., Schneller, J. and Seith, M. (1974) Water Plants of the World. Dr. W. Junk, The Hague, The Netherlands. 561 pp.

Corrick, A. H. (1981) Wetlands of Victoria II. Wetlands and waterbirds of south Gippsland. Proceedings of the Royal Society of Victoria 92: 187–200.

Corrick, A. H. (1982) Wetlands of Victoria III. Wetlands and waterbirds between Port Phillip Bay and Mount Emu Creek. Proceedings of the Royal Society of Victoria 94: 69–82.

Corrick, A. H. and Norman, F. I. (1980) Wetlands of Victoria I. Wetlands and waterbirds of the Snowy River and Gippsland Lakes catchment. Proceedings of the Royal Society of Victoria 91: 1–15.

Costin, A. B. (1954) A Study of the Ecosystems of the Monaro region of New South Wales. N.S.W. Government Printer, Sydney, Australia. 860 pp.

Costin, A. B. (1957) High mountain catchments in Victoria in relation to land use. Soil Conservation Authority of Victoria, Melbourne, Australia. 85 pp.

Costin, A. B., Gray, M., Totterdell, C. J. and Wimbush, D. J. (1979) Kosciusko Alpine Flora. Collins, Sydney, Australia. 408 pp.

Cowardin, L. M., Carter, V., Golet, F. C. and La Roe, E. T. (1979) Classification of Wetlands and Deepwater Habitats of the United States. Fish and Wildlife Service, United States Department of the Interior, Washington, DC, USA. 103 pp.

Cowie, I. D., Finlayson, C. M. and Bailey, B. J. (1988) Alien plants in the Alligator Rivers

Region, Northern Territory, Australia. Supervising Scientist for the Alligator Rivers Technical Memorandum 23, Australian Government Publishing Service, Canberra, Australia. 34 pp.
Croome, R. L. (1986) Biological studies of meromictic lakes. p. 113–130. *In*: P. De Deckker and W. D. Williams (eds.). Limnology in Australia. Dr. W. Junk, Dordrecht, The Netherlands.
Davis, C. (1936) Plant ecology of the Bulli district. Part I. Stratigraphy, physiography and climate; general distribution of plant communities and interpretation. Proceedings of the Linnean Society of New South Wales 61: 285–297.
Davis, C. (1941) Plant ecology of the Bulli district. Part II. Plant communities of the plateau and scarp. Proceedings of the Linnean Society of New South Wales 66:1–19.
De Deckker, P. (1982) Australian aquatic habitats and biota: Their suitability for paleolimnological investigations. Transactions of the Royal Society of South Australia 106: 145–153.
De Deckker, P. (1983) Australian salt lakes: Their history, chemistry and biota – a review. Hydrobiologia 105: 231–244.
De Deckker, P. and Williams, W. D. (1986) Limnology in Australia. Dr. W. Junk, Dordrecht, The Netherlands. 671 pp.
den Hartog, C. (1970) The Sea-grasses of the World. North Holland Publishing Co., Amsterdam, The Netherlands. 275 pp.
Dodson, J. R. (1974) Vegetation history and water fluctuations at Lake Leake, south-eastern South Australia. I. 10 000 B.P. to present. Australian Journal of Botany 22: 719–741.
Dodson, J. R. (1975) Vegetation history and water fluctuations at Lake Leake, south-eastern South Australia. II. 50 000 B.P. to 10 000 B.P. Australian Journal of Botany 23: 815–831.
Dodson, J. R., Greenwood, P. W. and Jones, R. L. (1986) Holocene forest and wetland vegetation dynamics at Barrington Tops, New South Wales. Journal of Biogeography 13: 561–585.
Doeg, T. J. (1984) Responses of the macroinvertebrate fauna of the Mitta Mitta River, Victoria, to the construction and operation of the Dartmouth Dam. 2. irrigation release. Occasional papers of the Museum of Victoria 1: 101–108.
Dowling, R. M. and McDonald, T. J. (1982) Mangrove communities of Queensland. p. 79–94, *In*: B. F. Clough (ed.). Mangrove Ecosystems in Australia: Structure, Function and Management. Australian National University Press, Canberra, Australia.
Ducker, S. C., Brown, V. B. and Calder, D. M. (1977) An Identification of the aquatic vegetation in the Gippsland Lakes. Ministry for Conservation, Victoria, Environmental Studies Series, Publication no. 136, Melbourne, Australia. 111 pp.
Duke, N. C., Bunt, J. S. and Williams, W. T. (1981) Mangrove litter fall in north-eastern Australia. I. Annual totals by component in selected species. Australian Journal of Botany 29: 547–553.
Dulhunty, J. A. (1945) On glacial lakes in the Kosciusko region. Journal and Proceedings of the Royal Society of New South Wales 79: 143–153.
Dulhunty, J. A. (1974) Salt crust distribution and lake bed conditions in southern areas of Lake Eyre South. Transactions of the Royal Society of South Australia 98: 125–133.
Durrington, L. R. (1977) Vegetation of Moreton Island. Botany Branch, Department of Primary Industries, Brisbane, Technical Bulletin no. 1, Brisbane, Australia. 44 pp.
Elsol, J. A. and Saenger, P. (1983) A general account of the mangroves of Princess Charlotte Bay with particular reference to zonation of the open shoreline. p. 37–46. *In*: H. J. Teas (ed.). Biology and Ecology of Mangroves. Dr. W. Junk, The Hague, The Netherlands.
Farrell, T. P. (1978) The spread and control of *Salvinia molesta* in Lake Moondarra, Mount Isa, Queensland. p. 179–187. *In*: Proceedings of the First Conference of the Council of Australia Weed Science Societies, Melbourne, Australia.
Farrell, T. P., Finlayson, C. M. and Griffiths, D. J. (1979) Studies of a tropical lake in north western Queensland. I. Seasonal changes in chemical characteristics. Australian Journal of Marine and Freshwater Research 30: 579–595.
Finlayson, C. M. (1979) Aquatic macrophytes in reservoirs – weed species, assessment and

control. p. 163–173. *In*: Proceedings of an Applied Field Workshop, Institute of Engineers, Brisbane, Australia.

Finlayson, C. M. (1980) Water birds recorded from five artificial lakes in north-west Queensland. Sunbird 11: 49–57.

Finlayson, C. M. (1984a) Salvinia, Water Hyacinth and Mimosa in the Alligator Rivers Region, Northern Territory. Australian Weeds 3: 83.

Finlayson, C. M. (1984b) Growth rates of *Salvinia molesta* in Lake Moondarra, Mt Isa, Australia. Aquatic Botany 18: 257–262.

Finlayson, C. M. (1988) Productivity and nutrient dynamics of seasonally inundated floodplains in the Northern Territory. p. 58–83. *In*: D. Wade-Marshall and P. Loveday (eds.). North Australia: Progress and Prospects, Vol. 2, Floodplain Research. Australian National University, North Australian Research Unit, Casuarina, Australia.

Finlayson, C. M. and Farrell, T. P. (1983) Chemical control of *Salvinia molesta*: Glasshouse and lake experiments. Australian Weeds 2: 90–93.

Finlayson, C. M. and Gillies, J. C. (1982) Biological and physiochemical characteristics of the Ross River Dam, Townsville. Australian Journal of Marine and Freshwater Research 33: 811–827.

Finlayson, C. M. and Mitchell, D. S. (1981) Aquatic weeds and impoundment limnology. p. 111–126. *In*: P. M. Fleming, R. J. Gunn, A. M. Reece and J. R. McAlpine (eds.). Burdekin Project Ecological Study. Australian Government Publishing Service, Canberra, Australia.

Finlayson, C. M. and Mitchell, D. S. (1982) Management of *Salvinia molesta* in Australia. Australian Weeds 2: 71–76.

Finlayson, C. M., Farrell, T. P. and Griffiths, D. J. (1980) Studies of the hydrobiology of a tropical lake in north-western Queensland. II. Seasonal changes in thermal and dissolved oxygen characteristics. Australian Journal of Marine and Freshwater Research 31: 589–596.

Finlayson. C. M., Farrell, T. P. and Griffiths, D. J. (1982) Treatment of sewage effluent using the water fern *Salvinia*. Water Research Foundation of Australia Technical Report No. 57, Kingsford, Australia, 35 pp.

Finlayson, C. M., Farrell, T. P. and Griffiths, D. J. (1984a) Studies of the hydrobiology of a tropical lake in north-western Queensland. III. Growth, chemical composition and potential for harvesting of the aquatic vegetation. Australian Journal of Marine and Freshwater Research 35: 525–536.

Finlayson, C. M., Farrell, T. P. and Griffiths, D. J. (1984b) The hydrobiology of five man-made lakes in north-western Queensland. Proceedings of the Royal Society of Queensland 95: 29–40.

Finlayson, C. M., Cullen, P., Mitchell, D. and Chick, A. (1986) An assessment of a natural wetland receiving sewage effluent. Australian Journal of Ecology 11: 33–47.

Finlayson, C. M., Bailey, B. J. and Cowie, I. D. (1988a) Macrophyte vegetation of the Magela Creek Floodplain, Alligator Rivers Region, Northern Territory. Supervising Scientist for the Alligator Rivers Region Research Report, Australian Government Publishing Service, Canberra, Australia. 18 pp.

Finlayson, C. M., Bailey, B. J., Freeland, W. J. and Fleming, M. R. (1988b) Wetlands of the Northern Territory. p. 103–126. *In*: A. J. McComb and P. S. Lake (eds.). The Conservation of Australian Wetlands. Surrey Beatty & Sons, Chipping Norton, Australia.

Finlayson, C. M., Cowie, I. D. and Bailey, B. J. (1990) Characteristics of a seasonally flooded freshwater system in monsoonal Australia. p. 141–162. *In*: D. F. Whigham, R. E. Good, and J. Kvet (eds.). Wetland Ecology and Management. Case Studies. Kluwer, Dordrecht, The Netherlands.

Fleming, P. M. (1981a) Introduction. p. 1–9. *In*: P. M. Fleming, R. H. Gunn, A. M. Reece and J. R. McAlpine (eds.), Burdekin Project Ecological Study. Australian Government Publishing Service, Canberra, Australia.

Fleming, P. M. (1981b) Hydrology and sediment transport. p. 143–181. *In*: P. M. Fleming, R. H. Gunn, A. M. Reece and J. R. McAlpine (eds.). Burdekin Project Ecological Study. Australian Government Publishing Service, Canberra, Australia.

Fleming, P. M. (1982) The water resources of the Murray-Darling Basin. p. 38–54. *In*: Murray-Darling Basin Development Study. CSIRO Division of Water and Land Resources, Canberra, Australia (Reference in Walker 1985).

Fletcher, A. R. (1986) Effects of introduced fish in Australia. p. 231–238. *In*: P. De Dekker and W. D. Williams (eds.). Limnology in Australia. Dr. W. Junk, Dordrecht, The Netherlands.

Fogarty, P. (1982) A preliminary survey of environmental damage associated with activity of feral buffalo. Conservation Commission of the Northern Territory, Technical Report, Darwin, Australia. 87 pp.

Fraser, L. and Vickery, J. W. (1939) The ecology of the upper Williams River and Barrington Tops District. III. The eucalypt forests and general discussion. Proceedings of the Linnean Society of New South Wales 64: 1–33.

Freeland, W. J. (1985) The need to control cane toads. Search 16: 211–215.

Freeland, W. J. and Martin, K. C. (1985) The rate of expansion by *Bufo marinus* in northern Australia 1980–1984. Australian Wildlife Research 12: 555–559.

Friend, G. R. and Taylor, J. A. (1984) Ground surface features attributable to feral buffalo, *Bubalus bubalis*. II. Their relationship to the abundance of small animal species. Australian Wildlife Research 11: 311–323.

Frith, H. J. (1977) Waterfowl in Australia. 2nd ed. Reed, Sydney, Australia. 328 pp.

Frith, H. J. and Davies, S. J. J. F. (1961) Ecology of the magpie goose, *Anseranas semipalmata* Latham (Anatidae). CSIRO Wildlife Research 6: 1–41.

Froend, R. H., Heddle, E. M., Bell, D. T. and McComb, A. J. (1987) Effects of salinity and waterlogging on the vegetation of Lake Toolibin, Western Australia. Australian Journal of Ecology 12: 281–298.

Fulton, W. (1983a) Macrobenthic fauna of Great Lake, Arthurs Lake and Lake Sorrell, Tasmania. Australian Journal of Marine and Freshwater Research 34: 775–785.

Fulton, W. (1983b) Qualitative and quantitative variation in the macrobenthic fauna of the original lake and new lake areas of Great Lake and Arthurs Lake, Tasmania. Australian Journal of Marine and Freshwater Research 34: 787–803.

Galloway, R. W. (1982) Distribution and physiographic patterns of Australian mangroves. p. 31–54. *In*: B. F. Clough (ed.). Mangrove Ecosystems in Australia: Structure, Function and Management. Australian National University Press, Canberra, Australia. 302 pp.

Garnett, S. T., Price, I. R. and Scott, F. J. (1985) The diet of the green turtle, *Chelonia mydas* (L.), in Torres Strait. Australian Wildlife Research 12: 103–112.

Geddes, M. C. (1986) Understanding zooplankton communities in farm dams: the importance of predation. p. 387–402. *In*: P. De Deckker and W. D. Williams (eds.). Limnology in Australia. Dr. W. Junk, Dordrecht, The Netherlands.

Glover, C. J. M. (1973) Adaptations of a central Australian gobiid fish. Bulletin of the Australian Society of Limnology 5: 8–10.

Glover, C. J. M. (1979) Studies on central Australian fishes: Further observations and records, Part 1. South Australian Naturalist 53: 58–62.

Glover, C. J. M. (1982) Adaptations of fishes in arid Australia. p. 241–246. *In*: W. R. Barker and P. J. M. Greenslade (eds.). Evolution of the Flora and Fauna of Arid Australia. Peacock Publications, Adelaide, Australia.

Glover, C. J. M. and Inglis, W. G. (1971) Freshwater fish of South Australia. p. 27–34 *In*: South Australian Yearbook 1971. South Australian Government Printer, Adelaide, Australia.

Glover, C. J. M. and Sim, T. C. (1978a) A survey of central Australian ichthyology. Australian Zoologist 19: 245–256.

Glover, C. J. M. and Sim, T. C. (1978b) Studies on central Australian fishes: A progress report. South Australian Naturalist 52: 35–44.

Goodrick, G. N. (1970) A survey of wetlands of coastal New South Wales. CSIRO Division of Wildlife Research Technical Memorandum No. 5, Canberra, Australia. 36 pp.

Goodrick, G. N. (1984) Wetlands of north-western New South Wales. National Parks and Wildlife Service [N.S.W.] Occasional Paper No. 6, Sydney, Australia. 16 pp.

Goodrick, G. N. (1985) Mapping and describing wetlands in arid New South Wales: An application of the landscape approach. p. 267–274. *In*: Survey Methods for Nature Conservation. CSIRO Division of Water and Land Resources, Canberra, Australia.

Gordon, D. M., Finlayson, C. M. and McComb, A. J. (1981) Nutrients and phytoplankton in three shallow freshwater lakes of different trophic status in Western Australia. Australian Journal of Marine and Freshwater Research 32: 541–553.

Graham, A., Begg, R., Graham, P. and Raskin, S. (1982) Buffalo in the Northern Territory. Conservation Commission of the Northern Territory, Technical Report, Darwin, Australia. 12 pp.

Greenslade, J., Joseph, L. and Reeves, A. (eds.) (1985) South Australia's Mound Springs. Nature Conservation Society of South Australia, Adelaide, Australia. 100 pp.

Grieve, A.M. (1987) Salinity and waterlogging in the Murray-Darling basin. Search 18: 72–74.

Grigg, G. C. and Taylor, J. A. (1980) An aerial survey of *Crocodylus porosus* nesting habitat in the Alligator Rivers Region, Northern Territory. Unpublished report to Australian National Parks and Wildlife Service, Canberra, Australia.

Hall, N., Wainwright, R. W. and Wolf, L. J. (1981) Summary of meteorological data in Australia. CSIRO Division of Forest Research Divisional Report No. 6, Canberra, Australia. 117 pp.

Harley, K. L. S. (1981) Biological control. p. 474–489. *In*: G. R. Sainty and S. W. L. Jacobs, Waterplants of New South Wales. New South Wales Water Resources Commission, Sydney, Australia.

Harris, M. McD., King, R. J. and Ellis, J. (1980) The eelgrass *Zostera capricorni* in Illawarra Lake, N.S.W. Proceedings of the Linnean Society of New South Wales 104: 23–33.

Hegerl, E. J. and Timmins, R. D. (1973) The Noosa River tidal swamps. Operculum 3: 38–43.

Heinsohn, G. E. and Birch, W. R. (1972) Foods and feeding habitats of the dugong, *Dugong dugon* Erxleben, in northern Queensland, Australia. Mammalia 38: 414–422.

Heinsohn, G. E. and Spain, A. V. (1974) Effects of a tropical cyclone on littoral and sublittoral biotic communities and on a population of dugongs (*Dugong dugon* Muller). Biological Conservation 6: 143–152.

Heinsohn, G. E., Wake, H., Marsh, H. and Spain, A. V. (1977) The dugong (*Dugong dugon* Muller) in the seagrass system. Aquaculture 12: 235–248.

Higginson, F. R. (1966) The distribution of submerged aquatic angiosperms in the Tuggerah Lakes system. Proceedings of the Linnean Society of New South Wales 90: 328–334.

Higginson, F. R. (1970) Ecological effects of pollution in Tuggerah Lakes. Proceedings of the Ecological Society of Australia 5: 143–152.

Hill, R. and Webb, G. J. W. (1982) Floating grass mats of the Northern Territory floodplains – an endangered habitat? Wetlands 2: 45–50.

Hillman, T. J. (1986) Billabongs. p. 457–470. *In*: P. De Deckker and W.D. Williams (eds.). Limnology in Australia. Dr. W. Junk, Dordrecht, The Netherlands.

Hodgkin, E., Birch, P. B., Black, R. E. and Humphries, R. B. (1981) The Peel-Harvey estuarine study (1976–1980). A report to the Estuarine and Marine Advisory Committee, Report No. 9, Department of Conservation and Environment, Western Australia. 72 pp.

Hutchings, P. A. and Recher, H. F. (1982) The fauna of Australian mangroves. Proceedings of the Linnean Society of New South Wales 106: 83–121.

Hutchings, P. A. and Recher, H. F. (1983) The faunal communities of Australian mangroves. p. 103–110. *In*: H. J. Teas (ed.). Biology and Ecology of Mangroves. Dr. W. Junk, The Hague, The Netherlands.

Hutchings, P. and Saenger, P. (1987) Ecology of Mangroves. University of Queensland Press, Brisbane, Australia. 388 pp.

Ivanstoff, W. and Glover, C. J. M. (1974) *Craterocephalus dalhousiensis* n. sp., a sexually dimorphic freshwater teleost (Antherinidae) from South Australia. Australian Zoologist 18: 88–98.

Jacobs, S. W. L. (1983) Wetlands vegetation. p. 14–19. *In*: C. Haigh (ed.). Wetlands in New South Wales. New South Wales National Parks and Wildlife, Sydney, Australia.

Jacobs, S. W. L. and Sainty, G. R. (1987) Water weeds into the third century. p. 148–151. *In*: D. Lemerle and A. R. Leys (eds.). Proceedings of the 8th Australian Weeds Conference, Sydney, Australia.

Jenkins, R. W. G. and Forbes, M. A. (1985) Seasonal variation in abundance and distribution of *Crocodylus porosus* in the tidal East Alligator River, Northern Territory. p. 63–70. *In*: G. Grigg, R. Shine and H. Ehmann (eds.). The Biology of Australasian Frogs and Reptiles. Surrey Beatty & Sons, Chipping Norton, Australia.

Jennings, J. V. and Mabbutt, J. A. (1977) Physiographic outlines and regions. p. 38–52. *In*: D. N. Jeans (ed.). Australia: A Geography. Sydney University Press, Sydney, Australia.

John, J. (ed.) (1987) The Swan River Estuary and Management. Curtin University Environmental Studies Group Report No. 1, Canberra, Australia. 327 pp.

Johnson, M. (1985) Progress report No. 2: hydrologic and organic fluxes within a coastal swamp system. Department of Mineral Resources New South Wales report No. 85/25, Lidcombe, Australia. 137 pp.

Julli, M. (1986) The taxonomy and seasonal population dynamics of some Magela Creek floodplain microcrustaceans (Cladocera and Copepoda). Supervising Scientist for the Alligator Rivers Region, Northern Territory, Australia, Technical Memorandum 18, Australian Government Publishing Service, Canberra, Australia. 72 pp.

Kabay, E. D. and Burbidge, A. A. (eds.) (1977) A biological survey of the Drysdale River National Park, north Kimberley, Western Australia, in August 1975. Wildlife Research Bulletin of Western Australia No. 6, 33 pp.

Keith, D. A. (1984) Report on the conservation status of O'Hares Creek catchment. Unpublished, available from library, National Herbarium of New South Wales, Sydney, Australia. 35 pp.

Keith, D. A. (1985) Diversified bad country: The swamps and forests of O'Hares Creek catchment. National Parks Journal 29: 13–15.

Kenneally, K. F. (1982) Mangroves of Western Australia. p. 95–110. *In*: B. F. Clough (ed.). Mangrove Ecosystems in Australia: Structure, Function and Management. Australian National University Press, Canberra, Australia.

Killigrew, L. P. and Gilkes, R. J. (1974) Development of playa lakes in south Western Australia. Nature 247: 454–455.

Kirkpatrick, J. B. and Glasby, J. (1981) Saltmarshes in Tasmania. Department of Geography, University of Tasmania, Occasional Paper No. 8, Hobart, Australia. 62 pp.

Kirkpatrick, J. B. and Harwood, C. E. (1983a) Plant communities of Tasmanian wetlands. Australian Journal of Botany 31: 437–451.

Kirkpatrick, J. B. and Harwood, C. E. (1983b) Conservation of Tasmanian macrophytic wetland vegetation. Papers and Proceedings of the Royal Society of Tasmania 117: 5–20.

Kirkpatrick, J. B. and Tyler, P. A. (1988) Tasmanian wetlands and their conservation. p. 1–16. *In*: A. J. McComb and P. S. Lake (eds.), The Conservation of Australian Wetlands. World Wildlife Fund, Chipping Norton, Australia.

Knights, P. (1980) Macquarie Marshes: Wetlands and Management Requirements. Total Environment Centre, Sydney, Australia. 87 pp.

Kratochvil, M., Hannon, N. J. and Clarke, L. D. (1973) Mangrove swamp and salt marsh communities in southern Australia. Proceedings of the Linnean Society of New South Wales 97: 262–274.

Kuo, J. and McComb, A. J. (1989) Seagrass taxonomy, structure and development. p. 6–73. *In*: A. W. Larkum, A. J. McComb and S. Shepherd (eds.). A Treatise on the Biology of Seagrasses With Special Reference to the Australian Region. Elsevier, Amsterdam, The Netherlands. 841 pp.

Lake, J. S. (1975) Fish of the Murray River. p. 213–224. *In*: G. V. Lawrence and G. K. Smith (eds.). The Book of the Murray. Rigby, Adelaide, Australia.

Lake, P. S., Barmuta, L. A., Boulton, A. J., Campbell, I. C. and St. Clair, R. M. (1985) Australian streams and Northern Hemisphere stream ecology: comparisons and problems. Proceedings of the Ecological Society of Australia 14: 61–82.

Lane, J. A. K. and Burbidge, A. A. (1978) Wetlands of the south-west of Western Australia, with special reference to the Busselton area. Department of Fisheries and Wildlife, Perth, Australia.

Lane, J. A. K. and McComb, A. J. (1988) Western Australian wetlands. p. 127–146. *In*: A. J. McComb and P. S. Lake (eds.). The Conservation of Australian Wetlands. Surrey Beatty & Sons, Chipping Norton, Australia.

Lanyon, J. (1986) Guide to the identification of seagrasses in the Great Barrier Reef region. Great Barrier Reef Marine Park Authority Special Publication Series (3), GBRMPA, Townsville, Australia. 54 pp.

Larkum, A. W. D., McComb, A. J. and Shepherd, S. (eds.) (1989) Biology of Seagrasses, a Treatise on the Biology of Seagrasses with Special Reference to the Australian Region. Elsevier, Amsterdam, The Netherlands. 841 pp.

Laut, P., Heyligers, P. C., Keig, G., Loffler, E., Margules, C., Scott, R. M. and Sullivan, M. E. (1977) Environments of South Australia. CSIRO, Canberra, Australia. 8 vols.

Lee, D. M. and Neal, A. B. (1984) The climate in northern Australia. p. 29–55. *In*: D. Parker (ed.). Northern Australia: The Arenas of Life and Ecosystems on Half a Continent. Academic Press, Sydney, Australia.

Lenanton, R. C. J. (1984) Life history strategies of fish in some temperate Australian estuaries. p. 119–137. *In*: E. P. Hodgkin (ed.). Estuarine Environments of the Southern Hemisphere. Department of Conservation and Environment, Western Australia, Bulletin No. 161.

Linacre, E. T., Hicks, B. B., Sainty, G. R. and Grauze, G. (1970) The evaporation from a swamp. Agricultural Meteorology 7: 375–386.

Linacre, E. and Hobbs, J. (1977) The Australian Climatic Environment. John Wiley & Sons, Brisbane, Australia. 354 pp.

Logan, B. W. and Cebulski, D. E. (1970) Sedimentary environments of Shark Bay, Western Australia. Memoirs of the American Association for Petrology and Geology 13: 1–37.

Lonsdale, W. M., Harley, K. L. S. and Gillett, J. D. (1988) Seed bank dynamics in *Mimosa pigra*, an invasive tropical shrub. Journal of Applied Ecology 25: 963–976.

Lothian, J. A. and Williams, W. D. (1988) Wetland conservation in South Australia. p. 147–166. *In*: A. J. McComb and P. S. Lake (eds.). The Conservation of Australian Wetlands. World Wildlife Fund, Chipping Norton, Australia.

Love, L. D. (1981) Mangrove swamps and salt marshes. p. 319–334. *In*: R. H. Groves (ed.). Australian Vegetation. Cambridge University Press, Cambridge, England. 449 pp.

Lyster, S. (1985) International wildlife law. Grotius Ltd., Llandysul, United Kingdom. 470 pp.

McComb, A. J. and Lake, P. S. (eds.) (1988) The Conservation of Australian Wetlands. World Wildlife Fund, Chipping Norton, Australia, 196 pp.

Mackay, N. J., Bennison, G., Hillman, T. J. and Waller, P. (1983) River Murray Commission Biological Monitoring Programme. Final report of development phase, River Murray Commission, Canberra, Australia. 84 pp.

Mclennan, P. M. and Moore, I. D. (1976) Review of Australia's water resources 1975. Australian Government Publishing Service, Canberra, Australia. 170 pp.

Macnae, W. (1966) Mangroves in eastern and southern Australia. Australian Journal of Botany 14: 67–104.

Macnae, W. (1968) A general account of the fauna and flora of mangrove swamps and forests in the Indo-West-Pacific region. Advances in Marine Biology 6: 73–270.

Magnusson, W. E., Webb, G. T. W. and Taylor, J. A. (1976) Two new locality records, a new habitat and a nest description for *Xeromys myoides* Thomas (Rodentia: Murridae). Australian Wildlife Research 3: 153–157.

Majer, K. (1979) Wetlands of the Darling system: Wetlands in Conservation Reserves and National Parks. Department of Conservation and Environment, Western Australia, Bulletin No. 61, Perth, Australia.

Marchant, R. (1982) Seasonal variation in the macroinvertebrate fauna of billabongs along Magela Creek, Northern Territory. Australian Journal of Marine and Freshwater Research 33: 329–342.

Marsh, H., Channels, P. W., Heinsohn, G. E. and Morrissey, J. (1982) Analysis of stomach contents of dugongs from Queensland. Australian Wildlife Research 9: 55–67.

Masini, R. J. (1986) Inland waters of the Pilbara. A report of a field study carried out in March–April 1983. Unpublished report to the Department of Conservation and the Environment, Perth, Australia.

May, V. (1970) A toxic alga in New South Wales and its distribution. Contributions from the New South Wales National Herbarium 4: 84–86.

May, V. (1972) Blue-green algal blooms at Braidwood, New South Wales (Australia). Science Bulletin of the New South Wales Department of Agriculture 82: 1–45.

Messel, H. and Vorlicek, G. C. (1986) Population dynamics and status of *Crocodylus porosus* in the tidal waterways of northern Australia. Australian Wildlife Research 13: 27–37.

Messel, H., Wells, A. D. and Green, W. J. (1979) Survey of tidal river systems in the Northern Territory of Australia and their crocodile population. The Alligator Region River system. Monograph No. 4, Pergamon Press, Sydney, Australia. 49 pp.

Michaelis, F. B. and O'Brien, M. (1988) Preservation of Australia's wetlands: a Commonwealth approach. p. 167–177. *In*: A. J. McComb and P. S. Lake (eds.). The Conservation of Australian wetlands. World Wildlife Fund, Chipping Norton, Australia.

Miles, J. M. and Burbidge, A. A. (eds.) (1975) A biological survey of the Prince Regent River Reserve, north-west Kimberley, Western Australia, in August 1974. Wildlife Research Bulletin of Western Australia 3: 1–116.

Miles, J.T. (1975) A review of literature and other information on wetlands in New South Wales. CSIRO Technical Memorandum 75/6, Canberra, Australia. 20 pp.

Mills, K. (1983) Evaluation and inventory of the wetland habitats associated with Lake Illawarra. Lake Illawarra Management Committee, Wollongong, Australia. 48 pp.

Mills, K. (1985) Lake Illawarra –a bibliography. Lake Illawarra Management Committee, Wollongong, Australia. 44 pp.

Milward, N. E. (1982) Mangrove-dependent biota. p. 121–139. *In*: B. F. Clough (ed.). Mangrove Ecosystems in Australia: Structure, Function and Management. Australian National University Press, Canberra, Australia.

Mitchell, B. D. (1985) Limnology of mound springs and temporary pools south and west of Lake Eyre. p. 51–63. *In*: Greenslade, P., Joseph, L. and Reeves, A. South Australian Mound Springs. Nature Conservation Society of South Australia, Adelaide, Australia.

Mitchell, B. D. (1987) Entomostracan zooplankton communities of Australian lakes and ponds. p. 369–388. *In*: P. De Deckker and W. D. Williams (eds.). Limnology in Australia. Dr. W. Junk, Dordrecht, The Netherlands.

Mitchell, B. D. and Williams, W. D. (1982a) Factors influencing the seasonal occurrence and abundance of the zooplankton in two waste stabilization ponds. Australian Journal of Marine and Freshwater Research 33: 989–997.

Mitchell, B. D. and Williams, W. D. (1982b) Dispersion patterns of zooplankton in two waste stabilization ponds. Australian Journal of Marine and Freshwater Research 33: 1123–1126.

Mitchell, D. S. (1978) Aquatic Weeds in Australian Inland Waters. Australian Government Publishing Service, Canberra, Australia. 189 pp.

Morley, A. W. (ed.) (1981) Review of Jabiluka Environmental Studies. Pancontinental Mining Pty. Ltd., Sydney, Australia. 7 vols.

Morton, S., Brennan, K. and Dostine, P. (1984) Diversity, abundance and assessment of avian links in food chains. p. 51–53. *In*: Alligator Rivers Region Research Institute Report 1983–84. Australian Government Publishing Service, Canberra, Australia. 74 pp.

Murray-Darling Basin Ministerial Council (1987) Murray-Darling basin environmental resources study. [N.S.W.] State Pollution Control Commission, Sydney, Australia. 426 pp.

Myerscough, P. J. and Carolin, R. C. (1986) The vegetation of the Eurunderee sand mass, headlands and previous islands in the Myall Lakes area, New South Wales. Cunninghamia 1: 399–466.

New South Wales Department of Water Resources (1987) Developing a water management strategy for the Barwon-Darling catchment. Issue paper September 1987 (copy held in library, National Herbarium, Sydney), Australia (incomplete pagination).

New South Wales Department of Water Resources and National Parks and Wildlife Service (1986) Water management plan for the Macquarie Marshes. Department of Water Resources and National Parks and Wildlife Service, Sydney, Australia. 18 pp.

Nicholls, N. (1981) Air-sea interaction and the possibility of long range weather prediction in the Indonesian Archipelago. Monthly Weather Review 109: 24–35.

Nicholls, N. (1984) A system for predicting the onset of the north Australian wet season. Journal of Climatology 4: 425–435.

Norman, F. I. and Corrick, A. H. (1988) Wetlands in Victoria: a brief review. p. 17–34. *In*: A. J. McComb and P. S. Lake (eds.). The Conservation of Australian Wetlands. World Wildlife Fund, Chipping Norton, Australia.

Paijmans, K. (1978) A reconnaissance of four wetland pilot study areas. CSIRO Division of Land Use Research Technical Memorandum 78/3, Canberra, Australia. 37 pp.

Paijmans, K., Galloway, R. W., Faith, D. P., Fleming, P. M., Haantjens, H. A., Heyligers, P. C., Kalma, J. D. and Loffler, E. (1985) Aspects of Australian wetlands. CSIRO Division of Water and Land Resources Technical Paper No. 44, Canberra, Australia. 77 pp.

Pen, L. (1987) Peripheral vegetation of the Swan-Canning estuary – past, present and future. p. 221–231. *In*: J. John (ed.). The Swan River Estuary Ecology and Management. Curtin University Environmental Studies Group Report No. 1, Perth, Australia.

Perry, R. A. (1953) The vegetation communities of the Townsville-Bowen region. CSIRO Australia Land Research Series 2: 44–54.

Pianka, E. R. (1983) Evolutionary Ecology. 3rd ed. Harper and Row, New York, U.S.A. 416 pp.

Pickard, J. and Jacobs, S. W. L. (1983) Vegetation patterns on the Sassafras Plateau. p. 54–65. *In*: R. W. Young and G. C. Nanson (eds.). Aspects of Australian Sandstone. Australian and New Zealand Geomorphology Group, Wollongong, Australia.

Pidgeon, I. M. (1938) The ecology of the Central Coast area of New South Wales. II. Plant succession on the Hawkesbury Sandstone. Proceedings of the Linnean Society of New South Wales 63: 1–26.

Poiner, I. R., Staples, D. J. and Kenyon, R. (1987) Seagrass communities of the Gulf of Carpentaria, Australia. Australian Journal of Marine and Freshwater Research 38: 121–131.

Pollard, D. A., Llewellyn, L. C. and Tilzey, R. D. J. (1980) Management of freshwater fish and fisheries. p. 227–270. *In*: W. D. Williams (ed.). An Ecological Basis for Water Resource Management. Australian National University Press, Canberra, Australia.

Ponder, W. F. (1985) South Australian mound springs, relict faunas in the desert. Australian Natural History 21: 352–355.

Ponder, W. F. (1986) Mound springs of the Great Artesian Basin. p. 403–420. *In*: P. De Dekker and W. D. Williams (eds.). Limnology in Australia. Dr. W. Junk, Dordrecht, The Netherlands.

Ponder, W. F. and Hershler, R. (1984) Hydrobiid studies. p. 5–1 to 5–16. *In*: Kinhill Stearns (ed.). Supplementary Environmental Studies – Mound Springs. Kinhill Stearns, Adelaide, Australia.

Porter, J. (1984) A survey of Blue Mountain wetlands. New South Wales Institute of Technology (Unpublished, copy held in library, National Herbarium, Sydney, Australia).

Potter, I. C., Loneragan, N. R., Lenanton, R. C. J., Chrystal, P. J. and Grant, C. J. (1983) Abundance, distribution and age structure of fish populations in a Western Australian estuary. Journal of Zoology 200: 21–50.

Powling, I. J. (1970) Phytoplankton of the Kosciusko region. Bulletin of the Australian Society of Limnology 3: 29–32.

Pressey, R. L. (1981) A survey of wetlands on the lower Hunter floodplain, N.S.W. New South Wales National Parks and Wildlife Service, Sydney, Australia. 97 pp.

Pressey, R. L. (1986) Wetlands of the River Murray below Lake Hume. River Murray Commission Report 86/1, Canberra, Australia. 134 pp.

Pressey, R. L. (1987a) A survey of wetlands on the lower Clarence floodplain, N.S.W. Report prepared for New South Wales National Parks and Wildlife Service, Sydney, Australia. 295 pp.

Pressey, R. L. (1987b) A survey of wetlands of the lower Macleay floodplain, N.S.W. Report prepared for New South Wales National Parks and Wildlife Service, Sydney, Australia. 244 pp.

Pressey, R. L. and Griffith, S. J. (1987) Coastal wetlands and associated communities in Tweed Shire, northern New South Wales. New South Wales National Parks and Wildlife Service, Sydney, Australia. 103 pp.

Pressey, R. L. and Harris, J. H. (1988) Wetlands of New South Wales. p. 35–57 *In*: A. J. McComb and P. S. Lake (eds.). The Conservation of Australian Wetlands. World Wildlife Fund, Chipping Norton, Australia.

Pressey, R. L., Bell, F. C., Barker, J., Rundle, A. S. and Belcher, C. A. (1984) Biophysical features of the Lachlan-Murrumbidgee confluence, south-western New South Wales. New South Wales National Parks and Wildlife Service, Sydney, Australia. 182 pp.

Raine, J. I. (1982) Dimictic thermal regime and morphology of Blue Lake in the Snowy Mountains of New South Wales. Australian Journal of Marine and Freshwater Research 33: 1119–1122.

Ramage, C. S. (1971) Monsoon Meteorology. Academic Press, New York, NY, U.S.A. 269 pp.

Riggert, T. L. (1966) A study of the wetlands of the Swan coastal plain. Department of Fisheries and Fauna, Western Australia, Perth, Australia. 36 pp.

Riley, S. J., Warner, R. F. and Erskine, W. (1984) Classification of waterbodies in New South Wales. Water Resources Commission of New South Wales, Sydney, Australia. 100 pp.

Room, P. M., Forno, I. W. and Taylor, M. J. F. (1984) Establishment in Australia of two insects for biological control of the floating weed *Salvinia molesta*. Bulletin of Entomological Research 74: 505–516.

Room, P. M., Harley, K. L. S., Forno, I. W. and Sands, D. P. A. (1981) Successful biological control of the floating weed Salvinia. Nature 294: 78–81.

Saenger, P., Specht, M. M. and Specht, R. L. (1977) Mangal and coastal salt-marsh communities in Australasia: Australia. p. 293–310. *In*: V. J. Chapman (ed.). Ecosystems of the World. Vol. 1, Wet Coastal Ecosystems. Elsevier, Amsterdam, The Netherlands.

Sainty, G. R. and Jacobs, S. W. L. (1981) Waterplants of New South Wales. New South Wales Water Resources Commission, Sydney, Australia. 550 pp.

Sanderson, N. T., Koontz, D. V. and Morley, A. W. (1983) The ecology of the vegetation of the Magela Creek Floodplain: Upper section from Oenpelli road crossing to Nankeen Billabong. Unpublished Report in Scientific Workshop, Environmental Protection in the Alligator Rivers Region, Jabiru, 17–20 May 1983 (Unpaginated).

Schodde, R., Mason, I. J. and Gill, H. B. (1982) The avifauna of the Australian mangroves – a brief review of composition, structure and origin. p. 141–150. *In*: B. F. Clough (ed.). Mangrove Ecosystems in Australia: Structure, Function and Management. Australian National University Press, Canberra, Australia.

Semeniuk, C. A. (1987) Wetlands of the Darling system [Western Australia] – a geomorphic approach to habitat classification. Journal of the Royal Society of Western Australia 69: 95–112.

Semeniuk, V. (1980) Mangrove zonation along an eroding coastline in King Sound, North Western Australia. Journal of Ecology 68: 789–812.

Semeniuk, V. (1981) Sedimentology and the stratigraphic sequence of a tropical tidal flat, north Western Australia. Sedimentary Geology 29: 195–221.

Semeniuk, V., Chalmer, P. N. and Le Provost, I. (1982) The marine environments of the Dampier Archipelago. Journal of the Royal Society of Western Australia 65: 97–114.

Shiel, R. J. (1980) Billabongs of the Murray-Darling system. p. 376–390. *In*: W. D. Williams (ed.). An Ecological Basis for Water Resource Management. Australian National University Press, Canberra, Australia.

Shiel, R. J. (1981) Plankton of the Murray-Darling River system with particular reference to the zooplankton. PhD. thesis, University of Adelaide, Adelaide, Australia. 287 pp.

Shiel, R. J. (1986) Zooplankton of the Murray-Darling system. p. 661–678. *In*: B. R. Davies and K. F. Walker (eds.). The Ecology of River Systems. Dr. W. Junk, Dordrecht, The Netherlands.

Shiel, R. J. and Koste, W. (1986) Australian Rotifera: ecology and biogeography. p. 141–150. *In*: P. De Dekker and W. D. Williams (eds.). Limnology in Australia. Dr. W. Junk, Dordrecht, The Netherlands.

Shiel, R. J. and Walker, K. F. (1985) Zooplankton of regulated and unregulated streams: The Murray-Darling river system, Australia. p. 263–270. *In*: A. Lillehammer and S. J. Saltveit (eds.). Regulated Rivers. University of Oslo Press, Oslo, Norway.

Shiel, R. J., Walker, K. F. and Williams, W. D. (1982) Plankton of the lower River Murray, South Australia. Australian Journal of Marine and Freshwater Research 33: 301–327.

Shine, R. (1986) Diets and abundances of aquatic and semi-aquatic reptiles in the Alligator Rivers region. Supervising Scientist for the Alligator Rivers Region Technical Memorandum 16, Australian Government publishing Service, Canberra, Australia. 57 pp.

Shine, R. and Lambeck, R. (1985) A radiotelemetric study of movements, thermoregulation and habitat utilization of Arafura file snakes (Serpentes: Acrochordidae). Herpetology 41: 351–361.

Shine, R., Ellway, C. P. and Hegerl, E. J. (1973) A biological survey of the Tallebudgera Creek estuary. Operculum 3: 59–83.

Shorthouse, D. J. (1984) The ecological resources of the ACT. National Capital Development Commission, Canberra, Technical Paper 42, Canberra, Australia. 141 pp.

Slatyer, R. O. (1962) Climate of the Alice Springs area. p. 109–128. *In*: Lands of the Alice Springs Area Northern Territory, 1956–57. CSIRO Australia, Land Research Series No. 6, 280 pp.

Sloane, R. D. (1984) Distribution and abundance of freshwater eels *Anguilla* sp. in Tasmania. Australian Journal of Marine and Freshwater Research 35: 463–470.

Smith, A. J. (1975a) A review of literature and other information on Victorian wetlands. CSIRO Division of Land Use Research Technical Memorandum 75/5, Canberra, Australia. Incomplete pagination.

Smith, A. J. (1975b) A review of literature and other information on the wetlands in South Australia. CSIRO Division of Land Use Research Technical Memorandum 75/7, Canberra, Australia. Incomplete pagination.

Smith, A. J. (1975c) A review of literature and other information on wetlands in Western Australia. CSIRO Division of Land Use Research Technical Memorandum 75/8, Canberra, Australia. Incomplete pagination.

Smith, A. J. (1975d) A review of literature and other information on Tasmanian wetlands. CSIRO Division of Land Use Research Technical Memorandum 75/9, Canberra, Australia. Incomplete pagination.

Smith, B. J. (1978) Molluscs of the Murray Darling river system. Proceedings of the Royal Society of Victoria 90: 203–209.

Smith, B. J. and Kershaw, R. C. (1979) Fieldguide to non marine molluscs of south east Australia. University of New South Wales Press, Sydney, Australia. 285 pp.

South Eastern Wetlands Committee (1984) Wetland resources of the south east of South Australia. Report of the South Eastern Wetlands Committee, Adelaide, Australia. 2 vols.

Spain, A. V. and Heinsohn, G. E. (1973) Cyclone associated feeding changes in the dugong (Mammalia: Sirenia). Mammalia 37: 678–680.

Specht, R. L. (1970) Vegetation. p. 44–67. *In*: G. E. Leeper (ed.). The Australian Environment. CSIRO, Melbourne, Australia.

Specht, R. L. (1981) Biogeography of halophytic angiosperms (salt-marsh, mangrove and seagrass). p. 575–590. *In*: A. Keast (ed.). Ecological Biogeography of Australia. Dr. W. Junk, The Hague, The Netherlands. 3 vols.

Spenceley, A. P. (1976) Unvegetated saline tidal flats in north Queensland. Journal of Tropical Geography 42: 78–85.

Stanton, J. P. (1975) A preliminary assessment of wetlands in Queensland. CSIRO Division of Land Use Research Technical Memorandum 75/10, Canberra, Australia. 42 pp.

Stoddart, D. R. (1980) Mangroves as successional stages, inner reefs of the northern Great Barrier Reef. Journal of Biogeography 7: 269–284.

Storrier, R. R. and Stannard, M. E. (eds.) (1980) Aeolian landscapes in the semi-arid zone of south eastern Australia. Australian Society of Soil Science Inc. Riverina Branch, Wagga Wagga, Australia. 250 pp.

Tait, R. D., Shiel, R. J. and Koste, W. (1984) Structure and dynamics of zooplankton communities, Alligator Rivers Region, N.T., Australia. Hydrobiology 113: 1–13.

Tamuly, A. (1970) Physical and chemical limnology of the Blue Lake of Mount Gambier, South Australia. Transactions of the Royal Society of South Australia 94: 71–86.

Taylor, J. A. and Friend, G. R. (1984) Ground surface features attributable to feral buffalo, *Bubalus bubalis* I. Their distribution relative to vegetation structure and plant life form. Australian Wildlife Research 11: 303–309.

Taylor, J. A. and Tulloch, D. (1985) Rainfall in the wet-dry tropics: extreme events at Darwin and similarities between years during the period 1870–1983. Australian Journal of Ecology 10: 281–295.

Thom, B. G. (1975) Mangrove ecology from a geomorphic viewpont. p. 464–481. *In*: G. Walsh, S. Snedaker and H. Teas (eds.). Proceedings of the International Symposium on Biology and Management of Mangroves, Vol. 2, University of Florida, Gainesville, Florida, U.S.A.

Thom, B. G., Wright, L. D. and Coleman, J. M. (1975) Mangrove ecology and deltaic-estuarine geomorphology: Cambridge Gulf –Ord River, Western Australia. Journal of Ecology 63: 203–232.

Thompson, M. B. (1983) The physiology and ecology of the eggs of the pleurodiran tortoise *Emydura macquarii* (Gray 1831). PhD. thesis, University of Adelaide, Adelaide, Australia. 193 pp.

Thompson, M. B. (1985) Reptiles of the mound spring area. p. 78–83. *In*: J. Greenslade, L. Joseph, and A. Reeves, (eds.). South Australian Mound Springs. Nature conservation Society of South Australia, Adelaide, Australia.

Thompson, M. B. (1986) River Murray Wetlands. Department of Environment and Planning, South Australia, Adelaide, Australia. 172 pp.

Timms, B. V. (1970) Chemical and zooplankton studies on lentic habitats in north-eastern New South Wales. Australian Journal of Marine and Freshwater Research 21: 11–33.

Timms, B. V. (1974) Morphology and benthos of three volcanic lakes in the Mt Gambier district, South Australia. Australian Journal of Marine and Freshwater Research 25: 287–297.

Timms, B. V. (1976) Morphology of Lakes Barine, Eacham and Euramoo, Atherton Tableland, north Queensland. Proceedings of the Royal Society of Queensland 87: 81–84.

Timms, B. V. (1980a) The benthos of Australian lakes. p. 23–39. *In*: W. D. Williams, (ed.). An Ecological Basis for Water Resource Management. Australian National University Press, Canberra, Australia.

Timms, B. V. (1980b) The benthos of the Kosciusko glacial lakes. Proceedings of the Linnean Society of New South Wales 104: 119–125.

Timms, B. V. (1980c) Farm dams. p. 345–359. *In*: W. D. Williams (ed.). Ecological Basis for Water Resource Management. Australian National University Press, Canberra, Australia.

Timms, B. V. (1982) Coastal dune waterbodies of north-eastern New South Wales. Australian Journal of Marine and Freshwater Research 26: 1–13.

Timms, B. V. (1986a) Reconnaissance limnology of some coastal dune lakes of Cape York Peninsula, Queensland. Australian Journal of Marine and Freshwater Research 37: 167–176.

Timms, B. V. (1986b) The coastal dune lakes of eastern Australia. p. 421–432. *In*: P. De Dekker and W. D. Williams (eds.). Limnology in Australia. Dr. W. Junk, Dordrecht, The Netherlands.

Toerien, D. F., Gary, P. R., Finlayson, C. M., Mitchell, D. S. and Weerts, P. G. J. (1983) Growth models for *Salvinia molesta*. Aquatic Botany 16: 173–179.

Tyler, M. J. (1976) Frogs. Collins, Sydney, Australia. 256 pp.

Tyler, M. J. and Crook, G. A. (1987) Frogs of the Magela Creek system. Supervising Scientist for the Alligator Rivers Region, Northern Territory, Technical Memorandum 19, AGPS, Canberra, Australia. 46 pp.

Tyler, M. J., Crook, G. A. and Davies, M. (1983) Reproductive biology of the frogs of the Magela Creek system, Northern Territory. Records of the South Australian Museum 18: 415–445.

Tyler, P. A. (1974) Limnological studies. p. 29–61. *In*: W. D. Williams (ed.). Biogeography and Ecology in Tasmania. Dr. W. Junk, The Hague, The Netherlands.

Wake, J. A. (1975) A study of the habitat requirements and feeding biology of the dugong, *Dugong dugon* (Muller). Thesis, James Cook University, Townsville, Queensland, Australia. (Cited in Heinsohn *et al.* 1977).

Walker, D. I. and Prince, R. I. T. (1987) Distribution and biogeography of seagrass species on the north-west coast of Australia. Aquatic Botany 29: 19–32.

Walker, K. F. (1981) Ecology of freshwater mussels in the River Murray. Australian Water Resources Council Technical Paper No. 63, Canberra, Australia. 119 pp.

Walker, K. F. (1985) A review of the ecological effects of river regulation in Australia. Hydrobiologia 125: 111–129.

Walker, K. F. (1986a) The Murray-Darling River system. p. 631–659. *In*: B. R. Davies and K. F. Walker (eds.). The Ecology of River Systems. Dr. W. Junk, Dordrecht, The Netherlands.

Walker, K. F. (1986b) The state of ecological research on the River Murray. p. 637–648. *In*: P. De Dekker and W. D. Williams (eds.). Limnology in Australia. Dr. W. Junk, Dordrecht, The Netherlands.

Walker, K. F. (1986c) The freshwater mussel *Velesunio ambiguus* as a biomonitor of heavy metals associated with particulate matter. p. 175–185. *In*: B. T. Hart (ed.). The Role of Particulate Matter in the Transport and Fate of Pollutants. Water Studies Centre, Chisolm Institute of Technology, Melbourne, Australia.

Walker, K. F. and Hillman, T. J. (1981) Phosphorus and nitrogen loads in waters associated with the River Murray near Albury-Wodonga and their effects on phytoplankton populations. Australian Journal of Marine and Freshwater Research 33: 223–243.

Walker, T. D. and Tyler, P. A. (1984) Tropical Australia, a dynamic limnological environment. Verhandlungen Internationale Vereinigung für Theoretische und Angewandte Limnologie 22: 1727–1734.

Walker, T. D., Waterhouse, J. and Tyler, P. A. (1984) Thermal stratification and distribution of dissolved oxygen in billabongs of the Alligator Rivers Region, Northern Territory. Supervising Scientist for the Alligator Rivers Region, Northern Territory, Australia, unpublished Open File Record 28, 79 pp.

Warcup, J. (1982) Wetlands of South Australia: A Selective Bibliography of Their Natural History. South Australian Department for Environment and Planning 4/82, Adelaide, Australia. 15 pp.

Warren, J. W. (ed.) (1975) The Westernport Bay symposium. Proceedings of the Royal Society of Victoria 87: 1–155.

Webb, G. J. W., Manolis, S. C. and Buckworth, R. (1982) *Crocodylus johnstoni* in the McKinlay River area, N.T. I. Variation in the diet, and a new method of assessing the relative importance of prey. Australian Journal of Zoology 39: 877–899.

Webb, G. J. W., Sack, G. C., Buckworth, R. and Manolis, S. C. (1983) An examination of *Crocodylus porosus* nests in two northern Australian freshwater swamps, with an analysis of embryo mortality. Australian Wildlife Research 10: 571–605.

Webb, G., Manolis, S., Whitehead, P. and Letts, G. (1984) A proposal for the transfer of the Australian population of *Crocodylus porosus* Schneider (1801), from Appendix I to Appendix II of C.I.T.E.S. Conservation Commission of the Northern Territory, Darwin, Australia. 82 pp.

Wells, A. G. (1982) Mangrove vegetation of Northern Australia. p. 57–78. *In*: B. F. Clough (ed.). Mangrove Ecology in Australia: Structure, Function and Management. Australian National University Press, Canberra, Australia. 302 pp.

Wells, A. G. (1983) Distribution of mangrove species in Australia. p. 57–76. *In*: H. J. Teas (ed.). Biology and Ecology of Mangroves. Dr. W. Junk, The Hague, The Netherlands.

West, R. J., Thorogood, C. A., Walford, T. R. and Williams, R. J. (1985) An estuarine inventory for New South Wales, Australia. Fisheries Bulletin 2, New South Wales Department of Agriculture, Sydney, Australia. 140 pp.

Westlake, D. F. (1975) Primary production of freshwater macrophytes. p. 189–206. *In*: J. P. Cooper (ed.). Photosynthesis and Productivity in Different Environments. Cambridge University Press, London, England.

White, J. M. (1986) Managing the New England lagoons for waterbirds. Master of Science Thesis, University of New England, Armidale, Australia. 228 pp.

Williams, A. R. (1979) Vegetation and stream pattern as indicators of water movement on the Magela Floodplain, Northern Territory. Australian Journal of Ecology 4: 239–247.

Williams, W. D. (1964) Some chemical features of Tasmanian inland waters. Australian Journal of Marine and Freshwater Research 15: 107–122.

Williams, W. D. (1967) The chemical characteristics of lentic waters in Australia. p. 53–58. *In*: A.H. Weatherly (ed.). Australian Inland Waters and Their Fauna: Eleven Studies. Australian National University Press, Canberra, Australia.

Williams, W. D. (1974) Biogeography and Ecology in Tasmania. Dr. W. Junk, The Hague, The Netherlands. 498 pp.

Williams, W. D. (ed.) (1980) An ecological basis for water resource management. Australian National University Press, Canberra, Australia. 417 pp.

Williams, W. D. (ed.) (1981a) Salt Lakes. Dr. W. Junk, The Hague, The Netherlands. 444 pp.

Williams, W. D. (1981b) The limnology of saline lakes in western Victoria. p. 233–259. *In*: W. D. Williams (ed.). Salt Lakes. Dr. W. Junk, The Hague, The Netherlands.

Williams, W. D. (1983) Life in Inland Waters. Blackwell, Melbourne, Australia. 252 pp.

Williams, W. D. (1984) Chemical and biological features of salt lakes on the Eyre Peninsula, South Australia, and an explanation of regional differences in the fauna of Australian salt lakes. Verhandlungen Internationale Vereinigung für Theoretische und Angewandte Limnologie 22: 1208–1215.

Williams, W. D., Walker, K. F. and Brand, G. W. (1970) Chemical composition of some inland surface waters and lake deposits of New South Wales, Australia. Australian Journal of Marine and Freshwater Research 21: 103–116.

Wood, E. J. F. (1959a) Some east Australian sea-grass communities. Proceedings of the Linnean Society of New South Wales 84: 18–26.

Wood, E. J. F. (1959b) Some aspects of the ecology of Lake Macquarie, N.S.W. with regard to an alleged depletion of fish. VI. Plant communities and their significance. Australian Journal of Marine and Freshwater Research 10: 322–340.

Woodroffe, C. D. (1985) Variability in detrital production and tidal flushing in mangrove swamps. p. 201–212. *In*: K. N. Barsdley, J. D. S. Davie and C. D. Woodroffe (eds.). Coasts and Tidal Wetlands in the Australian Monsoon Region. Australian National University North Australia Research Unit, Mangrove Monograph No. 1, Casuarina, Australia.

Woodroffe, C. D., Chappell, J. M. A., Thom, B. G. and Wallensky, E. (1985) Geomorphology of the South Alligator tidal river and plains, Northern Territory. p. 3–16. *In*: K. N. Barsdley, J. D. S. Davie and C. D. Woodroffe (eds.). Coasts and Tidal Wetlands in the Australian Monsoon Region. Australian National University North Australia Research Unit, Mangrove Monograph No. 1, Casuarina, Australia. 375 pp.

Yassini, I. (1985) Foreshore vegetation of Lake Illawarra. Wetlands 5: 97–117.

Young, A. R. M. (1986) The geomorphic development of dells (upland swamps) on the Woronora Plateau, N.S.W., Australia. Zeitschrift für Geomorphologie 30: 317–327.

Young, P. C. and Kirkman, H. (1975) The seagrass communities of Moreton Bay, Queensland. Aquatic Botany 1: 191–202.

Yugovic, J. Z. (1985) The vegetation of the Lake Conneware State Game Reserve. Deparment of Conservation and Forests, Victoria, Arthur Rylah Institute for Environmental Research Technical Report Series no. 18, Heidleberg, Victoria, Australia. 53 pp.

Wetlands of Papua New Guinea

P. L. OSBORNE

Abstract

Over 80% of the 5,000 lakes in Papua New Guinea lie below 40 m and most are associated with large rivers and surrounded by extensive wetlands. Mangroves, brackish swamps, freshwater swamps, and alluvial plains account for 7.5% of the total land area of the country but these regions are sparsely populated (2–4 persons per km^2). The island of New Guinea is geologically young and the region is seismically active with the north coast undergoing quite rapid uplift while the southern side is sinking slowly. The relief of Papua New Guinea is dominated by a central cordillera, north of which is a depression which is occupied by the Sepik River in the west and the Ramu River in the east. To the south, in the western part of Papua New Guinea, is a huge tract of low-lying land drained by the Fly and Strickland Rivers.

Thirty-one taxa of mangroves have been recorded and extensive communities occur in the tidal reaches of the larger rivers, particularly those draining into the Gulf of Papua. The utilisation of mangroves has, so far, been largely of a subsistence nature and away from urban areas they are mostly unspoiled. Lowland freshwater wetlands are a mosaic of open water, herbaceous swamp and swamp savanna, and woodland. Two aquatic weeds (*Salvinia molesta*, *Eichhornia crassipes*) are now widespread but biological control of *Salvinia* in the Sepik wetland has been successful. Wetland invertebrates have been poorly studied and work on fish has been largely taxonomic. The native fish fauna is derived directly from the marine fauna and it has been supplemented by the introduction of twenty-one exotic species. Two species of crocodile are found throughout the low-lying wetlands and crocodile farming is an important village-based activity. Of the 708 species of birds listed for New Guinea, 115 are waterbirds and six of these are endemic. Of the native mammals, only four water rats are regarded as wetland species but introduced Rusa deer occur in large numbers in the Fly River area. Most of the wetlands in Papua New Guinea are in pristine condition but one has been markedly

D.F. Whigham et al. (eds.), Wetlands of the World I, 305–344.

altered from sewage disposal and others are under threat from mine tailings disposal. Two case studies are described. While there is no specific legislation for wetland conservation, protection of them is afforded under a number of general Acts of Parliament.

Introduction

Papua New Guinea is predominantly a wet country with rainfall in some areas in excess of 10,000 mm per year and most of the country receives annual rainfall of between 2,500 and 3,500 mm (McAlpine *et al.* 1983). The relief is generally rugged and mountainous except in the south-west and along the banks of the lower reaches of the larger rivers (Löffler 1977). The central ranges are interrupted in places by large, elongate intermontane valleys. As a result of the high rainfall and the rugged topography, rivers are mostly fast-flowing with very high discharges. However, in their lowland reaches, some of them form the central feature of extensive wetlands. Chambers (1987) recorded a total of 5,383 freshwater lakes with a surface area greater than 0.1 ha. The lakes are mostly small, with only 22 having a surface area greater than 1,000 ha. Over 80% of the lakes lie below 40 m altitude and most of these are associated with large rivers. Although only 4% of the lakes are over 2,000 m in altitude, some have wetlands of conservation significance associated with them.

Biologically, the country is extremely diverse with environments ranging from lowland swamps and tropical rain forests to alpine grasslands and frost-covered mountain peaks. The main environments in Papua New Guinea and their approximate percentage are (Paijmans 1976):

Coastal beach ridges and flats	0.5
Coastal saline and brackish swamps	1.5
Lowland freshwater swamps	11.0
Lowland alluvial plains and fans	15.0
Foothills and mountains below 1000 m a.s.l.	43.0
Lower montane zone, 1000–3000 m a.s.l.	25.0
Upper montane zone, 3000–4000 m a.s.l.	4.0

Papua New Guinea (Fig. 1) has a large land area of some 462,000 km^2 and a human population of only 3,006,799 (1980 census) although this is estimated to be growing at a rate of 2.1% per annum and 47% of the population is under 15 years of age. Consequently, the environment is relatively unspoilt and many natural habitats could still be conserved. This is particularly true of the larger areas of wetlands where the population density is very low (2–4 persons per km^2). However, the environment, in general, is coming under increasing pressure and aquatic ecosystems, in particular, in

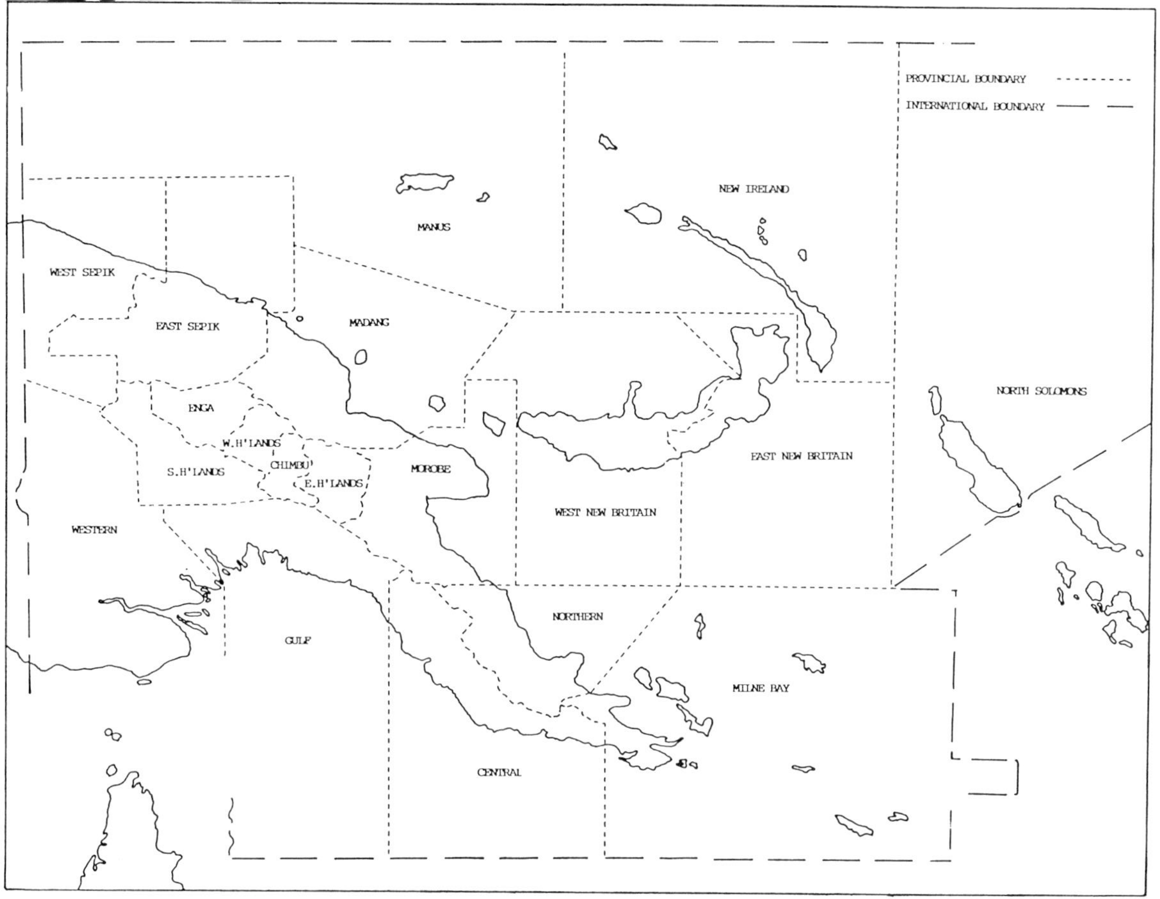

Figure 1. Map of Papua New Guinea showing international and provincial boundaries (redrawn from King, D. and Ranck, S. (eds.) PNG Atlas, Robert Brown and Associates, Australia).

Figure 2. The major geomorphic regions of Papua New Guinea (redrawn from Löffler 1977).

some parts of the country are being degraded through the disposal of mining, agricultural, and urban wastes. Furthermore, some of the pristine forests have been replaced by woody regrowth as the result of widespread shifting agriculture and in parts of the country, the capacity of the land to support the growing population is being exceeded. Over 95% of the land is held under customary tenure and most of the rural people rely to some degree on natural resources for their livelihood.

Physical environment

The island of New Guinea is geologically young and together with its associated smaller islands is situated between the stable land mass of Australia to the south and the deep ocean basin of the Pacific to the north. Young folded and faulted mountain chains, island arcs, and recent volcanic and seismic activity are characteristic features of this mobile region of the earth's crust. New Guinea lies in the zone of interaction between the northward-moving Australian continental plate and the north-westward-moving Pacific plate (Löffler 1977). The orthogonal movement of these plates results in a structurally complex geology. The main geomorphologic regions and altitudinal zones are shown in Figs. 2 and 3 respectively. Episodic uplift and faulting occurred during the late Oligocene to early Miocene and is still in progress. The southern half of the island was formed from uplift of the Australian continen-

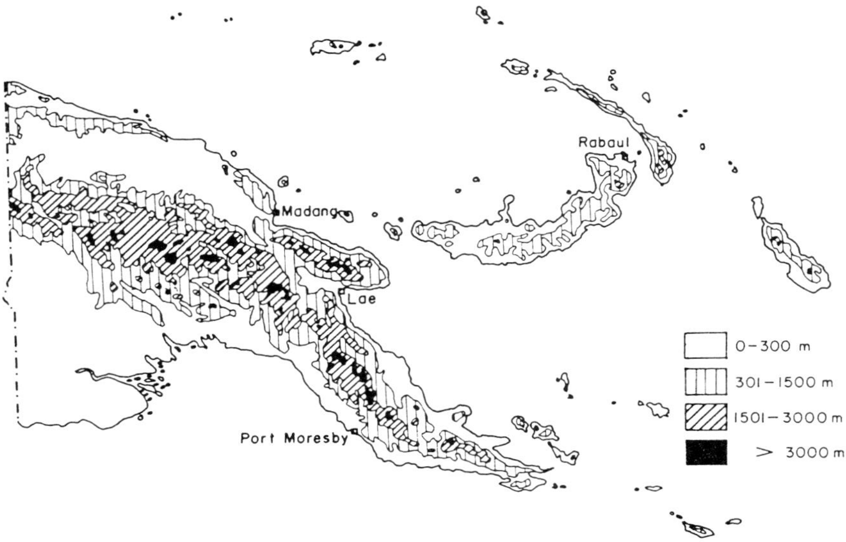

Figure 3. Altitudinal zones of Papua New Guinea (redrawn from Löffler 1977).

tal plate as it collided with the Pacific plate, these collisions also resulting in the formation of a chain of island arcs running from Sulawesi out into the Pacific. The north coast of Papua New Guinea is, for much of its length, undergoing quite rapid uplift, whilst in contrast, the southern side, particularly around the Gulf of Papua is sinking slightly (Löffler 1977). Seismic and volcanic activity have resulted in large areas being covered by volcanic deposits, while weathering and erosion of the steeply-sloping mountains has created extensive alluvial plains. Löffler (1977) subdivided Papua New Guinea into five major landscape regions and these are summarised below.

The central cordillera

The southern fold mountains, the highlands and the northern and eastern metamorphic ranges are often referred to as the central cordillera (Figs. 2 and 3). On the border with Irian Jaya the main mountain range is about 100 km wide, but it increases in width in the central highlands region to 300 km. From there the cordillera narrows towards Milne Bay. These highlands are a complex system of ranges and valleys with several huge stratovolcanoes and because of this heterogeneity, they can not be treated as one geomorphological region. The principal uplands of this region are the Star Mountains in the west, through the Hindenburg, Muller, Kubor, Schrader, Bismarck to the Owen Stanley Ranges, in the east. The highest peaks are Mt Wilhelm (4,509 m), Mt Giluwe (4,368 m), Mt Albert Edward (3,990 m)

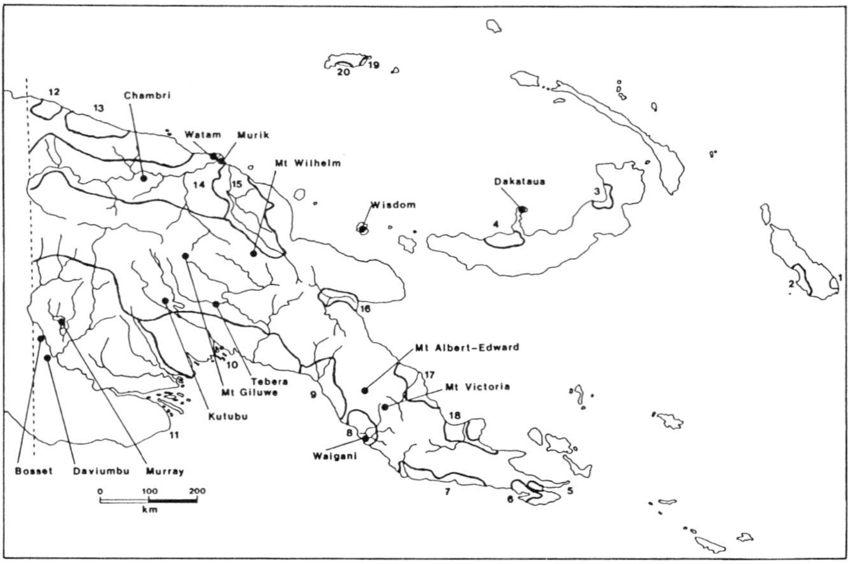

Figure 4. Major wetland areas in Papua New Guinea and geographical position of mountains and lakes mentioned in the text. 1. Abia, 2. Empress Augusta Bay, 3. Toriu, 4. Namo, 5. Rakua, 6. Mullin's Harbour, 7. Kemp Welch and Mori, 8. Vanapa, 9. Lakekamu, 10. Kikori, 11. Fly River, 12. Vanimo, 13. Arnold River, 14. Sepik, 15. Ramu, 16. Markham, 17. Mambare, 18. Musa, 19. Kelaua, and 20. Malai.

and Mt Victoria (4,035 m) (see Fig. 4). All these mountains were covered by glaciers during the Pleistocene. The most widespread landforms are mountains characterised by irregular branching ridges, with a relatively widely spaced pattern of steep-sided valleys. In the centre of the region is a succession of intermontane plains, broad upland valleys, and a number of large extinct volcanoes. Much of this central area is densely populated and covered by anthropogenic grassland but there are a number of small lakes and wetlands and, on the mountains, tarns, bogs and fens.

The intermontane trough

This huge structural depression is made up of plains, lowlands, and swamps and is, in most parts, flanked by steep mountains. The Markham and Ramu Rivers (Fig. 5) flow through the eastern part of the trough which is a narrow graben zone occupied by a series of alluvial fans. These fans are formed of coarse debris derived from the tectonically active Finisterre and Saruwaged Ranges which rise steeply along the north-eastern margin of the trough. The Markham River is only 170 km long but it is very wide and has a high discharge. The braided river channel is 3 km wide at the Ramu divide and

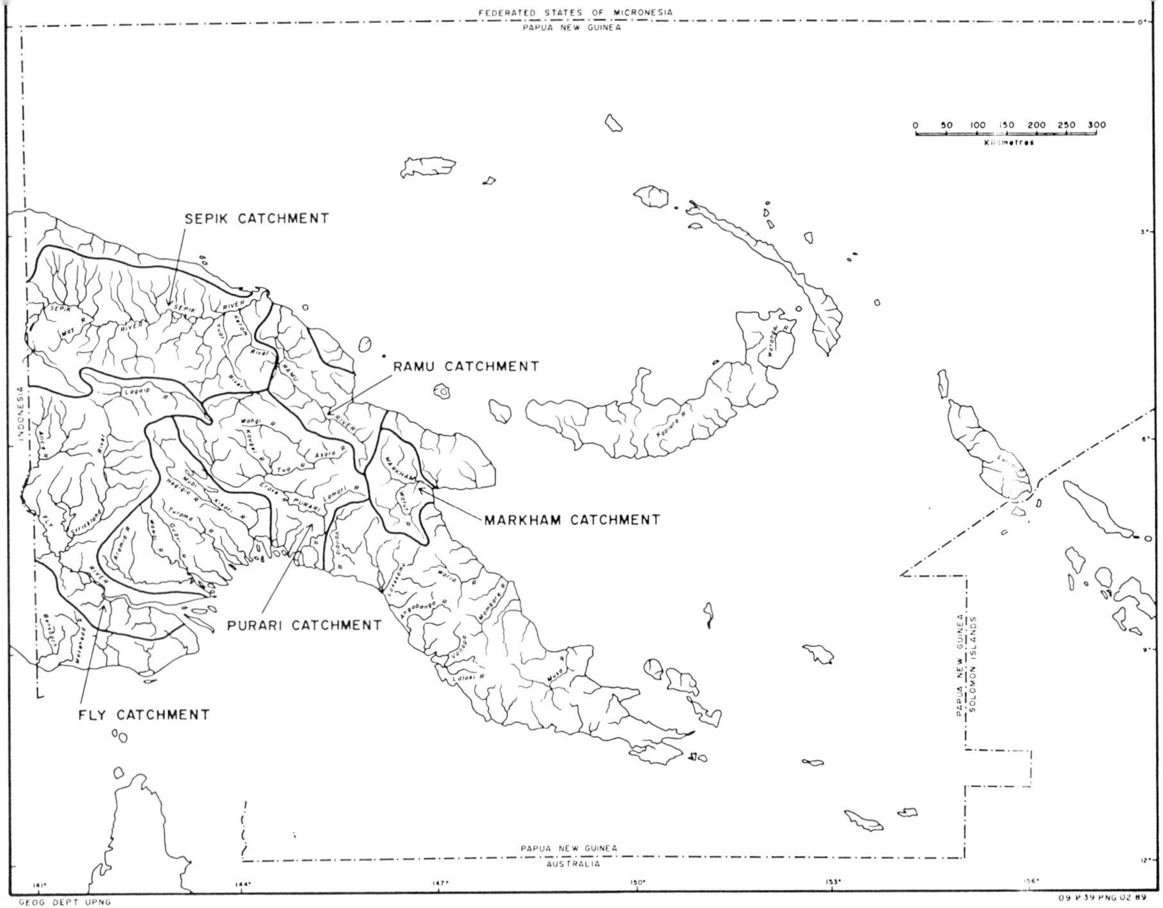

Figure 5. Map of the major river catchments of Papua New Guinea (redrawn from King, D. and Ranck, S. (eds.) PNG Atlas, Robert Brown and Associates, Australia.).

reaches its maximum width at its confluence with the Leron River. The Ramu River is approximately 720 km long but it has a relatively small catchment area. In its lower reaches, the floodplain terrain is very flat and swampy and the fall is imperceptible for a distance of 250 km from its mouth. Westwards, the narrow Markham-Ramu trough opens into the much more extensive Sepik depression. The Sepik River (Fig. 5) has a catchment area of 78,000 km^2, is approximately 1,100 km long and is navigable for about 500 km. It is Papua New Guinea's largest river in terms of catchment area but it has a lower discharge than the Fly River (Fig. 5). There are numerous (around 1,500) ox-bow and other lakes associated with the Sepik floodplain; the largest of these is Chambri Lake (Fig. 4) which is shallow (maximum 6 m) and has a highly variable area up to 250 km^2 in the flood season.

The main river channel is deep (over 35 m at Angoram) and consequently so are the more recently-formed ox-bow lakes in the lower floodplain. Mangrove areas associated with the river are negligible since the river discharges directly into the sea through a single outlet. This contrasts markedly with rivers in the south which invariably have extensive deltas and much larger estuarine zones. Modest mangrove areas occur to the west (Murik Lakes) and east (Watam Lakes) of the Sepik mouth (Fig. 4). These are separate from the main river but subject to the influence of its flood regime.

The fly platform and the southern plains and lowlands

The Fly Platform is the largest tract of low-lying land in Papua New Guinea, occupying nearly a third of the mainland and it constitutes a series of superimposed alluvial plains on structural plainlands. In the far west it is more than 400 km wide but it narrows towards the east. In the south, the area is gently undulating and flat areas are poorly drained and swampy. The Platform is drained by the Fly and Strickland Rivers (Fig. 5) and swamps and lakes occur mainly along the rivers occupying blocked valleys and along the coast behind beach ridges. The north-eastern part of the Platform is formed by extensive volcano-alluvial fans (see section on the Wetlands of Western Province).

In contrast to the relatively uniform Fly Platform, the southern plains and lowlands form a region of great heterogeneity, with hilly lowlands and intervening depositional plains at the mouths of the main rivers. The region is a narrow belt 15–20 km wide between the coast and the steeply rising central ranges, from the Kikori and Purari Rivers (Fig. 5) in the west to Mullin's Harbour in the east. The Purari River has a large catchment area (33,670 km^2), draining the central highlands. It is 630 km long and has an

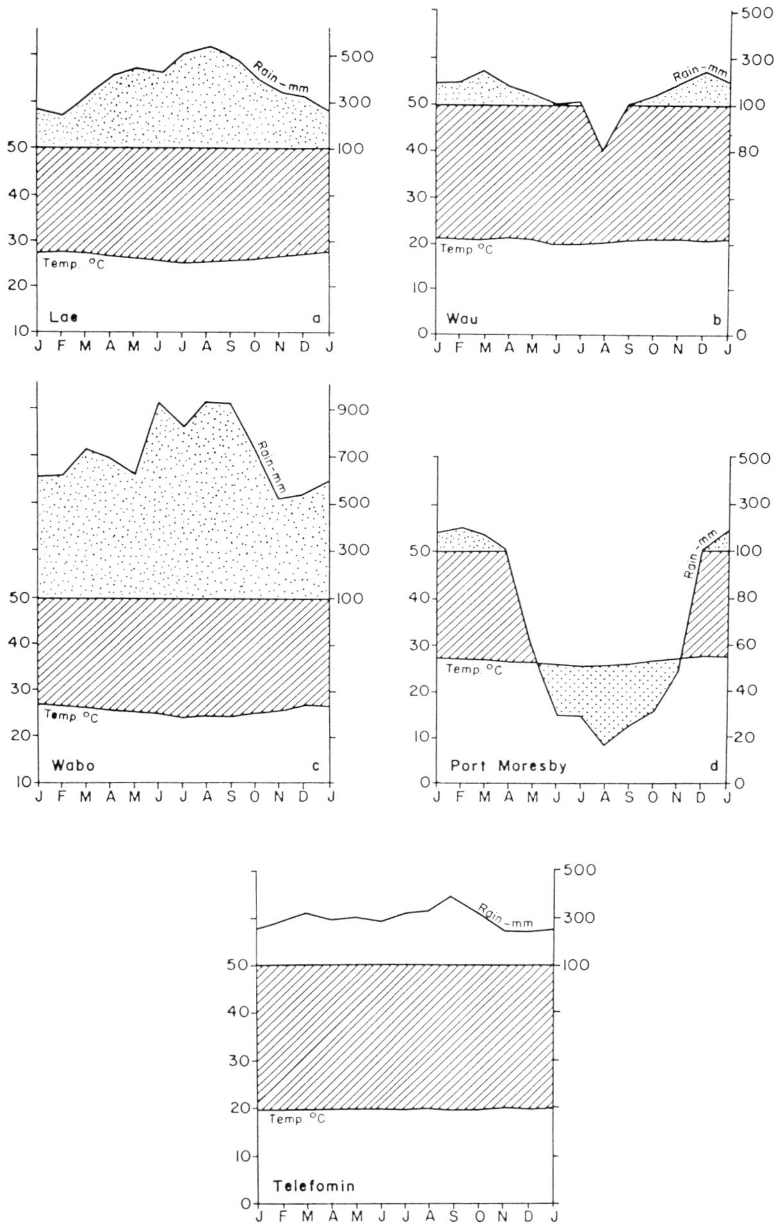

Figure 6. Climate diagrams showing mean monthly temperatures (° C lower half; scale at left) and mean monthly rainfall (mm; upper half; scale at right) from: Lae 10 m altitude; 6°44′ S, 147°00′ E; Wau 1150 m altitude; 7°10′ S, 146°4′ E; Wabo 40 m altitude; 7°00′ S, 145°03′ E; Port Moresby 10 m altitude; 9°29′ S, 147°09′; Telefomin 1520 m altitude; 5°09′ S, 141°38′ E (figures redrawn from Gressitt 1982.).

extensive delta and the discharge is 2667 $m^3 s^{-1}$ (Petr 1983). Other major rivers draining this area include the Vailala, Lakekamu, Angabanga, Vanapa, Laloki and Kemp-Welch (Fig. 5).

Northern coastal ranges

The northern ranges run parallel to the central ranges and consist of several individual ranges separated by lowlands such as the lower Sepik-Ramu plain and the Gogol depression (Figs. 2 and 3). They are generally much lower and narrower than the central ranges and consist predominantly of dendritic mountain ridges and steep-sided valleys. The ranges fall steeply down to a narrow, discontinuous coastal plain consisting of beach ridges, mangrove flats, alluvial plains and swamps, and coral platforms and terraces (Löffler 1977).

Islands

Numerous islands ranging in area from about 30,000 km^2 to a few square kilometres lie off the north coast of Papua New Guinea and its easternmost tip. The Bismarck Archipelago comprises all the larger islands which form an oval ring about the Bismarck Sea. The southern Bismarck Island Arc includes New Britain and the chain of volcanic islands off the north coast of New Guinea. It is a typical island arc with an arcuate shape, a deep ocean trench and a belt of volcanoes along its concave northern coast. Most of the volcanoes are active or potentially so and two large lakes occupy calderas within this region: Lake Wisdom on Long Island and Lake Dakataua on New Britain (Fig. 4). Lake Wisdom is surrounded by steep crater walls rising to 300 m a.s.l. and the lake is 360 m deep, making it one of the deepest lakes in the south-east Asia/Australia region (Ball and Glucksman 1978). Lake Dakataua was also formed by the post-eruptive collapse of a volcanic crater and the maximum depth of the lake is 120 m (Ball and Glucksman 1980).

Climate

New Guinea is a humid tropical island with moderate to very high rainfall (Fig. 6). Although the local climate is strongly related to topography, two surface pressure systems are the major climatic controls in Papua New Guinea. From about May to August the country is influenced by south-easterly trade winds which originate from the sub-tropical high pressure system located 25°–30° S. From January to April the inter-tropical convergence zone influences the climate and north-westerly winds predominate and, because the heaviest and most frequent rainfalls are associated with these winds, they are called monsoons. Most of Papua New Guinea experiences relatively high annual rainfall of 2,500–3,500 mm. Some lowland areas are drier but annual rainfall of less than 1,000 mm is restricted to Port Moresby and surrounds (Fig. 6). Large areas of the highlands receive in excess of 4,000 mm per year and in some places over 10,000 mm per year has been recorded. Three basic rainfall patterns are evident: Type A (Lae and Wabo in Fig. 6): rainy season in the middle of the year (May–August); Type B (Telefomin in Fig. 6): relatively uniform distribution throughout the year; Type C (Wau and Port Moresby in Fig. 6) rainy season from December–March (Abeyasekera 1987). Temperature seasonality is minimal but the diurnal cycle is important. Air temperatures are high throughout the year with daily mean maximum temperatures on the coast around 30–32° C and minima around 23° C. The most marked characteristic of temperature is the drop associated with increasing altitude (Fig. 6). Average temperature declines 0.6° C with each 100 m increase in altitude. Above 2,200 m altitude frosts occur but only rarely. Frosts are more common at over 3,000 m altitude and snow occasionally falls on the higher mountains. The combination of high rainfall and temperature results in high humidity, cloudiness and only moderate rates of evaporation. For further information on the climate of Papua New Guinea, see McAlpine *et al.* (1983).

Biological environment

The vegetation of New Guinea is among the most diverse in the world. Paijmans (1976) classified the vegetation on the basis of seven major environments as listed in the Introduction. Within these categories he listed wetland types based on vegetation form and species present. The geographic position of the major wetlands in Papua New Guinea is given in Fig. 4 and the distribution of mangroves, herbaceous and wooded freshwater swamps, and montane areas is shown in Fig. 7. Taylor (1959) from his study of lowland swamps in northeastern Papua New Guinea recognised eight groups of lowland swamp communities. The main factors differentiating his groups were the type of water, (whether fresh, brackish, or salt), and the depth, duration, and frequency of flooding. The wetland classification schemes of both Paijmans (1976) and Taylor (1959) are compared in Table 1.

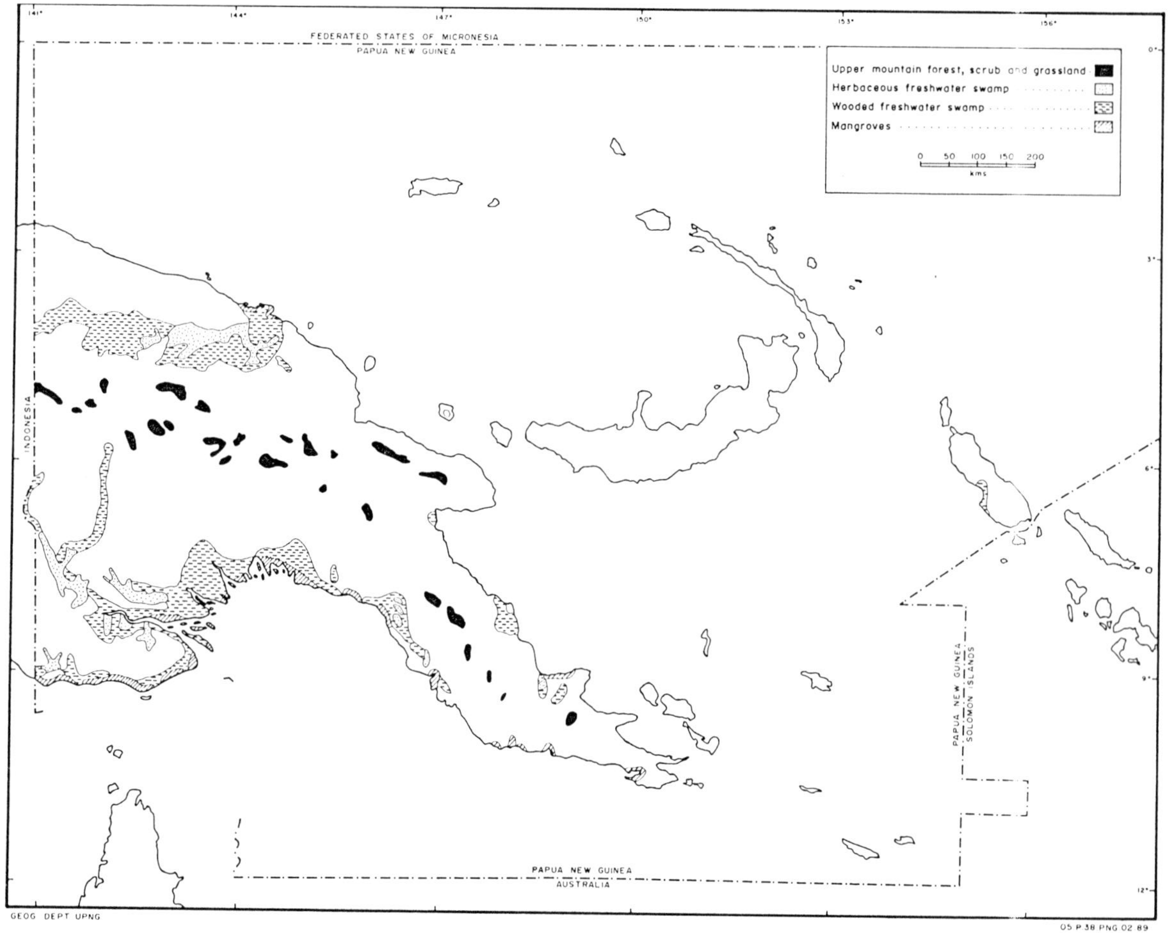

Figure 7. Map showing the distribution of the wetland vegetation in Papua New Guinea (redrawn from King, D. and Ranck, S. (eds.) PNG Atlas, Robert Brown and Associates, Australia).

Table 1. Major wetland types and communities as recognised by Paijmans (1976) and Taylor (1959).

	Paijmans	Taylor
MAJOR ENVIRONMENT	VEGETATION TYPE	SALT WATER SWAMPS
Saline and brackish swamps	Mangrove scrub	Tidal mangrove sequence
	Low mangrove forest	Herbaceous swamp
	Mature mangrove forest	Mangrove forest
	Avicennia scrub and woodland	Mangrove woodland
	Excoecaria scrub and woodland	Mangrove marsh sequence
	Nypa palm woodland	Mangrove woodland
		Mangrove thicket
		BRACKISH WATER SWAMPS
		Brackish water sequence
		Mangrove fern savanna
		Fern-tall tree savanna
		Estuarine sequence
		Swamp thicket
		FRESHWATER SWAMPS
Lowland freshwater	Herbaceous swamp vegetation	Permanent swamp sequence
		Swamp forest
	Leersia grass swamp	Swamp woodland
	Saccharum-Phragmites grass swamp	Swamp savanna
	Pseudoraphis grass swamp	Herbaceous swamp
	Mixed swamp savanna	Fluctuating swamp sequence
	Melaleuca swamp savanna	Seasonal swamp forest
	Mixed swamp woodland	Open seasonal swamp forest
	Sago swamp woodland	Seasonal swamp woodland
	Pandan swamp woodland	Herbaceous swamp
	Mixed swamp forest	Semi-seasonal swamp sequence
	Campnosperma swamp forest	Seasonal swamp forest
	Terminalia swamp forest	Open seasonal swamp forest
	Melaleuca swamp forest	Seasonal swamp woodland
		Herbaceous swamp
		Seasonal swamp sequence
Lower montane zone	Sedge-grass swamp	
	Phragmites grass swamp	
	Swamp forest	
Upper montane zone	Herbaceous swamp vegetation	

Mangroves

Mangrove flora

Mangrove swamps occupy large sections of the coast of Papua New Guinea but the most extensive communities occur in the tidal and brackish reaches of the estuaries of the larger rivers (Fig. 7). The largest areas of mangroves occur in the south, especially in the Gulf of Papua into which several large rivers (Fly, Kikori, Purari) flow (Fig. 5). Extensive stands also occur along the north coast around the Murik Lakes (Fig. 4) near the mouth of the Sepik and around the mouth of the Ramu River (Fig. 5).

The mangrove flora of Papua New Guinea is very rich with thirty-one obligate taxa listed by Johnstone and Frodin (1982). Percival and Womersley (1975) in their taxonomic treatment listed thirty-seven species that occur in mangrove swamps and indicated that most of these are restricted to this environment. Frodin and Leach (1982) can be used to identify mangroves of the Port Moresby region. The flora includes representatives of all genera currently recorded for Australia and South-East Asia with the exception of the genus *Kandelia*.

All species are not uniformly distributed throughout the region. The north coast of Papua New Guinea is not as rich in mangroves as the south coast but *Avicennia alba* and *Sonneratia caseolaris* (broad leaved-form) have been recorded there and not on the south coast. Conversely, *A. officinalis* has only been found on the south coast. A full list of mangrove species and their distribution is given in Table 2.

Where environmental gradients are present, mangrove species occur in characteristic patterns or zones. Zonation studies in Papua New Guinea have been carried out by Taylor (1959) (Mambare and Musa wetlands); Floyd (1977) (Purari Delta); Green and Sander (1979) (Manus Island) and Johnstone (1978) (Papa near Port Moresby). Taylor (1959) recognised three floristic communities of mangrove forest arranged in well-defined bands. Proceeding from rainforest, a narrow zone dominated by *Heritiera littoralis* was followed by a broad zone dominated by *Bruguiera gymnorhiza*. The third community was a broad zone dominated by *R. mucronata*. Percival and Womersley (1975) provided two zonation diagrams apparently based on transects. The actual species present varies considerably throughout Papua New Guinea and this complexity is probably due to a limited set of environmental factors that vary in their relative importance from site to site. These factors include: inundation, wave action, drainage, salinity, and substrate (Johnstone and Frodin 1982).

Mangrove fauna

Several studies have shown that there is a rich mangrove fauna in Papua New Guinea. The aquatic invertebrate fauna is dominated by crustaceans

Table 2. Obligate mangrove species of the Papuan subregion and their distribution in New Guinea (after Johnstone and Frodin 1982).

Mangrove species	Family	Distribution
Acanthus ilicifolius	Acanthaceae	Ubiquitous
Aegialites annulata	Plumbaginaceae	South Coast
Aegiceras corniculatum	Myrsinaceae	Ubiquitous
A. floridum	Myrsinaceae	Irian Jaya
Avicennia alba	Verbenaceae	North Coast
A. eucalyptifolia	Verbenaceae	Ubiquitous
A. officinalis	Verbenaceae	South Coast
A. rumphiana	Verbenaceae	Ubiquitous
Bruguiera cylindrica	Rhizophoraceae	Ubiquitous
B. exaristata	Rhizophoraceae	South Coast
B. gymnorhiza	Rhizophoraceae	Ubiquitous
B. parviflora	Rhizophoraceae	Ubiquitous
B. sexangula	Rhizophoraceae	Ubiquitous
Camptostemon schultzii	Bombacaceae	South Coast
Ceriops decandra	Rhizophoraceae	Ubiquitous
C. tagal		
var. *australis*	Rhizophoraceae	South Coast
var. *tagal*	Rhizophoraceae	Ubiquitous
Excoecaria agallocha	Euphorbiaceae	Ubiquitous
Heritiera littoralis	Sterculiaceae	Ubiquitous
Lumnitzera littorea	Combretaceae	Ubiquitous
L. racemosa	Combretaceae	Ubiquitous
Osbornia octodonta	Myrtaceae	South Coast
Rhizophora apiculata	Rhizophoraceae	Ubiquitous
R. lamarckii	Rhizophoraceae	S. Coast and New Hanover
R. mucronata	Rhizophoraceae	Ubiquitous
R. stylosa	Rhizophoraceae	Ubiquitous
Scyphiphora hydrophyllacea	Rubiaceae	Ubiquitous
Sonneratia alba	Sonneratiaceae	Ubiquitous
S. caseolaris	Sonneratiaceae	Ubiquitous
Narrow form "Claudie"	Sonneratiaceae	South Coast
Wide form "Tully"	Sonneratiaceae	North Coast
S. ovata	Sonneratiaceae	South Coast
Xylocarpus australasicus	Meliaceae	Ubiquitous
(*X. mekongensis*)		
X. granatum	Meliaceae	Ubiquitous

and molluscs (Liem and Haines 1977). In the Purari and Kikori mangroves Liem and Haines (1977) recorded 143 species of fish from 58 families and 95 species of tetrapods which included 12 mammals, 59 birds, 20 reptiles, and 4 amphibians. Some bird species appear to be confined to mangroves and include: Mangrove fantail (*Rhipidura phasiana*), Broad-billed flycatcher (*Myiagra ruficollis*), and Red-headed honeyeater (*Myzomela erythrocephala*) (Coates 1985). Species which are characteristic of mangroves but which also occur in other habitats include the mangrove heron (*Butorides striatus*), Mangrove warbler (*Gerygone levigaster*), Rufous fantail (*Rhipidura rufi-

frons), Mangrove robin (*Eopsaltria pulverulenta*), and Mangrove golden whistler (*Pachycephala melanura*) (Coates 1985, Beehler *et al*. 1986).

Mangrove utilisation

It is estimated that the mangrove forests in the Gulf of Papua occupy an area of between 162,000 and 200,000 ha (including *Nypa* palm stands). Rhizophoraceae dominate with 121,500 ha (56%) while *Bruguiera* and *Camptostemon* comprise 18% and 14% respectively. In the Purari Delta there are about 134,000 ha of mangroves (Liem and Haines 1977). The Central Province including the National Capital District has an estimated mangrove area of 57,770 ha (Rau 1984). These extensive stands of mangroves constitute an important resource and those near urban areas and in proximity to development schemes are subject to varying degrees of degradation. *Ceriops* spp., *Rhizophora* spp., and *Avicennia marina* are commonly used as firewood and mangroves, particularly members of the family Rhizophoraceae, are used for building houses and canoes. Mangroves provide a habitat for a range of animals which constitute an important source of protein for local inhabitants. These include the mud crab (*Scylla serrata*), bivalves (*Barbatia lima*, *Gelonia coaxans*) and the common cockle (*Anadara granosa*) (Rau 1984). Mangroves also provide raw materials for clothing, tools, dyes, tannins, and traditional medicines.

Freshwater wetlands

Flora

The freshwater flora of Papua New Guinea comprises eight species of Characeae, twenty-one species of ferns and fern-allies, and 130 flowering plants. Leach and Osborne (1985) provides a taxonomic treatment with keys to families, genera, and species; species descriptions and notes on taxonomy, habitat, and distribution. Freshwater algae in Papua New Guinea have been poorly studied and most work has been taxonomic (Thomasson 1967, Watanabe *et al*. 1979a, 1979b, Yamagishi 1975, Yamagishi and Watanabe 1979)

The herbaceous vegetation of lowland wetlands. The vegetation of the lowland freshwater swamps forms a continuous sequence from open water to tall mixed swamp forest, depending on the depth and quality of the water and drainage and flooding conditions (Fig. 7). The aquatic vegetation consists of free-floating, floating-leaved, and submerged plants. These either form a mixture or are arranged in concentric zones. They occupy the shallow margins between open water and grass swamp, and in places, cover entire lakes that have a uniform depth.

Herbaceous communities consisting of sedges, herbs, and ferns are characteristic of stagnant, permanent, relatively deep swamps. Common species include: *Thoracostachyum sumatranum*, *Scleria* sp., *Hanguana malayana*, and the fern *Cyclosorus interruptus*. *Phragmites karka* often dominates on poorly drained swamp margins whereas *Pseudoraphis spinescens* and *Ischaemum polystachyum* form narrow bands along more steeply sloping, wet-dry margins. Grasses such as *Leersia hexandra*, *Echinochloa praestans*, *Oryza* spp., *Panicum* sp., and *Hymenachne amplexicaulis* occupy permanently swampy river plains that may be under 3 m of water in the flood season. Conn (1983) records large floating islands composed of *L. hexandra* covering extensive areas of Lake Tebera, Gulf Province (Fig. 4). A sudd community of sedges (*Gahnia* and *Cyperus odoratus*), grasses (*Isachne* and *Ischaemum*), and herbs (*Limnophila indica*, *Ludwigia adscendens*, and *Polygonum*) developed on these islands. Herbs such as *L. adscendens* and *Ipomoea aquatica* are commonly present along the open water fringes of lowland swamps and may reach out over open water.

Tall swamp grasses, mainly *Saccharum robustum* and *P. karka* grow in swamps that are shallower than those described above and may be intermittently dry. *Pseudoraphis spinescens* is a low creeping swamp grass that is most extensive in south-west Papua New Guinea. Here it forms dense, almost pure stands on flood plains that are seasonally dry. In the south-west of Papua New Guinea these grasses are heavily grazed by deer and wallabies.

Two aquatic weeds, *Salvinia molesta* and *E. crassipes*, are now widespread in the low-lying wetlands of Papua New Guinea. *Salvinia* was first recorded in Papua New Guinea in 1977 at Wau and on the Sepik River where it was probably introduced in 1971/2 (Mitchell 1978/1979, 1979). By 1979 *Salvinia* covered 80 km^2 and the physical impact of the weed was reflected in the decline in fish catches, crocodile hunting, and sago gathering and also in the disruption to the lives of Sepik villagers. People in a number of villages were unable to reach markets to sell produce and children were prevented from attending school (Mitchell *et al.* 1980, Coates 1982, 1987a). A control programme was instituted in 1979 (Thomas 1979) and physical, chemical, and biological control methods were tested (Thomas 1985). Biological control using the South American weevil *Cyrtobagous salviniae* was spectacularly successful (Thomas 1985, Laup 1985, Thomas and Room 1986). The initial development of the weevil population introduced in 1982 was nitrogen limited (Room and Thomas 1985, 1986a, 1986b) but once-established on plants enriched with fertilizer, the populations became self-sustaining. By June 1985, the weevil had destroyed an estimated 2 million tonnes of weed which had covered 250 km^2. The local people have now resumed their former lifestyles (Thomas and Room 1986).

Eichhornia crassipes was first recorded in 1962 when it was found near Bulolo in old gold mining dredge ponds. It has, despite warnings (Mitchell

1978/1979, Osborne and Leach 1984), become widespread throughout lowland areas of Morobe and Madang Provinces and has recently spread to the wetlands west of Port Moresby, Central Province. It has also been reported from Manus, New Ireland, North Solomons, West New Britain, Eastern Highlands, and Western Province (Laup 1986a). Attempts at biological control using the weevil *Neochetina eichhorniae* have produced some promising results.

Pistia stratiotes is widespread throughout the lowlands of Papua New Guinea and although it is a weed in other parts of the world it has not reached weed proportions here (Laup 1986b). The same is true of *Hydrilla verticillata* which has only been recorded near Wau, Madang, Port Moresby, and at Lake Kutubu (Fig. 4) (Leach and Osborne 1985).

Lowland swamp savanna and woodland. Mixed swamp savanna is a transitional vegetation type between purely herbaceous swamps and swamp woodland and it occurs in permanent, stagnant swamps (Fig. 7). In addition to an herbaceous cover, there is an open layer of trees: *Nauclea*, *Campnosperma*, *Syzygium*, and *Melaleuca*. *Melaleuca* swamp savanna is characteristic of the fluctuating backswamps of the middle Fly and Strickland Rivers and also occurs along parts of the monsoonal south and south-west coasts. *Melaleuca* trees form an even, open almost pure, canopy. The main species is *M. cajuputi*. In the wet season *Melaleuca* swamp savanna is inundated and colonised by aquatic plants. In permanent swamps the tree storey of mixed swamp woodland is generally open and ranges from low to tall. Common trees are *Campnosperma*, *Nauclea coadunata*, *Mitragyna ciliata*, and *Timonius*. Palms and pandans fill in much of the space below the trees and *Hanguana*, sedges, and *Cyclosorus* from a dense ground cover. Sago palm, *Metroxylon sagu*, is widespread throughout more or less permanent swampy woodland. All gradations occur from stands of pure sago to woodland with a dense layer of trees and an open lower tier of sago. The palm grows best where there is a regular influx of freshwater. Swamp pandans occupy a habitat similar to that of sago palm but have a wider range. They form open to quite dense, pure stands in shallow, fresh to brackish, stagnant to frequently flooded swamps. Mixed swamp forest is the most common type of swamp forest. It generally has an open, but occasionally dense canopy. Some of the trees often found are: *Campnosperma*, *Terminalia canaliculata*, *Nauclea coadunata*, *Syzygium*, *Alstonia scholaris*, *Bischofia javanica*, and *Palaquium*.

Dense stands of *Campnosperma* (*C. brevipetiolata* and *C. coriacea*) are found in permanently flooded backswamps. Sago may form a dense understorey. *Terminalia* swamp forest is mainly found in North Solomons Province where *T. brassii* grows together with *Campnosperma* and locally dominates in the canopy of open swamp forest. Low-lying, frequently flooded, bouldery, and sandy rivers, and peat swamps with flowing waters are the habitats.

Lower montane wetlands. Communities dominated by sedges and grasses occur above about 1,800 m in swamps occupying intermontane basins, local depressions in valley floors, and seepage slopes. Many different sedges are present and they commonly make up most of the ground cover. Characteristic grasses are *Arundinella furva*, *Isachne* spp., and *Dimeria* spp. *Phragmites karka* commonly forms pure stands in seepage areas on slopes and on flat valley floors to over 2,500 m and it also occurs associated with *Miscanthus floridulus* along river banks and swamp margins. The vegetation associated with eight lakes within this zone, near Mount Giluwe in the Southern Highlands was described by Conn (1979). *Miscanthus floridulus* formed an almost monospecific community above the high water level, while the rooted emergent, semi-aquatic species, *Eleocharis sphacelata*, closely parallelled the minimum water level and was, in places, replaced by *Scirpus mucronatus* or *Hydrocotyle sibthorpioides*. *Nymphoides indica* and *Chara* species were the only true aquatic species found by him. Further work was carried out on some of these lakes by Chambers *et al.* (1987). The profiles obtained by them are reproduced in Fig. 8.

Upper montane wetlands. Herbaceous communities consisting of a mixture of low herbs, sedges, grasses, and mosses occupy depressions, fringe open water and in the higher parts of the zone, also occur on slopes. Common grasses include: *Anthoxanthum angustum*, *Agrostis reinwardtii*, and *Monostachya oreoboloides*. The sedge *Carpha alpina* and the fern *Gleichenia vulcanica* locally form pure stands.

Fauna

Invertebrates. The freshwater invertebrate fauna of Papua New Guinea has been poorly studied. MacArthur (1965) noted that numbers of species of freshwater invertebrates do not increase in the tropics. Lakes in Papua New Guinea feature low zooplankton diversities, with a notable paucity of cladoceran and copepod species. Twelve zooplankton species have been recorded from the lakes on Mount Wilhelm (Bayly and Morton 1980, Löffler 1973, McKenzie 1971). Low species diversity was attributed to the youthfulness of these lakes compared with the diversity of zooplankton in older, high altitude, tropical lakes. Taxonomic works on freshwater invertebrates include: Decopoda (Holthius 1974, 1982), leeches (Richardson 1977), *Macrobrachium* (Robertson 1983), Mollusca (Benthem-Jutting 1963), and mussels (McKenzie 1956, McMichael and Hiscock 1958).

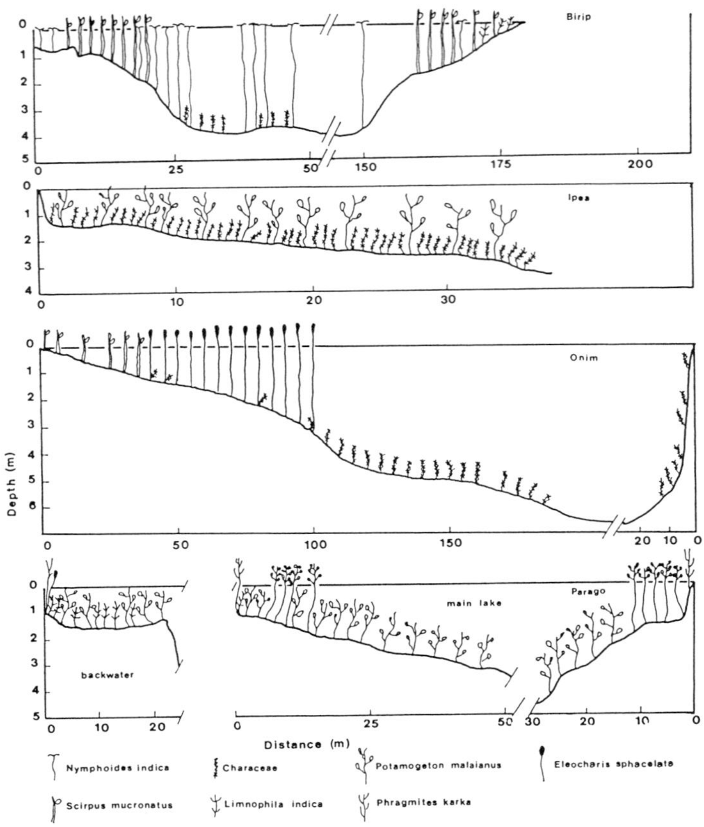

Figure 8. Aquatic plant profiles showing the dominant rooted plants for Lakes Birip, Ipea (Enga Province), Onim (Southern Highlands Province), and Parago (Enga Province), Papua New Guinea (redrawn from Chambers *et al.* 1987).

Freshwater fishes. The island of New Guinea lies east of "Wallace's Line" and thus forms part of the Australasian zoogeographic region. The inland and native freshwater fish fauna is derived directly from the marine fauna and primary freshwater ostariophysian fishes are absent with the exception of *Scleropages* spp. (Roberts 1978, McDowell 1981). All species naturally occurring in freshwater are either diadromous or descendants of marine families. This has been noted to have an adverse effect on species diversity and fish stock abundance resulting in low fisheries yields (Coates 1985).

The ichthyofauna can be clearly divided into two zoogeographic regions. Freshwater bodies to the south of the central cordillera have an ichthyofauna closely allied with that of northern Australia. The two land masses were originally joined. Whilst several of those species with diadromous habits can be found in both southern and northern rivers, the fish permanently inhabiting freshwater in the north are invariably different species from those in southern water bodies. Apart from the land barrier formed by the central cordillera, it is also known that northern rivers are much younger than southern rivers. Of those fish families common to both northern and southern rivers, species diversity is invariably lower in the north (Coates 1987b). Several fish families are also absent from the north whilst all those present there, are also present in southern rivers. A list of some freshwater fish taxonomy papers is given in Osborne (1988).

Twenty-two species of freshwater fish representing nineteen genera, eleven families and all six continents have been introduced into Papua New Guinea for various reasons (West 1973, West and Glucksman 1976, Glucksman *et al.* 1976). The effects of some of these introductions are not known but others have had significant positive impacts on fisheries (e.g. common carp (*Cyprinus carpio*) and tilapia (*Oreochromis mossambica*) in the Sepik (Coates, 1985)). Carp were released into lakes in the highlands which were previously without fish and Lakes Dakataua and Wisdom still lack a fish fauna. Robertson and Baidam (1983) found only one species (*Oxyeleotris fimbriata*) in Lake Wangbin, a remote lake in the Star Mountains of Western Province. West (1973) questioned the value of these introductions and the importation of exotic freshwater fishes is now strictly controlled (Coates 1987b).

Inland fisheries are invariably for subsistence purposes with the exception of a small commercial fishery for tilapia in the Sepik and a more important commercial fishery based on the migratory barramundi or sea bass (*Lates calcarifer*) in southern rivers. Despite their often low yields, inland fisheries are considered of major importance because of the large percentage (over 85%) of local inhabitants involved (Coates 1985, see also Schuster 1957, Haines 1979, 1983, Haines and Stevens 1983).

The southern rivers of Papua New Guinea have abundant fish and compared with the Sepik and highland rivers are richer in species. The Fly River ichthyofauna comprises 103 species in thirty-three families and has much in common with that of northern Australia (Roberts 1978). Barramundi (*L. calcarifer*) is the most abundant large commercial species in the waterways of southern New Guinea and it reaches its maximum abundance in rivers, swamps, and lagoons discharging into the Gulf of Papua. The fish is catadromous and spawning occurs from November to early March in the bays and estuaries of the Gulf. Following spawning, the fish migrates away from coastal waters into the rivers and lakes of Gulf and Western Provinces (Dunstan 1962, Moore 1980, 1982, Moore and Reynolds 1982, Reynolds 1978, Reynolds and Moore 1982). Barramundi from the Lake Murray (Fig. 4) region have been shown to accumulate mercury and the average mercury concentration recorded is close to the limit set by the Australian Government (Sorentino 1979, Petr 1979, Kyle and Ghani 1982a, 1984). Barramundi from other freshwaters in Papua New Guinea have low levels of mercury and it would appear that the source of the mercury is the catchment area of the Strickland River (Natural Systems Research 1988).

The ichthyofauna of the Sepik River appears to be less diverse than that of the Fly River and many of the species recorded are endemic to rivers of northern New Guinea and some are only known from the Sepik (Coates 1985, 1987b). Of twenty-nine native species recorded by Coates (1983) only *Anguilla bicolor* and *Lutjanus argentimaculatus* are of any importance to the subsistence fishery. Almost all the other native species are small. Two exotic species: *Oreochromis mossambica* and *Cyprinus carpio* are more productive and it has been recommended that further exotic species be introduced (Petr 1984, Coates 1983, 1984, 1986, 1987b). *Tilapia rendalli* was introduced to the Sepik and Ramu rivers in 1991 but their success can not yet be judged (D. Coates, personal communication).

Indigenous fish species in the upper Purari belong to the Families Anguillidae, Arridae, Plotosidae, Melanotaeniidae, Terapontidae, and Gobiidae but a complete list is not available. The freshwater fish fauna of the lower Purari has forty-nine species from twenty-four families. A notable feature of the Purari fish fauna is the paucity of herbivores and a lack of planktonic feeders; detritus forming the base of the food chain (Haines 1983).

Berra *et al.* (1975) recorded forty-three native species belonging to nineteen families and six exotic species from the Laloki, Brown, and Goldie Rivers draining the western half of Central Province (Fig. 5). The Laloki fauna is closely related to that of Northern Australia but about half the species are endemic to New Guinea. The seven small highland lakes studies by Chambers *et al.* (1987) contained only introduced carp.

The most up to date published list of freshwater species and their distribution is that of Allen (1991) which includes Irian Jaya and exotic introductions.

Amphibians. Of the three orders in this group neither caecilians nor salamanders occur in Papua New Guinea. Frogs (Anura: five families) are well represented, with over 200 species described at present and new species being recognised as current research proceeds. The five families are: Bufonidae (the introduced *Bufo marinus*), Hylidae (about 70 species), Leptodactylidae (5–10 species), Microhylidae (about 90 species) and Ranidae (10–15 species) (Menzies 1975). Not all species are aquatic: a large number are forest dwellers which burrow beneath the surface, or live beneath leaf litter. The hylids (*Litoria* and *Nyctimystes*) have aquatic larvae but in the majority of cases, the association with water is only temporary. Tadpoles are often present in large numbers and are often the only vertebrates present in mountain streams. One ranid, *Limnonectes grunniens*, is more permanently aquatic but occurs no further east than the Fly and Sepik River systems. It probably feeds on insects with aquatic larvae. One group of very small hylids (the *Litoria bicolor* group) is characteristic of open grassy swamps from Seram to New Britain and Australia. These frogs probably live permanently in such places and often in very large numbers so they must have a substantial impact on the swamp ecosystem. They probably feed on small insects that come to lay eggs in the water.

The majority of species are endemic to either Papua New Guinea, or the island of New Guinea. A southern group having its origins from Australia can be recognised, as can a group of species originating from the Solomon Islands to the south-east of Papua New Guinea. The surrounding islands have, in general, a depauperate amphibian fauna in comparison with the adjacent mainland. Much taxonomic work remains to be done and the rich variation in specialised habitat requirements already known, suggest that many more species are yet to be described.

Distribution and conservation of crocodiles. Two species of crocodile are found in Papua New Guinea. These are the New Guinea or Freshwater Crocodile (*Crocodylus novaeguineae*) and the Saltwater Crocodile (*Crocodylus porosus*). Both species are still found in relatively large numbers and are heavily exploited for hides and meat. Distribution maps of these two species are given in Osborne (1987). The distribution of the endemic New Guinea crocodile is not fully documented. It prefers a freshwater environment but is occasionally found in brackish waters such as the Fly delta. It is more often found in sluggish, shallow water rather than swifter-flowing or deeper areas, and hides, by day, in thick grass or under fallen trees. At night this crocodile moves into deeper water (see Burgin 1980a, Niell 1946, Pernetta and Burgin 1980, 1983).

Characteristically, the saltwater crocodile occurs in brackish areas such as estuaries and mangroves. Although once thought to be restricted to the coastal tidal areas, the species is now known to occur well inland. Inland populations are generally associated with freshwater pools and deep rivers. The species is also recorded from areas of shallow water including fast-flowing rocky streams up to 1,000 km inland (Burgin 1981). The saltwater crocodile is dangerous to man and livestock but it is relatively easy to hunt as its nests are easy to locate. This species is now rare in the large mangrove areas of Gulf and Western Provinces and also in East and West Sepik Provinces where it was once apparently common.

Numbers of both species declined during the late 1950s and 1960s through indiscriminate hunting. In 1969 the Crocodile Trade (Protection) Act (Chapter 213) was implemented which, to protect breeders, placed a ban on trade in skins greater than 51 cm belly width. This halted further decline in crocodile numbers, indicated by a steady level of export during the 1970s. The Act also allows for the control of the crocodile industry on a systematic basis. Regulations under this Act control crocodile farming and the purchase and export of skins. In 1981 a ban was placed on trade in skins smaller than seven inches. This ban was established because PNG was in a position to ranch crocodiles on a large scale. By 1984, although the number of skins exported was the same as in previous years, 30% were from ranched animals and consequently were of higher grade and greater size. Both species of crocodile are listed on Appendix 2 of CITES which means they are regarded as vulnerable but trading is allowed to continue. Papua New Guinea for many years has been the only country allowed by CITES to trade in *C. porosus*. In 1982 extensive monitoring of both species commenced especially in the Ambunti District of the East Sepik Province. From 1982–1985 the number of saltwater nests virtually doubled in this area – an indication of the effectiveness of the management policy (see Burgin 1980b, 1980c, Bolton 1978, Bolton and Laufa 1979).

Lizards, snakes, tortoises, and turtles. Between 150–200 species of lizard (Reptilia: Lacertilia) in five families occur in PNG. Only certain members of the families of dragon lizards (Agamidae) and Monitors (Varanidae) are habitually associated with water, but it does not appear to be an essential habitat for their survival. The water monitor (*Varanus indicus*) and Gould's monitor (*V. gouldii*) appear to be equally at home in water as on land, though their food habits show them to be primarily land animals.

Approximately 90–95 species of snakes from six families are recorded from Papua New Guinea. Three families are typically aquatic and can be found in the still and slower-flowing waters of the lowlands. The file-snake family (Acrochordidae) includes one or two genera with two or three species according to taxonomic opinion; the water-snakes of the family Colubridae (genera *Amphiesma*, *Cerberus*, *Enhydris*, *Fordonia*, and *Myron*) are regular inhabitants of wetlands, though the family also includes many species restricted to land. Sea-snakes (Hydrophiidae) are represented by just over twenty species in eight genera. These are all marine (although *Enhydris* and *Schistosa* have been recorded in north coast rivers away from the sea) and some species are frequently seen in shallow water over reefs and presumably in mangrove waters, while others seek their food at greater depths from the sandy bottom.

The six species of marine turtles in Papua New Guinea are principally pelagic, coming to breed on sand beaches rather than wetland or mangrove-mud areas. However, two species of turtle, the pit-shelled Turtle (*Carettochelys insculpta*) and the soft-shelled Turtle (*Pelochelys bibroni*) are found in freshwaters south of the central cordillera: the latter is found also in the freshwaters of the Sepik wetlands (and through Indonesia to India).

Five species of tortoises (Family: Chelidae) are either the same, or closely related, species to those found in eastern Australian freshwaters. Four of these are found only south of the central cordillera and are inhabitants mainly of still or slowly-flowing freshwater. *Elseya novaeguineae* and *Chelodina siebenrocki* are endemic to New Guinea and *C. parkeri* is endemic to the Fly River Basin and coastal areas (Goode 1967).

Birds. The avifauna of Papua New Guinea is relatively well-documented (Beehler *et al.* 1986, Coates 1985, Gould 1970). Of the 708 species of birds listed for New Guinea, 115 are waterbirds, and all but three of these occur in Papua New Guinea. Osborne (1987) provides a full list of wetland birds of Papua New Guinea and also lists them for the major wetlands in the country. Six species of wetland bird are endemic to New Guinea: *Zonerodius heliosylus*, *Anas waigiuensis*, *Rallina rubra*, *R. forbesi*, *R. mayri* and *Megacrex inepta*. *Zonerodius heliosylus* frequents streams, pools, and swamps and *Megacrex inepta* inhabits mangrove forests, river-edge bamboo thickets and watery ditches. *Anas waigiuensis* is restricted to the highlands, up to at least 3,850 m altitude, and is commonly the only duck encountered there, but never in large numbers. The low abundance and diversity of ducks in the highlands may be due to the small size and widespread distribution of the lakes (Diamond 1972). Chambers (1987) recorded pelicans (*Pelecanus conspicullatus*) on Lake Papapli (2,420 m altitude) and suggests that the expansion of its range (this bird was regarded as a rare visitor to southern New Guinea (Iredale 1956) but it has been more regularly recorded recently) may be due to the introduction of fish into highland lakes. Coates (1985), however, indicates that the distribution of this bird within New Guinea is more scattered during drought years in Australia. He also records that a major eruption into the New Guinea area occurred in 1977–78. MacKay (1970) suggested that the little pied cormorant (*Phalacrocorax melanoleucos*) increased in the Port Moresby area following introductions of exotic fish.

Waterbirds breeding in Papua New Guinea include two grebes, two cormorants, *Anhinga melanogaster*, twelve species of herons and egrets, *Ephippiorhynchus asiaticus*, *Anseranas semipalmata*, nine species of ducks, fifteen species of Rallidae, *Grus rubicunda*, *Irediparra gallinacea*, and six species of shorebirds. The great majority of passage migrants and winter visitors from Asia are shorebirds (thirty regular species and seven vagrants). Several of these occur in very large numbers en route to and from wintering areas in Australia. Furthermore, Rapson (1968) observed flocks of up to 100,000 magpie geese (*A. semipalmata*) on shallow Suki Lagoon in Western Province. Birds in such large numbers must have significant effects on the wetland ecology through feeding and nutrient cycling. Regular migrants from Australia include: *Pelecanus conspicillatus*, several herons and egrets, *Threskiornis aethiopicus*, *T. spinicollis*, *Plegadis falcinellus*, *Platalea regia*, *Haematopus longirostris*, *Stiltia isabella*, *Erythrogonys cinctus*, *Larus novaehollandiae*, *Chlidonias hybridus*, and *Hydroprogne caspia*. There are no less than twenty-two species of kingfishers in New Guinea, but many of these are forest and savanna birds and only the collared kingfisher (*Halcyon chloris*) can strictly be listed as a wetland bird. However, the common kingfisher (*Alcedo altthis*) and the little kingfisher (*A. pusilla*) are often found along river banks and in swamp forests or mangroves respectively. The twelve-wired bird of paradise (*Seleucidis melanoleuca*) inhabits sago swamps.

Waterbirds observed at Lake Kutubu (Fig. 4) by Schodde and Hitchcock (1968) included fish-eating birds of the open water (*Phalacrocorax sulcirostris*, *P. melanoleucos*, *Anhinga melanogaster*), predators operating at the lake margins (*Egretta alba*, *E. intermedia*, *Nycticorax caledonicus*), black duck (*Anas superciliosa*), and two waders (*Tringa hypoleucos*, *T. brevipes*). Ball and Glucksman (1978) recorded black duck on Motmot Island in the centre of Lake Wisdom (Fig. 4) and suggested that they were responsible for effecting the colonisation by plants of this recently formed volcanic island. Ball and Glucksman (1978) saw the little grebe (*Tachybaptus ruficollis*) on Lake Wisdom. Ball and Glucksman (1980) recorded the following birds on Lake Dakataua (Fig. 4): *Anas superciliosa*, *Tachybaptus ruficollis*, *Ixobrychus flavicollis*, *Nycticorax caledonicus*, *Phalaropus lobatus*, *Himantopus leucocephalus*, and *Sterna* sp.

Mammals. Some 190–200 species of mammals occur in Papua New Guinea. Active and critical analysis of the various genera currently taking place indicates that this figure is subject to revision: new species, and taxonomically hidden species are presently being described. Of this number of species only four can be clearly tied to the presence of water: *Hydromys chrysogaster*, *H. habbema*; and *Crossomys moncktoni*. *Hydromys chrysogaster* is a widespread lowland water rat (from Tasmania to New Guinea) whereas *H. habbema* and *Crossomys moncktoni* are montane species.

Mention should also be made of the introduced deer, the Javan Rusa, *Cervus timorensis*, which occurs in large numbers in the seasonally-flooded trans-Fly area, and lesser numbers in wetland areas near Port Moresby. Small isolated populations are known at scattered localities where they do not appear to be tied to water. The larger populations, however, are typically swamp-dwellers and frequently graze with their head submerged.

Wetland conservation

Legislation

There is no specific environmental legislation directed primarily towards the conservation of wetlands. Protection is afforded however under a number of Acts. The Conservation Areas Act (1978) allows for the establishment of "Conservation Areas", these being areas of land deemed worthy of legal protection for a variety of reasons. Whereas the National Parks Act 1982 permits the establishment, on state-owned or long-leased land, of a series of National Parks, protecting areas of outstanding scenic and scientific value.

The purpose of the Fauna (Protection and Control) Act 1974–1982 is to allow the systematic management, use, and conservation of the fauna of Papua New Guinea. Wildlife Management Areas may be established under this Act. These are similar to Conservation Areas but their purpose is restricted to the management of wildlife resources, whereas a Conservation Area may be established for its scenic, aesthetic, or historic values. The International Trade of Endangered Species of Fauna and Flora Act (1979) controls the export and import of certain wildlife species among countries which are signatories to the international agreement.

Two Acts control the disposal of a wide range of pollutants which could be detrimental to the environment (Dumping of Wastes at Sea Act 1981 and Environmental Contaminants Act 1978). The Environmental Planning Act 1978 controls the exploitation of environmental resources, particularly in regard to the preparation of environmental impact assessments for projects which may have massive and long-term effects upon the quality of the environment after the project has finished.

The Water Resources Act can regulate land drainage, river/stream channels (diversion and damming) and the disposal of wastes to land, swamps and water courses and bodies. It encompasses both fresh and saline waters to the territorial boundaries of Papua New Guinea. It is specifically designed for public water supply protection but could be used to include environmental conservation.

Wetland degradation and pollution

Most of the wetlands in Papua New Guinea are still in pristine condition. The human population density in wetland areas is mostly low and their utilisation has been largely at a subsistence level. There are a number of instances where wetlands have been abused or threatened and two case histories are presented below.

Waigani Swamp

Waigani Swamp comprises a number of small, shallow lakes near Port Moresby (Fig. 4). It is a small part of the extensive swamp/river system dominated by the Brown and Laloki Rivers (Fig. 5). The main Waigani Lake is shallow (1–2 m deep) with an open water area of 120 ha, situated in a valley with much recent urban development. The lake is surrounded by a *Phragmites karka* and *Typha domingensis* swamp and is heavily fished for two introduced species: *Oreochromis mossambica* and *Cyprinus carpio*. In 1965 sewage settling ponds were established near by; subsequently they were expanded, and now about 80% of the sewage from the capital city of Port Moresby (population approximately 200,000) enters the swamp/lake system. Major changes in the aquatic flora of this system have occurred over the last thirty years and these have been described from a series of aerial photographs by Osborne and Leach (1983). From 1942 to 1956, Waigani Lake was dominated by emergent vegetation and there was very little open water. Between 1956 and 1966 this emergent vegetation was replaced by dense stands of floating-leaved plants: *Nymphaea* spp. and *Nymphoides indica*. Following this period, the increase in sewage effluent disposal correlates with a decline in floating-leaved plants and by 1974 only a few small stands remained. By 1978 no floating-leaved plants could be found in the main lake and their decline in Waigani Lake was accompanied by a regression of the surrounding reedswamp. This latter decline has continued subsequent to the work by Osborne and Leach (1983) (Osborne unpublished information). Osborne and Leach (1983) suggested that the nutrient enrichment of the lake was the most likely cause of these vegetation changes and concluded that much more research is required before tropical wetlands are used for the purification of wastewaters.

Osborne and Polunin (1986) were able to trace changes in the recent ecological history of Waigani Lake through the analysis of short (c. 1 m) sediment cores. Their major findings are summarised in Fig. 9. In the lowermost sections of the sediment cores, seeds of emergent plants predominated. These sediments had low nitrogen and phosphorus concentrations, undetectable levels of plant pigments, and high densities of epiphytic diatom frustules. These characteristics are indicative of swamp and correlate with the aerial photographs described by Osborne and Leach (1983) taken in 1943 and 1956. Following this phase, major changes in the vegetation and chemistry of Waigani Lake occurred. Nitrogen and phosphorus concentrations in the sediments increased and probably indicate the start of sewage disposal into the lake. It would appear both from the analysis of aerial photographs and the density of seeds in sediments laid down during this period, that initially this nutrient enrichment may have stimulated the growth of the aquatic

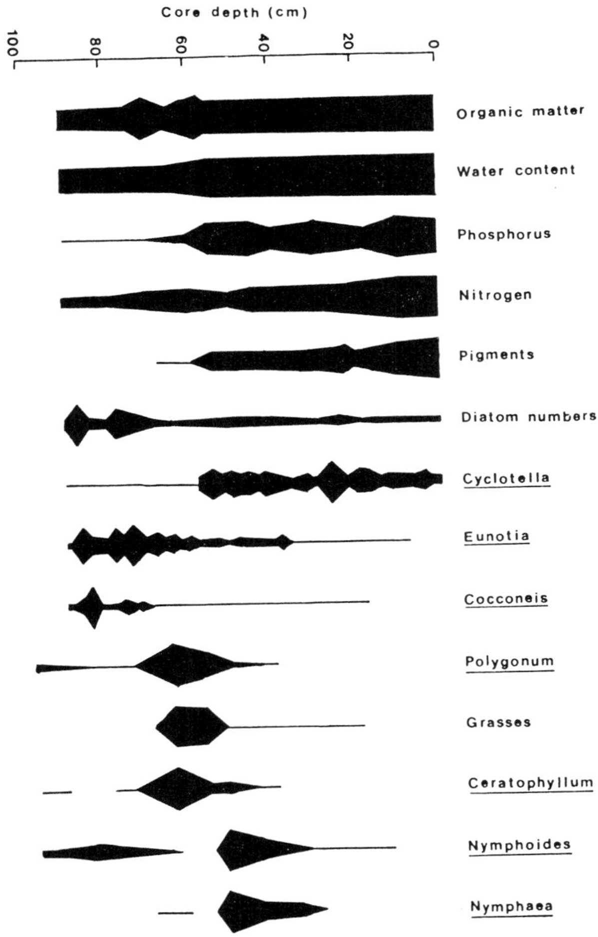

Figure 9. Profiles of organic matter, water content, total phosphorus, total nitrogen, plant pigment content, total diatom numbers, and numbers of the three major diatom genera expressed as percentages of maximum values from a sediment core collected from Waigani Lake. Also depicted are the profiles of seed numbres of *Polygonum attenuatum*, grasses, *Ceratophyllum demersum*, and *Nymphoides indica* from another core from the same lake (redrawn from Osborne and Polunin 1986).

vegetation. Further nutrient enrichment with the expansion of the sewage system in 1970 probably resulted in the higher nutrient deposition in the sediments and at around this time the switch from emergent vegetation to the one dominated by floating-leaved plants occurred. The increase in the area of open water was detected in the cores through the declining density of *N. indica* seeds and the increasing concentration of frustules of the planktonic diatom *Cyclotella meneghiniana*.

Osborne and Polunin (1986) pointed out that the changes in the aquatic vegetation of Waigani Lake were the reverse of the normal successional changes that occur in shallow lakes. They argued that although an increase in water-level is an obvious cause it can not explain all the features of the vegetation changes observed. The timing of the changes in the features measured in the cores suggested sewage effluent disposal was a possible cause but from information available it was not possible to rule out the effects of water level changes.

Further work by Polunin *et al.* (1988) has shown that timing of sewage effluent disposal into Waigani lake correlates with elevated levels in the sediments of lead, sulphur, manganese, barium, calcium, sodium, and strontium. The accelerated loadings of these elements may help to explain the disappearance of the swamp vegetation and it also appears that the different vegetation stages may have influenced patterns of sediment geochemistry. On the basis of the analysis of sediment cores from this and another shallow, lowland lake in the tropics, Osborne and Polunin (1988) concluded that sediment deposition in these lakes is sufficiently chronological to permit interpretations of their ecological history.

Wetlands of Western Province

The central feature of this area is the Fly River (Fig. 5) which is the second largest in Papua New Guinea with a discharge of $6000\,m^3\,s^{-1}$. The river is over 1,200 km long and is navigable as far as Kiunga, 800 km from its mouth. The gradient in its lower reaches is extremely gentle as Kiunga is only 20 m above sea level and the river is tidal 240 km from the sea. The middle Fly is characterised by an extensive floodplain 15–20 km wide with a mosaic of lakes, alluvial forest, swamp grassland, and swamp savanna. The river flows on an alluvial ridge formed by deposition of material eroded from the upper catchment and as this deposition was more rapid than that of tributary streams, the tributaries became blocked forming numerous tributary lakes (e.g. Bosset Lagoon and Lake Daviumbu). The river meanders extensively in this region, and in addition to tributary lakes there are numerous backswamps and ox-bows of variable depth depending on age (Fig. 10).

In the lower Fly region, mangroves and sago swamps dominate the vegetation of the estuary. The mouth of the Fly River is best described as a tidal delta. It can not be characterised as a true estuary, nor as a river delta because of its transient polyhaline structure and strong current regime.

Between its confluence with Fly River near Obo and the Strickland Gorge, the Strickland River floodplain contains tributary lakes, backswamps and oxbow lakes. Lake Murray (Fig. 4) is the largest of these habitats and lies in a shallow depression (maximum depth 7 m) at the confluence of the Strick-

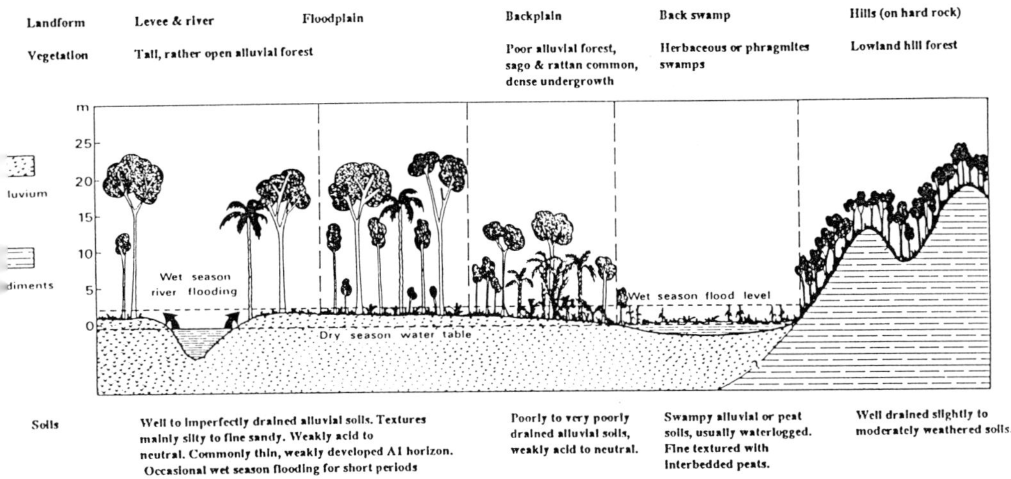

Figure 10. Schematic cross section through a lowland floodplain system showing river channel, swamp forest, and herbaceous swamp in a backswamp (redrawn from Bleeker 1983).

land and Fly Rivers. The lake is drained by the Herbert River, a tributary of the Strickland River.

This enormous wetland system is of high conservation value, particularly as it forms a refuge during drought years for water-fowl from Australia. Consequently, the environmental impacts on these wetlands resulting from the construction of the Ok Tedi and Porgera mines in the upper catchment areas of the Fly and Strickland Rivers respectively has been a cause for some concern (Pernetta 1988). No part of this wetland has been set aside as a conservation area but a Wildlife Management Area has been established at Tonda, on the Bensbach River, south-east of the Fly Delta.

The Ok Tedi mine is located in the upper catchment of the Ok Tedi, a tributary of the Fly River, and is now one of the largest open-cut copper and gold mines in the world. Waste rock and tailings are currently dumped into the Ok Tedi and Pickup *et al.* (1981) identified two potential, detrimental effects on the middle Fly wetlands. First, increased sediment loads in the Fly River could lead to increased siltation in the river system; and second, increased heavy metal concentrations in the water and sediments may result from mining activities.

Natural Systems Research (1988) indicated that the principal environmental threat from the Porgera copper and silver mine is through mine waste disposal into the Laiagap-Strickland River system. This mine went into production in 1990. The Porgera valley is in Enga Province and is drained by the Porgera River, a small, fast-flowing tributary of the Lagaip River, which in turn joins the Strickland. Environmental consequences similar to those identified for the Ok Tedi mine are listed by Natural Systems Research (1988) with the additional complication that the Porgera mine tailings will be higher in mercury than the natural suspended sediment load of the Strickland River. High levels of mercury in fish of the Fly-Strickland River systems have been known for some time (Lamb 1977; Kyle and Ghani 1982a; 1984). Furthermore, Kyle and Ghani (1982b) recorded elevated concentrations of mercury in the hair of the people living around Lake Murray. The Herbert River, which drains Lake Murray into the Strickland, has been shown to reverse flow on occasions, a result of the flatness of the topography and marked seasonal fluctuations in the water level of Lake Murray and the Strickland River (Natural Systems Research 1988; Osborne *et al.* 1987). This discovery indicated a pathway for mine wastes to accumulate in Lake Murray.

Both mining developments have been subjected to environmental impact assessments (Maunsell and Partners 1982, Natural Systems Research 1988) but our ecological understanding of tropical wetlands is so poor we are unable to predict accurately long-term effects of increased sediment and heavy metal loadings on them. Mowbray (1988a, 1988b) predicted that under conditions of low river flow, the effects of tailings from the Ok Tedi mine would be confined to the upper reaches of the Ok Tedi but chronic effects could extend down the Fly River. Even the mining company projected that the impact on the Middle Fly would be reduced numbers of fish species and lower populations (Townsend 1988). Osborne *et al.* (1988) showed that sediments transported by the Fly River are deposited in Lake Daviumbu, a backswamp of the Middle Fly region (Fig. 4). This contrasts with the environmental impact assessment which predicted that little river-borne sediment was expected to reach these backswamps (Maunsell and Partners 1982).

So far, apart from a number of critical incidents in the construction and operation of the Ok Tedi mine (see Townsend 1988), it has not been possible to show that either of these mines is having or will have long term effects on the middle Fly region, but the threat is real. They will, however, have major impacts in their upper catchments but the distance downstream that these will extend is largely unknown. Papua New Guinea faces the dilemma of choosing between potential environmental degradation from such projects and the large monetary rewards that they bring. This dilemma indicates the urgent need for research into the general ecological functioning of wetlands in the tropics, and, more specifically, into the effects of increased heavy metal and sediment loadings on the middle Fly wetlands.

Wetland research

Osborne (1988) lists 174 publications in a bibliography of freshwater ecology in Papua New Guinea. Very few of these publications relate to wetlands and most of them have a strong applied bias (e.g. *Salvinia*, fish introductions, mercury pollution, environmental impact assessments) or are taxonomic. A number of papers (Hope 1973, 1976, Hope *et al.* 1988, Walker 1972, Garrett-Jones 1979, Worsley and Oldfield 1988) report the results of analyses of sediment cores from a number of the small lakes and swamps in the highlands. These papers contain some incidental information on wetlands but are primarily biogeographical, studying changes in catchment use and climate. An inventory compiled by Osborne (1987) indicates the paucity of information and demonstrates that, apart from a few selected cases, (e.g. Waigani Swamp, Fly, and Sepik wetlands) our information is restricted to, probably incomplete, species lists.

Recommendations

1. Papua New Guinea is not yet party to the World Heritage Convention or the Ramsar Convention. The World Heritage Convention provides for the designation of areas of "outstanding universal value" as World Heritage sites, in order to promote their significance at local, national, and international levels. The Convention has provision for aid and technical cooperation to be offered to contracting parties for the protection of their World Heritage sites. The Ramsar Convention provides the framework for international cooperation to conserve wetlands. Contracting parties accept an undertaking to promote the wise use of all wetlands and to designate one or more wetlands for inclusion in a "List of Wetlands of International Importance" Papua New Guinea should sign both these conventions.

2. Two large areas of wetlands should be set aside as Conservation Areas. One of these should be the wetland surrounding the confluence of the Fly and Strickland Rivers, including Lake Daviumbu (Fig. 4). Lake Daviumbu supports very large populations of waterbirds and has a diverse aquatic flora with one species endemic to Western Province and another endemic to Papuasia. The wetland surrounding the Chambri Lakes (Fig. 4), should also be gazetted as a Conservation Area. This area also supports large bird populations and the fish fauna is completely different from that in Lake Daviumbu. Following the signing of the conventions described above, both these sites should be listed under the Ramsar Convention as wetlands of international importance and Lake Daviumbu should be included in the World Heritage List.

3. Twenty-one species of exotic fish have been introduced to the freshwaters of Papua New Guinea for a range of reasons. Further introductions should not be made without a thorough analysis of their impact and prior detailed study of areas into which they are to be introduced. A major problem with fish introductions is that, once released in the country, intra-country

transfers are hard to control. That each river basin in the country has a distinctive fauna (and in some cases, with no species overlap) means that a fish approved for release in the Sepik will probably not be suitable for release into the Fly River.

4. Two exotic aquatic plants are now widespread throughout the wetlands of Papua New Guinea. Biological control of *Salvinia molesta* has proved efficacious but its spread still needs to be carefully monitored. Water hyacinth (*Eichhornia crassipes*) has, despite warnings, spread throughout the north coast wetlands and has recently been recorded along the south coast. This plant could seriously disrupt the use of wetlands for subsistence fishing and transport and its control and eradication is urgently needed. Strict enforcement of import controls on aquatic plants is required so that further aquatic weeds are not brought into the country and education programmes need to be instituted to reduce intra-country movements of all plants and animals.

5. Further research into wetland ecology is needed. This should include detailed monitoring programmes to increase the data base of environmental conditions in pristine wetlands and a thorough study of the effects of heavy metals on tropical wetlands is urgently needed. Assistance from developed countries is needed and this should be in the form of trained manpower with financial support for capital equipment and maintenance costs.

Acknowledgments

I thank James Menzies, Helen Fortune-Hopkins, and Marjorie Sullivan for constructive criticism of the manuscript. Vagoli Bouauka drew Figures 1, 2, 3, 5, 6 and 7.

References

Abeyasekera, S. (1987) A classification of seasonal rainfall in Papua New Guinea. Science in New Guinea 13: 1–14.

Allen, G. R. (1991) Freshwater Fishes of New Guinea. Christensen Research Institute, Publication 9, Madang, Papua New Guinea.

Allen, G. R. and Boeseman, M. (1982) A collection of freshwater fishes from western New Guinea with descriptions of two new species (Gobiidae and Eleotridae). Records of the Western Australian Museum 10: 67–103.

Ball, E. and Glucksman, J. (1978) Limnological studies of Lake Wisdom, a large New Guinea caldera lake with a simple fauna. Freshwater Biology 8: 455–468.

Ball, E. and Glucksman, J. (1980) A limnological survey of Lake Dakataua, a large caldera lake on West New Britain, Papua New Guinea, with comparisons to Lake Wisdom, a younger nearby caldera lake. Freshwater Biology 10: 73–84.

Bayly, I. A. E. and Morton, D. W. (1980) A note on zooplankton from four Papua New Guinea lakes (altitudinal range 538–3630). p. 3–5. *In*: T. Petr (ed.). Purari River (Wabo)

Hydroelectric Scheme Environmental Studies, Vol. 11: Aquatic Ecology of the Purari River Catchment. Office of Environment and Conservation, Waigani, Papua New Guinea.

Beehler, B. M., Pratt, T. K. and Zimmerman, D. A. (1986) Birds of New Guinea. Princeton University Press, Princeton, NJ, USA. 293 pp.

Benthem-Jutting, W. S. S. van (1963) Non-marine molluscs of West New Britain, Part I, Mollusca from fresh and brackish waters. Nova Guinea, Zoology 20.

Berra, T.M., Moore, R. and Reynolds, L.F. (1975) The freshwater fishes of the Laloki River system of New Guinea. Copeia 1975: 316–326.

Bleeker, P. (1983) Soils of New Guinea. ANU Press, Canberra, Australia. 352 pp.

Bolton, M. (1978) Crocodile Farming in Papua New Guinea. Oryx 14: 365–369.

Bolton, M. and Laufa, M. (1979) The crocodile project in Papua New Guinea. Biological Conservation 22: 169–79.

Burgin, S. (1980a) The status of the biology and ecology of Papua New Guinea's crocodile, *Crocodylus novaeguineae* (Schmidt). Science in New Guinea 7: 163–171.

Burgin, S. (1980b) A review of crocodile farming in Papua New Guinea. Science in New Guinea 7: 73–88.

Burgin, S. (1980c) Crocodiles and crocodile conservation in Papua New Guinea. p. 295–300. *In*: L. Morauta, J. Pernetta and W. Heaney (eds.). Traditional Conservation in Papua New Guinea: Implications for Today, Monograph 16. Institute of Applied Social and Economic Research, Papua New Guinea.

Burgin, S. (1981) The biology of *Crocodylus porosus* (Schneider). Science in New Guinea 8: 9–37.

Chambers, M. R. (1987) The freshwater lakes of Papua New Guinea: An inventory and limnological review. Journal of Tropical Ecology 3: 1–23.

Chambers, M. R., Kyle, J. H., Leach, G. J., Osborne, P. L. and Leach, D. (1987) A limnological study of seven highlands lakes in Papua New Guinea. Science in New Guinea 13: 51–81.

Coates, B.J. (1985) The Birds of Papua New Guinea. Vol. 1. Dove Publications, Australia. 464 pp.

Coates, D. (1982) On the biological problems caused by the introduced water-weed *Salvinia molesta* Mitchell in the Sepik River. Regional workshop on limnology and water resource management in developing countries of Asia and the Pacific, Kuala Lumpur.

Coates, D. (1983) Raw data and preliminary analysis of gill-net catches from the Sepik River, roundwaters and floodplain. Report No. 83–16, Department of Primary Industry, Fisheries Research and Surveys Branch, Port Moresby, Papua New Guinea.

Coates, D. (1984) The fisheries and fish fauna of the Sepik River System: Recommendations for species introductions. Report No. 84–10, Department of Primary Industry, Fisheries Research and Surveys Branch, Port Moresby, Papua New Guinea.

Coates, D. (1985) Fish yield estimates for the Sepik River, Papua New Guinea, a large floodplain system east of "Wallace's Line". Journal of Fish Biology 27: 431–443.

Coates, D. (1986) Fisheries development of the Sepik River, Papua New Guinea: Proposed fish introductions. p. 367–370. *In*: J. L. Maclean, L. B. Dizon and L. V. Hosillos (eds.). The First Asian Fisheries Forum. Asian Fisheries Society, Manila, Philippines.

Coates, D. (1987a) On the biological problems caused by the introduced water weed *Salvinia molesta* Mitchell in the Sepik River, Papua New Guinea. Archive für Hydrobiologie 28: 205–208.

Coates, D. (1987b) Consideration of fish introductions into the Sepik River, Papua New Guinea. Aquaculture and Fisheries Management 18: 231–241.

Conn, B. J. (1979) The vegetation of the lakes of Mt Giluwe area (Southern Highlands Province), Papua New Guinea. p. 37–62. *In*: T. Petr (ed.). Purari River (Wabo) Hydroelectric Scheme, Vol. 10: Ecology of the Purari River Catchment, Office of Environment and Conservation, Waigani, Papua New Guinea.

Conn, B. J. (1983) Aquatic and semi-aquatic flora of the Purari River System. p. 283–293. *In*: T. Petr (ed.). The Purari – Tropical Environment of a High Rainfall River Basin. Dr. W. Junk, The Hague, The Netherlands.

Diamond, J. M. (1972) Avifauna of the Eastern Highlands of New Guinea. Nuttall Ornithological Club, Harvard University Club 12: 413–419.

Dunstan, D. J. (1962) The barramundi in New Guinea waters. The Papua and New Guinea Agricultural Journal 15: 23–31.
Floyd, A. G. (1977) Ecology of the tidal forests in the Kikori-Romilly Sound area, Gulf of Papua. Ecology Report 4, Office of Forests, Lae, Papua New Guinea.
Frodin, D, and Leach, G. J. (1982) Mangroves of the Port Moresby region. Occasional Paper 3 (revised edition), Biology Department, University of Papua New Guinea.
Garrett-Jones, S. E. (1979) Evidence for change in Holocene vegetation and lake sedimentation in the Markham Valley, Papua New Guinea. Ph.D. Thesis, Australian National University, Canberra, Australia. 420 pp.
Goode, J. (1967) Freshwater Tortoises of Australia and New Guinea (in the Family Chelidae). Lansdowne, Melbourne, Australia. 154 pp.
Gould, J. (1970) Birds of New Guinea. Methuen, London, England. 321 pp.
Glucksman, J., West, G. and Berra, T. M. (1976) The introduced fishes of Papua New Guinea with special reference to *Tilapia mossambica*. Biological Conservation 9: 37–44.
Green, W. and Sander, H. (1979) Manus Province tuna cannery environmental study. Office of Environment and Conservation, Waigani. 123 pp.
Gressitt, J. L. (1982) Biogeography and Ecology of New Guinea. Dr W. Junk, The Hague, The Netherlands. 983 pp.
Haines, A. K. (1979) The subsistence fishery of the Purari delta. Science in New Guinea 6: 80–95.
Haines, A. K. (1983) Fish fauna and ecology. p. 367–384. *In*: T. Petr (ed.). The Purari – Tropical Environment of a High Rainfall River Basin. Dr. W. Junk, The Hague, The Netherlands.
Haines, A.K. and Stevens, R.N. (1983) Subsistence and commercial fisheries. p. 385–408. *In*: T. Petr (ed.). The Purari – Tropical Environment of a High Rainfall River Basin. Dr W. Junk, The Hague, The Netherlands.
Hope, G. S. (1973) The vegetation history of Mt Wilhelm, Papua New Guinea. Ph.D. Thesis, Australian National University, Canberra, Australia. 460 pp.
Hope, G. S. (1976) The vegetational history of Mt Wilhelm, Papua New Guinea. Journal of Ecology 64: 627–664.
Hope, G., Gillieson, D. and Head, J. (1988) A comparison of sedimentation and environmental change in New Guinea shallow lakes. Journal of Biogeography 15: 603–618.
Holthuis, L. B. (1974) Notes on the localities, habitats, biology, colour and vernacular names of New Guinea freshwater crabs (Crustacea, Decapoda, Sundathelpusidae). Zoologische Verhandelingen 13: 3–47.
Holthuis, L. B. (1982) Freshwater crustacea Decapoda of New Guinea. p. 603–619. *In*: J. L. Gressitt (ed.). Biogeography and Ecology of New Guinea. Dr. W. Junk, The Hague, The Netherlands.
Iredale, T. (1956) Birds of New Guinea. Vols. I and II. Georgian House, Melbourne, Australia. 230 pp and 261 pp.
Johnstone, I. M. (1978) The mangroves and mangrove ecosystems of Papua New Guinea. 2nd International Congress of Ecology, mimeogr. 26 pp.
Johnstone, I. M. and Frodin, D. G. (1982) Mangroves of the Papuan subregion. p. 513–528. In J. L. Gressitt (ed.). Biogeography and Ecology of New Guinea. Monographiae Biologicae 42, Dr. W. Junk, The Hague, The Netherlands.
Kyle, J. H. and Ghani, N. (1982a) Methylmercury in ten species of fish from Lake Murray. Science in Guinea 9: 48–59.
Kyle, J. H. and Ghani, N. (1982b) Methylmercury in human hair. A study of a Papua New Guinean population exposed to methylmercury through fish consumption. Archives of Environmental Health 37: 266–270.
Kyle, J. H. and Ghani, N. (1984) Mercury in barramundi (*Lates calcarifer*) from the Gulf of Papua. Science in New Guinea 11: 105–113.
Lamb, K. P. (1977) Mercury levels in nine species of fish from the Ok Tedi and upper Fly River. Science in New Guinea 5: 7–11.

Laup, S. (1985) The Sepik *Salvinia* problem is beaten. Harvest 11: 49–52.
Laup, S. (1986a) Biological control of water hyacinth: Early observations. Harvest 12: 35–40.
Laup, S. (1986b) Biological control of water lettuce: Early observations. Harvest 12: 41–43.
Leach, G. J. and Osborne, P. L. (1985) Freshwater Plants of Papua New Guinea. UPNG Press, University of Papua New Guinea, Papua New Guinea. 254 pp.
Liem, D. and Haines, A. K. (1977) The ecological significance and economic importance of the mangrove and estuarine communities of Gulf Province, Papua New Guinea. p. 1–35. *In*: T. Petr (ed.). Purari River (Wabo) Hydroelectric Scheme, Vol. 3: Ecology of the Purari River Catchment. Office of Environment and Conservation, Waigani, Papua New Guinea.
Löffler, H. (1973) Tropical high mountain lakes of New Guinea and their zoogeographical relationship compared with other tropical high mountain lakes. Arctic and Alpine Research 5: 193–198.
Löffler, E. (1977) Geomorphology of Papua New Guinea. ANU Press, Canberra, Australia. 195 pp.
MacArthur, R. H. (1965) Patterns of species diversity. Biological Reviews 40: 510–533.
McAlpine, J. R., Keig, G. and Falls, R. (1983) Climate of Papua New Guinea. CSIRO and ANU Press, Canberra, Australia. 200 pp.
McDowell, R. M. (1981) The relationships of Australian freshwater fishes. p. 1253–1273. *In*: A. Keast (ed.). Ecological Biogeography of Australia, Dr W. Junk, The Hague, The Netherlands.
MacKay, R. D. (1970) The birds of Port Moresby and District. Nelson, Melbourne, Australia. 74 pp.
McKenzie, K. G. (1956) Notes on the freshwater mussels of New Guinea. Nautilis 70: 38–48.
McKenzie, K. G. (1971) Ostracoda from lake Piunde, near Mt Wilhelm, New Guinea. Zoologische Anzeiger 186: 391–403.
McMichael, D. F. and Hiscock, I. D. (1958) Monograph of the freshwater mussels (Mollusca: Pelecypoda) of the Australian Region. Australian Journal of Marine and Freshwater Research 9: 372–508.
Maunsell and Partners. (1982) Ok Tedi Environmental Study. Vols. 1–6. Ok Tedi-Fly River Aquatic Survey. Ok Tedi Mining Limited, Port Moresby, Papua New Guinea.
Menzies, J. I. (1975) Handbook of common New Guinea frogs. Wau Ecology Institute, Handbook 1, Wau, Papua New Guinea. 75 pp.
Mitchell, D. S. (1978/1979) Aquatic weeds in Papua New Guinea. Science in New Guinea 6: 154–160.
Mitchell, D. S. (1979) The incidence and management of *Salvinia molesta* in Papua New Guinea. Office of Environment and Conservation, Waigani, Port Moresby. 51 pp.
Mitchell, D. S., Petr, T. and Viner, A. B. (1980) The water fern *Salvinia molesta* in the Sepik River, Papua New Guinea. Environmental Conservation 7: 115–122.
Moore, R. (1980) Migration and reproduction in the percoid fish *Lates calcarifer* (Bloch). Ph.D. Thesis, University of London, London, England.
Moore, R. (1982) Spawning and early life history of barramundi, *Lates calcarifer* (Bloch) in Papua New Guinea. Australian Journal of Marine and Freshwater Research 33: 647–661.
Moore, R. and Reynolds, L. F. (1982) Migration pattern of barramundi *Lates calcarifer* (Bloch) in Papua New Guinea. Australian Journal of Marine and Freshwater Research 33: 671–682.
Mowbray, D. L. (1988a) Assessment of the biological impact of Ok Tedi mine tailings, cyanide and heavy metals through toxicity testing. p. 45–74. *In*: J. C. Pernetta (ed.). Potential Impacts of Mining on the Fly River. UNEP Regional Seas Reports and Studies 99 and SPREP Topic Review 33, UNEP.
Mowbray, D. L. (1988b) Evaluation of the Ok Tedi mine environmental monitoring programme, and the impact of tailings derived heavy metal residues in the Ok Tedi and Fly River system, 1981–1985. p. 75–98 *In*: J. C. Pernetta (ed.). Potential Impacts of Mining on the Fly River. UNEP Regional Seas Reports and Studies 99 and SPREP Topic Review 33, UNEP.
Natural Systems Research (1988) Porgera gold project environmental plan, Vols. A, B and C. Porgera Joint Venture, Report CR 257/13.

Niell, W. T. (1946) Notes on *Crocodylus novae-guineae*. Copeia 1: 17–20.
Osborne, P. L. (1987) A draft inventory of wetlands in Papua New Guinea. Department of Environment and Conservation, Waigani, Papua New Guinea. 163 pp.
Osborne, P. L. (1988) Bibliography of freshwater ecology in Papua New Guinea. Biology Department Occasional Paper 9, University of Papua New Guinea. 65 pp.
Osborne, P. L., Kyle, J. H. and Abramski, M. (1987) Effects of seasonal water level changes on the chemical and biological limnology of Lake Murray, Papua New Guinea. Australian Journal of Marine and Freshwater Research 38: 397–408.
Osborne, P. L. and Leach, G. (1983) Changes in the distribution of aquatic plants in a tropical swamp. Environmental Conservation 10: 323–329.
Osborne, P. L. and Leach, G. (1984) The spread of water hyacinth in Papua New Guinea –a second warning or is it too late? Harvest 10: 51–53.
Osborne, P. L. and Polunin, N. V. C. (1986) From swamp to lake: Recent changes in a lowland tropical swamp. Journal of Ecology 74: 197–210.
Osborne, P. L. and Polunin, N. V. C. (1988) Progress in elucidating the sediment record of two shallow lakes in Papua New Guinea. p. 31–37 *In*: J. C. Pernetta (ed.). Potential Impacts of Mining on the Fly River. UNEP Regional Seas Reports and Studies 99 and SPREP Topic Review 33, UNEP.
Osborne, P. L., Polunin, N. V. C. and Nicholson, K. (1988) Geochemical traces of riverine influence on a tropical lateral lake. Verhandlungen internationale der Vereinigung theoretische und angewandte Limnologie 23: 207–211.
Paijmans, K. (1976) New Guinea Vegetation. ANU Press, Canberra, Australia. 213 pp.
Percival, M. and Womersley, J. S. (1975) Floristics and ecology of the mangrove vegetation of Papua New Guinea. Botany Bulletin 8, Office of Forests, Papua New Guinea. 96 pp.
Pernetta, J. C. (1988) The Ok Tedi mine: Environment, development and pollution problems. p. 1–8. *In*: J. C. Pernetta (ed.). Potential Impacts of Mining on the Fly River. UNEP Regional Seas Reports and Studies 99 and SPREP Topic Review 33, UNEP.
Pernetta, J. C. and Burgin, S. (1980) Census of crocodile population and their exploitation in the Purari area (with an annotated list of of the herpetofauna). p. 1–44. *In*: T. Petr (ed.). Purari River (Wabo) Hydroelectric Scheme Environmental Studies, 14: Ecology of the Purari River catchment. Office of Environment and Conservation, Waigani, Papua New Guinea.
Pernetta, J. C. and Burgin, S. (1983) The status of crocodiles in the Purari. p. 409–428. *In*: T. Petr (ed.). The Purari – Tropical Environment of a High Rainfall River Basin, Dr W. Junk, The Hague, The Netherlands.
Petr, T. (1979) Mercury in the Papua New Guinea environment. Science in New Guinea 6: 161–176.
Petr, T. (ed.) (1983) The Purari – Tropical Environment of a High Rainfall River Basin. Dr. W. Junk, The Hague, The Netherlands. 624 pp.
Petr, T. (1984) Technical report on the possibilities of Sepik River fish stock enhancement. FAO Fisheries Travel Report 2505, FAO, Rome, Italy.
Pickup, G., Higgins, R. J. and Warner, R. J. (1981) Erosion and sediment yield in Fly River drainage basins, Papua New Guinea. p. 438–456. *In*: T. R. Davies and A. J. Pearce (eds.). Erosion and Sediment Transport in Pacific Rim Steeplands. International Association of Hydrological Sciences Publication 132, Canberra, Australia.
Polunin, N. V. C., Osborne, P. L. and Totome, R. G. (1988) Environmental archive: Tropical urban development reflected in the sediment geochemistry of a flood-plain lake. Archiv für Hydrobiologie: 114, 199–211.
Rapson, A. M. (1968) A brief history of fisheries of Papua and New Guinea. Annual Report and Proceedings of the Papua and New Guinea Scientific Society 19: 27–44.
Rau, M. T. (1984) Traditional uses of mangroves in Central Province of Papua New Guinea. Harvest 11: 5–11.
Reynolds, L. F. (1978) The population dynamics of barramundi *Lates calcarifer* (Pisces: Centropomidae) in Papua New Guinea. M.Sc. Thesis, University of Papua New Guinea, Papua New Guinea.

Reynolds, L. F. and Moore, R. (1982) Growth rates of barramundi, *Lates calcarifer* (Bloch) in Papua New Guinea. Australian Journal of Marine and Freshwater Research 33: 663–670.

Richardson, L. R. (1977) The zoological importance of the freshwater leeches of the Papuan subregion. Science in New Guinea 5: 115–120.

Roberts, T. R. (1978) An ichthyological study of the Fly River in Papua New Guinea with descriptions of new species. Smithsonian Contributions to Zoology 281. Smithsonian Institution Press, Washington, DC, USA. 72 pp.

Robertson, C. H. (1983) Aspects of the biology of various *Macrobrachium* species found in the Sepik River. Report No. 83–05. Fisheries Research and Surveys Branch, Department of Primary Industry, Port Moresby, Papua New Guinea. 61 pp.

Robertson, C. H. and Baidam, G. (1983) Fishes of the OK Tedi area with notes on five common species. Science in New Guinea 10: 16–27.

Room, P. M. and Thomas, P. A. (1985) Nitrogen and establishment of a beetle for biological control of the floating weed *Salvinia* in Papua New Guinea. Journal of Applied Ecology 22: 139–156.

Room, P. M. and Thomas, P. A. (1986a) Nitrogen, phosphorus and potassium in *Salvinia molesta* in the field: Effects of weather, insect damage, fertilizers and age. Aquatic Botany 24: 213–232.

Room, P. M. and Thomas, P. A. (1986b) Population growth of the floating weed *Salvinia molesta*: Field observations and a global model based on temperature and nitrogen. Journal of Applied Ecology 23: 1013–1028.

Schodde, R. and Hitchcock, W. B. (1968) Report on the birds of the Lake Kutubu area, Territory of Papua and New Guinea. CSIRO Division of Wildlife Research Technical Paper 13: 1–73.

Schuster, W. H. (1957) A survey of the inland fisheries of the territory of New Guinea and Papua. Australian Journal of Marine and Freshwater Research 2: 226–236.

Sorentino, C. (1979) Mercury in marine and freshwater fish of Papua New Guinea. Australian Journal of Marine and Freshwater Research 30: 617–623.

Taylor, B. W. (1959) The classification of lowland swamp communities in north-eastern Papua. Ecology 40: 703–711.

Thomas, P. A. (1979) Proposals for the management of *Salvinia molesta* in Papua New Guinea. FAO Report to the Department of Primary Industry, Port Moresby, Papua New Guinea. 60 pp.

Thomas, P. A. (1985) The management of *Salvinia molesta* in Papua New Guinea. FAO Plant Protection Bulletin 32: 50–56.

Thomas, P. A. and Room, P. M. (1986) Successful control of the floating weed *Salvinia molesta* in Papua New Guinea: A useful biological invasion neutralises a disastrous one. Environmental Conservation 13: 242–248.

Thomasson, K. (1967) Phytoplankton from some lakes on Mt Wilhelm, East New Guinea. Blumea 15: 285–296.

Townsend, B. (1988) Giving away the river: environmental issues in the construction of the Ok Tedi mine, 1981–84. p. 101–119. *In*: J. C. Pernetta (ed.). Potential Impacts of Mining on the Fly River. UNEP Regional Seas Reports and Studies 99 and SPREP Topic Review 33, UNEP.

Walker, D. (1972) Vegetation of the Lake Ipea Region, New Guinea Highlands. II. Kayamanda Swamp. Journal of Ecology 60: 479–504.

Watanabe, M., Prescott, G. W. and Yamagishi, T. (1979a) Freshwater algae of Papua New Guinea (2) Desmids from Woitape, Central District. p. 49–66. *In*: S. Kurokawa (ed.). Studies on Cryptogams of Papua New Guinea. Academia Scientific Book Inc., Tokyo, Japan.

Watanabe, M., Prescott, G. W. and Yamagishi, T. (1979b) Freshwater algae of Papua New Guinea (3) Blue-green algae from Mt Wilhelm. p. 67–85. *In*: S. Kurokawa (ed.), Studies on Cryptogams of Papua New Guinea. Academia Scientific Book Inc., Tokyo, Japan.

West, G. J. (1973) The establishment of exotic aquarium fish in Papua New Guinea. Papua New Guinea Agricultural Journal 24: 30–32.

West, G. J. and Glucksman, J. (1976) Introduction and distribution of exotic fish in Papua New Guinea. Papua New Guinea Agricultural Journal 27: 19–48.

Worsley, A. T. and Oldfield, F. (1988) Palaeoecological studies of three lakes in the Highlands of Papua New Guinea. II. Vegetational history over the last 1600 years. Journal of Ecology 76: 1–18.

Yamagishi, T. (1975) The plankton algae from Papua New Guinea. p. 43–74. *In*: Y. Otani (ed.), Reports on the Cryptogams in Papua New Guinea. National Science Museum, Tokyo, Japan.

Yamagishi, T. and Watanabe, M. (1979) Freshwater algae of Papua New Guinea (1) On some filamentous green algae. p. 49–66. *In*: S. Kurokawa (ed.). Studies on Cryptogams of Papua New Guinea. Academia Scientific Book Inc., Tokyo, Japan.

Wetlands of South Asia

BRIJ GOPAL AND K. KRISHNAMURTHY

Abstract

The paper reviews the available information on the distribution, vegetation, associated fauna, important ecological characteristics, use, management, and conservation status of wetlands in South Asia. The region, better identified as the Indian subcontinent, is a natural biogeographic region isolated from the remaining Asian landmass by nearly continuous mountain chain on its northwest, north, northeast, and eastern sides. The climate ranges from tropical through subtropical to montane temperate in different parts but is governed by the monsoons which cause large spatial and temporal variations in precipitation. Further, it is the most densely populated region of the world, and hence all natural ecosystems are under great stress from human exploitation and other human activities.

Wetlands occur in all climatic zones of south Asia but large majority of them are seasonal in nature due to long dry summer periods. A scheme of wetland classification is proposed to group them into saline and freshwater wetlands and categories them further according to hydrological factors (duration of flooding and nature of water body) and vegetation types (herbaceous and woody). Among saline wetlands, mangroves which are coastal and estuarine forested wetlands, are among the world's largest such formations. Mangroves are also most investigated wetlands in the region and considerably detailed information on their flora, fauna, ecological characteristics and zonation is summarised. There are also saline and estuarine shallow lakes and temporarily flooded scrublands. Mangroves are heavily exploited for timber, fuelwood, fisheries (including shrimps and prawns) and a number of minor forest products. These are being converted into paddy fields and for aquaculture. Frequent cyclones arising in the Bay of Bengal and reduced freshwater flows due to river regulation in mainland are also important factors causing degradation of these important wetlands.

Freshwater wetlands are dominated by shallow lakes, ponds, and tempor-

D.F. Whigham et al. (eds.), Wetlands of the World I, 345–414.

ary water bodies which become periodically dry. Permanent herbaceous wetlands and forested swamps occur mostly in the sub-Himalayan tract but very little is known of them. These are fast disappearing due to human exploitation. Seasonal wetlands have a large variety of submerged, floating leaved and emergent vegetation. These wetlands have become infested with exotic free-floating water hyacinth which was introduced in the region about a century ago. Extensive floodplain wetlands occur along all major and medium rivers but have not been investigated yet. Most of the freshwater wetlands have been converted into paddy fields and fish ponds, and the remaining are being rapidly lost or degraded by filling, drainage, disposal of sewage, and industrial effluents. Floodplains are also lost by construction of dams.

Whereas National Mangrove Committees have been formed in all countries to prepare appropriate management plans for mangrove wetlands, freshwater wetland management is not yet considered important. Though some freshwater wetlands were conserved long ago as important habitats of wildlife, in recent years steps have been taken to conserve more areas of both saline and freshwater wetlands.

Introduction

South Asia, comprising mainly of the Indian subcontinent, and lying north of the Equator, is characterized by a nearly continuous chain of mountain ranges which separates it from the rest of Asian landmass (Fig. 1). The greatest of these, the Himalayan ranges buttressing the high plateau of Tibet on its southern side, form an unbroken arc over 2,412 km. The Himalayas comprise three principal ranges in the west but only two ranges in the east. The Himalayan ranges continue on the east curving southeastward as the Patkai and Arakan ranges. This Burmese arc continues even southward remaining submerged in the Bay of Bengal and reappearing as Andaman and Nicobar Islands. On the western side, the Himalayas are continued as Hindukush ranges and also turn southwest as Suleiman and Kirthar ranges. The region is thus a roughly diamond shaped landmass surrounded on the north by the mountains and on the south by the oceans. Geologically, it forms one unit which once formed a part of the Gondwanaland whose splitting and drift towards the north-east resulted in the obliteration of the former Tethys sea and the uplift of the mountain ranges. Politically, it includes India, Pakistan, Nepal, Bhutan, Bangladesh, Burma, and Sri Lanka.

This subcontinent is the second most populous region of the world and has witnessed a long history of utilisation by man. The earliest civilization of Mohanjodaro and Harappa developed in the western part of this landmass.

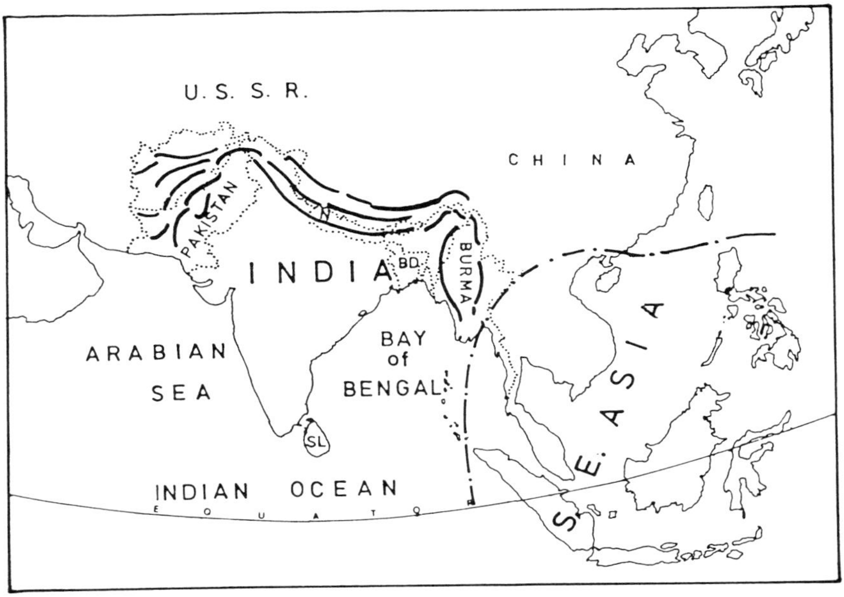

Figure 1. The South Asian region and its political constituents. The chains of mountain ranges in the north, Arabian Sea, and Bay of Bengal define its boundary with the remaining Asian landmass. BD – Bangladesh, N – Nepal, SL – Sri Lanka.

The past history of the region has been very tumultous over past several thousand years and included the migration of Aryans from the Middle East, several invasions from the western Asian region and later colonization by different European people, mainly the British. All this history had a great impact on the natural vegetation of the region and it is nearly impossible to say what and where the natural vegetation still exists, if any. During the past few decades, the rapid pace of economic development and the fast growth of human population has only added to the rate of decimation of the natural vegetation and natural habitats. About a century ago Hooker (1872–97) published the first comprehensive account of the vegetation of this region. Since then much effort has been spent in supplementing this account but as yet no thorough analysis has been made and there are no detailed accounts for the whole region. Forests and grasslands have received by far the greatest attention (Misra 1983, Puri *et al.* 1984, Singh and Singh 1987, Pandeya 1988, Melkania and Singh 1989). Wetlands, especially freshwater wetlands have received very little attention (Anonymous 1987a, Gopal 1982a, 1990b, Kaul and Handoo 1989). There has been no systematic survey or attempt to classify wetland vegetation in the region, and an account of the region's wetlands can only be sketchy. In this paper we propose a classification scheme and try to summarise the available knowledge on the distribution and vegetation of different kinds of wetlands, and also describe their traditional values and present conservation status.

Regional characteristics

Physiography

Physiographically, the Indian subcontinent may be divided into a northern alluvial plain formed by the basins of the Indus, Ganges and their tributaries; and a southern peninsular plateau (Fig. 1). The Tropic of Cancer roughly forms the boundary of the two regions. The Indo-Gangetic plain is traversed by relatively low hill ranges, the Siwaliks and the Aravallis extending in northeast to southwest direction. The greater part of the region lying west of these ranges is almost arid and continues westwards as the Great Indian Desert merging later with the desert of the middle East and the Asian desert.

The peninsular Plateau (450–600 m altitude) comprises of (a) Malwa Plateau bounded by the Aravalli hills on the northwest and the Vindhyan hills on the northeast, (b) Satpura range running west to east, and Chhota Nagpur Plateau, and (c) the Deccan proper lying south of Satpura range and flanked on the east and west by Eastern and Western Ghats. The Western Ghats are an area of relatively high mountain ranges (up to 900 m elevation) while the Eastern Ghats are a series of disconnected topographically lower hills, with gentle slopes and wide coastal plains along the eastern border of the land mass. The Eastern and Western Ghats meet in the south just north of Nilgiri Hills which rise up to 2,700 m.

Sri Lanka, a pear-shaped island, once connected to the main Indian peninsula, is essentially a detached fragment of the Deccan (Robinson 1976). It consists of a central mass of mountains (300–2,500 m elevation) flanked by broad coastal plains.

Much of north-western parts of Pakistan are covered by Hindukush ranges which form the north-west frontiers. The remaining part is desertic plain. The eastern and southeastern part of Pakistan are formed by the alluvial plains of river Indus and its tributaries. Nepal and Bhutan are land-locked states in the central and eastern region of the Himalayan ranges.

Bangladesh is primarily on an alluvial plain formed by the deltas of the Ganges and Brahmaputra rivers. In the east, Burma is also largely a hilly country covered by the Arakan ranges on the west and the Shan plateau (average altitude 900 m) on the east. Between the Arakan ranges and the Shan plateau lie the valleys of Irrawady and Sittang rivers which are separated by the Pegu Yoma hill range.

Climate

The climate of the region is highly variable but most often described as tropical to subtropical and monsoonic (Blanford 1889, Subrahmanyam 1965). Mountain ranges in the north exert the major influence in controlling the climate. The cold Arctic wind does not have any influence and therefore the most northerly parts of the region are still much warmer at the lower altitudes than similar areas elsewhere with the same latitude and altitude. The high altitude in the Deccan plateau around Nilgiris which lies otherwise in the tropical region, has a relatively cool climate that is often classified as montane temperate. The greater part of the southern India and northeast parts of the region (Bangladesh, Assam, and Burma) have typically tropical climates. Central and northern India are nearly subtropical with strong seasonality. In the foothills of the Himalayas and northward in Kashmir, the climate may be called warm or montane temperate. The western part of the region is semi arid to arid and relatively very hot.

There are, however, hardly any other regions on the earth with such great contrasts in the climate as in this south Asian region where the highest mountain peaks, the rainiest as well as the hottest place on the earth are found. The average climatic data are presented in Figs. 2–4. The summer temperature in most of the region rises to above 38°C (Fig. 2). The minimum winter season temperatures in southern India rarely go below 21°C and as one moves northwards the winter mean minimum temperature falls lower and lower, with about 4°C in the foothills of Himalayas and in Kashmir valley (Fig. 3). The lowest recorded temperature in the northwestern parts of the region is about −8°C. Winter frost is only rarely observed in parts of Punjab, Himachal Pradesh, and northwards.

The rainfall is most variable as it depends on the monsoon (Das 1968, Fein and Stephens 1987). The periodic failure and outbursts of the monsoon are a common feature of the region. The average rainfall (Fig. 4) decreases in general from east to west. Maximum rainfall occurs over Burma, Assam, and Bangladesh, and along the western Ghats. From the eastern part of the Indian subcontinent it decreases fast towards the west until around the Aravalli ranges there is very little precipitation. Much of the western part of the Deccan plateau east of the Western Ghats lies in the rain-shadow and receives relatively little rainfall. The eastern part of Sri Lanka also receives much lower rainfall than its southwestern part (Fernando 1985). The most important characteristic of the climate is the seasonality of rainfall and its high year to year variability (Fig. 5) which affect the vegetation. In the northern part the rainfall is received only from the southwest monsoon (June–September) and very little rain occurs during the period of retreating

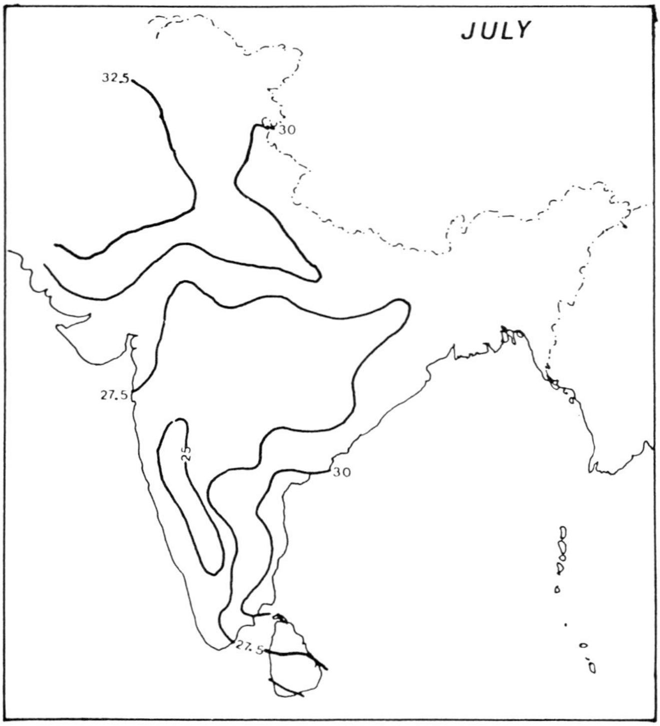

Figure 2. Isotherms of average monthly temperature during summer (July) in south Asia. Values are in°C.

monsoon (northeast monsoon) which brings larger amounts of rain to the southern part.

Soils

Alluvial, black and lateritic soils are the most widespread soil types in the region (Raychaudhuri *et al.* 1963). Alluvial soils often contain large amounts of calcium in the form of calcium carbonate granules (kankar) which form a hard pan which leads to waterlogging. Most alluvial soils also have a high clay content. Black soils are also subject to waterlogging. Saline soils are common in the northern and western parts of the region. Organic soils are nearly non-existent and only in forest soils and soils of the hilly areas is the organic matter content high. There are no significant peat deposits in the region.

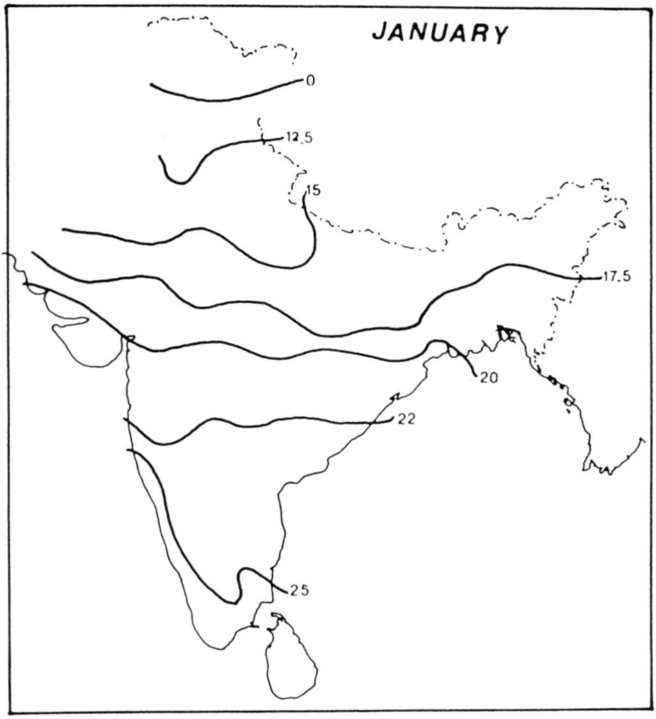

Figure 3. Isotherms of average monthly temperature during winter (January) in south Asia. Values are in°C.

Water resources

The whole region, except for the Great Indian Desert, is rich in water resources as there are numerous small and large rivers within the subcontinent (K. L. Rao 1975, Fig. 6). The major rivers in the northern plains arise in the Himalayas and derive their water largely from the melting snow and monsoon rainfall. This region can be divided into three river basins. The rivers of the Ganges basin carry water eastwards to the Bay of Bengal while those of the Indus basin discharge into the Arabian Sea. A few tributaries of Ganges flow northwards from central India before meeting Ganges. The river Brahmaputra flows in the Himalayas eastwards before entering India in the northeastern corner and flows southwest before meeting the Ganges in the deltaic region in Bangladesh.

Rivers in the plateau region (Godavari, Krishna, and Cauvery) arise on the eastern side of the Western Ghats and traverse the plateau flowing eastwards to the Bay of Bengal. The river Mahanadi arises in the Chhota Nagpur plateau and also flows eastwards to Bay of Bengal. Only two major

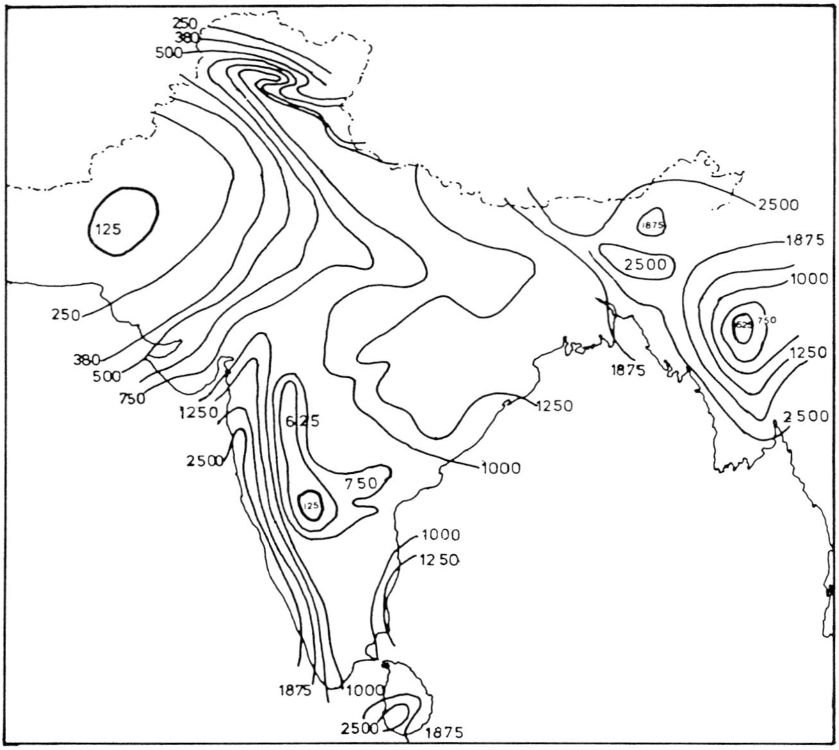

Figure 4. Isolines of average annual rainfall (mm) in south Asia.

rivers (Narmada and Tapti) originate in the eastern part of the Vindhyan range and flow westwards into the Arabian sea. A few small rivers in the south (Kerala State) arise in the Western Ghats and flow into the Arabian Sea.

In Burma, the river Irrawady and its major tributary river Chindwin originate in the eastern part of Himalayan range and flow southwards to the Bay of Bengal. The river Salween arises in Tibet and meanders its way southwards to the Bay of Bengal. In Sri Lanka several small rivers arise from the central plateau and flow towards coastal plains before discharging into the Indian Ocean.

Though natural lakes are few and mostly confined to the Himalayan range, a large number of small and big man-made reservoirs dot the entire landscape, particularly in the semi-arid region (Table 1). Innumerable small shallow depressions have been formed by the excavation of earth and they become filled with water during the monsoon. These areas result in the formation of many wetlands (see below).

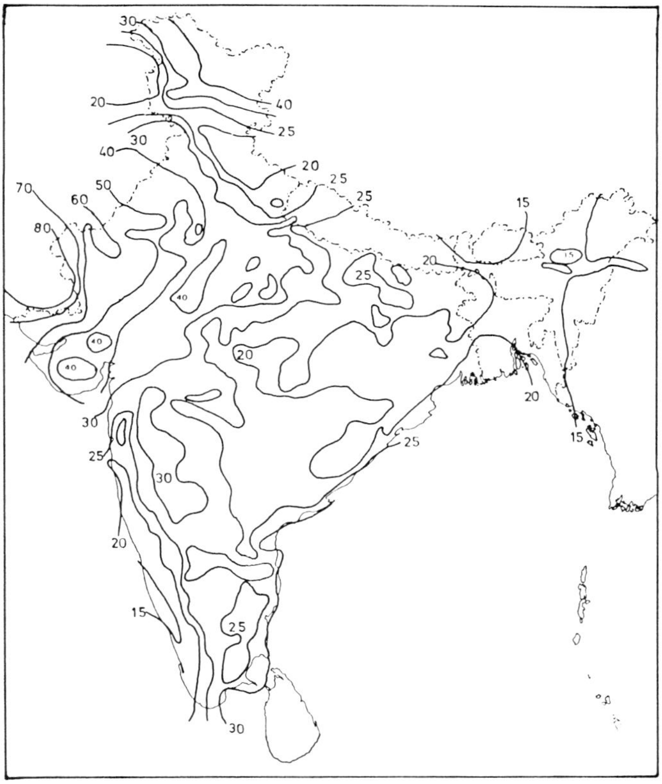

Figure 5. Isolines of coefficient of variability (%) in annual precipitation in south Asia.

Wetland habitats

A region rich in water resources also has a wide variety of wetlands. Herbaceous vegetation dominates wetlands in shallow waters and along rivers. The seasonality of rainfall and a relatively flat terrain results in wide floodplains on most of the major and medium rivers. There are also numerous man-made wetland habitats that have developed in areas influenced by human activities. For example, seepage of water from reservoirs or spill-over from the irrigation channels create substantial waterlogged areas which support wetland vegetation. Innumerable fish ponds and extensive paddy fields are important man-made wetlands in the region.

The deltas of many rivers along the eastern coastline include the world's largest area of mangroves. On the western coast, all rivers except the Indus, do not form deltas and therefore have fewer coastal wetlands. There are also extensive areas of marine wetlands (Silas *et al.* 1985, Hussain *et al.* 1985, Hussainy and Azariah 1985) dominated by marine algae and submersed

Figure 6. Major river systems of south Asia. Some important areas of freshwater wetlands are also marked by letters as described below. Forested wetlands are shown by solid triangles. Legend: b – Keoladeo Ghana National Park, Bharatpur; e – Kolleru lake (between delta of Krishna and Godavari); k – Kaziranga National Park, Assam; l – Loktak lake, Manipur; m – Manas sanctuary, Assam; s – Dal lake and other wetlands near Srinagar; w – Wular lake, Kashmir.

Table 1. Area (million ha) under lakes, reservoirs, ponds, and other shallow freshwater habitats in the South Asia (from Fernando 1984).

	Reservoirs		Natural lakes	Ponds and marshes	Paddy fields
	Existing	Planned			
India	2.00	5.00	0.20	0.50	38.60
Bangladesh	0.16	0.80		0.30	9.70
Burma	0.20	1.50		0.40	5.10
Nepal	0.02	0.10	0.01		1.20
Sri Lanka	0.13	0.25		0.20	0.63

angiosperms, in the eulittoral regions on both the east and west coast of India (Venkataraman *et al.* 1974). Coral reefs are common along the coast and around the numerous islands in both the Arabian Sea and the Bay of Bengal (Wafar 1986). Marine wetlands are not, however, discussed in this paper.

It is important to stress that majority of the wetland habitats in the region dry out almost completely during the summer. This makes it difficult to distinguish between different vegetation components whose presence and development depends upon the depth, duration, and frequency of flooding, and in turn, to identify which habitat should be considered a wetland. What might be recognized as a wetland in a year of excess rain may turn out to be dry grazing land in dry year and the survey in two different years would yield different results.

Wetland types

There are two ways of developing a detailed typology of wetlands. One involves detailed survey and data analysis to identify different wetlands. The limited amount of information on wetlands in the region makes this approach impossible. Alternatively, one must rely on more general information and develop a system of categories that then can be used to conduct field surveys. This approach has been more widely used and has resulted in several classification systems such as the one in the USA (Cowardin *et al.* 1979). However, this approach is still too elaborate for a region as large as south Asia. We do not have enough information on the wetlands in different parts in the region much less a detailed account of habitats and vegetation. Nevertheless, vegetation scientists have proposed their own schemes of classification where wetland communities have been given different status by different workers. In general, the climax vegetation under the tropical climate of the Indian subcontinent has been recognized as a forest and all variations are considered to be induced by edaphic or biotic factors. Thus, Champion (1936) developed a classification of forests that included some wetlands:

1. Beach forests (not truly wetland forests)
2. Tidal forests: Low mangrove, tree mangrove, salt water *Heritiera* and fresh water *Heritiera*
3. Delta freshwater swamps
4. Tropical valley freshwater swamps
5. Moist riparian fringing swamps
6. Khair-Sissoo forests (not truly wetland forests)
7. Riverine moist deciduous forests (not truly wetland forests): Southern

Table 2. Classification of Littoral and Swamp Forests recognised as a major category of vegetation in Indian subcontinent (adapted from Champion and Seth 1968).

A. Littoral forests
B. Tidal swamp forests
    Mangrove scrub
    Mangrove forest
    Saltwater mixed forest (*Heritiera*)
    Brackishwater Mixed forest (*Heritiera*)
    Palm swamp
C. Tropical freshwater swamp forests
    Myristica swamp
    Submontane hill valley swamp
    Creeper swamp
D. Tropical seasonal swamp forests
    Eastern seasonal swamp
    Barringtonia swamp
    Syzygium cumini swamp low forest
    Eastern seasonal swamp low forest (*Cephalanthus*)
    Eastern *Dillenia* swamp
    Eastern wet alluvial grassland
E. Tropical riparian fringing forests
    Riparian fringing forests

moist deciduous riverine, Gangetic moist deciduous riverine, Upper Assam moist deciduous riverine

8. Riverine semi-evergreen forests (not truly wetland forests)

Champion and Seth (1968), in their revised system of classification of Indian forest types, recognised "Littoral and Swamp Forests" which were further subdivided into a number of subtypes (Table 2). Besides the subtypes given in Table 2, Wet bamboo brakes in Assam and Alder forests and riverine blue pine forests in Himalayas may also be considered as wetlands. Unfortunately, detailed descriptions of these forests are not available as the categories in Table 2 were recognised on the basis of reports from different forest divisions at the beginning of this century.

The two above classifications just presented do not take into account herbaceous wetlands which are more widespread and are not associated with specific forest types. We suggest a classification scheme (Table 3) which emphasises the hydrological characteristics of the habitat and which would be useful guide for surveying wetlands in the region. Wetlands are first grouped into saline and freshwater types which are then subdivided on the basis of their hydrological regimes, particularly the duration of flooding. Further distinction is made between the wetlands with herbaceous and woody vegetation.

We wish to point out that a further distinction should be made between natural and man-made wetlands. As has been said earlier, there are more man-made aquatic habitats than natural ones in south Asia. Besides large

Table 3. Suggested scheme of classification of wetlands in south Asia.

I. SALINE WETLANDS
  A. Wetlands associated with permanent flooding
    i. Herbaceous vegetation (coastal beds of kelp and angiosperms)
    ii. Woody vegetation (mangroves)
  B. Wetlands associated with temporary flooding
    (Inland saline habitats with 1–3 months of flooding)
    i. Herbaceous vegetation – halophytes
    ii. Woody vegetation – saline scrubs
II. FRESHWATER WETLANDS
  A. Wetlands associated with permanent waterbodies
    a. Lotic habitats (rivers and streams)
      i. Herbaceous vegetation – reeds (*Phragmites*, *Arundo*), bamboos (*Bambusa*), canes (*Calamus*), other grasses
      ii. Woody vegetation – riparian fringing forests
    b. Lentic habitats (lakes/reservoirs)
    i. Herbaceous vegetation
      * Submerged, floating, and floating leaved macrophytes only
      ** Short emergent vegetation (sedges mainly)
      *** Tall emergent vegetation – reeds (*Phragmites*) and cattails (*Typha*)
    ii. Woody vegetation
  B. Wetlands associated with temporary waterbodies
    (3–9 months of flooding)
    a. Lotic habitats (floodplains of streams and rivers)
    a. Herbaceous vegetation – reeds (*Phragmites*), cattails (*Typha*), grasses (*Saccharum*, *Erianthus*, *Paspalum*), sedges (*Cyperus*, *Scirpus*) and other herbs
      ii. Woody vegetation – floodplain forests
    b. Lentic habitats (ponds, tanks, reservoirs)
      i. Herbaceous vegetation
        * Submerged, floating and floating leaved macrophytes only
        ** Short emergent vegetation (sedges mainly)
        *** Tall emergent vegetation – reeds (*Phragmites*) and cattails (*Typha*)

areas of paddy fields, there are numerous fish ponds (often modified from the marshes) and shallow reservoirs. Most of these anthropogenic wetlands are managed for specific economic gains while some are incidental to other forms of water resource utilisation (e.g., irrigation). Such habitats make it difficult to define and estimate the areal extent of wetlands in the region.

The available information on the distribution, area, salient habitat characteristics and vegetation of saline and freshwater wetlands is summarised in the following pages.

Saline wetlands

Saline wetlands are generally confined to tide influenced coastal, estuarine, and deltaic areas and most are dominated by arborescent or shrubby formations of mangroves. Other halophytic wetlands are associated with permanent waterbodies like saline lakes, coastal lagoons, and backwaters or temporarily

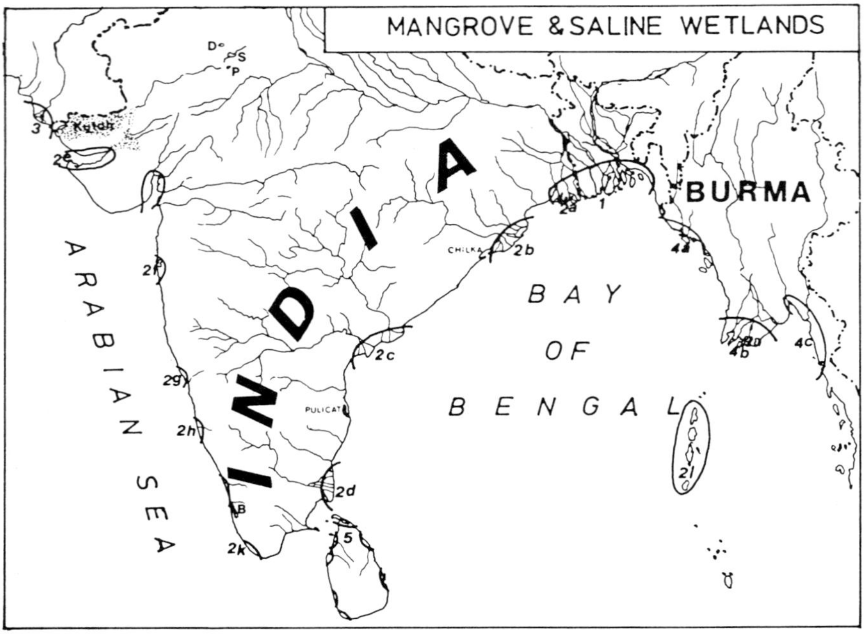

Figure 7. Distribution of mangroves and important saline wetlands in south Asia. Areas indicated by numbers and/or letters are: 1. Bangladesh: Sunderbans in Ganga-Brahmaputra delta; 2. India: a – Ganges delta, b – Mahanadi delta, c – Krishna-Godavari delta, d – Cauvery delta, e – Saurashtra coast, f – Maharashtra coast, g – Goa coast, h – Karnataka coast, k – Kerala coast, i – Andaman Nicobar Islands; 3. Pakistan: Indus delta; 4. Burma: a – Arakan coast, b – Irrawady delta, c – Tenasserim coast; 5. Sri Lanka: several coastal areas. Other saline wetlands shown here are Kutch, Chilka, Pulicat, Kerala backwaters (B), and salt lakes in Rajasthan (D – Deedwana, P – Pachpadra and S – Sambhar).

flooded inland areas. Typical coastal salt marshes are practically unknown in this region.

Saline wetlands other than mangroves have been rarely studied. There are two coastal lagoons (Lake Chilka and Lake Pulicat) on the eastern seacoast of India (Fig. 7). Lake Chilka covers more than 1,165 km^2, is very shallow (maximum depth between 2.5 and 3.6 m) and exhibits large spatial and seasonal variations in salinity (0.1 to 3.6‰; Jhingaran 1982). The vegetation consists primarily of submerged macrophytes, *Potamogeton pectinatus*, *Halophila ovalis*) and macroalgae (*Gracilaria*, *Enteromorpha*, *Chaetomorpha*). Lake Pulicat is approximately 777 km^2 and has a mean depth of only 1.5 m (Jhingaran 1982). Salinity in the lake ranges from nearly freshwater in the north to highly saline in the south. The vegetation is dominated by macroalgae (e.g., *Gracilaria confervoides*, *Enteromorpha intestinalis*, *E. compressa*, *Chaetomorpha indica*). Backwaters in Kerala (Fig. 7) on the southwestern coast of India comprise a system of interconnected lagoons, bays, and mangrove swamps which support a large diversity of fish and prawn species for which they are highly exploited (Jhingaran 1982).

Among the seasonally flooded saline wetlands, most important is the Rann of Kutch in western India (Fig. 7). Vast areas of Rann are inundated during the monsoon season but remain dry for most of the year. The soils are halomorphic and a white crust of salts forms on the surface during the dry season. The halophytic vegetation is chiefly composed of *Suaeda fruticosa*, *Salsola foetida*, *Salicornia brachiata*, *Haloxylon salicornicum*, *Atriplex stocksii*, *Cressa cretica*, and grasses like *Aeluropus lagopoides*, *Sporobolus helvolus*, and *Halopyrum mucronatum*. More details are given by Blasco (1975, 1977) and Puri *et al.* (1989).

There are also several salt lakes (Sambhar, Pachpadra, Deedwana, and Lunkaransar, and Mansar in Pakistan), referred to as salt playas by Puri *et al.* (1989), in western arid and semi-arid region of the subcontinent (Saxena and Gupta 1973). These shallow lakes collect the runoff during rainy season and support halophytic vegetation comprising of species of *Suaeda*, *Salsola*, *Salicornia*, and *Atriplex*, and the grass *Aeluropus lagopoides*.

Mangroves

Mangroves are among the best investigated wetlands and have been studied for more than a century. Among the earliest noteworthy reports are those on the Sunderbans (Roxburgh 1814, Schimper 1891, Clarke 1896, Prain 1903), Bombay Presidency (Blatter 1905), and Indus delta (Blatter *et al.* 1927–28). Mangroves were the subject of a national symposium in India in 1957 (Anonymous 1959) and interest in them has grown over the past few years. A number of regional and national reviews, varying in their scope and coverage have appeared (Mathauda 1959, Waheed Khan 1959, Rao and Sastry 1972, 1974, Navalkar 1973, Chapman 1970, 1974, 1975, Blasco 1975, 1977, Krishnamurthy *et al.* 1975, Sen and Raj Purohit 1982, Bhosale *et al.* 1983, Snedaker 1984, Kogo 1985, Pinto 1986, Ansari 1987, A.N. Rao 1987). A state-of-the-art report has been prepared by the Government of India (Anonymous 1987a) and a mangrove bibliography has been compiled (Untwale 1982). UNESCO in cooperation with United Nations Development Programme launched a long-term programme of research and training in mangroves of Asia and the Pacific. The program has organised many workshops, courses and conferences at which mangrove studies have been reviewed from time to time. Two important recent publications are by Soepadmo *et al.* (1984) and Umali *et al.* (1987).

Distribution and area. Mangroves in south Asia are part of the Indo-Pacific mangrove forests which form the world's most extensive and diverse mangrove system (Macnae 1968, Snedaker 1984). The term "mangal" is often

Table 4. Areal extent (sq. km) of mangroves in south Asia based on Blasco (1977) and field work of Krishnamurthy. Considerably different estimates are given by other authors.

Country and Location	Blasco	Krishnamurthy
1. Bangladesh: Ganges-Brahmaputra delta	6,000	6,000
2. India:		
a. Ganges delta (W. Bengal)	2,000	4,222
b. Mahanadi delta	50	150
c. Godavari and Krishna deltas	100	200
d. Cauvery delta	15	150
e. Saurashtra and Kutch coast	200	260
f. Bombay coast		330
g. Goa	200	200
h. Karnataka coast		60
i. Andaman and Nicobar Islands	1,000	1,190
3. Pakistan: Indus delta		2,495
4. Burma:		
a. Arakan coast		1,002
b. Irrawady delta		2,796
c. Tennasserim coast		1,842
5. Sri Lanka		320–400

used in reference to the living natural communities of organisms on coastal mudflats and waterways. Mangroves can be divided regionally into three zones (Fig. 7). In the Bay of Bengal, rivers such as Irrawaddy, Ganges (known as Padma in Bangladesh), Brahmaputra (Meghna in Bangladesh and Tsangpo in Tibet), Mahanadi, Godavari, Krishna, and Cauvery discharge enormous quantities of silt and freshwater, and form extensive deltas that are dominated by mangroves. The Arabian Sea coast is characterised by typical funnel-shapped estuaries of major rivers (Indus, Narmada, Tapti) or backwaters, creeks, and neritic inlets that are dominated by the estuarine and backwater type mangroves. A third type of mangrove occurs in the Bay of Bengal on islands (Andaman, Nicobar) which are in the "epicentre" of the tropical cyclone storms. On these islands, there are many tidal estuaries, small rivers, neritic islets, and lagoons which support a rich mangrove flora. Mangroves in Sri Lanka are of nearly similar nature.

Estimates of the area covered by mangroves differ widely because there is no agreed definition of the term "mangrove". Some authors include coastal saline areas without any significant vegetation in their estimates (e.g., Sidhu 1963). Blasco (1975, 1977) only considered forested areas. One of us (KK) has estimated that mangroves occupy an area of 21216 km^2 of which 6,760 km^2 are in India alone. The distribution of major mangrove areas is shown in Fig. 7 and the area estimates for different countries are given in Table 4.

Mangroves in the deltas of the Ganges and Brahmaputra in India and Bangladesh and Irrawady delta in Burma, and around the Andaman and

Nicobar Islands in the Bay of Bengal are among the largest in the Indo-Pacific region. Smaller patches of mangroves are associated with the deltas on rivers Mahanadi, Godavari, Krishna, and Cauvery on the eastern coast of India. In the western part of the region, large mangroves occur in the Indus delta of Pakistan, and smaller areas along the Indian coast near Bombay, Goa, and in Kutch (Saurashtra).

Habitat and vegetation. More than one hundred fifty species of angiosperms and ferns, often grouped into major and minor components and mangrove associates (A. N. Rao 1987), occur in the south Asian mangroves. Of these about eighty species are more common (Table 5). Dominant species belong to the genera *Rhizophora*, *Avicennia*, *Bruguiera*, *Kandelia*, *Ceriops*, *Excoecaria*, *Sonneratia*, *Lumnitzera*, *Nypa*, *Aegiceras*, *Heritiera*, *Aegialitis*, and *Xylocarpus*. Associated and usually less abundant species belong to the genera *Sesuvium*, *Suaeda*, *Salicornia*, *Acrostichum*, *Brownlowia*, *Thespesia*, *Clerodendron*, *Hibiscus*, *Derris*, *Salvadora*, *Phoenix*, *Porterasia*, *Aeluropus*, and *Urochondra*. Many species are found only in mangroves of Sri Lanka (Table 5). There are significant differences in species composition among mangroves of the east coast and also in different parts of the same coast depending upon the hydrological, edaphic, and biotic factors. Major vegetational features of different areas listed in Table 4 are therefore described below separately.

1. *Ganges-Brahmaputra delta*. Mangroves of the Ganges-Brahmaputra delta, commonly known as Sunderbans, are contiguous between India and Bangladesh (Fig. 8) and form the largest mangrove complex in the world. The undivided Sunderbans include the major portions of the Bakarganj and Khulna districts of Bangladesh and the 24–Parganas district of West Bengal (India). They occupy an area of more than 10,000 km^2 of which 4,222 km^2 is in India.

These mangroves were the first to receive botanical attention in the region (Clarke 1896, Prain 1903). During the past few years many studies have been made of these mangroves on the Indian side (Banerjee 1964, Blasco 1975, Mukherjee 1975, Mukherjee and Mukherjee 1978, Mukherjee 1984, Naskar and GuhaBakshi 1987) whereas very little is still known about them in Bangladesh (Ahmad 1984, Ismail 1989).

Various distributaries of the Ganges carry large amounts of freshwater which causes a distinct gradient in salinity in the eastern Sunderbans (LaFond 1966). The salinity differences result in three distinct areas: (a) a northeast area that is almost always fresh, (b) an area of moderate salinity east of the Raimangal river, and (c) an area of high salinity west of the Raimangal river. Vegetation and its zonation in the three areas were described by Curtis

Table 5. Distribution of mangrove species of South Asia. Sunderbans (I) of India and Bangladesh are contiguous, and hence placed together. Other mangrove areas are: II. India (a. Andaman and Nicobar Islands, b. Mahanadi delta, c. Godavari and Krishna deltas, d. Cauvery delta, e. Saurashtra and Kutch coast, f. Bombay coast, g. Goa, h. Karnataka coast, i. Kerala Coast), III. Burma, IV. Pakistan, and V. Sri Lanka. Trees are indicated with an * (adapted from Blasco 1975, 1977, A. N. Rao 1987).

Family	Species	Distribution												
		I	II									III	IV	V
			a	b	c	d	e	f	g	h	i			
Acanthaceae	*Acanthus ebracteatus* Vahl.		+											+
	Acanthus ilicifolius Lour.	+	+	+	+	+	+	+	+	+	+			+
	Acanthus volubilis Wall.		+											
Aizoaceae	*Sesuvium portulacastrum* L.		+											+
Apocynaceae	*Cerbera manghas* L. *		+											+
	Cerbera odollam Gaertn. *													+
	Ervatamia pandacagui Pichon.													+
Asclepiadaceae	*Sarcolobus carinatus* Wall.		+											
	Sarcolobus globulus Wall.		+											
	Finlaysonia obovata Wall.													+
Avicenniaceae	*Avicennia officinalis* L. *	+	+	+	+	+	+	+	+	+	+			
	Avicennia alba Blume *	+	+		+	+	+	+		+				+
	Avicennia marina (Forsk.) Vierh. *	+	+	+	+	+	+						+	+
Bignoniaceae	*Dolichandrone spathacea* (L.f.) Schum.													+
Boraginaceae	*Heliotropium curassavicum* L.													
Caesalpiniaceae	*Caesalpinia crista* L.													+
	Cynometra ramiflora Willd. *													+
Chenopodiaceae	*Arthrocnemum indicum* (Willd.) Moq.													+
	Atriplex stocksii Boiss.													
	Salicornia brachiata Roxb.													+
	Suaeda fruticosa Forsk.													
	Suaeda maritima (L.) Dum.													+
	Suaeda monoica Forsk.													+

Combretaceae	*Lumnitzera racemosa* Willd. *	+	+	+	+	+		+	+					+
(= Terminaliaceae)	*Lumnitzera littorea* (Jack.) Voigt		+			+								+
Convolvulaceae	*Stictocardia tiliaefolia* Hallier f.													
Cyperaceae	*Scirpus littoralis* Schrad.													
Euphorbiaceae	*Excoecaria agallocha* L. *	+	+	+	+	+		+	+	+				+
Lythraceae	*Pemphis acidula* J. R. & G. Fors.													
Malvaceae	*Hibiscus tiliaceus* L. *													+
Meliaceae	*Amoora cucullata* Roxb. *												+	+
	Xylocarpus granatum (L.) Koenig *	+	+	+		+		+						+
	(= *Xylocarpus obovatus* Grewe)													
	(= *Carapa obovata* (Bl.) Grewe)													
	Xylocarpus moluccensis (Lamk) Roem. *	+		+	+	+								+
	(= *Carapa moluccensis* Lamk.)													
	Xylocarpus mekongensis Pierre					+								
	Xylocarpus gangeticus (Prain) Parkinson *													
Myrsinaceae	*Aegiceras corniculatum* (L.) Blume	+	+	+	+	+	+	+	+	+			+	+
	(= *Aegiceras majus* Gaertn).													
	Ardisia littoralis Dryand.													+
	Myrsine umbellata Wall.													+
	Rapanea porteriana Merr.													+
	Rapanea umbellata Elm.													+
Myrtaceae	*Melaleuca leucodendra* L.													+
Palmae	*Nypa fruticans* Wurmb.	+	+											+
	Phoenix paludosa Roxb.	+	+	+										+
	Phoenix pusilla Gaertn.													
Pandanaceae	*Pandanus tectorius* Soland.													
Papilionaceae	*Dalbergia spinosa* Roxb.													
	Derris heterophylla Willd.													
	Derris trifoliata Lour.		+	+	+	+	+		+		+			
	Derris uliginosa Benth.					+								

Table 5. Continued.

Family	Species	Distribution I	II a	b	c	d	e	f	g	h	i	III	IV	V
Plumbaginaceae	*Aegialitis rotundifolia* Roxb.													+
Poaceae	*Aeluropus lagopoides* (L.) Trin.													+
	Myriostachya wightiana Hk. f.													
	Porteresia coarctata (Roxb.) Takeaka													
	Sporobolus virginicus Kunth													
	Urochondra setulosa (Trin.) Hubbard													
Rhizophoraceae	*Bruguiera gymnorhiza* (L.) Lamk. *	+	+	+	+	+	+	+	+		+		+	+
	(=*Bruguiera conjugata* Merr.)													
	Bruguiera cylindrica (L.) Blume *	+		+	+	+		+						+
	(= *Bruguiera caryophylloides* Bl.)													
	(= *Bruguiera caryophylloides* Burm.)													
	Bruguiera parviflora (Roxb.) Wight & Arn.	+	+	+			+	+	+	+				+
	Bruguiera sexangula (Lour.) Poir *	+	+	+	+	+								+
	(= *Rhizophora eriopetala*													
	Ceriops decandra (Grittith) Ding Hou *	+	+	+	+	+		+					+	+
	(= *Ceriops roxburghiana* Arn.)													
	Ceriops tagal (Perr) C. B. Robins *	+	+	+	+	+	+	+	+	+			+	+
	(= *Ceriops candolleana* Arn.)													
	Kandelia candel (L.) Druce *	+		+		+	+	+		+			+	
	(= *Kandelia rheedi* W. & A.)													
	Rhizophora mucronata Lamk. *	+	+	+	+	+	+	+	+	+	+			+
	Rhizophora stylosa Griff. *			+		+								
	Rhizophora apiculata Blume *	+	+	+	+	+	+	+		+	+			+
	(= *Rhizophora conjugata* (non L.) Arn.													
	(= *Rhizophora candelaria* DC.)													
	Rhizophora lamarckii Montrouz *					+								

Rubiaceae	*Guettarda speciosa* L.													+
	Scyphophora hydrophyllacea Gaertn.													+
Salsolaceae	*Salsola foetida* Delile													
	Salsola kali L.													
Salvadoraceae	*Salvadora persica* L. *													
	Salvadora oleoides Dcne *													
Sonneratiaceae	*Sonneratia apetala* Buch-Ham. *	+	+	+	+	+		+	+		+			+
	Sonneratia alba J. Sm. *		+					+	+					+
	Sonneratia griffithi Kurz. *													
	Sonneratia caseolaris (L.) Engler *		+	+			+	+	+	+			+	+
	(= *Sonneratia acida* (L.) Back.)													
Sterculiaceae	*Heritiera fomes* Buch-Ham. *	+		+										
	(= *Heritiera minor*)													
	Heritiera littoralis Dryand ex. Ait. *	+	+	+			+	+	+					+
	Kleinhovia hospita Linn.													+
Tamaricaceae	*Tamarix gallica* L.													+
Tiliaceae	*Brownlowia lanceolata* Benth.													+
Verbenaceae	*Clerodendron inerme* (L.) Gaertn.													+
Filicopsida (ferns)	*Acrostichum aureum* L.													+
	Acrostichum speciosum Willd.													+
	Stenochlaena palustre (Burm.) Bedd.													

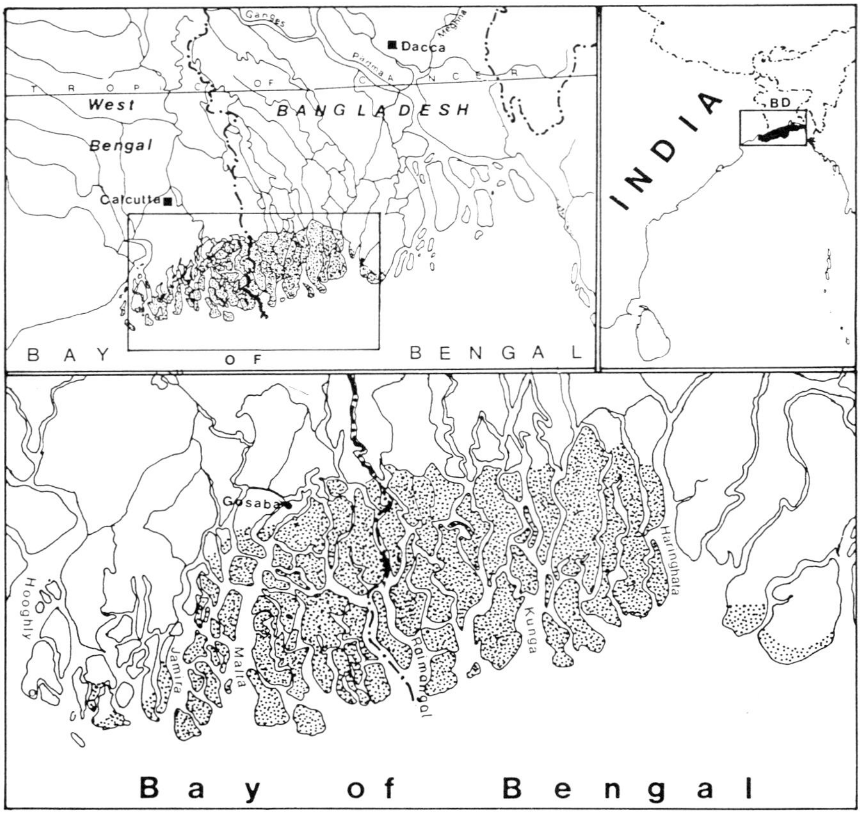

Figure 8. Sunderbans (stippled areas) in Ganges-Brahmaputra delta in India and Bangladesh.

(1933), Champion and Seth (1968), and Blasco (1977). Four or five zones are generally recognised in relation to the salinity gradient.

In the most saline inland areas, a scrub vegetation (referred to as back mangroves) is composed chiefly of *Salicornia brachiata*, *Heliotropium curassavicum*, *Suaeda maritima*, and *Sesuvium portulacastrum*. Occasionally, bushy growth form of *Aegialitis rotundifolia* and *Aegiceras corniculatum* occur on river banks. In areas regularly leached by freshwater, tall and dense forests (dense mangroves) that are dominated by *Heritiera fomes* ("Sunderi" in Bengali from which these forests derive their name), *Excoecaria agallocha*, *Xylocarpus moluccensis*, *Bruguiera cylindrica*, and *Sonneratia apetala*. *Ceriops decandra*, *Avicennia officinalis* and *A. corniculatum* are other important species but *Nypa fruticans* is now rare. In moderately saline water, *Rhizophora* and *A. rotundifolia* are abundant while saltwater areas are dominated by *Avicennia alba*, *A. marina*, *X. granatum*, and *Kandelia candel* (Ahmad 1966). A palm (*Phoenix paludosa*) occurs throughout the mangroves and forms dense stands in tidal zones near the edges of water courses. Other common species in the palm swamps are *E. agallocha* and *S. apetala*.

In the Sunderbans of West Bengal, freshwater contribution through the Ganga river is practically negligible and over the last few centuries major flow of the Ganges has been diverted from Bhagirathi to Padma. Considerable changes have subsequently occurred in the morphology of Ganges delta. This is supposed to have been the result of tectonic and morphogenetic uplift of the western part together with a eastward shift of the river (Gupta 1957, Chowdhury 1966). The lack of freshwater flow has affected the mangrove species such as *H. fomes* and *N. fructicans* which have practically disappeared from Indian part of Sunderbans.

The flow of sufficient freshwater through the Ganges-Brahmaputra riverine system throughout the year is essential for the deltaic mangroves of both India and Bangladesh. It may be pointed out that recent efforts in India to augment this flow (which would help restore the ecological balance, save species from extermination and also reduce siltation in Indian ports) have resulted in intergovernmental disputes.

2. *Mahanadi delta*. The River Mahanadi forms a delta somewhat southwest of the Sunderbans on the east coast. Mangroves here cover a relatively small area and very little is known of them although they are highly disturbed. These mangroves are floristically very similar to the Sunderbans as *H. fomes*, *H. littoralis*, and *P. paludosa* occur here as well. Recent studies show that many species, particularly *Kandelia rheedii* and *Rhizophora conjugata* have disappeared from this region during recent decades (Anonymous 1987a). *Aegialitis rotundifolia*, now restricted to Sunderbans, also occurred earlier in these mangroves (Rao and Sastry 1974). Species like *R. mucronata*, *A. majus*, and *P. paludosa* are now rare. Though earlier studies recognised a deltaic swamp forest zone and a littoral scrub fringe, the vegetation is highly degraded with stunted growth due to soil erosion and increasing salinity. Rao and Mukherjee (1972) recognised seven vegetation zones along Burabalanga estuary (Balasore district) and related them to differences in soil texture, moisture gradient, and soil chemistry. However, detailed information on these mangroves is not available.

3. *Godavari and Krishna deltas*. Southwards on the Andhra coast, the deltas of the Godavari and Krishna Rivers lying adjacent to each other support the second largest mangrove complex in the region. Dense mangrove forests also exist at Yanam on the banks of the Coringa river near Kakinada and in the Gautami-Godavari deltaic system. These mangroves are relatively better known floristically and ecologically.

The region is characterised by large seasonal variations in salinity. During the monsoon, the river Godavari carries large amounts of freshwater and salinity remains very low for at least half a year from July, and especially between October–November when rainfall is high (exceeding 200 mm a day). On the other hand, the development of an off-shore sand bar in the Coringa region has reduced the influence of sea water. However, during the dry season the salinity increases considerably.

Floristically, Godavari delta forms the dividing line between the mangroves of Mahanadi and Sunderbans on one hand and those of the peninsular India on the other. Venkateswarlu (1944) reported some 26 species of mangroves from the mouths of Godavari and Gautami rivers. More detailed studies in the Godavari delta were made by R. S. Rao (1959), Sidhu (1963), Venkatesan (1966), and T. A. Rao *et al.* (1972). The mangroves differ from those elsewhere in India in the dominance of *Avicennia* (represented by all the three species, *A. marina*, *A. alba*, and *A. officinalis*), *S. apetala*, and a grass (*Myriostachya wightiana*) which occurs otherwise only in Sunderbans. Members of the Rhizophoraceae are very rare except near river banks where *Bruguiera gymnorhiza* and *R. mucronata* are common. *Avicennia officinalis* and *Hibiscus tiliaceus* are common along rivers. Among other species, *E. agallocha*, *Dalbergia spinosa*, and *Stictocardia tiliaefolia* are common in somewhat inland areas.

4. *Cauvery delta*. Further south in Tamil Nadu, mangroves occur in the Cauvery delta Fig. 9). The mangrove forests in the region of Pichavaram (Vellar estuary) and Muthupet-Chattram (Cauvery proper) are also among the best studied wetlands (Rajagopalan 1952, Venkatesan 1966, Blasco and Caratini 1973, Caratini *et al.* 1973, Krishmamurthy *et al.* 1981, Krishnamurthy 1983, Lakshmanan *et al.* 1984). These mangroves are rich in species, and exhibit a clear zonation but occupy a very small area.

Near the shores, on constantly wet soils, there is a narrow belt of dense forest (Fig. 9) dominated by *R. apiculata*, *R. mucronata* and other Rhizophoraceae including *B. cylindrica* and *S. apetala*. Other common species are: *Lumnitzera racemosa*, *Aegiceras corniculatum*, *C. decandra*, and *Derris trifoliata*. Behind this belt is a belt of smaller trees of *A. marina* with shrubby undergrowth of *S. maritima* and *E. agallocha*. Further inland only halophytic shrubs and herbs like *S. brachiata*, *Acanthus ilicifolius*, *S. portulacasturm*, and *A. indicum* are found on periodically flooded highly saline soils.

Small patches of mangroves occur also on Pamban, Rameshwaram and other islands in the Gulf of Mannar but these have rarely been investigated (T. A. Rao *et al.* 1963 a,b). Common species are: *Rhizophora conjugata*, *A. alba*, *C. tagal*, *E. agallocha*, and *Arthrocnemum indicum*.

5. *West coast mangroves*. The west coast is characterized by the funnel

Figure 9. Mangrove forest dominated by *Rhizophora apiculata* and *Rhizophora mucronata* at Pichavaram in Cauvery delta.

shaped estuaries and typical deltas with alluvial deposits are almost totally absent. Thus, the mangroves on the west coast are of estuarine and backwater type as compared to deltaic type on the east coast. They are not extensive and are rapidly disappearing under anthropogenic pressure. Further, these mangroves differ markedly from those of the east coast by the absence of palms, and species of *Heritiera* and *Xylocarpus* (Table 5) whereas some species like *Sonneratia caseolaris* and *Urochondra setulosa* occur only on the west coast. An overview of these mangroves is provided by Untawale (1984).

In Kerala (most southern part of the Peninsula) only small mangrove area are now left near Quilon and Cochin. The mangroves at Veli near Trivandrum disappeared only about two decades ago. An important species still found in Kerala is *Cerbera manghas* (Blasco 1975).

North of Kerala, small areas of fringing mangroves occur on the Karnataka coast (Untawale 1984, Radhakrishnan 1985, Untawale and Wafar 1986). Fourteen species have been recorded with the dominants being *A. marina*, *A. officinalis*, *S. caseolaris*, *R. conjugata*, *R. mucronata*, *A. corniculatum*, *E. agallocha*, *H. littoralis*, *Cynometra numosoides*, and *Acanthus ilicifolius*.

The distribution, zonation, and ecology of mangroves around Goa have been studied in detail by Dwivedi *et al.* (1975), Bhosale (1978), Untawale *et al.* (1973, 1982), and Jagtap (1985, 1986). Most important mangroves occur along the Mandovi and Zuari estuaries. There are about twenty species of which *S. caseolaris*, *K. candel*, *R. mucronata*, *R. apiculata*, *S. alba*, *A. officinalis*, *B. parviflora*, *A. ilicifolis*, and *Derris heterophylla* are important. Untawale *et al.* (1982) described zonation of vegetation in relation to salinity (oligohaline to polyhaline) and sediments. *Sonneratia caseolaris* and *Acrostichum aureum* occur in oligohaline areas with silty substrata whereas polyhaline zones with sandy clay substrate are occupied by *R. mucronata*, *B. parviflora*, *A. marina*, and *S. alba*.

Further north, in the state of Maharashtra, mangroves occur around Bombay (Fig. 7) in small patches along creeks and on small islands. These are among the best studied mangroves in the country and many systematic and eco-physiological studies have been made (Cooke 1901–1908, Blatter 1905, Navalkar 1951, 1956, 1973, Qureshi 1959, Patil 1959, Kumar and Chaphekar 1985). The vegetation is dominated mainly by shrubby *R. apiculata* (Joshi and Bhosale 1982). *Rhizophora mucronata*, *B. cylindrica*, and *C. tagal* occur along the sea shore but inland the vegetation consists mainly of dense growth of *Avicennia* spp. and *E. agallocha*. *Salicornia brachiata* and *Derris scandens* are also found. *Sonneratia apetala*, and *S. alba* occur on some islands (Blasco 1977). Navalkar (1956) reported *K. candel* which is not found here any more. The presence of Salvadoraceae (*Salvadora persica* and *S. oleoides*) in Bombay is of interest as these are considered to represent old mangroves (Qureshi 1959). *Salvadora oleoides* does not occur at latitudes south of Bombay (Blasco 1975).

The halophytic formation in the Saurashtra region (Fig. 7) are often classified as mangroves but do not have characteristic mangrove plants (Rao and Aggarwal 1964, Rao *et al.* 1966, Rao and Shanware 1967). The area has been already described above as the seasonally flooded wetland. Only in the Gulf of Kutch, stunted woody growth of *A. marina* is obtained.

6. *Mangroves of Andaman-Nicobar islands.* The Andaman and Nicobar group of islands (Fig. 7) in the Bay of Bengal (6 to 14° N and 92 to 94° E) have an irregular coastline deeply indented with numerous tidal creeks and sheltered bays which provide excellent habitats for mangroves. These more or less virgin mangroves, due to the remoteness of the islands and low population density, account for about 17% of India's total mangrove area (Table 4). There are number of floristic surveys (Parkinson 1923, Chengapa 1944, Banerji 1958a,b, Sahni 1953, Thothathri 1960a,b, 1962) of this region, and in recent years, Mall *et al.* (1982, 1986) have made ecological studies of these mangroves. Floristically, the mangroves of Andaman-Nicobar islands stand in great contrast with those of the peninsular India (Thothathri 1981). *Nypa fruticans*, absent from peninsular India, is most abundant here. Other dominant species are: *R. mucronata*, *R. stylosa*, *R. apiculata*, *B. gymnorhiza*, *B. parviflora*, *C. tagal*, and *A. corniculatum*. *Xylocarpus granatum* and *Lumnitzera littorea* are also abundant but rare on the coasts of peninsular India.

Other important species found only in the island mangroves are *Guettarda speciosa*, *Hernandia ovigera*, *Brownlowia lanceolata*, and *Scyphiophora hydrophyllacea*. However, *Aeluropus lagopoides*, and *Porterasia coarctata* are absent from these islands. *Sonneratia apetala* reported absent earlier by Blasco (1975) has been recorded recently by Mall *et al.* (1986). Epiphytes like *Hydrophytum formicarum* and *Dischidia major* and the orchid *Papilionanthe teres* are also very common.

The vegetation zones based on the frequency and duration of inundation have been recognised by Mall *et al.* (1986). These are: a. Proximal zone with prolonged and most frequent inundation, b. Middle zone lying inwards and less frequently flooded, and c. Distal zone on the landward fringe with higher salinity. Species of *Rhizophora* dominate the proximal zone whereas species of *Bruguiera*, *C. tagal*, and *L. littorea* occur in the middle zone. *Excoecaria agallocha*, *Nypa fruticans*, and *H. littoralis* commonly occupy the distal zone.

7. *Mangroves of Pakistan*. The mangroves of Pakistan are confined to coastal Sind, particularly in the Indus delta and cover approximately 2,495 km^2 (Khan 1966, Saenger *et al.* 1983). They are nearly similar to those on the West coast of India but are floristically poor, being represented by only eight species (Table 5, Nasir and Ali 1970–85). There are several reports on the distribution and general ecological problems (Champion *et al.* 1965, Khan 1965, 1966, Kogo 1985, Ansari 1987) of these mangroves but detailed ecological studies have not been done.

Avicennia marina is the most dominant species forming dense mangroves along the creeks on recent alluvium. It is often associated with *C. tagal*. In the Indus delta, the normal mangrove vegetation is composed of *R. mucronata*, *R. apiculata*, *B. gymnorhiza*, *A. corniculatum*, and *S. caseolaris*. The associated vegetation in the sheltered areas include many other halophytes.

During recent years, many dams and barrages have been constructed on the river Indus for agriculture. Therefore, freshwater discharge into the coastal areas is small for about 9 months of the year. As a result, mangroves of the Indus delta are becoming decadent and their growth is retarded (Saenger *et al.* 1983, Ansari 1987).

8. *Mangroves of Sri Lanka*. Mangroves in Sri Lanka occur along the sea coast throughout the island (Fig. 10). In Jaffna peninsula (in the region of Gulf of Mannar) mangroves extend to seafront while in other places they are confined to estuaries and lagoons. The northern lagoons are in permanent communication with the sea (Raphael 1977) but those in the south are partially closed by sandbars for most part of the year and hence, experience lesser influence of sea. The tidal amplitude is also small (about one metre) and therefore, variation in inundation levels is not significant. The estimates

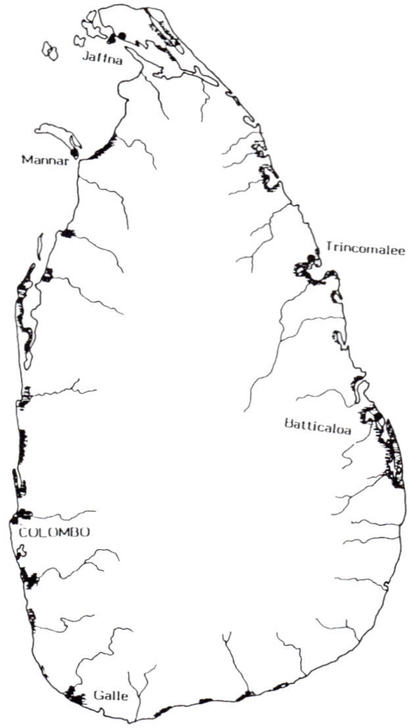

Figure 10. Distribution of mangroves in Sri Lanka (adapted from Jayewardene 1987).

of area under mangroves in Sri Lanka vary considerably: Seneviratne (1978) estimated the cover between 320 and 400 km^2 whereas Arulchelvam (1968) and Saengar *et al.* (1983) put the estimate at only 40 and 36 km^2 respectively. More recent estimates using remote sensing techniques show that about 63 km^2 mangrove area lies in the six coastal districts (Jayewardene 1987). The mangrove definitely occupied in the past a much larger area, most of which has been now reclaimed.

Thirty seven species of mangrove plants, including the associates, occur in Sri Lanka (Jayewardene 1987). *Rhizophora mucronata* or *R. apiculata* dominate near water's edge or on steep shores or river banks. Behind them, *B. gymnorhiza*, *S. caseolaris*, *A. officinalis*, *A. marina*, *C. tagal*, *C. decandra*, *A. corniculatum*, *Scyphiophora hydrophyllacea*, and *L. racemosa* are abundant. *Nypa fruticans* occurs on the southeastern coast and in some lagoons (Abeywickrama 1966). Though peats are not known from Indian mangroves, Abeywickrama (1966) has reported large peat deposits at Muturajawela in Sri Lanka.

Mangroves of Sri Lanka have been grouped into five types (de Silva 1985) namely: a. riverine mangroves in the estuaries of major rivers on south and southwest coast, b. fringing mangroves along shallow lagoons, c. basin mangroves associated with Vadamarachchi lagoon, d. scrub mangroves with stunted growth along lagoons on the east and west coasts, and e. overwash mangroves on small islands in Puttalam and Negombo lagoons.

9. *Mangroves of Burma*. The mangroves of Burma are distributed between 20 and 10° N Latitude and 94 and 98° E Longitude. They occur in estuarine areas and deltas wherever tidal action provides suitable conditions for growth of mangroves. Many islands along southern coastline have extensive mangrove forests. There are no accurate data on the mangroves of Burma. Wolker (1966) reported an area of about 5,200 km^2 in the Irrawaddy delta alone, leaving large areas on Arakan and Tennasserim coasts. Recently Than Htay and Saw Han (1984) put the estimate at over 5,700 km^2 of which 2,750 are in Irrawady delta, 1,863 on the Tenasserim coast, and 1,020 on the Arakan coast.

Moodie (1924–25) and Stamp (1925) referred to mangroves in the Irrawaddy delta and suggested that they are similar to mangroves in the Ganges-Brahmaputra delta. Like eastern Sunderbans, the deltaic mangroves in Burma are also dominated by *H. fomes* in freshwater dominated zones and *R. mucronata* in areas flooded with seawater. Other common taxa are: *B. gymnorhiza*, *A. officianalis*, *X. moluccensis*, *S. apetala*, *S. acida*, *E. agallocha*, *Ceriops roxburghiana*, and *N. fruticans*. The Arakan coastline is also dominated by members of Rhizophoraceae (*R. mucronata*, *B. gymnorrhiza*).

Associated fauna. Mangroves support a large diversity of both vertebrate and invertebrate fauna (A. N. Rao 1987) which are adapted to different salinity and hydrological gradients. In south Asia, the fauna of Indian Sunderbans have been investigated in more detail than of other mangroves (Chaudhuri and Chakrabanti 1973, Choudhury *et al.* 1984, Kurian 1984, Sarkar *et al.* 1984, Chakraborty and Choudhury 1986, Samant 1986, Kasinathan and Shanmugam 1986, Palaniappan and Baskaran 1986, Naskar and Guha Bakshi 1987, Mandal and Nandi 1989). There are also a few reports on the mangrove fauna of Sri Lanka (Pinto 1984, 1986, Jayewardene 1987) and Pakistan (Kogo 1985, Ansari 1987). Estuarine fisheries of India have been described in detail by Jhingaran (1982). Among invertebrates, more than 500 species of insects and Arachnida, 229 species of crustacea, 212 species of molluscs, 50 species of nematodes, and 150 species of planktonic and benthic organisms are known from Indian mangroves (Anonymous 1987a). Vertebrate fauna are represented by 300 species of fish, 177 species of birds, 36 species of mammals and 22 species of reptiles. Many of these are economically exploited.

Whereas some animals are only temporary visitors and move in and out of the mangroves at different times of the year, many are characteristic of these habitats. Important invertebrates include prawns (*Penaeus indicus*, *P. merguiensis*, *P. monodon*, *Macrobrachium rosenbergii*), crabs (*Uca lactea*, *Scylla serrata*, *Thalassina* sp., *Sesarma fascinata*, *Canosesarma minuta*, *Telescopicum telescopicum*, *Cerithidea alata*, *Clibanarius longitarsus*), molluscs, and oysters (*Crassostrea cucullata*, *Mytilus* sp.) and many insects especially honey bee (*Apis dorsata*, *Apis mellifera*), weaver ants (*Oecophylla* sp.), and mosquitoes (*Anopheles sundericus*, *Anopheles indigo*, *Culex fatigans*, *Aedes butleri*, *Aedes niveus*). Common fishes are mudskippers (*Periophthalmus* sp.), carangids, cluepeids, serranids, sciaenids, mullets, hilsa, seabass, and milkfish. Avifauna includes herons, storks, sea eagles, egrets, kingfishers, sandpipers, tits, and whistlers. Flamingoes are abundant in most of the areas (particularly Kutch). Frogs (*Rana hexadactyla*) and toads (*Rhocophorus maculatus*) are also common in Sunderbans. Sunderbans are well known for the Royal Bengal Tiger (*Panthera tigris*). In view of the rapid decline in tiger population, about 200 km^2 area of Indian Sunderbans (in 24–Paragnas district) had been protected as a Tiger Reserve in 1973. Chital deer (*Axis axis*), another mammal found only in Sunderbans, has also been recently protected to save it from extinction. Another important animal in Indian mangroves is crocodile (*Crocodylus porosus*) which occurs only in Mahanadi delta (Orissa) and in Andaman Nicobar islands. Excessive exploitation in the past reduced its populations to small number but the trend has now been reversed by breeding them in crocodile farms in coastal areas. The Pacific Ridley turtle (*Lepiodochelys olivaceae*) also nests on adjacent beaches. Other noteworthy animals are: dolphins (*Platenista gangetica*), mangrove monkey (*Macaca mulatta*), and otter (*Lutra perspicillata*).

Mention must also be made here of wild ass (*Asinus hemionus*) which occurs only in Kutch and feeds on saline scrub and grass. It is also an endangered species and efforts are being made to conserve it.

Important animals in mangroves of Sri Lanka (Pinto 1986) are: Portunid crabs (*Thalamita crenata*, *Portunus pelagicus*, *Scylla serrata*), Fiddler crabs (*Macrophthalmus depressus*, *Uca lactea*, *Uca dussumieri*), Graspid crabs (*Neosermatium malabaricum*, *Metaspograpsus messor*, *Chiromantes indiarum*), mud lobster (*Thalassina anomala*), prawns (*P. indicus*, *Metapenaeus dobsoni*, *M. rosenbergi*), molluscs (*Nerita polita*, *Littorina scabra*, *Gaffrarium tumidum*, *Geloina coaxans*), oysters (*Saccostrea* sp. and *Crassostrea* sp.) and mud skipper (*Periophthalmus sorbinus*).

In Pakistan, mangrove fauna include about 100 species of fish of which perciformes (46 species) and clupeiformes (15 species) are dominant groups (Ansari 1987). Many species of prawns, crabs and other crustaceans are abundant and form a major component of mangrove fauna. Lizards (*Stenodactylus orientalis*, *Acanthodactylus cantoris*, *Ophiomorus tridactylus*), and sea snakes (species of *Hydrophis*, *Microcephalopphis*, *Pelanis*) are also common.

Freshwater wetlands

Freshwater wetlands associated with both lentic and lotic waterbodies are widely distributed throughout the subcontinent from sea level to about 2,000 m in the Himalayan ranges. Because of the distinct seasonality in rainfall and a prolonged dry summer season, there are few permanently flooded natural areas. There are numerous man-made reservoirs that are generally small and shallow and often dry up completely during the summer. The large reservoirs also exhibit such large water level changes that their relatively large shallow littoral zones are subject to periodic drying. Thus, permanently flooded wetlands are rather rare, and most freshwater wetlands are only seasonal. Further, the long dry period is not conducive to the establishment and growth of woody species and most of the wetlands are, therefore, dominated by herbaceous vegetation. Forested or shrub dominated wetlands are confined to areas adjacent to perennial lotic water bodies. The herbaceous wetlands of temporary or permanent and lentic or lotic habitats exhibit only small differences in their floristic composition.

The relationship between the vegetation of different wetlands and their hydrological regimes has received little attention, and therefore a detailed account of wetland types suggested earlier (Table 3) is not possible. For the purpose of this review, freshwater wetlands are simply grouped into forested and herbaceous wetlands.

Forested wetlands

Forested wetlands occur primarily along rivers and are adapted to periodic flooding that is associated with the monsoonal rainy season. Wetland forests are, thus, best designated as floodplain or riparian forests. Forested wetlands of the Indian subcontinent are among the least investigated ecosystems. Besides a few preliminary studies, the only account of these wetlands is that by Champion and Seth (1968) whose classification is shown in Table 2. They emphasized that "ecologically they may be viewed as stages in natural succession or as edaphic preclimaxes". Following their scheme of classification, some important features of forested wetlands are given below.

Freshwater swamp forests. These forests occur on wet alluvium on the floodplains of rivers where soils are waterlogged throughout the year. These are subdivided into two categories.

1. Myristica *swamp forests.* They are distributed only in Travancore (Kerala) along streams (below 300 m altitude) on sandy alluvium rich in humus (Krishnamoorthy 1960). The soils are inundated during the latter half of the year. The dense evergreen, 15–30 m high forests are dominated by *Myristica mag-*

nifica, *Myristica canarica*, *Lophopetalum wightianum*, *Carallia brachiata*, *Pandanus furcatus*, *Calamus tenuis* and numerous species of Araceae, Cyperaceae and Scitaminae in the undergrowth.

2. *Tropical hill valley swamp forests*. They occur along streams on gravelly and sandy beds in submontane tracts of the Himalayas (in states of Uttar Pradesh, W. Bengal and Assam) and at few places in the Western Ghats. The swamps in Dehra Dun valley (western Himalaya) were first described by Kanjilal (1901). Detailed vegetation studies have been made by Dakshini (1960a,b, 1965), Som Dev and Aswal (1974), and Som Dev and Srivastava (1978). *Bischofia javanica* is the most dominant tree attaining a height of 12–15 m. In Golatappar swamp, *Shorea robusta* forms an emergent layer. The other common tree species are: *Ficus glomerata*, *Alstonia scholaris*, *Trewia nudiflora*, *Syzygium cumini*, *Toona ciliata*, *Pterospermum acerifolium*, *Machilus gamblei*, *Salix tetrasperma*, *Pyrus pashia*, *Elaeagnus conferta*, *Carallia brachiata* and *Phoebe lanceolata*. Shrubs and the cane, *Calamus tenuis* are common in the understorey. The herb stratum is rich in grasses (*Vetiveria zizanioides*, *Echinochloa crus-galli and Coix lachryma-jobi*), sedges (species of *Scirpus*, *Cyperus*,and *Carex*), ferns (*Dryopteris prolifera*) and other herbs (*Floscopa scandens*, *Acorus calamus*, *Polygonum stagnium*, and *Limnophila rugosa*).

In south India, these forested wetlands occur in Wynaad forest division in Nilgiris (Kerala). Here the important taxa are: *Anthocephalus cadamba*, *Elaeocarpus tuberculatus*, *Machilus macrantha*, *Pandanus tectorius*, *Mallotus albus*, and species of *Alpinia* and *Eugenia*.

3. *Creeper swamp forests*. These forests are found in Brahmaputra valley in low lying areas on heavy soils. The forests are dense, up to 10 m high, and have many vines. Important tree species are: *Magnolia griffithii*, *Drimycarpus racemosa*, *Machilus gamblei*, *Vatica lancaefolia*, and *Eugenia formosum*. Creepers and vines include *Calamus leptospadix*, *Calamus tenuis*, and species of *Cissus* and *Urcaria*. Among other aquatic plants, *Phragmites karka* is very common.

Seasonal swamp forests. These wetlands occur on floodplains of major rivers on alluvium and are subjected to only seasonal submergence. They are subdivided into:

1. *Eastern seasonal swamp forests*. Restricted to the Brahmaputra valley, these wetlands occur on black soils and are submerged for larger part of the year. Forests are dense and up to 20 m high. Abundant tree species are

Altingia excelsa, *Machilus gamblei*, *Elaeocarpus varunna*, and *Syzygium cumini*. Canes (*Calamus tenuis*, *C. latifolius*) and ferns are also abundant.

2. Barringtonia *swamp forests*. Evergreen forests dominated by *Barringtonia acutangula* occur in Ganges-Brahmaputra valley under conditions of deep flooding for prolonged periods. Silt deposits are deep and soils are rich in humus. The canopy averages 10–20 m. Other common taxa are: *Pongamia pinnata*, *Streblus asper*, *Ficus glomerata*, and *Salix tetrasperma*. *Calamus tenuis* is often abundant. Most of these wetland forests have been cleared and are now rarely found.

3. Syzygium cumini *swamp forests*. Along streams in the Gangetic alluvium and in the sub-Himalayan tract, there are seasonally flooded areas dominated by *Syzygium cumini*. Chaudhuri (1969) has described these forests from Bagdogra and Moraghat forests in north Bengal. Common tree species are *Syzygium formosum*, *Carallia integerrima*, and *Elaeocarpus* sp. *Pandanus furcatus*, and a tree fern (*Alsophila glabra*) are also abundant. Several shrubs (*Ardisia neriifolia*, *Fagraea obovata*, *Psychotria flava*), climbers, and palms (e.g. *Pinanga gracilis*) also occur. The ground flora consists of many herbaceous species of which most common are: *Pyrenium capitatum*, *Lasia heterophylla*, *Monochoria hastata*, *Arundo donax*, and *Pteris biaurita*.

4. *Low swamp forests*. In eastern Uttar Pradesh and Brahmaputra valley, these forests dominated by *Cephalanthus occidentalis* (only 5 m tall) occur on black heavy soils (Rowntree 1954). *Glochidion hirsutum* is co-dominant. Occasionally, *Shorea robusta* and *Syzygium cumini* are also found. Other common species are *Glycosmis pentaphylla*, *Saccolepis interrupta*, *Clinogyne dichotoma*, and *Hymenachne pseudo-interrupta*. *Leersia hexandra* and *Phragmites karka* are also abundant.

5. *Eastern* Dillenia *swamp forests*. These are evergreen to semi-evergreen forests with a dense but low canopy. They occur on heavy soil with seasonal flooding in northeastern India (mainly Assam). Important taxa are: *Dillenia indica*, *Bischofia javanica*, *Vatica lancaefolia*, *Altingia excelsa*, *Castanopsis indica*, *Mesua ferrea*, and *Premna bengalensis*.

Moist temperate alder forests. All along the Himalayan range (except Kashmir), at somewhat higher altitudes (between 1,000 and 3,000 m), under colder climate, the riparian forests are dominated by alders (*Alnus nitida* and *Alnus nepalensis*). These are generally confined to sites with permanent water supply. The common taxa associated with alders are *Ulmus ciliata* and *Populus ciliata*.

Herbaceous wetlands

Herbaceous wetlands (often referred to as marshes) are widely distributed throughout south Asia in a variety of habitats. However, there are few detailed studies of these wetlands in India (see reviews by Gopal 1982, Gopal and Sharma 1982, Zutshi 1989, Kaul and Handoo 1989, Vyas and Garg 1989) whereas practically nothing is known about them in other countries of the region. In India also most of the studies relate to the Indo-Gangetic Plain only. Further, there is little information on the wetlands associated with rivers and streams. Therefore, only a brief account of the distribution and vegetation of different wetlands is possible.

Distribution and area. The distribution and extent of herbaceous wetlands is difficult to quantify chiefly because there has not been any detailed survey. One of the earliest accounts of the freshwater vegetation was provided by Biswas and Calder (1936) who mention the widespread occurrence of lakes, ponds, pools, puddles, and other waterlogged depressions throughout British India. This account was prepared primarily in light of the malaria epidemic since those areas were the breeding grounds for the mosquitoes. They gave a detailed account of the algae and other aquatic plants but the geographical distribution and area of different kinds of habitats were not provided. There are numerous published reports of aquatic vegetation in different districts. They almost all have an identical style of presentation with a very brief account of the location, a few casual remarks on the water and soil characteristics and lists of species. In some cases, vegetation is grouped into emergent, floating, floating leaved, and submerged species but no further ecological information is given. Few reports contain information on the duration of flooding, frequency, and magnitude of water level changes.

Further difficulty arises from what is considered to be a wetland. It has been emphasised earlier as well that vast majority of wetland habitats in south Asia are only periodically flooded and often remain dry for larger part of the year. Thus, the wetland estimates would vary according to the period of survey. Floodplains or riparian areas also are often not included in wetland surveys. Varshney and Singh (1976) conducted in 1973 a questionnaire survey of aquatic weeds in India. Though such a survey has the potential of providing valuable information, the information provided is more often not accurate as most questionnaires are returned by people relying on indirect sources. The weed survey was also not fully relevant to the wetland survey because all weed infested areas are not necessarily wetlands (e.g. large reservoirs and rivers infested with water hyacinth or *Salvinia*) and all wetlands do not have important weed infestations. The mere presence of a weed in a region also does not mean that the region has significant wetlands. A small roadside depression may support the growth of *Hydrilla* or *Typha* but it can hardly be considered as a wetland of any significance.

More recently, Biswas (1983) reported some figures for wetlands in India. He reported 1,193 wetlands from 274 districts out of a total 385 in the country, covering an area of 39,045 km^2. Of this, 22,508 km^2 is reported brackish and 3,663 km^2 marine. Areas with more than 5 m water depth have also been included. Most wetlands are reported to be permanent. In absence of information on the data base or the methodology that was used for the estimates we do not believe that they are very accurate.

We wish to reiterate that the vast majority of herbaceous wetlands is man made as these are associated with small and large reservoirs or develop in low lying areas waterlogged by seepage from irrigation channels (for example, wetlands in the Damodar Valley and the Chambal Command Area). In India alone, there are more than 1,550 large reservoirs covering a total area of more than 1.45 million ha and more than 100,000 small and medium reservoirs cover another 1.1 million ha (Sharma 1985). Numerous small reservoirs have been built in the drier regions of India, Pakistan, and Sri Lanka (Fig. 11, Fernando 1973, 1984). Some of the largest natural freshwater wetlands are of fluvial origin (oxbow type) and occur in the floodplains of large rivers. Important examples are Wular, Dal and other valley lakes in Kashmir, Kaziranga in Assam, Suraha tal (Singh and Swarup 1980), Gujar tal (Ambasht and Ram 1976), Ramgarh lake (Sinha 1969), and Chilwa lake (Srivastava 1973) in Gangetic plain, and lake Kolleru in Andhra Pradesh on the east coast of India (Seshavatharam 1978). Similar shallow water wetlands occur elsewhere also and are known as Tals in Uttar Pradesh and Madhya Pradesh, Chaurs in Bihar, and Bils in Assam, West Bengal, and other eastern states of India.

Habitat and vegetation. The diversity of wetland habitats and their vegetation in south Asia has recently been reviewed in detail by Gopal (1990). More than 400 species of vascular plants occur under permanent or periodic flooding in different climatic zones. Of these, grasses and sedges outnumber all other aquatic plants which are otherwise generally dominant.

1. *Wetlands asssociated with lotic waters.* The majority of the larger rivers are perennial but exhibit very large variation in their discharge during the year. The alluvial plain of the river Ganges has only a negligible slope (from Aligarh to Farakka, less then 10 cm km^{-1}) and therefore, the river waters flood large areas of land, up to a distance of more than 200 km on either side. In the lower reaches, almost the whole of deltaic region of Bangladesh is flooded. Likewise, extensive areas in other river basins are also flooded during the rainy season. However, in the mountainous areas of Himalaya,

Figure 11. Distribution of man-made reservoirs in Sri Lanka. Each dot represents at least one reservoir. Hatched coastal areas are mangroves (redrawn from Fernando 1985).

Western Ghats, and in Sri Lanka, floodplains of rivers and streams are narrow.

Along the rivers and streams on clayey humus-rich soils, in northeast India, Kerala, Andaman Islands, and some parts of sub-Himalayan region, riparian fringes subjected to continuous waterlogging are formed by thick stands of *Calamus tenuis*, *Neohouzeaua dullooa*, *Ochlandra wightii*, *Ochlandra travancorensis*, *Bambusa schizostachyoides*, and *Bambusa arundinacea*. These are referred to as "wet bamboo brakes" by Champion and Seth (1968).

Relatively small low lying areas only of the floodplains in the zones of high rainfall (e.g. "Terai" belt [foothills] of Himalaya, northeast India, Bangladesh, and parts of Western Ghats in peninsular India) are flooded throughout or most part of the year. Extensive reed marshes occur in these areas. Champion and Seth (1968) included these marshes as a subtype "Eastern Wet Alluvial Grasslands" under Freshwater Swamp Forests. A typical example of these wetlands lies in the Kaziranga National Park (Assam). Chaudhuri (1965) gave a brief account of these marshes in a report on grasslands of West Bengal. The common species were: *Phragmites karka*,

Vetiveria zizanioides, *Saccharum procerum*, *Saccharum spontaneum*, *Arundo donax*, *Coix lachryma jobi*, *Typha elephantina*, *Erianthus arundinaceus*, *E. munja*, *Themeda arundinacea*, *Imperata cylindrica*, *Narenga prophyrocoma*, and *Sclerostachya fusca*.

On the river banks subjected to short-term flooding during the rainy season, many herbaceous annual species develop after the floods recede. No detailed account of this vegetation is available. Species of *Ranunculus*, particularly *R. sceleratus*, are typical of such habitats. Most other plants (e.g. *Rumex dentatus*, *Polygonum glabrum*, *Phyla nodiflora*, *Alternanthera sessilis*, *Eclipta alba*, *Eclipta prostrata*, *Aeschynomene aspera*, *Cynodon dactylon* and many members of Cyperaceae) are similar to those of temporary standing water bodies.

2. *Wetlands associated with lentic waters.* Permanent waterbodies (lakes and reservoirs) are almost exclusively restricted to the Himalayan belt. Noteworthy among them are the high altitude and forest lakes in Kashmir Himalayas (Zutshi *et al.* 1980, Zutshi 1989, Kaul and Kandoo 1989), the lakes in the Kumaun region (Singhal and Singh 1978), and the lakes of Everest region and Pokhra valley in Nepal (Löffler 1969, Swar and Fernando 1979, Swar 1980). Two important lakes in Burma are lake Inle and lake Indawgyi. There are no natural lakes in Pakistan and Sri Lanka.

There are no macrophytes in lakes above about 2,500 m (Zutshi 1975). The vegetation in shallow littoral zones of lakes at lower altitudes comprises mostly of submerged macrophytes like *Potamogeton pectinatus*, *Potamogeton crispus*, *Hydrilla verticillata*, *Ceratophyllum demersum*, *Myriophyllum spicatum*, and *Vallisneria spiralis* (Purohit and Singh 1985). On the marginal waterlogged soils, emergents like *Polygonum hydropiper*, *Polygonum glabrum*, *Polygonum amphibium*, *Eleocharis palustris*, and *Acorus calamus* are often present. Occasionally, a floating leaved species (*Nymphoides cristatum*) is found.

Some shallow floodplain lakes are among the best known wetlands in India. Large water level changes result in annual drying of a greater part of these wetlands. They are very rich in their floristic diversity and many growth forms of wetland vegetation occur in different zones of these wetlands (Fig. 12). Plants common to all wetlands include emergents like *Typha angustata*, *Phragmites karka* and *Eleocharis palustris*, submerged species like *Hydrilla verticillata*, *Ceratophyllum demersum*, *Potamogeton pectinatus*, *Potamogeton crispus*, and *Utricularia flexuosa*, floating leaved species such as *Nymphaea stellata*, *Nymphoides peltatum*, and *Trapa bispinosa*, and free floating *Lemna paucicostata* and *Spirodela polyrhiza*. Other species vary in different climatic zones. In Kashmir wetlands, common species are *Phragmites australis*, *Acorus calamus*, *Scirpus lacustris*, *Sparganium erectum*, *Phalaris arundinacea*,

Figure 12. Freshwater marsh dominated by *Phragmites karka*. *Alternanthera philoxeroides* is seen in the foreground.

Typha latifolia, *Scirpus palustris*, *Sagittaria sagittifolia*, *Myriophyllum spicatum*, *Salvinia natans*, and *Lemna trisulca* (Kaul *et al.* 1978, Handoo and Kaul 1982). In the Gangetic plain and eastern India, more common species include *Eleocharis plantaginea*, *Cyperus corymbosus*, *Scirpus alopecuroides*, *Scirpus roylei*, *Paspalum distichum*, *Monochoria vaginalis*, *Monochoria hastata*, *Ipomoea aquatica*, *Potamogeton nodosus*, *Azolla pinnata*, *Vallisneria spiralis*, *Ottelia alismoides*, and *Najas major*. It is noteworthy that several exotic plants have invaded these wetlands and are often a dominant component of their vegetation. Most significant of these is water hyacinth (*Eichhornia crassipes*; Fig. 13). *Ipomoea fistulosa* is spreading rapidly in many areas, and *Salvinia molesta* causes problem in lake Kolleru.

There are also numerous large reservoirs whose shallow littoral zones (Fig. 14) have beds of similar submerged, floating leaved and short emergent macrophytes (especially grasses and sedges). Tall emergents like *Phragmites karka* and *Typha angustata* are often present. An interesting feature of many reservoirs is the frequent development of "sudds" or floating islands which are initiated by explosive growth of exotic weeds like water hyacinth. As

Figure 13. A dense stand of water hyacinth (*Eichhornia crassipes*) which often replaces natural vegetation in freshwater marshes.

large amounts of dead organic mass accumulates in between the dense floating stands, it is gradually colonised by emergent aquatic plants (e.g. *Rumex dentatus*, *Polygonum glabrum*, *Typha angustata* and *Phragmites karka*). These floating islands are grounded during the low water level period and float again with the rise in water level. An important example of these sudds is Keibul Lamjao (2,160 ha) in Loktak lake (Manipur) where rural people live and cultivate vegetables on them (Yadav and Varshney 1981). The floating islands in Dal Lake of Kashmir are however different as their development is aided by man by using cattails and reeds (Kaul and Zutshi 1966).

Many thousand small and large temporary ponds (including small irrigation reservoirs in arid and semi-arid regions) are a common feature of the landscape in south Asia. These temporary water bodies are flooded during the rainy season and remain under water for three to five months (occasionally longer) and support a wide variety of herbaceous vegetation. Large areas of low lying lands associated with irrigation systems such as those in the Damodar Valley (Kachroo 1956) and Chambal Command Area (Brezny 1970, Reeders *et al.* 1986) are also waterlogged for varying periods. Physiognomically it exhibits great uniformity all over the region though small differences occur in species composition. In general, the dominant taxa are *Typha elephantina*, *Saccharum spontaneum*, *Saccharum bengalense*, *Vetiveria ziz-*

Figure 14. A shallow pond during rainy season. Two zones of floating leaved and short emergent vegetation can be noticed.

anioides, *Paspalum distichum*, *Paspalidium geminatum*, *Echinochloa colona*, *Hygrorhiza aristata*, *Eleocharis plantaginea*, several species of *Cyperus*, *Scirpus* and *Juncus*, *Alternanthera sessilis*, *Bacopa monnieri*, *Phyla nodiflora*, *Polygonum glabrum*, *Polygonum amphibium*, *Veronica anagallis*, *Ludwigia scandens*, *Hygrophila auriculata*, *Limnophila heterophylla*, *Equisetum arv-*

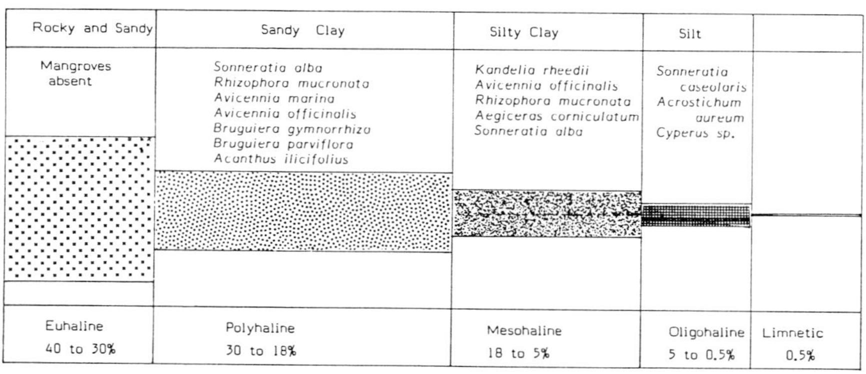

Figure 15. Zonation in mangroves in relation to texture of sediments and salinity gradients (redrawn from Untawale 1987a).

ense, *Equisetum ramossissimum*, *Potamogeton pectinatus*, *Potamogeton crispus*, *Potamogeton nodosus*, *Ceratophyllum demersum*, *Ceratophyllum echinatum*, *Hydrilla verticillata*, *Najas major*, *Najas graminea*, *Zanichellia palustris*, *Utricularia stellata*, *Utricularia flexuosa*, *Nymphoides indicum*, *Nymphoides cristatum*, *Marsilea minuta*, *Azolla pinnata*, and various lemnids (*Spirodela polyrhiza*, *Lemna paucicostata* and *Wolffia microscopica*).

In areas remaining under water for longer periods, *Phragmites karka*, *Typha angustata*, *Salvinia cucculata* and *Pistia stratiotes* are also common. *Eichhornia crassipes* is widespread throughout the region except in cold northern parts (Kashmir, Himachal Pradesh, Nepal, Bhutan and Sikkim).

Lastly, mention should be made of numerous fish ponds and extensive rice "paddy" fields which are important man-managed wetlands transformed from natural temporary wetlands (Jhingaran 1982). The total area under paddy cultivation is estimated at 38.6 million ha in India, 9.7 million ha in Bangladesh including more than two million ha under deep water rice (Whitton *et al.* 1988), and about 7.0 million ha in other countries in south Asia (Fernando 1984). A large variety of aquatic plants, often considered as weed, occurs in these habitats.

Associated fauna. Like mangroves, freshwater wetlands are also rich in faunal diversity. They are well known for their avifauna and fish many of which reside in wetlands only temporarily. Large number of insects belonging to different taxonomic groups depend on freshwater wetlands for completing their life cycle though the adults are not dependent on them. Further, numerous planktonic and benthic organisms represent all groups of invertebrates. Several amphibians and reptiles use wetlands as feeding or breeding grounds but few mammals only depend significantly on wetlands.

In south Asia, a total faunal inventory of any wetland has never been completed and the most investigated wetlands in the region are those around Srinagar in Kashmir (Kaul *et al.* 1978a, 1980a, Pandit 1980) and the Keoladeo Ghana National Park at Bharatpur (Salim Ali and Vijayan 1986, Vijayan 1988). Though there are many reports of zooplankton and benthic fauna of shallow water bodies (Adoni 1985, Fernando 1985), detailed studies have rarely been made. Gastropod molluscs are often abundant in freshwater wetlands and they are sometimes economically exploited as duck feed (Seshavatharam 1978). Among the vertebrates, the avifauna have been studied exhaustively by Salim Ali and Ripley (1983) and the studies on fishes are reviewed in detail by Jhingaran (1982), Fernando and Indrasena (1969), and Fernando (1985). In south Asia, 318 species of birds (about 26% of the total avifauna) representing about 40 families are dependent on different wetland habitats (Vijayan 1986). Besides these, more than 150 species of migratory birds also overwinter in wetlands of the region. The wetland at Bharatpur alone supports more than 350 species of birds of which more than 120 species are migratory.

Among other wetland animals, the rhinoceros (*Rhinoceros unicornis*) in alluvial marshes of Assam, swamp deer (*Cervus duvaucelli duvaucelli*) in terai region of upper Gangetic plains (sub-Himalayan tract), brown antler deer (*Cervus elde elde*) in marshes of lake Loktak (Manipur), wild buffalo (*Bubalus bubalis*) in Assam are important. The domesticated water buffaloes commonly wade into and often feed on herbaceous vegetation in wetlands.

General ecological features

Mangroves

Mangroves lie in the tropical and subtropical belt at the interface between land and sea. The characteristic mangrove vegetation owes its existence to the nutrient rich alluvium brought by the rivers, a perennial supply of freshwater along the deltaic coast, and its mixing with tidal sea water resulting in a gradient of salinity. The alluvial soils of mangroves are fine-textured unconsolidated loose mud or silt, rich in humus and sulphides (A. N. Rao 1987). Repeatedly flooded but well drained soils support good growth of mangroves but impeded drainage is detrimental. Geomorphology of the coastline is also important factor in their growth as the gently sloping areas, being less susceptible to erosion by waves favour mangroves. The balance between tidal supply of salts and their flushing with freshwater is, however, the major factor controlling the nature of the vegetation. The salinity gradient along which species with different tolerances to salinity levels are distributed, also contributes to species richness and variegated canopy of the mangrove forests. Frequency of flooding also affects the species composition through changes in salinity levels.

Among all wetland types in south Asia, mangroves have been ecologically investigated more than freshwater wetlands. An important climatic factor influencing the mangroves bordering the Bay of Bengal, particularly in the Sunderbans, are the frequent violent cyclones which result in upto 8 m high tidal waves (Blasco 1975). Thus salinity keeps on increasing in the area and much depends upon the inflow of freshwater which is uncertain depending upon the rainfall. The importance of nature and chemistry of soils has been shown in some studies (Blasco *et al.* 1986). Navalkar (1940, 1951), based on his studies around Bombay, showed the influence of soil nutrients on the occurrence of different species. He observed that *Avicennia alba* dominated on calcium-magnesium rich soils while *Acanthus ilicifolius* occurs on calcium-sodium rich soils and presence of potassium gives way to *Suaeda fruticosa*. In a field study, Karim *et al.* (1984) observed that the growth of *Avicennia*

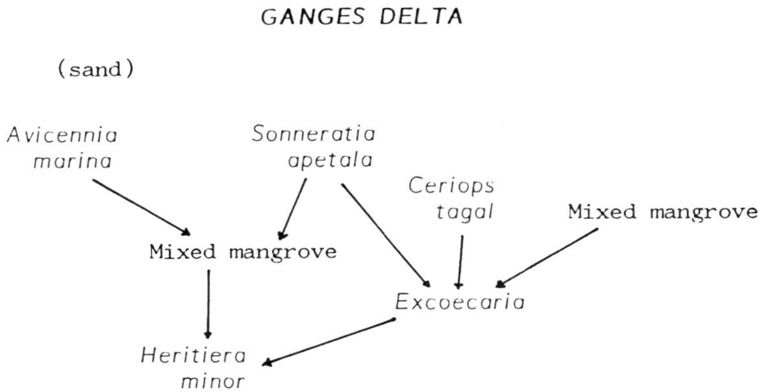

Figure 16. Successional trends in mangroves in Ganges delta (redrawn from Chapman 1970).

alba and *Avicennia officinalis* was directly correlated with the amount of silt deposition irrespective of its nutrient status.

The natural zonation of species along salinity gradient, salinity tolerance and its mechanism in different species, the successional changes in relation to salinity and tidal influences have received most attention. These studies have been reviewed in detail by Chapman (1970, 1976), Navalkar (1973), Walsh (1974), Sen and Rajpurohit (1982), Bhosale *et al.* (1983), Teas (1983, 1984), and Naskar and Guha Bakshi (1987). Recently, zonation in mangroves of Andaman and Nicobar islands has been described in detail by Garge *et al.* (1989).

Important aspects of zonation along salinity gradients in different mangrove areas of south Asia are described elsewhere in this chapter. In this section we focus on variables which correlate with zonation patterns. Zonation can be correlated with the frequency of flooding by tides as it directly affects salinity. For example, in Sunderbans, on the seaward more frequently flooded areas occur *Sonneratia apetala*, *Aegiceras corniculatum* and *Avicennia marina* whereas behind them under drier and more saline conditions the common taxa are *Xylocarpus moluccensis* and *Bruguiera gymnorhiza*. Untawale (1987a) has related the zonation with texture and salinity of the substrata (Fig. 15). Such zonation is often considered to represent a successional sequence and several workers have described successional trends in different mangroves (Figs. 16–18; Stamp 1925, Macnae 1968, Chapman 1970, 1976, Untawale 1987). However, arguments have been advanced in recent years to show that mangroves are indeed steady state "climax" communities (Lugo and Snedaker 1974, Lugo 1980).

Despite many reports on the distribution of mangrove species in different habitats with varying levels of salinity, the tolerance ranges of these species have not been investigated. The ecological responses of different plant and

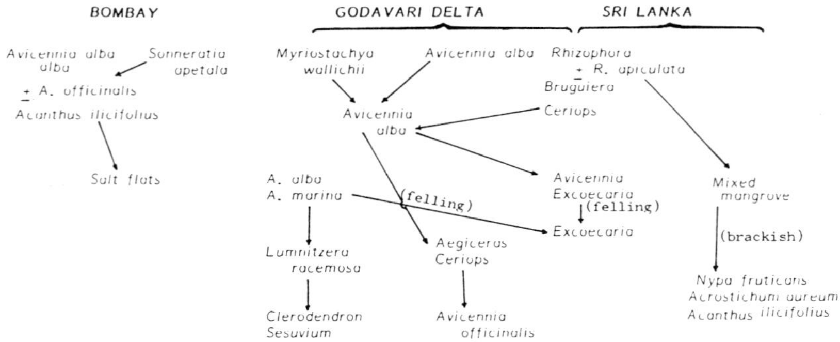

Figure 17. Successional trends in mangroves in Godavari delta, and on the coasts near Bombay and in Sri Lanka (redrawn from Chapman 1970).

animal species to other habitat factors such as frequency and duration of flooding, sedimentation, nutrients, and various stresses like grazing and pollution have also not been studied.

Mangroves are among the highly productive ecosystems (Lugo 1980, Ong 1986). However, there is no information on their biomass and primary production in south Asia though some studies have brought out that mangroves have many photosynthetic characteristics of the C4 plants (Joshi *et al.* 1975, Bhasale and Karadge 1975, Bhosale *et al.* 1983). Few observations have been made on chemical composition of several plant species (Bhosale *et al.* 1983), litter production (Yadav and Choudhury 1986), decomposition and microbial communities (Agate *et al.* 1988) but there is no detailed study of nutrient dynamics.

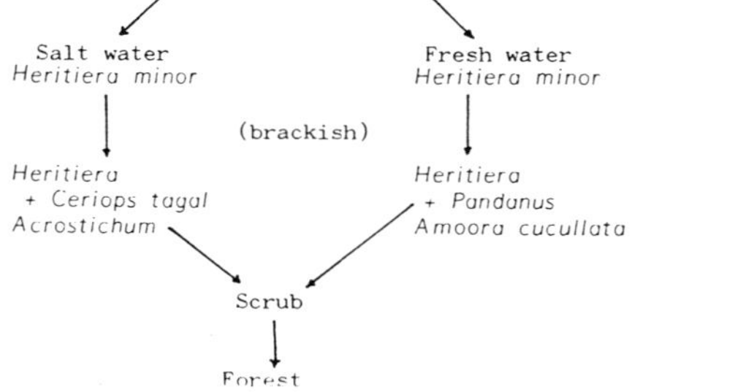

Figure 18. Successional trends in mangroves in Irrawady Delta (redrawn from Chapman 1970).

Freshwater wetlands

Our understanding of the ecology of freshwater wetlands is limited to standing water marshes, mostly temporary and in India. We do not know of any important studies from countries other than India, except for some observations on aquatic weeds, particularly water hyacinth. There is practically no ecological information on swamps or riparian wetlands, except for some comments on successional stages based on observed zonation (Dakshini 1960a,b, 1965). Ecological studies on marsh vegetation provide some insight into the ecological requirements of common species, and dry matter production, and nutrient dynamics of the herbaceous wetlands. These studies have recently been reviewed in detail (Gopal 1990).

The species composition, phenology and growth are largely controlled by the seasonality of climate, and the water level. The seeds and propagules of the aquatic and marsh plants sprout soon after the rains set in. After the first set showers one can find young sprouts and seedlings in all shallow depressions. Their survival however is very much dependent upon the intensity of rain in the following weeks. Heavy rains for long periods cause stirring of the substratum and the resultant turbidity does not allow the growth of submerged species. The lemnids and nymphaeids are most common in such cases. The shallow margins are quickly colonized by the sedges and grasses or other emergent species. Towards the end of the rainy season, the species diversity increases as more species establish themselves. The amount of precipitation and the shape of the basin of the waterbody influence to some extent the duration of the vegetation. Depending upon the water availability, the wetland vegetation may continue upto the early summer season or may die out even during the winter season.

Very few experimental or detailed field studies have been made on plant responses to different environmental factors. The distribution of both submerged and emergent species is governed to some extent by the magnitude and duration of water depth. *Phragmites karka* and *Typha angustata* require continuous waterlogging or submergence of the substrate though *P. karka* may tolerate drying for short periods only. On the other hand *Typha elephantina* colonizes even dry uplands near water bodies as it can draw its water requirement through deep seated rhizome system (Gopal 1982b). Water regimes have been observed to affect both growth and tuber formation in *Eleocharis palustris* (Pandey 1969). Maximum growth of *Paspalum distichum* is obtained on waterlogged soils and it declines progressively with increase in water depth (Davis and van der Valk 1989). The importance of alternate dry and submerged phases for germination and sprouting has been demonstrated in many species (Mall 1955). Generally, this is related to the requirement of high light intensity and/or alternating high and low temperatures for germination. For example, in *Typha angustata* seeds germinate under high light intensities and seedlings establish only under shallow water (Sharma and Gopal 1979a,b). Water level changes also induce the formation of vegetative propagules like turions in many submerged plants like *Hydrilla verticillata* and *Potamogeton crispus*.

Only preliminary reports are available on the effect of other habitat factors on wetland plants. Misra (1946) emphasized the importance of edaphic factors (exchangeable bases, redox potential, C:N ratio) in distribution of wetland plants, and Ramakrishnan (1965) grouped wetland communities on the basis of water depth and soil characteristics. Gopal (1969) suggested that the C:N ratio of soils may even affect reproduction in some wetland plants. The effects of temperature and photoperiod on growth and flowering have been reported in several species (Choudhuri 1966, Tripathi 1969, Gopal 1987). Flowering in *Typha angustata* occurs at different times of the year at different latitudes and this may be correlated with the temperature regimes. While in the northern latitudes flowering occurs over a short period of summer months, the plants flower almost throughout the year in southern latitudes (B. Gopal, personal observations).

There is hardly any information on the biotic interactions in wetland communities (Pandit 1980). It has been reported that growth and dry matter production are adversely affected at high densities in emergents like *Eleocharis palustris* and *Typha angustata*. On the other hand, in free-floating plants like *Salvinia molesta*, *Eichhornia crassipes*, and *Pistia stratiotes*, growth is not affected by high densities (Gopal 1976, 1987) as the plants compensate by vertical growth. Mehta and Sharma (1975) reported that *Brachiaria mutica* outcompetes *Typha angustata* in mixed cultures. Often wetland vegetation is grazed upon by cattle, especially water buffaloes but there are no reports on the influence of such grazing. Salim Ali and Vijayan (1986) have, however reported the increasing spread of *Paspalum distichum* after grazing was completely stopped in the marshes of Keoladeo National Park.

There are many reports on dry matter production and nutrient dynamics in herbaceous wetlands in India (for reviews, see Gopal 1990). Most of the studies report on standing crops or production based on maximum standing crop without taking into consideration the mortality during growth period, and translocations between below- and above-ground organs. Emergents like *Phragmites karka*, *Eleocharis plantaginea*, and *Typha* species are most productive of all wetland plants. Their annual production may be as high as 8–10 kg m^2. Among other highly productive species noteworthy is water hyacinth which may also produce upto 10 kg m^2 in highly eutrophic water bodies. The submerged macrophytes have very low annual production (200–600 g m^2) and other species fall in between these two limits (Gopal *et al.* 1978).

Studies on nutrient dynamics reveal that different wetland species accumulate large amounts of macro- and micro-nutrients from the water and/or sediments (Kaul *et al.* 1980b, Gopal and Kulshreshtha 1980, Kulshreshtha and Gopal 1982a). Following death and submergence, significantly high amounts of nutrients are leached out into water and further decomposition is rapid and the rate is influenced to some extent by the C:N ratio of the decomposing litter (Kulshreshtha and Gopal 1982b, Murty 1987, Davis and van der Valk 1989). There is also some evidence to suggest that a part of nutrients, particularly N and P, is cycled internally between the below- and above-ground organs of emergent plants like *Typha elephantina* (Gopal and Sharma 1984). Further, water depth and its seasonal changes also influence the rate of decomposition (Sharma and Gopal 1982, Davis and van der Valk 1989).

Utilization and human impacts

Traditional uses

Wetland habitats have been an integral part of the life of man in the region. The folklore and old Indian literature are full of references to the freshwater wetland plants. The floodplains of river Yamuna were once extensive enough where the herds of cows grazed in company of Lord Krishna. The reeds (*Phragmites karka*) were and are still widely used for thatching roofs and making screens. Cane (*Calamus tenuis*) well known for its strength, is used variously in making furniture. Sacred lotus (*Nelumbo nucifera*), held in reverence for being the seat of Gods and Godesses, was once abundant in shallow water bodies. The seeds of lotus and *Euryale ferox* are still a delicacy in India and thousands of kilograms are harvested every year. Besides seeds, the petioles and rhizomes are used as vegetable. Wild rice and other cereal grains from wetland habitats (*Panicum* sp., *Paspalum scrobiculatum*, *Echinochloa colona*, *Coix lachryma-jobi*) have been traditionally food of not only the poor but even the rich on certain religious days. *Trapa bispinosa* is cultivated throughout the region for its starchy fruits (Fig. 19). *Cyperus esculentus* and *Scirpus grossus* are cultivated for the tubers. *Typha elephantina* and *Typha angustata* are extensively used for thatching, mats, ropes, fodder (Fig. 20) and other decorative articles. *Ipomoea aquatica* is common vegetable. Roots of *Vetiveria zizanioides*, a grass, yield an aromatic oil and are also used in making screens to keep the houses cool with fragrance. Several other grasses and sedges are removed at the end of the growing season for fodder or are otherwise grazed by cattle, particularly buffaloes. Many wetland plants (e.g. *Acorus calamus*, *Bacopa monnieri*, *Eclipta alba Hygrophila auriculata*) are used extensively in indigenous medicine. Besides vegetation, wetlands have been valued also for fish and other animals. Even today, in several parts of the country, people live almost entirely on the

Figure 19. Cultivation of *Trapa bispinosa* in freshwater marshes.

Figure 20. Typha angustata harvested from a floodplain wetlands is used as fodder for elephants and for thatching and mat making.

wetland harvests; for example, around lake Kolleru, thousands of people depend on fisheries and duck farming, and use huge quantities of snails (*Pila globosa*) harvested from the wetland as feed for ducks.

Mangroves have been similarly exploited since ages though there is little information about their traditional uses (Walsh 1977). A significant section of human population in south Asia is settled in coastal areas near deltas of all major rivers. Lives of these people depend entirely on marine and estuarine fisheries which surely owe their existence to mangroves. These people depend on mangroves also directly for their requirements of fuelwood, building material, fodder, and minor products (Navalkar 1961, 1962). The leaves of *Nypa fruticans* are extensively used for thatching roofs. Fruits of *Sonneratia caseolaris* are eaten and the pneumatophores are used in cork in fishing. The wood from *Heritiera minor* and *Excoecaria agallocha* has since long been used for ship and boat making (Blasco 1977). The use of several mangrove species for medicine is mentioned in folklore (Naskar and Guha 1987, Bakshi 1987). However the "inhospitable environment of deltas" (loose muddy soil, frequent cyclones, abundance of stilt roots and pneumatophores, and crocodiles) greatly restricted human exploitation of mangroves in the past.

Direct and indirect human impacts

During recent decades, the rapid increase in human population and demand for natural resources for food (Fig. 21), fuel, and fodder have resulted in rapid deterioration and decline of all kinds of wetlands throughout the south Asian region. Diverse human activities on adjacent land and in water further aggravate the problem of wetland decline. It is interesting to point out also the impacts of introduced plants and animals on natural biota and processes in both mangroves and freshwater wetlands. Mukherjee and Tiwari (1984) listed many plants and animals which were introduced or immigrated into the Sunderbans mangroves and which caused changes in the natural plant and animal populations. Similarly, water hyacinth is the most significant exotic in freshwater wetlands throughout south Asia, and has drastically influenced the native biota (Gopal 1987, Mitchell and Gopal 1990).

Overexploitation and wetlands resources

Mangrove forests have been exploited most during the past 200 years for timber and fuelwood. The process of decimation of mangroves in the Indian subcontinent started with the official permission to clear Sunderban forests for human habitation and cultivation in 1771, by the Collector General of the East India Company. It was actually meant to collect valuable timber. Many species like *Sonneratia alba* and *Xylocarpus moluccensis* yield high quality timber for furniture. The tree trunks of *Heritieria fomes* and *Bruguiera*

Figure 21. Vegetables are cultivated on the floodplains after floods recede. The cultivators live on the floodplains in huts made from bamboos, *Typha elephantina*, *Phragmites karka* and *Saccharum spontaneum*.

gymnorhiza make excellent transmission poles for telegraph and electricity lines (Banerji 1958a, b). *Excoecaria agallocha* provides good pencil wood.

Mangroves are an important source of fuelwood and charcoal in all countries of south Asia. In Bangladesh mangroves have been exploited as a major source of revenue which increased about seven fold from 1970 to 1979 (Ahmad 1984). In Burma, nearly 350,000 ton of charcoal is made and about 40,000 ton is exported annually from the deltaic mangrove forests (Than Htay and Saw Han 1984). The bark of *Rhizophora mangle* has been used for extraction and export of tannin for leather industry since early nineteenth century. It is reported that the bark contains more tannin than the leaves and fruits. Several other species like *Ceriops candolleana*, *Kandelia candel*, and *Excoecaria agallocha* are also used for tannin (Mudaliar and Kamath 1952, Navalkar 1961, 1962, Venkatesan 1966) though Blasco (1977) says that these are not used any more. In West Bengal moiety of Sunderbans alone, the forest area has been reduced to almost half during the past 50 years.

Another important product from the mangrove forests is honey (Krishnamurthy 1990). The honey bee (*Apis dorsata*) prefers, in decreasing order of importance, *Excoecaria agallocha*, *Avicennia alba*, *Ceriops tagal*, *Rhizophora mucronata*, and *Heritieria fomes* for its combs. It is estimated that in Indian Sunderbans alone 60 to 180 ton of honey and 4–13 ton of bees wax is extracted every year (Chakrabarti and Chaudhuri 1972).

Figure 22. Cattle, sheep and goats grazing in seasonal wetlands during dry season.

Fishing and grazing by domestic cattle are other important human activities within wetlands, both saline and freshwater (Fig. 22). Enormous quantities of fish, shrimps, and prawns are harvested every year from coastal mangrove areas and estuarine lakes like lake Chilka and Pulicat (Jhingaran 1982, Krishnamurthy and Prince Jeyaseelan 1984). Numerous shallow lakes and marshes are important fishing areas (see Jhingaran 1982). Mangroves as well as the saline wetlands in the arid region of Saurashtra (Kutch) in western India are extensively used for cattle grazing (Kulkarni and Jungadd 1959, Blasco 1977, Krishnamurthy and Prince Jeyaseelan 1984). Floodplains and temporary marshes are favoured grazing lands especially during the long dry season. Large amounts of forage is also harvested, and often even the subterranean parts are scraped from these wetlands. Excessive grazing and forage removal not only depletes the herbaceous vegetation but also has impacts on other biota and soil properties which may be beneficial or harmful for the wetland. For example, grazing by domestic cattle has been regarded as the most serious problem in the wetland at Bharatpur (Keoldeo Ghana National Park, known for migratory birds) but their exclusion caused significant changes in vegetation.

Reclamation and conservation of wetlands

Most natural wetland habitats, both saline and freshwater, have been wholly reclaimed for different reasons or have been converted into fish ponds and paddy fields (Fig. 23). Regular cultivation in mangroves started in late 18th century, east of Calcutta in Matlah-Bidyadhari basins (Mukherjee and Tiwari 1984). Further reclamation of land by deforestation in 1830 around Bakarganj

Figure 23. Natural wetland areas in lake Kolleru (India) are dredged for conversion into fishponds.

paved the way for human colonization. Later, the forests were removed and land reclaimed for paddy cultivation and pisciculture (Rao 1959). These reclaimed areas are the mainstay of fisheries in the Sunderbans today (Ahmad 1966, Rabanal 1977). Ponnamperuma (1984) has indeed advocated the cause of conversion of mangroves into paddy fields to meet the growing good requirements in south and southeast Asia. In other saline wetlands brackish water fisheries and shrimp culture are being promoted (Jhingaran 1982). The most affected natural freshwater wetlands are perhaps those in the areas of the present urban centres. These have been filled or drained to bring land under habitation. This has happened to the floodplains of all major and medium rivers where the most important urban settlements are located today and which are preferred sites for landfill from urban solid wastes.

Indirect impact of human activities

Equally serious impacts on wetlands are made indirectly through activities which cause changes in their hydrological regimes. Most important of these are the river flow regulation projects aimed at flood control and water storage for irrigation or hydropower, and the denudation of catchment areas of the waterbodies by deforestation, excessive grazing, agriculture, and similar land based activities which increase soil erosion and the consequent inflow of silt and nutrients into the wetlands.

As pointed out earlier, in the Indian part of Sunderbans freshwater contribution through the Ganges is relatively small and is becoming still lower. In

Figure 24. Floodplains and other wetlands are used as landfill sites for disposal of solid wastes.

absence of flushing, the tributaries in the deltaic region become silt-laden and clogged, further reducing freshwater flow and increasing salinity of the soils. The strain and stress on mangroves has led to changes in species composition, their dominance, lower production, and diminished luxuriance. For example *Nypa fruticans*, *Heritiera fomes*, and *Phoenix paludosa* have almost disappeared from the Indian part of Sunderbans, and are replaced by *Excoecaria* species. The increasing salinity has affected also the paddy yields on reclaimed soils. The Dampiar-Hodges Line (based on salinity values) which once formed the natural frontier between the agricultural lands and the mangroves, has been pushed southwards and has lost its significance. The impacts of Mettur dam in Cauvery river basin and the consequent reduced freshwater supplies on the vegetation and fisheries in Pichavaram mangroves have been discussed by Krishnamurthy and Prince Jeyaseelan (1984).

Shallow lakes and marshes have filled up rapidly and the characteristic wetland vegetation has disappeared together with the dependent fauna. The floodplains are directly affected through construction of dams and barrages as the flooding regimes (frequency, duration, amplitude, and timing) are altered both upstream and downstream. Channelisation of rivers and streams has the same effect on floodplain wetlands.

Further important impacts on wetlands are made by discharge into the waterbody (or wetland itself) of domestic sewage and industrial effluents, and input of pesticides, and herbicides with the runoff from surrounding landscape (Fig. 24). Other human activities like washing, bathing, and recreation in water also adversely affect the wetland biota. Though the impact of sewage disposal and recreational activities on the decline of wetlands in Kashmir valley, particularly those around Srinagar is well known (Pandit and Fotedar 1982), effects of different human activities have rarely been investigated. Ramamurthy (1984) reported large scale fish mortality in mangroves on the Arabian seacoast of India, due to pollution by chlorinated hydrocarbons and organochlorine residues from the surrounding agricultural lands.

Oil spills from tankers and smaller transport boats as well as offshore oil exploration are a growing pollution problem in coastal areas, estuaries, and lower stretches of major rivers. Field and laboratory studies show that petroleum products cause mortality of seedlings, damage leaves and roots, block lenticels and retard growth of mangrove plants like *Avicennia officinalis* and *Rhizophora mucronata* (Jagtap and Untawale 1980, Jagtap 1985).

Management and conservation

Wetlands have provided for millenia throughout human history food, fodder, fiber and fuel. More than half of the world's population still depends on modified and intensively managed wetlands which provide rice and fish. However, growing dependence of mankind on land and terrestrial resources resulted in focus on wetlands as wastelands and their treatment as impediments to economic development. After large areas of natural wetlands were totally lost by drainage and land fills or were highly degraded by other human activities, their values and functions are being rediscovered. There is resurgence of interest in natural and semi-natural wetlands since the Ramsar convention (IUCN 1971) highlighted their value as habitats for wildlife, especially waterfowl, and called for conservation of internationally important wetlands. It is now well realised that wetlands support a large diversity of biota of which many are also economically important to mankind and that they perform many important functions like trapping of silt and nutrients, and regulating water flows, and have several ecological and aesthetic values which are related to their characteristic hydrology and structural attributes (Mitsch and Gosselink 1986, Patten *et al.* 1990). In this context both management and conservation of wetlands have received considerable attention also in the countries of south Asia.

Wetland management

Management is defined as a deliberate interference in a system to achieve desired organisation of its structural components and to optimise certain desirable functions and eliminate or minimise the undesirable functions. Thus, specific objectives are required to be laid down before developing appropriate management strategies. Wetlands considered valuable for maintaining biotic diversity or recreation cannot be used for wastewater treatment as the two values are not compatible. Management for maximising yields of biological resources like plant and fish necessarily alters the natural functions and values. It is important to understand the ecosystem processes in a wetland and the factors responsible for maintaining characteristics which impart it particular values. Further, it is essential to understand as well the impact of overexploitation and specific management activities (e.g. artificial regulation of water levels, manipulation of structure, introduction of plant and animal species) and animal activity (e.g. grazing). Very little information is available on these impacts on both saline and freshwater wetlands of south Asia. Therefore, most management practices are based on knowledge of wetland ecosystems outside the region, and there is much ad-hocism in policies and decisions concerning wetland management.

Mangroves have received most attention for their management as National Mangrove Committees were set up in concerned countries of the region (except Burma) under the UNESCO's programme on Integrated Management of Coastal Systems (COMAR). These Committees have evolved strategies and national plans for management of mangroves in respective countries (Untwale 1987a,b, Ansari 1987, Jayewardene 1987). There is great consciousness about the need to "integrate all ecologic, environmental and socio-economic components" in alternative schemes for mangrove development. Suggestions have been made to allocate specific areas for preservation, sustained yield, and conversion to different land uses (aquaculture, paddy cultivation, port and harbour development, and human settlements). In some countries, afforestation of degraded mangrove areas has been undertaken. However, emphasis continues to be laid on "development" for economic gains from paddy, coconut, fishery and, aquaculture (shrimp and prawn). Intrusion of salinity from sea is considered undesirable but little attention is paid to ensuring natural regimes of freshwater flow which is regulated upstream for other uses.

Management of freshwater wetlands has not received serious attention in any country of the region. This is mostly due to the fact that large majority of freshwater wetlands are seasonal; and their areal extent, biotic resources and ecological functions have not been adequately documented. Nature plays its own role in their degradation as large temporal and spatial variability in rainfall causes periodic droughts and floods. Wetlands become the first casuality when measures are taken to ensure supply of water (for domestic and agricultural needs) and safety of human lives. The management is therefore confined to a few wetland areas which have been identified for their value as wildlife habitats and are protected from human exploitation. However, there is no management policy or plan for freshwater wetlands in general.

Conservation status

Though the conservation movement is relatively recent, protection of wildlife, and natural landscapes has been a part of culture of the people in the region. People have always realised the value of conservation and nature reserves have been set up since ages. Many wetland areas were conserved as wildlife sanctuaries, with appropriate legislative measures as well, long before wetlands drew attention elsewhere. Sunderbans of Khulna district (now in Bangladesh) were declared a reserve in 1875. The area was prohibited by law for human settlement or colonization except for a few fishermen and forest guards. Keoladeo National Park at Bharatpur which is the most famous wetland known for migratory birds, especially Siberian crane (*Grus leucogeranus*) had been created by the former ruler, about 200 years ago, out of the floodplains of two seasonal rivers which were dammed to control flooding in the region. Kaziranga National Park in northeast India, the only habitat of unicorned rhino (*Rhinoceros unicornis*), representing an alluvial marsh dominated by *Phragmites karka*, was declared a rhino sanctuary long ago. Distributed throughout India there are 53 national parks and 247 wildlife sanctuaries of which many are important saline and freshwater wetland sites.

During recent years, great interest has been shown in conservation of wetland areas in south Asia. A part of Indian Sunderbans has been fully protected as Tiger Reserve since 1973. Indian Sunderbans have recently been designated by the Department of Environment, Govt. of India, as Biosphere Reserve together with three other saline wetlands (Anonymous 1987b) namely, North Andamans (1,375 km^2), Rann of Kutch (5000 km^2), and Gulf of Mannar (10,500 km^2). Two freshwater wetlands, Kaziranga (760 km^2) and Manas (600 km^2) have also been named as Biosphere Reserves.

The Ramsar Convention of the International Union for Conservation of Nature and Natural Resources (IUCN 1971) has also furthered the cause of wetland conservation in this region. Pakistan and India joined the Ramsar Convention in July 1976 and October 1981 respectively. Pakistan placed nine wetland sites (Keenjhar, Thanedarwala, Haleji, Khabbeki, Drigh, Kheshki, Tanda, Kandar and Malugul Dhand) totalling an area of 210 km^2 under the convention (Anonymous 1984a). India designated two wetlands, lake Chilka (1,165 km^2) and Keoladeo National Park, Bharatpur (29 km^2) under the convention (Anonymous 1984b). The Department of Environment, Government of India, has since then identified some more wetland sites for conservation in different states (Anonymous 1989). These include (State in parenthesis): lake Wular (Kashmir), lake Ujni (Maharashtra), Ashtamudi estuary and Sasthamkotta (Kerala), Lake Kolleru (Andhra Pradesh), lake Loktak (Manipur), Upper lake, Bhopal (Madhya Pradesh), Pichola and Sambhar lakes (Rajasthan), lakes Harike, Sukhna and Kanjli (Punjab), Nal Sarovar (Gujarat), lake Renuka (Himachal Pradesh) and Kabar lake (Bihar). Of these, Wular, Harike, Sambhar and Loktak have recently been placed under Ramsar Convention (Anonymous 1990).

Except for some areas of Sunderbans in Bangladesh, mangroves have not yet been conserved in Sri Lanka and Pakistan. There is no other information available concerning the conservation status of wetlands in the Indian subcontinent.

Before concluding this survey, it is important to record that whereas the magntiude of loss and conversion of natural wetlands is difficult to be estimated, large areas of freshwater wetland have been created through water resources development programmes. Innumerable reservoirs in arid and semi-arid regions have created wetlands out of totally dry landscapes. Construction of dams on major rivers also result in development of wetlands in the shallow littoral areas of reservoirs upstream. The example of Ujni reservoir (near Poona, Maharashtra) is also noteworthy. Rapid silting of the shallow littoral area of the reservoir within a few years of its filling, produced a wetland which attracts large populations of avifauna. It is now recommended for conservation. Most of the wetlands brought under conservation are man-made. It appears therefore that wetlands can be readily created and possibly restored. However, the floodplain wetlands cannot be created or restored once they are lost by regulating waterflows. These wetlands have so far been ignored and deserve most urgent attention.

Finally, all natural and semi-natural wetlands in south Asia are under great stress from both natural and human forces. Though mangrove wetlands are better understood than the freshwater wetlands, the management and conservation of all wetlands requires better understanding of the ecosystem processes and their responses of various impacts. Detailed investigations are therefore urgently needed.

Acknowledgments

We are grateful to several colleagues and friends for providing important literature on studies in this region, and wish to single out Dr. M. Vannucci, Dr. L. J. Bhosale, Dr. A. G. Untawale, and Dr. V. S. Vijayan for their help. We are thankful to Dr. Dennis F. Whigham and Dr. D. Dykyjova for their valuable suggestions on earlier drafts of this paper. Thanks are also due to our respective institutions for providing facilities.

References

Abeywickrama, B. A. (1966) The estuarine vegetation of Ceylon. p. 207–210. *In*: Humid Tropics Research. Scientific Problems of the Humid Tropical Zone Deltas and Their Implications. Proceedings of the Dacca Symposium. UNESCO, Paris, France.

Adoni, A. D. (ed.) (1985) Pure and Applied Limnology: Retrospect and Prospect. Proceedings of a National Symposium. Botanical Society, H. S. Gaur University, Sagar, India. 243 pp.

Agate, A. D., Subramanian, C. V. and Vannucci, M. (eds.) (1988) Role of Microorganisms in Nutrient Cycling of Mangrove Soils and Waters. UNDP/UNESCO Regional Project on Research and its Application to the Management of the Mangroves of Asia and the Pacific (RAS/86/120), New Delhi, India. 118 pp.

Ahmad, N. (1966) Fish and fisheries of the Sunderbans. p. 271–275. *In*: Humid Tropics Research. Scientific Problems of the Humid Tropical Zone Deltas and Their Implications. Proceedings of the Dacca Symposium. UNESCO, Paris, France.

Ahmad, N. (1984) Some aspects of economic resources of Sunderban mangrove forest of Bangladesh. p. 644–651. *In*: E. Soepadmo, A. N. Rao, and D. J. Macintosh (eds.). Proceedings of the Asian Symposium on Mangrove Environment: Research and Management. Percetakan Ardyas, Kuala Lumpur, Malaysia.

Anonymous (1959) Proceedings of the Mangrove Symposium, Government of India Press, Faridabad, India.

Anonymous (1984a) National Report of Pakistan. p. 381–390. *In*: Convention on Wetlands of International Importance Especially as Waterfowl Habitat. Proceedings of the Second Conference of the Contracting Parties. Ramsar Convention Bureau, International Union for Conservation of Nature and Natural Resources, Gland, Switzerland.

Anonymous (1984b) National report of India. p. 337–343. *In*: Convention on Wetlands of International Importance Especially as Waterfowl Habitat. Proceedings of the Second Conference of the Contracting Parties. Ramsar Convention Bureau, International Union for Conservation of Nature and Natural Resources, Gland, Switzerland.

Anonymous (1987a) Mangroves in India: Status Report. Ministry of Environment and Forests. Government of India, New Delhi, India. 150 pp.

Anonymous (1987b) Biosphere Reserves. Ministry of Environment and Forests. Government of India, New Delhi, India. 250 pp.

Anonymous (1989) Conservation of Wetlands in India. Ministry of Environment and Forests. Government of India, New Delhi, India. 67 pp.

Anonymous (1990) India names four new Ramsar sites. Asian Wetland News 3(2): 4.

Ansari, T. A. (1987) Country reports: Pakistan. p. 151–174. *In*: R. M. Umali, P. M. Zamora, R. R. Gotera, R. S. Jara, and A. S. Camacho (eds.). Mangroves of Asia and the Pacific: Status and Management. Technical Report of the UNDP/UNESCO Research and Training Pilot Programme on Mangrove Ecosystems in Asia and the Pacific (RAS/79/002). National Mangrove Committee of Philippines, Quezon City, Manila, Philippines.

Arulchelvam, K. (1968) Mangroves. Ceylon Forester 8: 59–62.

Banerji, I. (1958a) The mangrove forests of the Andamans. World Forestry Congress 3: 425–430.

Banerji, I. (1958b) The mangrove forests of the Andamans. Tropical Silviculture 20: 319–324.

Banerjee, A. K. (1964) Forest of Sunderbans. p. 166–175. *In*: Centenary Commemoration Volume. West Bengal Forests. Divisional Forest Office, Planning and Statistical Cell. Calcutta, India.

Bhosale, L. J. (ed.) (1986) The Mangroves. Proceedings of the National Symposium on Biology, Utilisation and Conservation of Mangroves. Shivaji University, Kolhapur, India. 558 pp.

Bhosale, L. J. and Karadge, B. A. (1975) Photosynthetic efficiency and productivity of mangroves of western India. Bulletin Department of Marine Science. University of Cochin 7: 205–212.

Bhosale, L. J., Waghmode, A. P. and Kotmire, S. Y. (1983) Biology of mangroves in Indian coastlands. Indian Review of Life Science 3: 265–286.

Biswas, B. (1983) India. p. 108–109. *In*: Proceedings of the Heilligenhafen Conference. IUCN, Gland, Switzerland.
Biswas, K. and Calder, C. C. (1936) Handbook of Common Water and Marsh Plants of India and Burma. Government Publications, Calcutta, India. 243 pp.
Blanford, H. F. (1889) A Practical Guide to the Climates of India, Ceylon and Burma. London, United Kingdom.
Blasco, F. (1975) Les mangroves de l' Inde (The mangrove of India). Institut Français de Pondichéry, Travaux de la Section Scientifique et Technique, Pondichéry 14: 1–175.
Blasco, F. (1977) Outlines of ecology, botany and forestry of the mangals of the Indian subcontinent. p. 241–260. *In*: V. J. Chapman (ed.). Ecosystems of the World. I. Wet Coastal Ecosystems. Elsevier Science Publishers, Amsterdam, The Netherlands.
Blasco, F., and Caratini, C. (1973) Mangrove de Pichavaram: Phytogeographie et palynologie. Travaux et Documentation Géographique Tropicale 8: 164–185.
Blasco, F., Kerrst, R. and Marius, C. (1986) Considerations on some ecological factors influencing the biology of Indian mangrove. p. 135–146. *In*: L. J. Bhosale (ed.). The Mangroves. Proceedings of National Symposium on Biology, Utilisation and Conservation of Mangroves. Shivaji University, Kolhapur, India.
Blatter, E. (1905) The mangrove of the Bombay Presidency and its biology. Journal of the Bombay Natural History Society 16: 644–656.
Blatter, E., McCann, C. and Sabnis, T. S. (1927) The flora of the Indus delta. Journal of the Indian Botanical Society 6: 31–47; 57–78; 115–132.
Blatter, E., McCann, C. and Sabnis, T. S. (1928) The flora of the Indus delta. Journal of the Indian Botanical Society 7: 22–43; 70–96.
Brezny, O. (1970) Final Technical Report on Aquatic Weed Control in Chambal Command Area. United Nations Development Program, New Delhi, India. 68 pp.
Caratini, C., Blasco, F. and Thanikaimoni, G. (1973) Relation between the pollen spectra and the vegetation of a south Indian mangrove. Pollen et Spores 15: 281–292.
Chakrabarti, K. and Chaudhuri, A. B. (1972) Wildlife biology of the Sunderbans forests: Honey production and behaviour pattern of the honey bee. Science and Culture 38: 269–276.
Chakraborty, S. K. and Choudhury, A. (1986). Distribution of fiddler crabs in Sunderbans mangrove estuarine complex, India. p. 467–472. *In*: L. J. Bhosale (ed.). The Mangroves. Proceedings of National Symposium on Biology, Utilisation and Conservation of Mangroves. Shivaji University, Kolhapur, India.
Champion, H. G. (1936) A preliminary survey of the forest types of India and Burma. Indian Forest Records New Series 1: 1–286.
Champion, H. G. and Seth, S. K. (1968) A Revised Survey of the Forest Types of India. Manager of Publications, Delhi, India. 404 pp.
Champion, H. G., Seth, S. K. and Khattak, G. M. (1965) Forest types of Pakistan. Pakistan Forest Institute, Peshawar, Pakistan.
Chapman, V. J. (1970) Mangrove phytosociology. Tropical Ecology 11: 1–19.
Chapman, V. J. (1974) Salt Marshes and Salt Deserts of the World. Cramer Lehre, 2nd ed., Interscience Publishers, New York, New York, U.S.A. 392 pp. (complemented with 102 pp.).
Chapman, V. J. (1975) Mangrove Vegetation. J. Cramer, Stuttgart, West Germany. 425 pp.
Chapman, V. J. (1976) Mangrove biogeography. p. 1–29. *In*: V. J. Chapman (ed.). Ecosystems of the World. I. Wet Coastal Ecosystems. Elsevier Science Publishers, Amsterdam, The Netherlands.
Chaudhuri, A. B. (1965) Grasses and grassland types of West Bengal and some aspects of their ecology. Bulletin of the Botanical Society of Bengal 19: 94–108.
Chaudhuri, A. B. (1969) A critical quantitative analysis and special ecological features of the vegetation of North Bengal. Bulletin of the Botanical Society of Bengal 23: 109–129.
Chaudhuri, A. B. and Chakrabarti, K. (1973) Wildlife biology of the Sunderbans forests – a study of the birds of Sunderbans with special reference to the breeding biology of open billed stork, little cormorant and large egrets. Science and Culture 39: 8–16.

Chengapa, B. S. (1944) Andaman forests and their regeneration. Indian Forester 70: 297–304.

Chowdhury, M. I. (1966) On the gradual shifting of the Ganges from west to east in delta building operations. p. 35–40. *In*: Humid Tropics Research, Scientific Problems of the Humid Tropical Zone Deltas and Their Implications. Proceedings of the Dacca Symposium, UNESCO, Paris, France.

Choudhuri, G. N. (1966) Seed germination and flowering in *Vallisneria spiralis*. Northwest Science 40: 31–35.

Choudhury, A., Das, A., Bhattacharya, S. and Bhunia, A. B. (1984) A qualitative assessment of benthic macrofauna in the intertidal mudflats of Sagar Island, Sunderbans, India. p. 298–310. *In*: E. Soepadmo, A. N. Rao, and D. J. Macintosh (eds.). Proceedings of the Asian Symposium on Mangrove Environment: Research and Management. Percetakan Ardyas, Kuala Lumpur, Malaysia.

Clarke, C. B. (1896) Presidential address of the Linnean Society on the Sunderbans of Bengal. Proceedings Linnean Society, London 32: 14–29.

Cooke, T. (1901–1908) The Flora of the Presidency of Bombay. Taylor and Francis, London, United Kingdom. (Reprinted ed. 1958), 3 vols.

Cowardin, L. M., Carter, V., Golet, F. C. and LaRoe, E. T. (1979) Classification of wetlands and deep water habitats of the United States. U. S. Fish and Wildlife Service, Washington, DC., USA. 103 pp.

Curtis, S. J. (1933) Working Plan for the Sunderbans Division (1931–1951). Forest Department of Bengal, Calcutta, India.

Dakshini, K. M. M. (1960a) The vegetation of Mothronwala swamp forest. A preliminary survey. Bulletin of the Botanical Survey of India 2: 57–59.

Dakshini, K. M. M. (1960b) The vegetation of Mothronwala swamp forest (Plant communities of swampy zone). Indian Forester 86: 728–733.

Dakshini, K. M. M. (1965) A Study of the vegetation of Mothronwala swamp forest, Dehra Dun, India. Journal of the Indian Botanical Society 44: 411–428.

Das, P. K. (1968) Monsoons. National Book Trust, New Delhi, India. 162 pp.

Davis, C. B. and van der Valk, A. G. (1989) Ecology of a semitropical monsoonal wetland in India: The Keoladeo Ghana National Park, Bharatpur, Rajasthan. Final Report. Iowa State University, Ames, Iowa, USA. 104 pp.

De Silva, M. W. R. N. (1985) Studies on Mangroves of Sri Lanka: A report. University of Peradeniya, Colombo, Sri Lanka. 38 pp.

Dwivedi, S. N., Parulekar, A. H., Goswami, S. C. and Untawale, A. G. (1975) Ecology of mangrove swamps of the Mandovi estuary, Goa, India. p. 115–125. *In*: G. E. Walsh, S. C. Snedaker, and H. J. Teas (eds.). Proceedings First International Symposium on the Biology and Management of Mangroves. Vol. I. Institute of Food and Agricultural Science, University of Florida, Gainesville, Florida, USA.

Fein, J. S. and Stephens, P. L. (eds.) (1987) Monsoons. John Wiley & Sons, Inc., New York, New York, USA. 631 pp.

Fernando, C. H. (1973) Man-made lakes of Ceylon: a biological resource. p. 664–671. *In*: W. C. Ackerman, G. F. White, and E. B. Worthington (eds.). Man Made Lakes: Their Problems and Environmental Effects. American Geophysical Union, Washington, D.C., USA.

Fernando, C. H. (1984) Reservoirs and lakes of southeast Asia (Oriental region). p. 411–446. *In*: F. B. Taub (ed.). Lakes and Reservoirs. Ecosystems of the World 23. Elsevier Science Publishers, Amsterdam, The Netherlands. 505 pp.

Fernando, C. H. (1985) Ecology and Biogeography of Sri Lanka. Dr. W. Junk Publishers, The Hague, The Netherlands.

Fernando, C. H. and Indrasena, H. H. I. (1969) The freshwater fisheries of Ceylon. Bulletin of the Fisheries Research Station Ceylon 20: 101–134.

Garge, A., Singh, V. P. and Mall, L. P. (1989) Zonational pattern of mangroves in Andaman Islands. p. 233–244. *In*: J. S. Singh and B. Gopal (eds.). Perspectives in Ecology. Jagmander Book Agency, New Delhi, India.

Gopal, B. (1969) Ecological studies on the genus *Marsilea*. II. Edaphic factors. Tropical Ecology 10: 278–291.
Gopal, B. (1976) A note on the growth and ecology of *Salvinia molesta* Mitch. p. 177–181. *In*: C. K. Varshney and J. Rzoska (eds.). Aquatic Weeds in South East Asia. Dr. W. Junk Publishers, The Hague, The Netherlands.
Gopal, B. (1982a) Ecology and management of freshwater wetlands in India. p. 127–162 *In*: Proceedings of the International Scientific Workshop (SCOPE-UNEP) on Ecosystem Dynamics in Freshwater Wetlands and Shallow Water Bodies 1. Centre of International Projects, GKNT, Moscow, USSR.
Gopal, B. (1987) Water Hyacinth. Elsevier Science Publishers, Amsterdam, The Netherlands. 471 pp.
Gopal, B. (ed.) (1990) Ecology and Management of Aquatic Vegetation of the Indian Subcontinent. Kluwer Academic Publishers, Dordrecht, The Netherlands. 257 pp.
Gopal, B. and Kulshreshtha, M. (1980) Role of aquatic macrophytes as reservoir of nutrients and in their cycling. International Journal of Ecology and Environmental Sciences 6: 145–152.
Gopal, B. and Sharma, K. P. (1982) Studies on wetlands in India with emphasis on structure, primary production and management. Aquatic Botany 12: 81–91.
Gopal, B. and Sharma, K. P. (1984) Seasonal changes in the concentration of major nutrient elements in the rhizomes and leaves of *Typha elephantina* Roxb. Aquatic Botany 20: 65–73.
Gopal, B., Sharma, K. P. and Trivedi, R. K. (1978) Studies on ecology and production in Indian freshwater ecosystems at primary producer level with emphasis on macrophytes. p. 349–376. *In*: J. S. Singh and B. Gopal (eds.). Glimpses of Ecology. International Scientific Publications, Jaipur, India.
Gupta, A. C. (1957) The Sunderbans: Its problems, its possibilities. Indian Forester 83: 481–487.
Handoo, J. K. and Kaul, V. (1982) Phytosociology and standing crop in four typical wetlands of Kashmir. p. 187–196. *In*: B. Gopal, R. E. Turner, R. G. Wetzel, and D. F. Whigham (eds.). Wetlands: Ecology and Management. National Institute of Ecology and International Scientific Publications, Jaipur, India.
Hooker, J. D. (1872–97) Flora of the British India. Vols. 1–7. L. Reeve and Company, Kent, United Kingdom.
Hussain, S. A., Sugathan, R. and Pandian, R. (1985) Importance of Vedaranyam swamp as a major ecological entity in the southeast coast of India. Papers for Symposium on Endangered Marine Animals and Marine Parks. Vol. 3. Marine Parks, Sanctuaries and Reserves. Mimeographed paper no. 38., Marine Biological Association of India, Cochin, India.
Hussainy, H. S. H. and Azariah, J. (1985) Point Calimere Marine Park: Conservation and management priorities. Papers for Symposium on Endangered Marine Animals and Marine Parks. Vol. 3. Marine Parks, Sanctuaries and Reserves. Mimeographed paper no. 39. Marine Biological Association of India, Cochin, India.
Ismail, M. (1989) Environment and ecology of forested wetlands of the Sunderbans of Bangladesh. *In*: A. E. Lugo, S. Brown, S., and M. M. Brinson (eds.). Ecosystems of the World, Forested Wetlands. Elsevier Science Publishers, Amsterdam, The Netherlands.
IUCN. (1971) The Ramsar Conference: Final act of the International Conference on the conservation of wetlands and waterfowl. Special Supplement, IUCN Bulletin 2: 1–4.
Jagtap, T. G. (1985) Ecological studies in relation to the mangrove environment along the Goa coast, India. Ph.D. Thesis, Shivaji University, Kolhapur, India.
Jagtap, T. G. (1986) Structure and composition of the mangrove forest along the Goa coast. p. 188–195. *In*: L. J. Bhosale (ed.). The Mangroves. Proceedings of National Symposium on Biology, Utilisation and Conservation of Mangroves. Shivaji University, Kolhapur, India.
Jagtap, T. G. and Untawale, A. G. (1980) Effects of petroleum products on mangrove seedlings. Mahasagar, Bulletin of the National Institute of Oceanography 13: 165–173.
Jayewardene, R. P. (1987) Country reports: Sri Lanka. p. 151–174. *In*: R. M. Umali, P. M.

Zamora, R. R. Gotera, R. S. Jara, and A. S. Camacho (eds.). Mangroves of Asia and the Pacific: Status and Management. Technical Report of the UNDP/UNESCO Research and Training Pilot Programme on Mangrove Ecosystems in Asia and the Pacific (RAS/79/002). National Mangrove Committee of Philippines, Quezon City, Manila, Philipppines.

Jhingaran, V. G. (1982) Fish and Fisheries of India. Hindustan Publishing Corporation, New Delhi, India. 660 pp.

Joshi, G. V. and Bhosale, L. J. (1982) Estuarine ecosystems of India. p. 21–33. *In*: D. N. Sen and S. K. Rajpurohit (eds.). Contributions to the Ecology of Halophytes. Dr. W. Junk Publishers, The Hague, The Netherlands.

Joshi, G. V., Bhosale, L. J., Jamale, B. B. and Karadge, B. A. (1975) Photosynthetic carbon metabolism in mangroves. p. 579–594. *In*: G. E. Walsh, S. C. Snedaker, and H. J. Teas (eds.). Proceedings International Symposium on Biology and Management of Mangroves, Vol. II, University of Florida Press, Gainesville, Florida, USA.

Kachroo, P. (1956) Plant types of the ponds of lower Damodar Valley. Journal of the Indian Botanical Society 35: 430–435.

Kanjilal, U. N. (1901) Swamp forest in Dehra Dun, N.W. Province. Indian Forester 27: 235–243.

Karim, A., Hossain, Z. and White, K. J. (1984) Study of the growth of mangrove plants in relation to edaphic factor in coastal afforestation plantation of Chittagong. p. 195–199. *In*: S. Soepadmo, A. N. Rao, and D. J. Macintosh (eds.). Proceedings of the Asian Symposium on Mangrove Environment: Research and Management. Percetakan Ardyas, Kuala Lumpur, Malaysia.

Kasinathan, R. and Shanmugam, A. (1986) Molluscan fauna of Pitchavaram mangroves, Tamilnadu. p. 438–443. *In*: L. J. Bhosale (ed.). The Mangroves. Proceedings of National Symposium on Biology, Utilisation and Conservation of Mangroves. Shivaji University, Kolhapur, India.

Kaul, V. and Handoo, J. K. (1989) Studies on the ecology of Kashmir Himalaya. p. 1–48. *In*: J. S. Singh and B. Gopal (eds.). Perspectives in Ecology. Jagmander Book Agency, New Delhi, India.

Kaul, V., Fotedar, D. N., Pandit, A. K. and Trisal, C. L. (1978a) A comparative study of plankton populations of some typical freshwater bodies of Jammu and Kashmir state. p. 249–269. *In*: D. N. Sen and R. P. Bansal (eds.). Environmental Physiology and Ecology of Plants. Bishensingh Mahendrapal Singh Publishers, Dehra Dun, India.

Kaul, V., Pandit, A. K. and Fotedar, D. N. (1980a) Ecology of freshwater snails (gastropod, mollusca) in Haigam, a typical wetland of Kashmir. Tropical Ecology 21: 32–46.

Kaul, V., Trisal, C. L. and Handoo, J. K. (1978a) Distribution and production of macrophytes in some water bodies of Kashmir. p. 313–334. *In*: J.S. Singh and B. Gopal (eds.). Glimpses of Ecology. International Scientific Publications, Jaipur, India.

Kaul, V., Trisal, C. L. and Kaul, S. (1980a) Mineral removal potential of some macrophytes in two lakes of Kashmir. Journal of the Indian Botanical Society 59: 108–118.

Kaul, V. and Zutshi, D. P. (1966) Some ecological considerations of the floating islands in Srinagar lakes. Proceedings of the National Academy of Sciences of India 36B: 273–281.

Khan, S. A. (1965) Mangrove forests, their past and present management in Hyderabad region. Science and Industry 3: 12–16.

Khan, S. A. (1966) Working plan of the coastal afforestation division from 1963–64 to 1983–84. Agriculture Department, Government of Pakistan, Lahore, Pakistan.

Kogo, M. (ed.) (1985) Mangroves of Pakistan. Proceedings of National Workshop on Mangroves held in August 1983. Pakistan Agricultural Research Council, Islamabad, Pakistan.

Krishnamoorthy, K. (1960) *Myristica* swamps in the evergreen forests of Travancore. Indian Forester 86: 256–263.

Krishnamurthy, K. (1990) The apiary of the mangroves. p. 135–140. *In*: D. F. Whigham, R. E. Good, and J. Kvet (eds.). Wetland Ecology and Management: Case Studies. Kluwer Academic Publishers, Dordrecht, The Netherlands.

Krishnamurthy, K. and Prince Jeyaseelan, M. J. (1984) Human impacts on the Pichavaram mangrove ecosystem: a case study from southern India. p. 624–632. *In*: E. Soepadmo, A. N. Rao, and D. J. Macintosh (eds.). Proceedings of the Asian Symposium on Mangrove Environment: Research and Management. Percetakan Ardyas, Kuala Lumpur, Malaysia.

Krishnamurthy, K., Kannan, L., Prince Jeyaseelan, M. J., Palaniappan, R. and Sultan Ali, M. A. (1981) A florstic study of halophytes of the Pichavaram mangroves. Bulletin of the Botanical Survey of India 23: 114–122.

Krishnamurthy, K. (1983) The conversion of mangrove lands and waters to other uses. p. 1–12. *In*: UNDP/UNESCO Workshop on Human Population, Mangrove Resources, Human Induced Stresses and Human Health. Bogor, Indonesia.

Krishnamurthy, K., Sundarraj, V. and Santhanam, R. (1975) Aspects of an Indian mangrove forest. p. 88–95. *In*: G. E. Walsh, S. C. Snedaker and H. J. Teas (eds.). Proceedings First International Symposium on Biology and Management of Mangroves, Vol. I. Institute of Food and Agricultural Science, University of Florida, Gainesville, Florida, USA.

Kulkarni, D. H. and Jungadd, C. F. (1959) Utilization of mangrove forests in Saurashtra and Kutch. p. 30–32. *In*: Proceedings of the Mangrove Symposium, Calcutta, 1957, Government of India Press, Faridabad, India.

Kulshreshtha, M. and Gopal, B. (1982a) Observations on nutrient removal by freshwater macrophytes under different habitat conditions in Jaipur (India). Internationale Revue der gesamten Hydrobiologie 67: 543–553.

Kulshreshtha, M. and Gopal, B. (1982b) Decomposition of freshwater vegetation. I. Submerged and free-floating macrophytes, and II. Aboveground organs of emergent macrophytes. p. 258–292. *In*: B. Gopal, R. E. Turner, R. G. Wetzel and D. F. Whigham (eds.). Wetlands: Ecology and Management. National Institute Ecology and International Scientific Publications, Jaipur, India.

Kumar, A. V. A. and Chaphekar, S. B. (1985) Ecology of estuaries – a case study of Ulhas river. p. 67–85. *In*: D. N. Rao, K. J. Ahmad, M. D. Yunus, and S. N. Singh (eds.). Perspectives in Environmental Botany, Vol. I. Print House (India), Lucknow, India.

Kurian, C. V. (1984) Fauna of the mangrove swamps in Cochin estuary. p. 226–230. *In*: E. Soepadmo, A. N. Rao, and D. J. Macintosh (eds.). Proceedings of the Asian Symposium on Mangrove Environment: Research and Management. Percetakan Ardyas, Kuala Lumpur, Malaysia.

La Fond, E. C. (1966) Bay of Bengal. p. 110–118. *In*: R. H. Fairbridge (ed.). The Encyclopedia of Oceanography. Reinhold Publishing Corporation, New York, New York, USA.

Lakshmanan, K. K., Ganapathy, A. and Balasubramanian, B. R. (1984) Mangrove phytosociology in Tamil Nadu. Journal Swamy Botanical Club 1: 35–48.

Löffler, H. (1969) High altitude lakes in Mt Everest region. Verhandlungen Internationale Vereinigung der Limnologie 17: 373–385.

Lugo, A. E. (1980) Mangrove ecosystems: successional or steady state. Biotropica 12: 65–72.

Lugo, A. E. and Snedaker, S. C. (1974) The ecology of mangroves. Annual Review of Ecology and Systematics 5: 39–64.

Macnae, W. (1968) A general account of the fauna and flora of mangrove swamps and forests in the Indo-West Pacific region. Advances in Marine Biology 6: 73–270.

Mall, L. P. (1955) Ecology of low-lying lands. Ph.D. Thesis. Sagar University, Sagar, India.

Mall, L. P., Billore, S. K. and Amritphale, D. (1982) Certain ecological observations on mangrove of the Andaman Islands. Tropical Ecology 23: 225–233.

Mall, L. P., Singh, V. P., Garge, A. and Pathak, S. M. (1986) Mangrove forest of Andamans and some aspects of its ecology. p. 25–38. *In*: L. J. Bhosale (ed.). The Mangroves. Proceedings of National Symposium on Biology, Utilization and Conservation of Mangroves. Shivaji University, Kolhapur, India.

Mandal, A. K. and Nandi, N. C. (1989) Fauna of Sundarban mangrove ecosystem, West Bengal, India. Fauna of Conservation Areas 3. Zoological Survey of India, Calcutta. 116 pp. and 15 plates.

Mani, M. S. (1974) Ecology and Biogeography of India. Monographiae Biologicae 23. Dr. W. Junk Publishers, The Hague, The Netherlands. 773 pp.

Mathauda, G. S. (1959) The mangroves of India. p. 666–87. *In*: Proceedings of the Mangrove Symposium, Calcutta, 1957. Calcutta, India. Government of India Press, Faridabad, India.

Mehta, I. and Sharma, R. K. (1975) Control of *Typha* by a competitive plant. Annals of Arid Zone 14: 175–182.

Melkania, N. P. and Singh, J. S. (1989) Ecology of Indian grasslands. p. 67–103. *In*: J. S. Singh and B. Gopal (eds.). Perspectives in Ecology. Jagmander Book Agency, New Delhi, India.

Misra, R. D. (1938) Edaphic factors in the distribution of aquatic plants in the English lakes. Journal of Ecology 26: 411–451.

Misra, R. (1946) A study in the ecology of low lying lands. Indian Ecology 1: 1–20.

Misra, R. (1983) Indian savannas. p. 151–166. *In*: F. Bourliere (ed.). Ecosystems of the World 13. Tropical Savannas. Elsevier Science Publishers, Amsterdam, The Netherlands.

Mitchell, D. S. and Gopal, B. (1990) Invasions of tropical freshwaters by alien aquatic plants. p. 139–154. *In*: P. S. Ramakrishnan (ed.). Ecology of Biological Invasions in the Tropics. International Scientific Publications, New Delhi, India.

Mitsch, W. J. and Gosselink, J. G. (1986) Wetlands. Van Nostrand Reinhold Company. New York, New York, USA. 536 pp.

Moodie, A. W. (1924–25) Working plan for the forests of the Irrawaddy delta. Forest Division, Moymyo, Rangoon, Burma.

Mudaliar, C. R. and Kamath, H. S. (1952) Distribution of *Rhizophora mucronata* Lam. in the backwater of the west coast and its economic importance. Madras Agricultural Journal 39: 610–615.

Mukherjee, A. K. (1975) The Sunderbans of India and its biota. Journal of the Bombay Natural History Society 72: 1–20.

Mukherjee, A. K. and Tiwari, K. K. (1984) Mangrove ecosystem changes under induced stress: The case history of the Sunderban, West Bengal, India. p. 633–643. *In*: E. Soepadmo, A. N. Rao, and D. J. Macintosh (eds.). Proceedings of the Asian Symposium on Mangrove Environment: Research and Management. Percetakan Ardyas, Kuala Lumpur, Malauysia.

Mukherjee, B. B. (1984) Mangrove forest types and their distribution in Sunderbans, India. p. 75–81. *In*: E. Soepadmo, A. N. Rao, and D. J. Macintosh (eds.). Proceedings of the Asian Symposium on Mangrove Environment: Research and Management. Percetakan Ardyas, Kuala Lumpur, Malaysia.

Mukherjee, B. B. and Mukherjee, J. (1978) Mangroves of Sunderbans, India. Phytomorphology 28: 177–192.

Murty, K. S. N. (1987) An ecological study of lake Kondakarla. Ph.D. Thesis. Andhra University, Waltair, India.

Nasir, E. and Ali, S. I. (eds.) (1970–85) Flora of Pakistan. Fascicles 1–175. Pakistan Agricultural Research Council, Islamabad, Pakistan.

Naskar, K. and Guha Bakshi, D. N. (1987) Mangrove swamps of the Sunderbans: An ecological perspective. Naya Prokash, Calcutta, India. 263 pp.

Navalkar, B. S. (1940) Studies in the ecology of mangrove. Journal of the University of Bombay New Series 8: 58–73 and 9: 78–91.

Navalkar, B. S. (1951) Succession of mangrove vegetation in Bombay and Salsette Island. Journal of the Bombay Natural History Society 5: 335–345.

Navalkar, B. S. (1956) Geographical distribution of the halophytic plants of Bombay and Salsette Islands. Journal of the Bombay Natural History Society 50: 157–160.

Navalkar, B. S. (1961) Importance of mangroves. Tropical Ecology 2: 89–91.

Navalkar, B. S. (1962) Mangrove habitat and its economic exploitation. Abstract. Symposium on Evaluation of Tropical Habitat for Production of Food, Fodder, Fuel and Fertilizers. International Society for Tropical Ecology, Cuttack, India. 6 pp.

Navalkar, B. S. (1973) Mangrove ecology. p. 101–109. *In*: R. Misra, B. Gopal, K. P. Singh, and J. S. Singh (eds.). Progress of Plant Ecology in India, Vol. 1. Today and Tomorrow's Printers and Publishers, New Delhi, India.

Ong, J. E. (1986) Partitioning of biomass and productivity in mangroves. p. 87–91. *In*: Workshop on Mangrove Ecoystem Dynamics, Port Moresby, Papua New Guinea. UNDP/UNESCO Research and Training Pilot Programme on Mangrove Ecosystems in Asia and the Pacific (RAS/79/002), New Delhi, India.

Palaniappan, R. and Baskaran, K. (1986) Distribution and abundance of zooplankton in Pichavaram mangrove, south India. p. 499–504. *In*: L. J. Bhosale (ed.). The Mangroves. Proceedings of National Symposium on Biology, Utilization and Conservation of Mangroves. Shivaji University, Kolhapur, India.

Pandey, S. B. (1969) *Eleocharis palustris*. p. 237–285. *In*: R. Misra (ed.). Ecological Studies of Noxious Weeds Common in India and America which are Becoming Increasing Problem in the Upper Gangetic Plain. Final Technical Report, PL480/Ecology Project, Banaras Hindu University, Varnasi, India.

Pandeya, S. C. (1988) Status of the Indian Rangelands. Presidential address to Third International Rangeland Congress, New Delhi, India. 215 pp.

Pandit, A. K. (1980) Biotic factor and food chain structure in some typical wetlands of Kashmir, Ph.D. Thesis. Kashmir University, Srinagar, India.

Pandit, A. K. and Fotedar, D. N. (1982) Restoring damaged wetlands for wildlife. Journal of Environmental Management 14: 359–368.

Parkinson, C. E. (1923) A Forest Flora of the Andaman Island. Reprinted ed. Bishen Singh Mahendra Pal Singh, Dehra Dun, India.

Patil, R. P. (1959) The mangrove in Salsette island near Bombay. p. 45–49. *In*: Proceedings of the Mangrove Symposium, Calcutta, 1957. Government of India Press, Faridabad, India.

Patten, B. C., Jorgensen, S. E., Dumont, H., Gopal, B., Koryavov, P., Kvet, J., Löffler, H., Sverizhev, Y. and Tundisi, J. G. (eds.) (1990) Wetlands and Continental Shallow Waterbodies. Vol. 1. Natural and Human Relationships. SPB Academic Publishing, The Hague, The Netherlands. 759 pp.

Pinto, M. L. (1984) Some ecological aspects of a community of mangrove crabs occurring within the islets of Negombo lagoon (Sri Lanka). p. 311–330. *In*: E. Soepadmo, A. N. Rao, and D. J. Macintosh (eds.). Proceedings of the Asian Symposium on Mangrove Environment: Research and Management. Percetakan Ardyas, Kuala Lumpur, Malaysia.

Pinto, L. (1986) Mangroves of Sri Lanka. Natural Resources, Energy and Science Authority of Sri Lanka, Colombo, Sri Lanka. 54 pp.

Ponnamperuma, F. N. (1984) Mangrove swamps in south and southeast Asia as potential rice lands. p. 672–683. *In*: E. Soepadmo, A. N. Rao, and D. J. Macintosh (eds.). Proceedings of the Asian Symposium on Mangrove Environment: Research and Management. Percetakan Ardyas, Kuala Lumpur, Malaysia.

Prain, D. (1903) Flora of the Sunderbans. Record of the Botanical Survey of India. 2: 231–370.

Puri, G. S., Gupta, R. K., Meher-Homji, V. M. and Puri, S. (1984) Forest Ecology. Vol. I. Phytogeography and Forest Conservation. Oxford and IBH Publishing Co., New Delhi, India. 549 pp.

Puri, G. S., Gupta, R. K., Meher-Homji, V. M. and Puri, S. (1989) Forest Ecology. Vol. 2, Plant Form, Diversity, Communities and Succession. Oxford and IBH Publising Co., New Delhi, India. 582 pp.

Purohit, R. and Singh, S. P. (1985) Submerged macrophytic vegetation in relation to eutrophication level in Kumaun Himalaya. Environmental Pollution 39: 161–173.

Qureshi, I. M. (1959) Botanical, silvicultural features of mangrove forests of Bombay State. p. 20–26. *In*: Proceedings of the Mangrove Symposium, Calcutta, 1957. Government of India Press, Faridabad, India.

Rabanal, H. R. (1977) Forest conservation and aquaculture development of mangrove areas. p. 145–148. *In*: Proceedings International Workshop on Mangrove and Estuarine Area Development for Indo-Pacific Region, Manila, Philippines.

Radhakrishnan, N. (1985) Studies on the mangroves along the central west coast: Achra (Maharashtra). p. 222–226. *In*: L. J. Bhosale (ed.). The Mangroves. Proceedings of National Symposium on Biology, Utilisation and Conservation of Mangroves. Shivaji University, Kolhapur, India.

Rajagopalan, V. R. (1952) Ecological Adaptations of the Mangrove Vegetation at Picharavam. Thesis. Annamalai University, India.

Ramakrishnan, P. S. (1965) Soil complex in relation to aquatic plant communities. Proceedings of the National Institute of Sciences India 30B: 348–353.

Ramamurthy, V. D. (1984) The effect of agricultural pollutants on fish and fish products in the mangrove swamps of southwest coast of India. p. 618–623. *In*: E. Soepadmo, A. N. Rao, and D. J. Macintosh (eds.). Proceedings of the Asian Symposium on Mangrove Environment: Research and Management. Percetakan Ardyas, Kuala Lumpur, Malaysia.

Rao, A. N. (1987) Mangrove ecosystems of Asia and the Pacific. p. 1–41. *In*: R. M. Umali, P. M. Zamora, R. R. Gotera, R. S. Jara, and A. S. Camacho (eds.). Mangroves of Asia and the Pacific: Status and Management. Technical Report of the UNDP/UNESCO Research and Training Pilot Programme on Mangrove Ecosystems in Asia and the Pacific (RAS/79/002). National Mangrove Committee of Philippines, Quezon City, Manila, Philippines.

Rao, K. L. (1975) India's Water Wealth: Its Assessment, Uses and Projections. Orient Longman, New Delhi, India. 255 pp.

Rao, R. S. (1959) Observations on the mangrove vegetation of the Godavari Estuary. p. 36–44. *In*: Proceedings of the Mangrove Symposium, Calcutta, 1957. Government of India Press, Faridabad, India.

Rao, T. A. and Aggarwal, K. R. (1964) Ecological studies of Saurashtra Coast and neighbouring Island. Bulletin of the Botanical Survey of India 6: 173–183.

Rao, T. A., Aggarwal, K. R. and Mukherjee, A. K. (1963a) Ecological studies on the soil and vegetation of Krusadi group of islands in the Gulf of Mannar. Bulletin of the Botanical Survey of India 5: 141–148.

Rao, T. A., Aggarwal, K. R. and Mukherjee, A. K. (1963b) An ecological account of the vegetation of Rameswaram Island. Bulletin of the Botanical Survey of India 5: 301–323.

Rao, T. A., Aggarwal, K. R. and Mukherjee, A. K. (1966) Ecological studies of Saurashtra coast and neighbouring Islands. I. Bulletin of the Botanical Survey of India 8: 60–67.

Rao, T. A. and Mukherjee, A. K. (1972) Ecological aspects along the shores of the Burabalanga tidal estuary. Balasore District. Proceedings of the Indian Academy of Sciences 76: 201–206.

Rao, T. A. and Sastry, A. R. K. (1972) An ecological approach towards classification of coastal vegetation of India. Indian Forester 98: 594–607.

Rao, T. A. and Sastry, A. R. K. (1974) An ecological approach towards classification of coastal vegetation of India. II. Estuarine border vegetation. Indian Forester 100: 438–452.

Rao, T. A., Sastry, A. R. K. and Shanware, P. G. (1972) Analysis of the basic patterns of an estuarine shore in the vicinity of Coringa Bay. Andhra Pradesh. Proceedings of the Indian Academy of Sciences 75: 40–50.

Rao, T. A. and Shanware, P. G. (1967) Ecological studies of Saurashtra coast and neighbouring islands. VI. An approach to the classification of Saurashtra coastland. Bulletin of the Botanical Survey of India 9: 240–248.

Rao, V. S. (1959) Mangrove forests and the problem of reclaiming saline blanks. p. 58–65. *In*: Proceedings of the Mangrove Symposium, Calcutta, 1957. Government of India Press, Faridabad, India.

Raphael, Y. I. (1977) Mangrove and estuarine development: Country statement – Sri Lanka. p. 119–120. *In*: Proceedings International Workshop on Mangrove and Estuarine Area Development, Manila, Philippines.

Raychaudhuri, S. P., Aggarwal, R. R., Datta Biswas, N. R., Gupta, S. P. and Thomas, P. K. (1963) Soils of India. Indian Council of Agricultural Research, New Delhi, India. 643 pp.

Reeders, H. H., van Schoubroeck, M. E. P., van Vierssen, W., Gopal, B. and Pieterse, A. H. (1986) Aquatic weeds and their implication for agriculture in the Chambal Irrigated Area, Kota (Rajasthan), India. p. 251–255. *In*: Proceedings EWRS/AAB 7th International Symposium on Aquatic Weeds, Loughborough, U.K. European Weed Research Society, Wageningen, The Netherlands.

Robinson, H. (1976) Monsoon Asia. 3rd ed. Macdonald and Evans, Estover, United Kingdom. 510 pp.

Rowntree, J. B. (1954) An introduction to the vegetation of the Assam valley. Indian Forest Records, New Series 9(1): 1–87.

Roxburgh, W. (1814) Hortus Bengalensis, or a Catalogue of Plants Growing in the Honourable East India Company's Botanical Garden at Calcutta. Mission Press, Serampore, India. 105 pp.

Saenger, P., Hegerland, E. J. and Davis, J. D. S. (eds.) (1983) Global status of mangrove ecosystems. IUCN Commission on Ecology paper no. 3. Environmentalist 3: 1–88.

Sahni, K. C. (1953) Botanical exploration in the great Nicobar. Indian Forester 79: 3–16.

Salim Ali and Ripley, S. D. (1983) Handbook of the Birds of India and Pakistan. 10 Vols. Oxford University Press, New Delhi, India.

Salim Ali and Vijayan, V. S. (1986) Keoladeo National Park: Ecology study, Summary report 1980–1985. Bombay Natural History Society, Bombay, India. 186 pp.

Samant, J. S. (1986) Avifauna of the mangroves around Ratnagiri, Maharashtra. p. 456–466. *In*: L. J. Bhosale (ed.). The Mangroves. Proceedings of National Symposium on Biology, Utilisation and Conservation of Mangroves. Shivaji University, Kohlapur, India.

Sarkar, S., Baidya, A., Bhunia, A. and Choudhury, A. (1984) Zooplankton studies in the Hooghly estuary around Sagar Island, Sunderbans, India. p. 286–297. *In*: E. Soepadmo, A. N. Rao, and D. J. Macintosh (eds.). Proceedings of the Asian Symposium on Mangrove Environment: Research and Management. Percetakan Ardyas, Kuala Lumpur, Malaysia.

Saxena, S. K. and Gupta, R. K. (1973) Vegetation of Pachpadra salt basin in Western Rajasthan. Journal of the Bombay Natural History Society 70: 104–127.

Schimper, A. F. W. (1891) Die indomalayische Strandflora. Botanishe Mitteilungen der Tropen 3: 32–67, 85–100, 158–184.

Sen, D. N. and Rajpurohit, K. S. (eds.) (1982) Contributions to the Ecology of Halophytes. Dr. W. Junk Publishers, The Hague, The Netherlands.

Seneviratne, E. W. (1978) The Sri Lanka mangroves. *In*: P. Kunstadter and S. C. Snedaker (eds.). Proceedings of Regional Seminar on Human Uses of Mangrove Ecosystems and Management Implications, Dacca, Bangladesh. Manuscript (unpublished).

Seshavatharam, V. (1978) Studies on ecology and weed biology of Kolleru lake. Final Technical Report of C.S.I.R. Project. Andhra University, Waltair, India. 92 pp.

Sharma, K. P. (1985) Impact of river basin development projects in India with reference to fisheries development. p. 25–32. *In*: Proceedings of the International Seminar on Environmental Impact Assessment of Water Resources Projects Vol. 1. University of Roorkee, Roorkee, India.

Sharma, K. P. and Gopal, B. (1979a) Effect of light intensity on growth and establishment of seedlings of *Typha angustata*. Polskie Archiwum Hydrobiologii 26: 495–500.

Sharma, K. P. and Gopal, B. (1979b) Effect of water regime on the growth and establishment of *Typha angustata* seedlings. International Journal of Ecology and Environmental Science 5: 69–74.

Sharma, K. P. and Gopal, B. (1982) Decomposition and nutrient dynamics in *Typha elephantina* Roxb. under different water regimes. p. 321–324. *In*: B. Gopal, R. E. Turner, R. G. Wetzel, and D. F. Whigham (eds.). Wetlands: Ecology and Management. National Institute of Ecology and International Scientific Publications, Jaipur, India.

Sidhu, S. S. (1963) Studies on the mangroves of India. I. East Godavari region. Indian Forester 89: 337–351.

Silas, E. G., Mahadevan, S. and Nayar, K. N. (1985) Existing and Proposed Marine Parks and Reserves in India. A review. Papers for Symposium on Endangered Marine Animals and

Marine Parks, Vol. 3. Marine Parks, Sanctuaries and Reserves. Mimeographed paper no. 36. Marine Biological Association of India, Cochin, India.

Singh, J. S. and Singh, S. P. (1987) Forest vegetation of the Himalaya. Botanical Review 53: 80–192.

Singh, S. R. and Swarup, K. (1980) Studies on the primary production in Suraha lake (Ballia, India). Internationale Revue der gesamten Hydrobiologie 65: 709–717.

Singhal, P. K. and Singh, J. S. (1978) Ecology of Nainital lakes; Morphometry and macrophytic vegetation. Tropical Ecology 19: 178–188.

Sinha, A. B. (1969) Investigations on the ecology of Ramgarh lake, Gorakhpur. Ph.D. Thesis. Gorakhpur University, Gorakhpur, India.

Snedaker, S. C. (1984) The mangroves of Asia and Oceania: Status and research planning. p. 5–15. *In*: E. Soepadmo, A. N. Rao, and D. J. Macintosh (eds.). Proceedings of the Asian Symposium on Mangrove Environment: Research and Management. Percetakan Ardyas, Kuala Lumpur, Malaysia.

Soepadmo, E., Rao, A. N. and Macintosh, D. J. (eds.) (1984) Proceedings of the Asian Symposium on Mangrove Environment: Research and Management. Percetakan Ardyas, Kuala Lumpur, Malaysia. 828 pp.

Som Dev and Aswal, B. S. (1974) Taxonomy and ecology of Mothronwala swamp. A reassessment. Indian Forester 100: 12–19.

Som Dev and Srivastava, M. M. (1978) An ecological study of the vegetation of Golatappar swamp, Dehra Dun. Indian Journal of Forestry 1: 44–52.

Srivastava, V. C. (1973) The limnology, primary production and energetics of Chilwa lake, Gorakhpur. Ph.D. Thesis. Gorakhpur University, Gorakhpur, India.

Stamp, D. L. (1925) The aerial survey of the Irrawady delta forests. Journal of Ecology 13: 262–276.

Subrahmanyam, V. P. (1965) Köppen and Thornthwaite systems of climatic classification as applied to India. Annals of Arid Zone 4: 45–55.

Swar, D. B. (1980) Present status of limnological studies and research in Nepal. p. 43–47. *In*: S. Mori and I. Ikusima (eds.). Proceedings of the First Workshop for the Promotion of Limnology in Developing Countries. Organising Committee of XXI SIL Congress, Kyoto, Japan.

Swar, D. B. and Fernando, C. H. (1979) Cladocera from Pokhra Valley, Nepal, with notes on distribution. Hydrobiologia 66: 113–128.

Teas, H. J. (ed.) (1983) Biology and Ecology of Mangroves. Dr. W. Junk Publishers, The Hague, The Netherlands.

Teas, H. J. (ed.) (1984) Physiology and Management of Mangroves. Dr. W. Junk Publishers, The Hague, The Netherlands.

Than Htay, U. and Saw Han, U. (1984) Mangrove forests of Burma. p. 82–85. *In*: E. Soepadmo, A. N. Rao, and D. J. Macintosh (eds.). Proceedings of the Asian Symposium on Mangrove Environment: Research and Management. Percetakan Ardyas, Kuala Lumpur, Malaysia.

Thothathri, K. (1960a) Botanical exploration in Car Nicobar and Nancouri Islands. Bulletin of the Botanical Survey of India 2: 341–346.

Thothathri, K. (1960b) Studies on the flora of Andaman Islands. Bulletin of the Botanical Survey of India 2: 357–373.

Thothathri, K. (1962) Contribution to the flora of the Andaman and Nicobar Islands. Bulletin of the Botanical Survey of India 4: 281–296.

Thothathri, K. (1981) Studies on the mangroves of peninsular India versus the Andaman and Nicobar Islands. Bulletin of the Botanical Survey of India 23: 151–154.

Tripathi, R. S. (1969) *Cyperus rotundus*. p. 583–647. *In*: R. Misra (ed.). Ecological Studies of Noxious Weeds Common in India and America Which are Becoming Increasing Problems in the Upper Gangetic Plain. Final Technical Report, PL480/Ecology Project, Banaras Hindu University, Varanasi, India.

Umali, R. M., Zamora, P. M., Gotera, R. R., Jara, R. S. and Camacho, A. S. (eds.) (1987) Mangroves of Asia and the Pacific: Status and Management. Technical Report of the

UNDP/UNESCO Research and Training Pilot Programme on Mangrove Ecosystems in Asia and the Pacific (RAS/79/002). National Mangrove Committee of Philippines, Quezon City, Manila, Philippines. 538 pp.

Untawale, A.G. (1982) Annotated Bibliography on Mangroves. National Institute of Oceanography, Goa, India. 110 pp.

Untawale, A. G. (1984) Present status of mangroves along west coast of India. p. 57–74. *In*: E. Soepadmo, A. N. Rao, and D. J. Macintosh (eds.). Proceedings of the Asian Symposium on Mangrove Environment: Research and Management. Percetakan Ardyas, Kuala Lumpur, Malaysia.

Untawale, A. G. (1987a) Country reports: India. p. 51–88. *In*: R. M. Umali, P. M. Zamora, R. R. Gotera, R. S. Jara, and A. S. Camacho (eds.). Mangroves of Asia and the Pacific: Status and Management. Technical Report of the UNDP/UNESCO Research and Training Pilot Programme on Mangrove Ecosystems in Asia and the Pacific (RAS/79/002). National Mangrove Committee of Philippines, Quezon City, Manila, Philippines.

Untawale, A. G. (1987b) Conservation of Indian mangroves – a national perspective. p. 85–104. *In*: Contributions in Marine Sciences. Prof. S. Z. Qasim Felicitation Volume. National Institute Oceanography, Goa, India.

Untawale, A. G., Dwivedi, S. N. and Singhal, S. Y. S. (1973) Ecology of mangroves in Mandovi and Zuari estuaries and interconnecting Cumbarjua canal of Goa. Indian Journal of Marine Sciences 2: 47–63.

Untawale, A. G. and Wafar, S. (1986) Coastal Karnataka and its ecodevelopment. p. 1–11. *In*: State of Environment Report 1984–1985. Government of Karnataka, Bangalore, India.

Untawale, A. G., Wafar, S. and Jagtap, T. G. (1982) Application of remote sensing techniques to study the distribution of mangroves along the estuaries of Goa. p. 51–76. *In*: B. Gopal, R. E. Turner, R. G. Wetzel, and D. F. Whigham (eds.). Wetlands Ecology and Management. National Institute of Ecology and International Scientific Publications, Jaipur, India.

Varshney, C. K. and Singh, K. P. (1976) A survey of aquatic weed problem in India. p. 31–42. *In*: C. K. Varshney and J. Rzoska (eds.). Aquatic Weeds in South East Asia. Dr. W. Junk Publishers, The Hague, The Netherlands.

Venkataraman, G. S., Goyal, S. K., Kaushik, B. D. and Roychoudhury, P. (1974) Algae: Form and Function. Today and Tomorrow's Printers and Publishers, New Delhi, India. 562 pp.

Venkatesan, K. R. (1966) The mangrove of Madras State. Indian Forester 92: 27–34.

Venkateswarlu, V. (1944) The esturial flora of the Godavari. Journal of the Bombay Natural History Society 44: 431–435.

Vijayan, V. S. (1986) On conserving the bird fauna of Indian wetlands. Proceedings of the Indian Academy of Sciences (Animal Sciences/Plant Sciences) Supplement: 91–101.

Vijayan, V. S. (1988) Keoladeo National Park Ecology Study – Annual report 1987. Bombay Natural History Society, Bombay, India. 163 pp.

Vyas, L. N. and Garg, R. K. (1989) Wetland Conservation. Agro-Botanical Publishers, Bikaner, India. 240 pp.

Wafar, M. V. M. (1986) Corals and coral reefs of India. Proceedings of the Indian Academy of Sciences (Animal Sciences/Plant Sciences) Supplement: 19–43.

Waheed Khan, M. A. (1959) Ecological studies of mangrove forests in India. p. 97–109. *In*: Proceedings of the Mangrove Symposium, Calcutta, 1957. Government of India Press, Faridabad, India.

Walsh, G. E. (1974) Mangroves: a review. p. 51–174. *In*: R. Reimold and W. H. Queen (eds.). Ecology of Halophytes. Academic Press, New York, New York, USA.

Walsh, G. E. (1977) Exploitation of mangal. p. 347–362. *In*: V. J. Chapman (ed.). Ecosystems of the World. I. Wet Coastal Ecosystems. Elsevier Science Publishers, Amsterdam, The Netherlands.

Whitton, B. A., Aziz, A., Francis, P., Rother, J. A., Simon, J. W. and Tahmida, Z. N. (1988) Ecology of deepwater rice fields in Bangladesh. I. Physical and chemical environment. Hydrobiologia 169: 3–22.

Wolker, D. C. (1966) The deltaic area of the Irrawaddy River in Burma. p. 373–378. *In*:

Humid Tropics Research. Scientific Problems of the Humid Tropical Zone Deltas and their Implications. Proceedings of the Dacca Symposium, UNESCO, Paris, France.

Yadav, V. K. and Choudhury, A. (1986) Litter production in mangrove forest of Lothian Island in Sunderbans, West Bengal. p. 227–229. *In*: L. J. Bhosale (ed.). The Mangroves. Proceedings of National Symposium on Biology, Utilisation and Conservation of Mangroves. Shivaji University, Kohlapur, India.

Yadav, P. S. and Varshney, C. K. (1981) Notes on the ecological and socioeconomic importance of wetlands of Manipur, N. E. India. International Journal of Ecology and Environmental Sciences 7: 149–150.

Zutshi, D. P. (1975) Associations of macrophytic vegetation in Kashmir lakes. Vegetatio 30: 61–65.

Zutshi, D. P. (1989) Twenty five years of ecological research on the lakes of northwestern Himalayas. p. 49–66. *In*: J. S. Singh and B. Gopal (eds.). Perspectives in Ecology. Jagmander Book Agency, New Delhi, India.

Zutshi, D. P., Subla, B. A., Khan, M. A. and Wanganeo, A. (1980) Comparative limnology of nine lakes of Jammu and Kashmir Himalaya. Hydrobiologia 80: 101–112.

Wetlands of Canada and Greenland

W. A. GLOOSCHENKO, C. TARNOCAI,
S. ZOLTAI AND V. GLOOSCHENKO

Abstract

Canada is a vast country characterized by a complex mosaic of climates and physiography. Climates range from cool temperate to cold arctic. In Canada, an estimated 127.2×10^6 ha of wetlands occur, some 14% of the surface area. Peatlands account for 88% of all wetlands. Five classes of wetlands are found in Canada: shallow open water, marshes (both freshwater and salt), swamps, fens, and bogs. Distinct regional differences occur in seven bioclimatic zones: Arctic, Subarctic, Boreal, Temperate, Prairie, Mountain, and Oceanic (both Pacific and Atlantic). Within these zones, major factors influencing wetland development include hydrology, water chemistry, time, nature of the terrain, and sedimentological processes.

Human activities including agricultural development, urbanization, peat extraction, forestry, and construction projects, are leading to depletion of wetlands in parts of Canada. The rates of such losses are being determined by wetland inventories in many parts of the country.

In Greenland, wetlands include shallow open water, saltmarshes, fens, and bogs. Limited data are available on their distribution and ecology.

Introduction

Canada is a vast country of 9,458,000 km^2 (Fig. 1). It is characterized by a range of climates from cool temperate to cold arctic with coasts moderated by oceanic influences. Physiographically, the landscape varies from the rugged western Cordillera to the Canadian Shield and adjacent lowlands to the Appalachian mountains on the east coast. Arctic conditions prevail on the northern mainland and the arctic islands. Approximately the northern one-third of the country is underlain by continuous permafrost while discontinuous permafrost can occur in another one-third of Canada. The continental

D.F. Whigham et al. (eds.), Wetlands of the World I, 415–514.

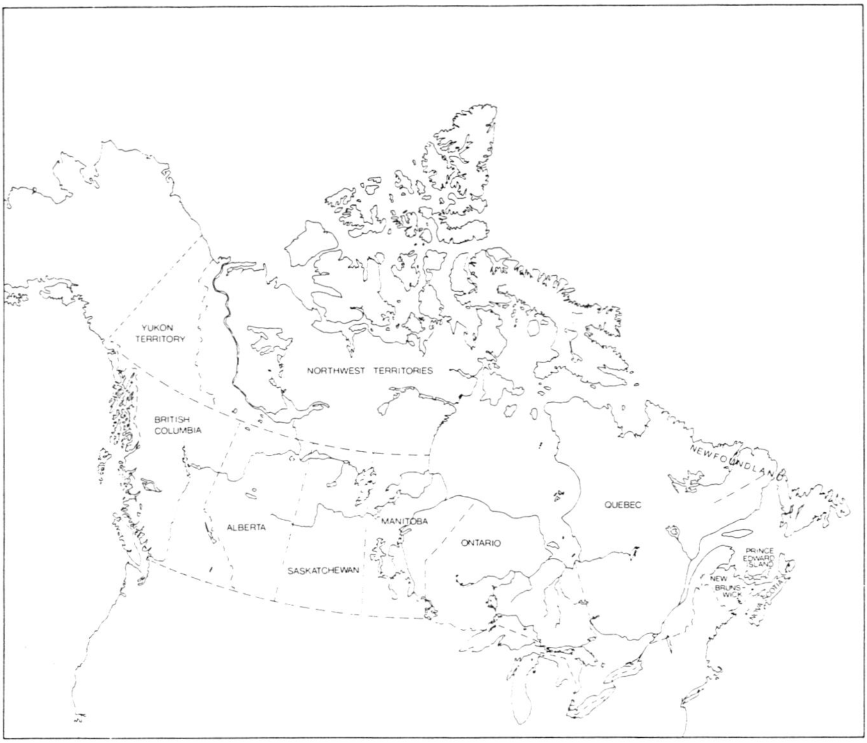

Figure 1. Map of Canada showing provinces and territories.

ice sheets of the Great Ice Age modified the landscape and their remnants retreated to their current position some 6,000 years ago.

With this climatic and geological diversity in mind, wetlands become an important part of Canada's landscape. An estimated 14% of the land area (127.2×10^6 ha) is classified as wetland. Of this, 88% is peatland with at least a 40 cm thickness of peat (Tarnocai 1983). Wetlands are unevenly distributed as shown in Table 1 and Fig. 2. Low mountains of western and eastern Canada and the hilly regions of Quebec and Ontario are too well drained to support extensive wetlands. Little precipitation is present in the prairies of interior British Columbia and the provinces of Alberta, Saskatchewan, and Manitoba. Here, wetlands are confined to shallow depressions. Yet, in wet areas such as coastal British Columbia and Newfoundland, peatland can occur on relatively steep slopes.

The major wetlands occur in an area extending from central Labrador to south of Hudson Bay and northwest to the Mackenzie River delta. This area is cool, moist and characterized by flat terrain. Other suitable areas for

Table 1. Distribution of wetlands in Canada (Tarnocai 1984).

Provinces and territories	Total wetland area (ha × 10^3)	% of land within designated area	% of total Canadian wetlands
British Columbia	3,100	3	2
Alberta	13,700	21	11
Saskatchewan	9,700	17	8
Manitoba	22,500	41	18
Ontario	29,200	33	23
Quebec	12,200	9	10
New Brunswick	550	8	1
Nova Scotia	200	3	1
Prince Edward Island	4	1	1
Newfoundland–Labrador	6,800	18	5
Yukon Territory	1,500	3	1
Northwest Territories	27,800	9	22
Canada	127,200	14	100

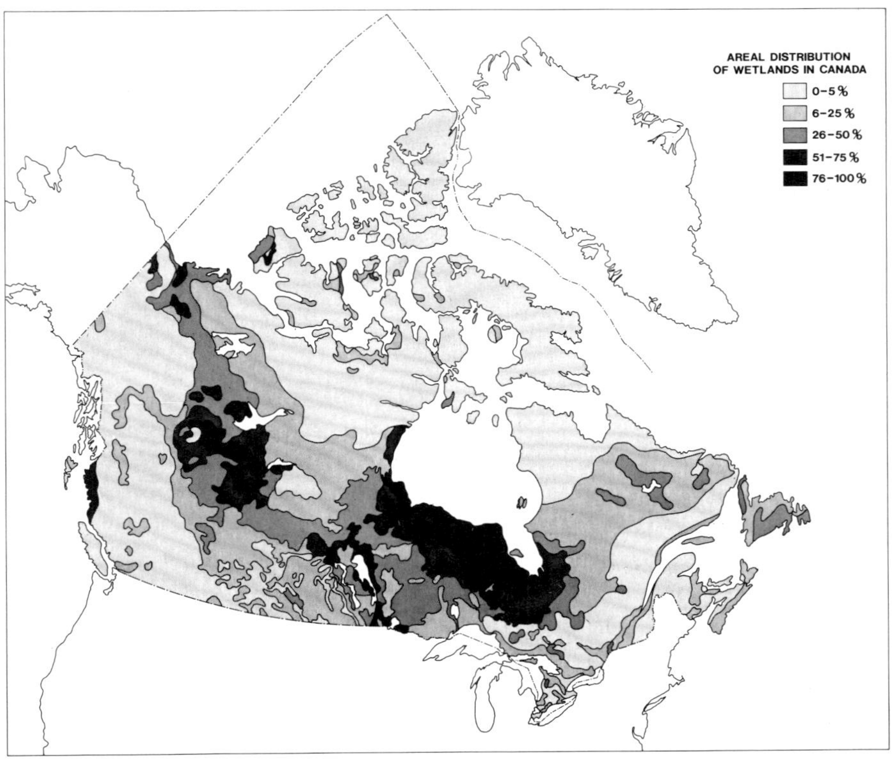

Figure 2. Map of Canada showing general distributions of wetlands.

wetland development include low, poorly-drained areas that were formerly occupied by glacial lakes.

This paper will cover the wetlands of Canada in a regional context. Twenty wetland regions have been delineated based on climate and similar topography, hydrology, and nutrient regime, as will be discussed in more detail later in this paper. The vegetation of the wetlands will be emphasized along with major factors controlling the vegetation including hydrology, water chemistry, and nutrient status. Peat formation will be stressed due to interest in this potential energy source and its value as a horticultural soil amendment (Rubec *et al.* 1988).

It is impossible to cover all the available literature on Canadian wetlands in a paper of this length. Previous work has been limited mainly to local and regional aspects as opposed to the country as a whole. For example, a review paper with a broad overview of Canadian wetlands is that by Zoltai and Pollett (1983). An example of a comprehensive review for a region of Canada is the paper by Wells and Pollett (1983) for Newfoundland, while an example of a comprehensive review of a wetland type, the salt marsh, are the papers of Glooschenko (1983) and Glooschenko *et al.* (1988). The National Wetland Working Group has prepared a book on the wetlands of Canada which covers many of these topics in a more comprehensive fashion (National Wetland Working Group 1988). Also, several proceedings have been published on Canadian wetlands. These include both Canada in general (Rubec and Overend 1988) and the Province of Ontario (Bardecki and Patterson 1989). Papers on the wetlands of Greenland are very limited; no comprehensive review was found.

Classification of wetlands

Definition of wetlands

Wetland is defined as land having the water table at, near, or above the soil surface or which is saturated for a long enough period to promote wetland or aquatic processes as indicated by hydric soils, hydrophilic vegetation, and various kinds of biological activity which are adapted to the wet environment (Tarnocai 1980).

Wetlands include peatlands, formed by the accumulation of plant materials. Peatlands have more than 40 cm of peat and are associated with organic soils, excluding Folisols (Canada Soil Survey Committee 1978). Wetlands also include areas that are influenced by excess water but which, for climatic, edaphic, or biotic reasons produce little or no peat. These wetlands are associated with Gleysolic or the peaty phase of Gleysolic soils.

Shallow open water, generally less than 2 m deep, is also included in wetlands. In certain types of wetlands, vegetation is lacking and soils are poorly developed as a result of frequent and drastic fluctuations of surface water levels or of wave action, water flow, turbidity, or high concentration of salts or other toxic substances in the water or in the soil. Such wetlands can be recognized by the presence of surface water or saturated soil at some time during each year. Wetland also include areas which are modified by water control structures or which are tilled and planted but which, if allowed to revert, again become saturated for long periods and are associated with wet soils (gleysols) and hydrophilic vegetation.

Wetland classification

The revised Canadian wetland classification (National Wetlands Working Group 1987) has been derived from several works (Jeglum *et al.* 1974, Tarnocai 1970, 1974 and 1980, Zoltai *et al.* 1973, Zoltai and Tarnocai 1975). Information concerning veneer bogs can be found in Mills *et al.* (1976, 1978), and the update of the marsh and shallow water classes are mainly from G. D. Adams (personal communication 1979).

This system of classification is hierarchical and has three levels. At the highest level, the class level, the wetlands are classified according to their genesis. At the wetland form level they are classified according to their surface morphology, surface pattern, morphology of the underlying mineral terrain, hydrology, and the type of water. At the lowest level, the wetland type, the wetlands are classified according to the general physiognomy of the vegetation cover.

Wetland classes

The definitions of the wetland classes (bog, fen, marsh, swamp, and shallow water) along with the associated wetland forms are given below.

Bog. A bog is a peatland which generally has a high water table. This water table is at or near the surface. The bog surface is either raised above or level with the surrounding wetlands and is virtually unaffected by the nutrient-rich ground waters from the adjacent mineral soils. Hence, the ground water of the bog is generally acid and low in nutrients. The dominant peat materials are undecomposed *Sphagnum* and moderately decomposed woody moss peat underlain, at times, by moderately to well decomposed sedge peat. The associated soils are Fibrisols, Mesisols, and Organic Cryosols (Canadian Soil Survey Committee 1978). Bogs may be treed with *Picea mariana* (Black Spruce) or tree-less and they are usually covered with *Sphagnum* spp. and feather mosses and ericaceous shrubs.

Fen. A fen is a peatland with a high water table, usually at or above the surface. The waters are mainly nutrient-rich, minerotrophic waters from adjacent mineral soils. The dominant peat materials are shallow to deep, well to moderately decomposed sedge or woody sedge peat. The associated soils are Mesisols, Humisols, and Organic Cryosols. The vegetation consists dominantly of sedges (Cyperaceae), grasses (Poaceae), reeds (Juncaceae), and brown mosses with some shrub cover and, at times, a scanty tree layer.

Marsh. A marsh is a mineral wetland or a peatland which is periodically inundated by standing or slowly moving waters. Surface water levels may fluctuate seasonally, with declining levels exposing drawn-down zones of matted vegetation or mud flats. The waters are nutrient-rich. The substratum usually consists of mineral material or moderately to well decomposed peat deposits. The associated soils are Humisols, Mesisols, and Gleysols. Marshes characteristically show a zonal or mosaic surface pattern of vegetation, comprised of unconsolidated grass and sedge sods, frequently interspersed with channels or pools of open water. Marshes may be bordered by peripheral bands of trees and shrubs, but the predominant vegetation consists of a variety of emergent non-woody plants such as rushes, reeds, reed-grasses, and sedges. Where open water areas occur, a variety of submerged and floating aquatic plants flourish.

Swamp. A swamp is a peatland or a mineral wetland with standing or gently flowing water in the form of pools and channels. The water table is usually at or near the surface. There is pronounced water movement from the margins or other mineral sources, hence the waters are nutrient-rich. If peat is present, it is mainly well decomposed woody or amorphous peat underlain, at times, by sedge peat. The associated soils are Mesisols, Humisols, and Gleysols. The vegetation is characterized by a dense tree cover of coniferous or deciduous species and by tall shrubs, herbs, and mosses.

Shallow water. Shallow water is semi-permanent to permanent standing or flowing water with relatively large and stable expanses of open water which are locally known as ponds, pools, sloughs, shallow lakes, bays, lagoons, oxbows, impoundments, reaches, or channels. Shallow waters are distinguished from deep waters by the upper 2 m limit, although depths may occasionally exceed this during periods of abnormal flooding. During droughts, low water or intertidal periods, drawn-down flats may be temporarily exposed. Included in this class are all basins in which summer open water zones exceed 8 ha in size, regardless of the extent of bordering wetlands. These shallow water units are delineated from wetland complexes by the outer border of floating vegetation mats or by mid-summer surface water levels, usually expressed by peripheral deep marsh emergents or shrubs. All other wetland basins less than 8 ha in area, with summer open water zones occupying 75% or more of the basin diameter, are classed as shallow water. The margins may be unvegetated or rooted emergent vegetation, including trees, confined to a narrow margin occupying no more than 25% of the basin diameter. Vegetation, if present in the open water zone, consists only of submerged and floating aquatic plant forms.

Bocher (1949) described and classified the lakes, including shallow water bodies, in the Sondre Stromfjord area of western Greenland. He identified: (1) lakes with acid water and low in salt (pH 5 to 6); (2) lakes with circumneutral water and low in salt (pH 7 to 7.5); (3) lakes with saline and alkaline water (pH 8 to 9); and (4) lakes with saline and highly alkaline water (pH 8.5 to 9.5). The lakes in groups 1 and 2 have well-vegetated shorelines; the lakes in groups 3 and 4 have sparsely vegetated or unvegetated shorelines.

Wetland forms and types. The wetland forms are determined primarily by the surface morphology of the wetlands, the morphology of the underlying mineral terrain, and the distribution of surface waters (Zoltai *et al.* 1973, Tarnocai 1970, 1980). Hydrotopographic features such as rivers and lakes and the type of water (e.g., fresh, brackish, or salt) also play an important role in determining the wetland forms.

In the classification, the wetland form terms are attached as modifiers to the wetland classes. The wetland forms recognized in Canada are indicated in Table 2. For example, the flat bog wetland form refers to an ombrotrophic peatland (a bog) having a relatively level surface with a relatively level underlying mineral interface and with connotations of hydrology and wetland dynamics. These wetland forms are readily identifiable on the ground, from the air, and on aerial photographs. Typical cross sections of fens (Fig. 3) and bogs (Figs. 4 and 5) demonstrates the great variability in wetland structure.

The term wetland type is used to describe the wetland based on the general physiognomy of the vegetation cover (Tarnocai 1980, National Wetland Working Group 1987). It is not a species description or vegetation community, but a term such as coniferous, hardwood, rush, or low shrub, to be used in connection with the wetland form. Eighteen types are recognized, but will not be discussed in this paper.

Regional aspects of Canadian wetlands

Wetlands exhibit regional differences across Canada both in terms of abundance and development. In general, there is a north-south temperature

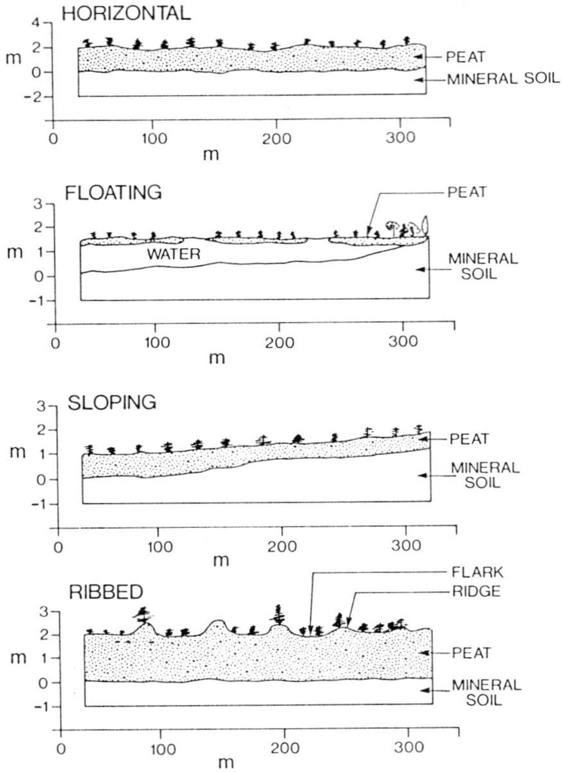

Figure 3. Cross-section of typical fens.

gradient and an east-west precipitation gradient with decreasing precipitation towards the west. Superimposed on this are the moderating influences of the Pacific and Atlantic Oceans upon their coasts.

The National Wetland Working Group has developed the concept of the "wetland region". This is defined as "Areas within which similar and characteristic wetlands develop in locations that have similar topography, hydrology, and nutrient regime. Subdivisions of these wetland regions are made based on the distribution of these wetlands, the relative abundance of the various kinds of wetlands (bogs, fens, swamps, marshes, and shallow water), or developmental trends somewhat divergent to those in the rest of the region" (National Wetland Working Group 1987). The distribution of the 20 wetland regions of Canada are given in Fig. 6 and Table 3. Some characteristics of subregions are presented in Table 4 including common wetlands, climate, and peat development.

Table 2. Classification keys to Canadian wetlands.

Part 1. Shallow water wetland forms

1. Inland; fresh to saline water bodies less than 2 m deep
   2. Associated with riverine systems
      3. Water continuously flowing in main water course .... Stream Water
      3. Water not continuously flowing
         4. Intermittent flowing water to discontinuous surface flow, confined to glacio-fluvial, eroded spillways .... Channel Water
         4. Intermittent flow or overbank flooding, impounded behind levees or ridges of alluvial deposits
            5. On river floodplains .... Oxbow Water
            5. On deltas .... Delta Water
   2. Not associated with riverine systems
      6. Surface catchment in topographically defined basin
         7. Basin not affected by permafrost
            8. Basin at terminus of drainage system .... Terminal Basin Water
            8. Basin not at terminus, water passes through the basin
               9. Shallow, gently sloping basin with relatively uniform depth .... Shallow Basin Water
               9. Relatively deep, bowl-shaped basin with moderately sloping sides .... Kettle Water
         7. Basin affected by permafrost
               10. Shallow basin with stable, steep shores .... Tundra Pool Water
               10. Shallow basin with unstable, collapsing shores .... Thermokarst Water
      6. Not in topographically defined catch basin, occupying the shallow shore zone of permanent open water bodies .... Shore Water

1. Coastal, estuarine, or marine water bodies less than 2 m deep
   11. Tidal water
      12. Estuarine channels or bays periodically inundated by fresh and brackish water .... Estuarine Water
      12. Coastal lagoons or bays primarily influenced by tidal action and marine salt water .... Tidal Water

   11. Non-tidal water
      13. Fresh to brackish water bodies located above mean high-tide zone .... Non-tidal Water

Table 2. Continued.

Part 2. Bog wetland forms	
1. Surface raised above surrounding terrain	
2. Surface convex	
3. Core frozen; abruptly domed; usually in fens	
4. Over 1 m high, diameter up to 100 m	Palsa Bog
4. Less than 1 m high, diameter up to 3 m	Peat Mound Bog
3. Core not frozen	
5. Convex surface small (1–3 m diameter); occurring in fens	Mound Bog
5. Convex surface often extensive; not occurring in fens	Domed Bog
2. Surface flat to irregular	
6. Core perennially frozen	
7. Surface with network of polygonal fissures	
8. Surface even	Polygonal Peat, Plateau Bog
8. Surface with high centres in a polygonal network	Lowland Polygonal Bog
7. Surface without polygonal fissures; surface about 1 m above the surrounding fen	Peat Plateau Bog
6. Core not frozen	
9. Bogs generally teardrop-shaped	Northern Plateau Bog
9. Bogs not teardrop-shaped; abundance of surface water	Atlantic Plateau Bog
1. Surface not raised above surrounding terrain	
10. Surface relatively level	
11. With abrupt marginal peat walls	Collapse Scar Bog
11. Without marginal peat walls	
12. Adjacent to water bodies	
13. Floating	Floating Bog
13. Not floating	Shore Bog
12. Not adjacent to water bodies	
14. Surface flat; topographically confined	
15. Basin deposit; depth greatest in centre	Basin Bog
15. Flat deposit; depth generally uniform	Flat Bog
14. Surface flat to undulating, often appreciably sloping	
16. Surface pattern of ridges and pools distinct	String Bog
16. Surface pattern of pools usually absent; extensive	Blanket Bog

10. Surface not level; appreciably sloping
    17. Core not frozen . . . . Slope Bog
    17. Core perennially frozen . . . . Veneer Bog

Part 3. Marsh wetland forms

1. Influenced by tidal water
  2. Water saline
    3. In river estuaries or connecting bays where tidal flats, channels, and pools are periodically inundated by water of varying salinity
      4. Located above mean high-water levels; inundated only at highest tides and/or storm surges . . . . Estuarine High Marsh
      4. Located below mean high-water levels; frequently inundated . . . . Estuarine Low Marsh
    3. On marine terraces, flats, embayments, or lagoons behind barrier beaches, remote from estuaries, where there is periodic inundation by tidal brackish or salt water including salt spray
      5. Located above mean high-water levels; inundated only at flood tides . . . . Coastal High Marsh
      5. Located below mean high-water levels . . . . Coastal Low Marsh
  2. Water fresh . . . . Tidal Freshwater Marsh

1. Not influenced by tidal water
  6. Located in topographically defined catch basins or valleys
    7. Associated with riverine or linear systems
      8. Adjacent to, or flooded by, flowing water
        9. Located on active fluvial floodplains adjacent to channels . . . . Floodplain Marsh
        9. Not on fluvial floodplains
          10. Occupying shorelines, bars, streambeds, or islands in continuously flowing water courses . . . . Stream Marsh
          10. Occupying abandoned glacial meltwater spillways, intermittent drainage courses, or open-ended, eroded channels . . . . Channel Marsh
      8. Located on river deltas
        11. Unrestricted water circulation, open connections to river channels and lakes, seasonally inundated . . . . Active Delta Marsh
        11. Restricted water circulation, inundated only during infrequent high river flows or wind tides

Table 2. Continued.

7. Associated with defined basins having poorly integrated surface drainage, fed by local runoff or groundwater	
12. Located at the terminus of an internal drainage system, may be flat or concave in topographically low areas, no outflow	Terminal Basin Marsh
12. Located along an internal drainage system; surface or underground water passes through the basin	
13. Shallow, gently sloping depressions that occur as natural swales or that occupy intervening areas between ridges or undulations on low-relief landforms	Shallow Basin Marsh
13. Sharply defined, bowl-shaped catch basin, usually located in high or intermediate topographic positions on moderate- to high-relief hummocky moraine, glacio-lacustrine or glacio-fluvial landforms	Kettle Marsh
6. Not located in topographically defined catch basins	
14. Occupying groundwater discharge sites, usually on or at the base of slopes	Seepage Track Marsh
14. Occupying the shores of semi-permanent or permanent lakes, receiving water from lake flooding or surface runoff	Shore Marsh

Part 4. Fen wetland forms

1. Surface not raised above surrounding terrain except in low hummocks and ridges	
2. Surface pattern of ridges and depressions	
3. Subparallel pattern of ridges and furrows	
4. Broad pattern; often very extensive	
5. Northern regions; lowland drainage; peat deep	Northern Ribbed Fen
5. Atlantic regions; mainly upland drainage; peat shallow	Atlantic Ribbed Fen
4. Narrow ladder-like pattern; along bog flanks	Ladder Fen
3. Reticulate pattern of ridges	Net Fen
2. Without pronounced surface pattern	
6. Featureless, adjacent to water bodies	
7. Floating	Floating Fen
7. Not floating	
8. Located in main channel or along banks of continuously flowing or semi-permanent streams	Stream Fen
8. Located along shores of semi-permanent or permanent lakes	Shore Fen
6. Depressed thaw hollows with high-water content peat; not adjacent to water bodies	Collapse Scar Fen

1. Surface raised above surrounding terrain
    9. Surface sloping appreciably
        10. With frozen core
            11. Mounds in patterned fen . . . . Palsa Fen
            11. Surface regular but sloping . . . . Snowpatch Fen
        10. Without frozen core
            12. Water from underground discharge . . . . Spring Fen
            12. Water from overland flow
                13. Surface with parallel drainage-ways . . . . Feather Fen
                13. Surface smooth or with irregular tracts . . . . Slope Fen
    9. Surface flat or depressional
        14. Core perennially frozen, surface with network of polygonal fissures . . . . Lowland Polygonal Fen
        14. Core not frozen, surface without pronounced surface pattern
            15. Basin part of regional drainage system
                16. Occupying broad depressions or plains . . . . Horizontal Fen
                16. Occupying well-defined, often eroded channels . . . . Channel Fen
            15. Basin does not receive regional drainage water . . . . Basin Fen

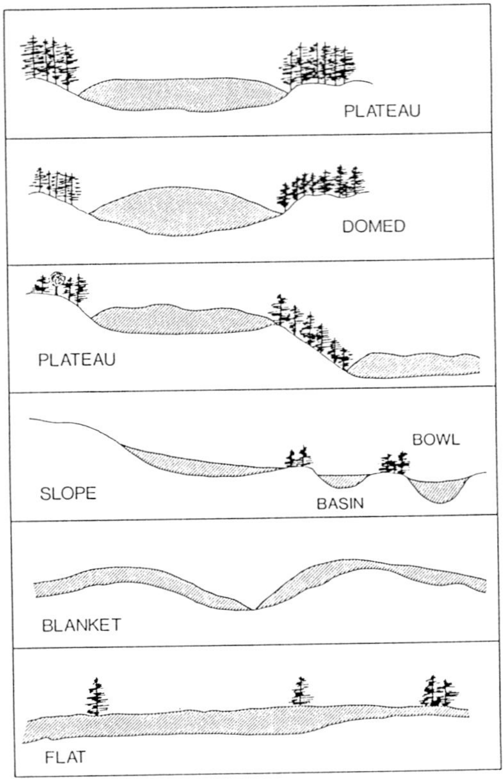

Figure 4. Cross-section of representative bogs in areas of no permafrost.

Wetland regions of Canada and Greenland

Low Arctic Wetland Region (AL)

This wetland region covers most of the continental arctic of Canada and part of the southern coast of Greenland (Fig. 6). Its southern limit coincides with the arctic tree line. The climate of this region is continental with the exception of areas along the arctic coast of mainland Canada and Greenland where the climate is marine-modified continental (Table 3). Bocher (1954) in Greenland differentiated between the oceanic and continental vegetation sub-regions using the 2,500 mm annual precipitation and the 25°C annual mean range of temperature as a boundary. The arctic continental climate is characterized by short cool summers, long cold winters, and low precipitation (Table 3). Permafrost is present under all land surfaces. The seasonal thawed active layer is approximately 40 cm deep under high centre polygons, 60 to 80 cm under wet fens, and 90 to 180 cm under marshes. The usual maximum

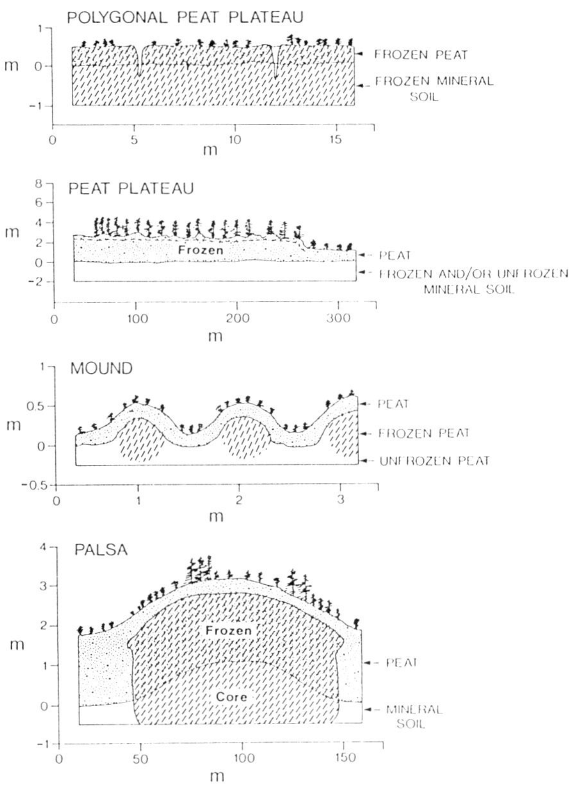

Figure 5. Cross-section of representative bogs in areas of permafrost.

thickness of peat is approximately 1.5 m on high centre polygons but only 50 cm on polygonal fens. Marshes have no surface peat layer. Fredskild (1961) reported that peat thicknesses up to 19 cm are associated with wetlands in the western part of Greenland.

In the Low Arctic Wetland Region, lowland polygon fens and bogs are by far the most widespread wetlands (Zoltai and Tarnocai 1975, Zoltai and Pollett 1983, Tarnocai and Zoltai 1988). Marshes are common along the coast and in the deltas while shallow water, a common phenomenon of the tundra, is also prevalent. Other wetlands occurring in this region are peat mound bogs (Fig. 5) and horizontal fens with peat cushions. Although peatlands are common in the extreme western parts of the region, they are scarce elsewhere. These peatlands occur mainly in depressional areas where the precipitation is concentrated either by runoff or by small creeks. Large expanses of tundra, covered with tussock-forming graminoid species such as *Carex bigelowii* and *Eriophorum vaginatum*, are not considered to be wetlands since they are not water-logged throughout the year (Zoltai and Pollett

Figure 6. Map of wetland regions of Canada. Abbreviations are described in the text.

1983). This view is in agreement with the definition of tundra bogs in Siberia used by Botch (1974).

A sequence of lowland polygon development is apparent in this wetland region. In the frozen fens polygonal cracks develop as a result of the intense cold. Ice wedges begin to form in these cracks, growing thicker and deeper year after year. The soil displaced by the ice wedges is elevated along the crack, forming a polygonal trough and shoulder pattern. In the initial low centre forms the shoulders enclose small, shallow ponds, giving it a paddy-like appearance. As peat is deposited in the centre of the polygon, it is gradually filled in with peat. This peat often contains mineral soil which has been mixed into it by cryoturbation. Once the centres are filled, the surface becomes relatively well drained and peat-forming plants can no longer grow there. In time, the polygonal troughs may become deeper as a result of thermal subsidence or erosion, leading to a high centre appearance. Such high centre polygons, if not protected by vegetation cover, are eroded by wind or by oxidation of the peat. At this stage, the high centre polygons are functionally no longer wetlands.

Table 3. Mean daily temperature and mean precipitation values in the wetland regions (Fig. 6) of Canada (Wetland W.G. 1984).

Wetland Region	Mean daily January temperature (°C)	Mean daily July temperature (°C)	Mean annual total precipitation (cm)	Mean annual snowfall (cm)
High Arctic-AH	−33	5	10	70
Mid Arctic-AM	−29	7	20	110
Low Arctic-AL	−29	9	15	80
High Subarctic-SH	−29	13	35	190
Low Subarctic-SA	−29	15	50	310
Atlantic Subarctic-SA	−5 to −15	12–15	85–130	270–460
High Boreal-BH	−23 to −25	13–16	40–70	160–260
Mid Boreal-BM	−20	17–19	50–80	120–280
Low Boreal-BL	−13	18	95	250
Atlantic Boreal-BA	−5 to −20	13–18	95–135	190–450
Continental Prairie-PC	−15 to −20	13–18	35–45	110–120
Intermountain Prairie-PI	−7	19		120
Eastern Temperate-TE	−9	20	35	210
Pacific Temperate-TP	−3	16	95	70
Atlantic Oceanic-OC	−3	14	155	190
Pacific Oceanic-OP	2 to 3	14	135	110
Coastal Mountain-MC	−4 to −19	13–17	195–305	130–270
Interior Mountain-MI	−8 to -27	13–18	30–105	120–250
Rocky Mountain-MR	−13 to −28	14–15	30–50	155–250
Eastern Mountain-ME	−23	12	70	370

Floodplain marshes occur in active floodplains adjacent to river channels. They usually occupy the low-lying alluvial islands which are flooded frequently and areas adjacent to river channels. Active delta marshes occur in river deltas with open drainage resulting from unrestricted connection to active channels.

The vegetation cover on low centre polygons is mainly sedges (*Carex* spp.) and cotton grasses (*Eriophorum* spp.) with mosses such as *Calliergon giganteum* and *Drepanocladus revolvens* also being present. The polygonal shoulders (along the trench) and the better drained portions of the high centre polygons are colonized by *Betula glandulosa* and ericaceous shrubs. The shoulders of the water-filled trenches, especially near the water level, are colonized by *Sphagnum* spp. Some high centre polygons are being eroded, mainly by wind, and have only a scattered vegetation cover. The vegetation associated with floodplain marshes is dominantly of the grass - *Equisetum* type. Flooded active delta marshes are associated with the grass-sedge and willow-sedge types of vegetation. The less frequently inundated portions of the active delta marshes are associated with low willows and patches of the sedge type of vegetation.

Table 4. Wetland regions of Canada.

Wetland region	Climate	Common wetlands	Peat development
High Arctic AH	Cold, short summers. Very cold winter. Very low precipitation. Permafrost under wetland.	Basin fen with or without lowland polygon development. Peat mound bog. Coastal marshes. Shallow water in low-lying areas.	Minimal, 0.5 m average thickness
Mid-Arctic AM	Cold, arid. Permafrost under wetlands.	Horizontal fens. Basin fens. Small, elevated peat mound bogs. Low centre polygonal bogs. Active high centre polygonal bogs rare. Coastal marshes. Shallow water in low-lying areas.	Peat 1.5 m in peat mounds and 0.5 m in fens.
Low Arctic AL	Cold, cool summers and low precipitation. Permafrost except close to lakes.	Lowland polygonal fens, both low and high centres. Spring fens. coastal marshes. Shallow water.	1.5 m in high-centre polygons, 0.5 m in polygonal fens.
High Subarctic SH	Cool summers. Low precipitation. Permafrost except close to lakes.	Polygonal peat plateau bogs. Basin and shore fens. Small palsa bogs. Coastal zone fens extensive. Stream fens. Shallow waters.	2 m average.
Low Subarctic SL	Cold winters and moderately cool summers. Low precipitation in west but higher in east. Permafrost under peats.	Peat plateau bogs. Palsa bogs mostly east of Nelson River. Ribbed and horizontal pens. Marshes and shallow water common in Hudson Bay lowland.	2–3 m average.
Atlantic Subarctic SA	Cool wet summers and relatively cool winters. High winds and extreme exposure. No permafrost.	Atlantic plateau bogs with palsa bogs in Labrador. Peat mound bogs and basin bogs. Ribbed fens at high elevations. Some slope fens.	1–2 m in bogs. Fens up to 2 m.
High Boreal BH	Cool winters, moderately cool summers. Low precipitation in west increasing eastwards. Sporadic permafrost.	Peat plateau bogs. Palsa bogs. String bogs. Ribbed fens. Veneer bogs in northern areas. Coastal marshes, horizontal fens and shallow water near Hudson Bay.	2–4 m, thickest peat in plateau bogs.

Mid-Boreal BM	Cold winters and warm summers in west to mild winters and cool summers in east. Precipitation highest in east. No permafrost.	Tree bogs and fens. Raised bogs in humid east. Floating fens and shore swamps bordering lakes and ponds. Marshes and fens locally common. Domed, flat and basin bogs in humid areas. Horizontal and ribbed fens in continental areas.	4–5 m except in areas transitional to Prairie Region in south where 2–3 m less peat in fens and minimal peat in swamps and marshes
Low Boreal BL	Cold winters and warm summers with high precipitation especially in east. No permafrost.	Bowl bogs, treed and often surrounded by peat margin swamps. Some basin swamps and fens in depressions.	Bog and fen peats average 5 m. Swamps seldom exceed 50 cm.
Atlantic Boreal BA	Relatively mild winters. Cool summers persistent fog. Precipitation ranges from 950 to 1,500 mm annually. No permafrost.	Domed bogs. Peat Plateau bogs. String bogs. Atlantic ribbed fens. Slope fens. Stream fens. Stream swamps. Basin swamps. Salt marshes on coast. Freshwater marshes along streams and floodplains.	5–10 m in domed bogs. 8–10 m in plateau bogs. 1–2 m in fens.
Continental Prairie PC	Cold winters and hot summers. Semi-arid. No permafrost.	Marshes including saline ones. Shallow water, both fresh and saline.	Mostly absent; if present, 50 cm.
Intermountain Prairie PI	Mild winters and hot summers. No permafrost.	Marshes. Emphemeral or semi-permanent shallow water, both fresh and saline.	Peat absent
Eastern Temperate TE	Mild winters and warm summers. High annual precipitation.	Basin and stream swamps with harbours basin and flag bogs bare. Shore marshes, stream marshes and fens along water bodies.	2 m in swamps and 3 m in bogs.
Pacific Temperate TP	Mild winters and warm summers. High annual precipitation.	Horizontal fens. Basin swamps. Flat and basin bogs. Marshes in floodplains and deltas. Coastal marshes.	4–5 m in swamps, 2 m in bogs and up to 4 m in fens.
Atlantic Oceanic BA	Cold winters and cool summers. High precipitation.	Blanket bogs. Small slope fens and slope bogs. Some ribbed fens.	Less than 2 m.

Table 4. Continued.

Wetland region	Climate	Common wetlands	Peat development
Pacific Oceanic OP	Mild winters and cool summers. High precipitation.	Flat bogs. Slope bogs. Horizontal and stream fens. Swamps. Coastal marshes.	1–4 m in bogs. 1–2 m in fens.
Coastal Mountain MC	Cool climate with moderate to high precipitation.	Flat bogs and horizontal fens in valleys. Small basin bogs and fens in alpine areas. Marshes along lakes and deltas.	1.5 m or less
Interior Mountain MI	Cool to cold with moderate to to low precipitation.	Flat and basin bogs. Horizontal fens in south. Palsa and peat plateau bogs in north. Marshes along lakes and deltas.	2–3 m in central to northern areas and 1–2 m in south.
Rocky Mountain MR	Cool to cold with moderate to low precipitation. Permafrost under wetlands in northern areas.	Flat and basin bogs in south. Palsa, peat plateau, and veneer bogs in north. Marshes along lakes and on delta.	Usually less than 1–15 m.
Eastern Mountain ME	Cool winters and cool summers. Low precipitation.	Slope fens. Ribbed fens.	10–20 cm.

Mid-Arctic Wetland Region (AM)

This wetland region covers the middle tier of arctic islands of Canada and the coast of central Greenland (Fig. 6). The climate of this region is a marine-modified continental type, characterized by short cool summers, long cold winters, and very low precipitation (Table 3). Permafrost is present under all land surfaces. The active layer is about 30 cm deep in peatlands and 40 cm deep under wet fens. The thickness of the peat is less than 150 cm on peat mounds and usually less than 50 cm on the fens. Marshes are not associated with peat.

In the Mid-Arctic Wetland Region the most common wetlands are lowland polygon fens. Lowland polygon bogs are rare except in eroding forms. Some small horizontal fens are also present, especially in depressions and on seepage slopes associated with snowbanks. These fens are often associated with elevated peat mounds (Zoltai and Pollett 1983). Horizontal fens are also common but are confined to poorly drained lowlands or to basins which receive runoff water from the surrounding mineral terrain. Salt marshes are common along the low-lying coastal lowlands.

Low centre lowland polygons occur locally but high centre lowland polygons are rare. The strongly eroded high centre polygons found in this region developed under a former climate and are considered to be relict phenomena. Horizontal fens and basin fens are similar, the major difference being that basin fens are confined and protected and receive more runoff from the surrounding mineral terrain. Peat mounds may be raised as much as 50 cm above the fen surface, elevated by ice accumulation in the peat, and may be regarded as palsas of the continuous permafrost zone (Washburn 1983a,b).

The vegetation cover on fens is *Carex* spp. and *Eriophorum* spp. with mosses such as *Aulacomnium turgidum* and *Drepanocladus revolvens* also occurring. Small isolated cushions of *Sphagnum fuscum* and *S. nemoreum* may occur on these fens. These sphagnum mosses often occur on rocks submerged in fens. In some cases the *Sphagnum* cushions coalesce to form peat mounds. The peat mounds are covered by lichens, sphagnum mosses, ericaceous shrubs, and dwarf birch (*Betula glandulosa*). High centre polygons are commonly eroded by wind and are entirely devoid of vegetation cover.

High Arctic Wetland Region (AH)

This wetland region covers most of the high arctic islands of Canada, including the northern and northeastern parts of Baffin Island as well as the north coastal strip of Greenland and higher elevation areas below the glacial ice on Greenland. The climate of this region is a marine-modified continental type, characterized by short cool summers, long cold winters, and very low precipitation (Table 3). Permafrost is present under all land surfaces. The active layer is about 20 to 30 cm deep in peatlands and 30 to 40 cm deep under wet fens. Peat development is minimal with the average thickness of peat being about 50 cm. None of the marshes are associated with peat. Because of the aridity of the high arctic, wetlands are scarce; they occur mainly in poorly drained lowlands and along the coastal lowlands.

The common wetland types in this wetland region are low centre lowland polygons, fens, often with shallow tundra ponds (shallow water), and peat mounds. Peat accumulation is very slow. The low centre lowland polygons often enclose shallow pools within the shoulders of the polygons. Active high centre lowland polygons are non-existent. The strongly eroded high centre polygons found on the high arctic islands developed under a former climate and are considered to be relict phenomena. Peat mounds similar to those occurring in the Mid-Arctic Wetland Region can be found here. As a result of ice accumulation in the peat these peat mounds may be elevated as much as 50 cm above the surrounding surface. Because of their internal morphology and since the ice occurs primarily in the form of segregated ice, these peat mounds may be regarded as palsas of the continuous permafrost zone (Washburn 1983a,b). Seepage fens occur on slopes where continuous moisture is available throughout the growing season. The source of the moisture is usually late-thawing snowbanks. Horizontal fens are also common but are confined to poorly drained lowlands or to basins which receive runoff water from the surrounding mineral terrain. Along the coast and in low-lying areas, marshes and shallow water are common.

The vegetation on basin and horizontal fens is mainly *Carex* spp. and *Drepanocladus revolvens*. The vegetation on seepage fens is similar to that found on basin and horizontal fens, with other mosses such as *Bryum cryophilum* and *Catoscopium nigritum* also being present. High centre polygons are unvegetated because of their actively eroding nature. The peat mounds are associated with the lichen and moss types of vegetation.

Although wet lowlands occupy only a small portion of the High Arctic, they constitute the most productive parts of the landscape (Muc 1977). The annual production of the bryophyte layer, based on *Messia triquetra* as marker species, was 77 g m^{-2} in a sedge meadow (Vitt and Pakarinen 1977). The same wetland, a type of fen, had an average total biomass of 202.7 g m^{-2} in the above-ground layer, and 1,366.8 g m^{-2} in the below-ground layer (Muc 1977). The total net primary production was 165.1 g m^{-2} (45.4 g m^{-2} above-ground and 128.7 g cm^{-2} below-ground) in this wetland community, the most productive of the habitats studied.

Arctic salt marshes

Salt marshes occur in various parts of the coastal areas of the arctic regions of Canada and Greenland. Since few major rivers influence these coastal areas, most marshes are of the coastal type as opposed to estuarine marshes. The marshes are characterized by the turf-forming grass species, *Puccinellia phryganodes*. Polunin (1948) has described salt marshes in northern Hudson Bay. There, he found *Puccinellia phryganodes*, several *Carex* species including *C. subspathacea*, *C. ursina*, and *C. glareosa*, various grasses (mainly *Dupontia fisheri*), and assorted forb species. Jefferies (1977) studied arctic salt marshes and found the pioneer community to be dominated by *Puccinellia phryganodes*, *Stellaria humifusa*, and *Cochlearia officinalis*. In more sheltered areas, common species were *Arctophila fulva*, *Dupontia fisheri*, *Hippuris tetraphylla*, and *Carex ramenskii*. Such marshes are of low productivity. The net annual primary production of *Puccinellia phryganodes* growing along the Arctic Ocean coast of Canada was estimated at 10 g m^{-2}.

Greenland has limited salt marsh development. Such marshes are restricted to protected coastal areas such as river mouths where fine-grained sediments occur, such as the Disko area on the west coast (Vestergaard 1978). Dominant species include *Puccinellia phryganodes*, *Carex glareosa*, *C. subspathacea*, *Potentilla egedii*, and *Stellaria humifusa*. These occur either in distinct zones or, more commonly, as patches. Freshwater marshes may be present behind the salt marshes with such species as *Carex rariflora*, *C. aquatilis* var. *stans*, *Equisetum arvense*, *Polygonum viviparum*, and *Salix glauca*.

Wetlands of the Subarctic Region (*SH*, *SL*, *SA*)

The uplands of the subarctic regions are characterized by open-canopied woodlands of coniferous trees with abundant ground lichen cover. The wetlands are mainly bogs and fens, many affected by permafrost. Subarctic wetland regions occupy about 1.5×10^6 km^2 of Canada, occurring as a broad belt between the treeless arctic regions to the north and the closed-canopy boreal forests to the south. In Greenland the subarctic zone is restricted to the southwestern coastal part of the island, generally south of Godthaab where there are remnants of the original treed vegetation (Bocher 1938).

The climate on the Canadian continent is characterized by cold winters and moderately warm, but short, summers, and low amounts of precipitation. In the coastal areas of eastern Canada the temperatures are more moderate, but the precipitation is higher (Table 3). In southwestern Greenland the summers are much cooler than on the continent and precipitation is moderate.

The physiography of this region is gently undulating in the Interior Plains and hilly on the Precambrian Shield. The entire region has been glaciated during the latest Pleistocene glaciation. The glaciers have removed most of the unconsolidated soils from the Precambrian bedrock area, leaving bare bedrock with only a thin morainic cover. Elsewhere, the till can be thick. Plains and broad valleys occur throughout the region, mainly in areas of post-glacial marine submergence or within the former beds of glacial lakes. Although wetlands are most extensive in such areas of low relief, they are abundant in all parts of the subarctic. It is estimated that about 40% of the land in the subarctic region is occupied by wetlands.

In the subarctic wetland regions the characteristic and common wetlands are bogs and fens (Zoltai *et al.* 1988). The bogs are nearly all affected by permafrost. As the permafrost develops, the bog surface is elevated above the regional water table. The surface is therefore well beyond the reach of the minerotrophic waters of the surrounding fen. The surface of the peatlands elevated by permafrost becomes dry and peat accumulation virtually ceases. Disturbances such as wildfires may destroy the insulating surface layer, causing the permafrost to thaw within the peat. This triggers a collapse of the elevated peatland as the surface bog levels sink into the surrounding fen.

Different kinds of wetlands occur in different portions of the subarctic region, and the subarctic has been divided into wetland regions and subregions on this basis. In the colder, more northerly belt, the High Subarctic Wetland Region (SH), all wetlands are underlain by permafrost and ice wedges have developed in some peatlands. In the continental part of the somewhat less cold Low Subarctic Wetland Region (SL), permafrost is widespread, but the fens are free of permafrost and ice wedges are rare in peatlands. In the coastal Low Subarctic Wetland Region (SL), as well as in southern Greenland, permafrost is sporadic in wetlands, and it is absent from the Atlantic Subarctic Wetland Region (SA).

The following is a description of the five most commonly occurring wetlands in the subarctic regions. Although these wetlands are characteristic of these regions, variations both in the configuration and vegetation of wetlands do occur.

Peat plateaus

Peat plateaus (Fig. 5) are peatlands that have a permafrost core. They are elevated about 1 m above the level of the surrounding fen. Their surface is generally level, but small damp depressions do occur. Peat depth is 2 to 3 m, frozen beneath the seasonally-thawed active layer. The permafrost extends into the underlying mineral soil. The water content of the frozen peat is similar to that of the unfrozen peat and ice lenses or layers are generally absent. Some ice accumulations may occur at the peat-mineral soil interface.

Peat plateaus vary in size from a few square metres to hundreds of square kilometers. The smaller ones occur as somewhat raised plateaus in wet fens, often with a melting, collapsing edge. The large peat plateaus often have small, unfrozen fen depressions on their surface. Peat plateaus, when occurring on a slight slope, have a well-defined surface drainage system of seasonal rivulets. Peat plateaus are the most common wetland form in the subarctic wetland regions, being especially widespread in the western part of the region.

Peat plateaus are characterized by scattered, stunted growth of *Picea mariana* trees (Horton *et al.* 1979). There is a well developed low shrub layer of *Ledum groenlandicum* or *L. palustre*. The sparse herb layer consists of *Vaccinium oxycoccus*, *V. vitis-idaea*, and *Rubus chamaemorus*. Most of the ground surface is covered by lichens, such as *Cladonia rangferina*, *C. mitis*, *C. alpestris*, and *C. uncialis*. *Sphagnum fuscum* and *S. nemoreum* occur in low cushions.

The peat plateaus often contain collapse scars which are surrounded on all sides by the peat plateau and are not connected to the surrounding fens. In such cases the collapse scars receive drainage from the peat plateau and are poor in nutrients. The dominant vegetation is *Carex paupercula*, *C. aquatilis*, *Sphagnum riparium*, *S. jensenii*, and *S. angustifolium* (Horton *et al.* 1979).

Fires often sweep across the dry, raised peat plateaus, destroying the living vegetation, but leaving the peat unburned (Jasieniuk and Johnson 1982). The vegetation after fire consists of species such as *Ledum groenlandicum*, *Polytrichum juniperinum*, *Pohlia nutans*, and *Ceratodon purpureus* that can readily regenerate from rhizomes and gemmae. These are followed by lichens which show a well defined successional sequence following the fire. The initial lichens are *Cladonia deformis* and *C. gracilis*, followed by *C. mitis*, *C. coccifera*, *C. amaurocrea*, *C. rangiferina*, *C. alpestris*, and *Cetraria nivalis*. The latter two species are found on peat plateaus that have not burned for a long time and therefore indicate a stable vegetation cover.

The development of a peat plateau-palsa complex was studied in northern Quebec (Couillard and Payette 1985). They found that peat deposition began about 3,700 years BP in an open fen that later progressed to a treed fen stage. Peat plateaus and palsas were formed as bog vegetation became established on parts of the fen after 2,700 years BP. Later the perennially frozen peatlands expanded during distinct periods at about 1,400, 1,100, 700, and 150 years BP. Collapse scars were formed after a fire around 1,100 years BP, and later at 340 years BP and at the present due to climatic warming.

Figure 7. Aerial view of a polygonal peat plateau, High Subarctic Wetland Region near Great Slave Lake. The diameter of polygons is approximately 15 m.

Polygonal peat plateaus

These perennially frozen peat landforms are similar to peat plateaus in being elevated above the fen levels and having a generally flat surface. They are distinguished from peat plateaus by the presence of a well-developed polygonal system of trenches (Fig. 5) which are underlain by ice wedges (Zoltai and Tarnocai 1975). The polygons are irregular in shape, with diameters in the 10 to 30 m range (Fig. 7). The polygonal trenches are 1 to 2 m wide and are about 50 cm below the plateau level. They are usually moist, but some may contain water. Corings and excavations revealed that there is an ice wedge under each polygonal trench. These ice wedges are about 1 m wide at the top, becoming narrower with depth. The ice wedges are usually in excess of 3 m long, extending well into the underlying mineral soil substrate.

Polygonal peat plateaus are common in the High Subarctic Wetland Region where they can occupy entire basins of several square kilometers in size. They may be associated with small fens in drainage tracks; in this region even the fens are underlain by permafrost.

The seasonally thawed active layer is thin, with a maximum thaw depth of 50 cm (Dredge 1979a), as the dry peat is an effective thermal barrier. However, in some instances the ice wedges may thaw, filling the trenches with water (Dredge 1979b). This allows the peat in the centre of polygons to become saturated with water, losing its insulating qualities. The surface

Figure 8. Ground aspect of polygonal peat plateau showing junction of polygon trenches marked by *Carex* tussocks. Site located near Great Slave Lake.

subsides as the permafrost thaws and thermokarst ponds are formed. This process can be accelerated by man-caused disturbances of the natural land surface.

The vegetation of polygonal peat plateaus is dominated by terricolous lichens. Trees are usually absent, although a few scattered individuals of stunted *Picea mariana* may occur. There is a low (20 cm high) shrub layer, consisting of *Ledum decumbens* and *Betula glandulosa*, along with *Vaccinium vitis-idaea* and *Rubus chamaemorus*. The dominant lichen layer consists mostly of *Cladonia mitis*, *C. alpestris*, *C. rangiferina*, *Cetraria nivalis*, *C. cucullata*, and *Alectoria ochroleuca*. In the trenches *Sphagnum balticum* and *Eriophorum russeolum* are usually present (Fig. 8).

The developmental history of a polygonal peat plateau, as shown by pollen, macrofossil and matrix composition, shows a hydroseral succession (Ovenden 1982). The initial deposits showed a progression from a pool to a marsh, then to a fen, and finally to a bog community. Permafrost affected the peatland some time after 9,600 yrs BP, after *Sphagnum* became the dominant vegetation.

Palsas

Palsas (Fig. 5) are peat mounds that have a perennially frozen core. The diameter of the mounds seldom exceeds 100 m, and their height may reach

Figure 9. Small palsa with unfrozen fen in foreground. Located in Low Subarctic Wetland Region, Mackenzie Valley.

4 m. They are situated as islands or peninsulas in very wet fens (Fig. 9). They appear to go through a period of growth, followed by erosion and final collapse (Railton and Sparling 1973, Kershaw and Gill 1979). They originate under *Sphagnum* cushions where small lenses of seasonal frost may persist. As the frost elevates the small moss cushion, it becomes drier and better insulating, resulting in accelerated growth. Water is drawn into the frozen peat and accumulates as ice in layers. After a certain size is reached, the peat may crack open and bare peat patches may appear, leading to accelerated wind and thermal erosion. Finally, the palsa may disappear completely. This process may take over 1,000 years (Kershaw and Gill 1979), but may occur in less than 200 years (Railton and Sparling 1973).

The palsas are composed of a frozen peat core covered by a seasonally thawed active layer of 40 to 70 cm thickness. The permafrost extends well into the underlying mineral soil. There may be some thin (10 to 20 cm) ice layers in the peat, but thick ice accumulations occur at the peat-mineral interface and in the upper part of the mineral soil. Thus, the height of the palsas is due to ice accumulation as well as to volume expansion as water changes to ice. Palsas occur in the Low Subarctic as well as in the High Boreal Wetland Regions. They may be locally numerous, but do not cover large areas.

The vegetation of palsas changes with the stage of their development

(Railton and Sparling 1973). In the youthful stage, *Sphagnum fuscum* dominates the vegetation, with low shrubs of *Kalmia polifolia* and occasional *Picea mariana* seedlings. The herb layer has *Vaccinium oxycoccus*, *Drosera rotundifolia*, and *Scirpus cespitosus*. In the mature stage there are low stunted *Picea mariana* trees on the palsa, with some *Ledum groenlandicum* shrubs. The surface, however, is dominated by lichens such as *Cladonia alpestris*, *C. rangiferina*, *C. arbuscula*, *C. gracilis*, and *C. pyxidata*. The moss *Polytrichum juniperinum* occurs on peat exposed by erosion.

Northern ribbed fens

These minerotrophic wetlands (Fig. 3) are characterized by low (70 cm high) and narrow (3 m wide) ridges that extend across the fens at right angles to the direction of drainage. The areas between the ridges (flarks) are usually wet, often with shallow standing water. The ridges (strings) are better drained and may contain permafrost, especially at the nodes where two or more ridges intersect. There is water movement through the fen and through the ridges, although the ridges impede the drainage to the extent that the level of successive flarks is lower than the level of the upslope flarks. Permafrost development at the nodes of ridge intersection or in small segments of the ridges does not interfere significantly with the drainage. However, permafrost development in the entire length of the ridges will stop the drainage through the peat and leads to the development of small open channels or creeks in the centre of the peatland that will provide surface drainage. This process leads to the gradual expansion of permafrost into the flarks and eventually the entire fen will become a peat plateau, perhaps with a few depressions marking the locations of the former flarks.

The vegetation of the flarks is dominated by *Carex aquatilis* and *C. limosa*. Other vascular species include *Drosera anglica* and *Menyanthes trifoliata*. The mosses, such as *Scorpidium scorpioides*, *Campylium stellatum*, *Drepanocladus revolvens*, and *D. exannulatus* are common. In the ridges that have no permafrost, low *Picea mariana* and occasional *Larix laricina* may occur, along with shrubs such as *Chamaedaphne calyculata*, *Andromeda polifolia*, and *Myrica gale*. The mosses *Sphagnum fuscum* and *S. magellanicum* are also common. On ridges with permafrost the vegetation resembles that of peat plateaus: low *Picea mariana* trees with *Ledum groenlandicum* and the dominant ground lichens, *Cladonia mitis*, *C. rangiferina*, and *C. alpestris*.

Slope bogs

These oligotrophic wetlands (Fig. 3) occur on slopes ranging up to 15% (Wells 1981) in the high rainfall area of the Atlantic Subarctic Wetland Region (SA), mostly in Newfoundland (Wells and Pollett 1983). Their development is related to climatic (large amounts of precipitation) and edaphic (soils derived from acidic bedrock) factors. In this area, small depressions on slopes initially develop into fens, both ribbed and slope, but the high rainfall and the nutrient-poor seepage water allow *Sphagnum* species to invade the fens and spread on the slope. The ability of this moss to retain moisture raises the permanent water table and bog conditions become prevalent along the entire slope. The peat thickness, however, seldom exceeds 2 m.

The slope bogs are usually treeless, but dwarf shrubs such as *Ledum groenlandicum*, *Kalmia angustifolia*, and *Empetrum nigrum* are common (Wells 1981). The main component of the bog is *Sphagnum fuscum*. On somewhat wetter depressions *Rhynchospora alba*, *Sphagnum tenellum*, *S. pulchrum*, and *S. cuspidatum* are common. Extensive carpets of wet *Sphagnum*, consisting of *S. magellanicum*, *S. rubellum*, and *S. papillosum* may occur. Sedges such as *Carex exilis*, *C. oligosperma*, and *Scirpus cespitosus* may be locally abundant.

Southwestern Greenland

Little information has been found on the wetlands of southern Greenland. Bocher (1938) indicated the presence of "bog" at his site #17. However, the dominant species, *Eriphorum scheuchzeri*, *Equisetum variegatum*, and *Saxifraga hirculus*, indicate that these were fens, resembling the seepage fens common in the low arctic in Canada.

Salt Marshes of the Low and High Subarctic Wetland Region

The Low and High Subarctic Wetland Regions (SL and SH) have a similar salt marsh vegetation. The Low Subarctic Wetland Region extends from approximately the Ontario/Quebec border at southern James Bay north to Cape Henrietta Maria on the Ontario coast where Hudson and James Bay come together. The High Subarctic Wetland Region extends along the Hudson Bay shoreline north from Cape Henrietta Maria to the Manitoba-Northwest Territory border (Fig. 6). The Hudson Bay Lowland extends from approximately the Ontario/Quebec border to Churchill, Manitoba thus favoring salt marsh development on extensive tidal flats. The salt marshes near Churchill have been studied by Jefferies *et al.* (1979). In river-influenced areas, *Hippuris tetraphylla* is the dominant species. On open tidal flats, *Puccinellia phryganodes* is the colonizing species and *Carex subspathacea*, *Cochlearia officinalis* var. *groenlandica*, and *Potentilla egedii* are common. Higher areas are characterized by forbs including *Chrysanthemum arcticum*, *Stellaria humifusa*, *Senecio congestus*, *Plantago maritima* var. *juncoides*, *Ranunculus cymbalaria*, *Triglochin maritima*, and *Salicornia europaea*. Fens occur landward of these marshes with *Carex aquatilis*, *C. glareosa*, *Eriophorum angustifolium*, *Calamagrostis neglecta*, and *Dupontia fisheri*. Kershaw (1976) described similar salt marsh vegetation further south at Pen Island at the Ontario/Manitoba border on Hudson Bay.

Glooschenko and Martini (1981) have studied the salt marshes of the Hudson Bay coast of Ontario. Glooschenko (1982) has described the vegetation on 26 transects along this coast. In areas of little river influence, *Puccinellia phryganodes* is the colonizing species with limited growth of other species such as *Triglochin palustris* in the low salt marsh. This gives way to a zone dominated by *Carex subspathacea* along with *Potentilla egedii*, *Stellaria humifusa*, and *Festuca rubra*. In areas where riverine influence is present, *Hippuris tetraphylla* is the dominant colonizing species. Landward of the salt marsh, fens are present if poor drainage occurs. Common species include *Calamagrostis neglecta*, *Carex glareosa*, *C. aquatilis*, *Eleocharis palustris*, *Eriophorum* spp., and *Salix* spp.

The Ontario coast of James Bay has been studied in detail (Glooschenko and Martini 1978). The coastline is ideal for wetland development as it is characterized by extensive tidal flats associated with a relatively rapid rate of isostatic rebound. The nearshore salinities are low but salt marsh vegetation occurs (Glooschenko and Clarke 1982). Glooschenko (1980a) has described three major marsh types in the area: 1) salt marshes, 2) river-influenced brackish marshes, 3) and estuarine marshes.

The salt marshes occur at a distance from the plumes of major rivers. The major colonizing species is *Puccinellia phryganodes*. Other species that occur are *Puccinellia lucida*, *Salicornia europaea* (in pans), *Glaux maritima*, *Scirpus maritimus*, *Triglochin maritima*, *Potentilla egedii*, *Plantago maritima*, *Festuca rubra*, *Juncus balticus*, *Cicuta maculata*, *Carex subspathacea*, *C. paleacea*, *Hordeum jubatum*, and *Atriplex patula*. The vegetation is arranged in distinct zones. The river-influenced brackish marshes lie south of major river mouths. Dominant species include *Carex paleacea*, *Hippuris tetraphylla*, and *Scirpus maritimus*. The salinity maximum can often occur up to 1 km inland of the shore and typical salt marsh vegetation lies in-shore of species more typical of brackish conditions. The estuarine marsh is found in, and adjacent to, river mouths. Major species include *Eleocharis palustris*, *E. acicularis*, *Sagittaria cuneata*, *Scirpus validus*, *S. americanus* and several species of the genera *Potamogeton*, *Carex*, *Juncus*, and *Equisetum*. Such wetlands have been described in detail for the Attawapiskat River estuary by Glooschenko and Martini (1985). These marshes are extremely important habitats for shorebirds and waterfowl (Martini *et al.* 1980). Riley and McKay (1980) have also discussed coastal wetland vegetation in some detail. Coastal marshes in the southernmost portion of James Bay have been studied by Ewing and Kershaw (1986) and Glooschenko and Martini (1987). Marshes here consist of brackish vegetation as previously described by Glooschenko (1980a).

The Quebec shoreline has little salt marsh development due to the lack of extensive tidal flats since the Canadian Shield extends to the shore. The marshes that do occur are estuarine and are found in river mouths and embayments. Dominant species include *Carex paleacea*, *Scirpus maritimus*, and *Hippuris tetraphylla* where brackish waters are present. Under freshwater conditions, common species include *Eleocharis* spp., *Scirpus validus*, *S. americanus*, *Calamagrostis* spp., *Potamogeton* spp., *Deschampsia caespitosa*, and *Carex glareosa* (Lamoureaux and de Repentigny 1972, Lamoureaux and Zarnovican 1972 and 1974, Laverdière and Guimont 1975).

The primary production of a James Bay salt marsh located at North Point north of the Moose River mouth was studied by Glooschenko and Harper (1982). The marsh consisted of six vegetation zones from a *Puccinellia phryganodes* – dominated intertidal zone to meadow-like zone with *Salix* thickets. Peak above-ground biomass occurred in early August. Net annual aerial primary productivity ranged from 119 to 384 g m^2 with lowest values in the high salt marsh and highest values in the landward *Salix* thicket. A mean value of 228 g m^{-2} was found which is low for salt marshes. Studies on the above-ground biomass of plants in the same marsh had been carried out one year earlier by Glooschenko (1978). Depending upon the vegetation, above-ground biomass in 1977 was 42.4% to 86.8% of 1976 which was attributed to the latter year having a cooler summer.

Further north in a Hudson Bay salt marsh located near Churchill, Manitoba, net above-ground primary production of 65 to 97 g m^{-2} was measured (Cargill and Jefferies 1984a). They also determined that inorganic nitrogen was limiting productivity of the marsh. Phosphorus was only limiting when nitrogen was previously supplemented. Another imporant factor in these marshes is the role of geese which are important grazers. Such geese influence the species composition, standing crop, and litter production of these subarctic salt marshes (Cargill and Jefferies 1984b, Bazely and Jefferies 1986).

Factors affecting subarctic wetland development

The wetlands of the subarctic regions are mainly bogs and fens. However, the characteristic wetlands are those in which permafrost conditions have developed.

Numerous corings (Reid 1974, Zoltai and Tarnocai 1975) have shown that in the peatlands presently affected by permafrost, the peat was initially deposited in non-permafrost envlronments of fens and bogs. The peat macrofossils show that the vegetation cover changed in the surface 50 to 100 cm. The peat in this layer shows much increased *Sphagnum* moss remnants, as well as roots and twigs associated with forested peatlands. The inference is that the establishment of *Sphagnum* cover initiated permafrost development, uplifting the peatand creating drier conditions which were suitable for tree growth. In the more northerly parts of the region this process was followed by ice wedge development in the peat.

The timing of the permafrost development is not known. In many areas there are newly developed, thin, frozen peat lenses (Reid 1974), indicating that permafrost development can be initiated under the present climatic conditions. At the same time, many instances of permafrost degradation were noted, often on the same peatland. This indicates a delicate balance where changes in permafrost can be initiated or destroyed by a slight alteration in the environment under the present climatic conditions.

Wetlands of the Boreal Region (BL, BM, BH)

The boreal region is characterized by closed forests of dominantly coniferous trees, or mixtures of coniferous and hardwood trees (Rowe 1972) on the uplands and by fen and bog peatlands in the waterlogged areas. The boreal wetland region covers approximately 3.1×10^6 km^2 of Canada, over one-third of the country (Fig. 6). The climate is characterized by cold winters and warm summers, with moderate to high amounts of precipitation (Table 3). There is a marked north-south gradient, with decreasing temperatures northward. An east-west precipitation gradient is also evident, with higher precipitation in the east, decreasing to much lower levels westward towards the centre of the continent. These climatic differences are reflected in the development and distribution of wetlands within the boreal area. Such ecological differences serve as a basis for the characterization of the boreal wetland regions in Canada (National Wetland Working Group 1985).

The physiography of this large region varies from the low hills of the Precambrian peneplain, through gently undulating or flat Interior Plains to the low mountains of the foothills of the Rocky Mountains. Within this area wetlands are most extensive in areas of low relief which have poor internal and external drainage, such as the extensive peatlands in the Hudson Bay Lowland which will be discussed later in this paper. Elsewhere, peatlands are extensive within glacial lake basins (central Manitoba, Ontario Clay Belt) or in other areas of low relief (northern Alberta). Within the boreal region it is estimated that about 25% of the land area is covered by wetlands.

The characteristic wetlands within the Boreal regions are bogs and fens, with swamps and marshes restricted to suitable areas (Zoltai *et al.* 1988b). Bogs develop chiefly in areas of high rainfall, where the precipitation occurs in sufficient quantities to maintain waterlogged conditions. In such areas the rapid accumulation of moisture raises the surface above the minerotrophic water table, leading to the development of raised bogs. Raised bogs are best developed in the humid climates of the eastern boreal region where they may attain a height of several metres above the regional water table. In the more arid west, the bog surfaces are raised only slightly (30 to 50 cm) above the minerotrophic water levels, but in many cases this is sufficient to establish true bog conditions.

Eight of the common and typical wetland types occurring within the boreal region are described below. These are not intended to be comprehensive descriptions of the wetlands of this large area, but rather to characterize the region, highlighting the differences and variabilities encountered.

Domed bogs

Domed bogs occur where the rapid accumulation of peat results in the elevation of the peat surface well above the regional water table. In cross section the surface is convex (Fig. 4), with the highest point near the centre, which may be several metres above the level of the regional water table. Domed bogs can be large, in the order of 1 km^2, and may occupy an entire wetland or a portion of a larger wetland complex. Domed bogs are characteristic of the humid portion of the boreal regions, and generally occur east of Lake Winnipeg.

The peat in the domed bogs can be several metres thick, composed mainly of *Sphagnum* peat (Bannatyne 1980). They often display drainage slots which radiate from the highest point of the domed bog. The vegetation of the raised portion consists of a dense cover of *Picea mariana*, with the trees often reaching heights in excess of 10 m. *Sphagnum fuscum* is the dominant ground cover, occurring in discrete cushions or as a carpet. Some minor *S. nemoreum* and *Pleurozium schreberi* may occur on some *Sphagnum* cushions, along with *Cladonia rangiferina* and *C. alpestris*. In openings low shrubs such as *Ledum groenlandicum*, *Kalmia polifolia* or *Kalmia angustifolia*, and *Chamaedaphne calyculata* may be present. *Sphagnum angustifolium* is usually found in wetter patches, together with *Rubus chamaemorus* and *Sarracenia purpurea*.

Northern plateau bogs

Northern plateau bogs have an ombrotrophic surface that is slightly (50 to 75 cm) above the minerotrophic water table of the surrounding fen. The surface topography is usually level, but there may be a sharp, sudden drop to the fen level at the edge of the plateau bog. Plateau bogs typically occupy part of a larger wetland complex, reaching several hundred hectares in size. They are surrounded on at least three sides by fen channels (laggs) where most of the drainage in the fen takes place. This imparts a somewhat streamlined shape to the plateau bogs when viewed from the air. Plateau bogs are common in the western parts of the Mid-Boreal and High Boreal Wetland Regions where precipitation is limited.

The thickness of the peat deposit is 2 to 4 m, with the surface 1 to 2 m composed of *Sphagnum* remains, and the rest usually composed of fen peat. The vegetation cover consists of a semi-open stand of stunted *Picea mariana*, where the trees seldom exceed 5 m in height. The shrub layer is composed of *Ledum groenlandicum*, with lesser amounts of *Chamaedaphne calyculata*, *Kalmia angustifolia*, and *Rubus chamaemorus*. The ground is covered by *Sphagnum fuscum* in cushions or in coalesced cushions. On the drier cushions *Vaccinium vitis-idaea* and *V. oxycoccus* grow with *Cladonia rangiferina*. In wetter hollows *Sphagnum angustifolium* and *Eriophorum vaginatum* may be found.

The primary productivity of three vegetation zones of a domed bog and its marginal fen lagg has been determined by Reader and Stewart (1972). They found that in a closed-canopied *Picea mariana* forest the primary production was 709.9 g m^{-2}, and the total biomass was 6,934.4 g m^{-2}. In the stunted *Picea mariana*-ericad woodland the net production was 992.6 g m^{-2} and the total biomass was 2,639.0 g m^{-2}. In a treeless ericaceous shrub bog the net production was 1,924.6 g m^{-2} and the total biomass was 2,516.0 g m^{-2}. In the fen lagg the net production was 1,631.0 g m^{-2} and the total biomass was 3,810.3 g m^{-2}. They measured the decomposition rates on these sites and estimated that less than 10% of the annual net primary production will accumulate as peat.

Basin bogs

Basin bogs are minerotrophic peatlands that develop with indiscrete basins of essentially closed drainage (Fig. 4), receiving water from precipitation and drainage from the surrounding slopes (Fig. 10). The surface is generally flat and featureless but there may be a narrow (up to 50 m) belt of treed or shrubby swamp along the margin where mineral-rich surface runoff water affects the vegetation. Basin bogs usually fill the entire topographic basins where the thickness of peat may exceed 3 m. Basin bogs are found throughout the boreal region. They occur in areas where the surrounding terrain is poor in nutrients or where the mineral-rich runoff water is insufficient to affect the ombrotrophic bog surface.

Basin bogs are usually treed with *Picea mariana*, but some may be treeless. In the treed form the trees are rarely more than 5 m tall and are widely spaced. The shrub layer is dominant in both the treed and treeless forms, composed of *Chamaedaphne calyculata*, *Kalmia polifolia* or *K. angustifolia*, and *Ledum groenlandicum*. The moss layer is dominated by *Sphagnum fuscum* and *S. magellanicum*.

Figure 10. Basin bog in BL Wetland Region near Atikokan, Ontario.

Wooded palsa
Peatlands affected by permafrost are encountered in the High Boreal Wetland Region. These peatlands consist of a perennially frozen core in the peat which may extend into the underlying mineral soil material. The permafrost peat landforms, peat plateaus and palsas, develop when the insulation provided by the living moss cover prevents the complete thawing of the seasonal frost. This process, repeated through the years, results in the formation of a lens of frozen peat. As the water in the peat changes into ice, its volume increases and the surface is uplifted. The result is a raised surface, about 1 m high, in the peat plateaus that may cover hundreds of hectares. Palsas are much smaller in aereal extent (up to 1 ha), but are considerably higher (up to 4 m) in Canada. This greater height of palsas is due to ice accumulation, usually at the peat-mineral soil interface.

The vegetation on palsas occurring within the High Boreal Wetland Region is characterized by dense forests of *Picea mariana*. The vegetation appears to be related to fire history (Zoltai and Tarnocai 1971). On the undisturbed palsas dense, but low (4 m) *Picea mariana* grows with a ground vegetation dominated by lichens such as *Cladonia rangiferina*, *C. mitis*, and *C. alpestris*. Shrubs (*Ledum groenlandicum*, *Chamaedaphne calyculata*) occur in small openings, along with *Aulacomnium palustre*. On palsas that have burned within the past 80 years, dense forests of *Picea mariana* are found,

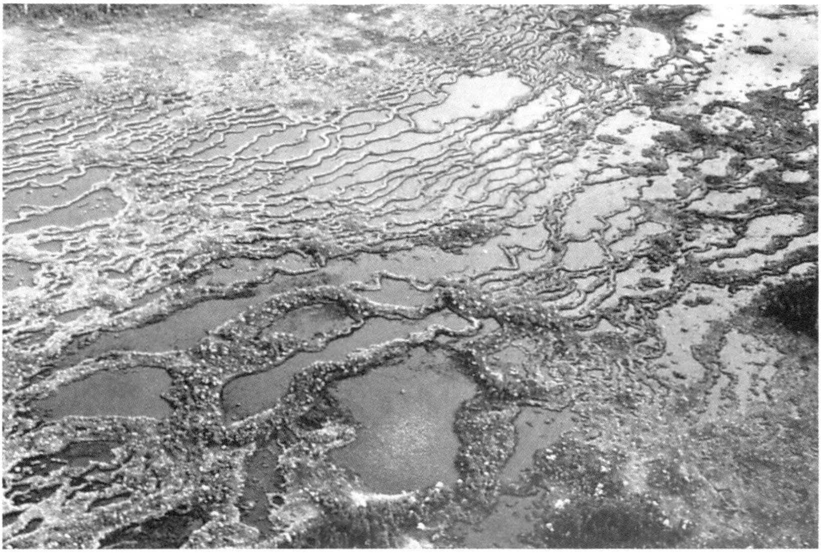

Figure 11. Aerial view of string fen, central Labrador (High Boreal Wetland Region).

with heights reaching 20 m. There is a nearly complete carpet of feather mosses, composed of *Ptilium crista-castrensis*, *Pleurozium schreberi*, and *Hylocomium splendens*.

Northern ribbed fens

These fens are characterized by the development of narrow (1 to 5 m wide), low (5 to 60 cm high) peat ridges that extend across the fen at right angles to the direction of water movement (Figs. 3 and 11). The ridges (strings) may loop across the fen in gentle arcs, or may link up with other ridges, enclosing small, wet depressions. The fens are usually gently sloping but drainage is by seepage through the fen rather than in defined surface water courses. The peat ridges act as impediments to drainage, as shown by the wet conditions in the depressions (flarks; Andersson and Hesselman 1907) along the upslope side of the ridges. The development of linear patterns is due to changing hydrological conditions and differential rates of peat accumulation (Foster *et al.* 1983), but the mode of initiation of the strings remains uncertain. Such ribbed fens are common in the BM and BH wetland regions and extend into the SL region. Permafrost is not associated with ridges here as in the subarctic.

The vegetation on the better-drained ridges and in the wet flarks are distinctly different (Slack *et al.* 1980). The ridges may be treed with *Larix laricina* and *Picea mariana*, with trees reaching a height of 10 m, although usually they are less than 4 m high. The shrub layer is generally well developed, consisting of *Betula pumila* var. *glandulifera*, *Salix pedicellaris*, and *Ledum groenlandicum*. The herb layer is usually sparse, represented by *Carex lasiocarpa* and *C. chordorrhiza*. The moss layer is nearly complete with *Tomenthypnum nitens* as the dominant species, with some *Sphagnum warnstorfii* and *Pleurozium schreberi*.

The vegetation of the flarks in the generally minerotrophic fens is influenced by the amount of water present (Slack *et al.* 1980). The deepest pools are dominated by *Scorpidium scorpioides*. Herbaceous plants such as *Carex limnosa*, *C. aquatilis*, *C. lasiocarpa*, *Triglochin maritima*, and *Menyanthes trifoliata* are typical components. In flarks with shallower pools, on firm root mats, *Carex* species dominate, mainly *C. aquatilis*, *C. limosa*, and *C. chordorrhiza*. The moss layer is dominated by *Drepanocladus revolvens*. In areas with increased water movement the flarks are characterized by *Scirpus caespitosus* and *S. hudsonianus*, with abundant *Campylium stellatum*.

Weakly minerotrophic fens that are nourished by groundwaters originating in nutrient-poor areas have a distinctly different vegetation (Vitt *et al.* 1975). The strings are usually well treed with *Picea mariana*, and only occasional occurrences of *Larix laricina*. The shrub layer consists mostly of ericaceous shrubs such as *Andromeda polifolia*, *Ledum groenlandicum*, and *Vaccinium vitis-idaea*. The mosses are dominantly *Sphagnum magellanicum* and *S. fuscum*, with some *Aulacomnium palustre* and *Tomethypnum falcifolium*.

The shallow-water flarks are characterized by *Carex limosa*, with a *Sphagnum jensenii* or *Drepanocladus exannulatus* moss layer. In flarks with deeper water *Menyanthes trifoliata* and *Eriophorum chamissonis* occur with lesser amounts of *Drepanocladus exannulatus*.

The process of patterned fen formation including Atlantic ribbed, ladder, and net fens was investigated in southeastern Labrador by Foster and King (1984). Surface pattern development was dependent upon topography and water movement. For example, the formation of string-flark features occurred on steeper slopes with large inputs of water. Three main parts characterize the process of surface pattern development. An irregular surface of hummocks and hollows takes places. This is followed by a gradual expansion and joining of depressions, a process controlled by differential rates of peat formation. Last, pools expand and coalesce by peat degradation. The developmental history of the fen was studied by peat stratigraphic investigations. Also in southern Labrador, studies have been carried out on bog vegetation and landform dynamics (Foster and Glaser 1986, Foster *et al.* 1988).

Horizontal fens

These minerotrophic peatlands occur on broad areas of low relief and slight slope gradient (Fig. 3). There is sufficient slope to allow a slow movement

Figure 12. Treed fen with small tamarack trees (*Larix laricina*) near Flin Flon, Manitoba (Mid-Boreal Wetland Region).

of minerotrophic waters through the fen but not enough to develop a pattern of ridges. The peat accumulation is seldom in excess of 3 m since deep basins, where thick peat accumulations could occur, are absent. The surface of the fen is usually featureless although "islands" of somewhat drier peatland may occur. Horizontal fens are common throughout the boreal region where poorly drained areas of broad, flat plains are found (Fig. 12).

The horizontal fens can be treed, shrubby, or open (graminoid, no trees or shrubs), according to the abundance and quality of the groundwater. The treed fens have an open stand of *Larix laricina*, usually with a well developed shrub layer of *Betula pumila*. In some cases scattered *Picea mariana* trees may be present. The herb layer is usually sparse, composed of various species of sedges, such as *Carex aquatilis*, *C. lasiocarpa*, or *Scirpus cespitosus*. The moss layer is prominent, composed mainly of *Campylium stellatum* and *Drepanocladus revolvens*. Some cushions of *Sphagnum warnstorfii* or *S. fallax* may develop in association with *Picea mariana*.

The shrubby fens are basically similar to the treed fens, but are without trees. In addition to *Betula*, *Myrica gale*, *Salix pedicellaris*, and *Salix candida* may be present. Lower shrubs may include *Ledum groenlandicum*, especially in the less minerotrophic fens.

The open fens are dominated by sedges and rushes, such as *Carex lasiocarpa*, *C. chordorrhiza*, and *C. aquatilis*, along with *Scirpus caespitosus*.

Mosses are abundant, including *Drepanocladus exannulatus*, *D. revolvens*, *Campylium stellatum*, and *Calliergon giganteum*. In particularly minerotrophlc fens *Scorpidium scorpioides* is abundant in consistently wet spots.

Conifer swamps

These mesotrophic wetlands occur in areas that receive overland or subsurface water flow of minerotrophic water. The peat is usually well decomposed under a fibrous cap. In many instances small (0.5 to 2 m diameter) openings up to 40 cm deep (sinkholes; Mueller-Dombois 1964) occur on the surface of the swamp. Most swamps develop on the margins of other wetlands where overland water flow reaches the wetland, on lakeshores, or on river floodplains where periodic inundations by mineral-enriched waters take place. Spruce (*Picea*) swamps occur throughout the boreal region, whereas cedar (*Thuja*) swamps are found in the Low Boreal and Eastern Temperate Wetland Regions.

The conifer swamps are characterized by dense growth of tall (> 10 m) trees. In spruce swamps the dominant species is *Picea mariana*, with some occasional occurrences of *Larix laricina*. Shrubs are usually present only in openings with such species as *Alnus rugosa*, *Salix* spp., and *Ledum groenlandicum* being common. In the dense stands feather mosses form a continuous carpet composed of *Pleurozium schreberi*, *Hylocomium splendens*, *Climacium dendroides*, and *Tomenthypnum nitens*. Some cushions of *Sphagnum fuscum* and *S. warnstorfii* may also be present.

Cedar swamps are similar in structure to spruce swamps. The main tree species is *Thuja occidentalis* growing in dense stands, with a few scattered occurrences of *Larix laricina*. Shrubs are infrequent, present only in small openings, and consist of *Alnus rugosa*, *Viburnum edule*, and *Salix* spp. In the openings, the herb layer may contain *Aralia nudicaulis*, *Clintonia borealis* and orchids such as *Habenaria hyperborea*, *H. obtusata*, *Orchis rotundifolia*, and *Cypripedium calceolus*. In the dense stands, mosses cover most of the ground and consist of *Pleurozium schreberi*, *Hylocomium splendens*, and *Climacium dendroides*.

Delta marshes (freshwater)

The major marshes occur where rivers reach large lakes and deposit their sediment load, gradually filling the proximal part of the lake. As the channels are silted up, new ones are cut, resulting in a maze of active and inactive channels, oxbow lakes, and basins enclosed by natural levees. This results in a variety of environments, depending on the proximity to ponds and active channels. The annual spring floods inundate much of the delta until some parts become sufficiently built up to escape all but the most severe floods. Because of shifting channels, some parts of the delta may become inactive, that is, no longer subject to frequent flooding. On such areas the fens and bogs develop that no longer reflect the delta influence. An example is the Cumberland wetlands of the Saskatchewan River (Dirschl and Dabbs 1969, Dirschl 1977, Dirschl and Coupland 1972). Other delta marshes may be formed in lagoons that are formed by barrier beaches. Such marshes are subject to fluctuating water levels of the lake. Large delta marshes are found on Lake Athabasca (Raup 1935, Dirschl *et al.* 1974), Lake Manitoba (Delta Marsh, Walker 1959) a site which will be discussed in more detail later in this paper, and Lake Winnipeg (Netley Marsh, Smith *et al.* 1967), along with numerous smaller marshes found on many lakes.

The wet meadows near ponds are dominated by *Carex atherodes* (Raup 1935), with lesser amounts of *C. aquatilis*, *Scirpus validus*, and *Eleocharis palustris*. At a greater distance from ponds large expanses of *Calamagrostis canadensis* occur in nearly pure stands. On slightly drier areas, scattered shrubs consisting mainly of *Salix planifolia* growing among *Calamagrostis canadensis* and *Poa palustris* invade the marsh meadows. On somewhat drier habitats, swamps formed by tall shrubs take over with *Salix bebbiana* shrubs forming nearly impenetrable thickets. The ground vegetation is sparse, consisting of *Equisetum palustre*, *Rubus idaeus*, and *Vicia americana*.

In the more southerly delta marshes, emergent vegetation dominates, growing in a soft muck (mud and decomposed organic matter) base. Such marshes are interspersed with pools of open water. The dominant species are *Phragmites australis*, *Typha latifolia*, *Scirpus validus*, and *Scirpus acutus*. Peat accumulation is largely absent from all marshes. This type of marsh is discussed in the Continental Prairie Wetland Region Section.

Another important area of marsh development occurs along the shores of oxbow lakes. Van der Valk and Bliss (1971) studied succession in marshes of the Pembina River floodplain in central Alberta. Twelve plant communities were present, ranging from submergents to shrub and forest communities. Water chemistry and waterlevel fluctuations caused by periodic flooding were felt to be the main factors controlling succession.

Studies have been conducted on ecological factors controlling the distribution of plant species on the shorelines of lakes located in boreal areas of the Provinces of Ontario and Nova Scotia. Keddy (1983) investigated zonation along exposure gradients of water depth and wave energy. Some species were found to reach their maximum distribution on exposed shores while others did so on sheltered areas, especially large, leafy plants. Maximum species richness occurred at intermediate exposures. In a further study, sediment particle size was found to be an important control of plant zonation; wave activity was found to influence particle size distribution and to disturb vegetation (Keddy 1984a). Water depth also is a control of zonation. The number of different species reaches a maximum at or just above the waterline depending upon the degree of esposure. The co-existence of species was not associated with increased specialization but appeared to be related to disturbance (Keddy 1984b). Organic content of sediments also appeared to control zonation (Wilson and Keddy 1985). Research was also performed upon species recruitment along lakeshore gradients. Water depth was found to be important in this respect to some species although most species exhibited broad tolerance limits (Keddy and Ellis 1985). In terms of sediment particle size, finer textured-sediments occurring in sheltered bays allowed for greatest recruitment (Keddy and Constabel 1986).

Many of these wetlands contain rare and endangered plant species. These species appear to be favoured by low fertility substrates and natural disturbance. Management of lakes must take such requirements into consideration (Keddy and Wisheu 1989, Moore *et al.* 1989). Keddy (1989) also studied the role of competition in controlling the growth of such species.

Forestry use of wetlands

Controlling water table levels in peatlands through drainage for the purpose of increasing wood production is a relatively new forest management technique in Canada (Hillman 1987). Although widely practiced in northern Europe, it is largely at an experimental stage in Canada. Early indications are that, by using the proper design and technique suited to different kinds of peatlands the yield of forests can substantially be increased and the time required for the trees to reach merchantable size can be shortened.

A number of drainage experiments were carried out in Newfoundland (Paivanen and Wells 1978). On the rarely treeless bogs of Newfoundland, the bogs had to be afforested with native and exotic species (*Larix laricina*, *L. kaempferi*, *Picea mariana*, *P. abies*, *Pinus contorta*, *P. sylvestris*). The results were variable both in terms of seedling survival and growth. The drainage design that afforded maximum protection to seedlings from exposure, coupled with adequate water table control, was the most successful.

In Quebec wetlands occur in some of the most accessible and productive forest areas (Bolghari 1986). As most peatlands are treed, planting a new tree crop is not a major concern. Forest drainage improvements are being implemented on an interim basis by the Quebec Wood Producers' Federation (Urgel Delisle et Associes 1983). Experimental forest drainage projects have been carried out, coupled with fertilization treatment with nitrogen, phosphorus, potassium, and copper (Stanek 1975). Periodic annual increment showed a five-fold increase over the control area. Trottier (1986) found that the growth of trees was always higher on the drained area than on the control plots; in some cases the increase was more than five times the pre-drainage rate.

In Ontario, drainage projects have been conducted on an experimental scale since 1929. Payandeh (1973) found that tree diameter and height growth showed significant increases after drainage. Stanek (1968) used dynamite to excavate drainage ditches. Although water flow in the resulting ditches was impeded by obstructions, they functioned well enough to increase both diameter and height growth in *Picea mariana* five times over the pre-drainage rates. A large experimental drainage project has been established recently (Hillman 1987). A number of treatments have been imposed on the drained area, such as different harvesting and regeneration techniques, planting different species, and fertilizing the original *Picea mariana* stand. These treatments will be monitored to evaluate the effect of drainage on them. Peatland forestry in Ontario has been reviewed by Haavisto and Wearn (1987).

A number of peatland drainage experiments have been conducted in Alberta since 1975 (Hillman 1987). The purpose of these experiments was to try various ditching equipment, ditch spacing and design, as well as fertilizing and thinning the original *Picea mariana* stands. Initial results show up to four-fold increase in leader growth. Another experimental area was established (Toth and Gillard 1984) where the drainage system design was based on synthetic groundwater hydrographs to optimize the spacing of ditches. Further experiments are in progress to test the use of synthetic groundwater hydrographs and the response of trees and vegetation to drainage in various peatland types under fully instrumented conditions (Hillman 1987).

Contaminant effects upon boreal peatlands

Recent concern has been made on the ecological impact of acid rain upon wetlands. Very little research is available on this topic (Gorham *et al.* 1984). Fens, particularly those with low alkalinities and pH values below 6, are felt to be especially vulnerable due to their low bicarbonate buffering capacity. These fens can have their buffering capacity overwhelmed by inputs of anthropogenic strong acids, both sulfuric and nitric, associated with acid rain. This is especially important if the fen has accumulated sufficient peat to reduce the input of minerotrophic ground waters which have buffering capacity. Other concerns related to acidification include 1) lower nutrient availability, 2) increased availability of metals, especially aluminum which may be toxic, 3) effects upon peat fauna, and 4) effects upon receiving waters from peatland drainage. Gorham *et al.* (1984) recommended more research be done upon the potential effect of acid rain upon individual plant and animal species, communities and ecosystems.

Recent research has been conducted upon a fen in northwestern Ontario which has been artificially acidified (Bayley *et al.* 1986). The vegetation and water chemistry of peatlands in the area was studied by Vitt and Bayley (1984). The fen acted as a sink for sulfate ion removing some 22 to 73%. During long, dry summers, reduced sulfur compounds could be oxidized and released to surface waters along with calcium and magnesium ions. Nitrate and sulfate ion associated with acid rain is also retained in such fens with up to 99% retention (Bayley *et al.* 1987). In the process of nitrate and sulfate retention, alkalinity is generated which resists lowering of pH. A short-term fertilization effect upon *Sphagnum* moss was found (Rochefort *et al.* 1990).

Smelting activities may have a negative impact upon peatlands located in boreal regions of Canada. *Sphagnum* mosses are killed in such ecosystems located at distances within 10 to 12 km from smelters such as those located at Sudbury, Ontario (Glooschenko *et al.* 1981, Gignac and Beckett 1986) and Rouyn-Noranda, Quebec (Glooschenko *et al.* 1986). This appears to be due to direct effects of sulfur dioxide gas emissions and heavy metal accumulation. A paper by Glooschenko (1986) discusses the accumulation of metals by bog vegetation and peats and reviews the role of peatlands in monitoring the atmospheric deposition of metals. Regional differences in contaminent deposition across Canada were found for arsenic and selenium (Glooschenko and Arafat 1988) and cadmium (Glooschenko 1989). Smelters were the main source of these potential contaminants. Other negative impacts on boreal peatland ecosystems include highway and powerline construction (Sims and Stewart 1981).

Atlantic Boreal Wetland Region

The Atlantic Boreal Wetland Region (BA) differs from the continental-influenced climates of the BH, BM and BL Wetland Regions (Fig. 6). Damman (1977) has shown that maritime-influenced climates have higher precipitation (including fog), lower summer temperatures, lower evapotranspiration, longer growing seasons, and an erratic snow cover. This leads to a higher moisture surplus and subsequent higher growth rate of *Sphagnum* mosses. This often results in large accumulations of peat in large domes. In other areas the high precipitation levels allow the formation of peat even on sloping land surface.

A study was made on primary productivity in peatlands located near Schefferville, Quebec located on the Quebec/Labrador border (Bartsch and Moore 1984). In the *Carex* dominated system, above-ground productivity ranged from 114 to 335 $g\,m^{-2}\,yr^{-1}$. Decomposition was also measured by litterbags and losses of 6.4 to 26.6% were measured over a year, 65% of which occurred over winter. Tissue quality was found to influence decomposition more than pH or temperature. Nutrient release was slow with the following decreasing order: $K > Mg > Ca > N,P$.

Figure 13. Two large domed bogs with crescent-shaped pools, central Newfoundland (Atlantic Boreal Wetland Region).

Domed bogs

Domed bogs are large (usually 500 m diameter) circular to elliptical bogs with a convex surface that rises several metres above the surrounding terrain. The centre drains in all directions. Several forms have been found, differing from one another in the presence and disposition of small pools, and in the shape of the dome. On concentric domed bogs, small crescentic pools occur around the centre which is the highest part of the dome. If the summit of the dome is off-centre, the pools form an eccentric pattern (Glaser and Janssens 1986, Foster and Glaser 1986). Atlantic plateau bogs are raised bogs that have a flat surface, often with a number of pools (Fig. 13).

It was found that the convexity of domed bogs increases with increased wetness of the climate (Damman 1986) thus differing from domed bogs elsewhere in Canada where drier climatic conditions prevail. For each climatic condition there is a maximum convexity (critical profile). If the convexity is below the critical profile, the water table will be near the surface of the bog centre. This situation exists on those domed and plateau bog that have a large number of pools.

Several studies have been made on Atlantic domed bogs. Damman (1977) studied raised bogs along the Bay of Fundy in New Brunswick and found four main groups of plant communities: 1) dwarf shrub heaths dominated by ericaceous shrubs including *Gaylussacia baccata*, *Kalmia angustifolia*, *Empe-*

trum nigrum, *Rubus chamaemorus*, and *Sphagnum*, mainly *S. fuscum*, 2) *Scirpus cespitosus* lawns and solid *Sphagnum* carpets occurring in the wet parts of bogs where the water table reaches above the surface, 3) extremely nutrient-poor fens with such species as *Eriophorum angustifolium*, *Smilacina trifolia*, *Myrica gale*, various species of *Sphagnum*, ericaceous shrubs, and occasionally *Larix laricina*, and 4) mud bottom communities that occupy wet depressions and contain *Sphagnum cuspidatum* and *Rhynchospora alba* or a *Utricularia cornuta–Cladopodiella fluitans* community. Damman and Dowhan (1981) also described the vegetation of a plateau bog on the southern Nova Scotia coast. This bog differed from the Bay of Fundy bogs mainly in the presence of coastal plain disjunct species.

The domed bogs along the St. Lawrence River (Gauthier and Gandtner 1975) are dominated by *Kalmia angustifolia*, *Ledum groenlandicum*, *Chamaedaphne calyculata*, and *Vaccinium angustifolium*. The main peat-forming mosses are *Sphagnum magellanicum*, *S. rubellum*, and *S. nemoreum*. In Labrador, the vegetation of domed bogs showed three distinct noda (Foster and Glaser 1986). The *Cladonia stellaris*, *Cladonia rangiferina–Kalmia angustifolia* nodum was characterized, in addition to these species, by *Chamaedaphne calyculata*, *Ledum groenlandicum* and some *Picea mariana*. *Sphagnum fuscum* was the dominant moss. The *Sphagnum rubellum–Scirpus cespitosus* nodum contained these species and *Chamaedaphne calyculata*, *Eriophorum spissum*, *Carex limnosa*, *Sphagnum tenellum*, and *Andromeda glaucophylla*. The *Sphagnum linderbergii* to *Scirpus cespitosus* nodum contained these species and *Eriophorum spissum*, *Cladopodiella fluitans*, *Andromeda glaucophylla*, and *Chamaedaphne calyculata*.

Slope bogs

Slope bogs are wetlands that occur on slopes (up to 15% slopes) in the maritime climate of Newfoundland (Wells 1981). These slope bogs have a relatively thin peat cover that seldom exceeds 2 m, but the water table is at or very close to the surface. The peat materials indicate fen conditions at the base, but later changing to bog as peat has accumulated. These slope bogs are similar in vegetation composition to those previously discussed in the Atlantic Subarctic Wetland Region.

Salt marshes of Atlantic Boreal Canada

Salt marshes occur along the coast of the Atlantic Ocean in the Atlantic Boreal Wetland Region. The Atlantic Oceanic Wetland Region (OA), which is found in parts of New Brunswick, Nova Scotia, and Newfoundland (Fig. 6), also has similar salt marsh vegetation and will be discussed in this section.

The salt marshes of the Atlantic Coast of Canada represent the northern limits of the *Spartina alterniflora* marshes of the east coast of the U.S.A. and

Figure 14. Salt marsh in southwestern Nova Scotia. Tall grass near tidal creek is *Spartina alterniflora* (Atlantic Boreal Wetland Region).

Gulf of Mexico (Chapman 1974). The Canadian salt marshes were previously reviewed by Glooschenko (1980b, 1982), Roberts and Robertson (1986) and Glooschenko *et al.* (1988). Figure 14 shows a typical *Spartina alterniflora* salt marsh. These marshes are quite different in species composition than those in the subarctic or Pacific Coastal areas.

The Bay of Fundy was one of the first areas of salt marsh in North America to be studied (Ganong 1903). The low marsh in this area is dominated by *Spartina alterniflora*, while the high marsh is characterized by *Spartina patens*, *Limonium carolinianum*, *Salicornia europaea*, *Suaeda maritimia*, *Atriplex patula*, *Plantago maritima*, *Puccinellia lucida*, *Triglochin maritima*, *Glaux maritima*, and *Hordeum jubatum*. At the landward edge, *Juncus gerardii* and *J. balticus* may be present, and this can give way to either freshwater marsh or bog. Chapman (1937) also studied the area. Further north at Cape Breton Island, Nichols (1918) noted similar vegetation, but also reported *Scirpus paludosus* and *Stellaria humifusa*. These two species are more common in boreal salt marshes.

Salt marshes occur along the St. Lawrence River in Quebec. The low marsh is dominated again by *Spartina alterniflora* which is replaced landward by *Spartina patens* and a *Juncus balticus* – *Juncus gerardii* zone. This can give way to a freshwater marsh (Reed and Moisan 1971). Pans can be found in the marsh with *Salicornia europaea* and ponds occur with *Ruppia maritima*.

Forbs also can be prevalent in the middle regions of the marsh. The Îles-de-la-Madeleine in the Gulf of St. Lawrence have a similar salt marsh vegetation, but *Spartina patens* is absent (Grandtner 1966).

The coastlines of New Brunswick and Prince Edward Island are characterized by salt marshes which develop on sandy sediments in the lee of barrier islands (Lucas 1980). She described three zones of vegetation. The first of these is the low marsh which colonizes the shoreline and occurs in wet depressions. *Spartina alterniflora* is the dominant species. At slightly higher elevations, which are protected from wave activity, *Salicornia europaea*, *Atriplex patula*, and *Suaeda* spp. are present. In barer areas, *Triglochin maritima* and *Plantago maritima* are also present. Above the height of daily flooding, *Spartina patens* is the main species which is found in a meadow-like, high marsh setting. Species including *Glaux maritima*, *Potentilla anserina*, *Carex paleacea*, *Scirpus americanus*, *Eleocharis* spp., and *Scirpus maritimus* are also present, the latter species occurring in pools. This gives way to a community with *Festuca rubra*, *Poa palustris*, *Agrostis alba*, *Puccinellia maritima*, and *Juncus gerardii*. *Juncus balticus* appears at the edge of this zone along with several other species. If freshwater inputs are high, a marsh with *Spartina pectinata*, *Typha latifolia*, and *Scirpus validus* is present.

Salt marshes in Newfoundland and Labrador have been described by Thannheiser (1981). *Spartina alterniflora*-dominated salt marshes occur as previously described for other parts of the Atlantic Coast of Canada. In other areas, where some freshwater influence occurs, the colonizing species are *Eleocharis halophila* and *E. parvula*. In the higher portions of these marshes, other species that occur include *Triglochin gaspense*, *Potentilla egedii*, *Carex paleacea*, *Carex subspathacea*, *C. salina*, *Plantago maritima*, *Ranunculus cymbalaria*, and *Puccinellia paupercula*. Some of these species are common in boreal salt marshes and it appears that Newfoundland represents a transition area in terms of salt marsh vegetation. The above-ground primary production of a *Spartina alterniflora* salt marsh in Nova Scotia was measured at 710 g m^{-2} by Hatcher and Mann (1975). This value is higher than measured in some Atlantic coast marshes located further south.

Continental Prairie Wetland Region (PC)

The climate of the Continental Prairie Wetland Region (Fig. 6) is semi-arid with cold winters and hot summers (Table 3). Marshes are the main wetland type and are usually associated with semi-permanent ponds. Two wetland districts are present - the Aspen Parkland Continental Prairie Wetland Region and the Grassland Continental Prairie Wetland District (Zoltai 1980). A comprehensive review of prairie wetlands can be found in Adams (1988) and the book of van der Valk (1989).

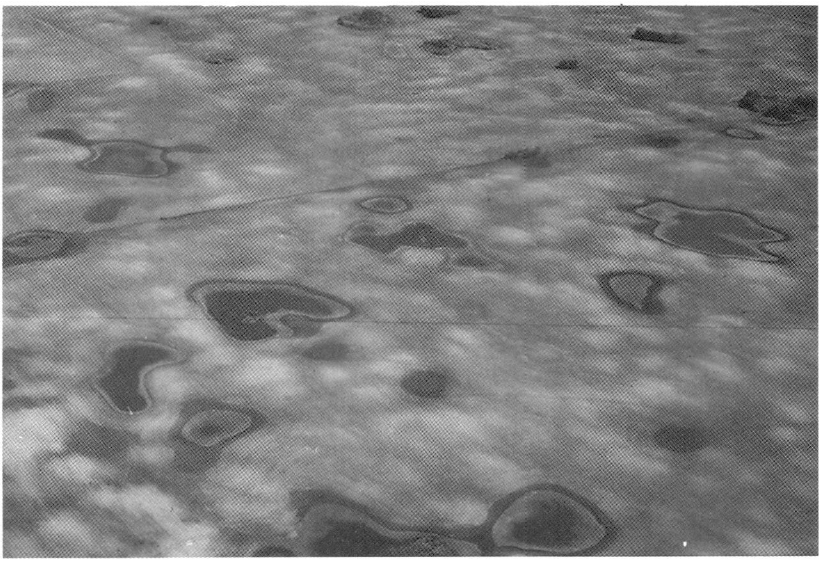

Figure 15. Aerial view of typical potholes surrounded by intensive agriculture in PC Wetland Region near Saskatoon, Saskatchewan.

In terms of relief, the Pleistocene glaciation left a legacy of millions of small depressions known as potholes or sloughs (Fig. 15). Sizes of these potholes can vary from several metres in diameter to small lakes of several hundred hectares. The area is known as the prairie pothole region and covers some 750×10^3 km^2 in south-central Canada and the adjacent north-central United States (Walker and Coupland 1970). Water chemistry in the area varies from freshwater to highly saline conditions. The vegetation of saline potholes will be discussed later in this section.

The main difference between the Aspen Parkland Continental Prairie and Grassland Continental Prairie districts is the type of vegetation that surrounds the potholes. In the former area, a surrounding border of thicket swamp can occur with such dominant species as *Salix bebbiana*, *S. discolor*, and *S. petiolaris*. In the latter area, various herbaceous wet meadow species form the surrounding vegetation. Prairie pothole vegetation usually occurs in bands or zones which can be correlated with depth and duration of submergence (Fig. 16). Millar (1969) describes four main zones in wetland basins. These are:

a. Wet meadow zone - submerged for a very short period of time in the spring with grasses such as *Hordeum jubatum* and *Poa palustris* and forbs.

Figure 16. Small marsh near Saskatoon, Saskatchewan. Note *Salix* tree on edge (PC Wetland Region).

b. Shallow marsh zone - usually flooded until June or July with grasses of intermediate height such as *Scolochloa festucacea*, *Glyceria grandis*, and *Carex atherodes*.
c. Deep marsh zone - normally flooded until late in the season or submerged through winter in wet years. Characterized by tall, coarse emergents such as *Typha latifolia*, *Scirpus validus*, and *S. acutus*.
d. Open water zone - normally flooded all year round with submerged aquatics such as species of *Potamogeton*. In drought years, with drawdown in the summer, a new temporary zone with weedy forbs can occur.

The vegetation ecology of potholes ("sloughs") of Saskatchewan was studied by Walker and Coupland (1968). They examined the relationship between herbaceous vegetation and environmental factors. In general, the vegetation of these ecosystems varied with water depth. The deeper potholes had submerged and floating vegetation in the centre with such species as *Potamogeton* ssp., *Lemna* spp., *Myriophyllum exalbescens*, and *Ranunculus circinatus* being present. The emergents in these deeper ponds included *Scirpus acutus*, *S. validus*, *Typha latifolia*, and *Scolochloa festucacea*. Shallower potholes were dominated by *Carex atherodes*, *Glyceria grandis*, *Sium suave*, and *Polygonum* spp. The marshes graded into a meadow zone composed of *Calamagrostis inexspansa*, *Carex lanuginosa*, *Poa palustris*, and

Juncus spp. Major factors in the environment influencing vegetation included disturbance and water level fluctuation, both within and between growing seasons, and salinity. Soil data showed very little influence on species distribution.

Walker and Coupland (1970) further investigated the vegetation associations in the aspen grove and grassland regions of Saskatchewan. They identified 27 vegetation groups; major factors of environmental interest were again water level, water salinity, and disturbance. Walker and Wehrhahn (1971) investigated vegetation-environment interactions in Saskatchewan wetlands in more detail. The most important factor in determining variation was salinity, followed in decreasing order by nutrient status, water relations of the substrate, and reducing conditions in the soil. Vegetational data analysis produced various indicators of environmental conditions. For example, the following species were indicators of stable, non-disturbed environments: *Carex atherodes*, *Lemna minor*, *L. trisulca*, *Carex rostrata*, and *Utricularia vulgaris*. On the other hand, *Eleocharis palustris*, *Glyceria grandis*, *Beckmannia syzigachne*, *Alopecurus aequalis*, and *Sium suave* were indicators of disturbed conditions.

Millar (1973) studied the response of wetland vegetation in Saskatchewan to water regime over a ten year period. The moisture regimes of such wetlands can vary from year to year depending upon variations in snow melt, summer precipitation, and evaporation. Shallow-marsh emergent species such as *Carex atherodes*, *Polygonum coccineum*, *Scolochloa festucacea*, and *Eleocharis palustris* decreased in abundance when greater than normal water depth was present at the beginning of the growing season. If this occurred for more than two successive years, i.e., continual flooding, the emergent cover was eliminated and open water resulted. This also required repeated autumnal reflooding. Basins subjected to grazing by cattle produced characteristic disturbance vegetation. In terms of indicator species, the presence of small amounts of deep water emergents in shallow marshes was not found to be an indicator of a wetter moisture regime. The presence of submergent rooted aquatics though did reflect the amount of year-round flooding. Here, *Potamogeton gramineus* occurred without year-round flooding, while *Utricularia vulgaris* usually required some year-round flooding. Two or more years of continuous flooding were required by *Ranunculus* spp. and *Potamogeton pusillus* while good, reliable indicators of several years of flooding were *Potamogeton pectinatus*, *Myriophyllum* spp., and *Ceratophyllum demersum*. Wetland size is also an important control on vegetation. Wetlands of less than 0.41 ha usually have a restricted moisture regime regardless of depth, and vegetation is limited to shallow water or drier conditions unless abnormally high surface or groundwater inflow occurs. Wetlands of larger area with depths shallower than approximately 90 cm are usually limited to shallow marsh vegetation. In addition, arid areas usually have more stable vegetation in large deep basins than in the aspen parklands.

The ecology of plant communities associated with shallow oxbow lakes in central Alberta was studied by van der Valk and Bliss (1971). Based upon phytosociological analyses, they identified 12 plant communities. Three communities consisted of submerged plants: 1) *Potamogeton pectinatus*, 2) mixed submerged and 3) *Potamogeton pectinatus – Ceratophyllum demersum*. Two communities were dominated by floating-leaved plants: 4) *Nuphar variegatum* and 5) *Potamogeton natans*. Four emergent communities were found: 6) *Equisetum fluviatile*, 7) *Eleocharis palustris*, 8) *Typha latifolia*, and 9) *Alisma plantago-aquatica*. Three meadow communities were present: including 10) *Carex-Acorus calamus*, 11) *Carex*-Bryoid and 12) *Acorus calamus-Sonchus uliginosus*. Water chemistry and water level fluctuations caused by periodic flooding were found to be the most important environmental factors controlling species succession.

Another major wetland area of the Canadian prairies is the Delta Marsh, a 15,000 ha system at the southern end of Lake Manitoba consisting of open water, channels, and emergent vegetation (Shay 1983). Dominant emergents in the area are *Phragmites australis*, *Typha* spp., and *Scirpus acutus*. Wet meadows are dominated by *Scolochloa festucacea* and *Carex atherodes* while submergent aquatics include mainly *Potamogeton pectinatus*, *P. vaginatus*, and *Myriophyllum exalbescens*. These marshes are very sensitive to water level. For example, in 1955, a very high natural water level killed thousands of hectares of vegetation, but water levels fell and four years later emergent dominants had returned.

One of the most important ecological processes in freshwater wetlands is that of succession. A qualitative model of this process was developed by van der Valk (1981), taking into account three major life history attributes: life-span, propagule longevity, and propagule establishment requirements. The author has applied this model to the Delta Marsh located at the southern end of Lake Manitoba. Three main vegetation types are present in this marsh: *Typha glauca* stands, *Phragmites australis* stands, and open water free of emergents. Thirteen additional species were represented in the seed bank. The author constructed a model of vegetational succession assuming that all of the vegetation would be completely destroyed by flooding, then a drawdown would take place followed by reflooding. During drawdown, mud-flat annuals such as *Atriplex patula*, *Lycopus asper*, *Mentha arvensis*, *Polygonum amphibium*, *Ranunculus sceleratus*, *Rumex maritimus* and seedlings of *Scirpus validus*, *Typha glauca*, *Sonchus arvensis*, and *Cirsium arvense* would form the vegetation. Reflooding would then eliminate mud-flat annuals and *Typha glauca* would be the only emergent remaining. If seed dispersal from adjacent areas occurred, then *Phragmites australis* could colonize mudflats during drawdown. Other related studies on the role of seed banks in prairie wetlands include van der Valk and Davis (1978), van der Valk (1986), and Kantrud *et al.* (1989). The management implications of seed banks was reviewed by Pederson and van der Valk (1985).

The ecology of prairie marshes was studied by Shay and Shay (1986), describing the habitats, propagation, and growth of five major species. They found that the mean August standing crop of *Phragmites australis* was 812 g m^{-2}, and the below-ground biomass accounted for 69 to 70% of the total biomass. *Typha latifolia* had an August above-ground biomass of 1,754 g m^{-2}, a standing dead mass of 1,224 g m^{-2}, of 233 g m^{-2}, and roots and rhizomes of 2,526 g m^{-2}. The above-ground biomass of *Scirpus glaucus* and *S. validus* varied from a low of 94 g m^{-2} to 570 g m^{-2} at five sites. *S. glaucus* has 82% of its total biomass in roots and rhizomes, but *S. validus* allocated 53% of its mass to roots and rhizomes. *S. maritimus* had a maximum standing crop of 625 g m^{-2}, with 41% of its biomass being below-ground.

A major classification for the prairie wetlands was developed by Millar (1976) to interpret the potential of different wetland types as wildlife habitat and as water resources. Previous classification schemes for the area were based on two common concepts: the permanence of water and the use of vegetation to determine water permanence. Millar included these two criteria plus guidelines for interpreting vegetation dynamics as related to wetland classification and use of physical features of wetlands to better predict long-term water regime. He proposed seven wetland zones as depicted in Fig. 17: 1. Wet meadow, 2. Shallow marsh, 3. Emergent deep marsh, 4. Transitional open water, 5. Shallow open water, 6. Open alkali, and 7. Disturbed - including cultivated, grazed, and drawdown. These zones appear as concentric bands around shallow open water. They follow basin contours and reflect the relative depth and duration of flooding. Such zones can be modified if water fluctuations occur too widely or if human disturbance takes place. They are also related in a successional sequence resulting from improving water regime.

In terms of physical features, Millar incorporates wetland size, basin or wetland depth, position of basin in the watershed, and origin and nature of alteration of the wetland. The system proposed by Millar is fairly involved and the reader is referred to the original paper for further details.

The prairies, both in Canada and in adjacent states of the U.S.A., are extremely important waterfowl breeding habitats. Over half of the waterfowl breeding in North America nest and raise their young in the grassland and aspen parkland potholes and lakes that are located in the southern portions of Alberta, Saskatchewan, and Manitoba (Kiel *et al.* 1972). It is beyond the scope of this paper to discuss wildlife and waterfowl production but the

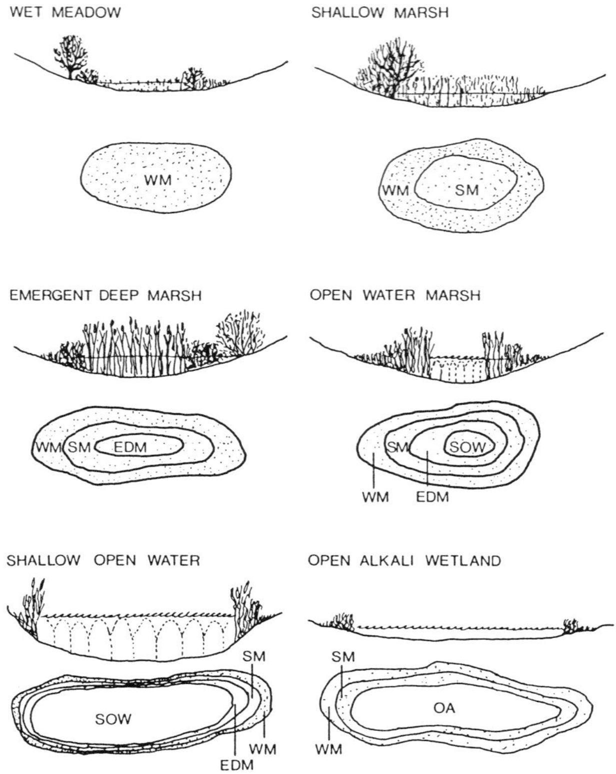

Figure 17. Wetland zonation in prairie marshes (figure redrawn from Millar 1976).

reader is referred to the Saskatoon Wetlands Seminar (Anonymous 1969) which is an excellent document covering waterfowl-wetland interactions in the prairies. More information on this topic can be found in the recent book by van der Valk (1989). Unfortunately, agricultural activities are threatening such wetlands as will be discussed later in this paper.

Intermountain Prairie Wetland Region (PI)

The Intermountain Prairie Wetland Region (PI) is characterized by a semi-arid climate with hot summers and mild winters (Table 3). Peat formation is limited. Marshes, both freshwater and saline, are the main wetland type found in the area and they surround both ephemeral and semi-permanent ponds. The ponds are characterized by shallow water wetland classes. Fens are also present.

Limited data is available on the wetlands of this region. The main study is that of Moon and Selby (1983) who characterized the wetlands and soils of the Cariboo-Chilcotin region of interior British Columbia. They described 11 vegetation types:

1. Aquatic – This includes areas of open water with submerged, rooted, and floating plants such as *Nuphar polysepalum*, *Potamogeton* spp., *Myriophyllum* spp., and *Utricularia vulgaris*.
2. Moss – Here, the soil surface is covered by brown mosses, mainly *Drepanocladus* spp. An open herbaceous stratum of *Eleocharis* spp., *Eriophorum* spp., *Carex lasiocarpa*, and *Menyanthes trifoliata* is present. Scattered woody shrubs including *Salix* spp. and *Betula glandulosa* can occur.
3. Cattail – The dominant cover is *Typha latifolia*, often occurring in a pure stand. It is found in standing or slow-moving water where it can be found with aquatic species.
4. Bulrush – The dominant cover is *Scirpus lacustris*. Some *Typha latifolia* and *Carex aquatilis* occur in a mixture.
5. Horsetail – This vegetation cover type is found in shallow standing water. The dominant species is *Equisetum fluviatile* with scattered *Sium suave*, *Potentilla palustris* and assorted aquatics such as *Lemna* spp.
6. Emergent grasses – This zone is characterized by hydrophytic grasses including *Glyceria* spp., *Alopecurus aequalis*, *Beckmania syzigachne*, and *Scolochloa festucacea*. Other plants, including *Potamogeton* spp., *Eleocharis* spp., and *Carex rostrata*, also occur but they are of low coverage.
7. Spike-rush – The dominant species is *Eleocharis palustris* with scattered *Polygonum amphibium*, *Beckmania syzigachne*, *Glyceria* spp., and *Hordeum jubatum*.
8. Sedge – This vegetation is found in shallow standing water, especially early in the growing season. Dominants include *Carex aquatilis*, *C. rostrata*, and *C. atherodes*. Mosses may also occur under the sedges.
9. Shrub-sedge – This area has both a herb and shrub (less than 6 m height) cover. The herbs consist mainly of *Carex aquatilis*, *C. atherodes*, and *C. rostrata*. Some mosses may be abundant, mostly *Aulacomnium* spp. and *Tomethypnum* spp. The shrubs are *Betula glandulosa* and *Salix* spp.
10. Water tolerant grass/forb – This area is dominated by a mixture of grasses, forbs, sedges, and rushes. There is both an alkaline and non-alkaline subclass. The alkaline component consists of *Distichlis spicata*, *Puccinellia* spp., *Spartina gracilis*, *Suaeda depressa*, *Triglochin maritima*, and *Hordeum jubatum*. The freshwater component includes *Juncus arcticus*, *Carex praegracilis*, *Hordeum jubatum*, *Poa pratensis*, *Muhlenbergia richardsoni*, *Potentilla anserina*, and *Taraxacum officinale*. Some of the species are more indicative of drylands.
11. Shrub-grass/forb – This area is dominated by shrubs, forbs, grasses, and grass-like species. The shrubs are *Betula glandulosa* and *Salix* spp. The

herbs are assorted grasses, *Juncus arcticus* and forbs more common of drylands. Mosses may occur in depressions.

Saline prairie wetlands

Another small but important group of wetlands in the prairie regions is the saline wetland. The source of salts is the discharge of saline groundwaters in contact with evaporite mineral deposits. Such lakes can be ephemeral and exhibit seasonal and annual changes in water chemistry. Dodd and Coupland (1966) studied the vegetation of saline wetlands in southern Saskatchewan. In terms of hydrophytic vegetation, submergents included *Potamogeton pectinatus* and *Chara* spp. while the main emergent was *Scirpus validus*. In slightly higher areas where some exposure of the soil surface occurs, *Scirpus paludosus* becomes common; with prolonged exposure these areas are invaded by *Hordeum jubatum*, *Atriplex* spp., *Chenopodium* spp., and *Sonchus uliginosus*. *Eleocharis palustris*, *Puccinellia airoides*, and *Carex* spp. are also abundant under slightly drier conditions. In slightly less depressed areas, where water seldom accumulates on the soil surface, halophytic vegetation occurs dominated by *Salicornia rubra*, *Triglochin maritima*, *Puccinellia airoides*, *Distichlis stricta*, *Hordeum jubatum*, and *Agropyron* spp.

Eastern Temperate Wetland Region (TE)

The Eastern Temperate Wetland Region (TE) is characterized by warm summers and mild winters. Precipitation is relatively high (Table 3). The dominant wetland types are hardwood and conifer-dominated swamps. Marshes are common along the shores of lakes, especially the Great Lakes. Fens may also occur near ponds and drainage ways, while coniferous bogs occur in flats and basins (Glooschenko and Grondin 1988).

Swamps dominated by hardwood species of *Acer* and *Fraxinus* are the most common wetland type found in the Eastern Temperate Wetland Region, although some conifers may be present (*Thuja occidentalis*, *Larix laricina*, and *Pinus strobus*). They are characterized by an irregular microtopography in which some portions are never inundated by water, some areas are always under standing or slowly moving water, and some portions are seasonally flooded. Swamps occur either adjacent to water bodies such as streams or lakes, or away from such water bodies in topographically-defined basins or in other areas such as flat deposits, floodplains, or associated with discharge areas such as springs. Peat deposits are often present with depths up to 2 m, but such peats do not appear to be forming at the present time (Zoltai and Pollett 1983, Eagle 1983). These swamps are particularly important as wildlife habitat (Glooschenko *et al.* 1987).

Vegetation consists mainly of woody plants including shrubs and trees

Figure 18. Hardwood swamp, TE Wetland Region, near Hamilton, Ontario.

(Fig. 18). Common shrubs are various species of *Salix*, *Alnus rugosa*, *Cornus stolonifera*, and *Cephalanthus occidentalis*. The most common trees are *Acer rubrum*, *A. saccharinum*, *A. negundo*, and *Thuja occidentalis*. Typical herbaceous species are *Symplocarpus foetidus*, *Caltha palustris*, and ferns (Pringle 1980). Little research has been done on these swamps.

Bogs occur in well-defined depressions and are of the bowl or basin type (Fig. 4). These localities are usually topographically low and trap cool air so they have a microclimate that is cooler than the surrounding area. Examples of such bogs are found near the northern shoreline of Lake Erie and near the cities of London, Hamilton, and Guelph, Ontario. They are characterized by a surface layer of *Sphagnum* moss, mainly *S. fuscum*. Ericaceous shrubs are present including *Chamaedaphne calyculata* and *Ledum groenlandicum*. The main tree species is *Larix laricina*. Fens, containing various species of *Carex*, *Eriophorum* and grasses, may occur where minerotrophic conditions are present.

One of the most common wetlands in the region are the marshes of the Great Lakes (Fig. 19). These wetlands were recently reviewed by Smith *et al.* (1991). The major marsh development occurs in low-energy environments along the shores of Lakes St. Clair, Erie, Huron, and Ontario (Fahselt and Maun 1979). Frequent dominants include *Typha latifolia*, *Sparganium eurycarpum*, *Phragmites australis*, *Spartina pectinata*, and *Scirpus validus*. Near Hamilton, Ontario, large marshes are dominated by *Glyceria maxima*.

Figure 19. Marsh on shoreline of Lake Ontario near Oshawa, Ontario in TE Wetland Region. Note intensive urban development in background. Major vegetation is *Typha latifolia*.

In some areas, *Lythrum salicaria* can be abundant (Pringle 1980). This species is displacing native marsh plants which is of concern.

The vegetation ecology of the Great Lakes wetlands has been discussed in a paper by Keddy and Reznicek (1986). These authors have pointed out that existing information on Great Lakes wetlands is quite limited and often beset with taxonomic inaccuracies. These authors describe the flora of Great Lakes wetlands as being rich with up to 450 species of vascular plants, the most important genera being *Carex*, *Cyperus*, *Eleocharis*, *Juncus*, *Polygonum*, *Potamogeton*, and *Scirpus*. They describe three major wetland types found in the Great Lakes. These are:

1. Wet meadow – This type is found most commonly on Lakes Huron and Michigan where slopes and substrates are neither too steep or rocky. They are quite rich in species numbers with dominant species being *Calamagrostis canadensis*, *Carex lanuginosa*, *C. lasiocarpa*, *C. sterilis*, *C. stricta*, *Cladium mariscoides*, *Deschampsia cespitosa*, *Equisetum variegatum*, *Eleocharis elliptica*, *Juncus acutus*, *S. americanus*, *S. cespitosus*, *Solidago ohioensis*, and *Spartina pectinata*. In areas subjected to calcareous seepage, fens may be present.
2. Marsh – Emergent species dominate the marsh ecosystem at depths up to 1.5 m. Species of *Typha* predominate with other important species including *Decodon verticillatus*, *Eleocharis smallii*, *Phragmites australis*,

Pontederia cordata, *Sagittaria latifolia*, *Scirpus acutus*, *S. fluviatilis*, and *Sparganium eurycarpum*. Other species dominate waters less than approximately 15 cm deep including *Carex aquatilis*, *C. atherodes*, *Leersia oryzoides*, *Lythrum salicaria*, and *Phalaris arundinacea*.
3. Aquatic – This ecosystem includes submerged and floating-leaf aquatics in shallow waters as an understory in emergent plant areas, and at depths of water greater than emergents can tolerate, up to at least 8 m. Important species can include *Ceratophyllum demersum*, *Elodea canadensis*, *Heteranthera dubia*, *Megalodonta beckii*, *Najas flexilis*, *Nymphaea odorata*, *Nuphar variegatum*, *Potamogeton* spp., *Ranunculus aquatilis*, *Utricularia vulgaris*, and *Vallisneria americana*.

As for important ecological factors influencing the vegetation, regular fluctuation in water level is very critical. Such fluctuations can increase the area of wetlands and also the diversity of vegetation and species composition. High water is important in that it can kill dominant emergents such as *Typha*, which would otherwise form extensive monocultures reducing diversity. High water also prevents woody vegetation and other terrestrial species from colonizing sites near the waterline. During low water levels, mudflat annual species, meadow and emergent marsh species can germinate from buried seed banks. The major management problem in such marshes is to stabilize water levels since high levels lead to reduced marsh areas, lower species diversity and less ecosystem diversity.

Lakeshore marshes in the vicinity of urban/industrial areas of the Great lakes are subjected to contaminant deposition. Glooschenko *et al.* (1981b) studied the sediment chemistry of a Lake Ontario marsh located near Toronto. They found elevated levels of metals in surficial sediments such as copper, zinc, cobalt, chromium, and nickel. Of special note was cadmium which was approximately four times elevated, and lead, eight times elevated than in sediments deposited before industrialization of the area. Organic contaminants also accumulated in the marsh including DDT and its degradation products, chlordane, mirex, HCB, and PCBs. Pathways of input of these contaminants included runoff from roads and urban areas and atmospheric deposition.

The impact of stress upon Great Lakes wetlands was reviewed by Patterson and Whillens (1985). Natural water level fluctuations were found to be a major control of wetland area. However, wetland expansion can be limited by geomorphic controls. Also, highly stressed wetlands tend to exhibit qualitative rather than areal changes.

Another important area of wetland development in the Eastern Temperate Wetland Region is along the shores of the St. Lawrence River. Lacoursière and Grandtner (1972) studied the intertidal freshwater marshes of Ile d'Orleans located in the St. Lawrence River near Quebec City. They determined ten wetlands associations. Characteristic vegetation of these associations is as follows with the first five associations being characterized by submergents and the last five by emergents: 1. *Potamogeton nodosus*, 2. *Elodea canadensis*, 3. *Nitella tenuissima*, 4. *Najas flexilis*, 5. *Sagittaria latifolia*, 6. *Scirpus americanus* including subassociations with *Sagittaria rigida*, *Sagittaria cuneata*, and *Sium suave*, 7. *Scirpus validus*, 8. *Spartina pectinata*, 9. *Sparganium eurycarpum*, and 10. *Typha latifolia*.

The vegetation ecology of Huntingdon Marsh, located on the St. Lawrence River in Quebec near the Ontario and New York, U.S.A. borders, was described by Auclair *et al.* (1973). This marsh, as do many of the wetlands located along the 1,200 km length of the St. Lawrence River, serves as an important breeding and staging area for migratory waterfowl and other forms of wildlife. Two major communities of vegetation are present, the emergent aquatic and the sedge meadow. The first of these communities has 14 species but only six of these are dominants: *Equisetum fluviatile*, *Scirpus fluviatus*, *Eleocharis palustris*, *Scirpus validus*, *Phragmites australis*, and *Typha angustifolia*. Present here also are floating and submerged plants including *Myriophyllum exalbescens*, *Lemna trisulca*, *Potamogeton zosteriformis*, *Ceratophyllum demersum*, *Elodea canadensis*, and *Vallisneria americana*. The sedge meadow community has 56 species, but few dominants. Those dominants included *Carex aquatilis*, *C. lacustris*, *C. lanuginosa*, *C. stricta*, *C. diandra*, *Calamagrostis canadensis*, and *Typha angustifolia*. Important factors influencing community structure in the sedge meadow community were found to be disturbance, water depth, and fire. In the emergent aquatic community, interaction between submerged and floating forms and competitive exclusion between dominants accounted for much of the variation.

Studies have been carried out upon primary production and nutrient dynamics in marshes located in southern Quebec. Auclair *et al.* (1967a) studied a *Scirpus–Equisetum* marsh. Annual above-ground production was estimated to be 845 g m^{-2}, and such productivity exhibited a bi-modal seasonal pattern with peaks in late-July and mid-September. Two-thirds of all litter was exported while the remaining litter decomposed in the following growing season. In terms of soil factors, potassium correlated highly with standing crop and nitrogen with productivity. Species diversity was negatively correlated with primary productivity and standing crop.

Similar research was carried out in a *Carex* – dominated meadow (Auclair *et al.* 1976b). Important ecological factors were found to be soil fertility, fire incidence, and topographic position. Productivity was closely associated with cations, especially calcium and phosphorus. Siltation associated with high nutrient levels accounted for the higher productivity of *Typha angustifolia* communities located near open water. Fire incidence was important as it influenced species diversity by scarification, reduction of littermass, and altered energy and nutrient budgets. Topographic gradient influenced soil fertility and controlled species composition and community structure. The effect of community and soil variables upon plant tissue nutrients was also studied (Auclair 1977). Important controls on nutrient uptake included water depth, fire influence, and soil nutrient concentration. Nutrient losses from the marsh occurred by a combination of volatilization, runoff, and leaching. Litter was found to be an active site for cation exchange. Season nutrient dynamics was also investigated (Auclair 1982).

Pacific Temperate Wetland Region (TP)

The Pacific Temperate Wetland Region occurs in southern British Columbia in the lower Fraser River valley and along a narrow coastal strip of eastern Vancouver Island adjacent to Georgia Strait (Fig. 6). Characteristic wetlands include conifer swamps, domed and flat bogs, and flat fens (Banner *et al.* 1988). Saline and brackish marshes are present on the coast while along and in the Fraser River, brackish and freshwater marshes are found but vegetation is quite different from other Canadian coastal marshes as previously discussed. The climate is characterized by high precipitation with mild winters and warm summers (Table 3).

Swamps are dominated by *Alnus oregona* and *Thuja plicata*. They have an understory of ferns such as *Athyrium filix-femina* and *Dryopteris austrica*. *Oplopanax horridum* and *Lysichiton camtschatcense* also occur (Zoltai and Pollett 1983). Bogs can have a lower layer of such *Sphagnum* species as *S. fuscum*, *S. nemoreum*, and *S. fallax*. Low ericaceous shrubs occur with the main species being *Empetrum nigrum*, *Ledum groenlandicum*, *Kalmia microphylla* spp. *occidentalis*, *Vaccinium myrtilloides*, *V. oxycoccus*, *V. uliginosum*, *V. ovalifolium*, *V. alaskaense*, *Menziesia ferruginea*, *Andromeda polifolia*, and *Gaultheria shallon*. Trees are mainly open stands of stunted *Pinus contorta*. Peat depths range from less than 1 m in swamps to 5 m in bogs (Hebta and Biggs 1981, Styan and Bustin 1983).

The estuarine marshes of the Fraser River delta have been reviewed in the papers by Glooschenko (1980b, 1982) and Glooschenko et al (1988). The lower portions of these marshes are dominated by *Scirpus americanus* with *S. paludosus* and *Carex lyngbyei* found at somewhat higher elevations in the marsh. This grouping of vegetation is indicative of brackish conditions due to the mixing of the fresh Fraser River waters and saline Georgia Strait waters. In the high marsh, *Typha latifolia* becomes the dominant species. In the areas where influence of the Fraser River plume is minimal, salt marshes occur. Here, dominant species include *Triglochin maritima*, *Salicornia virginica*, and *Distichlis spicata*. The estuarine marshes at the head of the fjord at Squamish north of Vancouver are dominated by *Carex lyngbyei* with scattered occurrences of *Eleocharis palustris*, *Deschampsia caespitosa*, *Festuca rubra*, *Hordeum brachyantherum*, *Potentilla pacifica* and other less common forbs (Lim and Levings 1973, Levings and Moody 1976). Similar vegetation occurs on the coastal marshes of Vancouver Island (Dawe and White 1982, Kennedy 1982).

Several studies have been made upon the primary productivity of these Pacific coast salt marshes. Yamanaka (1975) investigated salt marshes in the Fraser river delta dominated by *Carex lyngbyei*, *Scirpus americanus*, and *S. paludosus*. He found an average yield of 490 g m^{-2} on a dry-weight basis. A comprehensive study was made in a *Carex lyngbyei* marsh in the same delta by Kistritz ant Yesaki (1979). An annual net primary protuction of 634 g m^{-2} on an ash-free dryweight basis (AFDW) was measured. They also found the below-ground biomass to be five times that of the above-ground biomass. Detritus was measured to be 435 g m^{-2} on an AFDW of which 62% disappeared between September and June, the balance being buried by alluvial sediments. They also investigated nutrient dynamics.

Atlantic Oceanic Wetland Region (*OA*)

This wetland region covers the southern half of the Avalon and Burin peninsulas in Newfoundland (Wells 1981). The oceanic climate of this region is characterized by cool summers, cold winters and high precipitation (Table 3).

Wetlands characteristic of this region are the plateau raised bogs, blanket bogs, small seepage fens, and slope fens. Marshes and swamps are localized in their distribution with coastal marshes common only along portions of the northwest coast of Newfoundland. These have been previously discussed in the section on the Atlantic Boreal Wetland Region.

Plateau raised bogs are the dominant peatland type in the region. They have flat to gently undulating surfaces with distinct sloping margins having a gradient of 20 to 25%. Although pools are a common surface feature, they are not patterned, but rather form an indeterminate scattered network. Small seepage fens with distinct pool patterns in ladder-like formations occur along the bog margins. Peat thicknesses in the plateau bogs vary from 2 to 20 m with *Sphagnum* moss to sedge peat underlain by sedge peat layers. This peat deposition originated in moist depressions in shrubby swamps or fens. The resulting bog is maintained by high amounts of precipitation distributed evenly throughout the year. Blanket bogs occur in areas of high rainfall and fog. Unlike raised bogs, blanket bogs are not confined to valleys or basins but cover extensive areas with peat 1 to 3 m in thickness.

The dominant vegetation in the bogs is *Sphagnum fuscum* in the form of

Figure 20. Slope bog near Prince Rupert, B.C. in OP Wetland Region.

drier carpets and hummocks with *Scirpus caespitosus* and the ericaceous shrubs *Chamaedaphne calyculata*, *Kalmia angustifolia*, and *Ledum groenlandicum* also being common. *Sphagnum imbricatum* hummocks are also common. The representative bog association is *Kalmia-Sphagnum fuscum*. In moist or wet hollows the moss carpet consists of *Sphagnum capillifolium*, *Sphagnum tenellum* and sedges. In wet carpets *Sphagnum magellanicum* is dominant. These bogs are discussed further in Pollett ant Bridgewater (1973), Wells (1981) and Wells and Pollett (1983).

Pacific Oceanic Wetland Region (OP)

This wetland region covers all of the Queen Charlotte Islands, the northern coast of British Columbia and the northern and western coasts of Vancouver Island (Fig. 6). The oceanic climate of this region is characterized by cool summers and mild winters with high precipitation, most of it falling as rain (Table 3). Two wetland districts are recognized in this region, the North Coast and the South Coast Pacific Wetland Districts.

Bogs are the most commonly occurring peatlands in this region with slope bogs and flat bogs being the most prevalent (Banner *et al.* 1988). Slope bogs occur on sloping terrain, often with a pattern of peat ridges or steps that confine small pools of water on the slope (Fig. 20). Dome bogs are poorly developed and rare. Although they are not very common, fens are found throughout this region and are located mainly along streams, in shallow basins, and at the heads of bays. Sedimentary peat materials rich in diatoms are very common in this region. This peat could occur in bands a few centimetres thick within the peat deposit or it could form a deposit several metres deep. The thickness of the peat is generally 1.5–4 m in bogs and 1–2 m in fens.

The development of bogs in the Pacific Oceanic Wetland Region is associated with soil formation. Ugolini and Mann (1979) studied peatland development on marine terraces in nearby southeastern Alaska. They found that as podzol formation took place, iron-cemented hardpans were formed in the soil. This impedes drainage and litter accumulates due to anaerobic conditions. The soil pH is also lowered. This leads to peat formation and subsequent bog development. The process of bog formation was studied in northern coastal British Columbia near Prince Rupert (Banner *et al.* 1983). Using a combination of pollen analysis, peat stratigraphy and carbon-14 dating, they found a succession to take place starting with a *Pinus contorta–Alnus rubra*-fern alluvial forest and ending with a bog having a dominant species *Pinus contorta*, *Chamaecyparis nootkatensis*, several species of ericaceous shrubs and several species of *Sphagnum* moss. The authors related this succession to paleoclimatic change and edaphic factors. The relationship between vegetation and water chemistry in British Columbia coastal peatlands was studied by Vitt *et al.* (1990).

The region has limited salt marsh development. The coastline is very mountainous and there are few areas of protected tidal flats where vegetation can colonize. These are at the heads of fjords. Here, *Carex lyngbyei* is the dominant species (Fig. 21). The Queen Charlotte Islands contains some areas of salt marsh in locations fronted by shingle beaches or mudflats (Calder and Taylor 1968). Besides *Carex lyngbyei*, other species occur including *Deschampsia cespitosa*, *Hordeum brachyantherum*, *Festuca rubra*, *Triglochin maritima*, *Plantago macrocarpa*, and *Stellaria humifusa*. Where river inputs are high, estuarine marshes are present with *Triglochin maritima*, *Puccinellia pumila*, *Scirpus cernuus*, and *Lilaeopsis occidentalis* occurring.

Mountain Wetland Regions

In mountainous areas, wetlands are found in valleys, on mountain slopes, and in alpine regions. They generally cover small areas with the exception of those occurring in some valleys. The type of wetland that will develop on a particular site depends on the elevation and latitude of the site. Thus, for example, wetlands typical of the southern part of the Boreal Wetland Region may occur in the lower valleys but at higher elevations wetlands resemble

Figure 21. Salt marsh dominated by *Carex lyngbyei*, Queen Charlotte Islands, B.C.

those found in the High Boreal and Subarctic Wetland Regions. There are four distinct mountain wetland regions (Fig. 6).

Coastal Mountain Wetland Region (*MC*)

This wetland region covers the mountainous central part of Vancouver Island, most of the mountains along the mainland coast of British Columbia, and the southwestern corner of the Yukon Territory (Fig. 6). This region is generally dominated by high mountainous areas but it also includes the valleys which are climatically very much affected by the mountainous topography. This region is divided into three subregions, the North, Central, and South Coastal Mountain Wetlands.

The climate of this region is characterizet by cool summers, cool to cold winters, and moderately high precipitation (Table 3). No permafrost associated with wetlands was recorded in this wetland region. The peat is generally 1 m thick, although some deeper deposits are found in the valleys.

In the Coastal Mountain Wetland Region, the most common wetlands are flat bogs and horizontal fens in valleys and small basin bogs and fens in alpine areas. Marshes are generally found along the shores of lakes and in deltaic areas. In addition, ribbed fens are found in valleys and at higher elevations throughout this region but are more common in the northern and central areas. No taxonomic studies have been made in these wetlands.

Interior Mountain Wetland Region (MI)

This wetland region covers the interior portion of the Cordilleran mountains (Fig. 6). It is generally dominated by high mountains and the associated valleys. Climatically, this wetland region is divided into three subregions, the North, Central, and South Interior Mountain Wetlands.

The climate of this region is characterized by cold to very cold winters, cool summers, and moderate to low precipitation (Table 3). Permafrost is associated with some of the wetlands in the central and northern parts of this region. The average thickness of peat is 2–3 m in the northern and central subregions and 1–2 m in the southern subregion.

In the Interior Mountain Wetland Region, the most common wetlands are flat and basin bogs. In addition, horizontal fens occur in the south and ribbed fens and palsa and peat plateau bogs occur in the north. Marshes occur along the shores of lakes and in deltaic areas. Horizontal fens are commonly found in valleys but small horizontal fens also occur in alpine areas. Ribbed fens are found in valleys, especially in the northern and central subregions. They do occur in the southern part of this region, but at higher elevations. Palsas were reported at the 1,000 m elevation in the Atlin area of British Columbia in the extreme northern part of the central subregion by Seppala (1980). These palsas had a frozen silty core. Their height ranged from 0.5 to 3 m but they contained only 7 cm of peat. Small basin bogs are found in alpine areas throughout the region. These basin bogs are generally associated with *Picea mariana* or are treeless. Species in these wetlands are similar to those found in other palsas in Canada.

Rocky Mountain Wetland Region (MR)

This wetland region covers the Rocky Mountains in Alberta and British Columbia and the Selwyn, Mackenzie, Richardson, and British Mountains in the Yukon and Northwest Territories (Fig. 6). This region is divided into three subregions, the North, Central and South Rocky Mountain Wetlands.

The climate of this region is continental to arctic and the precipitation is moderate to low (Table 3). Permafrost is discontinuous in most of these areas except in the extreme northern portions of the Richardson and British Mountains. The peat is generally 1 m thick although some deeper deposits are found in the valleys.

In the Rocky Mountain Wetland Region the most common wetlands are flat and basin bogs and horizontal and ribbed fens. These are found mainly in the southern and central areas. The northern areas are associated with palsas, peat plateaus, and veneer bogs, which occur in the valleys and at lower elevations. Marshes are found along the shores of lakes and in deltaic areas. Palsas and peat plateaus were reported at Macmillan Pass in the Selwyn Mountains by Kershaw and Gill (1979). These peat landforms were found in bog and fen depressions at elevations between 1,285 and 1,690 m. Both of these peatlands were vegetated by *Cladonia–Betula glandulosa*, *Cladonia-Polytrichum – Cetraria*, and lichen – *Polytrichum* plant communities.

Eastern Mountain Wetland Region (ML)

This wetland region covers the Mealy Mountains in Central Labrador. The area is dominated by strongly glaciated mountains. The climate of this region is characterized by cold winters, cool summers, and low precipitation (Table 3). The peat thickness is usually between 10 and 50 cm.

In the Eastern Mountain Wetland Region the most common wetlands are slope and ribbed fens with the slope fens being associated with the southern portions of the mountains. Species in these wetlands are similar to those discussed in parts of Atlantic Canada characterized by lower elevations as previously discussed.

Development of wetlands

The development of mineral wetlands generally begins when conditions, especially hydrological conditions, create an environment suitable for wetland development. Little or no organic deposition takes place in these wetlands since the organic matter produced is able to decompose at a relatively fast rate because of the favourable oxygen conditions associated with these systems.

The development of peatlands begins when the basal peat is deposited and continues to the present. Peatlands, in a sense, represent a high energy balance system, where a great deal of energy is stored and very little is released by degradation. The energy which is released comes mainly from the surface layers with an increasingly smaller amount being released from the lower layers.

Organic material is continuously being added to the surface by vegetation litter. Thus, the peat deposit reflects the succession of vegetation, characterized by layers differing not only as to their degree of decomposition but also as to the nature of the parent materials.

Peatlands, in most cases, are composed of more than one peat layer. These peat layers are a reflection of the type of vegetation contributing to the organic layer rather than of the later decomposition processes.

Mineral wetlands are associated with a thin surface peat layer or an organic-rich mineral surface layer. Their development could thus be expressed using the equation Jenny (1941) developed for mineral soils. This equation is as follows:

$$S = f(cl, o, r, p, t \ldots)$$

where the soil development (S), depends on climate (cl), organisms (o), relief (r), parent material (p), and time (t). The dots indicate that additional forcing factors may have to be included. It is difficult to use Jenny's equation for peatlands, which are composed of several contrasting layers. The approach taken here is thus to show that the genesis of wetlands, especially those associated with peat, is greatly time-dependent and began at time zero, when the basal organic matter was deposited. Time is the dominant forcing factor and all other forcing factors are time-dependent (Tarnocai 1978).

A single, homogeneous peat layer is the result of the interaction of biological and physical forcing factors. Thus, the equation for a single peat layer (Sp) can be written:

$$Sp = f(cl, w, r, v, c, t \ldots)$$

where the variables are defined as: climate (c); water properties (w); relief or landform (r); vegetation (v); organisms (o); and time (t). The dots stand for unspecified components such as permafrost and dust or water pollution.

Peatlands, as indicated above, are composed of several peat layers, and all forcing factors are time-dependent. Thus, the equation for an organic soil (So) is:

$$So = Sp1 + Sp2 + Sp3 + \cdots + Spn = f(t)_{cl,w,r,v,o\ldots}$$

where Sp1, Sp2, Sp3 . . . Spn represent the individual layers of peat materials. Time (t) is the dominant factor and stands for the total time. The subdominant factors can change with time and are listed as subscripts.

This relationship is probably better shown in the schematic model suggested by Tarnocai (1978). This model, shown in Fig. 22, illustrates the relationship between the various forcing factors and time. The interaction of these factors can produce various peat layers, as in the case of peatlands, or very little peat material, as in the case of mineral wetlands. For example, a change of hydrology, especially water chemistry or climate, during time can give rise to various types of wetland development. Peatlands can start as mineral wetlands then develop into fens and finally into bogs. In coastal areas along the Hudson Bay Lowland coastal marshes represent the initial stage of development. Because of glacial rebound these change into fresh water marshes, then fens, and finally develop into bogs (Tarnocai 1982).

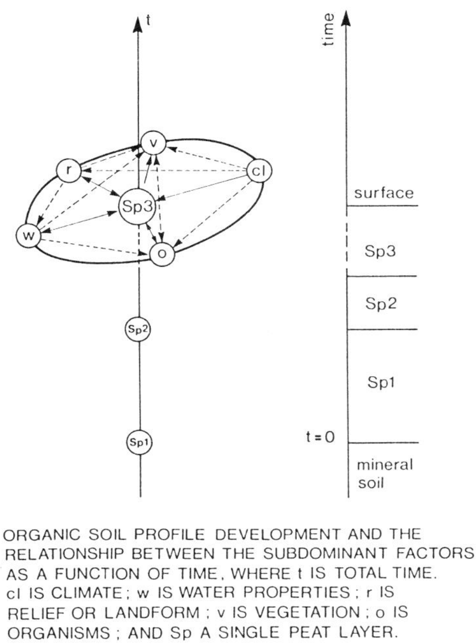

Figure 22. Relationship between forcing factors and time.

Forcing factors affecting wetland development

Time

As has already been indicated, time is the dominant factor in the genesis of wetlands. Time zero is the point at which wetland development began. In Canada, this is 4–6 $\times 10^3$ years B.P. in the continental regions and 8.5–9 $\times 10^3$ years B.P. in the arctic islands (Tarnocai 1978). From time zero onwards the interaction of the various factors produces the different layers of peat materials, the most recent one being on the surface. Any of the factors, however, can change with time, producing different organic layers, changing the rate of peat deposition or producing a different wetland. In the extreme case, the change (e.g., in climate) can be so drastic that wetland development ceases, as was the case with the arctic islands. Illustrations of the change in peatland vegetation over time in western Canada includes the paleoecological studies of Kubiw *et al.* (1989), Nicholson and Vitt (1990), and Zoltai and Vitt (1990).

Climate

There is very little peat development in either the prairie region, because the climate is dry and warm, or on the high arctic islands, because the climate is dry and cold. The most favourable climate for the development of peat in Canada coincides with that of the boreal and subarctic regions. Here, the climate is cool and moist, the optimum climatic conditions for peatland development (Terasmae 1972).

Many wetlands in Canada are affected by permafrost. Permafrost in the subarctic and boreal regions developed in some peatlands after the peat was deposited in a non-permafrost environment. This implies that a general cooling of the climate may be responsible for this phenomenon. However, it was found (Reid 1974, Zoltai 1972) that incipient permafrost as well as thawing permafrost may occur in the peatlands of these regions under the present climatic regime. It is possible, however, that a slightly cooler temperature was responsible for the initiation of most permafrost in these regions, as was found to have occurred in central Manitoba between 600 and 200 years B.P. (Thie 1974).

The influence of latitude, thus climate, was investigated upon the primary production of shoots of the bog shrubs *Chamaedaphne calyculata*, *Kalmia polifolia*, and *Ledum groenlandicum* along a gradient extending from southern to northern Ontario (Reader 1982). Most of the variability in shoot growth was explained by a combination of heat sum (degree-days), water-table depth, and water conductivity.

Water properties

The chemical composition of the ground waters, especially the cation content, is a very important factor influencing the floristic composition of wetland vegetation, hence, the type of peat development. This hypothesis is well demonstrated in several studies including Sjörs (1963), Heinselman (1970), Jeglum (1971), Tarnocai (1973), Mills *et al.* (1976), Karlin and Bliss (1984), Vitt and Bayley (1984), Vitt (1990), Gignac and Vitt (1990), and Vitt and Chee (1990).

In Table 5 the chemistry of surface waters from various peatlands and organic soils of northern Minnesota, southeastern Manitoba, the Hudson Bay Lowland, and the upper Mackenzie River area is given. These data indicate that bogs have developed in areas where the water is low in pH (3.3–4.7), low in calcium (0.5–4.0 ppm) and magnesium (0.1–0.7 ppm), and very low in sodium and potassium. The anion content of these waters is also very low. The waters associated with the fen type of peatland, on the other hand, are much higher in pH (5.3–7.8) and calcium (5.0–42 ppm), higher in magnesium (0.1–22.2 ppm) and sodium (6.0–7.6 ppm), low in potassium, and medium in anion content. Any changes in the hydrology of the wetland (excessive drainage or inundation) will also affect the vegetation and hence the formation of the wetland. The paper by Shotyk (1988) presents a review

Table 5. Chemical composition of surface waters from various types of peatlands and organic soils.

Peatland type	pH	C^{+2}	Mg^{+2}	Na^{+} mg/L	K^{+}	Cl^{+}	SO_4^{-2}	HCO_3^{-}	Source
Bog-pool	4.6	0.5	0.2	0.3	0.1	1.2	1.1	0	Sjörs 1963
Ombrotrophic bogs (bog-pool)	3.3–3.8	1.0–1.6	0.1–0.4	–	–	–	–	–	Heinselman 1970
Domed bog	4.5	1.9	0.6	–	–	–	–	–	Mills *et al.* 1974
Domed bog	4.1	4.0	0.7	–	–	–	–	–	Mills *et al.* 1974
Bog plateau	4.7	2.0	0.7	Tr[a]	–	Tr	Tr	Tr	Tarnocai 1973
Polygonal peat plateau	3.9	1.4	0.1	Tr	–	Tr	Tr	Tr	Tarnocai 1973
	4.0	2.8	0.1	Tr	–	Tr	Tr	Tr	Tarnocai 1973
Fen-flark	6.8	8.9	1.8	0.9	0.3	2.4	Tr	24.1	Sjörs 1963
Patterned fen	5.3–6.4	5.0–10.6	0.1–2.8	–	–	–	–	–	Heinselman 1970
Lowland fen[b]	6.9	36.0	4.6	–	–	–	–	–	Mills *et al.* 1974
	7.2	42.0	6.0	–	–	–	–	–	Mills *et al.* 1974
Patterned fen	7.8	18.6	22.2	7.6	–	17.7	29.7	97.0	Tarnocai 1973
Flat fen	6.9	37.1	3.3	6.0	–	3.5	1.9	164.1	Tarnocai 1973

[a]Traces, less than 0.1 pp.m.
[b]Periodically burned.

of the inorganic chemistry of peat and peatland water. Bourbonniere (1987) has reviewed the organic geochemistry of bog waters.

A recent paper by Glaser and Janssens (1986) discussed the relationship between the geographic patterns of bog landforms and peat stratigraphy. Autogenic bog proceses such as changes in the hydraulic properties of accumulating peat were found to be as important as climate on a regional basis.

Relief or landform

Landform types associated with peatlands in Canada have been studied and described by Tarnocai (1970) and Zoltai *et al.* (1975). They recognized three basic types of peatland classes: bogs, fens, and swamps. Subdivisions of these are based on the surface morphology (e.g., domed, plateau, flat, sloping, and patterned).

The landform type determines the moisture regime and the water source for the peatland and, thus, the type of vegetation growing on the peat deposit. In general, fen and swamp types of peatland forms are associated with a minerotrophic environment. They are characterized by saturated conditions and the water table is above or just at the surface for most of the growing season. The water supply of the fen and swamp types of landforms is mainly from mineral-rich ground waters. On the other hand, bog-type peat landforms are ombrotrophic (water supply is mainly from rain). In bogs, the water table is below the surface and, in the extreme case of the domed bogs, especially those associated with permafrost, there is a very dry surface peat cover.

Vegetation

Vegetation plays a very important part in the development of wetlands since the organic material originates from vegetation and reflects the succession of vegetation by its peat layers. The properties of the deposit (e.g., degree of decomposition and chemistry) are largely related to the type of vegetation from which the organic material was derived.

The peatland vegetation communities in the boreal region have been studied and described according to species composition by Heinselman (1970), Dansereau and Segadas-Vianna (1952), and Moss (1953). They delineated the floristic composition of the peatland environment and the successional stages resulting from environmental changes.

Organisms

The decomposition rates of organic materials associated with wetlands are frequently much slower than those of organic materials associated with uplands. This is basically due to low oxygen content associated with waterlogging, low nutrient content and pH, and low soil temperatures.

Latter *et al.* (1967) studied the microbiological activity in organic (peat) soils and compared the results with those obtained from mineral soils in the grassland region. They found that the total number of bacteria is approximately half as much in the peat soil ($14–35 \times 10^8$ cm^{-3}) as in the grassland soil ($16–79 \times 10^8$ cm^{-3}). They also estimated the total length of living fungal hyphae and found that the peat soil contained 15–180 m cm^{-3} and the grassland soil contained 160–580 m cm^{-3}. The ratio of bacteria to length of fungal mycelium is 1:300 in grassland soil ant 1:1,300 in peat soil. They also indicated that nitrogen-fixing bacteria, both aerobic and nitrifying, are virtually absent in peat soil.

Decomposition occurs most rapidly in the surface layer of the organic soil profile. The studies of Clymo (1965) show that the greatest loss in dry weight, which indicates the rate of decomposition, occurs in the surface 20 cm and becomes very low or disappears completely below this depth. This is due to the anaerobic conditions under which very few organisms can operate.

There are also indications that the rate of decomposition differs depending on the botanical origin of the organic soil material. *Sphagnum papillosum* decomposes at only about half the rate of *S. cuspidatum* (Clymo 1965). The more easily metabolized compounds will be used up by the organisms most rapidly, leaving less palatable compounds to be degraded more slowly. This means that the decomposition rate is rapid in the initial stages and becomes slower as time proceeds (Waksman and Stevens 1929, Theander 1954).

Soil animals also play an important role in the decomposition process (Cragg 1961, Macfadyen 1963). There is, however, too little data available for a complete assessment of their role in the decomposition of organic soils.

The Hudson Bay Lowland as an example of wetland development

The Hudson Bay Lowland lies on the western coast of James and Hudson Bays between the Quebec-Ontario border and Churchill, Manitoba. It is a large, flat, poorly-drained lowland with an area of 324×10^6 km^2. The area is underlain by Paleozoic sedimentary rocks, mainly carbonates, and the average slope is only 0.5–1.0 m per km. It lies adjacent to the Canadian Shield. The region is dissected by major rivers originating south on the Shield.

The lowland was deglaciated 7,400–8,000 years B.P. It was then invaded by the Tyrrell Sea which left a glacio-marine clay deposit which restricts drainage, promoting wetland formation. The area is undergoing isostatic rebound at a rate of 0.7–1.0 m per 100 years (Hunter 1970). Old coastal features such as beach ridges and dunes can be found inland from the coast at distances up to 350 km (Pala and Weischet 1982). The area represents an excellent chrono-sequence for pedological studies as time zero is represented by actively-forming beach ridges and dunes with well-developed podzols of ages up to 5,000 years being present (Protz *et al.* 1984). This also allows for a study of wetland development and succession processes. A climatic gradient is present. The southern portion of the lowlands lies in the Mid-Boreal Wetland Region while the northern portion lies in the High Subarctic Wetland Region (Fig. 23).

The wetland vegetation is of particular interest as it ranges from coastal salt marshes and brackish or freshwater marshes near major rivers to inland peatlands consisting of fens, bogs, and swamps. This will be discussed in some detail for two areas, southern James Bay and Hudson Bay near the northern limit of the lowlands.

The system starts out as salt marshes on the coast. These have been already discussed in this paper in the sections on the High and Low Subarctic and the High and Mid-Boreal wetland regions. Towards the landward-edge of these marshes, the fresh water influence becomes greater than the saline influence and wetlands in the southern portions of the lowland are characterized by both typical freshwater marsh species such as *Typha latifolia* and fen species including *Carex paleacea*, *C. diandra*, *Calamagrostis neglecta*, *Potentilla palustris*, *Myrica gale*, *Menyanthes trifoliata*, and species of *Eriophorum* (Glooschenko 1983). More important, peat begins to form in this location. Thus, the salt marsh wetland ecosystem gives way to a fen. Peat within 5 km of the coast is usually shallow, with depths around 40 cm. Peat depth increases inland and depths of 2 m are found 30–40 km inland (Sims *et al.* 1982a). Fresh water marshes disappear at approximately the High Boreal-Low Subarctic Wetland Region boundary (Fig. 6) and only fens are present. Treed fens with *Larix laricina* are particularly common in the southern portion of James Bay (Riley 1982, Sims *et al.* 1982a, b).

Away from the coast, increased peat thicknesses lead from minerotrophic to ombrotrophic ecosystems, i.e., bogs. Fens are still present where minerotrophic waters occur. The vegetation of inland peatlands was first described by such authors as Hustich (1957) and Sjörs (1961) for the Hawley Lake area in Ontario (Low Subarctic), and Sjörs (1963) for bogs and fens along the Attawapiskat River in the High Boreal Wetland Region. The succession of these wetlands has been discussed by Jeglum and Cowell (1982) for the Kinoje Lakes area located in the Mid-Boreal Wetland Region of the Hudson Bay Lowland, some 85 km inland NNW of Moosonee, Ontario located at the mouth of the Moose River. They discussed succession for three wetlands: those located along flowage lakes and streams, peatlands isolated from flow but under minerotrophic conditions, and ombrotrophic bogs. The flow-dominated wetlands begin as shallow marshes with thin peat. Major species here include *Carex rostrata*, *Eleocharis palustris*, and *Potentilla palustris*. This evolved through a meadow marsh stage to a thicket swamp with *Alnus*

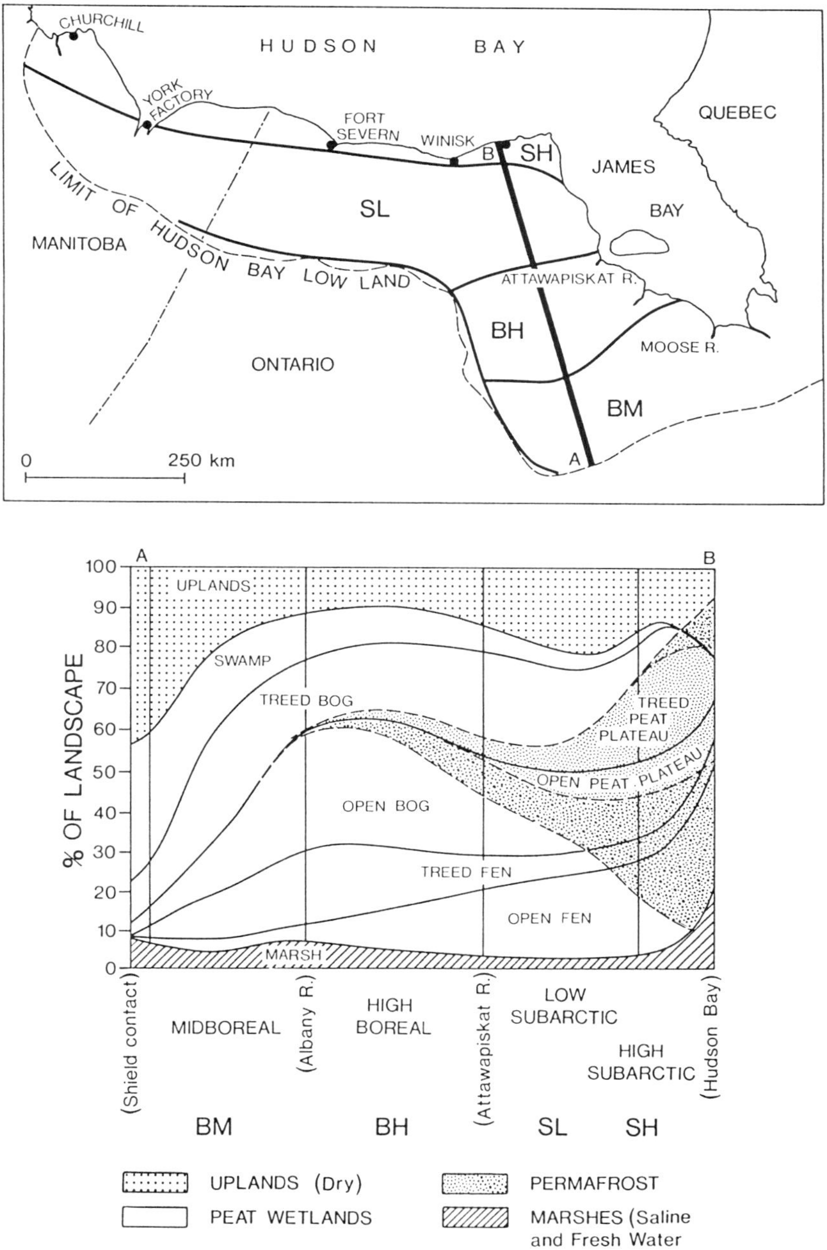

Figure 23. Map of Hudson Bay Lowland with Wetland Regions (figure redrawn from Riley 1982).

Figure 24. Bog near Winisk, Ontario in Hudson Bay Lowland (SL Wetland Region). Trees are stunted *Picea mariana*.

rugosa, *Betula pumila* var. *glandulifera*, and *Salix planifolla*. As peat deepens, a conifer swamp forms with *Picea mariana*, *P. glauca* and mosses. This can change with time to either an upland forest or a low-shrub treed bog. The minerotrophic peatlands isolated from flow start as a fen pool with mosses such as *Scorpidium scorpioides* and *Drepanocladus exannulatus*. This matures as a graminoid fen with *Menyanthes trifoliata*, *Rhynchospora alba*, *Scirpus caespitosus*, *C. interior*, *C. lasiocarpa*, and *Equisetum fluviatile* and mosses. Next in the sequence, a low shrub fen occurs with *Chamaedaphne calyculata*, *Myrica gale*, *Potentilla fruticosa*, and *Salix pedicellaris*. This is followed by a treed fen dominated by either *Larix laricina* or *Thuja occidentalis*. This then evolves into either a conifer swamp or a bog (Fig. 24). The ombrotrophic bog sequence evolved from a bog pool with *Sphagnum rubellum*, to a *Sphagnum* and graminoid bog with *Sphagnum magellanicum*, *S. pulchrum*, *S. tenellum*, *Carex oligosperma*, *Andromeda glaucophylla*, and *Oxycoccus microcarpus*. This successional sequence is terminated by a low shrub bog vegetated with *Kalmia angustifolia*, *Ledum groenlandicum*, *Sphagnum fuscum*, *Picea mariana*, *Rubus chamaemorus*, and *Cladonia* spp. This successional sequence has also been discussed in terms of formations, subformations, and physiognomic groups by Riley (1982).

In terms of climatic gradients, vegetation changes in the Hudson Bay Lowland are noted as one proceeds from the southern end of the lowland at the Canadian Shield (Mid-Boreal Wetland Region) north to the High Subarctic Wetland Region on Hudson Bay as depicted in Fig. 6 (Riley 1982). Swamps, treed bogs, and fens decrease in abundance. Open bogs dominate in High Boreal Wetland Region then decrease, while open fens increase to the north. In the Low Subarctic Wetland Region, permafrost becomes more important and peat plateau bogs, both open and treed, become the major wetland. An example of wetland succession at York Factory on the Hudson Bay shore in Manitoba is given by Tarnocai (1982).

The York Factory peninsula is located between the estuaries of the Hayes and Nelson Rivers in Manitoba. The seaward progression of Marsh Point (the tip of the peninsula) due to active glacial rebound is approximately 6 cm per year (Simpson 1972). This causes new surfaces to be continuously exposed and provides a gradient for wetland development with relatively recent salt marshes occurring close to the Hudson Bay shore and progressively older peatlands occurring further inland. The wetland, soil and vegetation development on the York Factory Peninsula was studied along an 18 km baseline (Tarnocai 1982). This distance represents a time span of approximately 2,100 years.

Low marshes influenced by fresh water occur along the coast below the mean high tide and are exposed during the low tide stage; tidal channels provide drainage. Their vegetation is composed of salt-tolerant species such as *Hippuris tetraphylla* and *Eleocharis* spp. High marshes are located above the mean high tide level and are inundated mainly during flood tides. These high marshes are well vegetated, dominantly by sedges, *Carex* spp., *Eriophorum* spp., *Triglochin maritima*, *T. palustris*, and occasional small clumps of *Salix* spp. This corresponds to the meadow zone described by Ritchie (1957).

Immediately inland from the high marshes is a transitional area composed of an alluvial plain with intermittent high marshes. This area, described as a shrub zone by Ritchie (1957), is covered with thick *Salix* spp. and inundated only during spring floods. When the flood waters recede, only depressional areas maintain the marsh characteristics. Peat development begins in this transitional area approximately 600 years after it emerges from Hudson Bay.

Large areas beyond this alluvial plain - high marsh unit are covered by horizontal fens, the most common terrain type on the peninsula. This area is subdivided into two units: the younger horizontal fen–palsa fen–beach unit and the older horizontal fen–palsa fen unit. The lack of beaches in the latter unit is accidental; the transect happened to cut through an area where no recognizable beaches were present. The vegetation can be divided into three zones on these units: *Salix* spp. (younger part), stunted *Larix laricina* (older part), and stunted *Picea glauca* and *Larix laricina* (oldest part). These vegetation zones correspond to the invading forest zone described by Ritchie (1957). It is in this horizontal fen zone that permafrost first develops, in association with beaches and palsa fens. Since most of the fens have a high water table, usually above the surface, permafrost develops on those areas which are slightly elevated above the water table. Flood waters from the Hayes River periodically inundate this area as is indicated by the alluvial layers within the peat and by occasional ice-rafted rocks and boulders found on the surface of the fens and palsas. This periodic inundation is very important for maintaining the minerotrophic characteristics of these peatlands.

When the peat surfaces are not affected by minerotrophic waters, the fens are slowly overtaken by bogs. Peat plateaus and palsa bogs are the dominant peatland types. They begin to develop on the Hayes River side of the peninsula approximately 2,000 years after the land emerges from Hudson Bay but on the Nelson River side development begins approximately 1,000 years after emergence. The earlier development of palsa and peat plateau bogs on the Nelson River side of the peninsula is related to the lack of flood waters and thus an ombrotrophic environment has been established much earlier. The vegetation in this unit is *Picea mariana*, ericaceous shrubs, and sphagnum mosses and corresponds to Ritchie's (1957) mound topography zone.

Thus, the Hudson Bay Lowland is a good example of wetland development as the important factors of time and climate can be separated in an area of uniform bedrock geology and physiography. Figure 25 summarizes the development of wetlands in the Hudson Bay Lowland starting with coastal marshes and ending with bogs or peat plateaus depending on the climate. This takes place in poorly-drained localities. In addition, upland forest can evolve into swamps if regional hydrology changes and water logging of the upland takes place.

Wetlands have been considered to be a major source of atmospheric methane, a gas that has been receiving increasing attention due to its functions as a greenhouse gas (Aselmann and Crutzen 1989). In order to understand the global budget of methane, studies are necessary on emissions from various wetlands, both in high and low latitudes. The Hudson Bay Lowland is an important area for studies of methane emissions as it is the second largest continuous wetland complex on earth next to the vast peatlands of western Siberia and shares a similar ecology to these wetlands.

The Northern Wetlands Study (NWS) was initiated under the auspices of the Canadian Institute for Research in Atmospheric Chemistry (CIRAC). The purpose of this study was to assess the importance of northern wetlands as a source/sink of biogenic gases to the atmosphere under current climatic conditions with emphasis on methane. This was in conjunction with intensive studies of the physical, chemical, hydrological, and biological processes influencing biogenic gas production (Roulet *et al.* 1991). A major field study was carried out during the summer of 1991 near Moosonee, Ontario involving

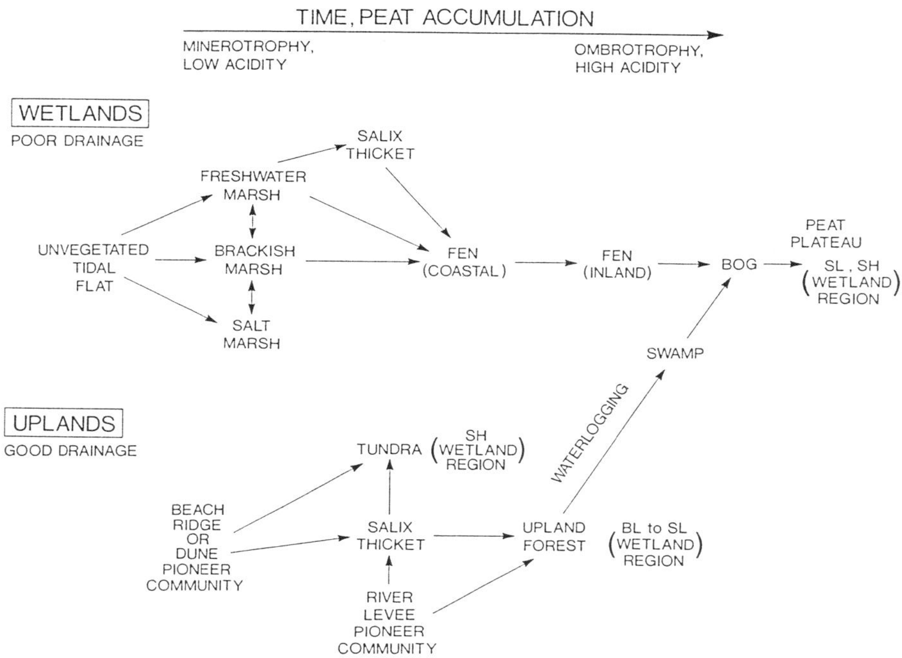

Figure 25. Wetland succession, Hudson Bay Lowland.

investigators from Canadian federal and provincial agencies, Canadian and U. S. universities, and NASA and the National Center for Atmospheric Research in the U.S. Preliminary results indicate that the Hudson Bay Lowland peatlands are a less important source of methane than anticipated (Roulet 1991).

Wetland inventories, change and evaluation in Canada

Wetland inventories

Wetland inventories for various purposes have been carried out throughout Canada in the past (Rubec *et al.* 1988). The majority of these inventories focussed on peatlands although some of the mineral wetlands, such as marshes, were also surveyed. The inventory methodologies varied. It became obvious from discussions during a workshop held in 1982 (Morgan and Pollett 1983) that no single data-gathering approach, system or methodology was used in Canada.

In the Newfoundland inventories an ecological division based on wetland types facilitated the gathering of data on peat volumes. In New Brunswick, the data gathered to provide a better understanding of the variation in peat and the data stored in a peat information bank can be used for resource planning and ecological interpretation. In Ontario an exploratory peatland inventory was conducted and detailed inventories were carried out in designated areas having potential for peat utilization. The Soil Survey of Canada, which is at present the only agency collecting data in a systematic fashion on wetlands across Canada, uses inventory methodology based on the wetland form (Tarnocai 1983). These landforms (e.g., flat bog, pattern fen, channel marsh, and basin swamp) provide an overall framework for establishing map units. Most of the inventories carried out by the Soil Survey of Canada rely heavily on remotely sensed data. Both LANDSAT data and various types of aerial photography are being used. Field data collection is carried out along a transect. This systematic data acquisition facilitates the generation of cross-sections of the wetland from which peat volumes and tonnages of peat can be calculated.

According to a recent estimate (National Atlas of Canada 1986) the total area of wetlands in Canada is about $127{,}194 \times 10^3$ ha, or 14% of the land area of Canada. The total area of peatlands (wetlands having greater than 40 cm of peat) in Canada is $111{,}328 \times 10^6$ ha, or 12% of the land area of Canada. Thus, mineral wetlands cover 2% of the land area of Canada. Table 1 shows the area of Canadian wetlands (both mineral wetlands and peatlands) while Table 6 (Tarnocai 1984) indicates only the peatland areas, volume, and tonnage by province and territory. No estimate was found in literature for the area of wetlands in Greenland.

The overall objective of the cooperative wetlands inventory in the Maritimes (Provinces of Nova Scotia, New Brunswick, and Prince Edward Island) is to provide information on the classification, size, distribution, and value to wildlife and other resources of wetlands in the Maritimes. Wetlands in the Maritime Provinces are being surveyed through a joint federal-provincial initiative known as the "Wetland Mapping and Designation Program", undertaken in 1980. No wetlands inventory exists for Quebec, but peatland inventories are in progress (Gerardin and Groudin 1986).

Inventories are being carried out in western Canada, mostly to monitor wetland changes (Rubec *et al.* 1988). In British Columbia, inventories are being carried out both in coastal estuarine wetlands and in interior British Columbia. A coastal inventory (five-year study) includes mapping, classifying in 35 major areas, and the evaluation of 200 to 300 wetlands (G. Adams, Canadian Wildlife Service, Saskatoon, personal communication).

The Alberta Fish and Wildlife Service is carrying out aerial survey work in cooperation with the U.S. Fish and Wildlife Service. This study provides spring pond index counts and as shown a steady erosion of wetlands from

Table 6. Peat resources of Canada (Tarnocai 1984).

Provinces and territories	Peatland areas			Indicated peat volumes		Indicated oven dry weight of peat	Indicated weights of peat with +50% water content	
	ha × 10^3	% of land area within designated areas	% of total Canadian peatlands	$m^3 \times 10^4$	%	Tonnes × 10^4	Tonnes × 10^4	%
Alberta	12,673	20	11	316,822	11	36,118	54,177	11
British Columbia	1,289	1	1	38,685	1	4,410	6,615	1
Manitoba	20,665	38	17	516,605	17	58,893	88,339	17
New Brunswick	120	2	19	4,800	#	466	698	#
Newfoundland–Labrador	6,429	17	#	257,160	8	24,945	37,417	8
Northwest Territories	25,111	8	6	577,553		65,841	98,762	19
Nova Scotia	158	3	23	6,320	#	613	920	#
Ontario	22,555	25	#	676,653	22	77,138	115,708	23
Prince Edward Island	8	1	20	312	#	30	45	#
Quebec	11,713	9	11	351,381	12	40,057	60,086	12
Saskatchewan	9,309	16	8	232,737	8	26,532	39,798	8
Yukon Territory	1,298	3	1	25,968	1	2,960	4,441	1
Canada	111,328	12		3,004,996		335,339	507,006	

#Less than 1%.
+ Oven dry weight basis.

agriculture, aggravated by drought. In the southern area of Alberta, wetland inventory work is concerned with the possible loss of wetlands which are maintained form seepage of old irrigation canals. Habitat loss was quantified with respect to upgrading of irrigation. The thrust of the present program is to retain habitat and includes landowner contact. Some wetland acquisition is carried out with Ducks Unlimited, a private organization, under the Wetlands for Tomorrow program.

Wetland monitoring and evaluation of change in Saskatchewan is being carried out by the Canadian Wildlife Service (CWS) in cooperation with Lands Directorate. They are using aerial photography provided by U.S. Fish and Wildlife Service to look at land use change, aimed at wetlands over a three Prairie Province area. This is to coincide with Ducks Unlimited work which is providing subsampling over three provinces or mapping work, also providing detailed groundtruthing. This will be redone every five years using Canada Land Data System.

The major federal wetland monitoring effort is the Canada Land Use Monitoring Program (CLUMP). This program established the Prime Wetlands Projects to provide a national overview of land-use change issues and dynamics on wetlands in southern Canada, improve federal wetland programs, and encourage provincial wetland initiatives. A national overview was obtained by monitoring wetland conversion trends around major Canadian cities and special regional studies (Lands Directorate 1986).

Wetlands loss in Canada

Many areas of southern Canada have declined in the areal extent of wetlands (Rubec *et al.* 1988). Lynch-Stewart (1983) has attempted to document this in southern Canada. Unfortunately, this is a hard task as she points out. There is a lack of interagency coordination and integration which leads to incompleteness, inconsistency, and duplication of data. She was able to find only 15 quantitative studies in all of southern Canada. Changes that occurred ran from less than 1% of wetlands being lost to over 70%. Some wetlands were altered while in other cases total destruction of the wetland occurred. In some areas, wetlands were created such as in parts of Alberta where Ducks Unlimited has created new artificial wetland habitat.

In the Province of British Columbia, major losses have occurred in wetlands in the Fraser River Delta due mainly to agriculture. For example, a 27% loss took place from 1967 to 1982, but much of this loss was from "natural" wetland to wetland use for recreation and conservation (Lands Directorate 1986). In the last century, however, marsh habitat has been lost by major dyke construction for flood control, landfills for urban and industrial development, and dredging for shipping. Past development also has led to wetland losses elsewhere in coastal British Columbia (Lynch-Stewart 1983). Near Vancouver, British Columbia, 70% of wetlands have been converted to agriculture and near Victoria, 58% of wetlands were lost to agriculture. In interior British Columbia, wetlands are being lost to increased ranging and grazing activities. The Prairie Provinces of Alberta, Manitoba, and Saskatchewan are characterized by an abundance of shallow "potholes" from less than one hectare to several hundred hectares in size. This "prairie pothole region" as it is called, provides habitat for rearing of approximately one-half of the population of North America waterfowl as well as many other migratory birds. These provinces are sites of major wetland loss due to agricultural activities such as drainage, in filling, and cultivation. No overall studies have been made of losses over the entire area, but estimates run up to a total of approximately 1.2×10^6 ha. Since settlement to 1976, this occurs directly in the loss of wetland basins to agriculture and indirectly through deterioration of marsh-edge vegetation which is essential for waterfowl habitat. In southwestern Manitoba, Rakowski *et al.* (1974) cited a 57% decline in total wetland areas from 1929–1974 and predicted further losses. Schick (1972) reported that only 39% of the original presettlement wetland area remained in the Alberta prairie parkland region. In southwestern Saskatchewan, Millar (1981) observed that by 1979, 84% of wetlands on sample transects had been affected by human activities. Major drainage projects account for about 20% of the total loss while the most significant cumulative losses are caused by drainage of small potholes by landowners or small drainage projects. Further details are given in Lynch-Stewart (1983).

Serious decline in wetland area in southern Ontario has been well documented. Snell (1982) has compared the extent of wetland area in 38 countries in pre-settlement (late 18th and early-mid 19th centuries) to that of the late 1960s. She estimated that over one million hectares (70% of area) has been converted to other uses. Agricultural reclamation, specifically drainage, is the major cause of such decline. A net decline of 1.8% occurred from 1967 to 1982. Similar findings on wetland loss were made by Bardecki (1981) who showed that 85% of wetlands loss in the period 1966–1970 was due to agricultural conversion. Three major areas affected: (1) southwestern Ontario, (2) eastern Ontario, and (3) the area south of Georgian Bay. In the three most southwestern Ontario counties, between 81 and 98% of wetlands have been lost. Another major area of wetland loss is the prime waterfowl habitat of the lower Great Lakes. Up to 35% of coastal wetlands there were lost to development by 1978 (Lands Directorate 1986). Wetland loss is significant along the Canadian shoreline of Lake Ontario (Whillans 1982). An estimated loss of 57% occurred with greater than 80% wetlands loss in some areas such as Toronto.

Quebec is another area of wetland loss. From November to March, this province is a major wintering ground for hundreds of thousands of aquatic birds. The wetlands here are thus of international importance. Important wetlands are also located along the lower St. Lawrence River valley where some 42% of wetlands were lost between 1950 to 1965, mostly due to agricultural conversion. An estimated 32% of salt marshes here have also been lost (Lands Directorate 1985). Urbanization in parts of Quebec has also led to wetlands decline. Factors causing decline in these urban centres include agriculture, urban growth, landfill, industry, and road construction (Lands Directorate 1986, Rubec *et al.* 1988).

The Atlantic provinces of Canada are an important area for coastal salt marshes. These marshes have declined in area by some 65% due to dyking and filling for agriculture. Tantramar Marsh on the Nova Scotia - New Brunswick border at the head of the Bay of Fundy is the largest single block of marshland in the Maritimes and contains some of the most productive habitat in eastern Canada (Jackson and Maxwell 1971). By 1920, 80% of the Tantramar Marsh was cultivated for hay and used for grazing, which severely limited waterfowl and wildlife habitat. At present, concern for the long-term management of competing interests has been acknowledged by the CWS and Lands Directorate of Environment Canada. The Bay of Fundy is also of international importance for seabird and shorebird species (Pearce and Smith 1974).

Inland wetlands in New Brunswick are not on a major flyway but do provide a dispersal route for waterfowl which travel up from the Bay of Fundy and the St. Johns River Valley. Staging areas for waterfowl are particularly active in wetlands south of Fredericton, New Brunswick. Other threats to Atlantic wetlands besides agriculture include urbanization and road construction. Potential threats are peat mining and tidal power projects (Wells and Hirvonen 1988).

Wetland evaluation and protection

This section will concentrate on wetland evaluation and protection in three areas of Canada: Ontario, Quebec, and the Maritime Provinces. This emphasis is due to a lack of wetland evaluation and conservation policy in other parts of Canada.

Ontario

Ontario is developing a wetland policy initiated by public and government concern about the future of wetlands. In 1981 the Government of Ontario released a discussion paper entitled "Towards a Wetland Policy for Ontario". Written by an inter-ministerial committee representing Ontario's resource ministries, this paper was designed to solicit public input concerning wetland management. Of the 520 responses which were received, 519 recognized the need to protect at least some wetlands. Ontario's wetlands are under much pressure due to rapid urbanization (see Bardecki and Patterson 1989).

The Guidelines for Wetlands Management in Ontario, released in the spring of 1984 and discussed earlier, are a political precursor to policy. They "represent the Province of Ontario's concern for wetlands and wetland management". Incorporated are the public's concern for the proper management of wetlands recognizing that other provincial and local interests including agriculture, housing, forestry, and recreation must also receive consideration in land-use planning.

To provide an objective base for many of the concerns with which the guidelines deal, southern Ontario's wetland evaluation system was incorporated into the decision-making process advocated by the guidelines. The evaluation system, which is now being used by the Ministry of Natural Resources and other agencies, ranks wetlands according to a point system based on their biological, social, hydrological, and special features values. The system is unique in southern Ontario. The evaluation system serves as a cornerstone of the guidelines in identifying valuable wetlands (EC/OMNR 1984, Glooschenko 1985).

Although the evaluation system pertains only to southern Ontario, the guidelines encompass all of Ontario's wetlands, both northern and southern. These guidelines represent the Province of Ontario's concern for wetland and wetland management in both southern and northern Ontario.

The wetland guidelines were designed to be incorporated by municipalities into their municipal planning process. As part of the government policy development, the guidelines were submitted to all 843 municipalities in Ontario. These were asked to comment on their appropriateness, applicability, and potential impact. Ultimately, the guidelines will be revised according to input by municipalities and other government and public agencies and will be incorporated into the Planning Act as official Government Policy. The wetland policy for Ontario is under development.

Federal interest in wetlands in Ontario has been centred on the St. Lawrence River wetlands. The St. Lawrence River lowlands region has been identified by the American-Canadian Planning Committee for the North American Waterfowl Management Plan as a high priority waterfowl staging and black duck area.

In 1985–86, the Canadian Wildlife Service (CWS) cooperated with the Ontario Ministry of Natural Resources in an initial study of St. Lawrence River wetlands. The objectives of this study were three-fold: (1) to take inventory and update size data for wetland areas along the St. Lawrence River; (2) to document the status of these wetlands with respect to aquatic vegetation, waterfowl habitat, significant wildlife species, waterfowl use; (3) to identify issues affecting these wetlands that are of relevance to CWS concerns. CWS also conducted spring and fall surveys of migrant waterfowl use of the St. Lawrence River-Ontario shorelines. Future work will include:

1. An initial survey of migrant waterfowl use of habitat on staging areas, during the spring and fall migration periods.
2. Identification of specific problems facing St. Lawrence River wetlands, such as degradation, disturbance, conversion, etc., and suggestions for amelioration.
3. Determination of wetland management needs and identification of appropriate habitat management techniques and management agencies.
4. Development of federal policy guidelines for the longterm protection and management of the St. Lawrence River wetland resource and exploration, with the provincial government, of alternative mechanisms for achieving preservation of selected wetland habitat.

Quebec

There is at present no evaluation or inventory of wetlands in Quebec with the exception of peatlands in Quebec below 54° N. The inventory stresses peat with potential economic use. They are protected, in theory at least, by the Environment Quality Act under the Quebec Department of the Environment (Environment Canada/Province of Quebec 1985) and legislative measures are in the planning stage.

Some degree of protection results from the acquisition of land by government or private conservation groups. The CWS administers a network of six National Wildlife Reserves, thus protecting 4,900 ha of habitat and many islands in the St. Lawrence Estuary. The Quebec Department of Recreation, Fish and Game also protects 9,700 ha of riparian land along the St. Lawrence and Ottawa Rivers. A number of migratory bird sanctuaries (14) have been established. At present, gaps in the Quebec Environment Quality Act, problems in applying the Act and the lack of a wildlife habitat protection act, leave most waterfowl habitats vulnerable.

The St. Lawrence Valley is the area where the need for habitat management is most urgent. Shoreline wetlands have been mapped here. Of the target areas, the marshes of Lake St. Frances and the Beauharnois region are especially significant. The wetlands of the lower Laurentians also call for rapid action because they are threatened or disturbed by farm reclamation and lumbering operations.

Maritime Provinces

Wetlands in the Maritime Provinces are being surveyed through a joint federal-provincial initiative known as "Wetland Mapping and Designation Program". The process began in 1980 and includes all wetlands over 0.25 ha in size. Similar to Ontario's system, the Maritimes wetland inventory is to be used in assisting federal, provincial, municipal, and town planning agencies in making land use decisions regarding wetland areas.

The end products of the inventory are a computer data base and a series of atlases. When completed, the wetlands inventory will be available to assist federal, provincial, municipal, and town planning agencies in making decisions relating to land use. Also, it will help the province to develop wetland policies and will provide a data base for a wide variety of wetland research and management programs.

High among the plans of this program is provision for the implementation of a federal-provincial agreement for wetland habitat protection. Under this agreement important wetlands could be designated for protection by both levels of government and neither would finance activity which would alter the natural habitats. Thus government assistance for agricultural drainage, industrial installations, sewage treatment plants and so on would not be approved for designated wetlands. It is hoped that such federal-provincial agreements can be developed soon after completion of the inventory. Recently, the government of Nova Scotia has used such wetlands information in the development of provincial water policies.

National and international aspects of wetlands protection

The long-term downward trends of migratory waterfowl, especially black duck, (*Anas rubripes*), goldeneye (*Bucephala clangula*), and green-winged teal (*Anas crecca*) populations due to loss of nesting habitat, wetland drainage, and degradation of migration and wintering habitat is of concern to both Canadian and U.S. federal governments (EC/USDI 1985). Designated areas of Key Priority Habitat listed in this document are: 1) the upper Atlantic coast, 2) the lower Great Lakes - St. Lawrence basin, and 3) Prairie potholes. It was recognized that efforts to maintain and enhance waterfowl habitat in North America are beyond the capability of public agencies alone. Thus long-term solutions will require the coordinated action of governments, private organizations and the involvement of landowners.

The program is intended to benefit both waterfowl and agricultural production by emphasizing land and water management and working to prevent soil erosion. Strategies are aimed at affecting small changes in land-use practice over a large area. These are planned to demonstrate that agriculture and wildlife production are compatible pursuits, and that wetlands are preserved. The plan was jointly signed by Canada and the U.S. in 1986. Its implementation will aid in protecting and improving 1.5×10^6 ha of Canada prairie wetlands, 28×10^3 ha of Great Lakes-St. Lawrence wetlands, and 4,000 ha of Atlantic wetlands important as waterfowl habitat.

The Canadian Wildlife Service is also acquiring waterfowl and wildlife habitats. To date, 44 national wildlife areas, many of them wetlands, have been acquired and more are being considered.

On an international basis, Canada is involved in the World Conservation Strategy which recognizes wetlands as of global significance. Canada also signed the Ramsar Convention on Wetlands of International Importance. Twenty-eight sites have been designated for protection (Rubec et al 1988) These sites are located mainly in marshes in southern Canada that are important waterfowl habitats.

There are factors impeding wetland protection in Canada. A major problem is fragmented jurisdiction for wetlands. No single federal agency exists to conduct wetland research. Federal, provincial, and municipal agencies share land-use management responsibilities and this can often cause coordination problems in wetland management and conservation. However, it appears that legislators and government agencies are beginning to recognize the importance of wetland conservation.

Recommendations for research, management and conservation of Canadian wetlands

In a large country such as Canada, which is characterized by cold climates and remote areas with limited access, it is difficult to carry out research because of such problems as limited access, short field seasons, and high costs of travel to conduct research. Limited funds tend to be spent in areas such as the populated areas of southern British Columbia and Ontario where threats exist to wetlands. Another factor that influences research is the economic importance of the wetland. For example, wetlands that are important waterfowl habitats, such as marshes in the prairies and salt marshes on the Pacific and Atlantic coasts, receive more research funds than arctic wetlands. Salt marshes in British Columbia have received greater emphasis than other salt marsh areas in Canada due to their importance as spawning areas for salmon, a commercially important species. In addition, until the rise in interest in peat as a potential energy source in the early 1980s, peatland research was limited. In general, wetlands research has not been given high priority in Canada and funding for such research has been minimal.

In terms of recommendations for future research, we see the following areas as priorities:

1. Wetland dynamics – Succession of wetland vegetation, both short- and long-term, is an important ecological process. It is necessary to understand how wetlands respond to various natural and anthropogenic perturbations in order to predict the impact of various management options. More research is needed on all aspects of this process in Canada, including

development of predictive models. This data is especially needed for wetland restoration and creation projects.

2. Hydrology – Little is known on the hydrology of Canadian wetlands, especially the potential impact of large peatland modifications upon regional hydrology. The current status of hydrological research on Canadian wetlands was reviewed by Roulet (1990).
3. The role of peatlands as fish and wildlife habitat – Very little research has been carried out on the ecological importance of peatlands and such research is necessary before large developments such as energy development and forest drainage take place in peatlands.
4. The impact of wetlands on water quality – Little research has been done on the influence of wetlands on the water quality of receiving water bodies such as streams and lakes. Research is also necessary on the role of wetlands as a sink for nutrients and contaminants.

It is not possible in this paper to go into problems relating to the management of wetlands in Canada. The country consists of many political jurisdictions including federal, provincial, and local-government levels. A federal policy is currently under government review. We do not intend to discuss needs relating to required wetland protection legislation. However, we do recommend several areas that need further consideration in order to assist government agencies in improved wetland protection:

5. Inventories – Improved wetland inventories are needed for all areas of Canada with emphasis on southern Canada. These then will serve to assist managers in determining specific wetlands or wetland areas that merit conservation.
6. Wetland evaluation – Improved methodologies are necessary in order to determine which wetlands must be preserved and which ones are not significant. This would include various considerations such as wildlife habitat, hydrological role, commonness or rarity of the wetland type in the locality, etc.
7. Impact of land use activities on wetlands – More research is necessary on how various activities impact on local wetlands. For example, how can agricultural practices be made more compatible with protection of pothole marshes in the prairies? How does water-level regulation in reservoirs and the lower Great Lakes affect wetland vegetation? How can a peatland be drained so as not to deleteriously impact on receiving waters?
8. Regional needs – There is little information available for some Canadian wetland types and geographic localities. These include the Mountain Wetland Regions, southern Ontario swamps, Great Lakes marshes and salt marshes on the Atlantic Ocean and St. Lawrence River. With the exception of the Mountain Wetlands, the other wetlands are in areas of development, including agriculture, urbanization, and recreation.

Acknowledgements

The authors wish to thank Elizabeth McCurdy and Dianne Crabtree of the National Water Research Institute, Burlington, Ontario, Canada and Carina Hernandez for the Faculty of Environmental Studies, York University, for their typing of the manuscript.

References

Adams, G. D. (1988) Wetlands of the Prairies of Canada. p. 155–198. *In*: National Wetlands Working Group (ed.). Wetlands of Canada. Ecological Land Classification Series No. 24. Sustainable Development Branch, Environment Canada, Ottawa. Polyscience Publishers, Montreal, Quebec, Canada.

Andersson, G. and Hesselman, H. (1907) Vegetation och flora i Hamra kronopark. (Vegetation and flora in the State Forest of Hamra). Meddelande Statens Skogsforsokanstalten 4: 35–102.

Anonymous (1969) Saskatoon wetlands seminar. Canadian Wildlife Service, Report Series No. 6, Ottawa, Canada. 262 pp.

Aselmann, I. and Crutzen, P. (1989) Global distribution of natural freshwater wetlands and rice paddies, their net primary productivity, seasonality and possible methane emissions. Journal Atmospheric Chemistry 8: 307–358.

Auclair, A. N. D. (1977) Factors effecting tissue nutrient concentrations in a *Carex* meadow. Oecologia (Berl.) 28: 233–246.

Auclair, A. N. D. (1982) Seasonal dynamics of nutrients in a *Carex* meadow. Canadian Journal of Botany 60: 1671–1678.

Auclair, A. N., Bouchard, A. and Pajaczkowski, J. (1973) Plant composition and species relations on the Huntingdon Marsh, Quebec. Canadian Journal of Botany 51: 1231–1247.

Auclair, A. N. D., Bouchard, A. and Pajaczkowski, J. (1976a) Plant standing crop and productivity relations in a *Scirpus – Equisetum* wetland. Ecology 57: 941–952.

Auclair, A. N. D., Bouchard, A. and Pajaczkowski, J. (1976b) Productivity relations in a *Carex* – dominated ecosystem. Oecologia (Berl.) 26: 9–31.

Bannatyne, B. B. (1980) *Sphagnum* bogs in southern Manitoba and their identification by remote sensing. Manitoba Department, Energy and Mines, Economic Geology Report, No. ER79–7. Winnipeg, Manitoba, Canada. 103 pp.

Banner, A., Pojar, J. and Rouse, G. E. (1983) Postglacial paleoecology and successional relationships of a bog woodland near Prince Rupert, British Columbia. Canadian Journal of Forest Research 13: 938–947.

Banner, A., Hebda, R. J., Oswald, E. T., Pojar, J. and Trowbridge, R. (1988). Wetlands of Pacific Canada. p. 305–346. *In*: National Wetlands Working Group (ed.). Wetlands of Canada. Ecological Land Classification Series No. 24. Sustainable Development Branch, Environment Canada, Ottawa. Polyscience Publishers, Montreal, Quebec, Canada.

Bardecki, M. J. (1982) The status of wetlands in southern Ontario. Wetlands 2: 262–270.

Bartsch, H. and Moore, T. R. (1984) A preliminary investigation of primary production and decomposition in four peatlands near Schefferville, Quebec. Canadian Journal of Botany 63: 1241–1248.

Bayley, S. E., Behr, R. S. and Kelly, C. A. (1986) Retention and release of S from a freshwater wetland. Water, Air, Soil Pollution 31: 101–114.

Bayley, S. E., Vitt, D. H., Newbury, R. W., Beaty, K. G., Behr, R. and Miller, C. (1981) Experimental acidification of a *Sphagnum*-dominated peatland: first year results. Canadian Journal Fisheries Aquatic Science 44 (Suppl. 1): 194–205.

Bazely, D. R. and Jefferies, R. L. (1986) Changes in the composition and standing crop of saltmarsh communities in response to the removal of a grazer. Journal of Ecology 74: 693–706.

Böcher, T. W. (1938) Biological distributional types in the flora of Greenland. Meddelelser om Gronland 106: 1–336.

Böcher, T. W. (1949) Climate, soil and lakes in continental west Greenland in relation to plant life. Meddelelser om Gronland 147: 1–63.

Böcher, T. W. (1954) Oceanic and continental vegetation complexes in southwest Greenland. Meddelelser om Gronland 148: 1–136.

Bolghari, H. (1986) Perspectives du drainage forestier au Quebec. p. 47–65. *In*: Proceedings, Symposium on Forest Drainage, Ordre des ingenieurs forestiers du Quebec. Sainte Foy, Quebec, Canada.

Botch, M. S. (1974) Bogs in the tundra zone of Siberia. (Bolota tundrovoi zony Sibiri [printsipy tipologii]). p. 146–154. *In*: Bog types of the U.S.S.R. and Principles of Their Classification. (Tipy bolot S.S.S.R. i printsipy ikh klassifikatsii. Nauka, Leningrad. English translation by National Research Council of Canada, Technical Translation No. 1925, Ottawa, Ontario.

Bourbonniere, R. A. (1987) Organic geochemistry of bog drainage waters. p. 139–145. *In*: C. D. A. Rubec and R. P. Overend (eds.). Proceedings, Symposium 1987, Wetlands/Peatlands. Edmonton, Alberta, Canada. Canadian National Committee, International Peat Society, Ottawa, Ontario.

Calder, J. A. and Taylor, R. L. (1968) Flora of the Queen Charlotte Islands. I. Systematics of Vascular Plants. Research Branch Canada Department Agriculture Monograph No. 4, Ottawa, Ontario, Canada. 659 pp.

Canada Soil Survey Committee (1978) The Canadian System of Soil Classification. Research Branch, Canadian Department Agriculture Publication No. 1646. Ottawa, Ontario, Canada. 164 pp.

Cargill, S. M. and Jefferies, R. L. (1984a) Nutrient limitation of primary production in a subarctic salt marsh. Journal of Applied Ecology 21: 657–668.

Cargill, S. M. and Jefferies, R. L. (1984b) The effects of grazing by lesser snow geese on the vegetation of a sub-arctic salt marsh. Journal of Applied Ecology 21: 669–686.

Chapman, V. J. (1937) A note on the salt marshes of Nova Scotia. Rhodora 39: 53–57.

Chapman, V. J. (1974) Salt Marshes and Salt Deserts of the World. 2nd Supplemented Reprint Edition, Verlag Von J. Cramer, Lehre, Germany. 392 pp.

Clymo, R. S. (1965) Experiments on the breakdown of *Sphagnum* in two bogs. Journal of Ecology 53: 747–758.

Couillard, L. and Payette, S. (1985) Evolution holocène d'une tourbière à pergelisol (Quebec nordique). Canadian Journal of Botany 63: 1104–1121.

Cragg, J. B. (1961) Some aspects of the ecology of moorland animals. Journal of Animal Ecology 30: 205–234.

Damman, A. W. H. (1977) Geographical changes in the vegetation patterns of raised bogs in the Bay of Fundy region of Maine and New Brunswick. Vegetatio 35: 137–151.

Damman, A. W. H. (1979) Geographic patterns in peatland development in eastern North America. p. 42–57. *In*: Proceedings International Symposium Classification Peat and Peatlands, Hyytiälä, Finland, 1979. International Peat Society, Helsinki, Finland.

Damman, A. W. H. (1986) Hydrology, development, and biogeochemistry of ombrogenous peat bogs with special reference to nutrient relocation in a Western New Foundland bog. Canadian Journal of Botany 64: 384–394.

Damman, A. W. H. and Dowhan, J. J. (1981) Vegetation and habitat conditions in western Head Bog, a southern Nova Scotia plateau bog. Canadian Journal of Botany 59: 1343–1359.

Dansereau, P. and Segadas-Vianna, F. (1952) Ecological study of the peat bogs of eastern North America. Canadian Journal of Botany 30: 490–520.

Dawe, N. K. and White, E. R. (1982) Some aspects of the vegetation ecology of the Little Qualicum River estuary, British Columbia. Canadian Journal of Botany 60: 1447–1460.

Dirschl, H. J. (1977) Geobotanical processes in the Saskatchewan River delta. Canadian Journal of Earth Science 9: 1529–1599.

Dirschl, H. J. and Coupland, R. J. (1972) Vegetation patterns and site relationships in the Saskatchewan River delta. Canadian Journal of Botany 50: 647–675.

Dirschl, H. J. and Dabbs, D. C. (1969) A contribution to the flora of the Saskatchewan River delta. Canadian Field-Naturalist 83: 212–228.

Dirschl, H. J., Dabbs, D. L. and Gentle, G. C. (1974) Landscape Classification and Plant Successional Trends: Peace-Athabasca Delta. Canadian Wildlife Service, Department Series No. 30, Edmonton, Alberta, Canada. 34 pp.

Dredge, L. A. (1979a) Thaw depths and permafrost in polygonal peat terrain, Hudson Bay Lowland, Manitoba. p. 27–30. *In*: Geological Survey Canada Paper 79–1C.

Dredge, L. A. (1979b) Thermal sensitivity and the development of tundra ponds and thermokarst lakes in the Manitoba portion of the Hudson Bay Lowland. p. 35–50. *In*: Geological Survey Canada Paper 79–1C, Ottawa, Ontario, Canada.

Dodd, J. D. and Coupland, R. T. (1966) Vegetation of saline areas in Saskatchewan. Ecology 47: 958–967.

Eagle, A. (1983) Peatlands of the Ontario portion of the St. Lawrence Lowland. pp 35–50. *In*: Agriculture Canada, Land Resource Research Institute, Contribution 83–03, Ottawa, Ontario, Canada.

EC/USDI (Environment Canada/U. S. Department of the Interior) (1985) North American Waterfowl Management Plan. Draft, Ottawa, Ontario, Canada. 37 pp.

Ewing, K. and Kershaw, K. A. (1986) Vegetation patterns of James Bay coastal marshes. I. environmental factors on the south coast. Canadian Journal of Botany 64: 217–226.

Fahselt, D. and Maun, M. A. (1979) A quantitative study of shoreline marsh communities along Lake Huron in Ontario. Canadian Journal of Plant Science 60: 669–678.

Foster, D. R. and Glaser, P. H. (1986) The raised bogs of south-eastern Labrador, Canada: Classification, distribution, vegetation and recent dynamics. Journal of Ecology 74: 47–71.

Foster, D. R. and King, G. A. (1984) Landscape features, vegetation and developmental history of a patterned fen in south-eastern Labrador, Canada. Journal of Ecology 72: 115–143.

Foster, D. R., King, G. A., Glaser, P. H. and Wright, H. E., Jr. (1983) Origin of string patterns in boreal peatlands. Nature 306: 256–258.

Foster, D. R., Wright, H. E., Jr. Thelalaus, M. and King, G. A. (1988) Bog development and landform dynamics in central Sweden and south-eastern Labrador, Canada. Journal of Ecology 76: 1164–1185.

Fredskild, B. (1961) Floristic and ecological studies near Jakobshavn, west Greenland. Meddelelser om Gronland 163: 1–82.

Ganong, W. F. (1903) The vegetation of the Bay of Fundy salt and diked marshes: an ecological study. Botanical Gazette 36: 161–186, 208–302, 340–367, 429–455.

Gauthier, R. and Grandtner, M. M. (1975) Étude phytosociologique des tourbières du Bas Saint-Laurent, Québec. Naturaliste canadien 102: 109–153.

Gignac, L. D. and Beckett, P. J. (1986) The effect of smelting operations on peatlands near Sudbury, Ontario, Canada. Canadian Journal of Botany 64: 1138–1147.

Gignac, L. D. and Vitt, D. H. (1990) Habitat limitations of *Sphagnum* along climatic, chemical and physical gradients in mires of western Canada. The Bryologist 93: 7–22.

Glaser, P. H. and Janssens, J. A. (1986) Raised bogs in eastern North America: transitions in landforms and gross stratigraphy. Canadian Journal of Botany 64: 395–415.

Glooschenko, V. (1985) Characteristics of provincially significant wetlands as assessed by the Ontario Wetland Evaluation System. p. 182–199. *In*: H. H. Prince and F. M. D'Itri (eds.). Coastal Wetlands. Lewis Publishers, Chelsea, Michigan, USA.

Glooschenko, V. and Grondin, P. (1988) Wetlands of eastern temperate Canada. p. 199–248. *In*: National Wetlands Working Group (ed.). Wetlands of Canada. Ecological Land Classification Series No. 24. Sustainable Development Branch, Environment Canada, Ottawa. Polyscience Publishers, Montreal, Quebec, Canada.

Glooschenko V., Archbold, J. H. and Weninger, J. (1987) The importance of swamp wetland habitat in Southern Ontario. p. 275–285. *In*: C. Rubec and R. Overend (eds.). Proceedings of Symposium '87, Wetlands/Peatlands. Edmonton, Alberta, Canada. 704 pp.

Glooschenko, W. A. (1978) Above-ground biomass of vascular plants in a subarctic James Bay salt marsh. Canadian Field-Naturalist 92: 30–37.

Glooschenko, W. A. (1980a) Coastal ecosystems of the James/Hudson Bay area of Ontario, Canada. Zeitschrift Geomorphologie N. F., Suppl-Bd. 34: 214–224.

Glooschenko, W. A. (1980b) Coastal salt marshes in Canada. p. 39–47. *In*: C. D. A. Rubec and F. C. Pollett (eds.). Proceedings Workshop on Canadian Wetlands. Environment Canada, Lands Directorate Ecological Land Classification Series No. 12, Ottawa, Ontario, Canada.

Glooschenko, W. A. (1982) Salt marshes of Canada. p. 1–8. *In*: B. Gopal, R. E. Turner, R. G. Wetzel, and D. F. Whigham (eds.). Wetlands, Ecology and Management. International Scientific Publications, Jaipur, India.

Glooschenko, W. A. (1983) Salt Marsh Vegetation of the Ontario Coast of Hudson and James Bay. Part 1. Transect Descriptions. Unpublished Report 83–23, National Water Research Institute, Environment Canada, Burlington, Ontario, Canada. 59 pp.

Glooschenko, W. A. (1986) Monitoring the atmospheric deposition of metals by use of bog vegetation and peat profiles. Advances Environmental Science Technology 17: 507–533.

Glooschenko, W. A. (1989) *Sphagnum fuscum* moss as an indicator of atmospheric cadmium deposition across Canada. Environmental Pollution 57: 27–33.

Glooschenko, W. A. and Arafat, N. (1988) Atmospheric deposition of arsenic and selenium across Canada using Sphagnum moss as a biomonitor. Science Total Environment 73: 269–275.

Glooschenko, W. A. and Clarke, K. (1982) The salinity cycle of a subarctic salt marsh. Naturaliste canadien 109: 483–490.

Glooschenko, W. A. and Harper, N. S. (1982) Net aerial primary production of a James Bay, Ontario, salt marsh. Canadian Journal of Botany 60: 1060–1067.

Glooschenko, W. A. and Martini, I. P. (1978) Hudson Bay Lowlands Baseline Study. p. 663–679. *In*: Proceedings Symposium Technical Environmental Socioeconomical Regulatory Aspects, Coastal Zone Management. American Society Civil Engineers, Minneapolis, Minnesota, USA.

Glooschenko, W. A. and Martini, I. P. (1981) Salt marshes of the Ontario coast of Hudson Bay, Canada. Wetlands 1: 9–18.

Glooschenko, W. A. and Martini, I. P. (1985) Wetlands of the Attawapiskat River mouth, James Bay, Ontario, Canada. Wetlands 3: 64–76.

Glooschenko, W. A. and Martini, I. P. (1987) Vegetation of river-influenced coastal marshes of the southwestern end of James Bay, Ontario. Wetlands 7: 71–84.

Glooschenko, W. A., Sims, R.A., Gregor, M. and Mayer, T. (1981a) Use of bog vegetation as a monitor of atmospheric input of metals. p. 389–399. *In*: S. J. Einsenreich (ed.). Atmospheric Pollutants in Natural Waters. Ann Arbor Science, Ann Arbor, Michigan, USA.

Glooschenko, W. A., Capobianco, J., Coburn, J. and Glooschenko, V. (1981b) Geochemical distribution of trace metals and organochlorine contaminants of a Lake Ontario shoreline marsh. Water, Air, Soil Pollution 15: 197–213.

Glooschenko, W. A., Holloway, L. and Arafat, N. (1986) The use of mires in monitoring the atmospheric deposition of heavy metals. Aquatic Botany 25: 179–190.

Glooschenko, W. A., Martini, I. P. and Clarke-Whistler, K. (1988) Salt marshes of Canada. p. 347–377. *In*: National Wetlands Working Group (ed.). Wetlands of Canada. Ecological Land Classification Series No. 24. Sustainable Development Branch, Environment Canada, Ottawa. Polyscience Publishers, Montreal, Quebec, Canada.

Gorham, E., Bayley, S. E. and Schindler, D. W. (1984) Ecological effects of acid deposition upon peatlands: A neglected field in "acid-rain" research. Canadian Journal Fisheries Aquatic Science 41: 1256–1268.

Grandtner, M. M. (1966) Première observation phytopedologiques sur les pies sales des Iles de la Madeleine. Naturaliste canadien 93: 361–366.

Hatcher, B. G. and Mann, K. H. (1975) Above-ground production of marsh cordgrass (*Spartina alterniflora*) near the northern end of its range. Journal Fisheries Research Board Canada 32: 83–87.

Hebda, R. J. and Biggs, W. G. (1981) The vegetation of Burns Bog, Fraser Delta, southwestern British Columbia. Syesis 14: 1–20.
Heinselman, M. L. (1970) Landscape evolution, peatland types, and the environment in the Lake Agassiz Peatlands natural area, Minnesota. Ecological Monographs 40: 235–261.
Hillman, G. R. (1987) Improving Wetlands for Forestry in Canada. Canadian Forest Service Information Report NOR-X-288, Northern Forest Center, Edmonton, Alberta, Canada. 64 pp.
Horton, D. G., Vitt, D. H. and Slack, N. G. (1979) Habitats of circumboreal-subarctic sphagna: 1. A quantitative analysis and review of species in the Caribou Montains, northern Alberta. Canadian Journal of Botany 57: 2283–2317.
Hunter, G. T. (1970) Postglacial uplift at Fort Albany, James Bay. Canadian Journal of Earth Science 7: 547–548.
Hustich, I. (1957) On the phytogeography of the subarctic Hudson Bay lowland. Acta Geografica 16: 1–47.
Jasieniuk, M. A. and Johnson, E. A. (1982) Peatland vegetation organization and dynamics in the western subarctic, Northwest Territories, Canada. Canadian Journal of Botany 60: 2581–2593.
Jefferies, R. L. (1977) The vegetation of salt marshes at some coastal sites in Arctic North America. Journal of Ecology 65: 661–672.
Jefferies, R. L., Jensen, A. and Abraham, K. F. (1979) Vegetational development and the effect of geese on vegetation at La Perouse Bay, Manitoba. Canadian Journal of Botany 57: 1439–1450.
Jeglum, J. K. (1971) Plant indicators of pH and water level in peatlands at Candle Lake, Saskatchewan. Canadian Journal of Botany 49: 1661–1676.
Jeglum, J. A. and Cowell, D. A. (1982) Wetland ecosystems near Kinoje Lakes, southern interior Hudson Bay Lowland. Naturaliste canadien 109: 621–635.
Jeglum, J. K., Boissonneau, A. N. and Haavisto, V. F. (1974) Toward a Wetland Classification for Ontario. Information Report 0–X-215. Canadian Forestry Service, Department of Environment, Sault Ste. Marie, Ontario, Canada. 54 pp.
Jenny, H. (1941) Factors of Soil Formation. McGraw-Hill, New York, New York, USA. 281 pp.
Karlin, E. F. and Bliss, L. C. (1984) Variation in substrate chemistry along microtopographical and water chemistry gradients in peatlands. Canadian Journal of Botany 62: 142–153.
Kautrud, H. A., Millar, J. B. and van der Valk, A. G. (1989) Vegetation of wetlands of the prairie pothole region. p. 132–187. *In*: A. van der Valk (ed.). Northern Prairie Wetlands. Iowa State University Press, Ames, Iowa, USA. 400 pp.
Keddy, P. A. (1983) Shoreline vegetation in Axe Lake, Ontario: Effects of exposure on zonation patterns. Ecology 64: 331–344.
Keddy, P. A. (1984a) Quantifying a within-lake gradient of wave energy in Gillfillan Lake, Nova Scotia. Canadian Journal of Botany 62: 301–309.
Keddy, P. A. (1984b) Plant zonation on lakeshores in Nova Scotia: A test of the resource specialization hypothesis. Journal of Ecology 72: 797–808.
Keddy, P. A. (1989) Effects of competition from shrubs on herbaceous wetland plants: a 4-year field experiment. Canadian Journal of Botany 67: 708–716.
Keddy, P. A. and Ellis, T. H. (1985) Seedling recruitment of 11 wetland plant species along a water level gradient: Shared or distinct responses. Canadian Journal of Botany 63: 1876–1879.
Keddy, P. A. and Reznicek, A. A. (1985) Vegetation dynamics, buried seeds, and water level fluctuations on the shorelines of the Great Lakes. p. 33–58. *In*: H. H. Prince and F. M. D'Itri (eds.). Coastal Wetlands. Lewis Publishers, Chelsea, Michigan, USA.
Keddy, P. A. and Constabel, P. (1986) Germination of ten shoreline plants in relation to seed size, soil particle size and water level: an experimental study. Journal of Ecology 74: 133–141.
Keddy, P. A. and Reznicek, A. A. (1986) Great Lakes vegetation dynamics: The role of fluctuating water levels and buried seeds. Journal Great Lakes Research 12: 25–36.

Keddy, P. A. and Wisheu, I. C. (1989) Ecology, biogeography and conservation of coastal plain plants: Some general principles from the study of Nova Scotian wetlands. Rhodora 91: 72–94.

Kennedy, K. A. (1982) Plant Communities and Their Standing Crops on Estuaries of the East Coast of Vancouver Island. M.Sc. Thesis, Department of Plant Science, University of British Columbia, Vancouver, British Columbia, Canada. 598 pp.

Kerbes, R. H., Kotanen, P. M. and Jefferies, R. L. (1990) Destruction of wetland habitats of lesser snow geese: A keystone species on the west coast of Hudson Bay. Journal of Applied Ecology 27: 242–258.

Kershaw, G. P. and Gill, D. (1979) Growth and decay of palsas and peat plateaus in the Macmillan Pass – Tsichu River area, Northwest Territories, Canada. Canadian Journal of Earth Science 16: 1362–1374.

Kershaw, K. (1976) The vegetational zonation of the East Pen Island salt marshes, Hudson Bay. Canadian Journal of Botany 54: 5–13.

Kiel, W. H. Jr., Hawkins, A. S. and Perret, N. G. (1972) Waterfowl Habitat Trends in the Aspen Parkland of Manitoba. Canadian Wildlife Series Report No. 18, Saskatoon, Saskatchewan, Canada. 62 pp.

Kistritz, R. U. and Yesaki, I. (1979) Primary Production, Detritus Flux, and Nutrient Cycling in a Sedge Marsh, Fraser River Estuary. Westwater Research Centre. University of British Columbia, Technical Report. No. 17, Vancouver, British Columbia, Canada.

Kubiw, H., Hickman, M. and Vitt, D. H. (1989) The developmental history of peatlands at Muskiki and Marguerite lakes, Alberta. Canadian Journal of Botany 67: 3534–3549.

Lacoursière, E. and Grandner, M. M. (1972) Les groupements végétaux ripariens entre Sainte-Famille et la Pointe d'Argentenaye, Ile d'Orleans, Québec. Naturaliste canadien 99: 469–507.

Lamoureaux, J. P. and de Repentigny, L. G. (1972) Contribution à la flore des marécages intertideaux de la Baie de Rupert (Que.). Report, Canadian Wildlife Service, Valleyfield, Quebec, Canada. 14 pp.

Lamoureaux, J. P. and Zarnovican, R. (1972) Approximation relative à la végétation riparienne de la Baie de Rupert, Buebec Report, Canadian Wildlife Service, Valleyfield, Quebec, Canada. 126 pp.

Lamoureaux, J. P. and Zarnovican, R. (1974) Les marécages côtiers de la Baie aux Oies, Quebec. Canadian Wildlife Service, James Bay Report Series No. 13, Quebec, Canada. 99 pp.

Latter P. M., Cragg, J. B. and Heal, O. W. (1967) Comparative studies of the microbiology of four moorland soils in the North Pennines. Journal of Ecology 55: 445–482.

Laverdiere, C. and Guimont, P. (1975) Le milieu bio-physique de la baie de Rupert. Environment Directorate, James Bay Development Corporation, Reconnaissance Report, Montreal, Quebec, Canada. 157 pp.

Lim, P. G. and Levings, C. D. (1973) Distribution and Biomass of Intertidal Vascular Plants of the Squamish Delta. Fisheries Research Board Canada, MS. Report 1219, Vancouver, British Columbia, Canada. 26 pp.

Levings, C. D. and Moody, A. I. (1976) Studies of Intertidal Vascular Plants, Especially Sedge (*Carex lyngbyei*), on the Disrupted Squamish River Delta, British Columbia. Fish Marine Service Technical Report No. 606, Vancouver, British Columbia, Canada. 51 pp.

Lucas, A. E. (1980) Phyto-geomorphic Sequences on Barrier Coastlines Within the Southern Gulf of St. Lawrence. Unpublished M.Sc. Thesis, Department of Geography, McMaster University, Hamilton, Ontario, Canada. 145 pp.

Lynch-Stewart, P. (1983) Land Use Change in Wetlands in Southern Canada: Review and Bbliography. Environment Canada, Lands Directorate, Working paper No. 26, Ottawa, Ontario, Canada. 115 pp.

Macfayden, A. (1963) The contribution of the microfauna to soil metabolism. *In*: J. Doeksen and J. Van der Drift (eds.). Soil Organisms, Amsterdam, The Netherlands.

Martini, I. P., Morrison, R. I. G., Glooschenko, W. A. and Protz, R. D. (1980) Coastal studies in James Bay, Ontario. Geoscience Canada 7: 11–21.

Millar, J. B. (1969) Observations on the ecology of wetland vegetation. p. 49–69. *In*: Saskatoon Wetlands Seminar. Canadian Wildlife Report Series No. 6, Ottawa, Ontario, Canada.

Millar, J. B. (1973) Vegetation changes in shallow marsh wetlands under improving moisture regime. Canadian Journal of Botany 51: 1443–1457.

Millar, J. B. (1976) Wetland classification in western Canada. Canadian Wildlife Report Series No. 37, Ottawa, Canada. 38 pp.

Mills, G. F., Hopkins, L. A. and Smith, R. E. (1976) Inventory and assessment for agriculture of the organic soils of the Roseau River Watershed in Manitoba. Canada-Manitoba Soil Survey, Canada Department of Agriculture, Winnipeg, Manitoba, Canada. 174 pp.

Mills, G. F., Forester, D. B and Veldhuis, H. (1978) A guide to biophysical land classification, Knee Lake 53M, Manitoba, Canada. Manitoba Soil Survey and Manitoba Department Natural Resources and Transport Service Technical Report, No. 78–2, Winnipeg, Manitoba, Canada. 38 pp.

Moon, D. E. and Selby C. J. (1983) Wetland systems of the Caribou-Chilcotin region of B.C. p. 54–74. *In*: S. M. Morgan and F. C. Pollett (eds.). Proceedings Peatland Inventory Methodology Workshop. Agriculture Canada, Land Resource Research Institute and Newfoundland Forest Research Centre, St. Johns's, Newfoundland, Canada.

Moore, D. R. J. and Keddy, P. A. (1989) The relationship between species richness and standing crop in wetlands: The importance of scale. Vegetatio 79: 99–106.

Morgan, S. M. and Pollett, F. C. (eds.). (1983) Proceedings of a Peatland Inventory Methodology Workshop. Land Resource Research Institute, Agriculture Canada and Newfoundland Forest Research Centre, Environment Canada, St. John's, Newfoundland, Canada. 121 pp.

Moss, E. H. (1953) Marsh and bog vegetation in northwestern Alberta. Canadian Journal of Botany 31: 448–470.

Muc, M. (1977) Ecology and primary production of sedge-moss meadow communities, Truelove Lowland. p. 157–184. *In*: L. C. Bliss (ed.). Truelove Lowland, Devon Island Canada: A High Arctic ecosystem. University of Alberta Press, Edmonton, Alberta, Canada.

Mueller-Dombois, D. (1964) The forest habitat types in southeastern Manitoba and their application to forest management. Canadian Journal of Botany 42: 1417–1444.

National Wetlands Working Group (1986) Distribution of Wetlands in Canada. National Atlas of Canada. Energy, Mines and Resources Canada, Map. No. MCR 4107. Ottawa, Ontario, Canada.

National Wetlands Working Group (1987) The Canadian Wetland Classification System. Lands Conservation Branch, Canadian Wildlife Service. Ecological Land Classification Series No. 21. Ottawa, Ontario, Canada. 18 pp.

National Wetlands Working Group (ed.). (1988) Wetlands of Canada. Ecological Land Classification Series No. 24. Sustainable Development Branch, Environment Canada, Ottawa. Polyscience Publishers, Montreal, Quebec, Canada. 452 pp.

Nicholas, G. E. (1918) The vegetation of northern Cape Breton Island, Nova Scotia. Transactions Connecticut Academy Arts and Sciences 22: 249–467.

Nicholson, B. J. and Vitt, D. H. (1990) The paleoecology of a peatland complex in continental western Canada. Canadian Journal of Botany 68: 121–138.

Ovenden, L. (1982) Vegetation history of a polygonal peatland, northern Yukon. Boreas 11: 209–224.

Päivänen, J. and Wells, E. D. (1977) Guidelines for the Development of Peatland Drainage Systems for Forestry in Newfoundland. Canadian Department Fisheries and Environment, Canadian Forest Service, Newfoundland Forest Research Centre, Information Report, St. John's, Newfoundland, Canada. 44 pp.

Pala, S. and Weischet, W. (1982) Toward a physiographic analysis of the Hudson Bay – James Bay Lowland. Naturaliste canadien 109: 637–651.

Patterson, N. and Whillens, T. H. (1985) Human interference with natural water regimes in the context of other cultural stresses on Great Lakes wetlands. p. 209–251. *In*: H. H. Prince and F. M. D'Itri (eds.). Coastal Wetlands. Lewis Publishers, Chelsea, Michigan, USA.

Payandeh, B. (1973) Analyses of a forest drainage experiment in northern Ontario. I. Growth analysis. Canadian Journal of Forest Research 3: 387–398.

Pearce, P. A., Smith A. D. (1974) The potential influence of waterfowl on port and tidal power development in the Bay of Fundy. Canadian Wildlife Service Report Series, Ottawa, Ontario 29: 91–100.

Pedersen, R. C. and van der Valk, A. C. (1985) Vegetation change and seed banks: ecological and management implications. Transactions North American Wildlife and Natural Resources Conference 9: 771–289.

Pollett, F. C. and Bridgewater, P. B. (1973) Phytosociology of peatlands in central Newfoundland. Canadian Journal of Forest Research 3: 433–442.

Polunin, N. (1948) Botany of the Canadian Eastern Arctic. Part III. Vegetation and ecology. Bulletin National Museum Canada, Ottawa, Ontario, Canada 104: 1–304.

Pringle, J. S. (1980) An Introduction to Wetland Classification in the Great Lakes Region. Technical Bulletin No. 10, Royal Botanical Gardens, Hamilton, Ontario, Canada. 11 pp.

Protz, R., Ross, G. J., Martini, I. P. and Terasmae, J. (1984) Rate of podzolic soil formation near Hudson Bay, Ontario. Canadian Journal of Soil Science 64: 31–49.

Railton, J. G. and Sparling, J. H. (1973) Preliminary studies on the ecology of palsa mounds in northern Ontario. Canadian Journal of Botany 51: 1037–1044.

Raup, H. M. (1935) Botanical Investigations in Wood Buffalo Park. Canadian Deptartment Mines, National Museum Canada, Ottawa, Ontario, Canada Bulletin 74. 174 pp.

Reader, R. J. (1982) Geographical variation in the shoot productivity of bog shrubs and some environmental correlates. Canadian Journal of Botany 60: 340–348.

Reader, R. J. and Stewart, J. M. (1972) The relationship between net primary production and accumulation for a peatland in south-eastern Manitoba. Ecology 53: 1024–1037.

Reed, A. and Moisan, G. (1971) The *Spartina* tidal marshes of the St. Lawrence estuary and their importance to aquatic birds. Naturaliste canadien 98: 905–922.

Reid, D. E. (1974) Permafrost in Peat Landforms in Northwestern Alberta. Canadian Arctic Gas Study Ltd., Biological Report Service Vol. 37, Chapter 2. Calgary, Alberta, Canada. 36 pp.

Riley, J. L. (1982) Hudson Bay Lowland floristic inventory, wetlands catalogue and conservation strategy. Naturaliste canadien 109: 543–555.

Riley, J. L. and McKay, S. M. (1980) The Vegetation and Phytogeography of Coastal Southwestern James Bay. Contribution Royal Ontario Museum, Life Sciences No. 124, Toronto, Ontario, Canada. 86 pp.

Ritchie, J. C. (1957) The vegetation of northern Manitoba. II. A prisere on the Hudson Bay Lowlands. Ecology 38: 429–453.

Rahefort, L., Vitt, D. H. and Bayley, S. E. (1990) Growth, production and decomposition dynamics of *Sphagnum* under natural and experimentally acidified conditions. Ecology 71: 1986–2000.

Roberts, B. A. and Robertson, A. (1986) Salt marshes of Atlantic Canada: their ecology and distribution. Canadian Journal of Botany 64: 455–467.

Rowe, J. S. (1972) The Forest Regions of Canada. Environment Canada Forest Service Publication No. 1300, Ottawa, Ontario, Canada. 172 pp.

Roulet, N. (1990) Focus: aspects of the physical geography of wetlands. The Canadian Geographer 39: 79–88.

Roulet, N. T. (1991) Role of the Hudson Bay Lowland as a source of atmospheric methane. EOS 72: 84.

Roulet, N. T., Barrie, L., Glooschenko, W. A., Harris, R. C., Higuchi, K., MacPherson, J. I., Moore, T. R., Newmann, H., Protz, R. and Schiff, H. I. (1991) Northern Wetlands Study (NWS). Ecological Bulletin (Sweden), in press.

Rubec, C. D. A., Lynch-Steward, P., Wickware, G. M., and Kessel-Taylor J. (1988) Wetland Utilization in Canada. p. 380–411. *In*: National Wetlands Working Group (ed.). Wetlands of Canada. Ecological Land Classification Series No. 24. Sustainable Development Branch, Environment Canada, Ottawa. Polyscience Publishers, Montreal, Quebec, Canada.

Seppala, M. (1980) Stratigraphy of a silt-cored palsa, Atlin region, British Columbia, Canada. Arctic 33: 357–365.

Shay, J. M. (1983) Prairie marshes in Western Canada. p. 40–41. *In*: Abstracts, Botanical Society America, Canadian Botanical Association Joint Meeting, Ecological Section.

Shay, J. M. and Shay, C. T. (1986) Prairie marshes in western Canada, with specific reference to the ecology of five emergent macrophytes. Canadian Journal of Botany 64: 443–454.

Shotyk, W. (1988) Review of the inorganic geochemistry of peats and peatland waters. Earth-Science Reviews 25: 95–176.

Simpson, S. J. (1972) The York Factory Area, Hudson Bay. Ph.D. Thesis, Department of Geography, University of Manitoba, Winnipeg, Manitoba, Canada. 293 pp.

Sims, R. A. and Stewart, J. M. (1981) Aerial biomass distribution in an undisturbed and disturbed subarctic bog. Canadian Journal of Botany 59: 782–786.

Sims, R. A., Cowell, D. W. and Wickware, G. M. (1982a) Use of vegetational physiognomy in classifying treed peatlands near southern James Bay, Ontario. Naturaliste canadien 109: 611–619.

Sims, R. A., Cowell, D. W. and Wickware, G. M. (1982b) Classification of fens near southern James Bay, Ontario using vegetational physiognomy. Canadian Journal of Botany 60: 2608–2623.

Sjörs, H. (1963) Bogs and fens on Attawapiskat River, northern Ontario. p. 45–133. *In*: Contributions to Botany, 1960–61, Bulletin No. 186. National Museum of Canada, Ottawa, Ontario, Canada.

Slack, N. G., Vitt, D. H. and Horton, D. G. (1980) Vegetation gradients of minerotrophically rich fens in western Alberta. Canadian Journal of Botany 53: 330–350.

Smith, R. E., Ehrlich, W. A. and Zoltai, S. C. (1967) Soils of the Lac du Bonnet Area. Manitoba Soil Survey, Soils Report No. 15, Winnipeg, Manitoba, Canada. 118 pp.

Smith, P. G. R., Glooschenko, V. and Hagan, D. A. (1991) Coastal Wetlands of three Canadian Great Lakes: inventory, current conservation initiatives and patterns of variation. Canadian Journal Fisheries Aquatic Science 48: 1581–1594.

Stanek, W. (1968) A forest drainage experiment in northern Ontario. Pulp and Paper Magazine Canada 69: 58–62.

Stanek, W. (1975) Growth of black spruce on drained peatland five years after fertilization. p. 69–75. *In*: G. P. Williams and J. Curran (eds.). Proceedings Sixteenth Muskeg Research Conference. Associate Committee on Geotechnical Research, National Research Council Canada, Technical Memorandum No. 116, Ottawa, Ontario, Canada.

Styan, W. B. and Bustin, R. M. (1983) Sedimentology of Fraser River delta peat deposits: A modern analogue for some deltaic coals. International Journal Coal Geology 3: 101–143.

Tarnocai, C. (1970) Classification of peat landforms in Manitoba. Canada Department of Agriculture, Research Station, Pedology Unit, Winnipeg, Manitoba, Canada. 45 pp.

Tarnocai, C. (1973) Soils of the Mackenzie River Area. Report No. 73–26. Environmental Social Program, Northern Pipelines, Edmonton, Alberta, Canada. 136 pp.

Tarnocai, C. (1974) Peat landforms and associated vegetation in Manitoba. p. 3–20. *In*: J. H. Day (ed.). Proceedings of the Canada Soil Survey Committee Organic Soil Mapping Workshop, Winnipeg, Manitoba, Soil Research Insititute, Ottawa, Canada.

Tarnocai, C. (1978) Genesis of organic soils in Manitoba and the Northwest Territories. p. 453–470. *In*: Proceedings of the Third York University Quaternary Symposium, Toronto, Ontario, Canada.

Tarnocai, C. (1980) Canadian wetland registry. p. 9–38. *In*: C. D. A. Rubec and F. C. Pollett (eds.). Proceedings of a Workshop on Canadian Wetlands, Environment Canada, Ecological Land Classification Series, No. 12, Ottawa, Ontario, Canada.

Tarnocai, C. (1982) Soil and terrain development on the York Factory peninsula, Hudson Bay Lowland. Naturaliste canadien 109: 511–522.

Tarnocai, C. (1983) Peatland inventory methodology used in soil surveys. p. 13–22. *In*: S. M. Morgan and F. C. Pollett (eds.). Proceedings of a Peatland Inventory Methodology Workshop, Land Resource Institute, Agriculture Canada and Newfoundland Forest Research Centre, Environment Canada, Ottawa, Ontario, Canada.

Tarnocai, C. (1984) Peat Resources of Canada. NRCC No. 24140, National Research Council

Canada and Land Resource Research Institute, Agriculture Canada, Ottawa, Ontario, Canada. 17 pp.

Tarnocai, C. and Zoltai, S. C. (1988) Wetlands of Arctic Canada. p. 27–53. *In*: National Wetlands Working Group (ed.). Wetlands of Canada. Ecological Land Classification Series No. 24. Sustainable Development Branch, Environment Canada, Ottawa. Polyscience Publishers, Montreal, Quebec, Canada. 452 pp.

Terasmae, J. (1972) Muskeg as a climate-controlled ecosystem. p. 147–158. *In*: Proceedings 14th Muskeg Research Conference, Kingston, 1971, NRC Technical Memo No. 102, Ottawa, Ontario, Canada.

Thannheiser, D. (1981) Die Küstenvegetation Ostkanadas. A. Mayr (ed.) Munster Geographical Series No. 10, Paderborn, Germany. 202 pp.

Theander, O. (1954) Studies on Sphagnum peat. 3. A quantitative study of the carbohydrate constituents of sphagnum mosses and sphagnum peat. Acta Chemica Scandnavica 8: 989–1000.

Thie, J. (1974) Distribution of thawing of permafrost in the southern part of the discontinuous permafrost zone in Manitoba. Arctic 27: 189–200.

Toth, J. and Gillard, D. (1984) Design of Drainage System in Forested Peatland with Special Reference to Water Regime Near Salteaux River, Slave Lake Forest: Phase II–III. University of Alberta, Edmonton, Alberta, Canada. 33 pp.

Trottier, F. (1986) Accroissement de certains peuplements forestiers attributable à la construction de cours d'eau artificiels. p. 66–84. *In*: Proceedings, Symposium on Forest Drainage. Ordre des ingenieurs forestiers du Quebec. St. Foy, Quebec, Canada.

Ugolini, F. C. and Mann, D. H. (1979) Biopedological origin of peatlands in Southeast Alaska, Nature 281: 366–367.

Urgel Delisle et Associes (1983) Measurement of the effectiveness of forest drainage on already drained sites in the Hemmingford Region of Quebec. Urgel Delisle et Associes, Quebec, Canada (Project Proposal). 20 pp.

van der Valk, A. G. (1981) Succession in wetlands: A Gleasonian approach. Ecology 62: 688–696.

van der Valk, A. G. and Bliss, L. C. (1971) Hydrarch succession and net primary production of oxbow lakes in central Alberta. Canadian Journal of Botany 49: 1177–1199.

van der Valk, A. G. (1986) The impact of litter and annual plants on recruitment from the seed bank of a lacustrine wetland. Aquatic Botany 24: 13–26.

van der Valk, A. G. (ed.). (1989) Northern Prairie Wetlands. Iowa State University Press, Ames, Iowa, USA. 400 pp.

van der Valk, A. G. and Davis, C. B. (1978) The role of the seed bank in the vegetation dynamics of prairie glacial marshes. Ecology 59: 322–335.

Vestergaard, P. (1978) Studies in vegetation and soil of coastal salt marshes in the Disko area, West Greenland. Meddelelser Om Gronland 204: 1–51.

Vitt, D. H. (1990) Growth and production dynamics of boreal mosses over climatic, chemical and topographic gradients. Botanical Journal Linnean Society 104: 35–59.

Vitt, D. H. and Bayley, S. E. (1984) The vegetation and water chemistry of four oligotrophic basin mires in northwestern Ontario. Canadian Journal of Botany 62: 1485–1500.

Vitt, D. H. and Pakarinen, P. (1977) The bryophyte vegetation, production, and organic components of Truelove Lowland. p. 225–244. *In*: Truelove Lowland, Devon Island, Canada: A High Arctic ecosystem. University of Alberta Press, Edmonton, Alberta, Canada.

Vitt, D. H. and Chee, W. L. (1990) The relationships of vegetation to surface water chemistry and peat chemistry in fens of Alberta, Canada. Vegetatio 89: 87–106.

Vitt, D. H., Achuff, P. and Andrus, R. E. (1975) The vegetation and chemical properties of patterned fens in the Swan Hills, north central Alberta. Canadian Journal of Botany 53: 2776–2795.

Vitt, D. H., Horton, D. G., Slack, N. G. and Malmer, N. (1990) *Sphagnum*-dominated peatlands of the hyperoceanic British Columbia coast: patterns in surface water chemistry and vegetation. Canadian Journal of Forest Research 20: 696–711.

Walker, B. H. and Coupland. R. T. (1970) Herbaceous wetland vegetation in the aspen grove and grassland regions of Saskatchewan. Canadian Journal of Botany 48: 1861–1878.

Walker, B. H. and Wehrhahn, C. F. (1971) Relationships between derived vegetation gradients and measured environmental variables in Saskatchewan wetlands. Ecology 52: 85–95.

Washburn, A. L. (1983a) Palsas and continuous permafrost. p. 1372–1377. *In*: Permafrost Fourth International Conference Proceedings, National Academy Press, Washington, DC, USA.

Washburn, A. L. (1983b) What is a palsa? p. 34–47. *In*: H. Poser and E. Schunke (eds.). Abhandlungen der Akademie der Wissenschaften in Gottingen, Sweden, Dritte Folge Nr. 35. Gottingen, Sweden.

Wells, E. D. (1981) Peatlands of eastern Newfoundland: distribution, morphology, vegetation and nutrient status. Canadian Journal of Botany 59: 1978–1997.

Wells, E. D. and Pollett, F. C. (1983) Peatlands. p. 207–265. *In*: R. G. South (ed.). Biogeography and Ecology of the Island of Newfoundland, Dr. W. Junk Publishers, The Hague, The Netherlands.

Wells, E. D. and Hirvonen, H. E. (1988) Wetlands of Atlantic Canada. p. 249–303. *In*: National Wetlands Working Group (ed.). Wetlands of Canada. Ecological Land Classification Series No. 24. Sustainable Development Branch, Environment Canada, Ottawa. Polyscience Publishers, Montreal, Quebec, Canada.

Whillens, T. H. (1982) Changes in marsh area along the Canadian shore of Lake Ontario. Journal Great Lakes Research 8: 570–577.

Wilson, S. D. and Keddy, P. A. (1985) Plant zonation on a shoreline gradient: physiological response curves of component species. Journal of Ecology 73: 851–860.

Yamanaka, K. (1975) Primary Productivity of the Fraser River Delta Foreshore: Yield Estimates of Emergent Vegetation. M.Sc. Thesis, University of British Columbia, Vancouver, British Columbia, Canada. 134 pp.

Zoltai, S. C. (1971) Southern limit of permafrost features in peat landforms, Manitoba and Saskatchewan. p. 305–310. *In*: Geological Association of Canada, Special Paper No. 9, St. John's, Newfoundland, Canada.

Zoltai, S. C. (1972) Palsas and peat plateaus in central Manitoba and Saskatchewan. Canadian Journal of Forest Research 2: 291–302.

Zoltai, S. C. and Pollett, F. C. (1983) Wetlands in Canada: Their classification, distribution and use. p. 245–268. *In*: A. J. P. Gore (ed.). Mires: Swamp, Bog, Fen and Moor. B. Regional Studies, Elsevier, Amsterdam, The Netherlands.

Zoltai, S. C. and Tarnocai, C. (1971) Properties of a wooded palsa in northern Manitoba. Arctic Alpine Research 3: 115–129.

Zoltai, S. C. and Tarnocai, C. (1975) Perennially frozen peatlands in the western arctic and subarctic of Canada. Canadian Journal Earth Science 12: 28–43.

Zoltai, S. C. and Vitt, D. (1990) Holocene climatic change and the distribution of peatlands in western interior Canada. Quaternary Research 33: 231–240.

Zoltai, S. C., Pollett, F. C., Jeglum, J. K. and Adams, G. D. (1973) Developing a wetland classification for Canada. p. 497–511. *In*: B. Bernier and C. H. Winget (eds.). Forest Soils and Forest Land Management, Proceedings of the Fourth North American Forest Soils Conference, Les Presses de l'Université Laval, uebec, Canada.

Zoltai, S. C., Tarnocai, C., Mills, G. F. and Veldhuis, H. (1988) Wetlands of subarctic Canada. p. 55–96. *In*: National Wetlands Working Group (ed.). Wetlands of Canada. Ecological Land Classification Series No. 24. Sustainable Development Branch, Environment Canada, Ottawa. Polyscience Publishers, Montreal, Quebec, Canada.

Zoltai, S. C., Taylor, S., Jeglum, J. K., Mills, G. F. and Johnson, J. D. (1988) Wetlands of boreal Canada. p. 97–154. *In*: National Wetlands Working Group (ed.). Wetlands of Canada. Ecological Land Classification Series No. 24. Sustainable Development Branch, Environment Canada, Ottawa. Polyscience Publishers, Montreal, Quebec, Canada.

Wetlands of the United States

BILL O. WILEN AND RALPH W. TINER

Abstract

This paper discusses the general types of wetlands found in the United States (including Alaska and Hawaii) as well as their classification, ecology, status and trends, and regional problem areas. It is based upon the work performed by the U.S. Fish and Wildlife Service's National Wetlands Inventory and also upon a review of existing information about the wetlands of the United States.

Introduction

The purpose of this chapter is to provide a general overview of the wetlands of the United States. It begins with a description of the study area and its major wetland types, followed by sections on wetland classification, characteristics (including wetland formation, hydrology, and key functions), use and conservation (including an overview of recent wetland trends and major threats and problem areas), and recommendations to improve wetland protection. This chapter is not intended to be exhaustive in its coverage (to do so would require a separate book), but the discussion should present the reader with a broad understanding of U.S. wetlands and identify sources for additional information.

Study area

The United States encompasses an area of approximately 8,625,000 km^2 extending from the Arctic Circle south to the Hawaiian islands, which lie just below the Tropic of Cancer at 23 1/2° N. Politically, the U.S. is comprised of 50 states. Figure 1 shows the location of each state, since numerous references to individual states are made throughout this chapter. Within this

D.F. Whigham et al. (eds.), Wetlands of the World I, 515–636.

Figure 1. Map of the United States.

broad area, regional variations in climate, topography, hydrology, geology, soils, and vegetation create a tremendous diversity of wetlands.

From an ecological standpoint, the U.S. has been divided into numerous "ecoregions" by the U.S. Forest Service. A map (Fig. 2) and descriptions of these ecoregions were prepared for the U.S. Fish and Wildlife Service's National Wetlands Inventory by Bailey (1976). Ecoregions are based on land-surface forms, regional climate, potential natural vegetation, and zonal soils. Land-surface forms are classified by Hammond (1964), climate by Koppen (1931), potential natural vegetation by Kuchler (1964), and zonal soils according to *Soil Taxonomy* (U.S. Department of Agriculture, Soil Conservation Service 1975). A general description of each ecoregion is presented in Table 1 (and a more detailed description in Bailey 1976).

Wetland Definition

The wetland concept used throughout this chapter follows the definition by the U.S. Fish and Wildlife Service (F.W.S.). This definition was developed for conducting an inventory of the wetlands of the United States. The Fish and Wildlife Service defines wetlands as follows:

> Wetlands are lands transitional between terrestrial and aquatic systems where the water table is usually at or near the surface or the land is covered by shallow water. For purposes of this classification wetlands must have one or more of the following three attributes: (1) at least periodically, the land supports predominantly hydrophytes; (2) the substrate is predominantly

undrained hydric soil; and (3) the substrate is nonsoil and is saturated with water or covered by shallow water at some time during the growing season of each year.(Cowardin *et al.* 1979).

All areas considered wetland must have enough water at some time during the growing season to stress plants and animals not adapted for life in water or saturated soils. Most wetlands, therefore, have hydrophytes and hydric soils present. The Fish and Wildlife Service has prepared a list of the nation's plants that occur in wetlands (Reed 1988). Approximately 31% of the 21,588 plant species found in the United States occur in wetlands and about half of these are restricted to or usually occur in wetlands (P. Reed, U.S.Fish and Wildlife Service, personal communication). The U.S. Department of Agriculture (USDA), Soil Conservation Service has published a list of the nation's hydric soils (U.S. Department of Agriculture, Soil Conservation Service 1987). A guide for identifying New England's hydric soils has been prepared (Tiner and Veneman 1987).

Wetland types

The United States possesses a wide variety of wetland types ranging from wet tundra in Alaska to tropical rain forests in Hawaii and desert wetlands in the arid Southwest region. Wetlands occur in every state of the country. Due to regional differences in climate, vegetation, soil, and hydrologic conditions, wetland diversity is tremendous. Vegetative communities are constantly being affected by three basic phenomena: succession, maturation, and fluctuation, or some combination of the three (van der Valk 1985).

The Fish and Wildlife Service's classification system (Cowardin *et al.* 1979) groups wetlands into categories sharing ecologically similar characteristics. It first divides wetlands and deepwater habitats into five ecological systems: (1) Marine, (2) Estuarine, (3) Riverine, (4) Lacustrine, and (5) Palustrine (Fig. 3). The Marine System generally consists of the open ocean and its associated coastline. It is mostly a deepwater habitat system, with marine wetlands limited to intertidal areas such as beaches, rocky shores, and intertidal coral reefs. The Estuarine System includes coastal wetlands (e.g. salt and brackish tidal marshes, mangrove swamps, and intertidal flats) as well as deepwater bays, sounds, and coastal rivers. The Riverine System is limited to largely freshwater river and stream channels and is mainly a deepwater flowing habitat system, although shallow or intermittent streams are considered wetlands. The Lacustrine System is dominated by deepwater habitats that include standing water bodies such as lakes, reservoirs, and deep ponds. The Palustrine System encompasses the vast majority of the country's inland marshes, wet meadows, bogs, swamps, bottomland hardwood forests, and

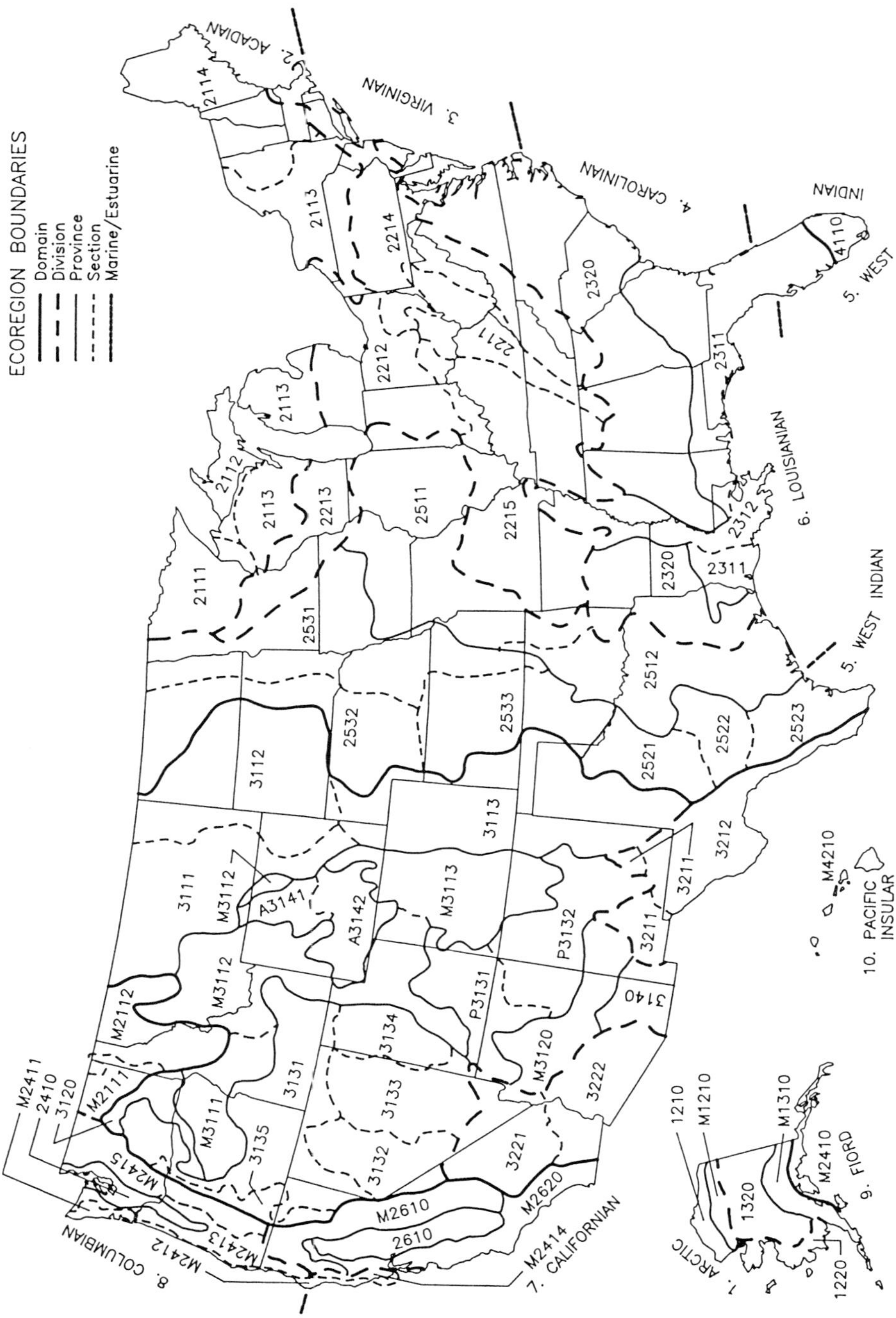
ECOREGION BOUNDARIES
Domain
Division
Province
Section
Marine/Estuarine
1. ARCTIC
2. ACADIAN
3. VIRGINIAN
4. CAROLINIAN
5. WEST INDIAN
6. LOUISIANIAN
7. CALIFORNIAN
8. COLUMBIAN
9. FIORD
10. PACIFIC INSULAR
2114
2113
2214
2212
2211
2112
2213
2111
2511
2215
2531
2320
4110
2311
2312
2512
2532
2533
2521
2522
2523
3112
3113
3111
M3112
A3141
A3142
M3113
3211
3212
M4210
P3132
P3131
3140
M2112
3134
3131
M3120
3222
M2111
M2411
2410
3120
M3111
3133
3135
M2415
3132
3221
M2610
M2620
2610
M2412
M2413
M2414
1210
M1210
1320
M1310
M2410
1220

- 1000 Polar
  - 1200 Tundra
    - 1210 Arctic Tundra
    - 1220 Bering Tundra
    - M1210 Brooks Range
  - 1300 Subarctic
    - 1320 Yukon Forest
    - M1310 Alaska Range
- 2000 Humid Temperate
  - 2100 Warm Continental
    - 2110 Laurentian Mixed Forest
      - 2111 Spruce-Fir Forest
      - 2112 Northern Hardwoods-Fir Forest
      - 2113 Northern Hardwoods Forest
      - 2114 Northern Hardwoods-Spruce Forest
    - M2110 Columbia Forest
      - M2111 Douglas-fir Forest
      - M2112 Cedar-Hemlock-Douglas-fir Forest
  - 2200 Hot Continental
    - 2210 Eastern Deciduous Forest
      - 2211 Mixed Mesophytic Forest
      - 2212 Beech-Maple Forest
      - 2213 Maple-Basswood Forest + Oak Savanna
      - 2214 Appalachian Oak Forest
      - 2215 Oak-Hickory Forest
  - 2300 Subtropical
    - 2310 Outer Coastal Plain Forest
      - 2311 Beech-Sweetgum-Magnolia-Pine-Oak
      - 2312 Southern Floodplain Forest
    - 2320 Southeastern Mixed Forest
- 2000 Humid Temperate
  - 2400 Marine
    - 2410 Willamette-Puget Forest
    - M2410 Pacific Forest (in conterminous U.S.)
      - M2411 Sitka Spruce-Cedar-Hemlock Forest
      - M2412 Redwood Forest
      - M2413 Cedar-Hemlock-Douglas-fir Forest
      - M2414 California Mixed Evergreen Forest
      - M2415 Silver fir-Douglas-fir Forest
    - M2410 Pacific Forest (in Alaska)
  - 2500 Prairie
    - 2510 Prairie Parkland
      - 2511 Oak-Hickory-Bluestem Parkland
      - 2512 Oak + Bluestem Parkland
    - 2520 Prairie Brushland
      - 2521 Mesquite-Buffalo Grass
      - 2522 Juniper-Oak-Mesquite
      - 2523 Mesquite-Acacia
    - 2530 Tall-Grass Prairie
      - 2531 Bluestem Prairie
      - 2532 Wheatgrass-Bluestem-Needlegrass
      - 2533 Bluestem-Grama Prairie
  - 2600 Mediterranean (Dry-summer Subtropical)
    - 2610 California Grassland
    - M2610 Sierran Forest
    - M2620 California Chaparral
- 3000 Dry
  - 3100 Steppe
    - 3110 Great Plains-Shortgrass Prairie
      - 3111 Grama-Needlegrass-Wheatgrass
      - 3112 Wheatgrass-Needlegrass
      - 3113 Grama-Buffalo Grass
- 3000 Dry
  - 3100 Steppe
    - M3110 Rocky Mountain Forest
      - M3111 Grand fir-Douglas-fir Forest
      - M3112 Douglas-fir Forest
      - M3113 Ponderosa Pine-Douglas-fir Forest
    - 3120 Palouse Grassland
    - M3120 Upper Gila Mountains Forest
    - 3130 Intermountain Sagebrush
      - 3131 Sagebrush-Wheatgrass
      - 3132 Lahontan Saltbush-Greasewood
      - 3133 Great Basin Sagebrush
      - 3134 Bonneville Saltbush-Greasewood
      - 3135 Ponderosa Shrub Forest
    - P3130 Colorado Plateau
      - P3131 Juniper-Pinyon Woodland + Sagebrush Saltbush Mosaic
      - P3132 Grama-Galleta Steppe + Juniper-Pinyon Woodland Mosaic
    - 3140 Mexican Highland Shrub Steppe
    - A3140 Wyoming Basin
      - A3141 Wheatgrass-Needlegrass-Sagebrush
      - A3142 Sagebrush-Wheatgrass
  - 3200 Desert
    - 3210 Chihuahuan Desert
      - 3211 Grama-Tobosa
      - 3212 Tarbush-Creosote Bush
    - 3220 American Desert
      - 3221 Creosote Bush
      - 3222 Creosote Bush-Bur Sage
- 4000 Humid Tropical
  - 4100 Savanna
    - 4110 Everglades
  - 4200 Rainforest
    - M4210 Hawaiian Islands

Figure 2. Map of the ecoregions of the United States with a key to the divisions (Bailey 1978).

Table 1. General environmental characteristic of second-order ecoregions (Bailey 1978). Area are given to nearest 1,000 km^2 and percentages to nearest tenth.

Division and extent of nation	Climate	Vegetation	Soils	Land-surface form
Tundra 541,000 km^2 6.4%	Very short, cool summers; long, severe winter; 55–188 days have a mean of 0°C; less than 200 mm precipitation; climate is humid due to low potential evaporation	Grasses, sedges and lichens with willow shrubs	Inceptisols with weakly differentiated horizons	Broad, level plain, less than 1,000 ft in elevation; thousands of lakes and wetlands along the coast Brooks Range is an extension of Rocky Mountain system 900 to 2,700 m
Subartic 745,000 km^2 8.8%	Summer warmth only thaws a few feet of permafrost; severe winter; less than 4 months have average temperature warmer than 10°C; average annual precipitation 425 mm; precipitation concentrated in 3 warm months	Needleleaf forest and open lichen woodland	Inceptisols with pockets of wet organic Histosols	Broad valleys, dissected uplands, and lowland basins, also includes Alaska and Aleutian Ranges; includes Mount McKinley at 6,193.5 m and 3,219 km volcanic arc of Aleutian Mountains
Warm continental 647,000 km^2 7.6%	Warm summer, cold snowy winter; 4 to 8 months temperature exceeds 10°C; precipitation ample all year 600–1,000 mm precipitation but substantially greater in spring	Needleleaf and mixed needleleaf-deciduous forest	Spodosols	Eastern portion has rolling hills and low mountains between 300 to 900 mm; western portion high, rugged mountains more than 2,700 m

Hot Continental 952,000 km^2 11.1%	Humid, hot summers; cool winters; 5 to 6 month frost free season; 4 to 8 month temperature exceeds 10°C; coldest month colder than 0°C; precipitation 900–1,500 mm; precipitation markedly greater in summer	Winter deciduous forest, dominated by tall, broadleaved trees	Inceptisols, Alfisols and Ultisols, rich in humus and moderately leached	Rolling, but some parts are nearly flat and up to 900 m in Appalachian Mountains
Subtropical 1,056,000 km^2 12.3%	Humid, rainy, hot summers; absence of really cold winters; average annual temperature 15 to 21°C; well distributed precipitation from 1,000 to 1,525 mm; driest summer month receives 30 mm	Sandy coastal region covered with longleaf, loblolly and slash pine forest; inland region deciduous forest	Strong leached Ultisols; rich in iron and aluminum oxides	Flat and irregular coastal plains; 50 to 80 percent gently sloping; relief less than 90 m on coast and up to 300 m on piedmont
Marine 369,000 km^2 2.7%	Temperate, rainy, warm summers, generally mild throughout year; annual temperature 2°C to 13°C; warmest month cooler 22°C, but at least 4 month average 10°C; precipitation from 380t o 1,525 mm in coast range and 750 to 3,800 mm in mountains; abundant throughout the year but markedly reduced in summer	Needleleaf forest; coast range has magnificent forest of Douglas-fir, redcedar, and spruce	Strongly leached, acid Inceptisols and Ultisols;	Valleys are nearly level to gently sloping flood plains with isolated hills; coastal plain mountains rise to 1,500 m, Cascade Range between 2,400 m to 2,700 m; one volcano mountain Rainier rises 4,300 m

Table 1. Continued.

Division and extent of nation	Climate	Vegetation	Soils	Land-surface form
Prairie 1,324,000 km^2 15.6%	Subhumid, summers are hot and winters range from cold in northern part to short and mild in southern part; average annual temperature ranges from 4°C in the north to 21°C in the south; 140 to more than 300 frost free days; evaportranspiration and precipitation balance; precipitation 500 to 1,000 mm	Tall grasses with subdominant broadleaved herbs; trees and shrubs are almost absent	Mollisols with black, friable, organic surface horizons and high content of bases	Northern portion was glaciated; mostly gently rolling plains and plateaus but steep bluffs border valleys; some areas nearly flat; others have high rounded hills; mesa and butte landscape is found in certain parts of Texas
Mediterrean 223,500 km^2 2.7%	Temperate rainy climate with dry, hot summers; winters are mild often foggy; wet winter followed by dry summer is typical; precipitation ranges from 150 mm in valleys, 750 mm on coast to 1,800 mm at highest elevations; 2 to 4 rainless months	Distinctive natural vegetation of broad-leaved evergreen trees and shrubs called sclerophyll	Generally Alfisols and Mollisols typical of semiarid climates	Ranges from a flat aluvial plain with elevations sea level to 150 m; to gently to steeply sloping, coast ranges with elevations to 760 m with 1,500 m peaks; to precipitous Sierra Nevada with elevations higher than 4,300 m

Steppe 2,360,300 km^2 27.9%	Summers warm to hot; winters cold and dry; average annual temperature 4°C to 15°C, except 2°C–7°C in mountains; frost free days range from less 100 to 200; precipitation ranges from 125 to 620 mm with over 1,000 mm in mountains; winters are dry	Short grasses that usually grow in bunches that are sparsely distributed, some scattered shrubs and low trees	Mollisols in steppe lands, Aridisols in semidesert lands, calcification, with salinization dominate poorly drained sites	Ranges from great basins with elevations under 200 m to rolling plains, to tableland, areas with isolated hills and mountains to 2,700 m, to steppe foothills, to high plateaus, to the Rockey Mountains with elevations of 4,300 m
Desert 366,700 km^2 4.4%	Continual desert, long summers with a high temperature of 57°C; average annual temperature 10°C–24°C; extreme variation between day and night, annual precipitation less than 200 mm and in some areas less than 100 mm	Xerophytic plants widely dispersed with negligible ground cover	Aridisols and dry Entisols; humus is lacking; dominant process is salinization	Mostly desert with extensive, gently undulating plains with isolated mountains and abruptly rising buttes; the plains range from 85 mm below sea level to 1,200 m above; mountains rise to 3,400 m
Savannah 20,200,000 km^2 0.2%	Tropical, average annual temperature is 21°C to 24°C between October thru February; no winter season; frost-free practically all year; precipitation 1,270 to 1,650 mm; wet and dry seasons, dry season last longer than 2 months	Tropical savannah characterized by open expanses of tall grasses, interspersed with hardy drought resistant shrubs and trees	Hislosols and Inceptisols, soils are excessively leached as a result of heavy rains and high temperatures	Almost flat, marl and limestone shelf covered with a few feet of muck and a little sand; elevation range from sea level to 7.6 m; contains massive areas of wetlands

Table 1. Continued.

Division and extent of nation	Climate	Vegetation	Soils	Land-surface form
Rainforest 17,400,000 km^2 0.3%	Tropical; average annual temperature 27°C, no perceptible seasons; rainfall heavy all year, no month average less than 60 mm	Rainforest or selva vegetatin type; evergreen broad-leaved trees	Ultisols and Oxisols especially rich in hydroxides of iron magnesium and aluminum	The five principal islands and 4 smaller ones are all volcanic in various stages of erosion; about $\frac{1}{4}$ rises less than 198 m, $\frac{1}{2}$ between 198 to 600 m, and $\frac{1}{4}$ higher than 600 m

shallow ponds, and does not include any deepwater habitat (The classification system is discussed in more detail later in this chapter). Characteristics of the major wetland types in the U.S. are described in the following sections which focus on estuarine and palustrine wetlands because they are the most abundant types. Examples are illustrated on pp. 632–635.

Estuarine wetlands

Estuarine wetlands are found along the entire U.S. coastline in association with tidal estuaries. These wetlands develop behind barrier islands and beaches or form along coastal rivers and embayments. From a salinity standpoint, estuaries can be divided into three distinct reaches: (1) polyhaline – strongly saline areas (18–30 parts per thousand, ppt), (2) mesohaline – moderate salinity areas (5–18 ppt), and (3) oligohaline – slightly brackish areas (0.5–5 ppt) (Cowardin *et al.* 1979). Large coastal rivers become increasingly fresher upstream from the river's mouth as salt water is diluted by the river's freshwater discharge. Since river discharge varies during the year, the salinity of coastal river systems vary on a seasonal basis. A variety of wetlands develop in estuaries largely because of differences in salinity and duration and frequency of tidal inundation. Major wetland types include: (1) emergent wetlands, (2) intertidal unconsolidated shores, and (3) scrub-shrub wetlands. Other coastal wetlands include intertidal coral and mollusk reefs, rocky shores, streambeds, and some forested wetlands. Submerged aquatic vegetation forming dense beds in permanent coastal waters are considered deepwater habitats.

Estuarine emergent wetlands
Estuarine emergent wetlands are usually dominated by grass or grasslike herbaceous plants. These wetlands, commonly called "salt marshes" and "brackish tidal marshes", are best represented along the coastlines of the Atlantic and the Gulf of Mexico. Differences in salinity and tidal flooding within estuaries have a profound effect on the emergent vegetation. Plant composition markedly changes from the more saline portions of estuaries to the slightly brackish areas. Even within areas of similar salinity, vegetation differs largely due to frequency and duration of tidal flooding. Two distinct zones can be observed based on hydrologic differences in frequency and duration of flooding: (1) regularly flooded marsh (low marsh) and (2) irregularly flooded (high marsh) (Fig. 4). The regularly flooded marsh is flooded and exposed at least once daily by the tides. In the northeastern U.S., this marsh is generally limited to tidal creek banks and the shores of coastal embayments and rivers, while in Georgia and South Carolina, it is the dominant coastal wetland type covering vast acreages. Along the coastlines

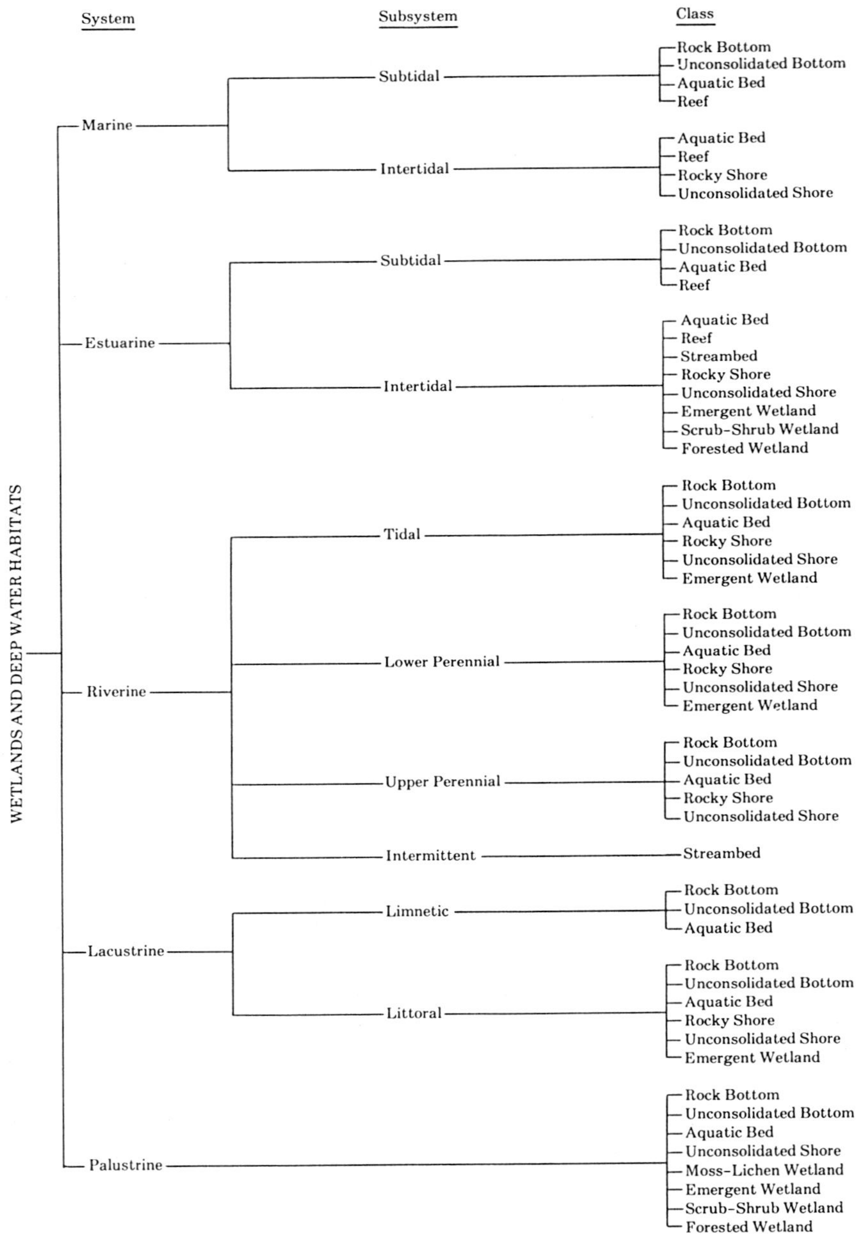

Figure 3. Classification hierarchy of wetlands and deepwater habitats, showing systems, subsystems and classes (Cowardin *et al.* 1979).

of the Atlantic Ocean and the Gulf of Mexico, smooth cordgrass (*Spartina alterniflora*) dominates the regularly flooded marsh, while on the coastline of the Pacific Ocean, California cordgrass (*Spartina foliosa*) prevails. These grasses are among the most productive marsh plants. Lying above the regularly flooded zone, the irregularly flooded marsh is exposed to air for long periods and flooded only at varying intervals. Vegetation in this zone is more diverse and includes salt marsh hay (*Spartina patens*), salt grass (*Distichlis spicata*), black grass (*Juncus gerardii*), alkali grasses (*Puccinellia* spp.) and baltic rush (*Juncus balticus*) along the North Atlantic coast, black needlerush (*Juncus roemerianus*), glassworts (*Salicornia* spp.), and sea ox-eye (*Borrichia frutescens*) along the South Atlantic and Gulf of Mexico coasts, high-tide bush (*Iva frutescens*), common reed (*Phragmites australis*), bulrushes (*Scirpus* spp.), asters (*Aster* spp.), and switchgrass (*Panicum virgatum*) on the Atlantic and Gulf coasts. de la Cruz (1979) discussed differences between South Atlantic and Gulf of Mexico coastal marshes. On the Pacific coast, common plants include *Spartina foliosa*, perennial glasswort (*Salicornia virginica*), common glasswort (*S. subterminalis*), salt dodder (*Cuscuta salina*), *Distichlis spicata*, California sea-blite (*Suaeda californica*), alkali heath (*Frankenia grandifolia*), California sea lavender (*Limonium californicum*), spreading alkali-weed (*Cressa truxillensis*), spiny rush (*Juncus acutus*), Lyngbye's sedge (*Carex lyngbyei*), tufted hairgrass (*Deschampsia caespitosa*), and fleshy jaumea (*Jaumea carnosa*). Salt marshes along the coast of Alaska are vegetated by *Puccinellia* spp., *Deschampsia caespitosa*, *Carex lyngbyei*, *C. cryptocarpa*, Ramensk's sedge (*C. ramenskii*), Hoppner's sedge (*C. subspathacea*), seaside arrow-grass (*Triglochin maritima*), MacKenzie water-hemlock (*Cicuta mackenziena*), and vetchling peavine (*Lathyrus palustris*).

Moving upstream in large coastal rivers where seawater is diluted by freshwater, brackish tidal marshes can be found. Salinity here fluctuates greatly with the tides, river flow, and the seasons. Nearest the salt marshes, *Juncus roemerianus* dominates brackish marshes along the South Atlantic and Gulf coasts, while big cordgrass (*Spartina cynosuroides*), wire grass (*Spartina patens*), *Panicum virgatum*, narrow-leaved cattail (*Typha angustifolia*), and *Scirpus* spp. are also dominant in brackish waters. As the upstream limit of salt water influence is approached, a highly diverse assemblage of emergent plants characterizes these marshes including *Spartina cynosuroides*, *Typha angustifolia*, pickerelweed (*Pontederia cordata*), southern wild rice (*Zizaniopsis miliacea*) (South Atlantic and the Gulf of Mexico), rose mallow (*Hibiscus moscheutos*), arrowheads (*Sagittaria* spp.), smartweeds (*Polygonum* spp.), sedges (*Carex* spp.), *Scirpus* spp., beggars-ticks (*Bidens* spp.), and *Phragmites australis*. Most of these plants, however, reach their maximum abundance in the inland wetlands. Numerous references on salt marsh vegetation in the United States are available including: Adamus (1963),

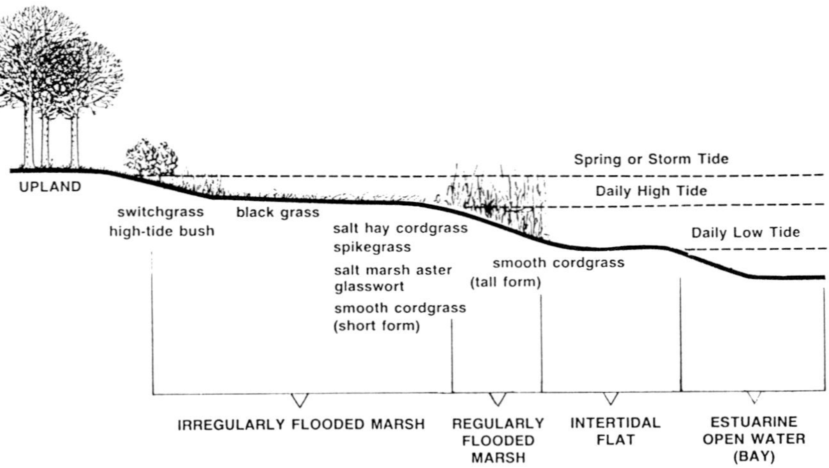

Figure 4. Cross sectional diagram of a northeastern salt marsh (Tiner 1984).

Carlton (1977), Chabreck (1972), Chapman (1938, 1940, 1960, 1976a), Conner and Day (1987), Copeland *et al.* (1983), Copeland *et al.* (1984), Day *et al.* (1973), de la Cruz (1981), Drew and Schomer (1984), Eleuterius and McDaniel (1978), Eleuterius (1972, 1980), Gosselink (1984), Hackney and de la Cruz (1982), Josselyn (1983), Kurz and Wagner (1957), Lewis and Estevez (1988), Livingston (1984), McCormick and Somes (1982), Niering and Warren (1980), Nixon (1982), Odum *et al.* (1984), Pomeroy and Wiegert (1981), Reimold and Queen (1974), Seliskar and Gallagher (1983), Stout (1984), Teal and Teal (1969), Tiner (1977, 1985a, 1985b, 1987, 1988), Zedler and Nordby (1986), and Zedler (1982).

Estuarine scrub-shrub wetlands

Estuarine scrub-shrub wetlands are characterized by salt-tolerant woody vegetation less than six meters in height. Common estuarine shrubs along the Atlantic and Gulf of Mexico coasts are *Iva frutescens*, groundsel tree (*Baccharis halimifolia*), and *Borrichia frutescens*. These shrubs occur at higher levels in the salt marshes. In particular, *Iva frutescens* is common along mosquito ditches where substrate material has been mounded and along the upper borders of many salt marshes. Estuarine scrub-shrub wetlands are perhaps best represented by mangrove swamps, which have a limited distribution in the U.S. (Fig. 5). Mangroves are generally found south of the 30° N. latitude and reach their maximum abundance in Florida, Puerto Rico, and the Virgin Islands. These wetlands are dominated by two forms of mangroves: (1) red mangrove (*Rhizophora mangle*) and (2) black mangrove (*Avicennia germinans*; Fig. 5). The former dominates the regularly flooded zone, while the

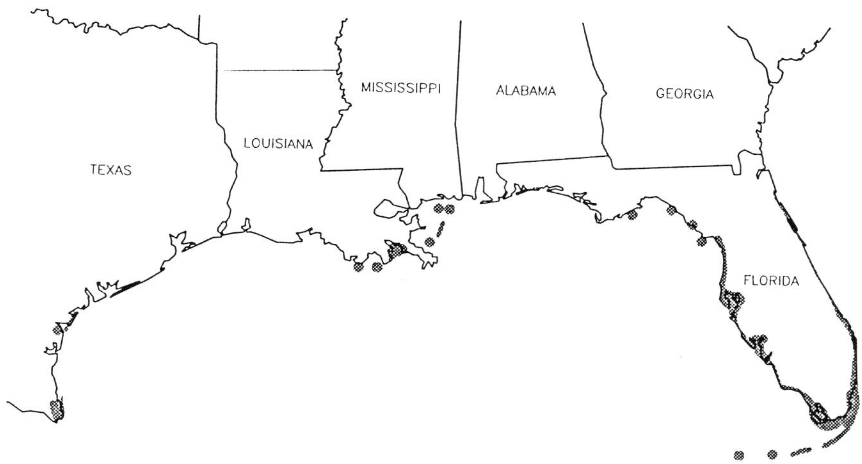

Figure 5. Distribution of *Avicennia germinans* (L.) L., black mangrove in Texas, Louisiana, Mississippi, and Florida (Little 1977).

latter species characterizes higher irregularly flooded areas. White mangrove (*Laguncularia racemosa*) may be intermixed (Chapman 1976b, Schomer and Drew 1982). Salt marshes of *Spartina alterniflora*, *Juncus roemerianus*, *Distichlis spicata*, woody glasswort (*Salicornia perennis*) and saltwort (*Batis maritima*) may be closely associated with Florida's mangroves swamps. Odum *et al.* (1982) have reported on the ecology of mangroves in South Florida.

Estuarine intertidal unconsolidated shores

Intertidal unconsolidated shores (also called "tidal flats") often lie seaward of tidal marshes and mangroves, at river mouths or along rocky shores. They also occur as barren areas within the high marsh in high salinity areas, especially along the South Atlantic and Gulf Coasts. At low tide, intertidal shores appear largely as unvegetated expanses of mud, sand, gravel, or cobbles or variations of these materials (Fig. 6). Microscopic plants like diatoms, bluegreen algae, and dinoflagellates may be extremely abundant. On occasion, macroscopic algae such as sea lettuce (*Ulva lactuca*) and *Enteromorpha intestinalis* may locally dominate these shores (Tiner 1987). These wetlands are particularly extensive in areas with high tidal ranges such as occur in Alaska and Maine.

Palustrine wetlands

Palustrine wetlands occur in the interior of the country and in coastal areas on the mainland and the interior of barrier islands. These wetlands are chiefly associated with river floodplains, topographic depressions, margins of lakes

Figure 6. Estuarine intertidal unconsolidated shore (Cowardin *et al.* 1979).

and ponds, limestone sinkholes, ground-water seepage slopes, and other areas where the water table is at or near the surface for a significant period during the growing season. In Alaska, they are also found in freeze and thaw basins, saturated permafrost areas, and below melting snow beds.

Hydrologic forces maintain wetlands in many ways according to local conditions. Most palustrine wetlands are flooded at some time during the year from only a week or two early in the growing season to the entire year (permanently flooded). Wetlands occurring on slopes may never be flooded, yet the soils remain saturated for all or most of the growing season. Along freshwater coastal rivers, some palustrine wetlands are flooded periodically by fresh tidal waters (less than 0.5 ppt). This occurs mainly along the Atlantic, Gulf, and Alaskan Coasts. Differences in local hydrology and other factors affect the plant composition of individual wetlands.

Palustrine wetlands are largely dominated by trees, shrubs, and persistent herbaceous plants that remain visible in wetlands through the winter and into the following spring. They mainly consist of freshwater types, although inland saline wetlands exist in arid and semiarid western regions of the country. Palustrine wetlands are represented by three major vegetated types: (1) emergent wetland, (2) scrub-shrub wetland, and (3) forested wetland.

Shallow open water bodies such as ponds and playa lakes (less than eight hectares in size and less than two meters deep), along with their aquatic beds also comprise palustrine wetlands.

Emergent wetlands

Palustrine emergent wetlands are dominated by erect, herbaceous vegetation, including many members of the Poaceae and Cyperaceae. These wetlands are commonly referred to by a host of terms including "marsh", "wet meadow", "fen", "inland salt marsh", and "alkali marsh", depending on the region of the country and individual characteristics. Many emergent wetlands are dominated by one or a few species of common marsh plants. These may include: cattails (*Typha* spp.), wild rice (*Zizania aquatica*), bluejoint (*Calamagrostis canadensis*), reed canary grass (*Phalaris arundinacea*), *Carex* spp., rushes (*Juncus* spp.), spikerushes (*Eleocharis* spp.), cottongrasses (*Eriophorum* spp.), *Scirpus* spp., rice cutgrass (*Leersia oryzoides*), water-willow (*Decodon verticillatus*), maiden-cane (*Panicum hemitomum*), *Phragmites australis*, bur-reeds (*Sparganium* spp.), *Sagittaria* spp., *Pontederia cordata*, and *Polygonum* spp.. The Everglades, located at the southern tip of Florida, contain the largest expanse of sawgrass (*Cladium jamaicense*) in North America; while the Prairie Pothole Region of the upper midwestern states of North Dakota, South Dakota, Minnesota, and Montana encompasses the highest density of emergent wetlands in the country. In central North Dakota, scientists have found an average of 11 wetlands basins per square kilometer and most of these were less than 0.4 hectares in size (Cowardin *et al.* 1981). This high density is related to past glacial events which left the landscape covered with numerous lakes, ponds, and undrained depressions. These pothole wetlands are exposed to a range of salinities from fresh (40–500 micromhos cm^{-1}) to saline (100,000 micromhos cm^{-1}). Stewart and Kantrud (1971) have described characteristic vegetation of different wetland types in this region. Inland salt marshes are found in Utah, Nevada, and other arid areas in adjoining states. Vegetation in these marshes includes red saltwort (*Salicornia rubra*), Utah glasswort (*S. utahensis*), iodine bush (*Allenrolfea occidentalis*), sea-blites (*Suaeda* spp.), *Distichlis spicata*, alkali sacaton (*Sporobolus airoides*) and *Cressa truxillensis* (Chapman 1960). Palustrine emergent wetlands are also prevalent in Alaska, often associated with saturated permafrost conditions. Important species include: cottongrasses (*Eriophorum angustifolium*, *E. scheuchzeri*, and *E. vaginatum*), sedges (*Carex aquatilis*, *C. bigelowii*, *C. limosa*, *C. lyngbyei*, *C. pluriflora*, and *C. rostrata*), *Calamagrostis canadensis*, pendant grass (*Arctophila fulva*), buckbean (*Menyanthes trifoliata*), Fisher's tundra grass (*Dupontia fisheri*), *Deschampsia caespitosa*, *Eleocharis* spp., *Juncus* spp., horsetails (*Equisetum* spp.), white beak-rush (*Rhynchospora alba*), and various mosses. Major reports on palustrine emergent wet-

Figure 7. Palustrine scrub-shrub (Pocosin) wetland in Brunswick County, North Carolina (Cowardin *et al.* 1979).

lands include the following: Batten and Murray (1982), Conner and Day (1987), Curtis (1959), Damman and French (1987), Drew and Schomer (1984), Eicher (1988), Geis and Kee (1977), Herdendorf *et al.* (1981), Herdendorf *et al.* (1986), Herdendorf (1987), Hobbie (1984), Hubbard (1988), Hubbard *et al.* (1988), Kantrud *et al.* (1989), Laessle (1942), Moore and Bellamy (1974), Nachlinger (1988), Nelson *et al.* (1983), Odum *et al.* (1984), Penfound (1952), Schomer and Drew (1982), Simpson *et al.* (1983), Stewart and Kantrud (1971, 1972), Tiner (1985a, 1985b, 1988, 1989), van der Valk (1985, 1989), Walker *et al.* (1989), Weller (1981), Windell *et al.* (1986), Zedler (1987).

Scrub-shrub wetlands

Inland wetlands dominated by woody vegetation less than six meters tall represent palustrine scrub-shrub wetlands (Fig. 7). Although not as abundant nationwide as palustrine emergent and palustrine forested wetlands, they occur widely throughout the nation and in some areas are a dominant type (e.g. boreal region). These shrub-dominated wetlands are commonly called "bogs", "pocosins", "shrub-carrs", or simply "shrub swamps".

Peat bogs are particularly interesting types of scrub-shrub wetlands. These wetlands are rarely flooded and are generally characterized by a saturated organic soil with the water table at or near the surface for most of the year. True bogs in Alaska are underlain by permafrost. Bogs in the northern part of the U.S. are prevalent in isolated depressions, along river courses, and along the margins of lakes in Alaska, Maine, Michigan, Minnesota, and Wisconsin. Typical northern bog plants include leatherleaf (*Chamaedaphne calyculata*), sweet gale (*Myrica gale*), *Eriophorum* spp., peat mosses (*Sphagnum* spp.), bog rosemary (*Andromeda glaucophylla*), Labrador tea (*Ledum groenlandicum*), bog laurel (*Kalmia polifolia*), blueberries and cranberries (*Vaccinum* spp.), as well as stunted trees of black spruce (*Picea marina*), larch (*Larix laricina*), and balsam fir (*Abies balsamea*). Alaskan bogs include many of these species plus others such as black crowberry (*Empetrum nigrum*), appleberry (*Rubus chamaemorus*), and lodgepole pine (*Pinus contorta*) (Batten and Murray 1982).

Bogs also occur along the southeastern Coastal Plain where they are called "pocosins". They are found on broad flat plateaus usually away from large streams. Pocosins are dominated by evergreen and deciduous shrubs, especially pond pine (*Pinus serotina*), sweet pepperbush (*Clethra alnifolia*), inkberry (*Ilex glabra*), fetterbush (*Lyonia lucida*), and swamp cyrilla (*Cyrilla racemiflora*). Other important scrub-shrub wetlands in the U.S. are characterized by buttonbush (*Cephalanthus occidentalis*), alders (*Alnus* spp.), willows (*Salix* spp.), dogwoods (*Cornus* spp.), and saplings of tree species like red maple (*Acer rubrum*) and poplars (*Populus* spp.). Examples of the variety of scrub-shrub wetland communities are presented in Table 2. Significant references addressing shrub wetlands include: Batten and Murray (1982), Conway (1949), Crum (1988), Curtis (1959), Damman (1977), Damman and French (1987), Dansereau and Segadas-Vianna (1952), Drury (1962), Gates (1942), Glaser (1987), Heinselman (1965, 1970), Johnson (1985), Kologiski (1977), Larsen (1982), Moore and Bellamy (1974), Osvald (1955), Richardson (1981a), Schomer and Drew (1982), Sjors (1959), Tiner (1985a, 1985b, 1988, 1989), and Windell *et al.* (1986).

Forested wetlands

Forested wetlands dominated by trees 6 meters or taller occur mostly in the eastern half of the United States and in Alaska. In the eastern U.S., they are the most abundant wetland type. They include such diverse types as black spruce bogs, cedar swamps, red maple swamps, pine swamps, and bottomland hardwood forests. In the Prairie Pothole Region of the upper midwestern states, forested wetlands are relatively scarce. As with other inland wetlands, flooding is extremely variable depending on regional climate, topographic position, and local hydrology. In the northern U.S., important trees of the wetter swamps include *Acer rubrum*, ashes (*Fraxinus*

Table 2. Examples of scrub-shrub wetland plant communities in the United States.

Wetland type and location	Dominant plants	Associated vegetation	Source
Buttonbush swamp; northern New Jersey	*Cephalanthus occidentalis*	*Spiraea tomentosa*, *Acer rubrum*, *Peltandra virginica*, *Juncus effusus*, *Scirpus cyperinus*, *Iris versicolor*, *Hypericum* sp., *Carex stricta*, *Boehmeria cylindrica*, *Polygonum sagittatum*, and *Lemnaceae*	Tiner (1985a)
Leatherleaf bog; southern New Jersey	*Chamaedaphne calyculata*	*Pinus rigida*, *Acer rubrum*, *Vaccinium corymbosum*, *Ilex glabra*, *Woodwardia virginica*, and *Sphagnum* spp.	Tiner (1986a)
Willow gravel bar thicket; interior Alaska	*Salix alaxensis*	*Salix richardsonii*, *S. reticulata*, *S. polaris*, *Poa alpina*, *Calamagrostis canadensis*, and *Trisetum spicatum*	Hanson (1958) as reported in Battan and Murray (1982)
Raised or blanket bog; southeastern Alaska	*Tsuga mertensiana*, *T. heterophylla*, and *Pinus contorta*	*Sphagnum spp.*, *Ledum* sp., *Empetrum* sp., *Kalmia* sp., *Carex pluriflora*, *Carex* spp., *Rubus chamaemorus*, *Vaccinium vitis-idaea*, and *Carex livida*	Neiland (1971) as reported in Battan and Murray (1982)
Pocosin; coastal North Carolina	*Pinus serotina*, *Cyrilla racemosa*, *Zenobia pulverulenta*, *Gordonia lasianthus*, and/or *Lyonia lucida*	*Clethra alnifolia*, *Kalmia angustifolia*, *Ilex glabra*, and *Chamaedaphne calyculata*	Christensen *et al.* (1981)
Northern coastal raised bog; eastern Maine	*Kalmia angustifolia*	*Empetrum nigrum*, *Sphagnum flaviocomans*, *S. imbricatum*, *Icmadophila ericetorum*, *Rubus chamaemorus*, *Picea mariana*, and *Larix laricina*	Damman (1977)

Table 2. Continued.

Wetland type and location	Dominant plants	Associated vegetation	Source
Northern bog; northern Minnesota	*Kalmia polifolia*, *Andromeda glaucophylla*, *Ledum groenlandicum*, *Chamaedaphne calyculata*, and *Sphagnum* spp.	*Carex oligosperma*	Glaser (1987)
Rich fen; northern Minnesota	*Betula pumila*, *Andromeda glaucophylla*, *Vaccinium oxycoccus*, and *Chamaedaphne calcyculata*	*Potentilla fruticosa* and *Carex cephalantha*	*Glaser* (1987)
Riparian sandbar thicket; Gila and San Francisco Rivers, New Mexico	*Populus fremontii* and *Salix gooddingii*	*Salix exigua*, *Baccharis glutinosa*, *Salsola kali*, *Conyza canadensis*, *Ambrosia artemisifolia*, *Sporobolus* spp., and others	Dick-Peddie *et al.* (1987)
Blueberry thicket; Rhode Island	*Vaccinium corymbosum*	*Ilex verticillata*, *Rhododendron viscosum*, *Acer rubrum*, *Eleocharis* sp., *Sphagnum* spp., *Carex stricta*, *Aronia* sp., *Amelanchier* sp., *Nyssa sylvatica*, *Pinus strobus*, *Osmunda cinnamomea*, *Maianthemum canadense*, *Iris versicolor*, *Betula populifolia*, *Kalmia angustifolia*, and *Spiraea latifolia*	Tiner (1989)
Meadowsweet thicket; western Maryland	*Spiraea alba*	*Calamagrostis canadensis*, *Carex* spp., *Scirpus cyperinus*, *Alnus* sp., and *Hypericum densiflorum*	Tiner (1988)
Shrub bog; southern part glaciated northeastern U.S.	*Sphagnum centrale* and *Chamaedaphne calyculata*	*Sphagnum fallax*, *S. fimbriatum*, *Carex stricta*, *Spiraea* spp., *Vaccinium corymbosum*, and *Rhododendron viscosum*	Damman and French (1987)

Table 2. Continued.

Wetland type and location	Dominant plants	Associated vegetation	Source
Riparian shrub wetland; Rio Arriba County, New Mexico	*Salix exigua*	*Populus wislizenii*, *Elaeagnus angustifolia*, *Conyza canadensis*, *Juncus* spp., *Apocynum cannabinum*, *Agrostis stolonifera*, *Elymus canadensis*, *Muhlenbergia asperifolia*, *Bromus japonicus*, *Sporobolus contractus*, and others	Dick-Peddie *et al.* (1984)

spp.), northern white cedar (*Thuja occidentalis*), *Picea mariana*, and *Larix laricina*. Bald cypress (*Taxodium distichum*), water tupelo (*Nyssa aquatica*), *Acer rubrum*, black gum (*Nyssa sylvatica*), Atlantic white cedar (*Chamaecyparis thyoides*), overcup oak (*Quercus lyrata*), sweet gum (*Liquidambar styraciflua*), and black willow (*Salix nigra*) are common in southern wet swamps. In the northwestern U.S., western hemlock (*Tsuga heterophylla*), red alder (*Alnus rubra*), and *Salix* spp. are important species. Swamps that flood only briefly during the growing season are characterized by silver maple (*Acer saccharinum*), pin oak (*Quercus palustris*), and sycamore (*Platanus occidentalis*) in northern areas and by *Liquidambar styraciflua*, loblolly pine (*Pinus taeda*), slash pine (*Pinus elliotti*), tulip poplar (*Liriodendron tulipifera*), beech (*Fagus grandifolia*), *Platanus occidentalis*, water hickory (*Carya aquatica*), pignut hickory (*C. glabra*), and oaks (e.g. *Quercus nigra*, *Q. laurifolia*, and *Q. phellos*) in the southern regions. Riparian wetlands along western streams are dominated by sugarberry (*Celtis laevigata*), *Liquidambar styraciflua*, willow oak (*Quercus phellos*), water oak (*Q. nigra*), overcup oak (*Q. lyrata*), *Carya aquatica*, Fremont's cottonwood (*Populus fremontii*), box elder (*Acer negundo*), *Salix* spp., red ash (*Fraxinus pennsylvanica*), and elms (*Ulmus* spp.). Major forested wetland species in Alaska are black spruce (*Picea mariana*), larch (tamarack) (*Larix laricina*), *Pinus contorta*, and *Abies balsamea*. Regional differences in composition of forested wetlands are illustrated by examples in Table 3. Major reports on forested wetlands include the following: Brabander *et al.* (1985), Brinson (1977), Brinson *et al.* (1981), Clark and Benforado (1981), Cohen *et al.* (1984), Conner and Day (1976, 1987), Crum (1988), Curtis (1959), Dabel and Day (1977), Damman and French (1987), Dick-Peddie *et al.* (1987), Drew and Schomer (1894), Duever *et al.* (1984), Erickson and Leslie (1988), Ewel and Odum (1984), Faber *et al.* (1989), Hall and Penfound (1943), Heinselman (1970), Hook and Lea

(1989), Jahn and Anderson (1986), Johnson *et al.* (1985), Kearney (1901), Laderman (1987, 1989), Laessle (1942), Larsen (1982), McCormick and Somes (1982), Metzler and Damman (1985), Monk (1966), Musselman *et al.* (1977), Penfound (1952), Rice (1965), Richardson (1981a), Schlesinger (1978), Schomer and Drew (1982), Shelford (1954), Tiner (1985a, 1985b, 1988, 1989), Veneman and Tiner (1989), Wharton *et al.* (1982), Wharton *et al.* (1976), Wilkinson *et al.* (1987), Windell *et al.* (1986), Wright and Wright (1932).

Current status of U.S. wetlands

Wetlands exist in every state in the U.S. but their abundance varies due to climate, soils, geology, land use, and other regional differences. Figure 8 shows the estimated extent of wetlands within each of the 50 states. Alaska, Florida, and Louisiana contain the most wetland area (listed in decreasing order). Other states with considerable wetland area include Minnesota, Texas, North Carolina, Michigan, Wisconsin, Georgia, and Maine (Tiner 1984, Dahl 1990). Smaller states like Delaware and New Jersey are also well represented by wetlands. Table 4 presents wetland area data for each state.

In the mid-1970s, an estimated 40 million ha of wetlands existed in the conterminous United States (Frayer *et al.* 1983). This amounts to an area equal to the size of California. Only five percent of the land surface of the lower 48 states contains wetland. Alaska and Hawaii, Puerto Rico or other U.S. territories are not included in these figures. Estimates of Alaska's wetland resource vary, but more than 68 million ha exist (Dahl 1990). Hawaii has approximately 20,972 ha of wetlands.

The abundance of major wetland types in the conterminous U.S. is shown in Table 5. In the mid-1970s, 37.9 million ha of palustrine wetlands were present, with over half of this acreage being forested wetland and about a third being emergent wetland. By contrast, only 2.1 million ha of estuarine wetlands existed by the mid-1970s, with nearly 75% being emergent wetlands and 10% either forested or scrub-shrub wetlands (mainly mangrove swamps). This amounts to an area representing only 0.3% of the land surface of all states except Alaska and Hawaii.

The distribution of palustrine wetland types by land-surface form in the conterminous U.S. is presented in Fig. 9 (a)–(d), by area in Table 6, and also by percentage in Table 7. The states of the Upper Midwest (Illinois, Indiana, Iowa, Minnesota, Ohio, Wisconsin, Michigan) have over 20% of the palustrine wetlands, while the combined Atlantic Coastal Flats (Delaware, Georgia, Maryland, New Jersey, New York, North Carolina, South Carolina, Virginia) and Gulf-Florida Coastal Flats (Alabama, Louisiana, Mississippi,

Table 3. Examples of forested wetland plant communities in the United States.

Wetland type and location	Dominant plants	Associated vegetation	Source
Red maple/white pine forested wetland; Rhode Island	*Acer rubrum* and *Pinus strobus*	*Betula alleghenesis*, *Quercus alba*, *Alnus* sp., *Vaccinium corymbosum*, *Ilex verticillata*, *Clethra alnifolia*, *Kalmia angustifolia*, *Viburnum recognitum*, *Osmunda cinnamomea*, *Aster* sp., *Sphagnum* spp., and *Vitis* sp.	Tiner (1989)
Coastal plain swamp; eastern Maryland	*Acer rubrum*, *Liquidambar styraciflua*, and *Fraxinus pennsylvanica*	*Magnolia virginiana*, *Vaccinium corymbosum*, *Smilax rotundifolia*, *Rhododendron viscosum*, and *Symplocarpus foetidus*	Tiner (1988)
Forested bog; northeastern Pennsylvania	*Larix laricina* and *Picea mariana*	*Acer rubrum*, *Vaccinium corymbosum*, *Ilex verticillata*, *Viburnum cassinoides*, *Vaccinium* sp., *Chamaedaphne calyculata*, *Ledum groenlandicum*, *Andromeda glaucophylla*, *Kalmia polifolia*, *Vaccinium oxycoccus*, *Carex trisperma*, *Osmunda cinnamomea*, *Sphagnum* spp., and others	Brooks *et al.* (1987)
Red maple swamp; northern New Jersey	*Acer rubrum*	*Ulmus rubra*, *Fraxinus americana*, *Quercus bicolor*, *Lindera benzoin*, *Sambucus canadensis*, *Rosa multiflora*, *Prunus pensylvanica*, *Ilex verticillata*, *Cornus amomum*, *Viburnum dentatum*, *Impatiens capensis*, *Geum* sp., *Solanum dulcamara*, *Carex stricta*, *Rumex* sp., *Aster novi-belgii*, *Eleocharis* sp., *Epilobium* sp., *Polygonum sagittatum*, *P. arifolium*, *Leersia oryzoides*, *Bidens* spp., *Arisaema triphyllum*, *Symplocarpus foetidus*, *Lysimachia ciliata*, *Toxicodendron radicans*, and *Parthenocissus quinquefolia*	Tiner (1985a)
Pitch pine lowland; southern New Jersey	*Pinus rigida*	*Sassafras albidum*, *Betula populifolia*, *Acer rubrum*, *Vaccinium corymbosum*, *Clethra alnifolia*, *Woodwardia virginica*, and *Smilax rotundifolia*	Tiner (1985a)

Loblolly pine forested wetland; Delaware	*Pinus taeda*	*Liquidambar styraciflua*, *Acer rubrum*, *Toxicodendron radicans*, *Vaccinium corymbosum*, *Ilex glabra*, *Magnolia virginiana*, *Parthenocissus quinquefolia*, *Smilax rotundifolia*, *Ilex opaca*, *Carpinus caroliniana*, *Nyssa sylvatica*, and *Fagus grandifolia*	Tiner (1985b)
Bottomland hardwood forest; eastern Texas	*Fraxinus pennsylvanica Ulmus crassifolia*, *Celtis laevigata*, *Quercus phellos*, *Q. nigra*, *Q. lyrata*, *Ulmus americana*, *Liquidambar stryaciflua*, and *Carya aquatica*	*Morus rubra*, *Carpinus caroliniana*, *Crataegus* spp., *Diospyros virginiana*, *Ilex opaca*, *I. decidua*, *Cornus drummondii*, *C. foemina*, *Sebastiania fruticosa*, *Halesia diptera*, *Ilex vomitoria*, *Callicarpa americana Toxicodendron radicans*, *Smilax rotundifolia*, *Berchemia scandens*, and *Vitis rotundifolia*	Wilkinson et al. (1987)
Bottomland hardwood forest; Reelfoot Lake area of Tennessee	*Salix nigra* and *Populus deltoides*	*Campsis radicans*, *Toxicodendron radicans*, *Vitis* sp., *Ampelopsis arborea*, *Ampelamus albidus*, *Brunnichia cirrhosa*, *Ipomoea lacunosa*, *Sambucus canadensis*, *Cornus drummondii*, *Echinochloa* sp., *Vernonia* sp., and *Rubus* sp.	Shelford (1954)
Bottomland hardwood forest; Louisiana	*Acer rubrum var. drummondii* and *Nyssa aquatica*	*Acer negundo*, *Populus heterophylla*, *Taxodium distichum*, *Cornus drummondii*, *Salix nigra*, *Ulmus americana*, *Carya ovata*, *Fraxinus tomentosa*, *Quercus nigra*, *Celtis laevigata*, *Diospyros virginiana*, *Ilex decidua*, *Carya cordiformis*, *Q. shumardii*, *Liquidambar styraciflua*, *Forestiera acuminata*, *Q. nuttallii*, *Persea palustris*, *Styra americana*, *Q. laurifolia*, *Sambucus canadensis*, *Toxicodendron radicans*, *Gelsemium sempervirens*, *Smilax* spp., *Ampelopsis* spp., and *Parthenocissus quinquefolia*	Conner and Day (1976)
Forested wetland; northern Minnesota	*Thuja occidentalis*, *Alnus rugosa*, *Fraxinus nigra*, *Larix laricina*, and *Picea mariana*	*Sphagnum* spp., *Gaultheria hispidula*, *Vaccinium oxycoccus*, *Ledum groenlandicum*, *Trientalis borealis*, *Smilacina trifolia*, *Calamagrostis canadensis*, *Impatiens biflora*, *Coptis trifolia*, *Mitchella nuda*, *Linnaea borealis*, *Cornus canadensis*, *Rubus pubescens*, and *Carex trisperma*	Heinselman (1970)

Table 3. Continued.

Wetland type and location	Dominant plants	Associated vegetation	Source
Cypress swamp; Okefenokee Swamp, Georgia	*Taxodium distichum*	*Tillandsia usneoides* (epiphyte). *Lyonia lucida*, *Nyssa sylvatica var. biflora*, *Clethra alnifolia*, *Itea virginica*, *Leucothoe racemosa*, *Cyrilla racemiflora*, *Ilex cassine*, *Pieris phillyreifolia*, *Decodon verticillatus*, *Smilax walteri*, and *Eriocaulon compressum*	Schlesinger (1978)
Forested fen; northern Minnesota	*Larix laricina* and *Picea mariana*	*Carex pseudocyperus*, *Aronia melanocarpa*, *Rubus pubescens*, and *Lonicera villosa*	Glaser (1987)
Pine flatwoods; northern Florida	*Pinus palustris*	*Pinus serotina*, *Pinus australis*, *Serenoa repens*, *Ilex glabra*, *Myrica cerifera*, *Rubus betulifolius*, *Aronia arbutifolia*, and *Smilax laurifolia*	Laessle (1942)
Bayhead; northern Florida	*Gordonia lasianthus*, *Tamala pubescens*, and *Magnolia virginiana*	*Ilex glabra*, *Lyonia lucida*, *Myrica cerifera*, *Smilax laurifolia*, *Osmunda cinnamomea*, *Anchestea virginica*, and *Sphagnum* spp.	Laessle (1942)
River swamp; coastal North Carolina	*Nyssa aquatica*	*Taxodium distichum*, *Fraxinus caroliniana*, *Saururus cernuus*, *Sagittaria* sp., *Peltandra virginica*, *Smilax* sp., *Ludwigia palustris*, *Nitella flexilis*, *Hydrocotyle* sp., *Fontinalis* sp., and algae	Brinson (1977)
Black gum swamp; Dismal Swamp in Virginia	*Nyssa sylvatica*	*Acer rubrum*, *Taxodium distichum*, *Nyssa aquatica*, *Chamaecyparis thyoides*, *Fraxinus caroliniana*, *Quercus phellos*, *Pinus taeda*, *Pinus serotina*, *Ilex opaca*, *Magnolia virginiana*, *Persea borbonia*, *Liriodendron tulipifera*, *Salix* sp., *Fagus grandifolia*, and *Ulmus* sp.	Whitehead (1972)
Black spruce bog; northern Minnesota	*Picea mariana*	*Larix laricina*, *Kalmia polifolia*, *Andromeda glaucophylla*, *Ledum groenlandicum*, *Chamaedaphne calyculata*, *Gaultheria hispidula*, *Sphagnum* spp., *Carex trisperma*, *Vaccinium vitis-idaea*, *Smilacina trifolia*, *Pleurozium schreberi*, *Dicranum* sp., and *Polytrichum strictum*	Glaser (1987)

Cypress swamp; Corkscrew Swamp, Florida	*Taxodium distichum*	*Acer rubrum*, *Ficus aurea*, *Fraxinus caroliniana*, *Annona glabra*, *Cephalanthus occidentalis*, *Persea borbonia*, *Ilex* sp., *Myrica cerifera*, *Salix carolinana*, epiphytic ferns, orchids and bromeliads	Duever *et al.* (1984)
Atlantic white cedar swamp; Alligator River, North Carolina	*Chamaecyparis thyoides*	*Clethra alnifolia*, *Vaccinium corymbosum*, *Gaylussacia frondosa*, *Ilex coriacea*, *Ilex glabra*, *Viburnum nudum*, *Lyonia lucida*, *Lyonia ligustrina*, *Myrica heterophylla*, *Mitchella repens*, *Peltandra virginica*, *Woodwardia areolata*, and *Sphagnum* spp.	Moore and Carter (1984)
Pocosin; coastal North Carolina	*Pinus serotina*, *Taxodium distichum*, *Acer rubrum*, and *Nyssa sylvatica var. biflora*	*Cyrilla racemiflora* and *Lyonia lucida*	Christensen *et al.* (1981)
Riparian forested wetland; Rio Grande Valley, New Mexico	*Populus fremontii*	*Elaeagnus angustifolia*, *Tamarix chinensis*, *Salix gooddingii*, *S. exigua*, and others	Freehling (1982)

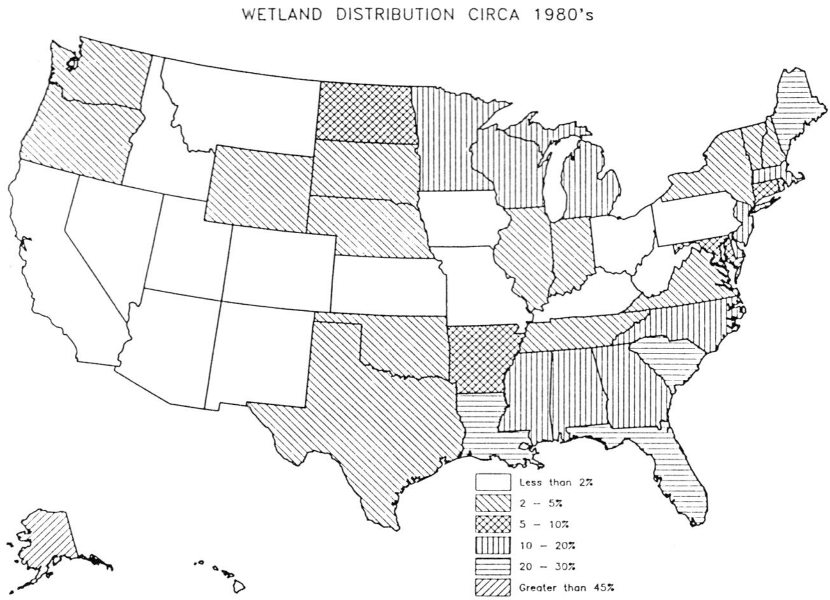

Figure 8. Distribution of wetlands in the conterminous United States circa 1980s (Dahl 1990).

Texas, Florida) totaled just over 25% of these wetlands. Forested wetlands are most abundant in the Gulf-Atlantic Rolling Plain (Alabama, Arkansas, Delaware, Florida, Georgia, Illinois, Louisiana, Maryland, Mississippi, Missouri, New Jersey, New York, North Carolina, Oklahoma, Pennsylvania, South Carolina, Tennessee, Texas, Virginia), Upper Midwest, Atlantic Coastal Flats, Gulf-Florida Coastal Flats, and Lower Mississippi Alluvial Plain (Arkansas, Florida, Kentucky, Louisiana, Mississippi, Missouri, Tennessee). Emergent wetlands predominate in the Upper Midwest and Gulf-Florida Coastal Flats, while scrub-shrub wetlands abound in the Upper Midwest. Ponds are most common in the Central Hills (Colorado, Iowa, Kansas, Missouri, Nebraska) and Plains (Colorado, Kansas, Nebraska, New Mexico, Oklahoma, South Dakota, Texas, Wyoming). Florida, Louisiana, and Minnesota had the greatest amount of palustrine wetlands compared to other states in the contiguous United States.

Of the 2.1 million ha of estuarine wetlands in the conterminous U.S., about 70% were found in the Gulf-Florida Coastal Zone and nearly 30% in the Atlantic Coastal Zone. Only a small percentage of the nation's estuarine wetlands were located along the Pacific Coast. Louisiana possessed nearly half (47%) of the estuarine emergent wetlands, followed by South Carolina (10%), Texas (9%), Florida (9%), Georgia (9%), New Jersey (4%), and North Carolina (4%).

Table 4. Wetland area of each state in hectares and percent of the land area of each state covered by wetlands (Dahl 1990).

State	Area (ha)	% of state
Alabama	1,531,903	11.5%
Alaska	68,825,910	45.3%
Arizona	242,915	0.8%
Arkansas	1,118,866	8.1%
California	183,806	0.4%
Colorado	404,858	1.5%
Connecticut	69,838	5.4%
Delaware	90,283	16.9%
Florida	4,862,687	29.5%
Georgia	2,145,020	14.1%
Hawaii	20,972	1.3%
Idaho	156,154	0.7%
Illinois	507,895	3.5%
Indiana	303,900	3.2%
Iowa	170,810	1.2%
Kansas	176,275	0.8%
Kentucky	121,457	1.2%
Louisiana	3,556,356	28.3%
Maine	2,104,939	24.5%
Maryland	178,138	6.5%
Massachusetts	238,253	11.1%
Michigan	2,260,486	15.0%
Minnesota	3,522,267	16.2%
Mississippi	1,646,559	13.3%
Missouri	260,324	1.4%
Montana	340,202	0.9%
Nebraska	771,457	3.9%
Nevada	95,688	0.3%
New Hampshire	80,972	3.4%
New Jersey	370,834	18.3%
New Mexico	195,101	0.6%
New York	414,980	3.2%
North Carolina	2,303,441	16.9%
North Dakoa	1,008,097	5.5%
Ohio	195,466	1.8%
Oklahoma	384,494	2.1%
Oregon	564,332	2.2%
Pennsylvania	202,030	1.7%
Rhode Island	26,378	8.4%
South Carolina	1,886,235	23.4%
South Dakota	720,648	3.6%
Tennessee	318,623	2.9%
Texas	3,081,948	4.4%
Utah	225,911	1.0%
Vermont	89,069	3.6%
Virginia	435,066	4.1%
Washington	379,757	2.1%
West Virginia	41,296	0.7%
Wisconsin	2,158,458	14.8%
Wyoming	506,073	2.0%

Table 5. Area (ha) of wetland types and selected combinations in the 1970's (symbols and types are described in Frayer *et al.* 1983).

Symbol	Type	Area	Percent standard error
M2	Marine intertidal	31,741	14.0
E2NV	Estuarine intertidal regularly flooded permanent – tidal	302,227	9.8
E2EM	Estuarine intertidal emergent	1,588,178	4.3
E2FO/SS	Estuarine intertidal forested/scrub-shrub	232,794	14.4
PUS	Palustrine unconsolidated shore	180,405	33.2
POW	Palustrine open water	1,778,583	7.7
PNV	Palustrine regularly flooded permanent – tidal	53,360	23.4
PFO	Palustrine forested	20,126,882	3.6
PSS	Palustrine scrub-shrub	4,295,992	12.5
PEM	Palustrine emergent	11,514,736	17.5
Combinations 1970's			
	Marine wetlands	31,741	14.0
	Estuarine wetlands	2,123,199	3.8
	Palustrine wetlands	37,949,958	6.8

Estimates of the original wetland area present at the time of European settlement vary, since the available information is scattered and largely incomplete. However, a very reliable account places this acreage at about 89 million ha for the contiguous United States (Tiner 1984, Dahl 1990). Today's wetland resource in the lower 48 states represents less than 47% of our original wetlands.

Wetland classification

Wetlands have been described in many ways for various purposes. Research studies focusing on small study areas often describe wetlands in great detail based on extensive collected data. While this approach may work well for intensive and site-specific studies, it is not suitable for inventorying wetlands at the state, regional, and national levels. There are far too many individual wetlands to reasonably undertake such an intensive effort and to do so would require expenditures of labor, time, and money well beyond the budget of any national resource agency. To determine the extent and distribution of wetlands in large geographical areas, wetlands are systematically combined into broad categories with similar ecological characteristics. In the U.S., numerous classification systems have been developed over time to inventory

Table 6. Area to the nearest thousands of hectares of palustrine wetlands for Hammond Physical Subdivisions. Reliable wetland area data not available for areas indicated with an *. The standard error of each entry is expressed as a percentage of the entry (SE%) and is given in parentheses.

Subdivision	Forested	Emergent	Scrub–shrub	Ponds
Atlantic Coastal Flats	3,579 (4.4)	292 (24.0)	389 (14.8)	27 (10.4)
Gulf and Florida Coastal Flats	2,867 (5.6)	2,006 (9.1)	397 (11.7)	78 (10.7)
Gulf-Atlantic Rolling Plain	4,941 (8.1)	371 (13.3)	235 (13.9)	344 (9.7)
Lower Mississippi Alluvial Plain	2,505 (6.0)	352 (15.9)	131 (15.0)	71 (30.2)
Eastern Highlands	1,363 (18.0)	384 (20.7)	362 (20.2)	239 (12.9)
Dakota-Minnesota Drift and Lake Bed Flats	413 (38.7)	1,152 (14.3)	143 (36.4)	81 (28.6)
Upper Midwest	3,752 (9.4)	2,486 (6.8)	1,418 (11.4)	269 (9.4)
Central Hills and Plains	*	1,313 (17.3)	158 (29.9)	546 (21.8)
Rocky Mountains	16 (23.0)	*	*	55 (43.4)
Intermontane	27 (47.7)	*	*	41 (44.7)
Pacific Mountains	148 (39.3)	148 (39.3)	77 (43.0)	23 (29.6)

Palustrine Emergent Wetlands: cover 11,514,736 hectares or 17.5% of the surface area of the 48 conterminous U.S.

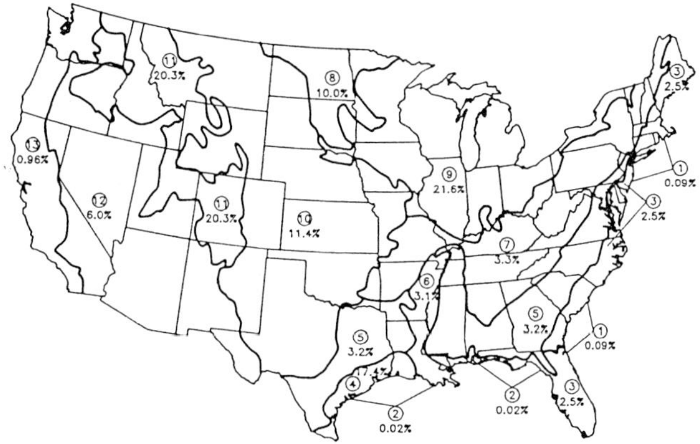

Palustrine Scrub/Shrub Wetlands: cover 4,295,992 hectares or 12.5% of the surface area of the 48 conterminous U.S.

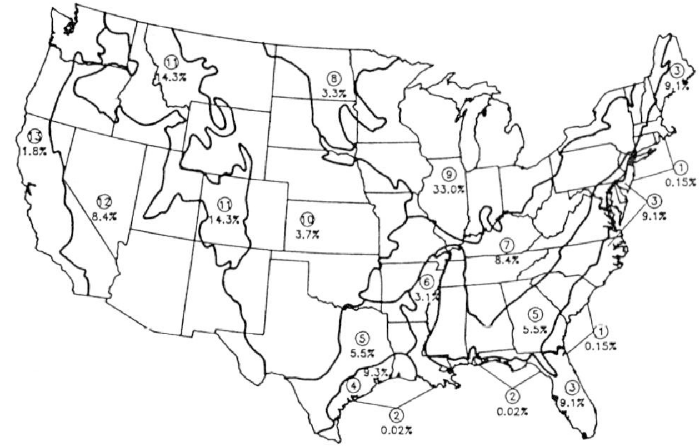

Palustrine Forested Wetlands: cover 20,126,882 hectares or 3.6% of the surface area of the 48 conterminous U.S.

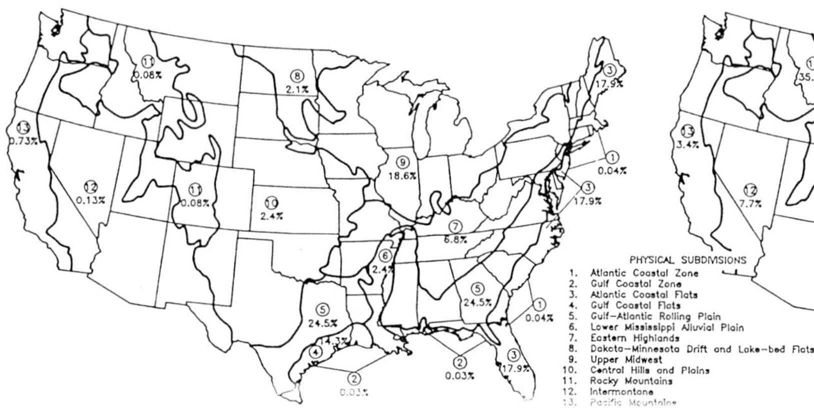

Lacustrine Wetlands: cover 1,778,583 hectares or 7.7% of the surface area of the 48 conterminous United States.

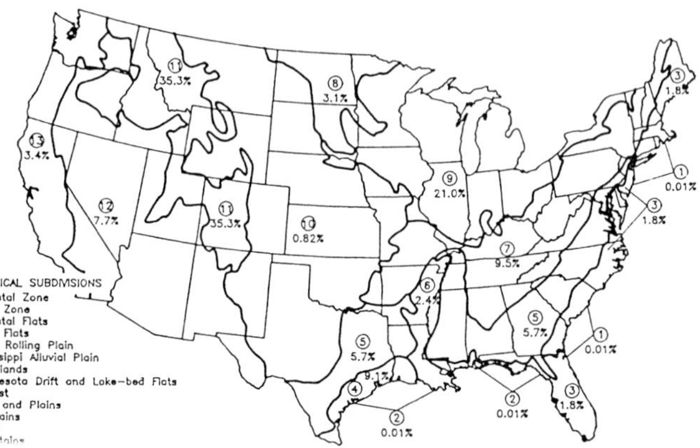

Figure 9a–d. Distribution of wetland types by land surface form

Table 7. Percentage of palustrine wetlands located in each Hammond Physical Subdivision. # indicates that the standard error is equal to or larger than the percentage estimate.

Subdivision	Forested	Emergent	Scrub–shrub	Ponds
Atlantic Coastal Flats	18	3	9	2
Gulf and Florida Coastal Flats	14	17	9	4
Gulf-Atlantic Rolling Plain	25	3	6	20
Lower Mississippi Alluvial Plain	12	3	3	4
Eastern Highlands	7	3	8	13
Dakota-Minnesota Drift and Lake Bed Flats	2	10	33	5
Upper Midwest	19	22	4	15
Central Hills and Plains	#	11	#	31
Rocky Mountains	>1	#	#	#
Intermontane	>1	#	2	2
Pacific Mountains	1	1		1

wetlands. The Federal government's first attempts at wetland classification were largely motivated by agricultural interests which sought to convert wetlands to cropland. The early notion was that wetlands in their natural state were worthless nuisances and that their primary value to the public and the nation could only be achieved through conversion to other uses, especially agriculture. The first classification systems put wetlands into a few general categories, such as river swamps (terrace and estuarine swamps), lake swamps (lake margin and quaking bogs), upland swamps (wet woods and climbing bogs), and ablation swamps or permanent swamps, wet grazing land, periodically overflowed land, and periodically swampy land as reported in Dachnowski (1920) and Wright (1907). Dachnowski (1920), with an interest in peat deposits, was dissatisfied by uses of certain common terms: "Progress in peat investigations has been severely checked by the widespread use of such terms as muck, overflowed land, swampy land, wetland and others". He proposed classifying peat deposits based on surface vegetation and recommended using such terms as marsh, fen, bog, heath, shrub, and forest.

Later wetland classifications developed from an ecological interest or a need to separate wetlands and other land cover types for regional and national planning purposes. These classification systems include the following: Akins and Jefferson (1973), Anderson *et al.* (1968, 1976), Braun (1950), Chabreck (1972), Chapman (1974, 1976a), Cowardin *et al.* (1979), Curtis (1959), Dansereau and Segadas-Vianna (1952), Dix and Smeins (1967), Eleuterius (1972, 1973), Golet and Larson (1974), Heinselman (1963, 1970), Jeglum *et al.* (1974), Kuenzler (1974), Martin *et al.* (1953), Millar (1976), Penfound (1952), Pestrong (1965), Redfield (1972), Stewart and Kantrud (1971), and Ungar (1974).

Most of these were regional systems, and only a few were nationally based. The system developed by Martin *et al.* (1953) was one of the few developed for a national survey. It was used to conduct an inventory of important waterfowl wetlands of the conterminous United States in 1954. The results of the inventory and an illustrated description of the 20 wetland types were published as U.S. Fish and Wildlife Service Circular 39 (Shaw and Fredine 1956). This Circular has been one of the most influential documents used in the continuous battle to preserve critically valuable, but rapidly diminishing wetlands (Stegman 1976). However, the shortcomings of this work are well known (Leitch 1966, Stewart and Kantrud 1971).

In attempting to simplify their classification, Martin *et al.* (1953) not only ignored ecologically critical differences, such as the distinction between fresh and mixosaline inland wetlands but also placed dissimilar habitats, such as forests of boreal *Picea mariana* and of southern *Taxodium distichum-Nyssa aquatica* in the same category, with no provisions in the system for distinguishing between them. Because of the central emphasis on waterfowl habitat, far greater attention was devoted to vegetated areas of known importance to waterfowl than to nonvegetated areas. Probably the greatest single disadvantage of the Martin *et al.* system was the inadequate definition of types, which led to inconsistencies in application across the country.

In 1974, the U.S. Fish and Wildlife Service began planning for a new national wetlands inventory and after examination of the existing classification decided to design a new one. The decision was made that the new system, unlike that described in Circular 39, should be hierarchical in structure, so that the users could select a level of detail appropriate to their needs. In addition, the new system had three primary objectives: (1) to group ecologically similar habitats so that value judgments can be made, (2) to furnish habitat units for inventory and mapping, and (3) to provide uniformity in concepts and terminology throughout the entire United States. An interim draft was prepared (Cowardin *et al.* 1976), reviewed, tested in the field, revised (Cowardin *et al.* 1977), and revised in 1979 and published as the Fish and Wildlife Service's official wetlands classification (Cowardin *et al.* 1979). The U.S. Fish and Wildlife Service developed this new wetland classification system in cooperation with other Federal state and other agencies.

The following discussion represents a simplified overview of the U.S. Fish and Wildlife Service's wetland classification system. Consequently, some of the more technical points have been omitted from this discussion. When actually classifying a wetland, the reader is advised to refer to the official classification document (Cowardin *et al.* 1979) and should not rely solely on this overview.

The wetland classification system is hierarchial in nature proceeding from general to specific, as noted in Fig. 3. In this approach, wetlands are first defined at a rather broad level the SYSTEM. The term SYSTEM represents "a complex of wetlands and deepwater habitats, that share the influence of similar hydrologic, geomorphologic, chemical, or biological factors". Five systems are defined: Marine, Estuarine, Riverine, Lacustrine and Palustrine. The Marine System generally consists of the open ocean and its associated coastline, while the Estuarine System encompasses salt and brackish marshes and waters of coastal rivers and embayments. Freshwater wetlands and deepwater habitat fall into one of the other three systems: Riverine (e.g. rivers and streams), Lacustrine, (e.g. lakes, reservoirs and large ponds) or Palustrine (e.g. marshes, bogs, swamps and small shallow ponds). Thus, at the most general level, wetlands can be defined as either Marine, Estuarine, Riverine, Lacustrine, or Palustrine.

Each system, with the exception of the Palustrine, is further subdivided into SUBSYSTEMS. The Marine and Estuarine Systems both have the same two subsystems, which are defined by tidal water levels: (1) Subtidal – continuously submerged areas and (2) Intertidal – areas alternately flooded by tides and exposed to air. Similarly, the Lacustrine System is separated into two subsystems based on water depth: (1) Littoral – wetlands extending from the lake shore to a depth of 2 meters below low water or to the extent of nonpersistent emergents [e.g. *Sagittaria* spp., *Pontederia cordata* or spatterdock (*Nuphar lutea*)] if they grow beyond that depth, and (2) Limnetic – deepwater habitats lying beyond the 2 meter depth at low water. By contrast, the Riverine System is further defined by four subsystems which represent different reaches of a flowing freshwater or lotic system: (1) Tidal – water levels subject to tidal fluctuations, (2) Lower Perennial – permanent, slow-flowing waters with a well-developed floodplain, (3) Upper Perennial – permanent, fast-flowing water with very little or no floodplain development, and (4) Intermittent – channels containing nontidal flowing water for only part of the year.

Below the subsystem, the CLASS level describes the general appearance of the wetland or deepwater habitat in terms of the dominant vegetative life form or the composition of the substrate (Table 8). Of the 11 classes, five refer to areas where vegetation covers 30% or more of the surface: Aquatic Bed, Moss-Lichen Wetland, Emergent Wetland, Scrub-Shrub Wetland, and Forested Wetland. The remaining six classes represent areas generally lacking vegetation, where the composition of the substrate and degree of flooding distinguish classes: Rocky Bottom, Unconsolidated Bottom, Reef (sedentary invertebrate colony), Streambed, Rocky Shore, and Unconsolidated Shore. Permanently flooded nonvegetated areas are classified as either Rocky Bottom or Unconsolidated Bottom, while exposed areas are typed as Streambed, Rocky Shore, or Unconsolidated Shore. Invertebrate reefs are found in both permanently flooded and exposed areas.

Each class is further divided into SUBCLASSES to better define the type of substrate in nonvegetated areas (e.g. bedrock, rubble, cobble-gravel, mud,

Table 8. Classes and subclasses of wetlands and deepwater habitats (source: Cowardin *et al.* 1979).

Class	Brief Description	Subclasses
Rock bottom	Generally permanently flooded areas with bottom substrates consisting of at least 75% stones and boulders and less than 30% vegetative cover.	Bedrock, rubble
Unconsolidated bottom	Generally permanently flooded areas with bottom substrates consisting of at least 25% particles smaller than stones and less than 30% vegetative cover.	Cobble-gravel, sand, mud, organic
Aquatic bed	Generally permanently flooded areas vegetated by plants growing principally on or below the water surface line.	Algal, aquatic moss rooted vascular, floating vascular
Reef	Ridge-like or mound-like structures formed by the colonization and growth of sedentary invertebrates.	Coral, mollusk, worm
Streambed	Channel whose bottom is completely dewatered at low water periods.	Bedrock, rubble, Cobble-gravel, sand, mud, organic, vegetated
Rocky shore	Wetlands characterized by bedrock, stones or boulders with areal coverage of 75% or more and with less than 30% coverage by vegetation.	Bedrock, rubble
Unconsolidated shore	Wetlands having unconsolidated substrates with less than 75% coverage by stone, boulders and bedrock and less than 30% vegetative cover, expect by pioneer plants.	Cobble-gravel, sand, mud, organic, vegetated
Moss-lichen wetland	Wetlands dominated by mosses or lichens where other plants have less than 30% coverage.	Moss, lichen
Emergent wetland	Wetlands dominated by erect, rooted, herbaceous hydrophytes.	Persistent, nonpersistent
Scrub–shrub wetland	Wetlands dominated by woody vegetation less than 6 m tall.	Broad-leaved deciduous, needle-leaved deciduous broad-leaved evergreen, needle-leaved evergreen, dead
Forested wetland	Wetlands dominated by woody vegetation greater than 6 m or taller.	Broad-leaved decidiuous, needle-leaved deciduous, broad-leaved evergreen, needle-leaved evergreen, dead

sand, and organic) or the type of dominant vegetation (e.g. persistent or nonpersistent emergents, moss, lichen, or broad-leaved deciduous, needle-leaved deciduous, broad-leaved evergreen, needle-leaved evergreen, and dead woody plants). Below the subclass level, DOMINANCE TYPE can be applied to specify the predominant plant or animal in the wetland community.

In order to allow better description of a given wetland or deepwater habitat in regard to hydrologic, chemical and soil characteristics and man's impacts, the classification system contains four types of specific modifiers: (1) Water Regime, (2) Water Chemistry, (3) Soil, and (4) Special. These modifiers may be applied to class and lower levels of the classification hierarchy.

WATER REGIME MODIFIERS describe flooding or soil saturation conditions and are divided into two main groups: (1) tidal and (2) nontidal. Tidal water regimes are used where water level fluctuations are largely driven by oceanic tides. Tidal regimes can be subdivided into two general categories, one for salt and brackish tidal areas and another for fresh tidal areas. This distinction is needed because of the specific importance of seasonal river overflow in fresh tidal areas. In contrast, nontidal modifiers define conditions where surface water runoff, ground-water discharge, wind effects, and/or lake seiches cause water level changes. Both tidal and nontidal water regime modifiers are presented and briefly defined in Table 9.

WATER CHEMISTRY MODIFIERS are divided into two categories which describe the water's salinity or hydrogen ion concentration (pH). Like water regimes, salinity modifiers have been further subdivided into two groups: halinity modifiers for tidal areas and salinity modifiers for nontidal areas. Estuarine and marine waters are dominated by sodium chloride, which is gradually diluted by fresh water as one moves upstream in coastal rivers. On the other hand, the salinity of inland waters is dominated by four major cations (calcium, magnesium, sodium, and potassium) and three major anions (carbonate, sulfate, and chloride). Interactions between precipitation, surface runoff, ground-water flow, evaporation, and sometimes plant evapotranspiration affect concentrations of these major ions. Table 10 shows ranges of halinity and salinity modifiers which are a modification of the Venice System (Remane and Schlieper 1971). The other set of water chemistry modifiers are pH modifiers for identifying acid (pH 5.5), circumneutral (pH 5.5–7.4), and alkaline (pH 7.4) waters. Some studies have shown a good correlation between plant distribution and pH levels in certain wetland habitats, see Table 11 (Jeglum 1971, Sjors 1950). Moreover, pH can be used to distinguish between mineral-rich and mineral-poor wetlands.

The third group of modifiers SOIL MODIFIERS are presented because the soil exerts strong influences on plant growth and reproduction as well as on the animals living in it. Two soil modifiers are given: (1) mineral and (2)

Table 9. Water regime modifiers of tidal and nontidal wetland groups (source: Cowardin *et al.* 1979).

Group	Type of water	Water regime	Definition
Tidal	Saltwater and brackish brackish areas	Subtidal	Permanently flooded by tides.
		Irregularly exposed	Exposed less often than daily by tides.
		Regularly flooded	Daily tidal flooding and exposure to air.
		Irregularly flooded	Flooded less often than daily and typically exposed to air.
	Freshwater areas	Permanently flooded-tidal	Permanently flooded by tides or exposed less often than daily by tides.
		Regularly flooded-tidal	Daily tidal flooding and exposed to air.
		Seasonally flooded-tidal	Flooded irregularly by tides and seasonally by river overflow.
		Temporarily flooded-tidal	Flooded irregularly by tides and for brief periods during growing season by river overflow.
Nontidal	Inland freshwater and saline areas	Permanently flooded	Flooded through the year in all years.
		Intermittently exposed	Flooded year-round except during extreme droughts.
		Semipermanently flooded	Flooded throughout the growing season in most years.
		Seasonally flooded	Flooded for extended periods in growing seasons, but surface water is usually absent by end of growing season.
		Saturated	Surface water is seldom present, but substrate is saturated to the surface for most of the season.
		Temporarily flooded	Flooded for only brief periods during growing season, with water table usually well below the soil surface for most of the season.
		Intermittently flooded	Substrate is usually exposed and only flooded for variable periods without detectable seasonal periodicity (not always wetland; may be upland in some situations).
		Artificially flooded	Duration and amount of flooding is controlled by means of pumps or siphons in combination with dikes or dams.

Table 10. Salinity Modifiers used in the classification system.

Coastal modifiers[a]	Inland modifiers[b]	Salinity (parts per thousand)	Approximate specific conductance (Mhos at 25°C)
Hyperhaline	Hypersaline	>40	>60,000
Euhaline	Eusaline	30.0–40	45,000–60,000
Mixohaline (Brackish)	Mixosaline[c]	0.5–30	800–45,000
Polyhaline	Polysaline	18.0–30	30,000–45,000
Mesohaline	Mesosaline	5.0–18	8,000–30,000
Oligohaline	Oigosaline	0.5–5	800–8,000
Fresh	Fresh	<0.5	<800

[a]Coastal modifiers are used in the Marine and Estuarine systems.
[b]Inland modifiers are used in the Riverine, Lacustrine, and Palustrine systems.
[c]The term Brackish should not be used for inland wetlands of deepwater habitats.

organic. In general, if a soil has 20% or more organic matter by weight in the upper 40 centimeters, it is considered an organic soil, whereas if it has less than this amount, it is a mineral soil. For specific definitions, refer to Appendix D of the Service's classification system (Cowardin *et al.* 1979) or to *Soil Taxonomy* (U.S.Department of Agriculture, Soil Conservation Service 1975).

The final set of modifiers SPECIAL MODIFIERS were established to describe the activities of man or beaver affecting wetlands and deepwater habitats. These modifiers include: excavated, impounded (i.e., to obstruct outflow of water), diked (i.e., to obstruct inflow of water), partly drained, farmed, and artificial (i.e. materials deposited by man to create or modify a wetland or deepwater habitat).

Ecological characteristics

Wetlands form in areas subjected to periodic flooding and/or to a seasonal high water table that creates saturated soil conditions. The presence of water for extended periods, either above the ground surface or in the soil, is related to an area's hydrology. Hydrology, therefore, is the primary forcing function responsible for creating and maintaining wetlands. Not only does it determine where wetlands develop and what types of plants become established, but hydrology drives other functions (e.g. nutrient cycling, primary and secondary productivity, fish and wildlife habitat, and flood storage) that make wetlands particularly valuable natural resources. The following discussion reviews wetland hydrology and other major functions. Sather and Smith (1984) present an overview of major wetland functions and values. Good *et al.* (1978), Greeson *et al.* (1979), Hook *et al.* (1988), Kusler and Brooks (1988), Majumdar *et al.* (1989), Mitsch and Gosselink (1986), and Sharitz and Gibbons (1989) are also among the useful references.

Table 11. Correlation of plant species with their associated pH levels for wetlands in northern Minnesota, USA (source: Gorham 1983).

Species	Median pH	Mean pH
Vaccintium vitis-idaea	4.0	3.9
Gaultheria hispidula	4.0	4.0
Eriophorum spissum	4.0	4.0
Carex trisperma	4.0	4.0
Carex oligosperma	4.1	4.1
Carex pauciflora	4.1	4.1
Ledum groenlandicum	4.1	4.3
Kalmia polifolia	4.1	4.3
Picea mariana	4.1	4.3
Smilacina trifolia	4.2	4.6
Chamaedaphne calyculata	4.2	4.6
Vaccinium oxycoccos	4.2	4.6
Carex paupercula	4.7	4.7
Larix laricina	4.4	4.7
Carex aquatilis	4.3	4.8
Drosera rotundifolia	4.8	4.9
Andromeda glaucophylla	4.9	5.0
Sarracenia purpurea	4.9	5.1
Juncus stygius	5.2	5.1
Carex tenuiflora	5.3	5.3
Scheuchzeria palustris	5.5	5.3
Eriophorum tenellum	5.8	5.4
Rhynchospora fusca	5.7	5 5
Carex limosa	5.8	5.6
Betula pumila	5.6	5.6
Pogonia ophioglossoides	6.1	5.8
Potentilla palustris	6.0	5.8
Carex lasiocarpa	5.9	5.8
Carex chordorrhiza	6.0	5.8
Salix pedicellaris	6.0	5.8
Equisetum fluviatile	6.2	5.9
Drosera intermedia	5.9	5.9
Rhynchospora alba	6.0	5.9
Menyanthes trifoliata	6.0	5.9

Wetland hydrology

Wetlands as the term implies are "wet" lands. They exist because the inflow of water exceeds the outflow for brief to extended periods of time during the growing season. Inland wetlands receive water from precipitation, snow melt, river overflow, surface overland flow, ground-water discharge, lake seiches and seepage from streams, lakes, ponds, and irrigation systems. Coastal wetlands receive water through tidal action, wave action, river overflow, storm surges, seiches, and ground-water inflows.

Table 11. Continued.

Species	Median pH	Mean pH
Carex exilis	6.2	5.9
Lysimachia thyrsiflora	6.0	6.0
Iris versicolor	6.0	6.0
Utricularia intermedia	6.1	6.0
Utricularia cornuta	6.0	6.0
Campanula aparinoides	6.1	6.1
Typha latifolia	6.2	6.1
Carex livida	6.2	6.1
Eleocharis compressa	6.2	6.1
Eriophorum gracile	6.3	6.2
Triadenum fraseri	6.3	6.2
Thelypteris palustris	6.2	6.2
Phragmites australis	6.2	6.2
Viola mackloskeyi	6.3	6.2
Galium labradoricum	6.3	6.3
Eriophorum chamissionis	6.4	6.3
Carex leptalea	6.3	6.3
Utricularia minor	6.3	6.3
Scirpus caespitosus	6.2	6.4
Drosera anglica	6.3	6.4
Triglochin maritima	6 3	6.4
Scirpus hudsonianus	6.4	6.4
Drosera linearis	6.6	6.4
Potentilla fruticosa	6.5	6.4
Carex interior	6.4	6.5
Thuja occidentalis	6.6	6.6
Muhlenbergia glomerata	6.4	6.6
Lobelia kalmii	6.6	6.7
Parnassia palustris	7.0	6.9

Most natural wetland functions are a result of, or closely related to, wetland hydrology. Wetland food chain, fish and wildlife habitat value, nutrient cycling, socio-economic values, heritage and even aesthetics values are tied to the source, velocity, frequency, timing, and quantity of water. Understanding wetland hydrology is the key to understanding wetland functions, but collecting hydrologic information is difficult. Since a given wetland's hydrology is reflected by its plants and soils, scientists study wetland plants and soils to obtain insights to a wetland's hydrology. Wetland plants often reflect the shorter term hydrologic regime, while soils tend to reflect the longer term.

Water budget

Water budgets or mass balance equations can be used for basic hydrolic analysis of wetlands (Carter *et al.* 1978, Heath 1975, Lichtler and Walker

1974, Rykiel 1977, Shjeflo 1968, Winner and Simmons 1977). The major components of a wetland water budget are expressed in the following equation:

$$P + SWI + GWI = ET + SWO + GWO + \Delta S$$

Where P is precipitation, SWI is surface water inflow (including overland runoff and tidal inundation), GWI is ground water inflow, ET is evapotranspiration (loss of water from the soil both by evaporation and as transpiration of the plants growing thereon), SWO is surface water outflow (discharge through aquifers and seepage), GWO is groundwater outflow, and ΔS is the change in storage. Monitored over long periods of time, inflow equals outflow. The close association between ground and surface water requires an understanding of both subjects to understand a given wetland's hydrology. Alterations to any part of the system will affect other components.

Determination of a wetland's water budget is not a simple matter (Carter 1989). All the inflows and outflows must be carefully measured. The hardest component of surface water inflow to measure is surface or overland sheet flow. It is affected by microtopography and vegetation existence and density. The ground water component of a wetland is also difficult to measure, even after a hydrologist has determined the configuration of the water table and its hydraulic potential. The hydrologist must understand the nature and properties of the soils in and surrounding a wetland, as well as in the underlying deposits to learn the path of the water movement between the aquifer and the wetland. Organic soils (Histosols) are common wetlands soils and our knowledge of the physical properties and the hydrology of Histosols is poor compared with that of mineral soils (Carter *et al.* 1978).

Evapotranspiration is affected by solar radiation, wind speed and turbulence, relative humidity, and available soil moisture. Bay (1966) cites several studies that indicate the evapotranspiration losses from bogs may also be affected by plant species, vegetation cover, climate, vegetation density. Estimates of evapotranspiration from different types of wetland vegetation during the growing season range from as low as 0.54 to as high as 5.3 times that of pan evaporation (Benton *et al.* 1978, Bonde *et al.* 1961, Heimburg 1976, Hughes and McDonald 1966, Penman 1963, Rogers and Davis 1972, Timmer and Weldon 1967, Wang and Heimburg 1976).

Wetland formation

Historical events and present hydrologic conditions have combined to create and maintain a great diversity of wetlands in the United States. Human activities have also exerted significant influence on wetland formation and hydrology. The following subsections address general differences between inland and coastal wetlands in their formation and hydrology.

The wetlands of the United States were created by five principal processes: (1) glaciers, (2) inundation of coastal lowlands, (3) accretion and erosion by rivers, (4) activities of man and beaver, and (5) special processes. The principal wetland creation process in the northern part of the United States including Alaska, was glaciation. The last major glaciation in the contiguous United States occurred more than 10–20,000 years ago. When the glaciers retreated, buried blocks of ice left by the receding glaciers created depressions in glacial till, outwash, and moraines. As this ice melted, wetlands, ponds, and lakes were formed where the depressions intersected the groundwater table or a confining layer of silt, clays, organic matter, dense basal till or a combination of all these formed to seal the bottom. The quality of the seal is related to the level of water in the soil profile, the hydrolic gradient, and the thickness of the confining layer. Glaciers also dammed river and stream outlets, changing drainage patterns and creating glacial lakes. When the ice retreated and the lakes partially drained, low-lying level areas with clay and silt bottoms conducive for wetland development remained. Glaciers also scooped out and scoured both large and small depressions in bedrock. Freeze-thaw basins formed just south of the glaciers, as can be observed today on the Arctic Plain. Wetlands have formed in these basins and have persisted for thousands of years.

Half the nation's wetlands are found in the southeastern United States, where historic sea level rise has inundated coastal lowlands forming extensive estuarine wetlands behind barrier islands and along tidal rivers. The land forms of this area are conducive to wetland formation. They include numerous drowned river valleys, a relatively flat coastal plain with elevations just above sea level, and extensive coastal flats. In fact the continuing rise of sea level is permanently inundating many coastal wetlands, especially in Louisiana. On the eastern shore of Chesapeake Bay, low-lying pine forests are slowly being converted to estuarine wetlands by rising sea level (Tiner 1984).

Wetlands are also formed by accretion and erosion of the broad floodplains of many rivers throughout the country. Examples include the Atchafalaya, Mississippi, Pearl, Tombigbee, Apalachicola, Altamaha, Savannah, Santee, and Roanoke in the Southeast and the Connecticut River in the northeastern United States. The annual precipitation of 100 to 150 centimeters in this region is especially conducive for maintaining wetlands. Seasonal river overflow creates and maintains these floodplain wetlands.

Many wetlands and deepwater habitats have been artificially created, while natural ones have been modified by man or beaver. Between the mid-1950s and mid-1970s, the number of small open water ponds created in the conterminous 48 states by man and beaver nearly doubled from 0.9 to 1.8

million ha. The acreage of deepwater habitats also increased by 600 thousand ha from 28.9 to 29.5 million ha. Most of these small ponds, lakes, and reservoirs were created on former uplands, but many involved wetland conversion (Tiner 1984).

Examples of wetlands created by special processes include: (1) the Sandhills of Nebraska where wind action scooped out depressions in the sandy plains, (2) the Florida Everglades where ground-water and surface water flow associated with a very shallow limestone bedrock and coastal inundation combined to form vast wetland expanses, (3) karst formations where "sink holes" are formed by percolating water dissolving holes in limestone formations (e.g. Kentucky, Indiana, Tennessee, North Carolina, South Carolina, Georgia, Texas, New Mexico, Florida, and other states), (4) earthquake formed-wetlands include Reelfoot Lake (Tennessee) and San Francisco Bay (California), and (5) permafrost which is a major factor in the formation of Alaskan tundra wetlands (Kusler 1983).

Hydrolic functions

Storm flows or floods are generated by snow melt and precipitation. The severity of the flood depends on the size of the drainage basin, antecedent-soil moisture, the quantity of water, infiltration rate of the soils in the basin, and types of land use (Anderson-Nichols and Co., Inc. 1971, Boelter and Verry 1977, Novitzki 1979, Verry and Boelter 1979). Wetlands reduce flood peaks and storm flows by temporarily storing surface water. The location and distribution of wetlands influence the flow distribution and the timing of the peak flow. In general, wetlands higher in the watershed have a greater effect on desynchronizing flood flows. Wetlands with no outlets or restricted outlets store more flood waters and have a greater impact on attenuating peak flood flows than wetlands with open outlets. Desynchronization potential increases directly with the number and size of obstructions (Novitzki 1979). Generally, the higher the percentage of wetland acreage in the watershed the lower the flood peaks. The storage capacity of wetland soils is related to the water table and whether the soil is organic or mineral. The lower the water table in combination with higher organic matter, the higher potential storage capacity is available.

Other factors that affect the flood storage and storm flow modification of wetlands are the steepness of the topography in the watershed, surficial and bedrock geology, whether the wetlands are part of the surface water drainage system, the coefficient of resistance and roughness coefficient of the wetland vegetation, and the depth of the soils. Kusler and Brooks (1988) provide a collection of recent investigations of wetland hydrology from a national symposium.

Flood storage

In their natural condition, most wetlands serve to temporarily store flood waters, thereby protecting downstream property owners from flood damage. This function becomes increasingly important in urban areas, where development has increased the rate and volume of surface water runoff and the potential for flood damage.

Floods are among the most significant natural disasters in the United States in terms of loss of life and property. Between 1960 and 1982, flooding caused an average of 151 fatalities annually. All states have flood-prone rural land. The Federal Emergency Management Agency (FEMA) has determined that about 20,000 out of 34,000 communities in the United States have some flood hazard areas. Two-thirds of the nation's flood damage occurs in rural areas. Flooding in upstream watersheds accounts for half of the flood damage, and over 80% of upstream damage occurs in rural areas. A little over 24 million hectares (14.4%) of the nation's cropland are in flood-prone areas and about 31% of all non-federal rural flood-prone land is used as cropland. Rural flood damages are expected to increase because existing cropland is being cropped more intensively, requiring greater investments. Flood-prone lands are the most productive, supplying as much as 20% of the nation's crop production. These lands are expected to stay in production because of their high productivity, and more flood-prone areas are likely to be farmed as some of the highly erodible soils are put into the conservation reserve. Upstream flood damages, apart from deaths and other social consequences, average more than $5 billion a year. By the year 2030, damages are projected to reach $9 billion or more. Damages in upstream watersheds account for half of all flood damages, and 80% of upstream damages occur in rural areas (U.S. Department of Agriculture 1989). Protection of wetlands is, therefore, an important means of minimizing flood damages in the future.

The U.S. Army Corps of Engineers has recognized the value of wetlands for flood storage in Massachusetts. In the early 1970s, the Corps of Engineers considered various alternatives to providing flood protection in the lower Charles River watershed near Boston, Massachusetts, including: (1) 68 million cubic meter reservoir, (2) extensive walls and dikes, and (3) perpetual protection of 3,440 hectares of wetlands (U.S. Army Corps of Engineers 1976). If 40% of the Charles River wetlands were destroyed, flood damages would increase by at least $3 million (U.S. dollars) annually. Loss of all basin wetlands would cause an average annual flood damage cost of $17 million (Thibodeau and Ostro 1981). The Corps of Engineers concluded that wetlands protection "Natural Valley Storage" was the least-cost solution to flooding problems. In 1983, they completed the acquisition of the wetlands for the flood control plan.

This flood storage value of wetlands has also been reported for other areas. In Pennsylvania, the 1955 floods washed out all but two bridges along one stream; the remaining bridges lay immediately downstream of a cranberry bog (Goodwin and Niering 1975). A Wisconsin study projected that floods may be lowered as much as 80% in watersheds with many wetlands compared with similar basins with little or no wetlands (Novitzki 1978). Prairie pothole wetlands in the Devils Lake Basin in North Dakota store nearly 75% of the total runoff that occurs in the watershed (Ludden *et al.* 1983).

Recent studies at National Wildlife Refuges in North Dakota and Minnesota have demonstrated the role of wetlands in reducing streamflow. Inflow into the Agassiz National Wildlife Refuge and the Thief River Wildlife Management Area was 142 m^3 sec^{-1} while outflow was only 39.6 cubic m^3 sec^{-1}. Storage capacity of those areas reduced flood peaks at Crookston, Minnesota, by 45 cm and at Grand Forks, North Dakota, by 15 cm (Bernot 1979). Drainage of wetlands was the most important land-use practice causing flood problems in a North Dakota watershed (Malcolm 1978, Malcolm 1979). Even northern peat bogs reduce peak rates of stream flow from snow melt and heavy summer rains (Verry and Boelter 1979). Destruction of wetlands through floodplain development and drainage has been partly responsible for recent major flood disasters throughout the country.

Recharge

There is considerable controversy about the role of wetlands in recharge in the United States (Carter 1989). Although most wetlands are discharge areas, some water may be recharged from depression wetlands that collected overland flow, such as cypress domes and some prairie potholes, but recharge may be less than from upland depressions (Odum *et al.* 1975, Stewart and Kantrud 1972, Winter and Carr 1980). Some wetlands may be alternately discharge or recharge areas (Lichtler *et al.* 1976). In general, the more temporary flooding of the wetland, the more likely the ground-water recharge will take place. Some recharge occurs on river flood plains (Mundorf 1950). Recharge may also take place through the bottoms of western streams, especially intermittently and temporarily flooded streams (Leopold and Miller 1961).

Water supply

In some parts of the country, a large number of wetlands may be underlain by aquifers with a high potential for water supply (Baker 1960, Heeley 1973, Heeley and Motts 1976, Motts and Heeley 1973). Heeley and Motts (1976) have shown that Massachusetts wetlands generally associated with stratified drift, alluvium, and lake bottom deposits are all considered favorable areas for ground water exploration and development.

Shoreline erosion control

Located at the interface of uplands with watercourses and waterbodies wetlands help protect upland soils from erosion. Wetland vegetation can reduce shoreline erosion in several ways, including: (1) increasing durability of the sediment through binding with its roots, (2) dampening waves through friction, and (3) reducing current velocity through friction and increased surface area (Dean 1979). This process also helps reduce turbidity and thereby improves water quality. Wetland vegetation has some inherent limitations: it can be undermined by waves and currents, damaged by ice and debris, covered by debris and silt during floods, killed by high salinity, ripped out by abnormally high flows, or destroyed by artificial water level fluctuations. Many wetland plants require relatively calm and sheltered waters to become established (Kadlec and Wentz 1974). Once established, however, they are effective in erosion control (Garbisch 1977). Temporary or permanent structures can be used to provide the shelter to help vegetation become established. Artificially-created wetlands are being constructed along many coastal areas to help reduce shoreline erosion. Obviously, trees are good stabilizers of river banks. Their roots bind the soil making it more resistant to erosion, while their trunks and branches slow the flow of flooding waters and dampen wave heights. The banks of some rivers have not been eroded for 100 to 200 years due to the presence of trees (Leopold and Wolman 1957, Sigafoos 1964, Wolman and Leopold 1957). Among the grass or grass-like plants, *Scirpus* spp. and *Phragmites* spp. have been regarded as the best at withstanding wave and current action (Kadlec and Wentz 1974, Seibert 1968). While most wetland plants need calm or sheltered waters for establishment, they will effectively control erosion once established (Garbisch 1977, Kadlec and Wentz 1974).

Wetland vegetation has been successfully planted to reduce erosion along United States waterways. *Salix* spp., *Alnus* spp., *Fraxinus* spp., cottonwoods and poplars (*Populus* spp.), maples (*Acer* spp.), and *Ulmus* spp. are particularly good stabilizers (Allen 1979). Successful emergent plants include *Phalaris arundinacea*, reed (*Phragmites australis*), *Typha* spp., and *Scirpus* spp. in freshwater areas (Hoffman 1977). Along the Atlantic and Gulf coasts, *Spartina alterniflora* and mangroves (*Avicennia germinans*, *Laguncularia racemosa*, and *Rhizophora mangle*) have been quite effective at controlling erosion (Lewis and Thomas 1974, Woodhouse *et al.* 1976).

Water quality

Water quality in natural wetlands reflects the quality of the inflow water and the interaction of the water with soils and vegetation. Salinity, pH, mineral and nutrient content, and pollutants in the inflow water all affect the wetlands. Wetlands affect water quality through element cycling, sediment deposition, ion and molecule absorption, and temperature modification (Carter *et al.* 1978, Hemond and Benoit 1988, Nixon and Lee 1986). High nutrient content of inflow waters can stimulate rapid growth and high productivity in aquatic macrophytes.

Wetlands with no outlets retain virtually all the inorganic sediments that enter them (Brinson 1988). Novitzki (1979) indicates that sediment yields in streams in north-central Wisconsin may be 90% lower in basins containing wetlands than in basins lacking wetlands. Dissolved materials may be retained in wetlands and water quality may vary seasonally or from year to year. Cycling of nutrients in wetlands includes breakdown or storage of plant material. Decomposition releases soluble materials and suspended breakdown products to the discharge water. CO_2, H_2S, CH_4, and other products of decomposition are released to the atmosphere. In coastal areas, wetlands dampen the effects of tidal storm surge. Large flat wetland areas dissipate wave energy as does the vegetation. They also prevent erosion and store water backed up by storm tides, while helping reduce or prevent salt water intrusion.

One topic that is still under examination is the effect of wetland evapotranspiration on climate moderation. The microclimate effect of wetlands on corn and wheat crops planted adjacent to small wetlands and the larger climate effect caused by vast wetlands such as Florida's Everglades is still being debated by the scientific community.

Nutrient cycling

Wetlands help maintain good water quality or improve degraded waters in several ways: (1) nutrient removal and retention, (2) processing chemical and organic wastes, and (3) reducing sediment loads of waters. Wetlands are particularly good water filters because of their location between land and water. Thus, they can both intercept runoff from land before it reaches the water and help filter nutrients, wastes and sediment from flooding waters. Clean waters are important to man as well as to aquatic life and other wildlife.

First, wetlands remove nutrients (especially nitrogen and phosphorus), particulates, and total biological oxygen demand from flooding waters for plant growth and help prevent eutrophication or over-enrichment of natural waters (Nixon and Lee 1986). It is, however, possible to overload a wetland and thereby reduce its ability to perform this function. Every wetland has a limited capacity to absorb nutrients and individual wetlands differ in their ability to do so.

Wetlands have been shown to be excellent removers of waste products from water. In fact, certain wetland plants are so efficient at this task that artificial wastewater treatment systems are using these plants. For example, the Max Planck Institute of Germany has a patent to create such systems, where a bulrush (*Scirpus lacustris*) is the primary waste removal agent (Carter *et al.* 1978). Numerous scientists have proposed that wetlands be used for domestic waste recycling, and some wetlands are already used for this purpose (Carter *et al.* 1978, Kadlec 1979, Sloey *et al.* 1978). The Brillion Marsh in Wisconsin has received domestic sewage since 1923. This cattail marsh on the average removed 80% of biological oxygen demand, 86% of coliform bacteria, 51% of nitrates, 40% of chemical oxygen demand, 44% of turbidity, 29% suspended solids and 13% of total phosphorus (Carter *et al.* 1978). Godfrey *et al.* (1985) discuss ecological considerations of using wetlands for treatment of municipal wastewater. Cypress domes in Florida are being used by some communities for wastewater disposal and using cypress may be feasible for about 35% of the wastewater treatment systems in Florida (Ewel and Odum 1984). Wetlands are being created in various areas to handle specific water quality problems. For example, cattail (*Typha latifolia*) marshes are being built in mining areas in Pennsylvania and Maryland to process acid mine drainage from both deep and surface coal mines.

One of the best examples of the importance of wetlands for water quality improvement is Tinicum Marsh (Grant and Patrick 1970). Tinicum Marsh is a 207 ha freshwater tidal marsh lying just south of Philadelphia, Pennsylvania. Three sewage treatment plants discharge treated sewage into marsh waters. On a daily basis, it was shown that this marsh removes 7,823 kg of biological oxygen demand, 4,978 kg of phosphorous, 4,369 kg of ammonia, and 62.6 kg of nitrate. In addition, Tinicum Marsh adds 20,320 kg of oxygen to the water each day.

A cooperative U.S. Environmental Protection Agency/U.S. Fish and Wildlife Service report entitled *The Ecological Impacts of Wastewater on Wetlands* (1984) is an annotated bibliography of over 1,000 articles on the subject (U.S. Environmental Protection Agency and U.S. Fish and Wildlife Service 1984). The role estuarine wetlands (salt marshes) in cycling nitrogen and phosphorus has received much attention (see Whitney *et al.* 1981 for a review). Nutrient dynamics in freshwater wetlands have been discussed by Good *et al.* (1978). Kadlec and Kadlec (1979) reviewed the relationship between wetlands and water quality – a fundamental societal concern. Wetlands have been proven to be important in reducing nutrient and heavy metal loading from urban runoff in the upper Delaware River (Grant and Patrick 1970, Simpson *et al.* 1983, Whigham and Simpson 1976). There is a seasonal pattern in nutrient cycling. During the growing season, wetland plants take up nutrients from the water and sediments and gradually store a portion of these nutrient in their underground root system or in wood (trees and shrubs). When the plants die back or woody plants shed their leaves at the end of the growing season, some nutrients are gradually released back into the water as leaves and herbaceous plant remains decompose. Consequently, there is a general net export of nutrients in fall and early spring. Several studies documenting this seasonality include: Klopatek (1978), Lee *et al.* (1975), Prentki *et al.* (1978), Simpson *et al.* (1983), Valiela *et al.* (1978), and Woodwell and Whitney (1977). This seasonal pattern has probably fueled much of the debate over whether wetlands are sources or sinks of various nutrients. van der Valk *et al.* (1979) and Mitsch and Gosselink (1986) provide overviews of nutrient cycling in wetlands.

Wetlands play a valuable role in reducing turbidity of flooding waters. This is especially important for aquatic life and for reducing siltation of ports, harbors, rivers, and reservoirs. Removal of sediment load is also valuable because sediments often transport absorbed nutrients, pesticides, heavy metals, and other toxins which pollute our nation's waters (Boto and Patrick 1979). In a southern Illinois alluvial swamp, phosphorus deposited with sediments by river flooding was ten times greater than the amount of phosphorus returned to the river during the rest of the year (Mitsch 1979). Depressional wetlands should retain all of the sediment entering them (Novitzki 1978). In Wisconsin, watersheds with 40% coverage by lakes and wetlands had 90% less sediment in water than watersheds with no lakes or wetlands (Hindall 1975). Creek banks along salt marshes typically support more productive vegetation than the marsh interior. Deposition of silt is accentuated at the water-marsh interface, where vegetation slows the velocity of water causing sediment to drop out of solution. In addition to improving water quality, this process adds nutrients to the creekside marsh which leads to higher plant productivity (DeLaune *et al.* 1978). The U.S. Army Corps of Engineers has investigated the use of marsh vegetation to lower turbidity of dredge disposal runoff and to remove contaminants. In a 20 ha impoundment near Georgetown, South Carolina, after passing through about 610 m of marsh vegetation, the effluent turbidity was similar to that of the adjacent river (Lee *et al.* 1976). Wetlands have also been proven to be good filters of nutrients and heavy metal loads in dredge material disposal effluents (Windom 1977).

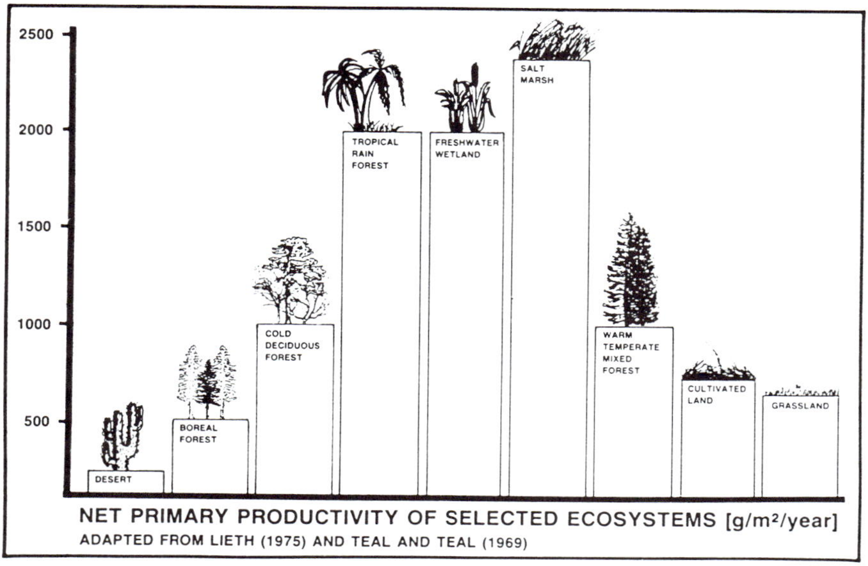

Figure 10. Net primary productivity of selected ecosystems.

Primary and secondary productivity

Wetlands are among the most productive ecosystems in the world rivaling the world's best agricultural fields (Fig. 10). Wetland plants are particularly efficient converters of solar energy. Through photosynthesis, plants convert sunlight into plant material or biomass and produce oxygen as a by-product. Wetlands vary in their productivity due to plant species composition, available nutrients, climate, hydrology, soils, and other factors. Examples of this variability are shown in Table 12.

Wetland biomass serves, in part, as food for a multitude of animals, both aquatic and terrestrial. For example, many waterfowl depend heavily on seeds of marsh plants, while muskrats (*Ondatra zibenthicus* and *Neofiber alleni*) eat *Typha* spp. rhizomes and young shoots. Moose (*Alces alces*), caribou (*Rangifer tarandus*), black bears (*Ursus americanus*), and brown bears (*Ursus arctos*) graze on marsh plants in Alaska (Crow and Macdonald 1979). Although direct grazing of wetland plants is generally limited, their major food value for aquatic animals is reached when plants fragment to form detritus. Water temporarily stored in wetland basins drops part of its nutrient and sediment load and picks up decomposition products and organic detritus for export. The major pulse of detritus from wetlands occurs in late winter through spring thus coinciding with the arrival of migrant species into estuaries for growth, spawning and the beginning of increased biological

Table 12. Examples of productivity estimates for wetlands in the United States (source: Mitsch and Gosselink 1986, Bradbury and Grace 1983). Values are either means or ranges.

Wetland types and location	Plant(s) or plant community	Annual net productivity $g\ m^{-2}$	Source
Salt marsh-Maine	*Spartina alterniflora* (short)	705	
	Spartina patens	2,740	
Salt marsh-Rhode Island	*S. alterniflora* (tall)	840	Nixon and Oviatt (1973)
	S. alterniflora (short)	432	
	S. patens	430	
Salt marsh-Georgia	*S. alterniflora* (tall)	2,000–3,700	Smalley (1959),
	S. alterniflora (short)	570–1,300	Gallagher *et al.* (1980)
	Juncus roemerianus	2,200	
Salt marsh-Louisiana	*S. alterniflora* (tall)	1,381	Hopkinson *et al.* (1980)
	S. patens	4,159	
	Juncus roemerianus	3,295	
Fresh tidal marsh-Mid-Atlantic Region	*Nuphar lutea*	780	Odum *et al.* (1984)
	Zizania aquatica	1,578	
	Typha angustifolia	1,420	
	Phragmites australis	1,872	
	Hibiscus moscheutos	868	
Nontidal marsh-New Jersey	*Typha latifolia*	1,904	Jervis (1969)
Nontidal marsh-South Carolina	*Typha latifolia*	530–1,132	Boyd (1971)
Nontidal marsh-North Dakota	*Typha latifolia*	404	McNaughton (1966)
Nontidal marsh-Oklahoma	*Typha latifolia*	703–1,527	Penfound (1956), McNaughton (1966)
Nontidal marsh-Florida	*Cladium jamaicense*	3,000	Stewart and Ornes (1973)
Nontidal marsh-South Carolina	*Pontederia cordata*	716	Polisini and Boyd (1972)
Nontidal marsh-Wisconsin	*Scirpus fluviatilis*	1,116	Polisini and Boyd (1972)
Nontidal marsh-Wisconsin	*Phalaris arundinacea*	1,253	Klopatek and Stearns (1978)
Nontidal marsh-New York	*Carex lacustris*	965	Bernard and Solsky (1977)
Nontidal marsh-Wisconsin	*Carex lacustris*	1,034	Klopatek and Stearns (1978)
Nontidal marsh-(sedge meadows) Alaska	*Carex* spp.	73–156	Haag (1974), Brown and West (1970)

Table 12. Continued.

Wetland types and location	Plant(s) or plant community	Annual net productivity g m^{-2}	Source
Shrub wetland-(fen) Michigan	*Chamaedaphne calyculata* and *Betula pumila*	341 Net	Richardson et al. (1976)
Shrub wetland-Alaska	*Salix planifolia*	459	Webber (1972)
Forested wetland-Minnesota	*Thuja occidentalis* and *Betula papyrifera*	1,032	Reiners (1972)
Forested wetland-Illinois	*Taxodium distichum* and *Nyssa sylvatica*	678	Mitsch (1979)
Floodplain forest-Illinois		1,250	Johnson and Bell (1976)
Floodplain forest-Kentucky		1,280–1,334	Taylor (1985)
Forested wetland-(nutrient poor) Georgia	*Taxodium distichum*	681	Schlesinger (1978)
Forested wetland-Florida	*Taxodium distichum* and *Nyssa sylvatica*	761	Mitsch and Ewel (1979)
Forested wetland-Florida		1,607	Brown (1978)
Forested wetland-Louisiana	*Taxodium distichum* and *Nyssa sylvatica*	1,120	Conner and Day (1976)

activity in fresh water bodies. The overwinter enrichment of wetland detritus by microbial action produces high quality food for detritus-based food chains both in wetlands and downstream areas.

This detritus forms the base of an aquatic food web which supports higher consumers, such as commercial fishes (Fig. 11). This relationship is especially well-documented for coastal areas. Animals such as shrimp, snails, clams, worms, killifish (*Fundulus confluentus* and *F. diaphanus*) and mullet (*Mugil cephalus*), eat detritus or graze upon the bacteria, fungi, diatoms, and protozoa growing on its surface (Crow and Macdonald 1979, de la Cruz 1979). Many of these animals are the primary food for commercial and recreational fishes. Salmon (*Oncorhynchus* spp.) are linked with wetlands and detritus. Juvenile salmon in Puget Sound, in the state of Washington, feed mainly on salt marsh midge larvae, which subsist on detritus (Crow and Macdonald 1979). Detritus from wetland vegetation along western rivers feeds aquatic insects important to the diet of resident fishes. Thus, wetlands can be regarded as the farmlands of the aquatic environment where great volumes of

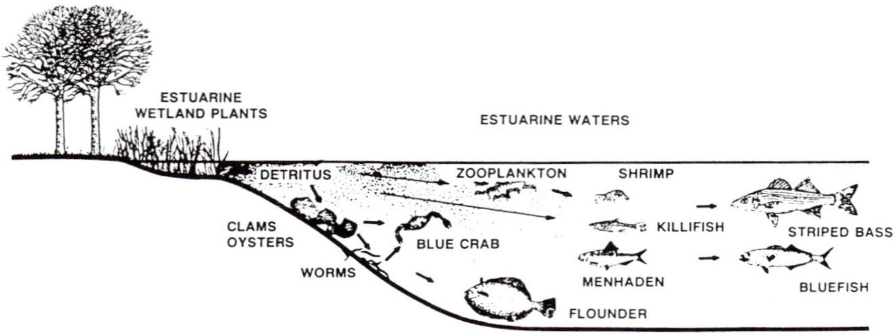

Figure 11. Simplified food pathways from estuarine wetland vegetation to commercial and recreational fishes (Tiner 1984).

food are produced annually. The majority of nonmarine aquatic animals depend, either directly or indirectly, on this food source.

Biodiversity

Approximately 6,728 (31%) of the 21,588 plant species that are found in the United States occur in wetlands. Wetlands are estimated to comprise about 5% of the land surface area of the conterminous United States. The proportion of species that have developed the ability to occur in wetlands is far greater than would be predicted if the flora occurring in wetlands was only proportional to the relative areal coverage of wetland habitats. Wetlands, therefore, include areas of extremely high diversity that have a disproportionate percentage of the flora of the United States. Approximately half of the species that occur in wetlands are restricted to or usually occur in wetlands. Wetlands thus provide critical habitat for the occurrence of a high percentage of the U.S. flora.

Fish and wildlife habitat

The variety of wetlands across the country creates habitats for many forms of fish and wildlife. Some animals spend their entire lives in wetlands, while others use wetlands primarily for reproduction and nursery grounds. Numerous fish and wildlife frequent marshes and swamps for feeding or feed on organisms produced in wetlands, while other animals visit wetlands for drinking water. Approximately 5,000 species of plants, 190 species of amphibians, and a third of all bird species in the nation occur in wetlands. Two-thirds of the 10 to 12 million waterfowl in the lower 48 states reproduce in the prairie potholes of the Midwest, and millions of ducks winter in the bottomland hardwoods of the south-central states. More than half of the marine sport fishes caught in the United States are dependent on wetland estuaries, and roughly two-thirds of the major U.S. commercial fish are dependent on estuaries and salt marshes for nursery or spawning grounds. Wetlands are also crucial for survival of numerous endangered animals and plants. Many plant and animal species which are federally listed as endangered or threatened are also dependent on wetlands for their survival. For example, 28% of the plant species and 50% of the animal species listed as endangered or threatened are wetland dependent. These animal species include 66% of the fish, 35% of the mammals, 25% of the birds, 30% of the mussels, 13% of the insects, 11% of the reptiles, and 8% of the amphibians (Niering 1988).

Fish and shellfish

Both inland and coastal wetlands are essential to maintaining fish populations. Estuarine wetlands are important producers of shrimp (*Penaeus* spp.), crabs (*Callinectes* spp.), oysters (*Crassostrea* spp. and *Ostrea* spp.), and clams (*Mya* spp. and *Mercenaria* spp.). Approximately two-thirds of the major U.S. commercial fishes depend on estuaries and salt marshes for nursery or spawning grounds (McHugh 1966). Among the more familiar wetland-dependent fishes are menhaden (*Brevoortia tyrannus* and *B. patronus*), bluefish (*Pomatomus saltatrix*), southern flounder (*Paralichthys lethostigma*), sea trout (*Cynoscion* spp.), spot (*Leiostomus xanthurus*), *Mugil cephalus*, Atlantic croaker (*Micropogonias undulatus*), striped bass (*Morone saxatilis*), and drum (*Pogonias cromis* and *Sciaenops ocellata*). Coastal marshes along the Atlantic and Gulf coasts are the most important in this regard. In the Pacific Northwest, coastal wetlands along spawning streams are vital to many salmon species (*Oncorhynchus* spp.) (Merrell and Koski 1979). Coastal wetlands are also essential for important shellfish like shrimp, blue crab, oysters, and clams. These areas serve as the primary nursery grounds for penaeid shrimp, whose young grow rapidly and reach adulthood here. Scientific studies have demonstrated a direct correlation between the amount of coastal marsh and shrimp production (Turner 1977).

Freshwater fishes also find wetlands important for survival. In fact, most freshwater fishes can be considered wetland-dependent because: (1) many species feed in wetlands or upon wetland-produced food, (2) many fishes use wetlands as nursery grounds and (3) almost all important recreational fishes spawn in the aquatic portions of wetlands (Peters *et al.* 1979). Marshes along Lake Michigan, for example, are spawning grounds for northern pike (*Esox lucius*), yellow perch (*Perca flavescens*), carp (*Cyprinus carpio*), smallmouth bass (*Micropteris dolomieui*), largemouth bass (*M. salmoides*), bluegill (*Le-*

pomis macrochirus), bullhead (*Ictalurus* spp.), and other fishes, including minnows (Jaworski and Raphael 1978). Prized gamefish spawn in flooded marshes as well as feed there. Bottomland hardwood forests of the southern U.S. serve as nursery and feeding grounds for young warmouth (*Lepomis gulosus*) and *Micropteris salmoides*, while adult bass feed and spawn in these wetlands. River swamps in Georgia produce 590 kg of fish per acre (Wharton 1970). The bottomlands of the Altamaha River in Georgia are used for spawning by hickory shad (*Alosa mediocris*) and blueback herring (*A. aestivalis*) (Wharton *et al.* 1982). Southern bottomland forested wetlands are also the home of the edible swamp crayfish ("crawdads") (*Procambarus* spp.) which burrow down to the water table when flood waters recede (Patrick *et al.* 1981). Wetland vegetation along western rivers is important to fishes in many ways, including providing cover, shade for water temperature regulation, and food for aquatic insects which are eaten by fishes. Crance (1988) provides a review of the relationship between riparian forested wetlands and fishery resources.

Waterfowl and other birds

In addition to providing year-round habitats for resident birds, wetlands are especially important as breeding grounds, overwintering areas, and feeding grounds for migratory waterfowl and numerous other birds. Both coastal and inland wetlands serve these valuable functions. Salt marshes along the Atlantic Coast of the United States are used for nesting by birds such as black ducks (*Anas rubripes*), laughing gulls (*Larus atricilla*), Forester's terns (*Sterna forsteri*), clapper rails (*Rallus longirostris*), blue-winged teals (*Anas discors*), willets (*Catoptrophorus semipalmatus*), northern harrier (*Circus cyaneus*), sharp tailed sparrows (*Ammodramus caudacuta*), and seaside sparrows (*A. maritima*). Wading birds like herons and egrets also feed and nest in and adjacent to coastal wetlands. Northeastern salt marshes are prime wintering grounds for *Anas rubripes* in the Atlantic Flyway. Atlantic coastal marshes are also important feeding and stopover areas for migrating snow geese (*Chen caerulescens*), peregrine falcons (*Falco peregrinus*), shorebirds, wading birds, and others. Intertidal mudflats along all coasts are principal feeding grounds for migratory shorebirds (e.g. oystercatchers (*Haematopus palliatus*), spotted sandpipers (*Actitis macularia*), yellowlegs (*Tringa melanoleuca* and *T. flavipes*), plovers (*Charadrius* spp.), and knots (*Calidris canutus*), while swallows can often be seen feeding on flying insects over the marshes.

As one moves upstream into the fresh coastal marshes, other birds can be observed nesting (Tiner 1985a). These include red-winged blackbirds (*Agelaius phoeniceus*), marsh wrens (*Cistothorus palustris*), least bitterns

(*Ixobrychus exilis*), and *Rallus longirostris* (Odum *et al.* 1984). Nesting birds of freshwater tidal marshes in New Jersey, for example, include these four birds, plus American goldfinch (*Carduelis tristis*), swamp sparrow (*Melospiza georgiana*), Indigo bunting (*Passerina cyanea*), common yellowthroat (*Geothlypis trichas*), yellow warbler (*Dendroica petechia*), Traill's flycatcher (*Empidonax traillii*), wood duck (*Aix sponsa*), green heron (*Butorides virescens*), and common gallinule (*Gallinula chloropus*) (Hawkins and Leck 1977). Many of these birds utilize nontidal wetlands as well for nesting.

The nation's inland wetlands are most noted for waterfowl production, although they also serve as important nesting, feeding and resting areas for other migratory birds. The Prairie Pothole Region located in North and South Dakota, and Minnesota is the principal breeding area for waterfowl in the United States. Waterfowl nesting in this region include 15 species, with mallard (*Anas platyrhynchos*), pintail (*Anas acuta*), and *Anas discors* being most abundant (Smith *et al.* 1964). Many of these nesters use different types of wetlands for mating and for rearing young. Individual mallard hens may use more than 20 different wetlands during the nesting season (Dwyer *et al.* 1979). Besides waterfowl, other birds also nest in these wetlands such as *Agelaius phoeniceus*, Brewer's blackbirds (*Euphagus cyanocephalus*), kingbirds (*Tyrannus* spp.), killdeer (*Charadrius vociferus*), *Actitis macularia*, sparrows, Wilson's phalaropes (*Phalaropus tricolor*), and black terns (*Chlidonias niger*). Potholes and other inland emergent wetlands also provide important winter cover and nesting habitat for ring-necked pheasant (*Phasianus colchicus*). In fact, the pheasant population in east-central Wisconsin is directly related to the amount and distribution of wetlands available (Gates and Hale 1974). Playa lake wetlands in the Texas Panhandle region are important nesting habitats for *P. colchicus*, mourning doves (*Zenaida macroura*), *Agelaius phoeniceus*, and others (Guthery 1981).

Bottomland forested wetlands of the southern U.S. are primary wintering grounds for waterfowl as well as important breeding areas for wood ducks, herons, egrets, and white ibises (*Eudocimus albus*). Wild turkeys (*Meleagris gallopavo*) even nest in bottomland hardwood forests. Other common bird inhabitants include barred owls (*Strix varia*), downy woodpeckers (*Picoides pubescens*), red-bellied woodpeckers (*Melanerpes carolinus*), cardinals (*Cardinalis cardinalis*), pine warblers (*Dendroica pinus*), wood peewees (*Contopus* spp.), common yellowthroats (*Geothlypis trichas*), and wood thrushes (*Hylocichla mustelina*) (Wharton *et al.* 1982).

In the northeastern U.S., red maple swamps are among the most abundant wetlands types. A study of breeding birds in swamps in Massachusetts revealed a total of 46 breeding species (Swift 1980). Most common breeders include Common yellowthroats, veery (*Catharus fuscescens*), Canada warbler (*Wilsonia canadensis*), ovenbird (*Seiurus aurocapillus*), northern waterthrush

(*S. noveboracensis*), and gray catbird (*Dumetella carolinensis*). The wood duck is another important resident of forested wetlands, primarily in the eastern half of the U.S., where it nests in cavities of dead trees or in man-made nesting boxes.

In the western U.S. riparian forested wetlands along rivers are valuable bird nesting and migratory stop-over areas. Wauer (1977) found 94 avian species nesting in riparian vegetation of the Rio Grande River including *Zenaida macroura*, verdin (*Auriparus flaviceps*), northern orioles (*Icterus galbula*), and brown-headed cowbirds (*Molothrus ater*). These riparian wetlands were very important to migratory birds in the spring and fall. In Arizona, the yellow-billed cuckoo (*Coccyzus americanus*) and blue-throated hummingbird (*Lampornis clemenciae*) are restricted to cottonwood-willow forested wetlands (Brown *et al.* 1977). Riparian wetlands may be more important to migratory birds in arid regions than in more humid areas. The availability of food, water, cover, and suitable north-south routing strongly influence migrants (Wauer 1977).

Alaskan and other tundra wetlands are prime breeding grounds for most shorebirds such as sandpipers, plovers, and their relatives. Nearly the entire Pacific Flyway populations of cackling Canadian goose (*Branta canadensis*) and the white-fronted goose (*Anser albifrons*) nest in Alaska's Yukon-Kuskokwin Delta. Alaska is also the most important production area for *Anas acuta* in the U.S. (U.S. Fish and Wildlife Service 1984). During droughts in the Prairie Pothole Region, Alaska's wetlands are heavily used by waterfowl for nesting.

Hawaii's wetlands are especially important to endangered birds. The Hawaiian stilt (*Himantopus himantopus knudseni*), Hawaiian coot (*Fulica americana alai*), Hawaiian gallinule (*Gallinula chloropus sandvicensis*), and Hawaiian duck (*Anas wyvilliana*) depend on Hawaiian wetlands for survival.

Wetlands are, therefore, crucial for the existence of many birds, ranging from waterfowl and shorebirds to songbirds. Some spend their entire lives in wetland environments, while others primarily use wetlands for nesting, feeding, or resting. The U.S. Fish and Wildlife Service prepared a report in August 1987 entitled *Migratory Nongame Birds of Management Concern in the United States: The* 1987 *List*. This report presented some important insight into the value of wetlands to those listed nongame birds. Compared to the nongame bird population at large, marsh/wading birds (representing 23% of the listed species) are overrepresented on the list. On a habitat basis, 47% of the listed species are associated with coastal and freshwater wetlands and beaches. A greater proportion of listed species (57%) forage in wetland substrate compared to the nongame bird population at large (34%).

Habitat loss is overwhelming and is the greatest threat facing the listed species, being mentioned in 85% of the references. Habitat loss is a major threat to all taxonomic/ecological groups and species-habitat associations. Human disturbance was viewed as the second greatest threat, being mentioned in 20% of the references. The species most impacted by human disturbance are the marsh/wading birds, birds of prey, and marine/shore birds, and species associated with coastal and freshwater wetlands and beaches. The species most often mentioned as suffering from human disturbance were the common loon (*Gavia immer*), trumpeter swan (*Cygnus buccinator*), snowy plover (*Charadrius alexandrinus*), and roseate tern (*Sterna dougallii*). It is clear that wetland-dependent birds are overrepresented on the list of nongame birds of management concern. These species need wetlands for cover and forage. If wetlands can be restored, it will help reduce the greatest threat facing these listed species.

Mammals and other wildlife

Many mammals live in wetlands or frequent wetlands in search of food, water, and cover. Furbearers like muskrats (*Ondatra zibenthicus* and *Neofiber alleni*), beavers (*Castor canadensis*), and nutria (*Myocastor coypus*) are common wetland mammals. Muskrats are the most wide-ranging of the three, inhabiting both coastal and inland marshes throughout the country. By contrast, beavers tend to be restricted to inland wetlands, with nutria limited to coastal wetlands of the southern United States. Other furbearers using wetlands including otters (*Lutra canadensis*), minks (*Mustela vison*), raccoons (*Procyon lotor*), skunks (*Mephitis* spp.), and weasels (*Mustela* spp.) Other mammals also frequent wetlands, such as marsh rabbits (*Sylvilagus palustris*), swamp rabbits (*S. aquaticus*), numerous mice, bog lemmings (*Synaptomys cooperi* and *S. borealis*), and shrews. Larger mammals may also be observed. *Ursus americanus* find refuge and food in forested and shrub wetlands in the northeastern states of Pennsylvania, West Virginia, and North Carolina (Richardson *et al.* 1981). In the northern states, white-tailed deer (*Odocoileus virginianus*) depend on Atlantic white cedar (*Chamaecyparis thyoides*) and other evergreen swamps for winter shelter and food. They are also common in red maple (*Acer rubrum*) swamps. By contrast, the extensive wetlands of Alaska's North Slope are used as summer range and calving areas by caribou (*Rangifer tarandus*).

Other forms of wildlife make their homes in wetlands. Turtles, reptiles, and amphibians are important residents. Turtles are most common in freshwater marshes and ponds. The more important ones nationally are the painted turtle (*Chrysemys* spp.), spotted turtle (*Clemmys guttata*), Blanding's turtle (*Emydoidea blandingi*), map turtle (*Graptemys* spp.), mud turtle (*Kinosternon* spp.), pond turtle, musk turtle (*Sternotherus* spp.), and snapping turtle (*Chelydra* spp.) (Clark 1979). The endangered Plymouth red-bellied turtle (*Pseudemys rubriventris*) and bog turtle (*Clemmys muhlenbergi*) are also wetland-dependent (Williams and Dodd 1979). Along the coast, the diamond-backed terrapin (*Malaclemys terrapin*) is a common inhabitant of salt marshes, while young loggerhead turtles (*Caretta caretta*) spend some time in estuaries after hatching before going out to sea.

The largest reptiles occurring in the U.S., the American alligator (*Alligator mississipiensis*) and the American crocodile (*Crocodylus acutus*), live in wetlands. The crocodile, an endangered species, is now found only in mangroves and coastal waters of Florida Bay, in the state of Florida, while the alligator lives in both brackish and freshwater wetlands, but is most abundant in the latter. Alligators create "gator holes" in the Florida Everglades, which persist through the dry season. Fishes and invertebrates concentrate in these holes which make them easy prey for birds and other animals. Gator holes with their abundance of food are, therefore, important to the breeding success of birds such as the wood ibis (*Mycteria americana*) (Williams and Dodd 1979).

Many snakes inhabit wetlands, with water snakes being most abundant throughout the U.S. (Clark 1979). Important wetlands snakes include cottonmouth moccassin (*Agkistrodon piscivorus*), garter (*Thamnophis sirtalis* and *T. cyrtopsis*), queen (*Regina septemvittata*), mud (*Farancia abacura*), and swamp snakes (*Liodytes alleni* and *Seminatrix pygaea*). In bottomland forested wetlands of the southern U.S., copperheads (*Agkistrodon contortrix*), and canebrake rattlesnakes (*Crotalus horridus atricaudatus*) can be found as well as northern brown (*Storeria dekayi dekayi*), garter (*Thamnophis sirtalis similis*), rough green (*Opheodrys aestivus*), and rat snakes (*Elaphe obsoleta*) (Wharton *et al*. 1982). The San Francisco garter snake (*Thamnophis sirtalis tetrataenai*), an endangered species, also requires wetlands for survival (Williams and Dodd 1979).

Nearly all of the approximately 190 species of amphibians in North America are wetland-dependent, at least for breeding (Clark 1979). Every freshwater wetland in the U.S., except in the Arctic tundra, probably has some frogs. Common frogs include the bull (*Rana catesbeiana*), green (*Rana clamitans melanota*), leopard (*Rana pipiens* and *R. utricularia*), mink (*R. septentrionalis*), pickerel (*R. palustris*), wood (*Rana sylvatica*), and chorus frogs (*Pseudacris* spp.), and spring peepers (*Hyla crucifera*). Many salamanders use temporary ponds or wetlands for breeding, although they spend most of the year in uplands. Numbers of amphibians, even in small wetlands, can be astonishing. For example, 1,600 salamanders and 3,800 frogs and toads were found in a small gum pond (less than 30 meters wide) in Georgia (Wharton 1978).

Wetland use and conservation

Review of major wetland inventories

The first attempt at a national wetlands inventory was conducted in 1906 by the U.S. Department of Agriculture (Wright 1907). The inventory was requested by the U.S. Congress to obtain information on the extent, character, and agricultural potential of the nation's wetlands. The inventory was conducted by mailing a questionnaire to one or more people in each county in the states east of the 115th meridian. The inventory was not a complete picture of the extent of the wetlands because eight of the public-land states in the arid western U.S. were excluded, as well as all intertidal wetlands. The purpose was to inventory the wetlands that probably could be easily drained and converted to agriculture. The study estimated that 32 million ha of wetlands could be profitably converted to agriculture.

The second attempt at a national wetlands inventory was conducted in 1922 by the Bureau of Agriculture Economics of the U.S. Department of Agriculture. It was based on data from the U.S. Bureau of Public Roads, soil-survey reports, topographic maps, various state reports and on the results of the 1920 census of drainage. Although not all inclusive, it was the most complete wetlands inventory conducted to that date and included intertidal wetlands. The 1922 inventory showed 37 million hectares of wetlands.

In 1940, the Soil Conservation Service of the U.S. Department of Agriculture estimated through the use of a drainage reconnaissance survey that there were 39.4 million ha of wetlands outside of organized drainage enterprises (Wooten *et al.* 1949). The U.S. Department of Agriculture, Technical Bulletin 1082 published in 1953, estimated 50 million hectares of wetlands.

Until that point in time, the purpose of all these national wetland inventories was to collect information on wetlands that could be drained and converted to other uses. In the early 1950s, it became apparent to the U.S. Fish and Wildlife Service that drainage of wetlands was having an adverse impact on wildlife habitat. Information was needed on the distribution, extent, and quality of the remaining wetlands in relation to their value as wildlife habitat. In order that the information would be most useful to the Fish and Wildlife Service, it was decided to place the primary emphasis on wetlands considered susceptible to drainage or other land use changes that destroyed wildlife habitat (Shaw and Fredine 1956). Through the use of aerial photographs, topographic maps, geodetic survey charts, U.S. Forest Service maps, soils maps, Federal and state land-use maps, county highway maps, and knowledge of state fish and game biologists, the inventory was completed in June, 1954. All the north central and southeastern states were inventoried. In the rest of the country the inventory efforts were restricted to physiographic regions where good waterfowl habitats were most abundant.

It was estimated that 90 percent or more of all wetlands used significantly by waterfowl were inventoried by this effort. The results of this inventory, *Wetlands of the United States*, were published as a national report (Shaw and Fredine 1956). It has been one of the most influential documents used in the protection of wetlands. It contained the gross average, general distribution and the relative importance to waterfowl of the 30.1 million hectares of wetlands that were included in the inventory.

Before the U.S. Fish and Wildlife Service's National Wetlands Inventory Project began in the mid-1970s, over 30% of the states did not have any state or local wetland inventories. An additional 20% only had narrative reports of gross maps at a scale of 1:250,000. The only extensive mapping at large scales had been conducted by coastal states and this was restricted primarily to their tidelands. These noted exceptions were Connecticut, Maryland, Massachusetts, New Jersey, New York, Delaware, Rhode Island, South Carolina, and West Virginia. In many other states wetland inventories of specific political units or specific study areas were conducted.

In general the detailed mapping of the tidelands was conducted as a result of state legislation. Most other mapping efforts were for the purposes of land use planning, wildlife management, environmental studies, wetland acquisition, or critical area surveys. For additional information refer to *Existing State and Local Wetlands Surveys* (U.S. Fish and Wildlife Service 1976).

In 1972 the U.S. Geological Survey published Circular 671 (Anderson *et al.* 1972) which was devised to provide a logical framework for land use and land cover information derived from remotely sensed data as well as to serve as the classification system for national inventory. The Land Use and Data Analysis (LUDA) Program was conceived in 1974 to fill the need for a baseline set of maps and data that could be used for several objectives related to national, regional, and state level resource management. The magnitude of the task required that the data had to be collected through the interpretation of data gathered by remote sensors. The land use and land cover mapping of the nation is complete at a scale of 1:250,000.

It became apparent by the early 1970s that wetlands had undergone many natural and man-induced changes. This coupled with the increased understanding of all wetland values, led the U.S. Fish and Wildlife Service to establish the National Wetlands Inventory Project (N.W.I.) in 1975. The U.S. Fish and Wildlife Service carefully reviewed the classification system procedures and products of the U.S. Geological Survey's Land Use and Data Analysis Program. Initially it was hoped that both agencies' goals could be achieved through a joint project. Although a lot of good information and insights were obtained for the review of the program, the Fish and Wildlife Service decided it needed more information and more detailed maps of the wetlands resources to meet its management needs. The N.W.I. is generating and disseminating scientific information on the characteristics and extent of the nation's wetlands. The purpose of this information is to foster wise use of the nation's wetlands and to provide data for making quick and accurate resource decisions. Decisionmakers cannot make informed decisions about wetlands without knowing how many of what type are where.

Two very different kinds of information are needed: (1) detailed maps at the scale of 1:24,000 and (2) status and trends reports. First, detailed wetlands maps for geographic areas of critical concern are needed for impact assessment of site-specific projects. These N.W.I. maps serve a purpose similar to the Soil Conservation Service's soil survey maps, the U.S. National Oceanic and Atmospheric Administration's coastal geodetic survey maps, and the Geological Survey's topographic maps. Detailed wetlands maps are used by local, State, and Federal agencies as well as by private industry and organizations for many purposes, including comprehensive resource management plans, environmental impact assessments, permit reviews, facility and corridor siting, oil spill contingency plans, natural resource inventories, and wildlife surveys. Secondly, national estimates of the current status and trends in losses and gains of wetlands are needed in order to provide improved information for reviewing the effectiveness of existing federal programs and policies, for identifying national or regional problems, and for general public awareness.

The N.W.I. has produced 1:24,000 scale maps for 60% of the contiguous United States, 16% of Alaska and all of Hawaii. Copies of wetland maps may be purchased through the United States Geological Survey or through 26 state run distribution centers. The first status and trends study, completed in 1982, was the first comprehensive, statistically valid effort to estimate the nation's wetlands. For the first time, a study had produced numbers on the total acreage of wetlands in the lower 48 states, their rate of disappearance over a 20–year interval, and the general cause for the losses. The results of this effort were reported in publications by Frayer *et al.* (1983) and Tiner (1984). The information generated from the analysis has been extremely useful, and has been cited by major reports that have discussed wetland loss rates (Office of Technology Assessment 1984). The data has also played a role in the review of some Federal policies regarding wetlands. Legislation language and Congressional reports make reference to these loss rates in both the Swampbuster Provision of the Food and Security Act of 1985 (P.L. 99–198) and the Emergency Wetlands Resources Act of 1986 (P.L. 99–645).

The U.S. Congress has recognized the need for monitoring the current status and trends of wetlands to provide information for making wise decisions, as well as the need for detailed wetland maps for impact assessment and site specific decisions. The Emergency Wetlands Resource Act of 1986 directs the Secretary of Interior, through the Director of the U.S. Fish and

Wildlife Service to produce by September 30, 1990, and at ten-year intervals thereafter, reports to update and improve the information contained in Frayer *et al.* (1983). The Act also requires the Fish and Wildlife Service to produce by September 30, 1998 the National Wetlands Inventory maps for the reminder of the contiguous United States and, as soon as practicable, wetland maps for Alaska and noncontiguous portions of the United States.

Recent national wetland trends

Information on historical wetland gains and losses is limited and often subjective. In 1983, the U.S. Fish and Wildlife Service completed a scientifically sound study of the current status and trends of U.S. wetlands between the mid-1950s and mid-1970s (Frayer *et al.* 1983). Although the results of this study are valid at the national level, few comparable statistics exist for individual states. Recently, a similar study was conducted for five states in the Mid-Atlantic region (Tiner and Finn 1986). The following discussions will summarize the results of the U.S. Fish and Wildlife Service's national study and other regional studies as reported by Tiner (1984). Specific problem areas where wetlands are in greatest jeopardy will be highlighted.

Slight net gains in deepwater habitats, manmade lakes and reservoirs, coastal waters, and in two wetland types (inland flats and ponds) took place between the mid-1950s and mid-1970s (Table 13). Lake acreage increased by 0.6 million hectares with 94% of this gain occurring in the eastern half of the country. These new lakes and reservoirs were mostly created from uplands, although vegetated wetlands were also destroyed. Some new wetlands, however, have formed along the edges of these new waterbodies. During the same period, coastal open water increased by 81 thousand hectares. Most of this gain came from Louisiana at the expense of coastal wetlands, which are being permanently flooded at an accelerating rate. Causes of this change from marsh to open water are numerous and complicated and include natural rise in sea level, subsidence of the coastal plain, levee construction, channelization, and oil and gas extraction (Turner 1987). Two wetland types experienced gains between the mid-1950s and mid-1970s: inland flats and ponds. Eighty-one thousand hectares of nonvegetated wetland flats and 0.9 million ha of ponds were created. Pond acreage nearly doubled from 0.9 million hectares to 1.8 million hectares, primarily due to farm pond construction in the central portions of the United States. Most of this pond acreage came from former upland, although 59 thousand hectares of forested wetlands and 156 thousand hectares of emergent wetlands were changed to open water.

Despite these modest gains, wetland losses were enormous. In the mid-1950s, there were an estimated 43.7 million hectares of wetlands in the lower

Table 13. Wetland area for the lower 48 states, 1950's–1970's. The standard error of each entry expressed as a percentage of the entry (SE%) is given in parenthesis (source: Frayer et al. 1983).

Wetland type	Hectares		Gains (+) and losses (−)	
Marine intertidal	31,740	(14.0)	−1,624	(57.5)
Estuarine intertidal emergent	1,588,164	(4.3)	−143,006	(8.3)
Estuarine forested scrub–shrub	231,991	(14.4)	−7,737	(93.2)
Estuarine nonvegetated	302,226	(9.8)	+2,175	(*)
Palustrine forested	20,126,692	(3.6)	−2,426,679	(3.7)
Palustrine scrub–shrub	4,295,976	(12.5)	−156,732	(56.7)
Palustrine emergent	11,514,441	(17.5)	−1,891,182	(5.2)
Palustrine unconsolidated shore	180,495	(33.2)	+61,637	(5.5)
Palustrine open water	1,778,564	(7.7)	+839,851	(2.5)
Palustrine other	53,365	(23.4)	+16,547	(39.9)
Total	40,103,654	(6.4)	3,707,328	(3.1)

*Standard error of estimate is equal to or larger than estimate.

48 states (Frayer *et al.* 1983). Just 20 years later, these wetlands were reduced to 40 million hectares, despite some gains in wetlands due to reservoir and pond construction, beaver activity, and irrigation and marsh creation projects. This loss of 3.7 million hectares equates to an area larger than Belgium, twice the area of Kuwait, more than three times the area of Gambia and Lebanon or nearly five times the area of Cypress. Actually, 4.5 million hectares of our most valuable natural wetlands were destroyed, but these acreage losses were reduced by gains of 0.8 million hectares of newly created wetlands, yielding a net loss of 3.6 million hectares. The average rate of wetland loss from the mid-1950s to the mid-1970s was 185 thousand hectares per year: 178 thousand hectares of palustrine losses and 7.3 thousand hectares of estuarine wetland losses. This annual loss equals an area about half the size of Rhode Island or to an area the size of the Caribbean island of Martinique.

Agricultural development involving drainage was responsible for 87% of recent national wetland losses, while urban development and other types of development caused only 8% and 5% of the losses, respectively. Agriculture had the greatest impact on forested and emergent wetlands, with losses of 2.3 and 1.0 million hectares, respectively. In addition, 162 thousand hectares of scrub-shrub wetlands were converted to agricultural use between the mid-1950s and the mid-1970s.

The most extensive wetland losses occurred in Louisiana, Mississippi, Arkansas, North Carolina, Florida, North Dakota, South Dakota, Minnesota, Texas, and Nebraska. Greatest losses of forested wetlands took place in the Lower Mississippi Valley with the conversion of bottomland hardwood forests to farmland. Large areas of shrub wetlands were lost in North Carolina where pocosin wetlands are being converted to cropland, pine plantations, or mined for peat. Inland marsh drainage for agriculture was most significant in the Prairie Pothole Region of North and South Dakota and Minnesota, Nebraska's Sandhills and Rainwater Basin, and Florida's Everglades. Between the mid-1950s and mid-1970s, estuarine wetland losses were heaviest in three states (Louisiana, Florida, and Texas) along the Gulf of Mexico. Most of Louisiana's coastal marsh losses were attributed to submergence by coastal waters. In other areas, urban development was the major direct man-induced cause of coastal wetland loss. Dredge and fill residential development in coastal areas was most significant in Florida, Texas, New Jersey, New York, and California.

While the national decline in wetlands is dramatic, losses in particular regions and states are even more startling. For example, California has lost over 90% of its original wetland resource (Dahl 1990). Less than 5% of Iowa's natural marshes exist and over 90% of the wetlands in Nebraska's Rainwater Basin have been destroyed (Bishop 1981, Farrar 1982). Only 20% of the original bottomland hardwood forests in the lower Mississippi alluvial plain remain (MacDonald *et al.* 1979). Other states with less than half of their original wetlands include Iowa, Indiana, Missouri, Illinois, and Connecticut (Dahl 1990). By 1955, Michigan had lost 3.2 million hectares of wetlands (Michigan Department of Natural Resources 1982). In selected areas of Illinois, wetland losses have been dramatic. For example, virtually all wetlands have been eliminated in the East-Central Region, Big Prairie Region, and Green River Watershed, while 98% of Illinois' southern bottomland swamps have been destroyed (Illinois Department of Conservation 1983).

In many areas, wetland destruction was greatest from the mid-1800s to the early 1900s due to passage of the Swamp Land Acts of 1849, 1850, and 1860. These acts granted all swamp and overflow lands to 15 states: Alabama, Arkansas, California, Florida, Illinois, Indiana, Iowa, Louisiana, Michigan, Minnesota, Mississippi, Missouri, Ohio, Oregon, and Wisconsin (Shaw and Fredine 1956). These states were to drain these wetlands for agriculture by constructing levees and drainage ditches. About 26.3 million hectares had been transferred from the federal government to the states by 1954.

The original 13 states had retained all lands within their borders when the federal government was established and Texas also kept all its land at the time of annexation. Interestingly, the extensive coastal wetlands of Texas, Georgia, South Carolina, North Carolina, Virginia, Maryland, Delaware, New Jersey, Connecticut, Rhode Island, and Massachusetts were never owned by the federal government and, by contrast, coastal wetland losses have been more recent. Between 1954 and 1978, the loss rate of coastal wetlands doubled due primarily to post-war urban and industrial development in the U.S coastal zone and to accelerated erosion and subsidence of Louisiana's vast coastal marshes (Gosselink and Baumann 1980).

While wetland losses in some states or regions may have been heaviest at the turn of the century, loss rates remain high in many areas. Between 1955 and 1978, Kansas lost 40% of its wetlands (C. Elliott, U.S. Fish and Wildlife Service, personal communication). In Illinois, an estimated 20% of its wetlands are destroyed every decade (Great Lakes River Basin Commission 1981). About 2.7 million hectares of Ohio's original wetlands have been drained, while over half of its wetlands along Lake Erie have been destroyed since 1954 (Weeks 1974). Kentucky's wetlands along the Mississippi and Ohio Rivers have been reduced by 37% in the past twenty years (Kentucky Department of Fish and Wildlife Resources 1983). Heavy annual losses are continuing in the bottomland hardwood forested wetlands of the Lower Mississippi Delta and accelerating in pocosin wetlands along the North Carolina coast (MacDonald *et al.* 1979, Richardson 1981a). Recent trends in Delaware, Maryland, and New Jersey illustrate the effect of state wetland protection. Before passage of the Wetlands Act of 1973, Delaware was losing almost 200 hectares of estuarine wetland each year. After the law, losses dropped to just 8 hectares annually (Hardisky and Klemas 1983). Coastal wetland losses in Maryland and New Jersey were also drastically reduced through wetland regulations. In addition to state laws, the Federal Clean Water Act added a level of governmental protection of these wetlands nationwide in the early 1970s. Effective implementation of similar laws in other states has probably reduced wetland losses substantially.

Major threats to wetlands

Wetlands represent dynamic natural environments which are subjected to both human and natural forces. These forces directly result in wetland gains and losses and affect their quality. Table 14 outlines major causes of wetland loss and degradation. Natural events influencing wetlands include rising sea level, natural succession, the hydrologic cycle, sedimentation, erosion, beaver dam construction, and fire. The rise in sea level, for example, both increases and decreases wetland acreage depending on local factors. Along the eastern shore of Chesapeake Bay of the Atlantic coastline, this is allowing coastal wetlands to replace pine forests, while permanently flooding wetlands at lowest elevations. Apparent sea level rise is one factor converting salt marshes to bay bottoms in Louisiana (Gosselink 1984). Natural succession and fire typically change the vegetation of a wetland, usually with no net loss or gain. However, fire in Alaska's permafrost wetlands may convert the area to nonwetland. Disturbance of the vegetation cover can cause the frostline to recede, and dry site plants may become established. The hydrol-

Table 14. Major causes of wetland loss and degradation (modified from Tiner 1984).

Human Threats

Direct:

1. Drainage for crop production, timber production, and mosquito control.
2. Dredging and stream channelization for navigation channels, flood protection, marinas, coastal housing developments, and reservoir maintenance.
3. Filling for dredged spoil and other solid waste disposal, roads and highways, and commercial, residential, and industrial development.
4. Construction of dikes, dams, levees, and seawalls for flood control, water supply, irrigation, storm protection, cranberry production, and muckland farming.
5. Discharges of materials (e.g. pesticides, herbicides, other pollutants, nutrient loading from domestic sewage and agricultural runoff, and sediments from dredging and filling, agricultural and other land development) into waters and wetlands.
6. Mining of wetland soils for peat, coal, sand, gravel, phosphate and other materials.

Indirect:

1. Sediment diversion by dams, deep channels, and other structures.
2. Hydrologic alterations by canals, spoil banks, roads, and other structures.
3. Subsidence due to extraction of groundwater, oil, gas, sulphur, and other minerals.

Natural Threats:

1. Subsidence (including natural rise of sea level)
2. Droughts
3. Hurricanes and other storms
4. Erosion
5. Sediment deposition (e.g. landslides, volcanic deposition and barrier beach migration)
6. Fire
7. Biotic effects (e.g. muskrat. nutria and goose "eat-outs")

ogic cycle refers to the natural cycle of wet and dry periods over time. Great Lakes water levels, for example, fluctuate drastically on a roughly 20–year cycle. This adds an important dimension to wetlands, making them vulnerable to drainage during dry periods. Similar conditions have resulted in wetland drainage in the Prairie Pothole Region. The activities of beavers create or alter wetlands by damming stream channels. Thus, natural forces act in a variety of ways to create, destroy and modify wetlands.

Human actions are particularly significant in determining the fate of wetlands. Unfortunately, many human activities are destructive to wetlands, either converting them to agricultural or other lands or degrading their quality. Key human impacts include drainage for agriculture, channelization for flood control, filling for housing, highways, industry, and sanitary landfills, dredging for navigation channels, harbors and marinas, impoundment and reservoir construction, timber harvest, peat mining, oil and gas extraction, strip mining and extraction of other minerals (e.g. sand and gravel), ground-water withdrawals and other hydrologic alterations, and various forms of water pollution and waste disposal (Kusler and Kentula 1989a, 1989b). Some activities do, however, create wetlands. Construction of farm ponds and, in some cases, reservoirs and irrigation projects may increase wetland acreage, although valuable natural wetlands may be destroyed in the process. Marsh creation projects and restoration of previously altered wetlands can also be beneficial, especially if not at the expense of natural systems. Federal and state fish and wildlife agencies and private organizations such as Ducks Unlimited traditionally manage wetlands to improve their value to waterfowl. Finally, wetland protection efforts by federal, state, and local governments serve to help maintain and enhance our wetland resources in the United States, despite mounting pressures to convert them to other uses.

In the Northeast, coastal wetlands are now well protected by state laws. Inland wetlands, however, continue to be vulnerable to development pressures in many areas, although they are protected to varying degrees by the federal government through the Clean Water Act and by a few states with wetland protection laws. Three states, Vermont, New Jersey, and Maryland recently passed laws to protect inland wetlands. Urbanization seriously threatens inland wetlands throughout the region. Peat mining and resort development are major causes of wetland losses in the Pocono Region of Pennsylvania. Agricultural impacts are greatest in the hardwood swamps of Delaware, Maryland, Virginia, and in New York's mucklands. Tiner and Finn (1986) reported on recent wetland losses in Delaware, Maryland, Pennsylvania, Virginia, and West Virginia.

Agricultural drainage of wetlands is continuing to destroy large tracts of wetlands in the southeastern U.S., especially in the Lower Mississippi Delta, Florida, and along the Coastal Plain of North Carolina. Bottomland hardwoods are being clearcut for timber, and then cleared and drained for crop production, chiefly soybeans. Pocosin wetlands are similarly used, as well as being mined for peat. Many inland wetlands are being converted to pine plantations throughout the Southeast. Phosphate mining in Florida and North Carolina is destroying considerable wetland acreage. Puerto Rico's inland marshes ("savannahs") are being transformed into sugar cane farms. Coastal wetland destruction has slowed in most states with passage of protection laws, but enforcement may present problems.

Agricultural development in the Midwest corn belt and Great Plains remains the greatest threat, by far, to the remaining inland wetlands. Coastal marshes along the Great Lakes are still impacted by industrial, residential, and agricultural development. Although several of the midwestern states have laws protecting certain wetlands or regulating certain activities in wetlands, agricultural drainage is still largely unregulated.

In the western states, agricultural development is still the primary threat to wetlands. With increased tension over water rights, remaining wetlands may be deprived of sufficient quantities of water to function properly. This is especially true in Colorado where high population growth has increased demand for water. Urban and industrial development is destroying wetlands near urban centers.

Along the West Coast, coastal wetlands are generally protected by state laws, yet they are still under heavy pressure for urban, residential, and industrial development. Inland wetlands remain subject to agricultural pressures, particularly in California's Central Valley and the Great Basin of Nevada, Oregon, and Idaho. Degradation of existing wetlands through urban and agricultural runoff remains a serious problem.

Alaska's wetlands were once subject to very few development pressures. With the discovery of significant deposits of oil and gas and the subsequent pipeline construction and energy development, many wetlands have recently been altered. The oil boom of the 1970s also increased human population densities, resulting in increased pressure on wetlands for urban development. Increases in timber harvest, mining, and agricultural activities are also threatening large areas of wetland in Alaska.

National problem areas

While wetland losses and degradation continue throughout the country, there are several areas where remaining wetlands are in greatest jeopardy from a national standpoint. These areas and their threatened wetland types include: (1) estuarine wetlands of the U.S. coastal zone, (2) Louisiana's coastal marshes, (3) Chesapeake Bay's submerged aquatic beds, (4) South Florida's palustrine wetlands, (5) Prairie Pothole Region's emergent wetlands, (6) wetlands of Nebraska's Sandhills and Rainwater Basin, (7) forested wetlands of the Lower Mississippi Alluvial Plain, (8) North Carolina's pocosins, (9) western riparian wetlands, and (10) urban wetlands. The following subsections summarize the nature of these national problems.

Estuarine wetlands of the U.S. coastal zone

Estuarine marshes and mangroves swamps are highly regarded for their commercial and recreational fisheries value. Protecting these wetlands has, however, only recently received national attention. In the past, coastal wetlands were viewed chiefly as potential sites for development. Between the 1950s and the mid-1970s, wetland losses were heaviest in Texas, Louisiana, and Florida (Table 15). The National Marine Fisheries Service (1983) estimated annual fishery losses at $208 million due to estuarine marsh losses from 1954 to 1978. Accelerating wetland destruction aroused much public concern which led to the passage of tidal wetland protection laws in many coastal states and to stricter enforcement of existing federal laws in the 1960s and the 1970s. Nonetheless, estuarine wetlands are still sought after by developers for residential and resort housing, marinas, and other uses.

Estuarine wetland losses have been greatest in five states: California, Florida, Louisiana, New Jersey, and Texas. Louisiana is losing them at a rate of 10 thousand hectares per year due to coastal subsidence and other causes (Fruge 1982; see the following subsection for discussion). Outside of Louisiana, coastal wetland losses are directly related to population density (Gosselink and Baumann 1980). Urbanization (i.e., residential home construction) has been responsible for over 90% of the losses directly attributed to human activities (Frayer *et al.* 1983). Accelerated urban development and increased ground-water withdrawals have resulted in salt water contamination of public water supplies in many coastal communities.

All coastal states, except Texas, have enacted special laws to protect estuarine wetlands. These laws vary considerably in the degree of protection, since a few exempt major activities that alter wetlands or apply only a state-owned lands. Section 10 of the Federal Rivers and Harbors Act of 1899 and Section 404 of the Federal Clean Water Act of 1977 mandate a strong federal role for protecting the nation's coastal wetlands. Federal permits are required for most types of construction in estuarine wetlands. While the regulatory tools to protect coastal wetlands are in place, continued enforcement of existing laws is required to maintain the integrity of the remaining wetlands. In addition to regulation, the Coastal Barrier Resources Act of 1982 removes federal subsidies and discourages development of approximately 1126 kilometers of designated coastal barriers and adjacent wetlands. Its greatest impacts in reducing coastal wetland loss should occur in Alabama, Florida, North and South Carolina, and Texas.

Louisiana's coastal marshes

Louisiana possesses roughly one-third of the coastal marshes in the conterminous U.S. (Turner and Gosselink 1975). The state's multi-million dollar commercial inshore shrimp fishery is directly proportional to the area of intertidal emergent wetland (Turner 1979). Along most coasts, salt marshes appear to be maintaining themselves through marsh building or accretion despite a rise in sea level. In Louisiana, however, this is not true as large expanses of coastal marshes are being permanently flooded by rising sea level. Vertical marsh accretion has not kept pace with coastal submergence over the past 30 years. The marsh is accreting at a rate of 8.4 millimeters yearly, while submergence is occurring at 12.7 millimeters per year (DeLaune *et al.* 1983). Currently, an estimated 116 square kilometers or nearly 12 thousand hectares of coastal marshes are lost each year. Besides direct losses, salt water intrusion is killing freshwater vegetation in tidal freshwater marshes

Table 15. Examples of recent wetland loss rates. Part A contains wetland acreage and percent of total wetland area lost. Part B contains annual rates of loss. * = Loss rate after passage of state coastal wetland protection laws. ** = Only 281,377 hectares of Pocosin remain undisturbed; the rest are partially drained, developed or planned for development (Source: Tiner 1984).
Part A.

State or region	Original wetlands (hectares)	Today's wetlands (hectares)	% of wetlands lost	Source
Iowa's natural wetlands	222,672	10,717	95	Bishop (1981)
California	2,024,291	182,186	91	U.S. Fish and Wildlife Service (1977)
Nebraska's Rainwater basin	38,057	3,425	91	Farrar (1982)
Mississippi alluvial plain	9,716,599	2,105,263	78	MacDonald *et al.* (1979)
Michigan	4,534,413	1,295,547	71	Michigan Dept. of Natural Resources (1982)
North Dakota	2,024,291	809,717	60	C. Elliott, U.S. Fish and Wildlife Service (personal communication)
Minnesota	7,449,393	3,522,267	53	University of Minnnesota (1981)
Louisiana's forested wetlands	4,574,899	2,281,377	50	Turner and Craig (1980)
Connecticut's coastal marshes	12,146	6,073	50	Niering (1982)
North Carolina's Pocosins	1,012,146	608,502**	40	Richardson *et al.* (1981)
South Dakota	809,717	526,316	35	C. Elliott, U.S. Fish and Wildlife Service (personal communication)

Table 15. Continued.
Part B.

State of region	Loss rate (hectares/year)	Source
Lower Mississippi alluvial plain	66,802	MacDonald *et al.* (1979)
Louisiana's forested wetlands	35,304	Turner and Craig (1980)
North Carolina's Pocosins	17,814	Richardson *et al.* (1981)
Prairie Pothole Region	13,360	Haddock and DeBates (1969)
Louisiana's coastal wetlands	10,121	Fruge (1982)
Great Lakes basin	8,097	Great Lakes River Basin Commission (1981)
Wisconsin	8,097	Wisconsin Dept. of Natural Resources (1976)
Michigan	2,632	Weller (1981)
Kentucky	1,457	Kentucky Dept. of Fish and Wildlife Resources (1983)
New Jersey's coastal wetlands	1,249	Ferrigno *et al.* (1973)
	20*	JACA Corporation (1982)
Palm Beach County, Florida	1,237	U.S. Fish and Wildlife Service (1982)
Maryland's coastal wetlands	405	Redelfs (1983)
	8*	
Delaware's coastal wetlands	193	Hardisky and Klemas (1983)
	8*	

and converting these types to more brackish wetlands or open water. It also has accelerated the advance of the predaceous oyster drill into productive oyster beds.

The causes of Louisiana coastal marsh loss are numerous and complicated (Boesch 1982, Craig *et al.* 1980). A combination of factors both natural and man-induced are responsible. Coastal subsidence, rise in sea level and the cyclical processes of Mississippi Delta growth represent the major natural forces. The Mississippi River is trying to shift its course into the Atchafalaya River, but the U.S. Army Corps of Engineers is only allowing 30% of the Mississippi and Red Rivers flows to be moved down the Atchafalaya. This is still enough to get some marsh building in Atchafalaya Bay. An estimated 49 thousand hectares of marsh will be created here in the next 30 to 50 years, but this will not offset heavy marsh losses in other areas of Louisiana (Louisiana State University 1983). Man's impacts include channelization and levee construction along the Mississippi River, canal dredging for navigation and energy operations, and subsidence from extraction of groundwater, minerals, oil and gas. Channelization and canal construction have increased marsh erosion and salt water intrusion along the coast. Man-made levees have disrupted the natural marsh building process by preventing overflow of sediment-rich waters.

Investigators have concluded that the rate of wetland loss in coastal Louisiana, for instance, has actually been increasing rapidly in recent years. The Louisiana Wetland Protection Panel (1987) noted average annual losses of 101 square kilometers of deltaic plain wetland between 1955 and 1978, increasing to 117 square kilometers in 1985. Efforts must be made to reduce man's adverse impacts on Louisiana's coastal marshes. Specific wetland preservation and restoration actions should be taken immediately. These actions include: diverting Mississippi and Atchafalaya River flows into areas experiencing salt water intrusion, creation of new marsh through careful placement of clean dredged material, improved water management in existing marsh areas, and reducing petroleum industry canal dredging through increased use of directional drilling. Future research studies should improve our understanding of the importance of causal factors and address mechanisms to improve the future for this rapidly diminishing resource.

Chesapeake Bay's submerged aquatic beds

Situated in eastern Maryland and Virginia, the Chesapeake Bay is the largest estuary in the contiguous United States. Many rivers drain into the Bay including the Susquehanna, Potomac, Patuxent, James, York, and Chester. The Bay once represented the primary overwintering area for canvasback ducks (*Aythya valisineria*) which fed on submerged aquatic vegetation. Fifty percent of the eastern population of canvasbacks were found in the Bay region (Stevenson and Confer 1978). While still among the more important overwintering areas for canvasbacks, the Chesapeake Bay is the single most important wintering ground in North America for the eastern population of tundra swans (*Cygnus columbianus*) (Bellrose 1976). Canada geese (*Branta canadensis*) and black ducks (*Anas rubripes*) also use the Bay area in winter. Aquatic grass beds provide spawning areas for estuarine-dependent fishes including striped bass (*Morone saxatilis*), shad (*Alosa sapidissima*), and herring (*Clupea harengus*, *C. pallasi*) and offer shelter for their young. Important submerged plants include pondweeds (*Potamogeton* spp.), redhead grass (*Panicum rigidulum*), eelgrass (*Zostera marina*), wild celery (*Vallisneria americana*), naiads (*Najas* spp.), waterweed (*Elodea* spp.), muskgrasses (*Chara* spp.), and Eurasian milfoil (*Myriophyllum spicatum*). According to Stevenson *et al.* (1979), submerged aquatic vegetation in Maryland decreased by almost 66% from 1971 to 1978 in selected areas. A similar decline has also been observed in Virginia waters. At the mouth of the Susquehanna River in Maryland, submerged grasses at a once prime waterfowl feeding area have virtually disappeared since 1971. Other areas have experienced declines in the numbers of plant species present. Since 1978, submerged aquatic vegetation appears to have stabilized, with a few areas even showing a slight increase (Orth and Moore 1981). Reductions in submerged vegetation have probably been the most important wintering habitat change which have led to declines in local populations of canvasbacks and redhead (Perry *et al.* 1981). These changes point to a stressed ecological system.

Although the causes for this vegetation decline are hard to pinpoint, researchers suggest a combination of natural and human-induced factors. Natural stresses include overgrazing by carp and cownose rays (*Rhinoptera bonasus*), Hurricane Agnes, a general warming of Bay waters, and natural diseases. In June 1972, Hurricane Agnes hit the Bay region. Its heavy rainfall lowered salinity in Chesapeake Bay and buried numerous grass beds with sediment carried by runoff. Human impacts on the submerged vegetation are largely from two general sources of water pollution: point and nonpoint sources. Point source pollution comes mainly from industrial and sewage treatment plant discharges, while nonpoint sources include failing septic systems, agricultural runoff, and urban runoff. These sources cause increased turbidity and sedimentation, nutrient overloading, and chemical pollution which have reduced or eliminated desirable aquatic beds from many areas. Channelization projects in bottomland hardwood forested wetlands have undoubtedly contributed to the problem by accelerating the discharge of agricultural runoff and eroded soil into the Bay.

The problem of the Bay's submerged aquatic vegetation is receiving special attention from the U.S. Environmental Protection Agency (EPA) and others.

EPA established a Chesapeake Bay Program to address this problem. Future studies should increase our understanding of the causes of the decline of desirable submerged aquatic vegetation and will hopefully lead to improved watershed management to restore and maintain a healthy Chesapeake Bay. Meanwhile, the governors of Maryland, Pennsylvania, and Virginia have joined together to address water quality problems in the Chesapeake Bay watershed. Only through interstate coordination and action can the Bay's problems be solved.

South Florida's palustrine wetlands

South Florida encompasses a 23 thousand kilometers area of lakes, rivers, and wetlands which extends from the central part of the Florida peninsula to the southernmost keys. While the Everglades dominates this region, Big Cypress Swamp, the Kissimmee River, and Lake Okeechobee are equally important. Freshwater runoff from this area helps maintain the salinity balance of estuaries which support 85% of South Florida's offshore fishery (Yates 1981). The wetlands are breeding grounds for many birds, notably wood ibises (*Mycteria americana*) and other ibises (*Plegadis falcinellus* and *Eudocimus albus*), roseate spoonbills (*Ajaia ajaja*), herons (*Ardea herodias*, *Egretta tricolor*, *E. caerulea* and *Butorides virescens*), and egrets (*E. rufescens*, *Casmerodius albus*, *E. thula* and *Bubulcus ibis*). They also support wintering populations of numerous waterfowl, especially lesser scaups (*Aythya affinis*), ring-necks (*A. collaris*), blue-winged teal (*Anas discors*), canvasbacks (*Aythya valisineria*), and American wigeons (*Anas americana*). Rare and threatened animals depend on these wetlands, including the Florida panther (*Felis concolor coryi*), American crocodile (*Crocodylus acutus*), West Indian manatee (*Trichechus manatus*), brown pelican (*Pelecanus occidentalis*), Everglades kite (*Rostrhamus sociabilis*), and southern bald eagle (*Haliaeetus leucocephalus*). The Everglades National Park was established to protect these natural resources.

South Florida's waters and wetlands have been subjected to various uses for many years (Yates 1981). In the 1920s, large wetland areas were drained and converted to sugar cane farms. Severe floods in 1928, 1947 and 1948 stimulated a massive flood control project in South Florida. The Central and Southern Florida Flood Control Project, authorized by the U.S. Congress, required the U.S. Army Corps of Engineers to construct a network of nearly 1,290 kilometers of new or improved levees and 805 kilometers of canals. This project completed drainage of the Kissimmee River wetlands, regulated Lake Okeechobee's water levels, and drained and irrigated the Everglades agricultural area. Channelization directly destroyed 16 thousand hectares of wetlands and facilitated drainage of more than 40 thousand hectares of contiguous wetlands (Thompson 1983). By reducing floods, the flood control project also accelerated filling of wetlands in some counties for urban expansion of coastal cities, as well as increasing agricultural conversion of wetlands in other areas. For example, between 1972 and 1980, Palm Beach County lost 9.6 thousand hectares of wetlands to agriculture and 265 hectares to urban development (U.S. Fish and Wildlife Service 1982) for a 6% wetland loss in just eight years.

Problems related to water supply have also resulted from this flood control project. Although three large impoundments called "conservation areas" were constructed to maintain recharge of the Biscayne Aquifer, salt water intrusion remains a constant threat. Urban growth and agricultural development increase demand for water. Public wells have been constructed further west which have lowered the Everglades water table and increased the flow of salt water into the Biscayne Aquifer. Besides public water supply problems, the flood control project has also seriously disrupted the natural hydrologic regime of the Everglades National Park. One levee, L-29, completely blocked sheet flow of freshwater into the Park in 1963. After much controversy and public debate, the Corps of Engineers in 1970 agreed to release a minimum of 388 million cubic meters of water annually (Yates 1981). Park officials estimate that at least twice this amount is needed and that the water must be distributed over a wider areas and be release on a more natural regime. These changes are necessary to help maintain the biological integrity of the Everglades National Park.

Wetland alterations in South Florida have created problems for many fish and wildlife species. Periodic discharges of freshwater from the conservation areas have disrupted fish nursery grounds in estuaries. Colonial wading bird populations have declined from about 1.5 million in 1935 to about 0.25 million today. Alligators have been eliminated from many areas and frog populations have been critically reduced from a commercial harvesting standpoint (Marshall 1981).

Possible effects of the Kissimmee River channelization and wetland drainage on local rainfall patterns have also been raised. Although quite controversial, some scientists have suggested that wetland drainage in South Florida has reduced the mist of evaporation and plant transpiration which triggers rainfall from sea breezes. This condition may be responsible for recent severe local droughts.

In 1976, the Florida legislature passed a mandate to restore the Kissimmee River. They recognized that channelization of this river among other things: increased the seriousness of water shortage and droughts, degraded water quality of Lake Okeechobee, eliminated vast acreages of wetlands, drastically reduced fish and wildlife populations, and destroyed a beautiful meandering river (Barada 1977). Ironically, the flood control project actually increased the potential for catastrophic floods and raised the costs to ranchers and farmers. Florida's Save Our River Act in 1981 created state funds to purchase threatened wetlands. Nongovernmental organizations such as The Nature Conservancy, the Richard King Mellon Foundation, and National Audubon Society have also been active in wetland acquisition. In 1983, the Governor of Florida announced a multimillion dollar "Save Our Everglades" program to restore the ecology of the Everglades, which includes acquisition of 101 thousand hectares of wetlands and improving hydrology (Thompson 1983). He also stressed the importance of federal-state cooperation in achieving this goal. These efforts should be instrumental in preserving these fragile wetlands and their associated values.

Prairie Pothole Region's emergent wetlands

Prairie potholes are the most valuable inland marshes for waterfowl production in North America. Although the Pothole Region accounts for only 10% of the continent's waterfowl breeding area, historically it has produced 50% of the duck crop in an average year and more than that amount in wet years (Smith *et al.* 1964). The Prairie Pothole Region extends from south-central Canada to the north-central United States, covering about 77 thousand square kilometers with roughly one-third in the United States. Due to glaciation thousands of year ago, the landscape is pock-marked with millions of pothole depressions. These pothole wetlands serve as primary breeding grounds for many kinds of ducks including: *Anas platyrhynchos*, *A. acuta*, *A. americana*, gadwall (*A. strepera*), northern shoveler (*A. clypeata*), teal (*Anas* spp.), *Aythya valisineria*, and redhead (*Aythya americana*). For example, in a study area in northeastern South Dakota, researchers found an average of 140 ducks produced in an area of 2.59 square kilometers per year (Evans and Black 1956).

In North and South Dakota, pothole wetlands originally covered 2.8 million hectares. Today, only slightly more than 1.6 million hectares remain. Over half have been destroyed by agriculture, irrigation, and flood control projects (C. Elliott, U.S. Fish and Wildlife Service, personal communication). Iowa has lost more than 95% of its natural marshes (Bishop 1981). Approximately 2.3 million hectares of potholes have been drained in western Minnesota. Since pothole wetlands are surrounded by farmland, they have been drained to create additional cropland, mostly for wheat and corn. Drainage in the Dakotas is largely done by open ditching in contrast to both open ditching and tile drainage in Minnesota and Iowa. These ditches drain into intermittent streams or highway right-of-way ditches. Highway ditches have been heavily used by local farmers to help drain wetlands. In western

Minnesota alone, an estimated 40.5 thousand hectares of wetland have been lost in this way. In addition, stream channelization sponsored by federal flood control projects, such as the small watershed protection and flood prevention program, have led to accelerated wetland drainage in the Pothole Region as they have elsewhere in the U.S. (Erickson *et al.* 1979). Drainage data for the Dakotas and Minnesota obtained from the U.S. Department of Agriculture's Production and Marketing Administration show that 76 thousand hectares were drained with federal assistance in 1949 and 1950 alone. Countless other acres were privately drained at the same time. Pothole wetland losses are estimated at more than 13.4 thousand hectares yearly (Haddock and DeBates 1969). Among the remaining wetlands, the drier ones (i.e., temporarily flooded) are often tilled during dry periods of the natural hydrologic cycle.

Each pothole drained leads to a further concentration of the breeding waterfowl population. This could result in decreased productivity, reduced size of the breeding population, and/or increased likelihood of diseases like avian cholera and botulism. Wetland drainage also destroyed habitats important to invertebrates used as food by breeding waterfowl such as *Anas acuta*, and *A. discors* (Krapu 1974, Swanson *et al.* 1974). Moreover, drainage eliminates the flood storage value of pothole depressions, thereby increasing flooding problems as in the James River Basin of North Dakota (Sidle 1983).

Agricultural activities on upland adjacent to potholes have also adversely impacted waterfowl production. Upland grasses bordering wetlands provide valuable nesting cover for mallard and other dabbling ducks. Conversion of rangelands to cropland, which destroys these nesting areas, has been accelerating. Between 1965 and 1975, approximately one half of the rangelands in the Coteau du Missouri counties of North Dakota were converted to cropland (U.S. Fish and Wildlife Service 1984). Excavation of ponds (dugouts) in pothole wetlands is also a problem. Out of an estimated 55,855 dugouts in eastern South Dakota, in 1976, 77% were in wetland basins or streambeds.

The Fish and Wildlife Service has been active in preserving Prairie Pothole wetlands through acquisition, easement, and other programs (Table 16). Recently, wetland acquisition in North Dakota was stopped for several years by state law. Due to a U.S. Supreme Court ruling against the state for this action, the Service's wetland acquisition is being resumed. The Clean Water Act generally regulates filling of pothole wetlands four hectares in size or larger, yet smaller isolated wetlands are largely unprotected. A 1984 settlement agreement between the Corps of Engineers and various environmental groups provided an opportunity to improve regulation of agricultural conversion of pothole wetlands. The Fish and Wildlife Service's acquisition and

Table 16. U.S. Fish and Wildlife Service wetlands in fee title ownership. MN = Minnesota, ND = North Dakota, SD = South Dakota. All water areas include wetland and deepwater basins in the prairie pothole region. * = Percent in fee title ownership exceeds percent natural distribution for semipermanent, intermittently exposed, and unknown wetland types.

Wetland type	States MN	ND	SD	% of all	Fee title	% of all
Temporary	931	12,528	2,252	15,711	4.9	13.6
Saturated	30,638	435	–	31,073	9.6	11.1
Seasonal	16,489	17,385	7,144	41,018	12.7	29.1
Semipermanent	25,216	67,328	20,336	112,880	35.0	21.6*
Intermittently exposed	3,324	100,577	4,946	108,847	33.8	10.4*
Permanent	2,942	–	–	2,942	0.9	13.0
Unknown	1,170	8,356	193	9,718	3.0	1.2*
Totals	80,710	206,609	34,871	322,189	100	100

easement program and improved federal regulations are needed to maintain valuable waterfowl producing wetlands, since pressures continue to convert such areas to agriculture.

The Food Security Act of 1985 contained a provision popularly referred to as the "Swampbuster" provision. It says that any person who in any crop year produces an agricultural commodity on converted wetland shall be ineligible for certain farm program benefits during that year. The list of programs consists of price and income support payments, farm storage facility loans, grain storage payments, Farmers Home Administration loans, crop insurance, disaster payments, and loans used to convert wetlands. Exemptions are included to make the law workable.

Wetlands of Nebraska's Sandhills and Rainwater Basin

Wetlands with the Sandhills and Rainwater Basin of south-central Nebraska are important to migratory sandhill cranes (*Grus canadenis*) and waterfowl that migrate along a central corridor of the country. About 2.5 million ducks and geese move through the Rainwater Basin each spring. Ninety percent of the mid-continent's white-fronted geese (*Anser albifrons*) stage in wetlands of the Basin and central Platte River each spring. Pheasants also depend on wetland vegetation for nesting and brood habitat (Farrar 1982). Eighty percent of the continent's population of sandhill cranes depend on wetlands along 113 kilometer of the Platte and North Platte Rivers as staging areas during spring migrations. Whooping cranes (*Grus americana*), an endangered species, also roost in broad reaches of the Platte River's channels (U.S. Fish and Wildlife Service 1981).

The Nebraska Sandhills Region is the largest sand dune formation in the western hemisphere covering approximately 52 thousand square kilometers. Formed primarily by wind action, the Sandhills consist of stabilized sand dunes, exposed ground-water lakes in the valley, and perched mineralized lakes on poorly drained soils. The grassland economy of the Sandhills is primarily one of cattle grazing. Large acreage of subirrigated meadows with water tables close to the surface offer great potential for increased grazing and hay production through development of level ditching. Wetland destruction in the Sandhills has accounted for over 11.3 thousand hectares or 15% of the original wetlands (Nebraska Game and Parks Commission 1972). Wetland loss has resulted from drainage, filling for pivot irrigation, and reduced ground-water levels from deep well irrigation.

Decreases in riverflows of the Platte River by upstream diversions for consumptive uses in the states of Colorado, Wyoming, and western Nebraska have reduced channel width by 80–90% in many areas. This condition has promoted growth of woody vegetation on former channel bars and islands. Sandhill cranes prefer roosting in shallows and sandbars where the channel is at least 152 meters wide and strongly avoid narrower channels. Reduction in natural channel width and increased growth of woody vegetation have caused crowding at remaining roost sites. This situation increases crane susceptibility to catastrophic losses due to severe storms and diseases. If the trend continues, sandhill cranes may shift to the Rainwater Basin where avian cholera is already a serious problem. Native grasslands along the rivers have also declined. These areas provide important food for the migrating cranes (U.S. Fish and Wildlife Service 1981).

The Rainwater Basin includes parts of 17 counties, roughly 11 thousand square kilometers in extent. Wetlands are formed in depressions underlain by clay on the rolling plain. Originally 4,000 marshes totaling 38 thousand hectares existed. Wetland destruction accelerated after World War II due to improved earth-moving equipment and deep well irrigation. Agriculture intensified in the Basin with the help of federal funds and technical assistance for wetland drainage. By the late 1960s, 18% remained and in 1981, less than 10% of the wetlands survived. Nine out of every ten wetlands have been drained or filled. Of those remaining, 43% are protected by state or federal wildlife agencies.

Losses of basin wetlands have forced ducks and geese to concentrate in the remaining wetlands. In 1980, about 80,000 waterfowl died due to avian cholera. This was the second largest cholera die-off reported in the country. During dry years with late winter storms, birds are forced to crowd into Basin wetlands, setting the stage for large die-offs. Waterfowl breeding populations have also been affected by wetland destruction. By 1975, the duck breeding population declined so much that the Nebraska Game and Parks Commission discontinued its aerial breeding bird survey.

Efforts to protect remaining wetlands have recently been weakened. The Federal Water Bank Program which provides payments to landowners preserving important waterfowl wetlands has been funded at lower levels. Wetland protection under the Clean Water Act of 1977 has been reduced through regulatory changes. Legal disputes between the U.S. Fish and Wildlife Service and others over water rights have affected management of 6.3 thousand hectares of waterfowl production areas in the Rainwater Basin. Along the Platte and North Platte Rivers, action is needed to protect native grassland near river channels and to maintain channel widths of 150 meters or more for suitable crane roost sites during migration. Acquisition and conservation easements are useful tools.

Forested wetlands of the Lower Mississippi Alluvial Plain

The bottomland hardwood forests of the lower Mississippi floodplain are among the nation's most important wetlands. They are prime overwintering grounds for many North American waterfowl, including 2.5 million of the 3 million mallards of the Mississippi Flyway, nearly all of the 4 million wood ducks, and many other migratory birds. Numerous finfishes depend on the flooded hardwoods for spawning and nursery grounds. These wetlands also support many other wildlife, including *Odocoileus virginianus*, squirrels (*Sciurus* spp.), *Procyon lotor*, *Mustela vison*, *Myocastor coypus*, *Castor canadensis*, foxes (*Vulpes* spp.), and rabbits (*Sylvilagus* spp.). They also play a vital role in reducing flooding problems by temporarily storing large quantities of water and by slowing the speed of flooding waters. In the process, these wetlands remove chemicals from the water such as fertilizers and pesticides and trap soil eroding from nearby farmlands.

Originally, the Mississippi Alluvial Plain included nearly 9.7 million hectares of bottomland forested wetlands. By 1937, only 4.8 million hectares or 50% of these remained. Today, there are less than 2.1 million hectares left, roughly 20% of the original acreage (MacDonald *et al.* 1979). Over half of this wetland is in Louisiana, with large amounts also in Arkansas and Mississippi. These forested wetlands have been cleared and drained for crop production. Federal flood control projects and small watershed projects have accelerated wetland conversion to cropland, especially from the 1950s to the present. An estimated 2.1% of the remaining bottomland forests are lost annually.

Historically, cotton and corn were the primary crops raised on former bottomlands, but since the mid-1950s, soybeans have dominated. In 1977, cropland acreage in soybeans amounted to more than the combined acreage of the four other principal crops of cotton, wheat, rice, and corn. Soybeans have major advantages over the other crops: (1) they have a very short growing season, so they can be planted in areas that are flooded till late June, and (2) they can be planted in a variety of soil conditions. Other crops, like cotton, require better drained soils than soybeans or rice. Heavy foreign demand for soybeans has made it the most lucrative cash crop. Traditionally, natural stands of bottomland hardwood forests were cut for timber. Recently, in an effort to maximize timber production, cottonwood and other silviculture plantations have been established to a limited extent. However, the economics of hardwood production cannot compete with farm crops, where they can be grown. The net economic return per acre is twice as high for farmland as for forest. Thus, conversion of bottomland hardwood forest to cropland can be expected to continue in the Mississippi Alluvial Plain as well as elsewhere in the Southeast. These losses seriously threaten some wildlife populations and increase the frequency of damaging floods like the April 1983 floods that caused millions of dollars of damage in the states of Louisiana and Arkansas.

The Federal Clean Water Act can be instrumental in regulating conversion of bottomland hardwood forests to agricultural uses. A 1979 U.S. District Court decision stated that a Section 404 permit is required for land clearing of wetlands for agriculture. Subsequently, the Corps of Engineers took a conservative position and regulated land clearing only in the Western District of Louisiana. On September 26, 1983, the Fifth Circuit Court of Appeals decision affirmed the district court's opinion by rejecting the contention that land clearing is a normal farming activity exempt from Section 404 permit requirements. This decision provides the legal framework for protecting remaining bottomland wetlands as well as other inland wetlands subject to agricultural conversion. In early 1984, an out-of-court settlement agreement on a U.S. District Court case among other things, ordered the Corps of Engineers to issue a regulatory guidances letter to be used in the permitting process. The future outcome of these decisions should lead to improved wetland protection under the Clean Water Act. Besides improved regulation, acquisition of bottomland hardwood forests in the Lower Mississippi Alluvial Plain is needed to protect the remaining wetlands. Accelerated acquisition efforts by the Fish and Wildlife Service, the State of Louisiana, the Nature Conservancy, and others are important steps to preserving these threatened forested wetlands.

North Carolina's Pocosins

Along the southeastern coastal plain, numerous evergreen forested and scrub-shrub wetlands called "pocosins" are found. Pocosins lie in broad, flat upland areas away from large streams. Their vegetation consists of a mixture of evergreen trees including *Pinus serotina*, loblolly bay (*Gordonia lasi-*

anthus), red bay (*Persea borbonia*), and sweet bay (*Magnolia virginiana*) with shrubs, including titi (*Cyrilla racemiflora*), zenobia (*Zenobia pulverulenta*), *Lyonia lucida*, wax myrtle (*Myrica* spp.), and leatherleaf (*Chamaedaphne calyculata*). Seventy percent of the nation's pocosins are in North Carolina, where they alone comprised about 890 thousand hectares or half of the state's freshwater wetlands in 1962 (Richardson *et al.* 1981).

Although pocosins are not essential for any wildlife species throughout its range, they do provide important habitat for many animals, especially black bear (*Ursus americanus*) along the coast (Monschein 1981). For example, the Dismal Swamp is reported to be the last refuge for black bears in coastal Virginia. More importantly, pocosin wetlands in coastal North Carolina are closely linked with the riverine and estuarine systems (Richardson 1981b, Street and McClees 1981). They help stabilize water quality and balance salinity in coastal waters. This is especially important for maintaining productive estuaries for commercial and recreational fisheries.

Historically, forestry and agriculture have had important influences on pocosins. During the past 50 years, forestry uses of pocosins have increased and today about 44% of North Carolina's pocosins are owned by major timber companies (Richardson *et al.* 1981). While some pocosins were drained and converted to pine plantations or agriculture prior to the early 1960s, most of the commercial development is more recent. Since 1970, timber companies have transferred nearly 203 thousand hectares to large-scale agriculture. Agricultural drainage has focused on the Albemarle-Pamlico peninsula where large corporate farms own 162 thousand hectares of pocosins. In addition to land clearing and extensive ditching, farming in these former wetlands requires adding fertilizers and lime. For example, 3.6 to 7.3 metric tons of lime must be added to new agricultural land, with 1.1 metric tons added every three years to keep former pocosin soils fertile (McDonald *et al.* 1983). Runoff from these farmlands degrades water quality of adjacent estuaries. Changes in nutrient loading and salinity patterns of adjacent estuaries have been observed (Barber *et al.* 1978). These changes may adversely impact fish nursery grounds.

Although forestry and agricultural uses of pocosins continue, peat mining represents a new threat to these wetlands. Peat deposits about four feet thick generally exist in coastal North Carolina. Interestingly enough, some of the large agricultural corporations which own many pocosins are already involved in peat mining operations. On December 22, 1982, the U.S. Synthetic Fuels Corporation endorsed federal subsidies for a $576 million synfuel project in North Carolina. This project would remove peat from 6,000 hectares of pocosins to produce methanol fuel and the land would subsequently be converted to farmland. The project was abandoned in early 1984 after the U.S. Synfuels Corporation formally rejected the proposal for loan and price guarantees. This practice of peat mining and agriculture has been conducted for years in northern states like Minnesota. Peat is very competitive with coal as a fuel for electric power generation.

About 1 million hectares of pocosins once existed in North Carolina (Richardson *et al.* 1981). Today, nearly 405 thousand hectares survive in their natural condition. Thirty percent of the original pocosins were converted to agriculture or managed forests, while another 30% was partially drained or cleared or planned for development. Federal wetland protection efforts through the Clean Water Act have been inconsistent to date. In September 1983, the U.S. Army Corps of Engineers was sued by various environmental groups over the Corps' failure to take jurisdiction over a large pocosin. The outcome of this court case may establish guidelines for future protection. If the present trend continues, however, we can expect that many pocosins will be lost in the near future. Moreover, a predicted change in estuarine salinity patterns may adversely affect valuable fish and shellfish nursery grounds and North Carolina's multi-million dollar commercial fishery.

Western riparian wetlands

Lands within the 100–year floodplain and along the margins of ponds and lakes in the arid and semiarid regions of the country (e.g. Arizona, New Mexico, Utah, Nevada, Colorado, California, and eastern Oregon and Washington) are commonly called riparian ecosystems. They include both wetlands along streams and other waterbodies, and uplands on floodplain terraces. Existing information on the extent of this resource does not make a clear distinction between wetlands and upland because the system as a whole is so important. However, loss of riparian habitats in general serves to reflect trends in associated wetlands.

Riparian ecosystems provide abundant food, cover, and water for resident and migrating animals. These thin ribbons of vegetation along streams and lakes support a disproportionately large variety of wildlife. Woody vegetation is used for nesting by birds and for food and shelter by various mammals. Mule deer (*Odocoileus hemionus*) migrate along streams between high elevation summer ranges and low elevation winter ranges (Thomas *et al.* 1979). Cottonwood (*Populus deltoides*) and willow (*Salix* spp.) wetlands are the prime bird habitats in the West (Anderson *et al.* 1977). Migrating birds follow the Rio Grande corridor in the spring and fall and riparian wetlands are very important to these birds (Wauer 1977). Along the Lower Verde River in Arizona, 166 bird species frequented riparian habitats, including the endangered bald eagle (*Haliaeetus leucocephalus*) and endangered Yuma clapper rail (*Rallus longirostris yumanensis*) (McNatt *et al.* 1980).

Unfortunately, riparian ecosystems have been grossly mistreated by man to the point where we can say that they represent the most modified land type in the western United States. Many riparian forests have been converted to cropland and grass prairie. Others have been badly overgrazed by livestock. Heavy grazing has destroyed understory vegetation and has prevented regeneration of riparian vegetation in many places. In Arizona, dam construction on rivers poses the greatest threat to remaining riparian lands (Todd 1978). Pumping of groundwater for irrigation, municipal, and industrial uses has lowered the water table in many areas, drying up riparian wetlands and/or changing plant species composition.

The magnitude of riparian forest losses is alarming. For example, cottonwood communities along the Colorado River in Arizona have been reduced by 44%, while in Colorado more than 90% of the river's riparian habitats were destroyed (Ohmart *et al.* 1973). Only 2% of the original riparian forest along the Sacramento River in California remains (McGill 1975, 1979). In Oklahoma, Rush and Wildhorse Creeks in the Washita watershed experienced a 93% and 84% reduction in bottomland forests between 1871 and 1969 (Barclay 1980). Today, no natural wetlands exist within their floodplain.

Flood control projects supported by public law have reduced flood frequency and magnitude. This in combination with channelization, has created drier conditions which may be the main factor for lower abundance of amphibians, reptiles, birds, and mammals on channelized sites (Barclay 1978). Besides direct losses of habitat, the quality of remaining riparian lands is changing due to water quality degradation, reduced stream flow, and the invasion of saltcedar (*Tamarix* spp.), an exotic tree of lower wildlife value (Ohmart *et al.* 1973). Because these riparian zones are of such tremendous value to wildlife, it is incumbent upon public agencies to treat them with a conservationist attitude. When a water project does extensive damage to a riparian area, there should be every effort made to mitigate that damage, either by planting of riparian species in nonvegetated riparian areas or acquisition and enhancement of existing riparian zones.

Urban wetlands

Wetlands near urban centers are under increasing development pressure for residential housing, industry, and commercial facilities. Rising population and economic growth create high demand for real estate in suburban localities. As suitable upland becomes exhausted, pressure intensifies to develop wetlands for residential housing, manufacturing plants, business office complexes and similar uses. In many communities, urban wetlands represent the last large parcels of open space. They often are also the final haven for wildlife in an increasing urban environment and these wetlands support many upland animals displaced by development.

With accelerating development of adjacent uplands, the role of urban wetlands in flood protection and water quality maintenance becomes critical. Urban and industrial development increases the amount of surface water runoff from the land after rainfalls. This raises flood heights and increases flow rates of the rivers, thereby increasing the risks of flood damages. Increased runoff brings with it various substances that degrade water quality, such as fertilizer chemicals, grease and oil, road salt, and sediment. Effluent from some sewage treatment plants built to handle the needs of growing communities also reduce water quality. By passing through wetlands, a type of cleansing action takes place as many pollutants are removed from the water and retained or utilized by the wetlands. Urban wetlands in certain instances function as recharge areas. This is especially true in communities where groundwater withdrawals are heavy. Thus, urban wetlands may be essential to preserving public water supplies.

Urban wetlands are vulnerable to development for several reasons including: (1) in many cases, they represent the last large tracts open land, (2) increased population in metropolitan areas has raised land values and demand for real estate, (3) relatively new interstate highways have improved access to many areas which has increased development opportunities, (4) wetlands may be zoned for light industry or residential housing by local governments, (5) the lack of any comprehensive state wetland protection for inland wetlands in most states, and (6) many inland wetlands do not meet specific requirements for federal jurisdiction under Section 404 of the Clean Water Act of 1977. Many of the more urbanized states have passed wetland protection laws. New Jersey and Maryland are among the latest states to enact laws to protect their remaining inland wetlands. Similar initiatives are needed in other states to reduce losses of inland wetlands to urbanization and industrial development. Moreover, federal regulation under the Clean Water Act is also vital to protecting these wetlands.

The future of U.S. wetlands

While predicting the future of the nation's wetlands is extremely difficult and complex, an examination of recent trends in population, agriculture, and wetland protection provides insight into what can be expected. Population growth and distribution and agricultural development greatly affect land-use patterns which impact wetlands. Government's wetland protection efforts are key to preserving wetland functions and values for today's public and for future generations. Ironically, once wetlands have been destroyed, humans

Table 17. Examples of potential manmade substitutes for wetland services lost due to wetland destruction (source: National Fish and Wildlife Foundation 1988).

Ground water recharge – artificial recharge pits, reservoir construction, induced recharge, sediment flushing to increase recharge.

Flood control – dams, floodways, dikes, levees, floodwalls, diversions, zoning, relocation of property, land acquisition, flood proofing, detention depressions, reservoirs, land treatment measures.

Shoreline anchoring – riprap, bulkheads, jetties, stream restoration, regulation of boat traffic, zoning of erosion-hazard areas, relocation of property, tax policies, land acquisition, flood proofing, flood forecasting, detention depressions, reservoirs, land treatment measures.

Sediment/toxic retention – sedimentation depressions, land treatment measures, dilutional flushing, buffer strip policies, zoning, tax policies, water treatment facilities, dredged removal of contaminants.

Nutrient retention – same as Sediment/Toxic Retention, plus chemical treatment, aeration/circulation.

Fishery habitat, aquatic diversity – creation of replacement habitat, diversion of fishing effort to unaffected species or non-fishing industries or recreational activities, improvement of habitat (e.g. stream restoration, placement of artificial shelters), stocking, predator management, modification of harvest restrictions, regulation of other limiting factor (e.g. pollutants).

Wildlife habitat, general diversity – similar to Fishery Habitat (above).

Active recreation – diversion of activities to alternate sites, construction of new sites (e.g. reservoirs, swimming pools), diversion to less water-dependent activities.

then attempt to replace their lost functions through numerous costly engineering projects and further landscape alterations (Table 17).

The U.S. population is growing by 1.7 million each year. In 1976, nearly 53% of Americans lived within 50 miles of a major coast. Population density in the coastal zone was six times that of the rest of the country (Council of Environmental Quality 1981). Pressures to develop estuarine and palustrine wetlands in coastal areas will remain intense, despite the existence of laws to protect estuarine wetlands. As adjacent upland becomes developed, public managers will be greatly challenged to protect wetlands from future development. A recent population shift from industrialized northeastern and north-central states to states of the Southeast and Southwest will increase urban and industrial development pressures on wetlands in these latter regions. This new growth will also heighten competition for water between agricultural and nonagricultural users, with fish and wildlife probably being the biggest losers.

Since 1970, non-metropolitan areas have grown faster than metropolitan areas. Suburban counties have grown most rapidly, threatening remaining wetlands with urban development. Since most states do not have wetland protection laws, federal regulation through the Clean Water Act is the key means to protecting these wetlands.

Increases in the world's population are likely to continue to have significant impacts on America's wetlands through agriculture. In the 1970s, U.S. export of grains and soybeans accelerated to help meet the worldwide rise in demand for food. This increased demand for U.S. farm products reversed a 40–year trend of declining cropland use (National Research Council 1982). It also led to conversion of vast acreages of bottomland forested wetlands to cropland, especially in the Mississippi Alluvial Plain. Increased demand for U.S. food will add more pressure to drain wetlands. Without adequate regulations, many palustrine wetlands will be converted to cropland in the near future. Other recent agricultural trends likely to increases wetland conversion include:

1. Increasing costs of production and declining net returns per unit of product force farmers to increase production.
2. Conversion of prime agricultural land to nonagricultural uses (e.g. urban) will lead to conversion of rangelands to pasture, and wetlands to cropland.
3. Increasing irrigation will lower water tables and dry up wetlands, especially in the West.

The Food Security Act of 1985 contains a provision popularly referred to as the "Swampbuster" provision which was described earlier. The statute does not provide sanctions for converting wetlands to agriculture or any other function. The sanctions are only for planting annual crops during the current year on wetlands converted after enactment. As a result, there are numerous instances in which wetlands can be lost, while the law provides no penalties. The intent of the law is to reduce federal subsidies which provide financial incentives to convert wetland to cropland. Depending upon how it is implemented and enforced, this provision could be an important deterrent to wetland conversion. The effectiveness of the Swampbuster provision in slowing the rate of wetlands conversion is yet to be determined. Although denying farm program benefits to operators who plant on converted wetlands will forestall conversion in many areas, the impact of the sanctions will vary greatly from region to region, being most effective in areas where participation in farm programs is high.

Agriculture will also continue to play a major role in degrading water quality, fish and wildlife habitat, and the quality of wetlands, unless improved management technique are employed. About 68% of water pollution in the U.S. is caused by agriculture, with soil erosion from cropland being the single greatest contributor to stream sediment (National Research Council 1982). Before considering conversion of wetlands and other lands to agricultural uses, improved soil management practices should be employed on existing farmland.

Conservation initiatives

A variety of techniques have been used in the United States to protect our remaining wetlands, including land-use regulations, direct acquisition, conservation easements, tax incentives, and public education. Kusler (1983) describes these techniques in great detail in *Our National Wetland Heritage: A Protection Guidebook*. Opportunities also exist for state private initiatives by individual landowners, groups, and corporations to help in conserving our Nation's wetlands. For a more detailed discussion refer to Burke *et al.* (1988), Kusler (1978), and Rusmore *et al.* (1982).

Wetland protection in the U.S. currently is accomplished by two primary techniques: (1) acquisition of priority wetlands and (2) regulation of wetland uses. Both federal and state governments are involved to varying degrees in wetland acquisition and regulation. The use of tax incentives to encourage preservation of wetlands by landowners, although not widely used to date, represents a potentially valuable tool for protecting wetlands. The removal of government subsidies which encourage wetland destruction would also benefit wetlands greatly.

Acquisition of wetlands to preserve fish and wildlife values is ongoing at both federal and state levels. The three key Federal programs are: (1) the Fish and Wildlife Service's National Wildlife Refuge System (2) the Soil Conservation Service's Water Bank Program, and (3) Migratory Bird Conservation Act. The Fish and Wildlife Service's acquisition efforts focus on wetlands important to migratory birds, especially waterfowl breeding and overwintering grounds. Wetlands are protected by direct purchase or by acquiring conservation easements which prevent wetlands from being drained, burned, leveled, or filled. The Migratory Bird Conservation Act of 1929, the Migratory Bird Hunting and Conservation Stamp Act of 1934, and the Land and Water Conservation Fund Act provide the authority and/or funds to purchase wetlands. The F.W.S. presently controls nearly 13 million hectares of palustrine wetlands and about 810 thousand hectares of estuarine wetlands. Most of this area (11.3 million palustrine hectares and 400 thousand estuarine hectares) is in Alaska.

A second important preservation program is administered through the Agricultural Stabilization and Conservation Service's (ASCS) Federal Water Bank Program. Under this program, wetland owners enter an agreement with ASCS promising not to drain, burn, fill, level, or use the wetland for 10 years, In exchange, the landowner receives a standardized annual payment that is determined for the entire country by the administration of the U.S. Department of Agriculture. If the land is also under a F.W.S. agreement, the annual payment is reduced by 20 percent. When accepting an area into the program, ASCS tries to maintain a 3:1 or 4:1 ratio of uplands to wetlands.

During 1989, the F.W.S. purchased 12,760 hectares and brought under easement or lease an additional 7,366 hectares of waterfowl production areas (Annual Reports of Lands Under Control of F.W.S. – 10/01/88 and 09/30/89; Migratory Bird Conservation Commission 1989 Annual Report). Water Bank has a nationwide appropriation of approximately $10 million per year, much of which is committed to the Prairie Pothole Region. However this funding level only allows about 32 thousand hectares of wetlands and 81 thousand hectares of adjacent upland to be enrolled in the program at any one time. Thus, the protected acreage is small relative to the developmental pressure and the annual amount of conversion.

State fish and game agencies are also active in wetland acquisition as part of fish and wildlife management areas. For example, within New Jersey, the state government possesses much more wetland acreage than the federal government. Its wildlife management areas, state parks and state forest contain numerous wetlands, ponds, lakes, and streams. The New Jersey Department of Environmental Protection is actively acquiring wetlands in the Pinelands National Reserve, using federal and state funds. Acquisition efforts have focused on several watersheds with a goal of acquisition of 41 thousand hectares of land (upland and wetland areas). Through a New Jersey program called the Green Acres Program, additional wetland and upland habitats are acquired for conservation and recreation purposes. This program also permits acquisition of conservation easements. County and municipal parks may hold wetlands in public ownership as well.

In May 1986 the United States and Canada signed the North American Waterfowl Management Plan, the most ambitious habitat protection agenda ever agreed to by the two nations. Negotiations are underway to secure Mexico's participation. The plan calls for the restoration of North American waterfowl populations to 1970s levels by the turn of the century, to 62 million breeding ducks with a fall flight of over 100 million – a 60 percent increase over current numbers. It identifies over 20 million hectares of wetlands and adjoining upland habitats in the U.S. and Canada for acquisition, restoration, and management. The plan's estimated total cost will be $1.5 billion (U.S.) over 15 years.

During 1987, the National Fish and Wildlife Foundation was asked by a consortium of state agencies and by members of the private conservation community to marshall public and private funds in support of the plan. Ducks Unlimited started the fund raising by providing a $1 million (U.S.) challenge, matched by $1 million from 12 states, the Fish and Wildlife Foundation obtained $2 million (U.S) in matching federal funds. This initial $4 million (U.S.) has been earmarked for wetland restoration and acquisition, primarily in Canada (National Fish and Wildlife Foundation 1988).

Beyond the $1 million (U.S.) already contributed to the Fish and Wildlife

Foundation fund raising effort in support of the North American Waterfowl Management Plan, the 600,000–member Ducks Unlimited, Inc. has protected and enhanced over 1.8 million hectares of wetlands and wetland complexes throughout North America. Ducks Unlimited intends to intensify its efforts by pledging an additional $299 million (U.S.) over the next 15 years toward implementation of the $1.5 billion, 28 million hectares of the North American Waterfowl Management Plan agreed to by the U.S. and Canada. In order to fully finance the plan, federal, state, provincial and other private sources need to participate. Ducks Unlimited has challenged these interested parties to make similar or larger contributions to the plan.

Many private organizations beyond Ducks Unlimited are involved in wetland acquisition and protection. The Audubon Society has a membership of 370,000 individuals in 394 local chapters. It owns, leases or patrols over 71 thousand hectares of sanctuaries, many of which include estuarine and palustrine wetlands. The Nature Conservancy has 35 chapters in 29 states with a membership of 130,000. It has preserved 728 thousand hectares of land and participated in approximately 3,000 preservation projects. The organization's success in land acquisition stems from its wealth of tax experience, a large revolving property acquisition fund, a well- established line of credit with institutional lenders, and an ability to act quickly and flexibly when natural areas are threatened. In 60% of its projects, the Conservancy retains ownership of the acquired land. In others, the land is transferred for management to the federal government, a state, a university or another conservation organization (Kusler 1983).

Private corporations are often in a position to work with private nonprofit conservation organizations and government agencies to protect wetlands. The Prudential Insurance Company of America, headquartered in Newark, New Jersey donated nearly 49 thousand hectares of prime wetlands and forest in coastal North Carolina to the F.W.S.'s National Wildlife Refuge System. The Nature Conservancy played an active role in this donation. This is an excellent example of private and public cooperation to achieve wetland protection goals. The Richard King Mellon Foundation gave the Nature Conservancy a $25 million (U.S.) grant towards its effort to conserve wetlands. The National Wildlife Federation is the largest nonprofit citizen organization in the world, with 3.5 million members in primary and affiliated organizations in all 50 states, Guam, Puerto Rico, and the Virgin Islands. It strongly advocates wetland protection and conducts widespread advertising campaigns to encourage protection at all levels of government. The Sierra Club has also shown a strong interest in wetlands protection through lobbying and public education efforts. It has not engaged in large-scale land acquisition programs comparable to those of the Nature Conservancy and the Audubon

Society but has assisted those organizations in their wetland acquisition efforts.

Many other national environmental organizations, while not directly managing wetland areas, carry out various activities (e.g. education) that help protect wetlands. Hundreds of other organizations on a local or regional level have been active in wetland protection, including fish and game clubs, hunting organizations, and general or special purpose environmental organizations.

Acquisition, although especially useful for preserving priority wetlands of a particular value, cannot be expected to provide protection for all of the nation's important wetlands. Wetland regulation at the federal and state levels are vital to preserving America's wetlands and saving the public values they provide.

The foundation of federal wetland regulations is Section 10 of the Rivers and Harbors Act of 1899 and Section 404 of the Clean Water Act of 1977, while twenty-four states have passed laws to regulate wetland uses. Federal permits for many types of construction in wetlands are required from the U.S. Army Corps of Engineers, although normal agricultural and silvicultural activities are exempt from permit requirements. Other federal conservation agencies play an active role in the permit process by reviewing permit applications and making recommendations based on environmental considerations, under authority of the Fish and Wildlife Coordination Act. The 1982 changes in the regulations reduced the federal government's role in protecting wetlands and generated much controversy and debate both within and outside of the government. Numerous lawsuits were filed against the U.S. Army Corps of Engineers by concerned environmental groups over these changes. Under a recent out-of-court settlement, the Army Corps of Engineers will propose new regulations requiring closer federal and state review of proposals to fill wetlands. This agreement should broaden federal protection of wetlands. Meanwhile, nearly half of the 50 states have laws in place which regulate wetland uses to varying degrees (Fig. 12). Most of these states protect estuarine wetlands, with palustrine wetlands being largely unprotected. For these latter wetlands, federal regulations are the principal means of protection. Unless these regulations are strengthened, extensive wetland acreages will be destroyed before the end of this century. Agriculture will continue to convert wetlands to cropland in the Mississippi Alluvial Plain, Prairie Pothole Region, South Florida, Nebraska's Sandhills and Rainwater Basin, California's Central Valley, and other areas. Urban development of wetlands will continue around urban centers throughout the country. Even if direct losses are controlled, the problem of degrading quality of wetlands must be addressed by government agencies to maintain the biological inte-

Figure 12. Current status of State wetland protection efforts. Shaded areas have enacted laws to regulate wetland use. States with only coastal wetland laws are shaded along their coastlines (modified from Tiner 1984).

grity of these valuable natural resources. There are several other statutes and other measures that may limit the impact of projects on wetlands.

The *Fish and Wildlife Act* of 1956 authorizes the development and distribution of fish and wildlife information to the public, as well as the Congress and the President of the U.S., and the development of policies and procedures that are necessary and desirable to carry out the laws relating to fish and wildlife. This act gives broad authority to the administration of the U.S. department of the Interior to take such steps as "may be required for the development, advancement, management, conservation and protection" of fish and wildlife resources. The strength of this authority is that it provides the F.W.S. through the U.S. Department of the Interior, with an avenue for developing new policy and making recommendations for new legislation that could be used to improve federal wetland protection alternatives.

The *Fish and Wildlife Coordination Act* of 1934 authorizes the F.W.S. to investigate and report on all water resource development projects that require a Federal permit or license. This Act specifies that fish and wildlife conservation shall receive equal consideration with other project purposes. This authority requires that all U.S. Government agencies will consult with the F.W.S. and state fish and wildlife agencies concerning the effect on fish and wildlife resources of all water resource development projects that require a permit or license.

The *Endangered Species Act* of 1973 requires that threatened and endangered species and the ecosystems on which these species depend be conserved. This is one of the stronger authorities the F.W.S. works under to protect the fish and wildlife resources, including wetland habitats for listed species.

The *National Environmental Policy Act* (NEPA) of 1975 requires that all officials responsible for major government actions shall consult with and obtain the comments of all government agencies which have jurisdiction by law or expertise with respect to any environmental impact involved.

The *Coastal Zone Management Act* of 1972 declares a national interest in the effective management, beneficial use, protection and development of the coastal zone. This act makes government funds available to encourage states to develop comprehensive management programs, in cooperation with the national and local governments. Opportunities for wetland protection are limited to the narrow, but ecologically important coastal planning in the coastal zone to protect important coastal resource values.

The *Coastal Barriers Resources Act* of 1982 established a network of 186 units in 15 states along the Atlantic and Gulf Coasts within which most federal expenditures are no longer available if they promote activities that are incompatible with protecting ecologically sensitive coastal areas, including wetlands.

The *Watershed Protection and Flood Prevention* Act of 1954 authorizes the F.W.S. to make surveys and investigations and prepare a report for the conservation of wildlife resources on Soil Conservation Service small watershed projects. This authority provides the F.W.S. with an opportunity to make recommendations to protect wetlands at proposed small watershed project sites.

The *Federal Power Act* of 1920, as amended, provided for cooperation between the Federal Energy Regulatory Commission and other federal agencies in the investigation of proposed power projects, and for other agencies to provide information to the Commission upon request. Section 4(e) of the Act requires coordination with the Secretary of the U.S. Department of the Interior regarding construction of fishways, and Section 30 (c) requires coordination with the F.W.S. for exceptions from licensing.

The *Emergency Wetlands Resources Act* (Wetland Act) was enacted in 1986 to promote the conservation of our nation's wetland in order to maintain the public benefits they provide. The intent was to intensify cooperative and acquisition efforts among private interest and local, state, and federal governments for the protection, management, and conservation of wetlands. The Wetlands Act contains a broad variety of measures available to the F.W.S. to promote wetland conservation and to offset or prevent wetland losses. These include new options for generating revenues for acquisition and protection of wetlands, establishing a National Wetlands Priority Conservation Plan for wetland acquisition, requiring that Statewide Comprehensive

Outdoor Recreation Plans specifically address wetlands, completing the mapping of the nation's wetlands, and studying the effects of federal programs on wetlands.

Executive Order 11990 – *Protection of Wetlands.* This executive order from the President of the U.S. directs each federal agency to provide leadership and take action to minimize the destruction, loss or degradation of wetlands, and to preserve and enhance the natural and beneficial values in carrying out agency responsibilities.

Executive Order 11988 – *Floodplain Management.* This executive order directs federal agencies to take floodplain management into account when formulating or evaluating water or land use plans. This order is applicable because of the strong interrelationship between wetland and floodplains.

Recent legislation has removed government subsidies which encouraged wetland drainage. The Coastal Barriers Resources Act does not allow most federal expenditures within the 186 designated units. The Food Security Act of 1985 denies farm program benefits to persons who produce agriculture commodities on converted wetlands. The Tax Reform Act of 1986 removes the deductions, credits, and preferential taxation that adversely affect wetlands.

Convention on Wetlands of International Importance. The Senate of the United States ratified the Convention and the Paris Protocol for the "Wetlands of International Importance" on 9 October 1986. The President signed the Instruments of Ratification on 10 November 1986, and the Instruments were deposited with UNESCO on 18 January 1987. The convention maintains a list of wetlands of international importance and works to encourage the wise use of all wetlands in order to preserve the ecological characteristics from which wetland values originate. Responsibility for implementation of the Convention rests with the U.S. Fish and Wildlife Service. The current data on wetlands designated for the List of Wetlands of International Importance in the U.S. is given in Table 18 (U.S. Fish and Wildlife Service 1990).

Recommendations

In an effort to halt or slow wetland losses and to enhance the quality of the remaining wetlands, many opportunities are available to both government agencies and the private sector. Their efforts will determine the future course of the nation's wetlands. The Environmental Law Institute's publication *Our*

Table 18. Wetlands designated for the list of wetlands of international importance (source: U.S. Fish and Wildlife Service 1990).

Site name	State	Size (ha)
Izembek Lagoon National Wildlife Refuge and State Game Range	Alaska	168,433
Forsythe National Wildlife Refuge	New Jersey	13,080
Okefenokee National Wildlife Refuge	Georgia/Florida	159,889
Ash Meadows National Wildlife Refuge	Nevada	9,509
Everglades National Park	Florida	585,867
Chesapeake Bay Estuarine Complex	Maryland/Virginia	13,425
Cheyenne Bottoms State Game Area	Kansas	8,036
Cache/Lower White Rivers	Arkansas	145,690
	Total	1,103,929

National Wetland Heritage discussed in detail public and private means of protecting wetlands (Kusler 1983). Major options have been outlined below.

Government options:

1. Develop a consistent national policy to protect wetland resources.
2. Strengthen federal, state, and local wetlands protection efforts.
3. Ensure proper implementation of existing laws and policies through adequate staffing, surveillance, and enforcement.
4. Continue recent efforts to remove government subsidies which encourage wetland drainage.
5. Provide tax and other incentives to private landowners and industry to encourage wetland preservation and remove existing tax benefits which encourage wetland destruction.
6. Increase wetland acquisition for conservation purposes.
7. Improve wetland management of federal and state-owned lands, including rangelands and forests.
8. Require that Federal water projects affecting wetlands be financed in accordance with the benefit principle of public finance, i.e., each beneficiary should bear the cost (including interest costs and any wetland opportunity cost) of generating his benefits.
9. Amend the Food Security Act of 1985 so that any agriculture activity conducted on a wetland converted after enactment triggers the Swampbuster provision, not just the production of a commodity crop produced by annual tilling of the soil and sugar cane.
10. Extend the provisions of Section 403 of the Tax Reform Act of 1986 to include gains from the sale of all converted wetlands not just those converted for farming.
11. Encourage the Secretary of the Army to develop and implement projects

for the creation, protection, restoration, and enhancement of wetlands in conjunction with authorized projects for navigation, flood control, and drainage in the Lower Mississippi Valley.

12. Extend the mitigation requirements of the Water Resources Development Act of 1986 to all projects which affect wetlands, not just those constructed by the U.S. Army Corps of Engineers.
13. Increase wetland restoration efforts.
14. Increase public awareness of wetland values, threats, and the need to protect them for the future.
15. Develop educational materials about wetlands for use in primary and secondary schools.

Private options:

1. Rather than drain or fill wetlands, seek compatible uses of those areas (e.g. waterfowl production, fur harvest, hay and forage, wild rice, hunting leases).
2. Donate wetlands or funds to purchase wetlands to private and public conservation agencies for tax purposes.
3. Work in concert with government agencies to educate the public on wetland values, etc.
4. Maintain buffers around wetlands and woody vegetation along streams and rivers.
5. Construct ponds in upland areas and manage for wetland and aquatic species.
6. Purchase federal and state duck stamps to support wetland acquisition.

Many of our current wetland problems have international, national and multi-state implications. For example, wetland drainage in one state may increase flood damages in another state. Cooperation between federal, state, and local governments and nongovernmental organizations is imperative to solving these problems. Opportunities also exist for the private sector to join with government in protecting wetlands. Large and small landowners can also contribute to this effort by managing their lands in ways that minimize wetland alterations. With over half of the wetlands in the conterminous U.S. already lost, it is imperative that appropriate steps be taken to protect our remaining wetlands. Wetland protection demands both public and private sector cooperation and action to ensure that Americans will continue to receive the many public benefits that wetlands provide.

Acknowledgments

The authors wish to express their appreciation to the National Wetlands Inventory's Regional Wetland Coordinators for their contributions to this paper as well as to all the rest of the National Wetlands Inventory staff, particularly Mary Bates for editing the numerous revisions that this paper has gone through. We would also like to thank Lajaun Randolph from the Division of Habitat Conservation who initially typed the manuscript, Warren Wilcox from the Division of Realty for providing the graphics, and others from the Division of Habitat Conservation staff who assisted in typing the tables. Lastly and most importantly, we would like to thank Dennis Whigham of the Smithsonian Environmental Research Center for his helpful editorial comments and his patience.

References

Adamus, D. A. (1963) Factors influencing vascular plant zonation in North Carolina salt marshes. Ecology 44: 445–456.

Akins, G. J. and Jefferson, C. Q. (1973) Coastal wetlands of Oregon. Oregon Coastal Conservation and Development Commission, Florence, Oregon, USA. 159 pp.

Allen, H. H. (1979) Role of wetland plants in erosion control of riparian shorelines. p. 403–414. *In*: P. E. Greeson, J. R. Clark, and J. E. Clark (eds.). Wetland Functions and Values: The State of Our Understanding. American Water Resources Association, Minneapolis, Minnesota, USA.

Anderson, B. W., Higgins, A. and Ohmart, R. D. (1977) Avian use of saltcedar communities in the Lower Colorado River Valley. p. 128–136. *In*: R. R. Johnson and D. A. Jones (eds.). Importance, Preservation and Management of Riparian Habitat: A Symposium. U.S. Forest Service General Technical Report RM-43, Ft. Collins, Colorado, USA.

Anderson, J. R., Hardy, E. E. and Roach, J. T. (1972) A land-use classification system for use with remote sensor data. U.S. Geological Survey Circular 671, Reston, Virginia, USA. 16 pp.

Anderson, J. R., Hardy, E. E., Roach, J. T. and Whitman, R. E. (1976) A Land Use and Land Cover Classification System for Use with Remote Sensor Data. U.S. Geological Survey Professional Paper 964, Reston, Virginia. USA. 28 pp.

Anderson-Nichols and Co., Inc. (1971) Neponset River Basin Flood Plain and Wetland Encroachment Study. Massachusetts Water Resources Commission, Boston, Massachusetts, USA. 61 pp.

Anderson, R. R., Brown, R.G. and Rappleye, R.D. (1968) Water quality and plant distribution along the upper Patuxent River, Maryland. Chesapeake Science 9: 145–156.

Bailey, R. G. (1976) Ecoregions of the United States. U.S. Forest Service, Intermountain Region, Ogden, Utah, USA. (Map only; scale 1:7,500,000.)

Bailey, R. G. (1978a) Ecoregions of the United States. U.S. Forest Service, Intermountain Region, Ogden, Utah, USA. 77 pp.

Bailey, R. G. (1978b) Description of the Ecoregions of the United States. U.S. Department of Agriculture Miscellaneous Publication 1391, Washington, DC, USA. 77 pp.

Baker, J. A. (1960) Wetland and Water Suppply. U.S. Geological Survey Circular 431, Reston, Virginia, USA. 3 pp.

Barada, W. R. (ed.) (1977) Restoring the Kissimmee River may be Florida's environmental armageddon. ENFO Newsletter, Florida Conservation Foundation, Winter Park, Florida, USA. 8 pp.

Barber, R. T., Kirby-Smith, W. W. and Parsley, P. E. (1978) Wetlands alterations agriculture. p. 642–651. *In*: P. E. Greeson, J. R. Clark and J. E. Clark (eds.). Wetland Functions and Values: The State of our Understanding. American Water Research Association, Minneapolis, Minnesota, USA.

Barclay, J. G. (1978) The effects of channelization on riparian vegetation and wildlife in south central Oklahoma. p. 129–138. *In*: R. R. Johnson, and J. F. McCormick (eds.). Strategies for Protection and Management of Floodplain Wetlands and Other Riparian Ecosystems. U.S. Forest Service General Technical Report WO-12, Washington, DC, USA.

Barclay, J. G. (1980) Impact of Stream Alterations on Riparian Communities in Southcentral Oklahoma. U.S. Fish and Wildlife Service, FWS/OBS-80/17, Washington, DC, USA. 91 pp.

Batten, A. R. and Murray, D. F. (1982) A Literature Survey on the Wetland Vegetation of Alaska. U.S. Army Corps of Engineers, Wetland Research Program, Technical Report Y-82–2, Vicksburg, Mississippi, USA. 22 pp.

Bay, R. R. (1966) Factors influencing soil-moisture relationships in undrained forested bogs. p. 335–343. *In*: Proceedings National Science Foundation Advanced Science Seminar, International Symposium on Forest Hydrology. Washington, DC, USA.

Bellrose, F. C. (1976) Ducks, Geese and Swans of North America. Stackpole Books, Harrisburg, Pennsylvania, USA. 544 pp.

Benton, A. R., Jr., James, W. P. and Rouse, J. W., Jr. (1978) Evapotranspiration from water hyacinth (*Eichhornia crassipes* [Mart.] Solms) in Texas reservoirs. Water Resources Bulletin 14: 919–930.

Bernard, J. M. and Solsky, B. A. (1977) Nutrient cycling in a *Carex lacustris* wetland. Canadian Journal of Botany 55: 630–638.

Bernot, C. (1979) Water bank: keeping wetlands wet. The Minnesota Conservation Volunteer 42(246):4.

Bishop, R. A. (1981) Iowa's wetlands. Proceedings of the Iowa Academy of Science 88: 11–16.

Boelter, D. H. and Verry, E. S. (1977) Peatland and Water in Northern Lake States. North Central Forest Experiment Station, U.S. Forest Service General Technical Report NC-31, St. Paul, Minnesota, USA. 22 pp.

Boesch, D. F. (ed.) (1982) Proceedings of the Conference on Coastal Erosion and Wetland Modification in Louisiana: Causes, Consequences, and Options. U.S. Fish and Wildlife Service, FWS/OBS-82/59, Washington, DC, USA. 256 pp.

Bonde, A. N., Ives, J. D. and Lawrence, D. B. (1961) Ecosystem studies at Cedar Creek Natural History Area: III – Water Use Studies. Proceedings of the Minnesota Academy of Science 29: 190–198.

Boto, K. G. and Patrick, W. H., Jr. (1979) Role of wetlands in the removal of suspended sediments. p. 479–489. *In*: P. E. Greeson, J. R. Clark and J. E. Clark (eds.). Wetland Functions and Values: The State of Our Understanding. American Water Resources Association, Minneapolis, Minnesota, USA.

Boyd, C. W. (1971) The dynamics of dry matter and chemical substances in a *Juncus effusus* population. American Midland Naturalist 86: 28–45.

Brabander, J. J., Masters, R. E. and Short, R. M. (1985) Bottomland Hardwoods of Eastern Oklahoma. U.S. Fish and Wildlife Service, Mimeo report, Washington, DC, USA. 143 pp.

Bradbury, I. K. and Grace, J. (1983) Primary production in wetlands. p. 285–310. *In*: A. J. P. Gore (ed.). Ecosystems of the World, 4A Mires: Swamp, Bog, Fen and Moor. Elsevier Scientific Publishing Company, New York, New York, USA.

Braun, E. L. (1950) Deciduous Forests of Eastern North America. Hafner Publishing Co., New York, New York, USA. 596 pp.

Brinson, M. M. (1977) Decomposition and nutrient exchange of litter in an alluvial swamp forest. Ecology 58: 601–609.

Brinson, M. M., Swift, B. L., Plantico, R. C. and Barclay, J. S. (1981) Riparian Ecosystems:

Their Ecology and Status. U.S. Fish and Wildlife Service, FWS/OBS-81/17, Washington, DC, USA. 154 pp.
Brinson, M. M. (1988) Strategies for assessing the cumulative effects of wetland alteration of water quality. Environmental Management 12: 655–662.
Brooks, R. P., Arnold, D. E. and Bellis, E. D. (1987) Wildlife and Plant Communities of Selected Wetlands: Pocono Region of Pennsylvania. U.S. Fish and Wildlife Service, NWRC Open File Report 87–02, Washington, DC, USA. 41 pp.
Brown, D. E., Lowe, C.H. and Hausler, J. F. (1977) Southwestern riparian communities: their biotic importance and management in Arizona. p. 201–211. *In*: R. R. Johnson and D. A. Jones (eds.). Importance, Preservation and Management of Riparian Habitat. U.S. Forest Service, General Technical Report RM-43, Ft. Collins, Colorado, USA.
Brown, J. and West, G. C. (1970) Tundra biome research in Alaska. p. 70–71. *In*: The Structure and Function of Cold-Dominated Ecosystems. U.S. IBP Tundra Biome Report, University of Alaska, Fairbanks, Alaska, USA.
Brown, S. L. (1978) A Comparison of Cypress Ecosystems in the Landscape of Florida. Ph.D. dissertation, University of Florida, Gainesville, Florida, USA. 569 pp.
Burke, D. G., Myers, E. J., Tiner, R. W., Jr. and Groman H. (1988) Protecting Nontidal Wetlands. American Planning Association, Planning Advisory Service Report No. 412/413, Chicago, Illinois, USA. 76 pp.
Carlton, J. M. (1977) A survey of selected coastal vegetation communities of Florida. Florida Department of Natural Resources, Marine Research Laboratory, Marine Research Publication No. 30, St. Petersburg, Florida, USA. 40 pp.
Carter, V., Bedinger, M.S., Novitzki, R.P. and Wilen, W.O. (1978) Water resources and wetlands. p. 344–376. *In*: P. E. Greeson, J. R. Clark and J. E. Clark (eds.). Wetland Functions and Values: The State of Our Understanding. American Water Resources Association, Minneapolis, Minnesota, USA.
Carter, V. (1989) An overview of the hydrologic concerns related to wetlands in the United States. Canadian Journal of Botany 64: 364–374.
Chabreck, R. H. (1972) Vegetation, water and soil characteristics of the Louisiana coastal region. Louisiana State University Agricultural Experiment Station Bulletin 664, Baton Rouge, Louisiana, USA. 72 pp.
Chapman, V. J. (1938) Studies in salt marsh ecology I–III. Journal of Ecology 26: 144–221.
Chapman, V. J. (1940) Studies in salt marsh ecology VI–VII. Journal of Ecology 28: 118–179.
Chapman, V. J. (1960) Salt Marshes and Salt Deserts of the World. Interscience Publishers, Inc., New York, New York, USA. 392 pp.
Chapman, V. J. (1974) Salt Marshes and Salt Deserts of the World. 2nd supplemented edition. J. Cramer, LeHre, Germany. 392 pp.
Chapman, V. J. (1976a) Coastal Vegetation. 2nd edition. Pergamon Press, Oxford, England. 292 pp.
Chapman, V. J. (1976b) Mangrove Vegetation. J. Cramer, Leuterhausen, Vaduz, Germany. 447 pp.
Christensen, N. L., Burchele, R. B., Liggett, A. and Simms, E. L. (1981) The structure and development of Pocosin vegetation. p. 43–61. *In*: C. J. Richardson (ed.). Pocosin Wetlands. Hutchinson Ross Publishing Co., Stroudsburg, Pennsylvania, USA.
Clark, J. E. (1979) Fresh water wetlands: habitats for aquatic invertebrates, amphibians, reptiles, and fish. p. 330–343. *In*: P. E. Greeson, J. R. Clark and J. E. Clark (eds.). Wetland Functions and Values: The State of Our Understanding. American Water Resources Association, Minneapolis, Minnesota, USA.
Clark, J. R. and Benforado, J. (eds.)(1981) Wetlands of Bottomland Hardwood Forests. Elsevier Scientific Publishing Co., New York, New York, USA. 401 pp.
Cohen, A. D., Casagrande, D. J., Andrejko, M. J. and Best, G. R. (eds.)(1984) The Okefenokee Swamp: Its Natural History, Geology, and Geochemistry. Wetlands Surveys, Los Alamos, New Mexico, USA. 709 pp.
Conner, W. H., and Day, J. W., Jr. (1976) Productivity and composition of a bald cypress –

water tupelo site and a bottomland hardwood site in Louisiana swamp. American Journal of Botany 63: 1354–1364.

Conner, W. H., and Day, J. W., Jr. (eds.)(1987) The Ecology of Barataria Basin, Louisiana: An Estuarine Profile. U.S. Fish and Wildlife Service, Biological Report 85(7.13), Washington, DC, USA. 165 pp.

Conway, V. M. (1949) The bogs of central Minnesota. Ecological Monographs 19: 173–206.

Copeland, B.J., Hodson, R. G., Riggs, S. R. and Easley, J. E., Jr. (1983) The Ecology of Albemarle Sound, North Carolina: An Estuarine Profile. U.S. Fish and Wildlife Service, FWS/OBS-83/01, Washington, DC, USA. 68 pp.

Copeland, B. J., Hodson, R. G. and Riggs, S. R. (1984) The Ecology of the Pamlico River, North Carolina: An Estuarine Profile. U.S. Fish and Wildlife Service, FWS/OBS-82/06, Washington, DC, USA. 83 pp.

Council of Environmental Quality (1981) Environmental Trends. U.S. Government Printing Office, Washington, DC, USA. 346 pp.

Cowardin, L. M., Carter, V., Golet, F. C. and LaRoe, E. T. (1976) Interim classification of wetland and aquatic habitats of the United States. Addendum to Sather, J. H. (ed.) Proceedings National Wetland Classification and Inventory Workshop. U.S. Fish and Wildlife Service, Washington, DC, USA. 110 pp.

Cowardin, L. M., Carter, V., Golet, F. C. and LaRoe, E. T. (1977) Classification of Wetland and Deep-water Habitats of the United States. U.S. Fish and Wildlife Service, Washington, DC, USA. 100 pp.

Cowardin, L. M., Carter, V., Golet, F. C. and LaRoe, E.T. (1979) Classification of Wetlands and Deepwater Habitats of the United States. U.S. Fish and Wildlife Service, FWS/OBS-79/31, Washington, DC, USA. 103 pp.

Cowardin, L. M., Gilmer, D. S. and Mechlin, L. M. (1981) Characteristics of central North Dakota wetlands determined from sample aerial photographs and ground study. Wildlife Society Bulletin 9: 280–288.

Craig, N. J., Turner, R. E. and Day, J. W., Jr. (1980) Wetlands losses and their consequences in coastal Louisiana. Zeitschrift fur Geomorphologie N.F. Supplement. Bdl. 34: 225–241.

Crance, J. H. (1988) Relationships Between Palustrine Wetlands of Forested Riparian Floodplains and Fishery Resources: A Review. U.S. Fish and Wildlife Service, Biological Report 88(32), Washington, DC, USA. 27 pp.

Crow, J. H. and MacDonald, K. B. (1979) Wetland values: Secondary production. p. 146–161. *In*: P. E. Greeson, J. R. Clark and J. E. Clark (eds.). Wetland Functions and Values: The State of Our Understanding. American Water Resources Association, Minneapolis, Minnesota, USA.

Crum, H. (1988) A Focus on Peatlands and Peat Mosses. University of Michigan Press, Ann Arbor, Michigan, USA. 306 pp.

Curtis, J. T. (1959) The Vegetation of Wisconsin. The University of Wisconsin Press, Madison, Wisconsin, USA. 657 pp.

Dabel, C. V. and Day, F. P., Jr. (1977) Structural comparison of four plant communities in the Great Dismal Swamp, Virginia. Bulletin of the Torrey Botanical Club 104: 352–360.

Dachnowski, A. P. (1920) Peat deposits in the United States and their classification. Soil Science 10: 453–465.

Dahl, T. E. (1990) Wetland losses in the United States 1780s to 1980s. U.S. Fish and Wildlife Service, Washington, DC, USA. 21 pp.

Damman, A. W. H. (1977) Geographic changes in the vegetation patterns of raised bogs in the Bay of Fundy Region of Maine and New Brunswick. Vegetatio 35: 137–151.

Damman, A. W. H. and French, T. W. (1987) The Ecology of Peat Bogs of the Glaciated Northwestern United States: A Community Profile. U.S. Fish and Wildlife Service, Biological Report 85 (7.16), Washington, DC, USA. 100 pp.

Dansereau, P. and Segadas-Vianna, F. (1952) Ecological study of peat bogs of eastern North America. I. Structure and evolution of vegetation. Canadian Journal of Botany 30: 490–518.

Day, J. W., Jr., Smith, W. G., Wagner, P. R. and Stowe, W. C. (1973) Community Structure and Carbon Budget of a Salt Marsh and Shallow Bay Estuarine System in Louisiana. Center for Wetland Resources, Louisiana State University, Publication No. LSU-SG-72–04, Baton Rouge, Louisiana, USA. 79 pp.

Dean, R. G. (1979) Effects of vegetation on shoreline erosional processes. p. 415–426 *In*: P. E. Greeson, J. R. Clark and J. E. Clark (eds.). Wetland Functions and Values: The State of Understanding. American Water Resources Association, Minneapolis, Minnesota, USA.

de la Cruz, A. A. (1979) Production and transport of detritus in wetlands. p. 162–174. *In*: P. E. Greeson, J. R. Clark and J. E. Clark (eds.). Wetland Functions and Values: The State of Our Understanding. American Water Resources Association, Minneapolis, Minnesota, USA.

de la Cruz, A. A. (1981) Difference between South Atlantic and Gulf Coast marshes. p. 10–20. *In*: R. C. Carey, P. S. Markovits, and J. B. Kirkwood, (eds.). Proceeding of the U.S. Fish and Wildlife Service Workshop on Coastal Ecosystems of the Southeastern United States. U.S. Fish and Wildlife Service, FWS/OBS-80/59, Washington, DC, USA.

DeLaune, R. D., Patrick, W. H., Jr., and Buresk, R. J. (1978) Sedimentation rates determined by 137Cs dating in a rapidly accreting salt marsh. Nature 275: 532–533.

DeLaune, R. D., Baumann, R. H. and Gosselink, J. G. (1983) Relationships among vertical accretion, coastal submergence, and erosion in a Louisiana Gulf Coast marsh. Journal of Sedimentary Petrology 53: 147–157.

Dick-Peddie, W. A., Meents, J. K. and Spellenberg, R. (1984) Vegetation Resources Analysis for the Velarde Community Ditch Project, Rio Arriba and Santa Fe Counties, New Mexico. U.S. Bureau of Reclamation, Amarillo, Texas, USA. 238 pp.

Dick-Peddie, W. A., Hardesty, J. V., Muldavin, E. and Sallach, B. (1987) Soil-Vegetation Correlations in the Riparian Zones of the Gila and San Francisco Rivers in New Mexico. U.S. Fish and Wildlife Service, Biological Report 87(9), Washington, DC, USA. 29 pp.

Dix, R. L. and Smeins, F. E. (1967) The prairie, meadow and marsh vegetation of Nelson County, North Dakota. Canadian Journal of Botany 45: 21–58.

Drew, R. D. and Schomer, N. S. (1984) An Ecological Characterization of the Caloosahatchee River/Big Cypress Watershed. U.S. Fish and Wildlife Service, FWS/OBS-82–58.2, Washington, DC, USA. 225 pp.

Drury, W. H., Jr. (1962) Patterned Ground and Vegetation on Southern Bylot Island, Northwest Territories, Canada. Harvard University Gray Herbarium Contributions 190, Boston, Massachusetts, USA. 111 pp.

Duever, M. J., Carlson, J. E. and Riopelle, L. A. (1984) Corkscrew Swamp: A virgin cypress strand. p. 334–348. *In*: K. C. Ewel and H. T. Odum (eds.). Cypress Swamps. University of Florida Press, Gainesville, Florida, USA.

Dwyer, T. J., Krapu, G. and Janke, D. (1979) Use of prairie pothole habitat by breeding mallards. Journal of Wildlife Management 43: 527–531.

Eicher, A. L. (1988) Soil-vegetation Correlations in Wetland and Adjacent Uplands of the San Francisco Bay Estuary, California. U.S. Fish and Wildlife Service, Biological Report 88(21), Washington, DC, USA. 35 pp.

Eleuterius, L. M. (1972) The marshes of Mississippi. Castanea 37: 153–168.

Eleuterius, L. M. (1973) The distribution of certain submerged plants in Mississippi Sound and adjacent waters. p. 191–197. *In*: J. L. Christmas (ed.). Cooperative Gulf of Mexico Estuarine Inventory and Study, Mississippi Gulf Coast Research Laboratory, Ocean Springs, Mississippi, USA.

Eleuterius, L. M. and McDaniel, S. (1978) The salt marsh flora of Mississippi. Castanea 43: 86–95.

Eleuterius, L. M. (1980) An Illustrated Guide to Tidal Marsh Plants of Mississippi and Adjacent States. Mississippi-Alabama Sea Grant Consortium, Publication No. MASGP-77–039, Ocean Springs, Mississippi, USA. 131 pp.

Erickson, N. E. and Leslie, D. M., Jr. (1988) Soil-Vegetation Correlations in Coastal Mississippi Wetlands. U.S. Fish and Wildlife Service, Biological Report 89(3), Washington, DC, USA. 48 pp.

Erickson, R. E., Linder, R. L. and Harmon, K. W. (1979) Stream channelization (P.L. 83–566) increase wetland losses in the Dakotas. Wildlife Society Bulletin 7: 71–78.

Evans, C. D. and Black, K. E. (1956) Duck Production Studies on the Prairie Potholes of South Dakota. U.S. Fish and Wildlife Service, Special Scientific Report – Wildlife No. 32, Washington, DC, USA. 59 pp.

Ewel, K.C. and Odum, H. T., (eds.)(1984) Cypress Swamps. University of Florida Press, Gainesville, Florida, USA. 472 pp.

Faber, P. M., Keller, E., Sands, A. and Massey, B. M. (1989) The Ecology of Riparian Habitats of the Southern California Coastal Region: A Community Profile. U.S. Fish and Wildlife Service, Biological Report 85(7.27), Washington DC, USA. 152 pp.

Farrar, J. (1982) The Rainwater Basin: Nebraska's Vanishing Wetlands. Nebraska Game and Parks Commission, Lincoln, Nebraska, USA. 15 pp.

Ferrigno, F., Widjeskog, L. and Toth, S. (1973) Marsh Destruction. New Jersey Pittman-Robertson Report. Project W-53–R-1, Job I-G, Trenton, New Jersey, USA. 20 pp.

Frayer, W. E., Monahan, T. J., Bowden, D. C. and Graybill, F. A. (1983) Status and Trends of Wetlands and Deepwater Habitats in the Conterminous United States, 1950s to 1970s. Department of Forest and Wood Sciences, Colorado State University, Ft. Collins, Colorado, USA. 32 pp.

Freehling, M. D. (1982) Riparian Woodland of the Middle Rio Grande Valley, Mexico: A Study of Bird Populations and Vegetation with Special Reference to Russian Olive (*Elaeagnus angustifolia*). U.S. Fish and Wildlife Service, Albuquerque, New Mexico, USA. 35 pp.

Fruge, D. W. (1982) Effects of wetland deterioration on the fish and wildlife resources of coastal Louisiana. p. 99–107. *In*: D. F. Boesch (ed.). Proceedings of the Conference on Coastal Erosion and Wetland Modification in Louisiana: Causes, Consequences, and Options. U.S. Fish and Wildlife Service, FWS/OBS-82/59, Washington, DC, USA.

Gallagher, J. L, Reimold, R. J., Linthurst, R. A. and Pfeiffer, W. J. (1980) Aerial production, mortality, and mineral accumulation-export dynamics in *Spartina alterniflora* and *Juncus roemerianus* plant stands in Georgia salt marsh. Ecology 61: 303–312.

Garbisch, E. W., Jr. (1977) Marsh development for soil erosion. pp. 77–94. *In*: Proceedings of the Workshop on the Role of Vegetation in Stabilization of the Great Lakes Shoreline. Great Lakes Basin Commission, Ann Arbor, Michigan, USA.

Gates, F. C. (1942) The bogs of northern Lower Michigan. Ecological Monographs 12: 213–254.

Gates, J. M., and Hale, J. B. (1974) Seasonal Movement, Winter Habitat Use and Population Distribution of an East-Central Wisconsin Pheasant Population. Wisconsin Deptartment of Natural Resources. Technical Bulletin No. 76, Madison, Wisconsin, USA. 56 pp.

Geis, J. W., and Kee, J. L. (1977) Coastal Wetlands Along Lake Ontario and St.Lawrence River in Jefferson County, New York. State University of New York, Institute of Environmental Program Affairs, Syracuse, New York, USA. 130 pp.

Glaser, P. H. (1987) The Ecology of Patterned Boreal Peatlands of Northern Minnesota: A Community Profile. U.S. Fish and Wildlife Service, Biological Report 85(7.14), Washington, DC, USA. 98 pp.

Godfrey, P. J., Kaynor, E. R., Pelczarski, S. and Benforado, J. (eds.)(1985) Ecological Considerations in Wetlands Treatment of Municipal Wastewater. Van Nostrand Reinhold Company, New York, New York, USA. 474 pp.

Golet, F. C. and Larson, J. S. (1974) Classification of Freshwater Wetlands in the Glaciated Northeast. U.S. Fish and Wildlife Service, Resource Publication 116, Washington, DC, USA. 56 pp.

Good, R. E., Whigham, D. F. and Simpson, R. L. (eds.)(1978) Freshwater Wetlands, Ecological Processes and Management Potential. Academic Press, Inc. New York, New York, USA. 378 pp.

Goodwin, R. H. and Niering, W. A. (1975) Inland Wetlands of the United States. Evaluated as Potential Registered Natural Landmarks. National Park Service, Natural History Theme Studies, No. 2, Washington, DC, USA. 550 pp.

Gosselink, J. G. and Bauman, R. H. (1980) Wetland inventories: wetland loss along the United States coast. Zeitschrift fur Geomorphologie N.F. Supplement Bdl. 34: 173–187.

Gosselink, J. G. (1984) The Ecology of Delta Marshes of Coastal Louisiana: A Community Profile. U.S. Fish and Wildlife Service, FWS/OBS-84/09, Washington, DC, USA. 134 pp.

Grant, R. R., Jr. and Patrick, R. (1970) Tinicum Marsh as a water purifier. p. 105–123. *In*: J. McCormick, R. R Grant and R. Patrick (eds.). Two Studies of Tinicum Marsh. The Conservation Foundation, Washington, DC, USA.

Great Lakes River Basin Commission (1981) Wetlands. Great Lakes Communicator 11(9). 8 pp.

Greeson, P. E., Clark, J. R. and Clark, J. E. (eds.)(1979) Wetland Functions and Values: The State of Our Understanding. American Water Resources Association, Minneapolis, Minnesota, USA. 674 pp.

Guthery, F. S. (1981) Playa basins and resident wildlife in the Texas Panhandle. p. 47–51. *In*: J. S. Barclay and W. V. White (eds.). Playa Lakes Symposium Proceedings. U.S. Fish and Wildlife Service, FWS/OBS-81/07, Washington, DC, USA.

Haag, R. (1974) Nutrient limitations to plant production in two tundra communities. Canadian Journal of Botany 52: 103–116.

Hackney, C. T. and de la Cruz, A. A. (1982) The structure and function of brackish marshes in the north central Gulf of Mexico: a ten year case study. p. 89–107. *In*: B. Gopal, R. E. Turner, R. G. Wetzel, and D. F. Whigham (eds.). Wetlands: Ecology and Management. International Scientific Publishers, Jaipur, India.

Haddock, J. L. and DeBates, L. W. (1969) Report on Drainage Trends in the Prairie Pothole Region of Minnesota, North Dakota and South Dakota. U.S. Fish and Wildlife Service, Minneapolis, Minnesota, USA. 8 pp.

Hall, T. F. and Penfound, W. T. (1943) Cypress-gum communities in the Blue Girth Swamp near Selma, Alabama. Ecology 24: 208–217.

Hammond, E. H. (1964) Classes of Land-Surface Form in the Forty-Eight States U.S.A. Annals of the Association of American Geographers volume 54, Map Supplement No. 4, scale 1:5,000,000.

Hanson, H. C. (1958) Analysis of Nelchina Caribou Range. U.S. Fish and Wildlife Service, Federal Aid in Wildlife Job Completion Report, Job 6, W-3-k-12, Juneau, Alaska, USA. 68 pp.

Hardisky, M. A. and Klemas, V. (1983) Tidal wetlands natural and human made changes from 1973 to 1979 in Delaware: mapping techniques and results. Environmental Management 7: 1–6.

Hawkins, P. and Leck, C. F. (1977) Breeding bird communities in a tidal freshwater marsh. Bulletin of the New Jersey Academy of Science 22: 12–17.

Heath, R. C. (1975) Hydrology of the Albemarle-Pamlico Region, North Carolina. U.S. Geological Survey Water Resources Investigation 9–75, Reston, Virginia, USA. 98 pp.

Heeley, R. W. (1973) Hydrogeology of Wetlands in Massachusetts. National Technical Information Service, PB236:945, Springfield, Virginia, USA. 129 pp.

Heeley, R. W. and Motts, W. S. (1976) A model for the evaluation of ground water resources associated with wetlands. p. 52–65. *In*: J. S. Larson (ed.). Models for Assessment of Freshwater Wetlands. Water Resources Research Center Publication No. 32, University of Massachusetts, Amherst, Massachusetts, USA.

Heimburg, K. (1976) Hydrology of Some North Central Florida Cypress Domes. Unpublished Master's Thesis, University of Florida, Gainesville, Florida, USA. 114 pp.

Heinselman, M. L. (1963) Forest sites, bog processes, and peatland types in the Glacial Lake Agassiz Region, Minnesota. Ecological Monographs 33: 327–374.

Heinselman, M. L. (1965) String bogs and other patterned organic terrain near Seney, Upper Michigan. Ecology 46: 185–188.

Heinselman, M. L. (1970) Landscape evolution, peatland types, and the environment in the Lake Agassiz Peatlands Natural Area, Minnesota. Ecological Monographs 40: 235–261.

Hemond, H. F. and Benoit, J. (1988) Cumulative impacts on water quality functions of wetlands. Environmental Management 12: 639–653.

Herdendorf, C. E., Hartley, S. M. and Barnes, M. D. (eds.)(1981) Fish and Wildlife Resources of the Great Lakes Coastal Wetlands within the United States. Vol. I: Overview. U.S. Fish and Wildlife Service, FWS/OBS-81/02, Washington, DC, USA. 469 pp.

Herdendorf, C. E., Raphael, C. N. and Jaworski, E. (1986) The Ecology of Lake St. Clair Wetlands: A Community Profile. U.S. Fish and Wildlife Service, Biological Report 85(7.7), Washington, DC, USA. 187 pp.

Herdendorf, C. E. (1987) The Ecology of the Coastal Marshes of Western Lake Erie: A Community Profile. U.S. Fish and Wildlife Service, Biological Report 85(7.9), Washington, DC, USA. 171 pp.

Hindall, S. M. (1975) Measurements and Prediction of Sediment Yields in Wisconsin Streams. U.S. Geological Survey, Water Resources Investigations 54–75, Reston, Virginia, USA. 27 pp.

Hobbie, J. E. (1984) The Ecology of Tundra Ponds of the Arctic Coastal Plain: A Community Profile. U.S. Fish and Wildlife Service, FWS/OBS-83/25, Washington, DC, USA. 52 pp.

Hoffman, G. R. (1977) Artificial establishment of vegetation and effects of fertilizer along shorelines of Lake Oahe and Sakakawea mainstem Missouri River reservoirs. p. 95–109. *In*: Proceedings Workshop on the Role of Vegetation in Stabilization of the Great Lakes Shoreline. Great Lakes Basin Commission, Ann Arbor, Michigan, USA.

Hook, D. D., McKee, W. H., Jr., Smith, H. K., Gregory, J., Burrell, V. G., Jr., DeVoe, M. R., Sojka, R. E., Gilbert, S., Banks, R., Stolzy, L. H., Brooks, C., Matthews, T. D. and Shear, T. H. (eds.)(1988) The Ecology and Management of Wetlands. Timber Press, Portland, Oregon, USA. Vol. 1 (Ecology of Wetlands), 592 pp., Vol. 2 (Management, Use and Value of Wetlands). 394 pp.

Hook, D. D. and Lea, R. (eds.)(1989) Proceedings of the Symposium. The Forested Wetlands of the Southern United States. U.S. Forest Service, Southeastern Forest Experiment Station General Technical Report SE-50, Washington, DC, USA. 168 pp.

Hopkinson, C. S., Jr., Gosselink, J. G. and Parrondo, P. T. (1980) Production of coastal Louisiana marsh plants calculated from phenometric techniques. Ecology 61: 1091–1098.

Hubbard, D. E. (1988) Glaciated Prairie Wetland Functions and Values: A Synthesis of the Literature. U.S. Fish and Wildlife Service, Biological Report 88(43), Washington, DC, USA. 50 pp.

Hubbard, D. E., Millar, D. D., Malo, D.D. and Higgins, K.F. (eds.)(1988) Soil-vegetation correlations in Prairie Potholes of Beadle and Deuel Counties, South Dakota. U.S. Fish and Wildlife Service, Biological Report 88(22), Washington, DC, USA. 97 pp.

Hughes, G. H. and McDonald, C. C. (1966) Determination of water use by phyreatophytes and hydrophytes. American Society of Civil Engineers, Journal of Hydraulics, Division 92 (Hy2:63–81).

Illinois Department of Conservation (1983) February 16 letter from D. Kenny to B. Wilen, U.S. Fish and Wildlife Service, Washington, DC, USA.

JACA Corporation (1982) A Case Study of New Jersey Wetland Trends and Factors Influencing Wetland Use. Prepared for the Office of Technology Assessment, U.S. Congress, Washington, DC, USA. p. 1–1 to R-2

Jahn, L. A. and Anderson, R. V. (1986) The Ecology of Pools 19 and 20, Upper Mississippi River: A Community Profile. U.S. Fish and Wildlife Service, Biological Report 85(7.6), Washington, DC, USA. 142 pp.

Jaworski, E. and Raphael, C. N. (1978) Fish, Wildlife, and Recreational Values of Michigan's Coastal Wetlands. Michigan Department of Natural Resources report, Lansing, Michigan, USA. 209 pp.

Jeglum, J. K. (1971) Plant indicators of pH and water level in peatlands at Candle Lake, Saskatchewan. Canadian Journal of Botany 49: 1661–1676.

Jeglum, J. K., Boissonneau, A. N. and Haavisto, V. G. (1974) Toward a Wetland Classification for Ontario. Canadian Forest Service Information Report 0–X-215, Quebec, Canada. 54 pp.

Jervis, R. A. (1969) Primary production in the freshwater marsh ecosystem of Troy Meadows, New Jersey. Bulletin of the Torrey Botanical Club 96: 209–231.

Johnson, C. W. (1985) Bogs of the Northeast. University Press of New England, Hanover, New Hampshire, USA. 269 pp.
Johnson, F. L. and Bell, D. T. (1976) Plant biomass and net primary production along a flood-frequency gradient in a streamside forest. Castanea 41: 156–165.
Johnson, R. R., Ziebell, C. D., Patton, D. R., Folliott, P. F. and Hainre, R. H., Technical coordinators (1985) Riparian Ecosystems and Their Management: Reconciling Conflicting Uses. First North American Riparian Conference. U.S. Forest Service, General Technical Report Rm-120, Washington, DC, USA. 523 pp.
Josselyn, M. (1983) The Ecology of San Francisco Bay Tidal Marshes: A Community Profile. U.S. Fish and Wildlife Service, FWS/OBS-83/23, Washington, DC, USA. 102 pp.
Kadlec, J. A. and Wentz, W. A. (1974) State-of-the-art Survey and Evaluation of Marsh Plant Establishment Techniques: Induced and Natural. U.S. Army Coastal Engineerign Research Center Contract Report D-74–9. U.S. Army Corps of Engineers Waterways Experiment Station, Vicksburg, Mississippi, USA. 231 pp.
Kadlec, R. H. (1979) Wetland for tertiary treatment. p. 490–504. *In*: P. E. Greeson, J. R. Clark and J. E. Clark (eds.). Wetlands Functions and Values: The State of Our Understanding. American Water Resources Association, Minneapolis, Minnesota, USA.
Kadlec, R. H. and Kadlec, J.A. (1979) Wetlands and water quality. p. 436–456. *In*: P. E. Greeson, J. R. Clark, and J. E. Clark (eds.). Wetland Functions and Values: The State of Our Understanding. American Water Resources Association, Minneapolis, Minnesota, USA.
Kantrud, H. A., Krapu, G. L. and Swanson, G. A. (1989) Prairie Basin Wetlands of the Dakotas: A Community Profile. U.S. Fish and Wildlife Service, Biological Report 85(7.28), Washington, DC, USA. 116 pp.
Kearney, T. H. (1901) Report on a botanical survey of the Dismal Swamp Region. U.S. Department of Agriculture, Contributions to the U.S. National Herbarium 5: 321–550
Kentucky Department of Fish and Wildlife Resources (1983) February 23 letter from C.E. Kays, Commissioner to B. Wilen, U.S. Fish and Wildlife Service. Washington, DC, USA. 2 pp.
Klopatek, J. M. (1978) Nutrient dynamics of freshwater riverine marshes and the role of emergent macrophytes. p. 195–216. *In*: R. E. Good, D. F. Whigham, and R. L. Simpson (eds.). Freshwater Wetlands: Ecological Processes and Management Potential. Academic Press, Inc., New York, New York, USA.
Klopatek, J. M. and Stearns, F. W. (1978) Primary productivity of emergent macrophytes in a Wisconsin freshwater marsh ecosystem. American Midland Naturalist 100: 320–332.
Kologiski, R. L. (1977) The Phytosociology of the Green Swamp, North Carolina. North Carolina Agricultural Experiment Station, Technical Bulletin 250, Raleigh, North Carolina, USA. 101 pp.
Koppen, W. (1931) Grudriss der limakunde. Walter de Gruyter & Co. Berlin, Germany. 388 pp.
Krapu, G. L. (1974) Feeding ecology of pintail hens during reproduction. The Auk 91: 278–290.
Kuchler, A. W. (1964) Potential Natural Vegetation of the Conterminous United States (map and manual). American Biographical Society, Special Publication 36, New York, New York, USA. 116 pp.
Kuenzler, E. J. (1974) Mangrove swamp systems. p. 346–371 of Vol. 1 (533 pp.). *In*: H. T. Odum, B. J. Copeland and E. A. McMahan (eds.). Coastal Ecological Systems of the United States. The Conservation Foundation, Washington, DC, USA. 4 Vols.
Kurz, H. and Wagner, K. (1957) Tidal Marshes of the Gulf and Atlantic Coasts of Northern Florida and Charleston, South Carolina. Florida State University, Florida State University Studies No. 24, Tallahassee, Florida, USA. 168 pp.
Kusler, J. A. (1978) Strengthening State Wetland Regulations. U.S. Fish and Wildlife Service, FWS/OBS-78/98, Washington DC, USA. 147 pp.
Kusler, J. A. (1983) Our National Wetland Heritage. A Protection Guidebook. Environmental Law Institute, Washington, DC, USA. 167 pp.

Kusler, J. A. and Brooks G. (eds.)(1988) Proceedings of the National Wetland Symposium: Wetland Hydrology. Association of State Wetland Managers, Berne, New York, USA. 339 pp.

Kusler, J. A. and Kentula, M. E. (1989a) Wetland Creation and Restoration: The Status of the Science. Vol. I: Regional Reviews. U.S. Environmental Protection Agency, EPA 60013–89/038a, Corvallis, Oregon, USA. 159 pp.

Kusler, J. A. and Kentula, M. E. (1989b) Wetland Creation and Restoration: The Status of the Science. Vol. II: Perspectives. U.S. Environmental Protection agency, EPA 60013–89/038b, Corvallis, Oregon, USA. 172 pp.

Laderman, A. D. (ed.) (1987) Atlantic White Cedar Wetlands. Westview Press. Boulder, Colorado, USA. 401 pp.

Laderman, A. D. (1989) The Ecology of Atlantic White Cedar Wetlands: A Community Profile. U.S. Fish and Wildlife Service, Biological Report 85(7.21), Washington, DC, USA. 114 pp.

Laessle, A. M. (1942) The Plant Communities of the Welaka Area With Special Reference to Correlation between Soils and Vegetational Succession. University of Florida, Biological Science Series IV(1.), Gainesville, Florida, USA. 143 pp.

Larsen, J. A. (1982) Ecology of the Northern Lowland Bogs and Conifer Forests. Academic Press, New York, New York, USA. 307 pp.

Lee, G. F., Bentley, E. and Amundson, R. (1975) Effects of marshes on water quality. p. 105–127. *In*: A. D. Hasler (ed.). Coupling of Land and Water Systems. Springer-Verlag, New York, New York, USA.

Lee, C. R., Hoeppel, R. E., Hunt, P. G. and Carlson, C. A. (1976) Feasibility of the Functional Use of Vegetation to Filter, Dewater, and Remove Contaminants from Dredged Material. U.S. Army Corps of Engineers, Waterways Experiment Station, Technical Report D-76-4, Vicksburg, Mississippi, USA. 80 pp. plus Appendicies.

Leitch, W. G. (1966) Historical and ecological factors in wetland inventory. Transactions of North American Wildlife and Natural Resources Conference 31: 88–96.

Leopold, L. B. and Wolman, M. G. (1957) River Channel Patterns Braided, Meandering, and Straight. U.S. Geological Survey Professional Paper 282 B. p. 39–85. *In*: Physiologic and Hydrolic Studies of Rivers 1956–1961. U.S. Geological Survey Professional Paper 282 A-G, Reston, Virginia, USA.

Leopold, L. B. and Miller, J. P. (1961) Ephemeral Streams – Hydraulic factors and their relation to the drainage net. U.S. Geological Survey Professional Paper 282 A. p. 1–37. *In*: Physiologic and Hydrologic Studies of Rivers 1956–1961. U.S. Geological Survey Professional Paper 282 A-G, Reston, Virginia, USA.

Lewis, R. and Thomas, J. (1974) Proceedings of the First Annual Conference on Restoration of Coastal Vegetation in Florida. Hillsborough Community College, Tampa, Florida, USA. 55 pp.

Lewis, R. R. III and Estevez, E. D. (1988) The Ecology of Tampa Bay, Florida: An Estuarine Profile. U.S. Fish and Wildlife Service, Biological Report 85(7.18), Washington, DC, USA. 132 pp.

Lichtler, W. F. and Walker, P. N. (1974) Hydrology of the Dismal Swamp, Virginia-North Carolina. U.S. Geological Survey, Water Resources Division, Reston, Virginia, USA. 50 pp.

Lichtler, W. F., Hughes, G. H. and Pfischner, F. L. (1976) Hydrologic Relations Between Lake and Aquifers in a Recharge Area Near Orlando, Florida. U.S. Geological Survey Water-Resources Investigations 76–65, Reston, Virginia, USA. 61 pp.

Little, E. L., Jr. (1977) Atlas of United States Trees, Vol. 4: Minor Eastern Hardwoods. U.S. Forest Service Miscellaneous Publication No. 1342, Washington, DC, USA. n.p.

Livingston, R. J. (1984) The Ecology of the Appalachicola Bay System: An Estuarine Profile. U.S. Fish and Wildlife Service, FWS/OBS-82/05, Washington, DC, USA. 148 pp.

Louisiana State University (1983) Exploring the Atchafalaya Delta. Aquanotes 12: 1–4.

Ludden, A. P., Frink, D. L. and Johnson, D. H. (1983) Water storage capacity of natural wetland depressions in the Devils Lake Basin of North Dakota. Journal of Soil and Water Conservation 38: 45–48.

MacDonald, P. O., Frayer, W. E. and Clauser, J. K. (1979) Documentation, Chronology, and Future Projections of Bottomland Hardwood Habitat Loss in the Lower Mississippi Alluvial Plain. Vol. I: Basic Report. U.S. Fish and Wildlife Service, Ecological Services, Vicksburg, Mississippi, USA. 133 pp.
Majumdar, S. K., Brooks, R. P., Brenner, F. J. and Tiner, R.W., Jr. (eds.)(1989) Wetlands Ecology and Conservation: Emphasis in Pennsylvania. The Pennsylvania Academy of Science, Lafayette College, Easton, Pennsylvania, USA. 395 pp.
Malcolm, J. (1978) Study of Wetland Drainage in Relation to Souris River Water uality as It Impacts J. Clark Salyer National Wildlife Refuge. U.S. Fish and Wildlife Service, J. Clark Salyer National Wildlife Refuge, Unpublished Special Report, Upham, North Dakota, USA. 24 pp. plus attachments.
Malcolm, J. (1979) Relationship of Wetland Drainage to Flooding and Water Quality Problems and the Impacts on J. Clark Salyer National Wildlife Refuge. U.S. Fish and Wildlife Service, J. Clark Salyer National Wildlife Refuge, Unpublished Special Report, Upham, North Dakota, USA. 29 pp. plus attachments
Marshall, A. R. (1981) Restoring the Florida Everglades. p. 153–160. *In*: R. C. Carey and P. S. Markovitis (eds.). Proceedings U.S. Fish and Wildlife Service Workshop on Coastal Ecosystems of the Southeastern United States. FWS/OBS-80/59, Washington, DC, USA.
Martin, A. C., Hotchkiss, N., Uhler, F. M. and Bourn, W. S. (1953) Classification of Wetlands of the United States. U.S. Fish and Wildlife Service, Special Scientific Report – Wildlife 20, Washington, DC, USA. 14 pp.
McCormick, J. and Somes, H. A., Jr. (1982) The Coastal Wetlands of Maryland. Maryland Department of Natural Resources. Coastal Zone Management Program. Annapolis, Maryland, USA. 243 pp.
McDonald, C. B., Ash, A. N. and Kane, E. S. (1983) Pocosins: A Changing Wetland Resource. U.S. Fish and Wildlife Service, FWS/OBS-83/32, Washington, DC, USA. 22 pp.
McGill, R. R., Jr. (1975) Land Use Changes in the Sacramento River Riparian Zone, Redding to Colusa. California Deptartment of Water Resources, Northern District, Sacremento, California, USA. 23 pp.
McGill, R. R., Jr. (1979) Land Use Change in the Sacramento River Riparian Zone, Redding to Colusa, an Update – 1972 to 1977. California Deptartment of Water Resources, Northern District, Sacremento, California, USA. 34 pp.
McHugh. J. L. (1966) Management of estuarine fishes. p. 133–154. *In*: S. Smith, A. H. Swartz and W. H. Massmann (eds.). Symposium on Estuarine Fishes. American Fisheries Society, Special Publication No. 3, Bethesda, Maryland, USA. 154 pp.
McNatt, R. M., Hallock, R. J. and Anderson, A. W. (1980) Riparian Habitat and Instream Flow Studies: Lower Verde River, Ft. McDowell Reservation, Arizona. U.S. Fish and Wildlife Service, Albuquerque, New Mexico, USA. 52 pp.
McNaughton, S. J. (1966) Ecotype function in the *Typha* community-type. Ecological Monographs 36: 297–325.
Merrell, T. J., Jr. and Koski, K. V. (1979) Habitat values of coastal wetlands for Pacific Coast salmonids. p. 256–266. *In*: P. E. Greeson, J. R. Clark and J. E. Clark (eds.). Wetland Functions and Values: The State of Our Understanding. American Water Resources Association, Minneapolis, Minnesota, USA.
Metzler, K. J. and Damman, A. W. H. (1985) Vegetation patterns in the Connecticut River floodplain in relation to frequency and duration of flooding. Le Naturaliste Canadien 112: 535–549.
Michigan Department of Natural Resources (1982) Michigan's Wetlands. Lansing, Michigan, USA. 47 pp.
Millar, J. B. (1976) Wetland Classification in Western Canada: A Guide to Marshes and Shallow Open Water Wetlands in the Grasslands and Parklands of the Prairie Provinces. Canadian Wildlife Service Report Series 37, Ottawa, Ontario. 38 pp.
Mitsch, W. J. (1979) Interactions between a riparian swamp and a river in southern Illinois. p. 63–72. *In*: R. R. Johnson and J. F. McCormick, (Technical eds.). Strategies for the Protection

and Management of Floodplain Wetlands and Other Riparian Ecosystems. U.S. Forest Service General Technical Report WO-12, Washington, DC, USA.

Mitsch, W. J. and Ewel, K. C. (1979) Comparative biomass and growth of cypress in Florida wetlands. American Midland Naturalist 101: 417–426.

Mitsch, W. J. and Gosselink, J. G. (1986) Wetlands. Van Nostrand Reinhold Co., New York, New York, USA. 539 pp.

Monk, C. D. (1966) An ecological study of hardwood swamps in North-central Florida. Ecology 47: 649–654.

Monschein, T.D. (1981) Values of pocosins to game and fish species in North Carolina. p. 155–170. *In*: C. J. Richardson (ed.). Pocosin Wetlands. Hutchinson Ross Publishing Co., Stroudsburg, Pennsylvania, USA.

Moore, J. H. and Carter, J. H. III. (1984) Habitats of white cedar in North Carolina. p. 177–190. *In*: A. D. Laderman (ed.). Atlantic White Cedar Wetlands. Westview Press, Boulder, Colorado, USA.

Moore, P. D. and Bellamy, D. J. (1974) Peatlands. Springer-Verlag, Inc., New York, New York, USA. 221 pp.

Motts, W. S. and Heeley, R. W. (1973) Wetlands and ground water. p. 5–8. *In*: J. S. Larson (ed.). A Guide to Important Characteristics and Values of Freshwater Wetlands in the Northeast. University of Massachusetts Water Resources Research Center Publication No. 31, Amherst, Massachusetts, USA.

Mundorff, M. J. (1950) Flood-Plain Deposits of North Carolina Piedmont and Mountain Streams as a Possible Source of Ground-Water Supply. North Carolina Division of Mineral Resources Bulletin 59. North Carolina Department of Conservation and Development, Raleigh, North Carolina, USA. 20 pp. plus maps.

Musselman, L. J., Nickrent, O. L. and Levy, G. F. (1977) A contribution towards a vascular flora of the Great Dismal Swamp. Rhodora 79: 240–268.

Nachlinger, J. L. (1988) Soil-vegetation Correlations in Riparian and Emergent Wetlands, Lyon County, Nevada. U.S. Fish and Wildlife Service, Biological Report 88(17), Washington, DC, USA. 39 pp.

National Fish and Wildlife Foundation (1988) Fish and Wildlife Service Five-Year Needs Assessment. Dec. 1 1988 (revised), Washington, DC, USA. 402 pp.

National Research Council (1982) Impacts of Emerging Agricultural Trends on Fish and Wildlife Habitat. National Academy Press, Washington, DC, USA. 303 pp.

National Marine Fisheries Service (1983) Unpublished coastal wetland (national and southeast) estimates. St. Petersburg, Florida, USA. 3 pp.

Nebraska Game and Parks Commission (1972) Survey of Habitat. Work Plan K-71. Pittman-Robertson Project W-15–R-28, Lincoln, Nebraska, USA. 78 pp.

Neiland, B. J. (1971) The forest-bog complex of Southeast Alaska. Vegetatio 22: 1–64.

Nelson, R. W., Logan, W. J. and Weller, E. C. (1983) Playa Wetlands and Wildlife on the Southern Great Plains: A Characterization of Habitat. U.S. Fish and Wildlife Service, FWS/OBS-83/28, Washington, DC, USA. 163 pp.

Niering, W. A. and Warren, R. S. (1980) Vegetation patterns and processes in New England salt marshes. Bioscience 30: 301–307.

Niering, W. A. (1982) Statement before the House Committee on Merchant Marine and Fisheries (August 10). Available from the author. Connecticut College, New London, Connecticut, USA.

Niering, W. A. (1988) Endangered, threatened and rare wetland plants and animals of the continental United States. p. 227–238. *In*: D. D. Hook, W. H. McKee, Jr., H. K. Smith, J. Gregory, V. G. Burrell, Jr., M. R. DeVoe, R. E. Sojka, S. Gilbert, R. Banks, L. H. Stolzy, C. Brooks, T. D. Matthews and T. H. Shear (eds.). The Ecology and Management of Wetlands. Vol. 1. Ecology of Wetlands. Timber Press, Portland, Oregon, USA.

Nixon, S. W. (1982) The Ecology of New England High Salt Marshes: A Community Profile. U.S. Fish and Wildlife Service, FWS/OBS- 81/55, Washington, DC, USA. 70 pp.

Nixon, S. W. and Oviatt, C. A. (1973) Ecology of a New England salt marsh. Ecological Monographs 43: 463–498.

Nixon, S. W. and Lee, V. (1986) Wetlands and water quality: a regional review of recent research in the United States on the role of freshwater and saltwater wetlands as sources, sinks, and transformers of nitrogen, phosphorus, and various heavy metals. U.S. Army Corps of Engineers, Technical Report Y-86–2, Vicksburg, Mississippi, USA. 228 pp.
Novitzki, R. P. (1978) Hydrology of the Nevin Wetland near Madison, Wisconsin. U.S. Geological Survey, Water Resources Investigations 78–48, Reston, Virginia, USA. 25 pp.
Novitzki, R. P. (1979) The hydrologic characteristics of Wisconsin's wetlands and their influence on floods, streamflow, and sediment. p. 377–388. *In*: P. E. Greeson, J. R. Clark and J. E. Clark (eds.). Wetland Functions and Values: The State of Our Understanding. American Water Resources Association, Minneapolis, Minnesota, USA.
Odum, H. T., Ewel, K. C., Mitsch, W. J. and Ordway, J. W. (1975) Recycling Treated Sewage Through Cypress Wetlands in Florida. University of Florida. Center for Wetlands, Occasional Publication No. 1, Gainesville, Florida, USA. 13 pp.
Odum, W. E., McIvor, C. D. and Smith, T. J. III. (1982) The Ecology of the Mangroves of South Florida: A Community Profile. U.S. Fish and Wildlife Service, FWS/OBS-81/24, Washington, DC, USA. 144 pp.
Odum, W. E., Smith. T. J. IV, Hoover, J. K. and McIvor, C. C. (1984) The Ecology of Tidal Freshwater Marshes of the United States East Coast: A Community Profile. U.S. Fish and Wildlife Service, FWS/OBS-83/17, Washington, DC, USA. 177 pp.
Office of Technology Assessment (1984) Wetlands: Their Use and Regulation. U.S. Congress, Office of Technology Assessment, OTA-O-206, Washington, DC, USA. 208 pp.
Ohmart, R. D., Deason, W. O. and Burke, C. (1973) A riparian case history: the Colorado River. p. 35–47. *In*: R. R. Johnson and D. A. Jones (eds.). Importance, Preservation and Management of Riparian Habitat: A Symposium. U.S. Forest Service. General Technical Report RM-43, Washington, DC, USA.
Orth, R. J. and Moore, K. A. (1981) Submerged aquatic vegetation of the Chesapeake Bay: past, present, and future. p. 271–283. *In*: K. Sabol (ed.). Transactions of the 46th North American Wildlife and Natural Resources Conference, Wildlife Management Institute, Washington, DC, USA.
Osvald, H. (1955) The vegetation of two raised bogs in Northeastern Maine. Srensk Botanisk Tidkrift 49: 110–118.
Patrick, W. H., Jr., Dissmyer, G., Hook, D. D., Lambou, V. M., Leitman, H. M. and Wharton, C. H. (1981) Characteristics of wetlands ecosystems of southeastern bottomland hardwood forests. p. 64–89. *In*: J. R. Clark and J. Benforado (eds.). Report on a Bottomland Hardwood Wetlands Workshop. Lake Lanier, Georgia. June 1–6 1980, National Wetlands Technical Council, Washington, DC, USA.
Penfound, W. T. (1952) Southern swamps and marshes. Botanical Review 18: 413–446.
Penfound, W. T. (1956) Primary production of vascular aquatic plants. Limnology and Oceanography 1: 92–101.
Penman, H. L. (1963) Vegetation and Hydrology. Commonwealth Bureau of Soils, Technical Communication No. 53, Commonwealth Bureau, Bucks, England. 124 pp.
Perry, M. C., Munroe, R. E. and Haramis, G. M. (1981) Twenty-five year trends in diving duck populations in Chesapeake Bay. p. 299–309. *In*: K. Sabol (ed.). Transactions of the 46th North American Wildlife and Natural Resources Conference. Wildlife Management Institute, Washington, DC, USA.
Pestrong, R. (1965) The development of drainage patterns on tidal marshes. Stanford University Publications in Geological Science 10: 1–87.
Peters, D. S., Ahrenholz, D. W. and Rice, T. R. (1979) Harvest and value of wetland associated fish and shellfish. p. 606–617. *In*: P. E. Greeson, J. R. Clark and J. E. Clark (eds.). Wetland Functions and Values: The State of Our Understanding. American Water Resources Association, Minneapolis, Minnesota, USA.
Polisini, J. M. and Boyd, C. E. (1972) Relationships between cell wall fractions, nitrogen and standing crop in aquatic macrophytes. Ecology 53: 484–488.
Pomeroy, L. R. and Wiegert, R. G. (eds.)(1981) The Ecology of a Salt Marsh. Springer-Verlag, New York, New York, USA. 271 pp.

Prentki, R. T., Gustafson, T. D. and Adams, M. S. (1978) Nutrient movements in lakeshore marshes. p. 169–194. *In*: R. E. Good, D. F. Whigham and R. L. Simpson (eds.). Freshwater Wetlands: Ecological Processes and Management Potential. Academic Press, Inc., New York, New York, USA.

Redelfs, A. E. (1983) Wetlands Values and Losses in the United States. M.S. Thesis, Oklahoma State University, Stillwater, Oklahoma, USA. 143 pp.

Redfield, A. C. (1972) Development of a New England salt marsh. Ecological Monographs 42: 201–237.

Reed, P. B. Jr. (1988) National list of plant species that occur in wetlands: 1988 National Summary. U.S. Fish and Wildlife Service, Biological Report 88(24), Washington, DC, USA. 244 pp.

Reimold, R.J. and Queen, W. H. (eds.)(1974) Ecology of Halophytes. Academic Press, New York, New York, USA. 605 pp.

Reiners, W. A. (1972) Structure and energetics of three Minnesota forests. Ecological Monographs 42: 71–94.

Remane, A. and Schlieper, C. (1971) Biology of Brackish Water. Wiley Interscience Division. John Wiley & Sons, New York, New York, USA. 372 pp.

Rice, E. L. (1965) Bottomland forests of North-central Oklahoma. Ecology 46: 708–714.

Richardson, C. J., Wentz, W. A., Chamie, J. P. M., Kadlec, J. A. and Tilton, D. L. (1976) Plant growth, nutrient accumulation and decomposition in a central Michigan peatland used for effluent treatment. p. 77–117. *In*: D. L. Tilton, R. H. Kadlec and C. J. Richardson (eds.). Freshwater Wetlands and Sewage Effluent Disposal. University of Michigan, Ann Arbor, Michigan, USA.

Richardson, C. J., ed. (1981a) Pocosin Wetlands. An Integrated Analysis of Coastal Plain Freshwater Bogs in North Carolina. Hutchinson Ross Publishing Co., Stroudsburg, Pennsylvania, USA. 364 pp.

Richardson, C. J. (1981b) Pocosins: ecosystem process and the influence of man on system response. p. 135–151. *In*: C. J. Richardson (ed.) Pocosin Wetlands. Hutchinson Ross Publishing Co., Stroudsburg, Pennsylvania, USA.

Richardson, C. J., Evans, R. and Carr, D. (1981) Pocosins: an ecosystem in transition. p. 3–19. *In*: C. J. Richardson (ed.). Pocosin Wetlands. Hutchinson Ross Publishing Co., Stroudsburg, Pennsylvania, USA.

Rogers, H. H. and Davis, D. E. (1972) Nutrient removal by water hyacinth. Weed Science 20: 423–425.

Rusmore, B., Swaney, A. and Spader, A. D. (eds.)(1982) Private Options: Tools and Concepts for Land Conservation. Island Press, Covelo, California, USA. 292 pp.

Rykiel, E. J., Jr. (1977) The Okeefenokee Swamp Watershed: Water Balance and Nutrient Budgets. Ph.D. Dissertation, University of Georgia, Athens, Georgia, USA. 246 pp.

Sather, J.H. and Smith. R. D. (1984) An Overview of Major Wetland Functions and Values. U.S. Fish and Wildlife Service, FWS/OBS- 84/18, Washington, DC, USA. 68 pp.

Schlesinger, W. H. (1978) Community structure, dynamics, and nutrient ecology in the Okefenokee Cypress Swamp-Forest. Ecological Monographs 48: 43–65.

Schomer, N. S. and Drew, R. D. (1982) An Ecological Characterization of the Lower Everglades, Florida Bay, and the Florida Keys. U.S. Fish and Wildlife Service, FWS/OBS-82/58.1, Washington, DC, USA. 246 pp.

Seibert, P. (1968) Importance of natural vegetation for the protection of the banks of streams, rivers, and canals. p. 35–67. *In*: Nature and Environment Series. Publication No. 2 (Volume on Freshwater), Council of Europe, Strasbourg, France.

Seliskar, D. M. and Gallagher, J. L. (1983) The Ecology of Tidal Marshes of the Pacific Northwest Coast: A Community Profile. U.S. Fish and Wildlife Service, FWS/OBS-82/32, Washington, DC, USA. 65 pp.

Sharitz, R. R. and Gibboons, J. W. (eds.)(1989) Freshwater Wetlands and Wildlife. U.S. Department of Energy, Washington, DC, USA. 1,265 pp.

Shaw, S. P. and Fredine, C. G. (1956) Wetlands of the United States. Their Extent and

Their Value to Waterfowl and Other Wildlife. U.S. Fish and Wildlife Service, Circular 39, Washington, DC, USA. 67 pp.

Shelford, V. E. (1954) Some Lower Mississippi Valley floodplain biotic communities: their age and elevation. Ecology 35: 126–142.

Shjeflo, J. B. (1968) Evapotranspiration and the Water Budget of Prairie Potholes in North Dakota. U.S. Geological Survey Professional Paper 585–B, Reston, Virginia, USA. 49 pp.

Sidle, J. G. (1983) Wetland drainage and the James River. North Dakota Outdoors (August): 10–14

Sigafoos, R. S. (1964) Botanical Evidence of Floods and Floodplain Deposition, Vegetation, and Hydrologic Phenomena. U.S. Geological Survey Professional Paper 485–A, Reston, Virginia, USA.

Simpson, R. L., Good, R. E. Dubinskil, B. J., Pasquale, J. J. and Philipp, K. R. (1983) Fluxes of Heavy Metals in Delaware River Freshwater Tidal Wetlands. Rutgers University, Center for Coastal and Environmental Studies, New Brunswick, New Jersey, USA. 79 pp.

Simpson, R. L., Good, R. E., Leck, M. A. and Whigham, D. F. (1983) The ecology of freshwater tidal wetlands. BioScience 33: 255–259.

Simpson, R. L., Good, R. E., Walker, R. and Frasco, B. R. (1983) The role of Delaware River freshwater tidal wetlands in the retention of nutrients and heavy metals. Journal of Environmental Quality 12: 41–48.

Sjors, H. (1950) On the relation between vegetation and electrolytes in north Swedish mire waters. Oikos 2: 241–258.

Sjors, H. (1959) Bogs and fens in the Hudson Bay lowland. Journal of the Arctic Institute of America 12: 2–10.

Sloey, W. E., Spangler, F. L. and Fetter, C. W. Jr. (1978) Management of freshwater wetlands for nutrient assimilation. p. 321–340. *In*: R. E. Good, D. F. Whigham and R. L. Simpson (eds.). Freshwater Wetlands. Ecological Processes and Management Potential. Academic Press, Inc., New York, New York, USA.

Smalley, A. E. (1959) The growth cycle of *Spartina* sp. and its relations to the insect population in the marsh. p. 96–100. *In*: Proceeding of the Salt Marsh Conference, Sapelo Island, Georgia, March 25–28, 1958. Marine Institute, University of Georgia, Athens, Georgia, USA.

Smith, A. G., Stoudt, J. H. and Gollop, J. B. (1964) Prairie potholes and marshes. p. 39–50. *In*: J. P. Linduska (ed.). Waterfowl Tomorrow. U.S. Fish and Wildlife Service, Washington, DC, USA.

Stegman, J. L. (1976) Overview of current wetland classification and inventories in the United States and Canada: U.S. Fish and Wildlife Service. p. 102–120. *In*: J. H. Sather (ed.). National Wetland Classification and Inventory Workshop Proceedings-1975, University of Maryland. U.S. Fish and Wildlife Service, Washington, DC, USA.

Stevenson, J. 0., Confer, N. and Pieper, C. B. (1979) The Decline of Submerged Aquatic Plants in Chesapeake Bay. U.S. Fish and Wildlife Service, FWS/OBS-79/24, Washington, DC, USA. 12 pp.

Stevenson, J. C. and Confer, N. M. (1978) Summary of Available Information on Chesapeake Bay Submerged Vegetation. U.S. Fish and Wildlife Service, FWS/OBS-78/66, Washington, DC, USA. 335 pp.

Steward, K. K. and Ornes, W. H. (1973) Assessing the Capability of the Everglades Marsh Environment for Renovating Wastewater. Final Report. Agricultural Research Service, Fort Lauderdale, Florida. USA. 28 pp.

Stewart, R. E. and Kantrud, H. A. (1971) Classification of Natural Ponds and Lakes in the Glaciated Prairie Region. U.S. Fish and Wildlife Service, Resource Publication 92, Washington, DC, USA. 57 pp.

Stewart, R. E. and Kantrud, H. A. (1972) Vegetation of Prairie Potholes, North Dakota, in Relation to Quality of Water and Other Environmental Factors. U.S. Geological Survey Professional Pap. 585–D, Reston, Virginia, USA. 36 pp.

Stout, J. P. (1984) The Ecology of Irregularly Flooded Salt Marshes of the Northeastern Gulf of Mexico: A Community Profile. U.S. Fish and Wildlife Service, Biological Report 85(7.1), Washington, DC, USA. 98 pp.

Street, M. W. and McClees, J. D. (1981) North Carolina's coastal fishing industry and the influence of coastal alterations. p. 238–251. *In*: C. J. Richardson (ed.). Pocosin Wetlands. Hutchinson Ross Publishing Co., Stroudsburg, Pennsylvania, USA.

Swanson, G. A., Meyer, M. I. and Serie, J. R. (1974) Feeding ecology of breeding blue-winged teal. Journal of Wildlife Management 38: 396–407.

Swift, B. L. (1980) Breeding Bird Habitats in Forested Wetlands of West-Central Massachusetts. M.S. Thesis, University of Massachusetts, Amherst, Massachusetts, USA. 90 pp.

Taylor, J. R. (1985) Community Structure and Primary Productivity in Forested Wetlands in Western Kentucky. Ph.D. Dissertation, University of Louisville, Louisville, Kentucky, USA. 139 pp.

Teal, J. and Teal, M. (1969) Life and Death of the Salt Marsh. Audubon/Ballantine Books, New York, New York, USA. 274 pp.

Thibodeau, F.R. and Ostro, B.O. (1981) An economic analysis of wetland protection. Journal of Environmental Management 12: 19–30.

Thomas, J. W., Maser, C. and Rodick, J. E. (1979) Wildlife Habitats in Managed Rangelands - the Great Basin of Southeastern Oregon: Riparian Zones. U.S. Forest Service, General Technical Report PNW-80, Washington, DC, USA. 18 pp.

Thompson, R. (1983) America's Disappearing Wetlands. Congressional Quarterly Inc. Editorial Research Reports (August 19): 615–631.

Timmer, D. E. and Weldon, L. W. (1967) Evapotranspiration and pollution of water by water hyacinth. Hyacinth Control Journal 6: 34–37.

Tiner, R. W., Jr. (1977) An Inventory of South Carolina's Coastal Marshes. South Carolina Marine Resources Center, Technical Report No. 23, Charleston, South Carolina, USA. 33 pp.

Tiner, R. W., Jr. (1984) Wetlands of the United States: Current Status and Recent Trends. U.S. Fish and Wildlife Service, National Wetlands Inventory, NWI Report, Washington, DC, USA. 59 pp.

Tiner, R. W., Jr. (1985a) Wetlands of New Jersey. U.S. Fish and Wildlife Service, National Wetlands Inventory, Newton Corner, Massachusetts, USA. 117 pp.

Tiner, R. W., Jr. (1985b) Wetlands of Delaware. U.S. Fish and Wildlife Service. National Wetlands Inventory, Newton Corner, Massachusetts and Delaware Deptartment of Natural Resources and Environmental Control, Wetlands Section, Dover, Delaware, USA. Cooperative Publication. 77 pp.

Tiner, R. W., Jr. (1987) A Field Guide to Coastal Wetland Plants of the Northeastern United States. University of Massachusetts Press, Amherst, Massachusetts, USA. 285 pp.

Tiner, R. W., Jr. (1988) Field Guide to Nontidal Wetland Identification. U.S. Fish and Wildlife Service. Region 5, Newton Corner, Massachusetts and Maryland Dept. of Natural Resources, Annapolis, Maryland, USA. Cooperative Publication. 283 pp. and color plates.

Tiner, R. W., Jr. (1989) Wetlands of Rhode Island. U.S. Fish and Wildlife Service, Region 5 Newton Corner, Massachusetts and U.S. Environmental Protection Agency, Region 1 Boston, Massachusetts, USA. Cooperative Publication. 71 pp. and Appendix.

Tiner, R. W., Jr. and Finn, J.T. (1986) Status and Recent Trends of Wetlands in Five Mid-Atlantic States: Delaware, Maryland, Pennsylvania, Virginia, and West Virginia. U.S. Fish and Wildlife Service, National Wetlands Inventory, Newton Corner, Massachusetts and U.S. Environmental Protection Agency, Philadelphia, Pennsylvania, USA. Cooperative Publication. 40 pp.

Tiner, R. W., Jr. and Veneman, P. L. M. (1987) Hydric Soils of New England. University of Massachusetts Cooperative Extension. Bulletin C-183, Amherst, Massachusetts, USA. 27 pp.

Todd, R. (1978) Where to find Arizona's birds. Arizona Wildlife Views 21: 6.

Turner, R. E. (1977) Intertidal vegetation and commercial yields of penaeid shrimp. Transactions of the American Fisheries Society 106: 411–416.

Turner, R. E. (1979) Louisiana's coastal fisheries and changing environmental conditions. p. 363–370 *In*: J. W. Day, Jr., D. D. Culley, Jr., R. E. Turner and A. J. Mumphrey, Jr. (eds.). Proceedings of the Third Coastal Marsh and Estuary Management Symposium. March 6–7, 1978. Louisiana State University, Baton Rouge, Louisiana, USA.

Turner, R. E. (1987) Relationship between canal and levee density and coastal land loss in Louisiana. U.S. Fish and Wildlife Service, Biological Report 85(14), Washington, DC, USA. 58 pp.

Turner, R. E. and Gosselink, J. G. (1975) A note on standing crops of *Spartina alterniflora* in Texas and Florida. Contributions to Marine Science 19: 113–118.

Turner, R. E. and Craig, N. J. (1980) Recent areal changes in Louisiana's forested wetland habitat. Proceedings of the Louisiana Academy of Science Vol. XL III: 61–68.

U.S. Army Corps of Engineers (1976) Natural Valley Storage: A Partnership with Nature Newsletter. New England Division, Waltham, Massachusetts, USA. 4 pp.

U.S. Department of Agriculture, Soil Conservation Service (1975) Soil Taxonomy: A Basic System of Soil Classification for Making and Interpreting Soil Surveys, Agriculture Handbook 436, Washington, DC, USA. 754 pp.

U.S. Department of Agriculture, Soil Conservation Service (1987) Hydric Soils of the United States. In cooperation with the National Technical Committee for Hydric Soils, Washington, DC, USA. n.p.

U.S. Department of Agriculture (1989) The Second RCA Appraisal. Soil, Water, and Related Resources on Nonfederal Land in the United States. Analysis of Condition and Trends, Washington, DC, USA. 280 pp.

U.S. Environmental Protection Agency and U.S. Fish and Wildlife Service (1984) The Ecological Impacts of Wastewater on Wetlands: An Annotated Bibliography. U.S. EPA, Region IV, EPA-905/3–84–002, Chicago, Illinois, USA.

U.S. Fish and Wildlife Service (1976) Existing State and Local Wetland Surveys (1965–1975). Vol. II. Narrative. Office of Biological Services Report, Washington, DC. 453 pp.

U.S. Fish and Wildlife Service (1977) Concept Plan for Waterfowl Wintering Habitat Preservation: Central Valley California. Region 1. Portland, Oregon, USA. 116 pp. + appendices.

U.S. Fish and Wildlife Service (1981) The Platte River Ecology Study. Northern Prairie Wildlife Research Center Special Resource Report, Jamestown, North Dakota, USA. 187 pp.

U.S. Fish and Wildlife Service (1982) Agricultural Resources and Wetland changes 1972–1980, Palm Beach County, Florida. National Wetlands Inventory, St. Petersburg, Florida, USA. Unpublished mimeo. 8 pp.

U.S. Fish and Wildlife Service (1984) Report of the Waterfowl Habitat Strategy Team (Draft), Atlanta, Georgia, USA.

U.S. Fish and Wildlife Service (1989) Annual Report of the Migratory Bird Conservation Commission. U.S. Fish and Wildlife Service, Washington, DC, USA. 35 pp.

U.S. Fish and Wildlife Service (1990) U.S. National Report to the Ramsar Secretariat, Washington, D.C, USA. 10 pp.

Ungar, I. A. (1974) Inland halophytes of the United States. pp 235–305. *In*: R. J. Reimold and W. H. Queen (eds.). Ecology of Halophytes. Academic Press, Inc., New York, New York, USA.

University of Minnesota, Center for Urban and Regional Affairs (1981) Thematic Maps: "Presettlement Wetlands of Minnesota" and "Available Wetlands for Bioenergy Purposes: Land Use and Drainage Constraints". Prepared under contract for Minnesota Energy Agency, Minneapolis, Minnesota, USA. 2 8 1/2" × 11" black and white maps.

Valiela, I., Teal, J. M., Volkmann, S., Shafer, D. and Carpenter, E. J. (1978) Nutrient and particulate fluxes in a salt marsh ecosystem: Tidal exchanges and inputs by precipitation and groundwater. Limnology and Oceanography 23: 798–812.

van der Valk, A. G. (1985) Vegetation dynamics of prairie glacial marshes. p. 293–312. *In*: J. White (ed.). The Population Structure of Vegetation. Dr. W. Junk Publishers, Dordrecht, The Netherlands.

van der Valk, A. G., ed. (1989) Northern Prairie Wetlands. Iowa State University Press, Ames, Iowa, USA. 400 pp.

van der Valk, A. G., Davis, C. B., Baker, J. L. and Beer, C. E. (1979) Natural freshwater wetlands as nitrogen and phosphorus traps for land runoff. p. 457–467. *In*: P. E. Greeson, J. R. Clark, and J. E. Clark (eds.). Wetland Functions and Values: The State of Our Understanding. American Water Resources Association, Minneapolis, Minnesota, USA.

Veneman, P. L. M. and Tiner, R. W. (1990) Soil-Vegetation correlations in the Connecticut River Floodplain of Western Massachusetts. U.S. Fish and Wildlife Service, Biological Report 90(6), Washington, DC, USA. 51 pp.

Verry, E. S. and Boelter, D. H. (1979) Peatland hydrology. p. 389–402. *In*: P. E. Greeson, J. R. Clark and J. E. Clark (eds.). Wetland Functions and Values: The State of Our Understanding. American Water Resources Association, Minneapolis, Minnesota, USA.

Walker, M. D., Walker, D. A. and Everett, K. R. (1989) Wetland Soils and Vegetation, Arctic Foothills, Alaska. U.S. Fish and Wildlife Service, Biological Report 89(7), Washington, DC, USA. 116 pp.

Wang, F. C. and Heimburg, K. (1976) Hydrologic budget model. p. 68–109. *In*: H. T. Odum and K. C. Ewel (eds.). Cypress Wetlands for Water Management, Recycling, and Conservation. Third Annual Report to the National Science Foundation (RAWN) and Rockefeller Foundation, Center for Wetlands, University of Florida. Gainesville, Florida, USA.

Wauer, R. H. (1977) Significance of Rio Grande riparian system upon the avifauna. p. 165–174. *In*: R. R. Johnson and D. A. Jones (eds.). Importance, Preservation and Management of Riparian Habitat. A Symposium. U.S. Forest Service, General Technical Report RM-43, Washington, DC, USA.

Webber, P. J. (1972) Comparative ordination and productivity of tundra vegetation. p. 55–60. *In*: S. Bowen (ed.). Proceedings of IBP Tundra Biome Symposium. University of Washington, Seattle, Washington, USA.

Weeks, J. L. (1974) Ohio Wetland Inventory. Ohio Department of Natural Resources, Division of Wildlife. Federal Aid Project W-104–R-16, Columbus, Ohio, USA.

Weller, M. W. (1981) Freshwater Marshes. Ecology and Wildlife Management. University of Minnesota Press, Minneapolis, Minnesota, USA. 137 pp.

Wharton, C. H. (1970) The Southern River Swamp -A Multiple Use Environment. School of Business Administration, Georgia State University, Atlanta, Georgia, USA. 48 pp.

Wharton, C. H. (1978) The Natural Environments of Georgia. Georgia Department of Natural Resources, Atlanta, Georgia, USA. 277 pp.

Wharton, C. H., Odum, H. T., Ewel, K., Duever, M., Lugo, A., Boyt, R., Bartholomew, J., DeBellevue, E., Brown, W. Bown, M. and Duever, L. (1976) Forested Wetlands of Florida – Their Management and Use. Center for Wetlands, University of Florida, Gainesville, Florida, USA. 421 pp.

Wharton, C. H., Kitchens, W. M., Pendleton, E. C. and Sipe, T. M. (1982) The Ecology of Bottomland Hardwood Swamps of the Southeast: A Community Profile. U.S. Fish and Wildlife Service, FWS/OBS-81/37, Washington, DC, USA. 133 pp.

Whigham, D. F. and Simpson, R. L. (1976) The potential use of freshwater tidal marshes in the management of water quality in the Delaware River. p. 173–186. *In*: J. Tourbier and R. W. Pierson, Jr. (eds.). Biological Control of Water Pollution. University of Pennsylvania Press, Philadelphia, Pennsylvania, USA.

Whitehead, D. R. (1972) Development and environmental history of the Dismal Swamp. Ecological Monographs 42: 301–315.

Whitney, D. M., Chalmers, A. G., Haines, E. B. Hanson, R. B. Pomeroy, L. R. and Sheer, B. (1981) The cycles of nitrogen and phosphorus. pp. 163–181. *In*: L. R. Pomeroy and R. G. Weigert (eds.). The Ecology of a Salt Marsh, Springer-Verlag, New York, New York, USA.

Wilkinson, D. L., Schneller-McDonald, K., Olson, R. W. and Amble, G. T. (1987) Synopsis of Wetland Functions and Values: Bottomland Hardwoods with Special Emphasis on Eastern Texas and Oklahoma. U.S. Fish and Wildlife Service, Biological Report 87(12), Washington, DC, USA. 132 pp.

Williams, J. D. and Dodd, C. K., Jr. (1979) Importance of wetlands to endangered and threatened species. p. 565–575. *In*: P. E. Greeson, J. R. Clark and J. E. Clark (eds.). Wetland Functions and Values: The State of Our Understanding. American Water Resources Association, Minneapolis, Minnesota, USA.

Windell, J. T., Willard, B. E., Cooper, D. J., Foster, S. Q., Knud-Hanson, C. F., Rink, L.P. and Kiladis, G. N. (1986) An Ecological Characterization of Rocky Mountain Montane and Subalpine Wetlands. U.S. Fish and Wildlife Service, Biological Report 86(11), Washington, DC, USA. 298 pp.

Windom, H. L. (1977) Ability of Salt Marshes to Remove Nutrients and Heavy Metals from Dredged Material Disposal Area Effluents. U.S. Army Corps of Engineers, Waterways Experiment Station, Technical Report D-77–37, Vicksburg, Mississippi, USA.

Winner, M. D., Jr. and Simmons, C. E. (1977) Hydrology of the Creeping Swamp Watershed, North Carolina, With Reference to Potential Effects of Stream Channelization. U.S. Geological Survey Water-Resources Investigation 77–26, Reston, Virginia, USA. 54 pp.

Winter, T. C. and Carr, M. R. (1980) Hydrologic Setting of Wetlands in the Cottonwood Lake Area, Stutsman County, North Dakota. U.S. Geological Survey, Water Resources Investigations 80–99, Reston, Virginia, USA. 42 pp.

Wisconsin Department of Natural Resources (1976) Wetland Use in Wisconsin: Historical Perspective and Present Picture. Division of Environmental Standards, Water Quality Planning Section, Madison, Wisconsin, USA. 48 pp.

Wolman, W. G. and Leopold, L. B. (1957) River Floodplains. Some Observations of Their Formation. U.S. Geological Survey Professional Paper 282–C, Reston, Virginia, USA.

Woodhouse, W. W., Seneca, E. D. and Broome, S. W. (1976) Propagation and Use of *Spartina alterniflora* for Shoreline Erosion Abatement. U.S. Army Coastal Engineering Research Center, Technical Report 76–2, Ft. Belvoir, Virginia, USA. 73 pp.

Woodwell, G. M. and Whitney, D. E. (1977) Flax Pond ecosystem study: exchanges of phosphorus between a salt marsh and the coastal waters of Long Island Sound. Marine Biology 41: 1–6.

Wooten, H. H. and Purcell, M. R. (1949) Farm Land Development: Present and Future by Clearing, Drainage, and Irrigation. U.S. Department of Agriculture, Circular 825, Washington, DC, USA. 67 pp.

Wright, A. H. and Wright, A. A. (1932) The habitats and composition of the vegetation of Okefenokee Swamp, Georgia. Ecological Monographs 2: 109–232.

Wright, J. O. (1907) Swamp and Overflowed Lands in the United States. U.S. Department of Agriculture, Office of Experiment Stations, Circular 76, Washington, DC, USA. 23 pp.

Yates, S. (1981) Florida's broken rain machine. The Amicus (Fall): 48–55.

Zedler, J. B. (1982) The Ecology of Southern California Coastal Salt Marshes: A Community Profile. U.S. Fish and Wildlife Service, FWS/OBS-81/54, Washington, DC, USA. 110 pp.

Zedler, P. H. (1987) The Ecology of Southern California Vernal Pools: A Community Profile. U.S. Fish and Wildlife Service, Biological Report 85(7.11), Washington, DC, USA. 136 pp.

Zedler, J. B. and Nordby, C. S. (1986) The Ecology of Tijuana Estuary, California: An Estuarine Profile. U.S. Fish and Wildlife Service, Biological Report 85(7.5), Washington, DC, USA. 104 pp.

Additional recent references

Dahl, T. E. and Johnson, C. E. (1991) Wetlands Status and Trends in the Conterminous United States Mid-1970's to Mid-1980's U.S. Fish and Wildlife Service, Washington, DC, USA. 28 pp.

Field, D. W., Reyer, A. J., Genovese, P. V. and Shearer, B. D. (1991) Coastal Wetlands of the United States. An Accounting of a Valuable National Resource. National Oceanic and Atmospheric Administration, Strategic Assessment Branch, Rockville, MD, USA. 59 pp.

Frayer, W. E. and Hefner, J. M. (1991) Florida Wetlands Status and Trends, 1970's to 1980's. U.S. Fish and Wildlife Service, Atlanta, GA, USA. 33 pp.

Frayer, W. E., Peters, D. D. and Pywell, H. R. (1989) Wetlands of California Central Valley: Status and Trends 1939 to Mid-1980's. U.S. Fish and Wildlife Service, Portland, OR, USA, 27 pp.

Hall, J. V. (1988) Alaska Coastal Wetlands Survey. U. S. Fish and Wildlife Service and National Oceanic and Atmospheric Administration Cooperative Report, Washington, DC, USA. 36 pp.

Metzler, K. J. and Tiner, R. W. (1992) Wetlands of Connecticut. Connecticut Department of Environmental Protection, State Geological and Natural History Survey of Connecticut. Report of Investigation No. 13. Hartford, CT, USA. 115 pp.

Tiner, R. W. (1992) Field Guide to Coastal Wetland Plants of the Southeastern United States. University of Massachusetts Press, Amherst, MA, USA. 285 pp.

Estuarine emergent wetland (salt marsh) in New England. (Photograph by Ralph Tiner).

Estuarine emergent wetland (brackish marsh) along the Gulf Coast. (Photograph by Ralph Tiner).

Estuarine scrub-shrub wetland (mangrove swamp) in Florida (Photograph by Ralph Tiner).

Palustrine emergent wetland in the western U.S. (Photograph by Ralph Tiner).

Palustrine emergent wetland (prairie pothole marsh). (Photograph by Charles Elliott).

Palustrine emergent wetland (wet meadow). (Photograph by Bill Zinni).

Palustrine scrub-shrub wetland (willow swamp) in Maine. (Photograph by Glenn Smith).

Palustrine scrub-shrub wetland (northern bog). (Photograph by Ralph Tiner).

Palustrine scrub-shrub wetland in Alaska. (Photograph by U.S. Fish and Wildlife Service).

Palustrine forested wetland (red maple swamp). (Photograph by Ralph Tiner).

Palustrine forested wetland (southern bottomland swamp). (Photograph by Ralph Tiner).

Palustrine forested wetland (riparian cottonwood forest) in the western U.S. (Photograph by David Cooper).

Wetlands of Mexico

INGRID OLMSTED

Abstract

Among the many diverse habitats in Mexico, wetlands vary with regard to hydrological, geomorphological, and biological factors. Humid tropical to temperate mountain climates over complex geological formations have contributed to the diversity. Estuarine and marine wetlands are the most extensive types along the 10,000 km long Mexican coastline. The Tabasco/Campeche system of the Usumacinta/Grijalva rivers is the largest watershed of the country. Palustrine habitats are described as floodplain marshes and savannas as well as forested wetlands in the form of riparian forests, palm thickets, and inundated low forests on the Yucatan Peninsula. Lacustrine wetlands are the least abundant, located in the mountainous inland areas.

The review of Mexican wetlands is based on a literature review and personal observations by the author for the Yucatan Peninsula. Wetland classification follows Cowardin *et al.* (1979). Deforestation, agricultural development and expansion, oil refineries, industrial production, and local tourist development over the last 100 years have reduced and contaminated or damaged the wetlands of the country. In spite of the detrimental effects on the coastal wetlands, large tracts of vegetation and animal populations remain in tact. Municipal, state, and federal wildlife reserves and refuges have been established and are being considered, though legislation for these areas is slow.

Introduction

Mexico stretches latitudinally (32° to 14° N) and longitudinally (88° to 117° W) over a very diverse area of habitats. Its peculiar shape and irregular outline, caused by the narrowing of the continent in a north-south direction, are

D.F. Whigham et al. (eds.), Wetlands of the World I, 637–677.

Figure 1. Map of Mexico with coastal lagoons. Letters and numbers represent sites as follows:

Region A: 1 = Estero San Miguel, 2 = Estero de Punta Banda, 3 = "Laguna Vicente Guerrero", 4 = Bahía de San Quintín, 5 = Laguna Manuela, 6 = Laguna Guerrero Negro, 7 = Laguna Ojo de Liebre, 8 = Bahía Tortuga, 9 = "Laguna Abreojos", 10 = "Estero Ballenas", 11 = Laguna San Ignacio, 12 = "Estero San Renito", 13 = Laguna San Gregorio, 14 = Laguna Santo Domingo, 15 = Bahía Magdalena, 16 = Bahía Almejas.

Region B: 1 = Estuario del Río Colorado, 2 = "Estero el Moreno", 3 = "Estero de San Lucas", 4 = Bahía de Concepcion, 5 = Ensenada de la Paz.

Region C: 1 = Bahía de Aduar, 2 = Bahía Cholla, 3 = "Estero Peñasco", 4 = Laguna Salada, 5 = "Estero de San Jorge", 6 = "Estero del Sargento", 7 = Laguna de la Cruz, 8 = Estero Tastiota, 9 = Bahía San Carlos, 10 = Laguna de Guaymas, 11 = "Laguna Vicicori", 12 = Estero Tortuga, 13 = Estero de Tecolote, 14 = Estero de Algodones, 15 = Estero de la Luna, 16 = Estero de Lobos, 17 = Estero Corga, 18 = Estero de Huivulay, 19 = Estero Ciaris, 20 = Estero de Santa Lugarda, 21 = Bahía de Yavaros, 22 = Estero de Agiabampo, 23 = Bahía San Esteban, 24 = Bahía de Topolobampo, 25 = Bahía Ohuira, 26 = Bahía de San Ignacio, 27 = Bahía de Navachiste, 28 = Bahía de Playa Colorada, 29 = Bahía de Santa María, 30 = Ensenada del Pabellon, 31 = "Ensenada de Quevedo".

Region D: 1 = Estero de Urias, 2 = Laguna de Huizache, 3 = Laguna de Caimanero, 4 = Laguna de Escuinapa, 5 = Laguna de Agua Brava, 6 = Laguna Mexcaltitlán, 7 = Boca Cegada, 8 = Estero del Pozo, 9 = Estero del Rey, 10 = Estero de San Cristobal, 11 = Laguna Agua Dulce, 12 = "Estero de Navidad", 13 = Laguna Cuyutlan, 14 = Laguna de Potosí, 15 = Las Salinas de Cuajo, 16 = Laguna de Nuxco, 17 = Laguna Mitla, 18 = Laguna Coyuca, 19 = Laguna Tres Palos, 20 = Laguna Tecomate, 21 = Laguna Chautengo, 22 = Laguna Apozahualco, 23 = Laguna de Alotengo, 24 = Laguna de Chacahua, 25 = Laguna de Pastoria, 26 = "Estero de Punta Conejo", 27 = Estero del Rio Tehuantepec, 28 = Laguna Superior, 29 = Laguna Inferior, 30 = Mar Muerto, 31 = Laguna La Joya, 32 = Laguna del Viejo.

emphasized by its extensive borders and the peninsulas of Baja California and Yucatan (Fig. 1).

Mexican wetlands are varied with regard to hydrological, geomorphological and biological factors. The diversity of climate from humid tropical to temperate mountain and the complex orogeny and geology have produced a large number of habitats. The amount of rainfall together with flat topography over poorly drained areas, modified by the influence of marine systems, determines Mexican wetlands. The most extensive wetlands (Fig. 1) occur along the Mexican coastline which extends 10,000 km along the Pacific Ocean, the Gulf of Mexico, the Gulf of California, and the Caribbean Sea (Lankford 1977).

Mexican wetlands as a whole have not been described, and therefore there is no classification system. Because of the extent of the coast line and its economic and ecological importance, coastal wetlands have been treated more often than those located inland. Work on marine and estuarine wetlands in Mexico abound in the literature, while riverine, lacustrine, and palustrine habitats have not been covered extensively. Recently, Lot and Novelo (1990) have published the first description of forested wetlands.

While the importance, especially of coastal wetlands, has been recognized for some time, it has not kept development from occurring at an ever faster rate. The conversion of wetlands to agricultural lands with accompanying dredging and contamination has reduced the available wetlands and changed the freatic level and inundation cycles (Rzedowski 1983).

This chapter will summarize the literature of marine and estuarine wetlands as well as palustrine wetlands, mostly in the coastal plains of Tamaulipas, Veracruz, Tabasco, the lowlands of the Yucatan Peninsula, and Pacific Coast wetlands from Baja California to Chiapas. Lacustrine wetlands are mentioned mostly for states other than those on the two peninsulas.

Region E: 1 = Laguna Madre de Tamaulipas, 2 = Laguna de San Andres, 3 = Laguna Chijol, 4 = Laguna de Pueblo Viejo, 5 = Laguna de Tamiahua, 6 = Laguna Tampamuchoco, 7 = Laguna Grande, 8 = Laguna Verge, 9 = Laguna Mandinga, 10 = Laguna Camaronero, 11 = Laguna Tlalixcoyan, 12 = Laguna de Alvarado, 13 = Laguna de Santecomapan, 14 = Laguna de Ostión, 15 = Laguna de Carmen, 16 = Laguna Machona, 17 = Laguna Tupilco, 18 = Laguna Mecoapan, 19 = Estero de Chiltepec, 20 = Laguna Pom, 21 = Laguna Atasta, 22 = Laguna de Términos, 23 = Laguna Sabancuy.

Region F: 1 = Laguna de Celestum, 2 = "Estero de Progreso", 3 = "Estero de Telchae", 4 = "Estero de Punta Arenas", 5 = Laguna Lagartos, 6 = Laguna de Yalahua, 7 = Cayo Arcas, 8 = Arrecifes Triángulos, 9 = Arrecife Alacrán.

Region G: 1 = Bahía Contoy, 2 = Laguna Nichupté, 3 = Laguna Chumyaxchac, 4 = Bahía de la Ascención, 5 = Bahía del Espiritu Santo, 6 = Bahía Chetumal, 7 = Banco Chinchorro, 8 = Unnamed reef lagoons. Redrawn from Lankford 1977.

Factors influencing wetland types

Climate

Apart from latitudinal controls, the major factors influencing the climate in Mexico are the characteristics of the oceans and coastal zones(of which the Carribbean and the Gulf coasts have a larger influence than the Pacific), the distribution of landmass to water, and the complex topography (Koppen 1936). The closeness of the seas influences the temperature and precipitation. The cold California current of the Pacific causes the temperature in Baja California to be lower on the west coast than on the east coast. This current also makes for air stability which in turn prevents summer rains. The warm current of the Gulf of Mexico has an effect on the temperatures in the east. It also produces humidity which falls as rain along the Gulf Coast, but diminishes somewhat over the Peninsula of Yucatan (Garcia 1976). The Pacific and Atlantic Oceans and the Caribbean also produce hurricanes which affect east and west coasts alike.

According to Köppen (1936) and as modified by Garcia (1976), there are four climate types in Mexico: A (hot and humid), B (dry), C (temperate humid), D (cold). Am, Af, and Aw climates are hot humid to subhumid. Figure 2 shows the distribution of the A climates. Af is hot humid with rains all year and Am is the same with rain in the summer. Aw is a subhumid climate with rains in the summer. Areas with A climates, receiving annually between 1500 mm and 4000 mm of precipitation, have most of the important wetlands. Most A climates are along the coast except where mountainous topography causes high rainfall in the southeast of Mexico. On the whole the Gulf of Mexico produces more precipitation than the Pacific, and the largest extent of wetlands is found along the Gulf Coast and the southeast of Mexico, including the Yucatan Peninsula. Hot humid and subhumid climates are encountered in areas with wetland vegetation: mangroves, palm swamps, fresh and saltwater marshes, savannas, low inundated forests, and riverine forests. The drier climates support few wetland types.

Topography and geology

Mexico extends over an area of 2 million km^2 about equally distributed on either side of the tropic of Cancer. Because of the extensive coastline there is no place in Mexico that is further than 500 km from the sea (Rzedowski 1983).

Mexico's surface is one of the most dissected and convoluted in the world. Geologically the country is rich in volcanic and sedimentary rock, the latter occurring mostly in the east, south and southeast, while the mountain systems of the Sierra Madre Occidental, Transverse Volcanic Axis, and the Sierra

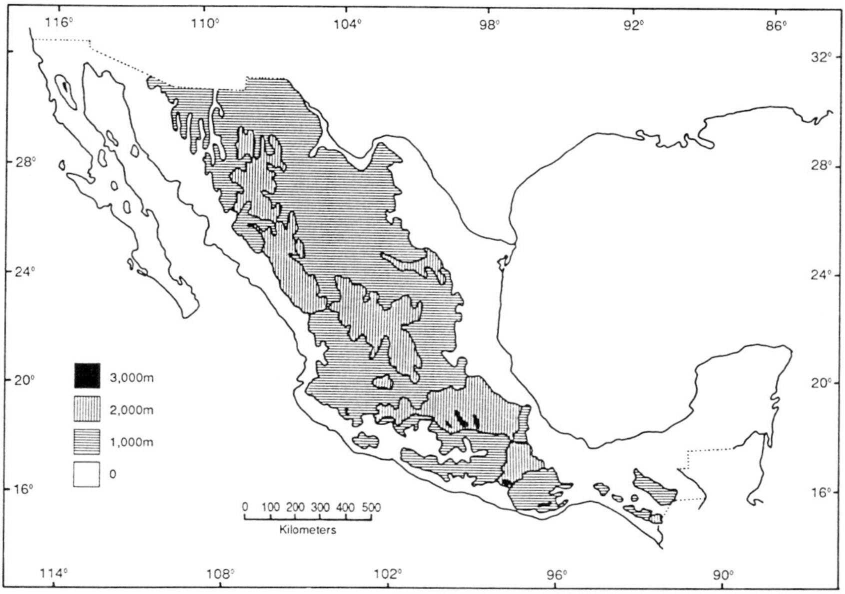

Figure 2. Altitudinal belts of Mexico.

Madre del Sur are volcanic in origin. The country has more mountains than plains, and more than half of the country's surface lies at elevations above 1000 m (Rzedowski 1983). Figure 3 shows the different altitudinal belts of Mexico, including the principal mountain ranges and plains or depressions. The dissected mountain systems provide powerful rivers whose waters are stored in reservoirs used for hydroelectric power, especially in the states of Mexico, Oaxaca, Chiapas, Chihuahua, Zacatecas, Sinaloa, and Durango. The Altiplano is a tremendously large plateau which reaches from Chihuahua to Hidalgo. The largest plain is the entire peninsula of Yucatan, which does not reach more than 300 m at its highest and has no surficial waterflow, with a highly permeable rock base. The Sierra are mostly made up of volcanic rock, other mountain systems of intrusive and metamorphic rock, the Altiplano of alluvium, the whole Gulf coastal plain and the Yucatan Peninsula of marine sedimentary rock, and the Tabasco lowlands partly of alluvium. The sediments put down by the sea are often limestone which produces karst landscapes, as is common on the Yucatan Peninsula.

Wetland types and their geomorphological setting

Since there is no Mexican classification for wetlands, wetlands are presented in this paper using the classification employed by the U.S. Fish and Wildlife

Figure 3. Distribution of A climates in Mexico according to Köppen.

Service for the United States (Cowardin *et al.* 1979). Because of lack of information on Mexican wetlands, I have modified the US system to fit Mexican wetlands. At the system level, I consider marine, estuarine, lacustrine, and palustrine wetlands. Riverine systems are included under estuarine and palustrine, since Cowardin *et al.* (1979) classification considers only the actual river. There are no descriptions of rivers in that sense in the current Mexican literature. Only one subsystem (tidal) is used, and the wetland classes are based on lifeforms as well as vegetative or dominance characteristics. Modifiers as used in the US system are considered in the section on "Ecological characteristics".

Marine

Bays

Bays are open marine systems that are exposed to ocean waves, receive no or little freshwater run-off from the land, and salinity is more than 30‰. In bays that receive some freshwater runoff, the distinction between marine and estuarine systems would be difficult, as in the case of Bahia del Espiritu Santo and Bahia de la Ascensiontwo which are located along the Caribbean Coast of Quintana Roo (Fig. 1).

Seagrass beds

This wetland type is called "ceibadales" in Mexico. They are abundant along the shores, except along the east and west coasts of Northern Baja California. The dominant seagrasses are all tropical in the Gulf of Mexico and Caribbean, and include *Syringodium filiforme*, *Halodule wrightii*, *Halophila decipiens* var. *pubescens*, *Thalassia testudinum*, and *Ruppia maritima*. Lot-Helgueras (1971) described the ecology and environmental conditions of these communities in Veracruz and gives references to ecology, taxonomy, anatomy, and productivity of the species. Two more species occur in the western and southern Gulf, in addition to the above, namely, *Halodule beaudettei* and *Halophila engelmanni*. *Thalassia*, *Halodule*, and *Syringodium* are dominant along the Caribbean and *Ruppia maritima* occurs along all Mexican coasts.

Along the Pacific coast of Baja California, eelgrass (*Zostera marina*) is very abundant (Lot-Helgueras 1977). *Phyllospadix scouleri* and *P. torreyi* are two dominant seagrasses found along the Pacific coast.

Estuarine

The estuarine systems include river deltas and coastal lagoons. Wetlands within these geomorphological settings are complex and numerous and share many common characteristics. Consequently, estuarine wetlands are presented together for the two systems.

Mangrove swamps

Mangrove swamps are the most common type of coastal wetland along the Pacific Coast, from Xcalak on the southern Caribbean coast of Quintana Roo to Matamoros along the Gulf Coast. The mangrove communities are dominated by four species. Red mangrove (*Rhizophora mangle*) has its northern limit at the 29th parallel (Rzedowski 1983), black mangrove (*Avicennia germinans*) and white mangrove (*Laguncularia racemosa*) occur in almost all areas over the entire range of this wetland type. Lot-Helgueras *et al.* (1977) describe the floristic changes near the northern limit of mangroves in the Gulf of Mexico, with *Rhizophora* and *Laguncularia* distributed to about 23.8 degrees N, while *Avicennia* extends to about 27.5 degrees N. The fourth species is buttonwood (*Conocarpus erectus*). While it most often occurs in mangroves, it also grows in other types of wetlands such as the low inundated forest (Olmsted and Duran 1986). Rico-Gray (1981) reports *Rhizophora harrisonii* from the Pacific coast.

Because of different combinations of tidal influence, substrate composition, wave action, salinity and disturbance history, the structure and productivity of mangrove vegetation varies widely (Flores-Verdugo *et al.* 1987).

The most productive mangroves occur along river deltas where alluvium provides an annual replenishment of nutrients (Table 5). The best developed riverine mangroves are found in the Usumacinta River delta near Laguna de Terminos in Campeche (Lot and Novelo 1990), the Atasta and Pom lagoons in Campeche, the Sontecomapan lagoon in southern Veracruz (Menéndez 1976), the Agua Brava lagoon in Nayarit (Rollet 1974), Teceapan Lagoon in Sinaloa (Pool *et al.* 1977), and Mezcalapa Delta in Tabasco (Thom 1967). Mangroves in all of these areas occur on deep organic soils and the canopy, at least in Veracruz, Tabasco, and Campeche, is approximately 30 m. Mangrove forests of lesser stature surround most coastal lagoons on the Pacific Ocean (Flores-Verdugo *et al.* 1990) from Sonora to Chiapas. Along coastal rocky areas of Oaxaca, Guerrero, Michoacan, and Jalisco the distribution of these mangrove forests is discontinuous (Rzedowski 1983).

Fringing mangrove forests are less extensive along the Gulf coast of Tamaulipas and Veracruz. On the sedimentary limestone coasts of the Yucatan peninsula, there is little freshwater runoff into the coastal lagoons and mangroves are restricted to narrow strips with the trees usually not more than 8–12 m tall. There is a shallow organic layer over the calcareous soil (Olmsted, personal observation).

There are also extensive areas of scrub-shrub mangroves on the Yucatan Peninsula along the Gulf and Caribbean Coast (Rico-Gray 1982, Olmsted *et al.* 1983, Olmsted and Duran 1988). All three mangrove species and buttonwood occur in the scrub mangroves, but red mangrove and buttonwood are most abundant where they occur in sparse as well as dense cover types. The plants vary from 2–5 m in height. Scrub-shrub mangroves are especially abundant in the biosphere reserve of Sian Ka'an in Quintana Roo (Olmsted *et al.* 1983). The scrub-shrub mangroves occur on calcareous marl with high sulfide content (Lot and Novelo 1990). Near the coast the organic layer may be as deep as 1 m over marl. These scrub mangroves can be found up to 20 km inland from the sea. Rico-Gray (1982) described this type of system for a large wetland area along the north coast of Campeche and Yucatan, called "Los Petenes". Vasquez-Yanes (1971) mentions the occurrence of scrub *Avicennia* marshes near Laguna Mandinga in Veracruz.

Hammocks

Hammocks (petenes) are common features in coastal wetlands of Campeche, Yucatan, and Quintana Roo. While only infrequently affected by tidal influence, these forested islands are common in the matrix of marsh and mangrove vegetation. In Campeche, Duran (1987) found hammocks dominated by *Rhizophora*, *Avicennia*, and *Laguncularia* up to 25 m in height. In one hammock, very large *Laguncularia* trees were codominant with *Manilkara zapota*, a common upland species in the Yucatan peninsula. These hammocks are inundated for part of the year, occur on deep organic soil that has developed over marl, and have some freshwater input, usually from springs which are called cenotes. The surrounding marshes or mangrove wetlands are saline, while even during the height of the dry season, the hammocks have some fresh water input. Olmsted *et al.* (1983) mention the hammocks of Sian Ka'an in Quintana Roo, and Olmsted and Duran (1988) compare hammocks over organic soil in Campeche, Quintana Roo, and Florida. Hammocks with semi-evergreen forests, growing on rock platforms in a marsh or prairie matrix occur in Sian Ka'an in Quintana Roo (Olmsted *et al.* 1983).

Saline and brackish marshes

In northern Mexico, saltmarshes are dominated by *Spartina* spp., *Sporobolus virginicus*, *Distichlis spicata*, *Monanthochloe littoralis*, and *Uniola* spp. (Rzedowski 1983). All of these marshes are to a greater or lesser extent influenced by tides. The saltmarshes are more abundant along the northern Pacific coasts than along the tropical Gulf or Caribbean. Zedler (1982) gave a community profile of southern California coastal saltmarshes which includes the estuary of the Tijuana River in Mexico. She includes the saltmarshes of northern Baja California in the southern California type, based on species composition which shows *Salicornia virginica* and *Spartina foliosa* as dominants and *Cordylanthus maritimus* as endangered (Zedler 1985).

In more humid and/or tropical conditions, Mexican saltmarshes are floristically different from those described in temperate regions. Brackish and saline marshes are commonly found in association with mangroves along the Pacific, Gulf, and Caribbean coasts, especially near coastal lagoons or near river deltas with low sediment load (Lankford 1977, West *et al.* 1969, Rzedowski 1983, Olmsted *et al.* 1983). In a study of Mexican wetlands important for waterfowl, Scott and Carbonell (1986) listed a number of coastal areas where brackish marshes are common. They range from the Yaqui delta in Sonora to Topolobampo Lagoon in Sinaloa, to Marismas Nacionales in Nayarit, and the coastal lagoons in Guerrero, Oaxaca and Chiapas. *Scirpus* spp., *Eleocharis* spp., *Typha domingensis*, *Cladium jamaicense*, and *Cynodon* spp. are the most common genera of emergent plants, but other grasses and sedges are common, as well as the seagrass *Ruppia maritima*.

Along the coast of Quintana Roo, brackish marshes are scattered, usually small in extent, occur on marl with a thin organic layer, and are adjacent to mangroves and/or buttonwood. The dominant species are *Typha domingensis*, *Phragmites australis*, *Cladium jamaicense*, and *Eleocharis cellulosa*. These species are all characteristic of brackish conditions, but cannot survive in highly saline areas. *Spartina spartinae*, *Distichlis spicata*, *Fimbristylis* spp., and *Fuirena* spp. occur sparingly throughout the Yucatan Peninsula (Olmsted *et al.* 1983).

Salt flats

Another estuarine wetland, small in extent, with high soil salinities, and periodically tidally influenced, is salt flats dominated by *Salicornia* sp., *Batis maritima*, *Suaeda linearis*, and *Sesuvium portulacastrum*. They are found along all coastlines, with the species composition being different on east and west coast. The four species mentioned above occur on the Yucatan Peninsula. González-Medrano (1972) found three associations on salt flats surrounding the Laguna Madre de Tamaulipas on the Gulf coast. *Suaeda nigra* and *Salicornia ambigua* formed one association. A second association was dominated by *Batis maritima*, *Borrichia frutescens*, *Clappia suaedifolia*, and *Maytenus phyllanthoides*, while a third had *Distichlis spicata* and *Monanthochloe littoralis*. Vazquez-Yanes (1971) described non-tidally influenced areas, "espartales", near Laguna Mandinga, Veracruz. *Salicornia* spp., *Suaeda* spp., and *Sesuvium* spp. were the dominant species. For the Gulf of California, Johnston (1924) describes a halophytic community with *Salicornia pacifica*, *Monanthochloe littoralis*, *Batis maritima*, *Salicornia europaea*, and *Frankenia grandifolia*. Knapp (1965) mentions *Salicornia virginiana*, *Suaeda californica*, *Distichlis spicata*, *Jaumea carnosa*, among others, for a saline marsh on the U.S. border in northwestern Baja California.

Lacustrine

Freshwater lakes, reservoirs, and their littorals

Mexico has quite a number of natural lakes as well as numerous man-made reservoirs (Fig. 1). Many of the reservoirs have been built for hydroelectric purposes or irrigation, and the most important ones are listed in Table 1. Deepwater lakes and reservoirs will not be considered here as they have been poorly studied. The materials that follow are primarily for shallow lakes, but they would also apply to the littoral areas of many deep lakes and reservoirs.

Submerged vegetation of shallow lakes consists mostly of the species *Cabomba palaeformis*, *Ceratophyllum demersum*, *Myriophyllum* spp., *Nitella* spp., and *Vallisneria americana* in quiet areas, and *Chara* spp., *Potamogeton* spp., *Ranunculus*, and *Utricularia* spp. in flowing water areas (Rzedowski 1983).

Floating vegetation in shallow lakes and seasonally inundated ponds can often be quite dense and has low diversity. The pan-tropical water hyacinth (*Eichhornia crassipes*) covers extensive areas, especially where man has intervened. The water lettuce (*Pistia stratiotes*) is another common plant on

Table 1. Some freshwater and saltwater lakes and reservoirs of Mexico by State. Based on SPP Carta Topografica 1985, Scott and Carbonell (1986), and Saunders and Saunders (1981).

State	Lake	Reservoir
Aguascalientes	–	3 small ones
Baja California Norte/Sur	Laguna Salada	–
Chiapas	Montebello	Chicoasan
		Netzahualcoyotl
		Angostura
Campeche	Misteriosa	–
Chihuahua	Encinillas	Luis L. Leon
	Bustillas	La Boquilla
	Palomas	Abraham Gonzalez
		Francisco Madero
Coahuila	Cuatro cienegas	Venustiano Carranza
		La Amistad (Rio Bravo)
Colima	–	–
Durango	Santiaguillo	Lazaro Cardenas
	La Mancha	Francisco Zarco
Guanajuato	Yuriria	Ignacio Allende
		Solis
		Corralejo
Guerrero	Tuxpan	Valerio
		Vicente Guerrero
		Hermenegildo Galeana
Hidalgo	Tecocomulco	Endho
	Metatitlan	Javier Rojo Gomez
		Vicente Aguirre
		Requena
Jalisco	Chapala	Santa Sosa
	Zapotlan	–
	La Vega	–
	San Marcos	–
Mexico	Zumpango	Danxho
	Texcoco	Huapango
	Xochimilco	Taxhimay
		Guadalupe
		Ignacio Ramirez
		Antonio Azate
		Villa Victoria
		Valle de Bravo
Michoacan	Cutzeo	Infiernillo
	Patzcuaro	Aristeo Mercado
	San Juanico	El Bosque
	Zitahuen	Tupextepec
Morelos	Laguna Zempoala	–
	Coatetelco	–
Nayarit	–	–
Nuevo Leon	–	Santa Rosa
Oaxaca	Coyuca	Miguel Aleman
		Benito Juarez
Puebla	Oriental	Tejocotal
		Manuel A. Camacho
Queretaro	–	Constitucion de 1917

Table 1. Continued.

State	Lake	Reservoir
Quintana Roo	Nobec	–
San Luis Potosi	some small ones	La Lajillas
Sinaloa	–	Miguel Hidalgo
		Josefina O. de Doming.
		Eustaquio Balbuena
		Adolfo Lopez Mateo
Sonora	Plutarco Elias Calles	Alvaro Obregon
	La Angostura	Adolfo Ruiz Cortines
	A. L. Rodriguez	–
Tabasco	lots of contact lakes	–
Tamaulipas	La Tortuga	Falcon
		Marte R. Gomez
		Vicente Guerrero
		Ramiro Caballero
Tlaxcala	–	Atlangatepec
Veracruz	Catemaco	–
	La Chila	–
	Pueblo Viejo	–
	Mandinga	–
	Maria Lisamba	–
Yucatan	–	–
Zacatecas	–	Leobardo Reynosa

water surfaces in humid, hot climates. Another group of floating species are the water ferns which can totally cover lake surfaces. This group includes the species *Salvinia* spp., *Lemna gibba*, *Spirodela polyrrhiza*, *Wolffia brasiliensis*, *W. columbiana*, and *Azolla* spp. *Salvinia* occurs in hot humid climates and all the rest in colder climates, especially higher altitudes (Rzedowski 1983). Floating-leaved species that are rooted in the substrate in areas of quiet waters include the species *Nymphaea ampla* (in some areas *Nymphaea gracilis*) and *Nymphoides fallax* or *N. indica*.

Large lakes such as Champala and Atotonilco in Jalisco or Patzcuaro and Cuitzeo in Michoacan have freshwater marshes along the littoral zone. They are mostly dominated by *Scirpus californicus*, *Typha latifolia* and species of *Polygonum*. Other grasses and sedges are common and *Eichhornia* is always present.

In a study of 6 crater lakes in the state of Puebla, Ramirez Garcia A. (1983) describes the vegetation as consisting mostly of the submersed species *Potamogeton pectinatus* and *Ruppia maritima* and emergents such as *Typha domingensis*, *Phragmites australis*, *Cyperus laevigatus*, *Juncus andicolus*, and *Scirpus californica*.

The aquatic flora and vegetation were presented for Lakes Coatetelco (Morelos), Yuriria (Guanajuato), Cuitzeo (Michoacan), and Mitla (Guerrero) (see Figs. 1 and 4) at the 10th Mex. Bot. Congress in Guadalajara in 1987.

Salt lakes
In arid regions of Baja California, Chihuahua, Coahuila, Durango, and Zacatecas (Figs. 1, 4) highly saline lakes occur. These areas only receive water seasonally and become highly saline. Vegetation is very sparse, except for stands of halophytes such as *Suaeda* and *Salicornia*. Examples of this type of wetland can be found in Lake Salada in Baja California Norte and Lake Mayran in Coahuila (Goldman 1951). In the same states endorheic lakes occur which have the same vegetation as saline lakes.

In the state of Coahuila in Cuatro Ciénagas, there is an area of gypsum which has a number of lakes and springs with attendant endemic fauna and flora. Because of certain characteristics of each spring or lake and their separation, endemic species of fish, snails, turtles, and other animals have evolved. The area is being disturbed by canalization and diversion of the waters for agricultural purposes and gypsum mining. Water levels have receded in some places, and waters have mixed in others, endangering the fauna. Receding lake shores are being invaded by *Typha* spp. and *Eleocharis rostellata* replacing the original submerged aquatic flora (Almeda Villela and Contreras Balderas 1984).

Palustrine

Emergent wetland
1. *Floodplain marshes*. In the states of Tabasco and Campeche there is a type of marsh not found along the Pacific coast nor on other parts of the Gulf Coast. The common name "popal" is given to this wetland, which occurs on floodplains on deep organic soils that are almost continuously flooded. The wetland is dominated by tall herbs with *Thalia geniculata* usually the dominant species. Another common dominant is *Typha latifolia*. Popal wetlands are most extensive in the deltaic regions of the rivers flowing to the Gulf in the states of Tabasco, Campeche and Veracruz (West *et al*. 1969). *Sagittaria* sp., *Echinodorus* sp., *Cyperus giganteus* and *C. articulatus* also occur in the popal marsh. In some places toward Campeche, *Heliconia* and *Calathea* species occur (Rzedowski 1983). In areas of Tabasco, shallow open water lakes are dominated by floating species such as *Eichhornia crassipes*, *Pistia stratiotes* or by rooted species such as *Nymphea ampla*.

2. *Freshwater marshes*. There is very little information on freshwater marshes of palustrine emergent wetlands in Mexico, although they are abun-

dant. On the Yucatan Peninsula, they are most extensive in the state of Quintana Roo where they usually occur over Cenozoic limestone. As part of the vast coastal lowland, they are usually adjacent to the slightly brackish wetlands described earlier. *Cladium jamaicense* is the dominant species although various species of *Eleocharis* (*E. cellulosa*, *E. caribea*, *E. interstincta*) and several species of *Rhynchospora* (*R. tracyi*, *R. microcarpa*, *R. nervosa*) occur. Other common emergents are *Typha domingensis*, *Phragmites australis*, *Sagittaria lancifolia*, and *Bacopa* sp. (Olmsted *et al.* 1983, Olmsted and Duran 1988).

This freshwater marsh type is also encountered around large lakes in Michoacan and Jalisco and some of the other Pacific states. It also extends from the head waters of some rivers, as in the case of the Lerma River in the state of Mexico (Goldman 1951). A major plant species of this wetland type is *Scirpus californicus*.

3. *Prairies*. Prairies are not very well known. In Mexico the common name for anything that looks like a prairie and may have a few trees is "savana". However, prairies and savannas should be distinguished. In the bisophere reserve Sian Ka'an, Quintana Roo on the Yucatan Peninsula, a prairie wetland type occurs over marl and limestone. It is physiognomically very similar to the *Muhlenbergia* Prairie of southern Florida, which also occurs over marl and limestone. Over extensive areas in the northern portion of the reserve the wetland prairies are dominated by *Schoenus nigricans* and *Cladium jamaicense*. *Andropogon glomeratus*, *Panicum virgatum*, and *Bletia purpurea* also occur. The diversity of this prairie is much lower than its equivalent in southern Florida. This type of vegetation has not been described elsewhere in Mexico, except by Olmsted and Duran (1988) for Quintana Roo. Shrubby trees of the genus *Conocarpus* do occur sparingly in a random manner, unlike the more even distribution of trees in savannas.

4. *Savanna*. The savanna, a type of prairie with trees, is most common in southeastern Mexico, where it occurs in Campeche, Tabasco, Veracruz, and Chiapas. It is also represented in other states along the Pacific from Sinaloa to Chiapas but is much reduced in area. Savannas are common in humid and subhumid hot climates and are included in this chapter because they are often inundated for 6 months or more. The vegetation is exposed to frequent fires during the dry season. Original savannas were edaphically and not climatically controlled (Beard 1955, Rzedowski 1983). Especially in Tabasco, savannas have become more extensive because of human manipulation of the terrain (Thom 1967). Savannas are dominated by Graminae, with a very open stratum of low trees. The most common tree species are *Byrsonima crassifolia*, *Curatella americana*, *Crescentia alata*, and *C. cujete*, as well as

species of *Coccoloba*, *Quercus*, and the paurotis palm, *Acoelorrhaphe wrightii* (Rzedowski 1983).

The best represented genera of the graminoids are *Paspalum*, *Aristida*, *Imperata*, *Leptochroyphium*, *Digitaria*, and *Axonopus*. They are usually tall and sometimes form bunchgrasses. Other frequent plants are in the Cyperaceae, Leguminosae, and Compositae (Rzedowski 1983). The authors most frequently mentioned with regard to savannas in the various states are Miranda (1952, 1958) for the Yucatan Peninsula; Sarukhan (1968) for Chiapas; Sousa (1968) for Veracruz; Vasquez (1963) for Campeche; Puig (1972) for Tabasco. Expansive grasslands, dominated by *Hilaria mutica* occur in Chihuahua and Coahulia on soils that are poorly drained. *Florestina tripteris*, *Viguiera phenax*, and *Xanthocephalum gymnospermoides* are associated species (Rzedowski 1983, Shreve 1942).

Shrub-scrub wetlands. The "mucal" described by Olmsted and Duran (1986) and Lot and Novelo (1990) for Quintana Roo and Tabasco, respectively, falls into this category. It is dominated by *Dalbergia glabra* in Quintana Roo and by *D. brownei* in Tabasco. Both species grow as viney shrubs and do not exceed 4 m in height. The vegetation is rather impenetrable. Lot and Novelo (1990) describe the "zarzal" which is dominated by *Mimosa pigra* and occurs in flooded pastures in Campeche and Tabasco (West 1966, Thom 1967). *Bravaisia tubiflora* ("julubal") is another shrubby species that occurs in pure stands of up to 3 m tall in the southeast of Mexico. The stands are usually small and occur along ecotones with mangroves and savannas.

Forested wetlands

1. *Riparian forests*. Riparian forests are also known as gallery forests (Rzedowski 1983). They occur along rivers that are seasonally flooded and are distributed from the mountains to the plains. The trees range from 4–40 m in height, and the most common genera are *Platanus mexicana*, *Populus fremontii*, *Salix humboldtiana*, and *Taxodium mucronatum* in temperate areas. *Inga spuria*, *Ficus* spp., *Bambusa* spp., *Pachira aquatica*, and *Astianthus viminialis* dominate riparian forests in tropical areas (Rzedowski 1983). The riparian forests of Veracruz, Tabasco, and Campeche (Lot and Novelo 1990) also consist of *Salix chilensis*, which attains 20 m in height, in periodically inundated lowland sites.

High to medium riparian forests, up to 30 m in height, dominated by *Pachira* and *Ficus* are found on alluvial soils. Other dominant species on river flood plains of the Gulf of Mexico are: *Lonchocarpus hondurensis*, *L. guatemalensis* var. *mexicanus*, *L. pentaphyllus*, *L. cruentus*, *L. unifoliolatus*, *Inga vera* subsp. *spuria*, *Pithecellobium belizense*, *P. calostachys*, *Machaer-*

ium falciforme, *M. lunatum*, *Vatairea lundellii*, *Muellera frutescens*, and *Combretum laxum* (Chavelas 1967, Sousa 1968, Menéndez 1976, West 1966).

Some seasonally inundated lagoon fringes or river fringes support tall evergreen forest (Lot and Novelo 1990). Species tolerant of such short-term flooding are *Voychisia guatemalensis*, *Andira galeottiana*, *Terminalia amazonica*, *Xylopia frutescens*, *Miconia argentea*, *Calophyllum brasiliense* and *Tabebuia rosea* (Gomez-Pompa 1965, Sarukhan 1968, Puig 1972, Orozco and Lot-Helgueras 1976).

In southern Tabasco and northern Chiapas, medium height evergreen wetland forests are dominated by *Bravaisia integerrima* (Lot and Novelo 1990). These wetlands occur in swampy areas with shallow calcareous soils. This forest type is locally named "canacoital", and it has been described by Miranda (1952), Miranda and Hernandez (1963), Puig (1972), and Rzedowski (1983). The latter author mentions that there are also patches of "canacoitales" on the Pacific slope. In those areas, *Bravaisia* has stilt roots of up to 2 m. Another type of floodplain forest, dominated by *Lonchocarpus sericeus* has been described by Sousa (1964) in Tuxtepec, Oaxaca.

2. *Palm thickets*. Monospecific aggregations of palms have been recognized as a distinct vegetation type by Rzedowski (1983). The name palm thicket was given by Lot and Novelo (1990) and stands for "palmar". They usually occur on soil different from other forest soils, or inundation regimes and occur equally along the Pacific, Gulf, and Caribbean coasts in hot, humid to subhumid climates. These wetlands can be divided into those that can tolerate continuous flooding and those intolerant of more than 6 months of inundation.

The "tasistal" association dominated by Paurotis palm (*Acoelorrhaphe wrightii*) is common in the southeast of Mexico. This association has been described by West 1966, Orozco and Lot-Helgueras 1976, and Olmsted and Duran 1986. Olmsted and Duran (1986) described this association in Quintana Roo where it occurs in association with *Cladium* marshes over calcareous marl with a variable depth of organic soil. *Dalbergia glabra* and *Jacquinia aurantiaca* occur in those marshes. Paurotis palm can grow to more than 8 m, but is usually 3–5 m tall. In an emergent herbaceous wetland associated with the tasistal palm thicket, Olmsted and Duran (1986) found *Rhynchospora tracyi* and *R. microcarpa*, which had not previously been reported for Mexico.

The flood-intolerant palm species occur where inundation occurs no longer than 6 months. *Sabal mexicana* up to 15 m high, may occur in upland and wetland associations in various parts of Mexico. According to Rzedowski (1983) this association is found along the Pacific slope of Oaxaca and Chiapas just beyond the littoral zone. The soils are deep and because of poor drainage they are annually inundated. Another species of the same genus, *Sabal morrissii*, grows in "botanales" in southern Quintana Roo (Miranda 1958). These associations are found in areas that are periodically inundated, mostly as transition zones between inundated vegetation and evergreen forests. *Sabal mexicana* associations occur on the coastal plain north of Veracruz and are derived from semi-evergreen and deciduous forests, and that the only original palm associations are located near the river Papaloapan in Veracruz. *Roystonea regia* may form palm associations along the littoral zone of northeastern Quintana Roo while *Roystonea dunlapiana* grows (Miranda and Hernández X. 1963) in inundated areas in Quintana Roo, Veracruz, and Tabasco. *Roystonea* also grows in tall and medium-sized evergreen forests in areas with slow drainage (Miranda and Hernández X. 1963, Rzedowski 1983).

Along the Gulf coast, in the southern portion of the Yucatan Peninsula (especially in Campeche and Quintana Roo), in Tabasco, Chiapas, and Veracruz one finds the *Scheelia liebmannii* association. It is well represented in the River Papaloapan region of Veracruz. This association occurs on deep clay soils subject to inundations (Miranda 1958). *Orbignya cohune*, a species of the Pacific slope reaching up to 30 m on well drained soil, occurs as an associate of *Scheelia* in Quintana Roo close to rivers and in areas occasionally inundated. Lot and Novelo (1990) mention a palm association of low stature that grows in shallow water throughout the Yucatan Peninsula. *Bactris balanoides* and *B. trichophylla* with spines, are the dominant species and grow in depressions in and around forests.

3. *Low inundated forests* ("*selva baja inundable*"). There are a variety of forested wetlands occupying areas with hydroperiods of 6–12 months. Most of them occur in southeastern Mexico in Tabasco, Campeche, and Quintana Roo (Fig. 4). Orozco and Lot-Helgueras (1976) described a vegetation type in southern Veracruz dominated by *Annona glabra* and *Chrysobalanus icaco*. Those wetlands are situated on fluvisols that are permanently flooded. Miranda (1958) described the inundated forests dominated by *Haematoxylon campechianum* from Campeche and Quintana Roo. They occur on shallow calcareous soils, are poorly drained, and are inundated during the rainy season and often dry during the dry season. *Bucida buceras*, *Eugenia lundellii*, *Metopium brownei*, *Coccoloba reflexiflora*, *Cameraria latifolia*, *Crescentia cujete*, and *Curatella americana* are associated species in this vegetation type. Similar forests have recently been described in the Sian Ka'an biosphere reserve in Quintana Roo by Olmsted and Duran (1986). The forests occur on calcareous marl, and a thin organic layer may be present. Olmsted and Duran recognize four variants (shown in parenthesis) of this vegetation type based on the dominant species: *Dalbergia glabra* ("mucal"), *Bucida spinosa*

Figure 4. Wetland distribution of Mexico.

("bucidal"), *Haematoxylon campechianum* ("tintal"), or *Bucida buceras* ("pucteal"). The "mucal" is more properly classified as a scrub-shrub wetland and is described in a previous section.

The four communities occur along a hydroperiod gradient associated with elevation differences. The longest hydroperiod occurs in the mucal and the shortest in the pucteal, which borders on low semi-evergreen upland forest. With decreasing hydroperiod, Olmsted and Duran observed an increase in species richness and% cover. The trees in the mucal cover an area of 400 m^2 ha^{-1}, which increases to 26,600 m^2 ha^{-1} in the pucteal. Species richness increases from 32 species in the mucal to 69 species in the pucteal. The forests are often disturbed by fire during the dry season. Fires oxidize the surface organic matter and the soil surface, causing a change in the hydroperiod and a subsequent shift in species. However, the authors did not find a progressive succession from the longest inundated to the shortest inundated forest as has been suggested by Miranda (1958). These forests are rich in epiphytes, each community having different abundances of different species.

Other inundated forests, "chechenales", are dominated by *Metopium brownei*, *Haematoxylon campechnianum*, and *Cameraria latifolia* in Quintana Roo and Campeche (Miranda 1958). *Metopium* seems to have a wide tolerance of conditions, as it grows everywhere in disturbed sites as well as in upland forests.

Two interesting species of the low inundated forests belong to the genus *Bucida*. *Bucida buceras* can grow to 40 m in evergreen forests on good soil, but in low inundated forest on calcareous marl it is usually between 5 m and 12 m tall depending on site conditions. It can withstand short inundations as a tall tree of the evergreen forests, as described by Rzedowski (1983). *Bucida spinosa* on the other hand occurs only in the low inundated forest and was until 1990 only known from the Sian Ka'an biosphere reserve in Quintana Roo (Olmsted and Duran 1986). However, Olmsted and Duran (personal communication) found populations in low inundated forests in the northernmost wetlands of Quintana Roo, a rather discontinuous distribution. The species does not grow anywhere else in Mexico, but occurs in Cuba and on the Bahama Islands. Lot and Novelo (1990) mention another restricted formation of *Pachira aquatica*, locally known as "apompal", which grows in mangrove ecotones with freshwater seeps (Vasquez-Yanes 1971, Chavelas 1967). All of these low forested wetlands are called "bosques espinosos" by Rzedowski (1983) because they contain a large number of spiny individuals, cacti (*Selenicereus donkelaari*) and vines. The legumes *Pithecellobium albicans*, *Dalbergia glabra*, *Mimosa bahamensis*, *Haematoxylon campechianum*, as well as *Bucida spinosa* are all armed with spines.

Another wetland type closely related to the savanna is one dominated by *Byrsonima crassifolia*, *Curatella americana*, and *Crescentia alata* (Miranda and Hernández X. 1963). In the savanna, the trees are widely spaced, while the canopy in these sites is closed (Rzedowski 1983, Sarukhan 1968). It occurs only from Veracruz to Tabasco in the humid hot climate zone, and Rzedowski and McVaugh (1966) describe a forest like this in Colima on the Pacific.

Geographical distribution and extent

Figure 4 indicates the distribution of the major Mexican wetlands (mangroves, saltmarshes, salt flats, low inundated forests, popales, savannas, freshwater marshes, prairies, and palm thickets). It is not possible to provide exact delimitations of each of these wetland types for all of Mexico as many of them are physically juxtaposed. This map, therefore, only demonstrates the locations of areas with wetlands. Other wetlands mentioned in the text are not in Fig. 4 because their distribution is very patchy and they are small in extent. These include the lake littoral habitats in the altiplano and mountains, and the palm associations. Major rivers, lakes, and reservoirs are also indicated in Fig. 4 with letters, and are cross-referenced on Table 1, giving names of lakes and reservoirs. Lake littoral habitats are associated with these lakes. For some larger lakes inland in Jalisco and Michoacan, surrounding freshwater wetlands are indicated. Halophytic wetlands are indicated in Coahuila and Chihuahua associated with semi-permanent salt lakes such as Lake Mayran in Coahuila.

Riverine forests are not specifically mapped, but they almost all occur along the major rivers shown in Fig. 4. The inundated low forests only occur in the southeast of Mexico and are situated around and within each wetland area indicated in Fig. 4, as with savannas, prairies, and freshwater wetlands in the same area.

The coastal lagoons are indicated in Fig. 1. According to Lankford (1977), there are 123 coastal lagoons in Mexico, covering approximately 12,500 km^2 according to Yañez-Arancibia (1986). They are habitats for the seagrass wetlands, and most have fringing mangrove forests.

Table 2 shows the wetland extent as indicated by Scott and Carbonell (1986), West *et al.* (1969), and certain maps (SPP, carta topográfica y Uso del suelo, Direccion General de Geografía 1980). This table shows the most important wetlands of Mexico, with a total area estimated to be of 3,318,500 ha. The largest continuous wetland (1,420,000 ha) is located in Tabasco and Campeche. Other important wetlands are located in Quintana Roo and Yucatan with 335,000 and 184,000 ha respectively; together with Tabasco, Veracruz, Campeche, and Chiapas, these make the southeast of Mexico the most significant wetland region of Mexico. On the Pacific coast, Sinaloa, Nayarit, Sonora, and Oaxacer have the majority of wetlands.

Wetland classification and inventory

The word "wetland" does not translate easily into Spanish. The importance of wetlands as a whole has not been addressed in Mexico, although there are two Federal Secretariats (SARH and Plan Nacional Hidraulico) responsible for wetlands. To my knowledge, a unifying classification system for all wetland types of Mexico has not been established or considered, and wetland descriptions have been quite diverse and variable depending on the objective of specific studies. Although a specific wetland classification system does not exist, wetlands have been included in a few classification systems.

Mexican vegetation types, including wetlands, have been described by Miranda and Hernández X (1963), Rzedowski (1983), and Lot and Novelo (1990). In Table 3, I have attempted to relate the terminology used in this paper to that used in the three papers cited above. The table shows only forested wetlands, as mentioned by Lot and Novelo. The names are rather similar, and the Spanish equivalents are easy translations. This classification is based on lifeform, inundation, and/or salinity. The Lot and Novelo system

Table 2. Extent and location of some important wetlands in Mexico.

Type of wetland	State	ha
Coastal lagoons and brackish marshes	Sonora	108,000
Coastal lagoons, brackish marshes and sand beaches, estuaries, mudflats, mangrove swamps	Sinaloa	377,000
Coastal lagoons, marshes, mangrove swamp	Nayarit	140,000
Coastal lagoons, mangrove swamps	Colima	9,500
Coastal lagoons, mangrove swamps	Guerrero	17,000
Coastal lagoons, mudflats, mangrove swamps	Oaxaca	149,000
Coastal lagoons, mangrove swamps	Chiapas	73,800
River delta, freshwater lakes and marshes (Rio Tamesi, R. Panuco, Tampico Lagoons)	Tamaulipas, Veracruz, San Luis Potosi	54,400
Freshwater lagoons and marshes, lakes R. Papaloapan	Veracruz	208,000
Brackish lagoon	Veracruz	105,000
Tidal saline lagoons, mangrove forests, gallery forests, freshwater lagoons, freshwater marshes, swamps	Tabasco	1,000,000
Tidal saline lagoons, mangrove forests, gallery forests, freshwater lagoons, freshwater marshes, swamps	Campeche	420,000
Coastal lagoons, mangrove forests, salt flats	Yucatan	184,000
Mangrove swamps, large coastal bays, brackish and freshwater marshes, savannas, prairies, low inundated forests and palm thickets	Quintana Roo	335,000
Freshwater Lakes with fringing marshes	Jalisco	120,000
Alkaline lakes	Jalisco	19,200
	Total	3,318,500

is somewhat different from that used in this paper, but all of the wetland types have been recognized (see Table 3). Different types of coastal lagoons, a distinct geomorphological entity, have been described by Lankford (1977).

An inventory of all Mexican wetland types does not exist. One may come across wetland inventories indirectly when considering inventories of waterfowl, animals, or plants. Saunders and Saunders (1981) did a survey of "Waterfowl and their Wintering Grounds in Mexico, 1937–64" which contains descriptions of wetlands from Baja California to Yucatan. Apart from the geographic breadth of the survey, it gives information more than 20 years old on areas which may be significantly altered today. A more recent inventory is that by Scott and Carbonell (1986) "Inventario de Humedales de la Region Neotropical". Their objective was to describe wetlands of importance for waterfowl warranting preservation or protection. Information on wetland distribution from these two books is included in the wetland distribution map (Fig. 4).

Table 3. Equivalence table of vegetation types based on Lot and Novelo (1990), Miranda (1958), and Rzedowski (1983).

Olmsted	Lot and Novelo	Miranda	Rzedowski
Low inundated forest	Low wetland forest	selva baja inundable	bosque espinosa
Riparian forest	Evergreen riparian forest	bosque perennifolia riparia	(bosque de galeria)
Scrub-shrub wetland	Wetland thornless scrub	matorral inerme inundable	
	Wetland thorn scrub	matorral espinosa inundable	
Palm thicket	Wetland palm thicket	palmar inundable	palmar
	High to medium riparian forest	selva alta-median riparia	bosque tropical perennifolio
Mangrove swamps	Mangrove forest	Manglar	Vegetacion acuatica y subacuatica

Table 4. Birds (breeding, passing, and wintering) reported from coastal wetlands by Scott and Carbonell (1986), and Perales and Contreras (1986). Birds noted with an * only occur on the Gulf and Caribbean coasts.

Pelecanus erythrorhynchos	White pelican
P. occidentalis	Brown pelican
Branta bernicle	Brant goose
Anas americana	
A. crecca	European teal
A. acuta	Pintail
A. clypeata	Red-breasted merganser Mergus serrator
Grus canadensis	Sandhill crane
Pandion haliaetus	Osprey
Anser albifrons	White-fronted goose
Oxyura jamaicensis	
Athya afinis	Lesser scaup
Fulica americana	Coot
Phalacrocoras sp.	Cormorant
Nyctinassa violacea	Yellow-crowned night heron
Bubulcus ibis	
Egretta caerulea	Little blue heron
E. tricolor	Tricolored heron
E. rufescens	
E. thula	
E. alba	
Eudocimus albus	White ibis
Ajaia ajaja	Roseate spoonbill
Recuvirostra americana	Avocet
Ardea herodias	Great blue heron
Mycteria americana	Wood stork
Dendrocygna autumnalis	
Nycticorax nycticorax	Black crowned night heron
Plegadis chihi	
Cairina moscata	
Branta canadensis *	Canada goose
Aramus guarauna *	Limpkin
Jacana spinosa *	Jacana
Anhinga anhinga *	Anhinga
Tigrisoma mexicanum *	Tiger heron
Cochlearius cochlearius *	Boat-billed heron
Agamia agami *	Agami heron
Jabiru mycteria *	Jabiru stork
Himantopus himantopus *	Black-necked stilt
Phoenicopterus ruber *	Flamingo

Ecological characteristics

Forcing functions

Marine – seagrass beds

There is little information on the subject of forcing functions for seagrass beds in Mexico, although Lot-Helgueras (1971) has studied them on the coast of Veracruz. Literature abounds on the subject, and seagrasses have been studied in many parts of the world. In this discussion, I refer to the paper by Lot-Helgueras, and there have been recent studies on the Laguna de Terminos in Campeche by a group of Mexican and U.S. scientists Yañez-Arancibia and Day 1982).

Lot-Helgueras divided coastal systems into four zones. Seagrass beds occur mostly in the third or infralittoral zone where there are stands dominated by *Thalassia testudinum*, mixed with *Syringodium*, at depths of 1.8 to 2.4 m. In shallower waters, he found beds of *Halodule* which can tolerate a greater range of temperatures and salinity. The limiting factors for all species are temperature, turbidity, and wave action. In Veracruz, northerly winds also seem to play an important role. *Halophila decipiens* var. *pubescens* always occurs in the deeper waters ranging from 3–5 m. Seagrass beds are best developed near protected reefs. *Ruppia maritima*, though mentioned as a marine seagrass, hardly ever grows offshore, but is usually limited to the waters of coastal lagoons where it tolerates a large range of salinities.

In an ecological characterization of Terminos Lagoon in Campeche, Yañez-Arancibia and Day (1982) mapped seagrass beds and related the occurrence of the most abundant species of *Thalassia testudinum* to conditions of circulation, water clarity and salinity. *Halodule wrightii* is a colonizing species in the delta region.

Estuarine (including riverine) wetlands

More ecological literature exists on estuarine, tidal wetlands than on any other wetland type in Mexico. Coastal lagoons are the most common setting for estuarine wetlands in Mexico, and they are very numerous. Lankford (1977) reviewed the origin of the 123 coastal lagoons of Mexico and classified them on the basis of geological origin. Analysis of physiographic and geologic controls, climatic conditions, and coastal oceanography according to their geographic distribution leads him to divide the Mexican coast into seven large regions (Fig. 1). Tides and waves influence the coastal lagoons dramatically. There is a difference of 7 m in daily tides between the head of the Gulf of California and the island of Cozumel in the Caribbean. Tidal currents are important energy sources in coastal lagoons and important to sediment erosion and suspension.

Coastal lagoons are clearly in a dynamic state and change continually. Flooding depth, flooding periodicity, salinity, water chemistry, nutrient loads, and oxygen content of the water determine together with the physiograpic, geologic, and climatic conditions mentioned which type of mangrove vegetation or salt marsh or salt flat will prevail in a certain area surrounding coastal lagoons or river deltas.

Ayala-Castañares's work (1969) on the geology of the Laguna Madre de Tamaulipas, Laguna de Tamiahua, and Laguna de Terminos of Campeche,

Table 5. Litterfall and structural components of mangrove wetlands in Mexico.

	Litterfall ($g\ m^2\ yr^{-1}$)	Stem density (stems ha^{-1})	Basal area ($m^2\ ha^{-1}$)	Height (m)
Laguna de Terminos[1]				
Estero Pargo (fringing)	834	7510	23.3	6
Boca Chica (riverine)	1252	3360	34.2	20
El Verde[2] (Fringing)	1100	1800	9.9	6
Aqua Brava[3]				
Laguna Grande (riverine)	1263	1567	16.4	–
Boca La Tigra (fringing)	1417	3203	14.0	–
Punta Raquel (basin)	1015	2022	12.5	7
Laguna de Mecoacan[4]	614	–	–	–
Laguna de la Manchas	905	–	–	12–10
Tecoapan[5]				
riverine (Roblit)	–	2240	29.6	8
riverine (Isla Palma)	–	2360	60.8	17
riverine (Rio Canas)	–	1790	57.8	16
overwash (Isla Ros.)	–	1480	28.5	8
basin (El Calc.)	–	3120	15.2	9

Sources of data are: [1]Day et al. in press, [2]Flores Verdugo et al. in press, [3]Flores Verdugo 1986, [4]Lopez Portillo and Ezcurra 1985, [5]Rico-Gray and Lot 1983, [6]Pool et al. 1977.

is the most important paper bearing on the forcing functions or modifiers of estuarine wetlands. It stresses in particular the influence of the dynamic state of the barriers on the sedimentation and salinity of the lagoons. In all cases, mangroves result as principal surrounding vegetation, except in the Laguna Madre which is almost completely filled and closed except for one inlet, where the vegetation consists of a few black mangroves (very salt tolerant) and other halophytes such as *Distichlis spicata*, *Sporobolus virginicus*, *Suaeda*, and *Salicornia*.

A very interesting history of a strand plain, lagoonal coast is portrayed by Curray *et al.* (1969) for the coastline of Nayarit, part of which encloses the Marismas Nacionales, an important mangrove-marsh wetland. The interesting part about this system is the 280 subparallel ridges formed by successive accretion to the shoreline of low narrow beach ridges, overlying longshore bars. It is a complex of beach ridges, inundated depressions, marshes, and mangrove swamps, and lagoons.

It was in the Marismas Nacionales of Sinaloa near Teacapan, that Pool *et al.* (1977) found especially riverine mangrove forests, which they consider typical of habitats that receive large quantities of freshwater runoff and nutrient-rich sediments from upland watersheds. Structure and productivity data of this study are compared with data for other Mexican mangrove forests in Table 5. The complexity index calculated for the various mangrove forest types (based on classification by Lugo and Snedaker (1974)) shows the basin forest as having the lowest index, which may be explained by the fact that it is not flushed by daily tides, as the riverine and overwash forests are. The lack of this tidal influence increases the soil salinity at certain times.

Another set of important lagoons is described by Yañez-Arancibia (1978) for the coast from Zihuatenejo to south of Acapulco in Guerrero. All of these lagoons have ephemeral inlets and therefore a rather varying cycle of salinities, water depths, and sedimentation rates. In a study of mangroves in El Verde Lagoon with an ephemeral inlet along the Sinaloa coast, Flores-Verdugo *et al.* (1987) report on a very different pattern of structure, composition and productivity of *Laguncularia racemosa* as compared with other mangrove forests. They found *Laguncularia* to be dominant in this and other lagoons with ephemeral inlets. Because of the particular conditions of such lagoons the mouth of the lagoon is only open during the wet season; as the water level rises due to river flow, water flows over the barrier and erodes it open. As the water level in the lagoon decreases, flow is confined to a channel; tidal currents enter the lagoon and flow is bidirectional. Flow through the inlet becomes almost solely tidal when river flow diminishes. Littoral currents close the inlet. Both tidal and riverine flooding give rise to strong flushing of the lagoon. The authors conclude that only *Laguncularia* may be established under these conditions because of faster germination and root establishment, which occur only during a very short period. During most of the wet season propagules would be flushed from the system, during the dry season the propagules either fall on the dry forest floor or in the water, and there is no flow between the two. The high productivity of 1,100 g m^{-2} yr^{-1} is in the upper range for mangroves, but the structural development is very poor. The authors suggest that the high freshwater turnover of the river permits the high productivity, as also reported by Pool *et al.* (1977), even though it is located in a semi-arid region. For an arid site with only 250 mm annual precipitation in Baja California, total red mangrove litterfall was between 948 and 1,631 g m^{-2} yr^{-1}. Data for El Verde forests are reported in Table 5.

In their papers on mangrove productivity in Laguna de Terminos (Campeche), Day *et al.* (1982, 1987) demonstrated that riverine mangrove forests were more productive than others because of high freshwater input, which accounts for lower salinity, higher nutrient inputs, and lower hydrogen sulfide levels.

In a study of relationships among physical characteristics, vegetation distribution, and fisheries yield in the Gulf of Mexico, Deegan *et al.* (1986) suggest that physical driving forces in estuaries act in time frames (i.e. geological events to seasonal variability), that intertidal and open water areas are determined by the geologic template of the estuary, and that the type and extent of intertidal vegetation are determined in turn by the intertidal area and by climate.

In his paper on mangrove ecology and deltaic geomorphology, Thom (1967) describes in detail the relationship between geomorphic processes occurring in the Grijalva-Usumacinta delta in Tabasco and the type of mangroves that grow there. In this vast Pleistocene-dated wetland one finds mangroves along river channels, coastal lagoons (as previously mentioned), interdistributary basins, or on narrow swales. As in many other areas, frequency of inundation and salinity are two important factors which influence mangrove distribution. At variance from other alluvial plains, the tidal influence in this area is not as important as seasonal variations in rainfall which affect stream discharge.

Habitats along some rivers (Rio Gonzalez and Rio Boca Grande) receive more freshwater discharge because of a major diversion of the Rio Mezcalapa in 1932. This has influenced the vegetation, which is composed not of mangroves but of grasses like *Panicum crus-galli*, *Paspalum fluitans*, and *Distichlis spicata*, along with sedges like *Scirpus maritimus*, and *Fimbristylis* sp., as well as *Phragmites australis* and *Typha* spp. Thom (1967) also describes the different stages of the establishment of a new distributary in this type of delta.

In addition to sedimentation, subsidence plays a major role in the formation of habitats on which mangroves become established. Thom concludes that "in short-term changes strong interaction between geomorphic and biotic processes takes place, so that the resulting change in vegetation is the effect of many contributing factors. But the long-range trends are dictated by physiographic processes in the deltaic plain, processes which are continually changing and which influence such phenomena as degree of water saturation of the soil, salinity of ground and surface water, soil type, and drainage of the surface. One may even interpret these long-term changes as cyclic in character with the freshwater communities of the active delta giving way to mangrove communities, when the centre of active sedimentation and discharge shifts elsewhere. Under continuous subsidence and compaction, the mangrove area becomes increasingly wetter, peats develop, more water-tolerant species spread laterally, and lagoons and ponds enlarge where the rate of organic sedimentation cannot keep up with subsidence".

Lopez-Portillo and Ezcurra (1989) relate distribution, abundance, and physiognomy of five halophytes, pertaining to mangrove and saltflat vegetation, at Laguna de Mecoacan (Tabasco), to edaphic conditions, especially physico-chemical characteristics.

Lacustrine

Very little information on forcing functions is available on Mexican lakes. Lot-Helgueras and Novelo (1978) describe Lake Tecocomulco in the state of Hidalgo at an elevation of 2,500 m as one of the least contaminated and with the best preserved vegetation. Lake Tecocomulco is the most important in the Valley of Mexico which, hundreds of years ago, used to be a large system of lakes. Average water depth is 1.5 m and it receives its water from small streams of the Mountain of Zempoala. At the end of the rainy season, water is diverted via canals to the Papalote River.

The vegetation is distributed with four associations according to the water level and topography of the lake. The extent and presence of each association depends on the dominant life forms. Emergent hydrophytes with *Scirpus lacustris* dominant is the rooted association in areas that are always inundated. At the lake margins occurs an association of emergent hydrophytes consisting of *Lilaeopsis schaffneriana*, *Elatine triandra*, *Polygonum punctatum*, and *P. lapathifolium*. The third association consists of rooted hydrophytes with floating leaves, such as *Nymphoides fallax*, *Potamogeton nodosus*, and *P. illinoensis*, located in open areas within the *Scirpus* community. Submerged hydrophytes grow in the deepest areas and consist of *Najas guadalupensis*, various species of *Potamogeton*, *Ranunculus aquaticus*, and *Sagittaria demersa*, endemic of Mexico and in danger of extinction.

The sump lakes (bolsones) or endorheic basins in the interior highlands of Chihuahua, Coahuila, Durango, and Zacatecas were more important wetlands 30 years ago than today. They are lake basins, fed by rivers, without outlets. Deforestation of the surrounding hills, agricultural activity, drainage, and reservoirs above the sumps have reduced most of these lakes to insignificant areas, and most of them have dried up. Lake Mayran in Coahuila was at the time of the conquest in the 16th century the largest lake in Mexico, but is mostly dry now (Goldman 1951).

Palustrine

The very large marshes in Tabasco and Campeche occur in the deltaic basin of the Grijalva and Usumacinta rivers. They do not carry as much sediment as some of the other rivers and therefore large bodies of water open up between the levees. The popal marsh, described earlier, so characteristic of Tabasco, makes up 3/4 of the marshes. High water tables and water-logged soils are responsible for the marsh formations.

The interdistributary and interdeltaic basins of the Mexcalapa and Usumacinta Rivers are recent fluvial surfaces which to the south contact Pleistocene terraces that are also inundated and have characteristic contact lakes. In many instances marshes and lakes mingle imperceptibly. The marshes are characterized by fine silts and clays interbedded with decaying plant material. Description of the whole Tabasco lowland is given by West *et al.* (1969). With regard to any kind of hydrograhic, geomorphic, or other environmental factor, the West *et al.* study gives the history of the processes responsible for today's habitats and vegetation.

Biota

Plants

Major plant species have been mentioned in the descriptive portions of each wetland type and efforts have been made and are being made by the Autonomous University of Mexico to study the floristics and ecology of various aquatic systems of the country (Tellez *et al.* 1982). The most recent studies were presented at the 10th Botanical Congress of Mexico in September 1987 and are listed here. A preliminary list of aquatic plants of Mexico was presented by M. and C. Antonio Lot. The following authors published in the Abstracts of the 10th Botanical Congress of Mexico (1987): Mijanos, C. Marco and A. Novelo R. (Inst. de Biologia, UNAM), Ramos Ventura, L. and A. Novelo (Inst. de Biologia, UNAM), Rojas, J. and A. Novelo (Inst. de Biologia, UNAM), Ramirez, P. and A. Lot (Inst. de Biologia, UNAM), Martinez Mahinda (Inst. de Ecologia y Alimentos, UAT) and A. Novelo (Inst. de Biologia, UNAM), Lozado Lucio (Facultad de Ciencias, UNAM), Siqueiros Delgado (Ma. Elena; Centro Basico, Dept. de Biologia Universidad Aut. de Aguascalientes). Ramirez, P. and A. Lot (Inst. de Biologia, UNAM).

Other major references which list plants of wetlands are:

1. Vegetacion de Mexico by Rzedowski (1983) contains various lists and references to floras of wetlands in Mexico.
2. Olmsted and Duran (1986) presents lists of plants of low inundated forests.
3. Duran (1987) contains floristic lists for Petenes region, Campeche.
4. González García, R. (1985) describes aquatic plants of the Tabasco wetlands.

The majority of wetland plants are widely distributed and do not seem to be in danger. *Sagittaria macrophylla*, which is endemic to the Valley of Mexico, could become endangered with the reduction of habitat and contamination going on in that region.

Animals

Animals in wetlands can be divided into those living in wetlands all year round, those coming to the wetland during the dry season, and those that are migratory. Birds (Table 4) comprise most of the last category. There are few data about animals in Mexican wetlands, and most of the information has been reported in connection with conservation projects or studies of species in danger of extinction. I shall treat animals associated with wetlands according to major groups, e.g., mammals, reptiles and amphibians, fishes, birds, insects.

Mammals. Ceballos and Galindo (1984) list seven endangered mammals which depend on riparian habitats in Mexico. The beaver (*Castor canadensis*) and muskrat (*Ondatra zibethicus*), both mostly temperate, are listed as endangered in Mexico. They are restricted to the Colorado River areas on the Sonora–Baja California border and to the Rio Bravo and its tributaries. The muskrat seems to best survive in Nuevo Leon along 860 km of riverine habitat (Bernal 1982). The other five listed are tropical mammals. The manatee (*Trichechus manatus*), once an abundant animal along the Gulf and Caribbean coasts, today can be found in small populations in coastal lagoons and slightly inland in rivers of Tabasco, Chiapas, Yucatan, Campeche, and Quintana Roo (Gallo 1982, Colmenero-R. and Zárate 1990). In a paper on the distribution and conservation of manatees in Mexico, Colmenero and Hoz Zavala (1985) estimated a population of 5,000 but thought it to be too high. The manatee feeds on various aquatic herbs such as manatee-grass (*Cymodocea filiformis*) and the exotic water hyacinth (*Eichhornia crassipes*) It is mostly endangered because of illegal hunting and reduction of habitat. A rat (*Rheomys mexicanus*), endemic to Oaxaca, lives on river banks and streams. *Chironectens minimus*, the water opossum, lives along rivers in tropical evergreen forest in Oaxaca, Chiapas, and Tabasco. The river otter (*Lutra longicaudis*) is still widespread in Mexico, but occurs in low densities from Sinaloa to the Yucatan Peninsula. One of the rarest animals of Mexico is the tapir (*Tapirus bairdii*), which lives in areas close to water, e.g. accessible riverine habitats or low inundated forests. Population figures for Mexico are not known, as it is a very shy animal. It has mostly been reported from Chiapas and Quintana Roo.

Daniel Navarro (personal communication) has found that other mammals depend on wetlands. The bulldog bat (*Noctilio leporinus*) feeds on fish in riverine habitats from Sinaloa to Yucatan, as does *Myotis vivesi*, the fish-eating bat of Baja California and Sonora. The marsh rice rat (*Oryzomys palustris*) is widely distributed in inundated areas close to rice plantations from the Pacific to the Gulf and Caribbean. The ocelot (*Felis paradalis*) the ocelot and some other felines like *F. wiedii* (tigrillo) and the jaguar (*Panthera onca*) are often seen in mangroves and swamps looking for food; besides they are excellent swimmers. They can also be seen around low inundated forests on the Yucatan Peninsula during the dry season. All of these animals are distributed in the tropical forests of Oaxaca, Chiapas, Veracruz, Tabasco, and the Yucatan Peninsula.

West *et al.* (1969) mention various animal species for each of the wetland types encountered in the Tabasco lowlands. Among mammals, the cottontail rabbit (*Sylvilagus floridanus yucatanensis*) and the gray fox (*Urocyon cinereoargenteus*) are found in the savanna wetland of Tabasco. The raccoon (*Procyon lotor hernandezii*), common all over Mexico, needs to live very close to water because it feeds on aquatic animals like frogs and land crabs.

In their study of the mammals of the Valley of Mexico, Ceballos and Galindo (1984) list the mouse *Reithrodontomys megalotis saturatus* as living in grass dominated vegetation close to water and the shrew *Soerex vargas orizabae* in riparian habitats. Other mammals living close to bodies of water or rivers mentioned by the same authors are the armadillo (*Dasypus novemcinctus mexicanus*) and the bat *Myotis yumanensis*; the former is also common in coastal plains and the extreme north of the Yucatan Peninsula.

Reptiles and amphibians. Coastal wetlands of the southeast, especially Tabasco and Quintana Roo, are important for the crocodiles (*Crocodylus morelettii* and *C. acutus*) and the rare caiman (*Caiman osclerops*). Crocodiles were formerly much hunted for their skins and are protected today, but illegal poaching continues. Reduction and contamination of their habitat are important reasons for establishment of a biosphere reserve in Centla, Tabasco, including the Grijalalva-Usumacinta delta. In Chiapas, Tabasco, and Quintana Roo projects are underway to raise crocodiles on farms.

Marine turtles are important inhabitants of the Pacific, Caribbean, and Gulf of Mexico; they come to the beaches to nest and use the bays and lagoons to feed in. *Chelonia mydas*, the green turtle, and the hawksbill (*Eretmochelys imbricata*) were at one time commonly found feeding on the *Thallassia* beds in Tabasco (West *et al.* 1969). *Chelonia mydas* and *Caretta caretta*, the white turtle, are common visitors to the Caribbean coast in Quintana Roo. The populations of all marine turtles, including *Dermochelys coriacea*, suffered great losses especially along the Pacific coast and the Gulf of Mexico until about eight years ago. Protection laws and vigilance as well as various studies during the last eight years have slowed the consumption of these animals and their eggs. Studies in Quintana Roo regarding protection of juveniles have been especially successful (Aviña-Carlin 1987).

Terrestrial turtles such as *Chelydra sepentina* are common in the Tabasco wetlands (West *et al.* 1969) and *Kinosternon leucostomun* (Lee 1980) is sought after for food. The marshes in Tabasco abound with amphibians including various species of *Rana*, *Hyla*, and *Bufo*. The same genera are found in the marshes of Quintana Roo. In the mangrove communities of Tabasco, West *et al.* (1969) reported the common iguana in *Rhizophora* along stream channels and the stenosaur (*Stenosaura acanthinura*) in the *Avicennia* forest. Boas and water snakes are found in mangrove swamps and other wetlands in the southeast of Mexico. The rattlesnake, *Crotalus durissus*, may be encountered at the water's edge in mangrove communities or in ecotones with upland forests in Quintana Roo.

There are three endemic salamanders threatened in three different wetland areas: *Ambystoma dumerili dumerili* is only known from Lake Patzcuaro (Michoacan); *Ambystoma lermaensis* is only known from the upper Rio

Lerma, and *Ambystoma mexicanum* is found only at Lake Xochimilco (Scott and Carbonell 1986).

Fishes, aquatic crustaceans, and mollusks. Emergent freshwater wetlands in the Tabasco lowlands provide daily food for local human inhabitants. Cichlids are the most plentiful fish and forms the basis of the diet for local residents. Six genera with 20 species were identified as being important, including catfish of the genus *Rhamdia*, mullet, and a needlefish which is considered to be a delicacy (West *et al.* 1969). The freshwater fish fauna of the marshes and scrub mangroves in Sian Ka'an, Quintana Roo, has been studied by Loftus, Ramo and Lopez Ornat (in prep.). Endemic and endangered species of fish of freshwater cenotes in Quintana Roo were described by Miguel Navarro (unpublished).

Contreras-Balderas (1986) has studied the fish populations of various springs, lakes, and lagoons of Cuatro Cienegas in Coahuila, of which 18 species are endemic to this particular area and depend totally on these wetlands for their survival. The same author has studied the ichthiofauna of other arid-zone lakes of Mexico (Contreras-Balderas 1969, 1977) and gives a list of threatened and endangered species of Mexican fishes (Contreras-Balderas, in press).

The saline lagoons of the Tabasco coast and other coasts of Mexico have a faunal assemblage very similar to that of marine waters. The marine fish population of the coastal lagoons consists of many commercial species such as mullet (*Mugil cephalus*), snapper (*Lutjanus*), snook (*Centropomus*), tarpon (*Megalops atlantica*) and pompano (*Trachinotus*) (West *et al.* 1969). The same genera are found in the Caribbean.

The common oyster (*Crassostrea virginica*) thrives abundantly in the coastal lagoons and is commercially very important, as is the marine shrimp, which uses the lagoons as nurseries. In Terminos Lagoon, the sessile fauna has been described by Espinoza (1980), the micromollusks by Garcia-Cubos (1963), the shrimp by Signoret (1984), ostracods by Morales (1966), and sponges by Nuñez (1979).

The literature on fishes of estuaries and other coastal areas in the Mexican tropics is quite abundant. I shall mention only a few that give information on species, their ecology, and their environment. Yáñez-Arancibia (1986) give information on the fish composition, community structure, and function in Terminos Lagoon (Campeche), the southern Gulf of Mexico, and the whole coastal zone (Yáñez-Arancibia 1981, 1978). Information on fishes in the Gulf of California may be obtained in Schwartzlose and Hendrickson (1983).

Mexico's most important lobster fishery is in the Gulf of California and along the Caribbean coast. Studies of recruitment and population dynamics of the spiny lobster (*Panulirus argus*) are underway in Sian Ka'an Reserve of Quintana Roo. Some life stages of the lobster depend on the mangrove-lagoonal system for survival. The importance of commercial fishing in coastal wetlands is described in Cardenas (1969) for all of Mexico.

Birds. Mexico is rich in waterfowl, especially since large numbers of migratory aquatic birds winter in or around Mexican wetlands. Since the 1930s the U.S. Fish and Wildlife Service has censused migratory aquatic birds here, so there are fairly good records of most important wintering grounds. These censuses also make it possible to assess the tremendous change in wetland habitat occurring during the last 50 years. Saunders and Saunders (1981) record the waterfowl seen in Mexican wetlands between 1937 and 1964. Starker Leopold (1977), who took part in some of the bird survey flights of the US Fish and Wildlife Service, describes many of the aquatic birds he saw. The bird surveys for the years 1964–85 were reported by Brazda (1986) for the Gulf coast wetlands of Mexico. The most recent update on the waterfowl and their habitat appeared in 1986 in an inventory of neotropical wetlands by Scott and Carbonell (1986), Saunders and Saunders (1981), and Brazda (1986). Table 4 lists the most important birds of the coastal wetlands, based on Scott and Carbonell (1986).

López-Ornat reports on the status of the threatened and endangered species of Ciconiformes in Mexico which will appear in Ceballos and Navarro (in press). The Jabiru stork (*Jabiru mycteria*) is in danger of extinction and still nests in Campeche and Quintana Roo in various wetland types. The following are endangered (vulnerable): the great blue heron (*Ardea herodias occidentalis*) nesting in Quintana Roo, the Flamingo (*Phoenicopterus ruber*) nesting in the Rio Lagartos refuge of Yucatan (the only nesting area of this species in all of Mexico), the wood stork (*Mycteria americana*) nesting in wetlands of the southeast and in Nayarit, and the reddish egret (*Egretta rufescens*) nesting in the southeast, Sinaloa, and Tamaulipas. The reasons for their threatened status are various, including habitat reduction, hunting for feathers, and contamination by pesticides. However, in arriving at the threatened status for these five species, López-Ornat used various criteria: the commonness and size of populations, restricted distribution as in the case of the Flamingo, where all 6,000 pairs nest only on the northern tip of Yucatan, and size of the bird (very large birds of more than 1 m length such as the Jabiru stork and the woodstork need a much larger undisturbed area in which to feed.

Insects. Insects are very abundant in almost all of the wetlands, and mosquitoes abound in all mangroves everywhere in Mexico. Deer flies and tabanus (greenheads) are also very common, as are ants and termites in wetlands, even though the ground is inundated most the time. In a study of the relationships between ants and epiphytes, Dejean and Olmsted (in prep.) found 26 different ant species in one *Tillandsia* species. The trees and epiphytes, often abundant in low inundated forests, contain the nests of these insects.

Wetland use and conservation

Mexican wetlands, especially in the Southeast, have traditionally been used for agriculture by the Indians. It is thought that today's "unoccupied" appearance of the regions in the Southern portion of the Yucatan Peninsula gave the opposite impression during Maya times. Raised fields in inundated "bajos" (low inundated forest, scrub shrub and savannas) are thought to have been very productive and intensively cultivated. The "chinampas" of the Valley of Mexico (now mostly Mexico City) and the raised fields of Veracruz (Siemens 1980, 1989) and Tabasco were similar.

The traditional uses of wetlands maintained them; modern use eliminates or degrades them. It is their use that makes their conservation so important. In Mexico, wetland use continues and increases without coordination between use and conservation. Responsibility for the conservation and management of wetlands lies with SEDUE, the Secretariat for Urban Development and Ecology, and SARH, the Secretariat for Agriculture and Hydrological Resources, as part of the land in general. There is no specific legislation for wetlands alone. In 1984, SEDUE issued the "National Program for Ecology" (Programa Nacional de Ecología) which sets forth the conservation priorities of the country. The Comisión Nacional de Ecología published "100 needed actions" (1987, 100 Acciones Necesarias, Ecología) which include the most important environmental problems facing Mexico. Quite a few wetland areas are mentioned: lakes, rivers, coastal lagoons, fauna and flora of wetlands, and whole watersheds. In a report on the state of the environment, SEDUE (1986) declared 20 watersheds between Baja California and Veracruz as being in the greatest danger of contamination. The most harmful contaminating industries are the food and drink, paper, textile, sugar, and chemical industries. In the southeastern tropical region, not included in the 20 watersheds mentioned, the most destructive industry is oil and petrochemical.

In their inventory of wetlands of Mexico, Scott and Carbonell (1986) found that only 4 wetland areas of 40 described and visited are under protection (Fig. 1, Table 1). Of 26 other areas in the nearctic region of Mexico, not described and visited by those authors, but included in this paper (see Fig. 4, Table 1), only one area is protected. These 26 wetland areas are all located in Baja California or in inland mountainous non-tropical regions.

However, SEDUE has published a list of all protected parks, reserves, and refuges in Mexico which include wetland areas, and names at least 20. In another list, other important wetland areas have been identified as important for conservation: watersheds of southern Sonora, the watershed of Rio Fuerte in Chihuahua and Sonora, Cuatro Ciénegas in Coahuila, El Morro de la Mancha in Veracruz, the Marismas Nacionales (Teacapan-Agua Brava) in Nayarit, the coastal addition of Estacion Chamela in Jalisco (administered by the Institute of Biology of UNAM, the National Autonomous University of Mexico), Delta Usumacinta-Grijalva, Laguna de Chaschoe, Laguna Mecoacan, and Rio Gonzalez in Tabasco, Los Petenes, and Laguna de Terminos in Campeche.

In all areas of Mexico the greatest pressure on wetlands has been agricultural use, which has caused the draining of extensive areas on the Pacific and Gulf coasts. Extensive drainage and irrigation systems have reduced the available reservoirs for various species of ducks in the Rio Grande Delta (Brazda 1986). The Tamesi and Panuco River delta (Tampico Lagoons) had substantially fewer waterfowl in 1985 than before; the reduced flow of the Tamesi River from impoundments upstream has been blamed along with oil and industrial pollution. Dams constructed on the major rivers have reduced the extent of flooding in Laguna Alvarado, Laguna Camaronera, and the deltas of the Rio Papaloapan and San Juan in Veracruz. The Yaqui Delta in Sonora supports intensive agriculture surrounding the remaining marshes. Overall, the resultant effects from agriculture are reduction of wetland area due to actual agricultural use, draining of wetlands for canals, and drying due to damming and irrigation upstream. Cacao, sugar cane, pineapples, bananas, and coconuts are now the main crops grown in Veracruz, Tabasco, and Campeche in drained wetland areas, as well as along river levees and coastal beaches.

Another closely related use is cattle ranching, which, especially in Tabasco and Campeche, uses most of the natural savannas and creates many new ones by draining the wetlands. Toledo (1984) considers this activity the second most destructive of Mexican tropical wetlands, after the oil industry.

In and around Tabasco, tintales (Olmsted and Duran 1986, West *et al.* 1969) were formerly important for the extraction of logwood, *Haematoxylon campechianum*. They were exploited from 1600 to 1900, especially by the English who logged primarily in Tabasco, Campeche, and Quintana Roo. According to West *et al.* (1969), many tintales in Tabasco were overcut and the tintales disappeared. In other areas in Campeche, tintales and other inundated forests were drained and used for cattle, mostly during this century. Most intact tintales can probably be found in Quintana Roo, though there are certainly fewer than before. The overuse may have favored other species.

"How to destroy paradise" (Toledo 1984) and "Petroleo y Desarrollo" (Beltran 1985) describe the destruction and contamination of wetlands occurring in Tabasco and Veracruz due to offshore and inland oil extraction. Oil pollution threatens the life of Laguna Tamiahua in Veracruz, all of the Usumacinta Delta, the Tabasco Lagoons, and Laguna de Terminos in Campeche (a total of 1,000,000 ha). Apart from the contamination and actual destruction of wetlands due to oil extraction, the oil industry brings urbanization and the petrochemical industry. The total effect of these processes are apparent in the areas of Coatzacoalcos and Minatitlan in Veracruz (Contreras 1986), with its many oil refineries. The effects of oil extraction, oil pollution, urbanization, and the petrochemical industry on rivers, lakes, coastal lagoons, mangrove swamps, and marshes in Veracruz and Tabasco have been studied by a group of scientists of the Centro de Ecodesarrollo from 1978 to the present. Ten volumes have appeared so far and five more are planned.

Many of Mexico's wetlands are located in coastal areas, potentially important for tourist development. This type of development brings destruction of mangrove areas for hotel construction and contamination of coastal lagoons and mangrove forests by hotel waste. Most threatened are the Marismas Nacionales near San Blas in Nayarit, Laguna Cuyutlan near Manzanillo in Colima, Laguna Coyuca and Papagayo near Acapulco in Guerrero, and the whole length of the Caribbean coast in Quintana Roo, especially between Cancun and Tulum. Associated with tourism is the use of powerboats on lakes and lagoons. The latter use endangers nesting of Flamingos in Rio Lagartos in Yucatan.

Wetlands in endorheic basins of the interior highlands of Mexico used to be important wintering areas for migratory waterfowl (Saunders and Saunders 1981). Many of the wetlands in Chihuahua, Coahuila, Guanajuato, Jalisco, and Michoacan were described by Goldman (1951) on his expedition between 1896 and 1906. The U.S. Fish and Wildlife Winter Surveys of the same areas give an indication of the changes that have occurred. Due to higher population densities in the interior highlands as compared to most coastal areas, the mountainous slopes surrounding many of the lakes and marshes were logged, and decreased runoff and excessive grazing together with drought often dried the marshes and lakes and altered the vegetation. In some places, diversion of water changed the water levels. Examples are Lake Chapala in Jalisco and Lake Mayran in Coahuila. Lake Chapala was formerly the most important lake to waterfowl in all of Mexico. Cultivation of the extensive marshes surrounding the lake, sediment load from Rio Lerma, and diversion of water upstream have reduced the attractiveness of this lake to waterfowl.

Numerous wetlands visited by Scott and Carbonell (1986) (Fig. 1, Table 1) are extremely important breeding grounds for waterfowl. Conservation of many of these areas is necessary. A small portion of the Laguna de Chacagua and Pastoria near Puerto Escondido in Oaxaca are protected in a National Park, and Isla Contoy in Quintana Roo is an ecological reserve. In January of 1986, the Mexican president established the MAB (Man and the Biosphere) reserve of Sian Ka'an in Quintana Roo, a 528,000 ha marsh-mangrove-tropical forest area. Reasearch has been conducted on wetland and forest vegetation, migratory and aquatic birds, ethnobotany, and marine turtles and preliminary studies were published in 1983 (CIQRO 1983); other studies have been referenced in wetland descriptions. Research is also underway in Celestún, Yucatan, on some coastal lagoons, a faunal refuge of 59,130 ha where hunting is prohibited. Major plans are underway by the government of Tabasco, supported by SEDUE, INIREB and various international conservation agencies, to protect portion of the Usumacinta-Grijalva delta and adjacent lands in Campeche. The Tabasco-Campeche wetlands of the delta are considered to be the most important wetlands of Mesoamerica. (Symposium Usumacinta Delta, Villahermosa 1987).

References

Almada Villela, P. and Contreras Balderas, S. (1984) El bolsón de Cuatro Ciénegas, Coah. Mex., p. 125–129. *In*: Proceedings, Reunión Regional de Ecología, Norte, Monterrey, N.L., SEDUE, Subsecretaría de Ecología, México, D. F., México.

Aviña-Carlin, R. (1987) Conservación de las Tortugas marinas de Quintano Roo. Unpublished report. Centro de Investigaciones de Quintana Roo. Cancun, Quintana Roo, México. 65 pp.

Ayala-Castañares, A. (1969) Datos Comparativos de la Geología Marina de tres lagunas litorales del Golfo de México. Anales del Instituto de Biología de la Universidad Nacional Autónoma de México 40, Serie Ciencias del Mar y Limnología 1: 1–116.

Beard, J. S. (1955) The classification of tropical American vegetation types. Ecology 36: 89–99.

Beltran, J. E. (1985) Petroleo y Desarrollo, La Politica Petrolera en Tabasco. Centro de Estudios e Investigación del Sureste. México, D. F., México. 346 pp.

Bernal, J. A. (1982) Estado actual del castor, *Castor canadensis* V. Bailey 1913, en el Estado de Nuevo León, México. Unpublished Thesis, Facultad de Ciencias Biológicas, Universidad Autónoma de Nuevo León, Monterrey, Nuevo Leon, México.

Brazda, A. R. (1986) Winterfowl Populations and Habitat Evaluation, Aerial Surveys, East Coast of Mexico. U.S. Fish and Wildlfe Service, Washington, DC, USA. 26 pp.

Cardenas, M. F. (1969) Pesquerías de las lagunas litorales de México. p. 645–652. *In*: Lagunas Costeras, un Simposio, Memorias del Simposio Internacional de Lagunas Costeras, UNAM-UNESCO, Nov. 1967, México, D. F., México.

Ceballos González, G. and Galindo Leal, C. (1984) Mamíferos silvestres de la cuenca de México. MAB, UNESCO e Instituto de Ecología y Museo de Historia Natural de la Ciudad de México. Editorial LIMUSA. México, D. F., México. 299 pp.

Chavelas P. J. (1967) Estudio florístico-sinecológico del Campo Experimental Forestal "El Tormento". Escárcega, Campeche. p. 130–221. *In*: Comisión de Estudios sobre la Ecología de Dioscóreas. V. Informe. Instituto Nacional de Investigación Forestal, México D. F., México.

CIQRO (1983) Sian Ka'an. Estudios Preliminares de una zona en Quintana Roo propuesta como Reserva de la Biósfera. Centro de Investigaciones de Quintana Roo, A. C. Puerto Morelos, Quintana Roo, México. 215 pp.

Colmenero-R., L. C. and Hoz Zavala, M. E. (1985) Distribución de los manatíes, situación y su conservación en México. Anales del Instituto de Biología, Universidad Nacional Autónoma de México 56, Serie Zoologia 3: 955–1020.

Colmenero-R., L. C. and Zárate, B. E. (1990) Distribution status and conservation of the West Indian manater in Quintana Roo, Mexico, Biological Conservation 52: 27–35.

Contreras, F. (1986) La Riqueza del pantano. Vol. V de Serie Medio Ambiente en Coatzacoalcos. Centro de Ecodesarrollo, México, D. F., México. 98 pp.

Contreras-Balderas, S. (1969) Perspectivas de la ictiofauna de las areas áridas del norte de México. Memorias del 1er simposio internacional del aumento de produccion de alimentos en Zonas Aridas, ICASAALS, Texas Technical Publication 3:293–304.

Contreras-Balderas, S. (1977) Speciation aspects and man-made community composition changes in Chihuahuan Desert fishes. p. 405–431. *In*: R. H. Wauer and D. H. Riskind (eds.). Transactions Biological Research, Chihuahuan Desert Region. U.S.– Mexico National Park Service Transactions 3.

Contreras-Balderas, S. (1986) Lista anotada de especies de peces Mexicanos amenazadas o en peligro de extinción. Simposio Nacional Areas Protegidas, México.

Cowardin, L. M., Carter, V., Folet, F. L. and La Roe, E. T. (1979) Classification of Wetlands and Deepwater Habitats of the United States. Fish and Wildlife Service. U. S. Department of the Interior. Superintendent of Documents, U. S. Government Printing Office, Washington, DC, USA. 103 pp.

Curray, J. R., Emmel, F. J. and Crampton, P. J. R. (1969) Holocene History of a strand plain lagoonal coast, Nayarit, Mexico. p. 63–100. *In*: A. Ayala-Castañares and B. Phleger (eds.). Lagunas Costeras, un Simposio. Memorias Simposio Internacional de Lagunas Costeras, UNAM-UNESCO November, 1967. Universidad Nacional Autónoma de México. México, D. F., México.

Day, J. W., Jr., Day, R. H., Barreiro, M. T., Ley-Lou, F. and Madden, C. J. (1982) Primary production in the Laguna de Terminos, a tropical estuary in the southern Gulf of Mexico. Oceanologica Acta 5: 269–276.

Day, J. W., Jr., Conner, W. H., Day, R. H., Ley-Lou, F., and Navarro, A. M. (1987) The productivity and composition of mangrove forests, Laguna de Términos, México. Aquatic Botany 27: 267–284.

Deegan, L. A., Day, J. W. Jr., Gosselink, J. G., Yáñez-Arancibia, A., Chavez, G. S. and Sanchez-Gil, P. (1986) Relationships among physical characteristics, vegetation distribution and fisheries yield in Gulf of Mexico estuaries. p. 83–100. *In*: D. A. Wolfe (ed.). Estuarine Variability, Academic Press, Inc., New York, New York, USA.

Duran Garcia, G. R. (1987) la vegetación de los Petenes del noroeste de Campeche. Biotica 12: 181–198.

Éspinoza, M. (1980) La fauna sesil intermareal del manglar relacionada con algunos parámetros ambientales de la Laguna de Términos, Campeche, México. p. 102–120. *In*: Estudio Científico e impacto humano en el Ecosistema de Manglares. Memorias Seminario UNESCO -Colombia, Cali, Nov. 27–Dic. 1, 1978.

Flores-Verdugo, F. J. (1986) Aporte de materia orgánica por los principales productos primarios a un ecosistema lagunar-estuarina de boca efímera. Ph.D. dissertation, Universidad Nacional Autónoma de México. México, D. F., México. 96 pp.

Flores-Verdugo, F. J., Day, Jr. W. Jr., and Briseño-Dueñas, R. (1987) Structure, litter fall, decomposition, and detritus dynamics of mangroves in a Mexican coastal lagoon with an ephemeral inlet. Marine Ecology – Progress Series 35: 83–90.

Flores-Verdugo, F., Gonzáles-Farías, F., Ramírez-Flores, O., Amexuca-Linares, F., Yáñez-Arancibia, A., Alvarez-Rubio, M. and Day, J. W., Jr. (1990) Mangrove ecology, aquatic primary productivity, and fish community dynamics in the Teacapan-Agua Brava lagoon-estuarine system (Mexican Pacific). Estuaries 13: 219–230.

Gallo, J. P. (1982) Notas sobre la distribución del manatí (*Trichechnus manatus*) en las costas de Quintana Roo. Anales del Instituto de Biología, Universidad. Nacional Autónoma de México 53, Serie Zoologia 1: 443–448.

García, E. (1976) Modificaciones al sistema de clasificación climática de Koeppen. Instituto de

Geografía, Universidad Nacional Autónoma de México. 2a. edicion. México, D. F. México. 246 pp.

García-Cubos, A. (1963) Sistemática y distribución de los micromoluscos de la Laguna de Términos, Campeche, México. Boletin Instituto de Geología, Universidad Nacional Autónoma de México. Publicacio Especial 5: 1–180.

Goldman, E. A. (1951) Biological Investigation in Mexico. Smithsonian Miscellaneous Collections. Vol. 115, Publication 4017. Smithsonian Institution, Washington, D.C., USA. 476 pp.

Gómez-Pompa, A. (1965) La vegetación de México. Boletin de Sociedad Botánica de México 29: 76–120.

Gonzalez Garcia, R. (1985) Plantas acuaticas. p. 79–105. *In*: Usumacinta, investigacion cientifica en la Cuenca del Usumacinta. Gobierno del Estado de Tabasco. Villahermosa, Tabasco, México.

González-Medrano, F. (1972) La vegetación del nordeste de Tamaulipas. Anales del Instituto de Biología de México Serie Botánica 42: 11–50.

Knapp, R. (1965) Die Vegetation von Nord- und Mittelamerika und der Hawaii-Inseln. Gustav Fischer Verlag. Stuttgart, West Germany. 373 pp.

Köppen, W. (1936) Das Geographische System der Klimate. p. 1–44. *In*: Handbuch der Klimatologie Band I, Teil C, West Germany.

Lankford, R. (1977) Coastal Lagoons of Mexico, their origin and classification. p. 182–215. *In*: M. Wiley (ed.). Estuarine Processes: Circulation, Sediments, and Transfer of Material in the Estuary. Academic Press, New York, New York, USA.

Leopold, A. S. (1965) Fauna Silvestre de México. Instituto Mexicano de Recursos Naturales Renovables, México, D. F., México. 600 pp.

Lopez-Portillo, J. and Ezcurra, E. (1985) Litter fall of *Avicennia germinans* L. in a one-year cycle in a mudflat at the Laguna de Mecoacán, Tabasco, Mexico. Biotropica 17:186–190.

Lopez-Portillo, J. and Ezcurra, E. (1989) Zonation in mangroves and salt marsh vegetation at Laguna Mecoacan, Mexico. Biotropica 21: 107–114.

Lot-Helgueras, A. (1971) Estudios sobre Fanerógamas Marinas en las Cercanías de Veracruz. Anales Instituto Biología. Universidad Nacional Autónoma de México 42, Serie Botánica 1: 1–48.

Lot-Helgueras, A. (1977) General status of research on seagrass ecosystems in Mexico. p. 233–245. *In*: C. P. McRoy and C. Helfferich (eds.). Seagrass Ecosystems, A Scientific Perspective. Marine Science, Vol. 4. Marcel Dekker, Inc., New York, New York, USA.

Lot-Helgueras, A. and A. Novelo Retana. (1978) Laguna de Tecocomulco, Hidalgo. Guias Botánicas de Excursiones en México. Sociedad Botánica de México, A. C. México, D. F., México. 19 pp.

Lot-Helgueras, A. and Novelo, A. (1990) Forested wetlands of Mexico. p. 287–298. *In*: A. E. Lugo, M. M. Brinson and S. Brown (eds.). Forested Wetlands. Elsevier Scientific Publication Co., Amsterdam, The Netherlands.

Lot-Helgueras, A., Novelo, A. and Ramirez-Garcia, P. (1986) Listados floristicas de México. V. Angiospermas Acuaticas Mexicanas I. Instituto de Biologia, Universidad Nacional Autónoma de México, México, D. F. México. 60 pp.

Lot-Helgueras, A., Vazquez-Yanez, C. and Menendez, L. F. (1975) Physiognomic and floristic changes near the northern limit of mangroves in the Gulf coast of Mexico. p. 52–61. *In*: G. Walsh, S. C. Snedaker and H. J. Teas (eds.). Proceedings of the International Symposium on Biology and Management of Mangroves, Vol. I. School of Forest Resources and Conservation. University of Florida, Gainesville, Florida, USA.

Lugo, A. E. and Snedaker, S. L. (1974) The ecology of mangroves. Annual Review Ecology and Systematics 5: 39–64.

Menéndez, L. F. (1976) Los Manglares de la Laguna de Sontecomapán, Los Tuxtlas, Veracruz. Estudio Florístico -ecológico. Tesis, Facultad de Ciencias. Universidad Nacional Autónoma de México, México, D. F., México. 115 pp.

Miranda, F. (1952) La vegetación de Chiapas. Ediciones del Gobierno del Estado, Tuxtla Gutiérrez. Vol. I., 334 pp.

Miranda, F. (1958) Estudio acerca de la vegetación. p. 215–271. *In*: E. Beltrán (ed.). Los Recursos Naturales del Sureste y su Aprovechamiento, 2. Instituto Mexicano de Recursos Naturales Renovables A. C., México, D. F. México.

Miranda, F. and Hernández, X. E. (1963) Los tipos de vegetación de México y su clasificación. Boletin Sociedad Botánica México 28: 29–178.

Morales, G. A. (1966) Ecology, distribution and taxonomy of recent ostracods of the Laguna de Términos, Campeche, México, Boletin Instituto Geología, Universidad Nacional Autónoma, México, 81: 1–103.

Nuñez, M. C. (1979) Estudio taxonómico de las esponjas de la Laguna de Términos, Campeche, Tesis profesional, Facultad Ciencias, Universidad Nacional Autónoma de México. Bulletin American Association of Petroleum Geology 55: 2130–2140.

Olmsted, I., Lopez-O. A., and Duran G. R. (1983) Vegetación de Sian Ka'an, Reporte Preliminar. p. 63–84. *In*: Sian Ka'an, Estudios Preliminares de una zona en Quintana Roo propuesta como Reserva de la Biósfera. Centro de Investigaciones de Quintana Roo, Puerto Morelos, Quintana Roo, México.

Olmsted, I. and Duran G. R. (1986) Aspectos ecológicos de la selva baja inundable de la Reserva Sian Ka'an, Q. Roo. Biotica 11: 151–179.

Olmsted, I. and Duran G. R. (1988) Aspectos ecológicos de los Petenes en Florida, Campeche y Q. Roo. p. 519–536. *In*: Proceedings Symposium on the ecology of the Usumacinta-Grijalva Delta, Villabermosa, Tab. 1987. INIREB, Division Regional Tabasco, Villahermosa, Tabasco, México.

Orozco, A. and Lot-Helgueras, A. (1976) La vegetación de las zonas inundables del sureste de Veracruz. Biotica 1: 1–44.

Perales, F. and Contreras Balderas, A. J. (1986) Aves acuaticas y semiacuticas de la Laguna Madre, Tamaulipas, México. Universidad y Ciencia 3: 39–46.

Pool, D. J., Snedaker, S. C. and Lugo, A. E. (1977) Structure of mangrove forests in Florida, Puerto Rico, Mexico and Costa Rica. Biotropica 9: 152–212.

Puig, H. (1972). La sabana de Huimanguillo, Tabasco, México. p. 389–411. *In*: Memorias del Prime Congreso Latinoamericano de Botánica. México, D. F., México.

Ramirez Garcia, J. P. (1983) Estudio de la Distribucion y Producción de la vegetación acuática en seis lagos crater de Puebla. Tesis profesional. Universidad Nacional Autónoma de México. México, D. F., México. 37 pp.

Rico-Gray, V. (1981) *Rhizophora harrisonnii* (Rhizophoraceae), un nuevo registro de las Costas de México. Boletin de Sociedad Botánica Mexicana 41: 163–165.

Rico-Gray, V. (1982) Estudio de la vegetación de la zona costera inundable del noreste del Estado de Campeche, México: Los Petenes. Biotica 7: 171–190.

Rico-Gray, V. and Lot, A. (1983) Producción de hojarasca del manglar de la laguna de la Mancha, Veracruz, México. Biotica 8: 295–301.

Rollet, B. (1974) Introduction àl'étude des mangroves du Mexique. Bois Forêts des Tropiques 156: 3–26; 157: 53–74.

Rzedowski, J. (1983) Vegetación de México. Limusa, México. 432 pp

Rzedowski, J. and McVaugh, R. (1966) La vegetación de Nueva Galicia. Contribution University of Michigan Herbarium 9: 1–123.

Sarukhan, K. J. (1968) Los tipos de vegetación arbórea de la zona cálida-húmeda de México. p. 3–47. *In*: T. D. Pennington and K. J. Surukhan (eds.). Manual para la identificación de los principales árboles tropicales de México. Instituto Nacional de Investigaciones Forestales, México, D. F., México.

Sarukhan, K. J. (1968) Análisis sinecológico de las selvas de *Terminalia amazonica* en la Planicie Costera del Golfo de México. Tesis. Colegio de Postgraduados, Escuela Nacional de Agricultura, Chapingo, México. 300 pp.

Saunders, G. B. and Saunders, D. C. (1981) Waterfowl and their Wintering Grounds in Mexico, 1937–1964. U. S. Department of the Interior, Fish and Wildlife Service, Resource Publication 138, Washington, DC, USA. 151 pp.

Schwartzlose, R. A. and Hendrickson, J. R. (1983) Bibliografía del Golfo de California: Ciencias Marinas (through 1981). Instituto de Ciencias del Mar y Limnología, Universidad Nacional Autónoma de México, Publicacion Especial 7: 1–212.

Scott, D. A. and Carbonell, M. (1986) A Directory of Neotropical Wetlands. IUCN Cambridge and IWRB Slimbridge, England. 665 pp.
SEDUE. (1986) Informe sobre el Estado del Medio Ambiente en México, D. F., México. 83 pp.
Sian Ka'an. Estudius Preliminares de una zona en Quintana Roo propuesta como Reserva de la Biosfera (1983) Centro de Investigaciones de Quintana Roo, A. C. Puerto Moreles, Quintana Roo, México. 215 p.
Siemans, A. H. (1980) Indicios de aprovechamiento agrícola prehispánico de tierras inundablas en el centro de Veracruz. Biotica 5: 83–92.
Siemans, A. H. (1989) Reducing the risk: Some indications regarding pre-hispanic wetland agricultural intensification from contemporary use of a wetland/terra firma boundary zone in central Veracruz. p. 233–250. *In*: S. R. Gliessman (ed.). Agroecology. Researching the Ecological Basis for Sustainable Agriculture. Springer-Verlag, New York, New York, USA.
Signoret, M. (1984) Abundancia, tamaño y distribución de Camarones (Crustacea, Peneidae) de la Laguna de Terminos, Campeche y su relación con algunos factores hidrológicos. Anales Instituto de Biología, Universidad Nacional Autónoma de México, Serie Zoología 45: 119–140.
Sousa, M. (1964) Estudio de la vegetación secundaria en la region de Tuxtepec, Oaxaca. p. 91–105. *In*: Instituto Nacional Investigaciones Forestales, Publicacion Especial No. 3. México, D. F., México.
Sousa, S. M. (1968) Ecologia de las leguminosas de Los Tuxtlas, Veracruz. Anales Instituto de Biología, Universidad Nacional Autónoma de México, Serie Botanica 39: 121–160.
Tellez, O., Sousa, S. M. y Cabrera-Cano, E. (1982) Imagenes de la Flora Quintanarroense. Centro de Investigaciones de Quintana Roo., e Instituto de Biologia de la UNAM, México, D. F., México. 224 pp.
Thom, B. G. (1967) Mangrove ecology and deltaic geomorphology, Tabasco, Mexico. Journal of Ecology 55: 301–343.
Toledo, A. (1984) Como destruir el paraiso -El desastre ecológico del sureste. Centro de Ecodesarrollo. México, D. F., México. 151 pp.
Vasquez-Yanez, C. (1971) La vegetación de la Laguna de Mandinga, Veracruz. Anales Instituto de Biología, Universidad Nacional Autónoma de México 42, Serie Botanica 49–94.
Vasquez, S. J. (1963) Clasificación de las masas forestales de Campeche. Instituto Nacional de Investigaciones Forestales. Boletin Tecnico 10, México, D. F., México. 30 pp.
West, R. C. (1966) The natural vegetation of the Tabascan lowlands, Mexico. Revista Geografica, (Rio de Janeiro) 64: 107–122.
West, R. C., Psuty, N. P. and Thom, B. G. (1969) The Tabasco Lowlands of Southeastern Mexico. Technical Report #70. Coastal Studies Institute, Louisiana, State University, Baton Rouge, Louisiana, USA. 193 pp.
Yañez-Arancibia, A. (1978) Taxonomia, Ecología y Estructura de las Comunidades de peces en Lagunas Costeras con Bocas Efímeras del Pacifico de México. Centro Ciencias Mar y Limnología, Universidad Autónoma de México. Publicacion Especial 2: 1–306.
Yañez-Arancibia, A. (1981) The occurrence, diversity and abundance of fishes in two tropical coastal lagoons with ephemeral inlets on the Pacific coast of Mexico. Proceedings, Symposium Duke University Marine Laboratory, Beaufort, NC, USA, UNESCO/IABO, 1978, UNESCO Technical Paper Marine Sciences 33: 233–260.
Yañez-Arancibia, A. and Day, J. W. (1982) Ecological characterization of Terminos Lagoon, a tropical lagoon-estuarine system in the Southern Gulf of Mexico, Oceanologia Acta, 1982. No. SP.
Yañez-Arancibia, A. (1986) Ecología de la zona costera – análisis de siete topicos. AGT (ed.). S. A., México, D. F., México. 189 pp.
Zedler, J. B. (1982) The Ecology of Southern California Coastal Salt Marshes: A Community Profile. FWS/OBS-81/54, Biological Services Program, Fish and Wildlife Service, U.S. Department of the Interior, Washington, DC, USA. 110 pp.
Zedler, J. B. (1985) Vegetación de la Marisma, Ejemplos del Estuario del Rio Tijuana. Una publicacion del "California Sea Grant College Program". Reporte No. E-CSGCP-004, Institute of Marine Resources, University of California. La Jolla, California, USA. 35 pp.

Wetlands of tropical South America

WOLFGANG J. JUNK

Abstract

The climate of tropical South America is characterized over large areas by a high annual precipitation, varying from 1,000 mm to more than 5,000 mm per year. A pronounced seasonality in rainfall results in the periodic flooding of large areas covered by forests or savanna vegetation. Therefore, most of the wetlands in this area belong to the category of seasonal wetlands with a pronounced dry period.

Flooding may occur by lateral overflow of rivers and streams or by sheet flooding due to excess rain and insufficient drainage. The floodpulse is monomodal and predictable in the savannas and the fringing floodplains along the large rivers whereas it is polymodal and unpredictable in the floodplains along small streams.

Plants and animals respond to this pulsing with a large set of morphological, anatomical, physiological, and ethological adaptations. Inspite of the physiological stress of the change between aquatic and terrestrial conditions, species diversity is comparatively high. Floodplains of tropical South America may be considered as areas of speciation, contributing to the great species diversity in the area. On the other hand, the floodpulse results in a periodic exchange of biological information between the wetlands and the drainage system, often over long distances. Therefore, many species have a wide range of distribution.

Nutrient status of the wetlands varies from extremely low levels in areas flooded by nutrient-poor water (e.g. from rains or black water rivers) to high levels in the fringing floodplains of white water rivers, rich in fertile sediments and electrolytes. Consequently, productivity varies from low to very high, reaching maximum values up to 100 t dry material per hectare per year in the floodplain of the Amazon River. There is a complex exchange of nutrients and energy between the terrestrial and the aquatic phase and between the floodplains and the connected river systems.

D.F. Whigham et al. (eds.), Wetlands of the World I, 679–739.

Further wetland types occur mainly along the coast of the Atlantic ocean, partly in the form of mangroves or salt marshes. Peat bogs, cushion bogs, and reed swamps occur, to a small extent, in the wet Paramos of the high Andes, salt pans occur in the dry Puna. There is no exact information about the total wetland area in tropical South America, partly due to the seasonal character of the wetlands which have been poorly studied and are often not recognized as wetlands. It is estimated, that more than 2,000,000 km^2 may belong to the wetland category corresponding to about 20% of the area.

In recent times, all wetland types are becoming increasingly influenced and modified by man. In floodplains agriculture and husbandry are the main anthropogenic factors, modifying natural vegetation by deforestation and use of fire for weed control. Some river-floodplains are becoming strongly affected by the construction of flood-control measures and hydroelectric power schemes. Water pollution due to the input of sediments, agro-industrial wastes, agrochemicals, and mercury is becoming a serious threat. Mangroves are probably the most endangered wetlands due to industrial pollution, colonization projects, timber extraction, and large scale fish- and shrimpculture.

Introduction

Tropical South America includes Venezuela, the Guianas, Columbia, Ecuador, Bolivia and the northern parts of Brazil, Peru, and Paraguay. The central part of tropical South America is covered by the largest rain forest on earth while areas to the north and south are dominated by savannas. The slopes of the Andes in the west are covered by different kinds of rain and cloud forest, while high elevation areas, the Altiplano, have different types of grassland (Puna and Paramo) vegetation. The human population density is generally low and is concentrated in a few urban centres. The highest population densities are in the Altiplano. Human impact has been slight in the past, however, it has been increasing sharply in the last decades.

A great variety of wetlands occur throughout the region. The Central Brazilian Shield and the Guyana Shield contain periodically flooded savannas. Periodically flooded woodlands and palm swamps are scattered over nearly all lowland areas. Most rivers and streams have fringing floodplains. In the high Andes, there are salt marshes, swamps, and peat bogs. The Tepuis of Venezuela are partly covered by a peat layer several meters thick. On the Atlantic coast there are extensive marine and freshwater wetlands. However, detailed knowledge about wetlands is still limited, and most information comes from a few areas, which have been rather well studied. Large areas remain virtually unknown scientifically. A summary on available literature and general information is given by Scott and Carbonell (1986).

The scarcity of information available on wetlands in the region is due to the difficulty in gaining access to them as well as to the relative unimportance presently attached to wetlands by planners and scientists (Junk 1980). This situation may change, as wetlands in tropical South America are becoming increasingly important as land resources for fast growing populations (Junk 1982a, 1989a). Inland fisheries in all large rivers are becoming increasingly important as suppliers of protein (Bonetto and Castello 1985). Furthermore, there are plans for the construction of large hydroelectric facilities along many large rivers, and there is considerable concern about their negative ecological side effects (Junk and Mello 1986). A major question which remains is whether or not the rate of increase in our knowledge about wetlands and their importance will be able to match the rate at which they will be damaged by development schemes.

An important first step is to summarize our current knowledge about wetlands in the region. Because classification systems do not exist for the region, I will describe wetlands on a geographical basis.

General aspects

Hydrologically, South America is characterized by large river systems, such as (Fig. 1) the Amazon, Orinoco, and Magdalena. The Paraná River basin lies to the south of the Amazon basin and is separated from it by the Central Brazilian Shield. The Orinoco basin in the Northeast is separated from the Amazon by the Guyana Shield. All three river systems have their origins in the Andes. Within the Andes, a large endorheic basin contains Lake Titicaca (Fig. 1) and its wetlands. The Caribbean Cordilleras, a branch of the Andes, forms the northern boundary of the Orinoco basin. The Magdalena River (Fig. 1) drains into the Caribbean Sea, while several smaller rivers emerge from the Guyana Shield and flow into the Atlantic Ocean. These include the Essequibo, Suriname, and Oiapoc Rivers. The drainage basin of the Paranaiba River is the largest, arising from the Central Brazilian Shield.

The river systems of the Amazon and Orinoco basins are connected by the famous anastomosis of the Cassiquiare, mentioned first by Acuña in 1639 and confirmed by Humboldt in 1800. Periodically, additional anastomoses exist between the Amazon and the Essequibo Rivers in the Northeast, and the Amazon and the Paraná Rivers in the South. The geological division of the region is similar to the geographic one. The shields of Guyana and Central Brazil are of Precambrian origin and belong to the oldest formations on the planet. They are, in part, covered by sediments of marine, lacustrine,

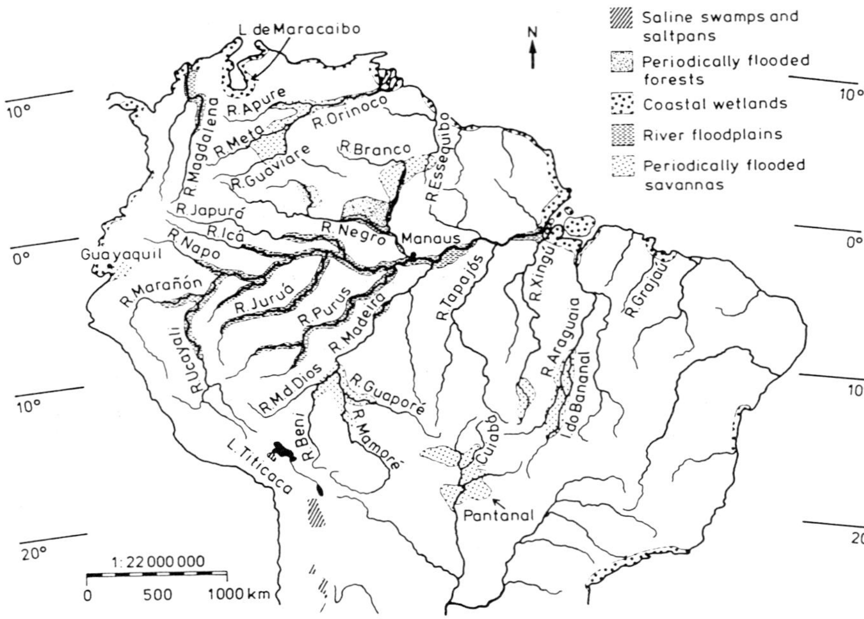

Figure 1. Map of the study area.

and riverine origin which have probably been deposited since the Cambrian. The Andes were formed at the end of the Tertiary by plate-tectonic processes (Putzer 1984). The Amazon basin is one of the largest Tertiary sedimentation basins in the world. Sedimentation and little tectonic activity explains the lack of large, deep lakes in the region in spite of abundant precipitation. The same geological processes explain the existence of large, flat plains which have insufficient drainage systems to quickly carry away surplus water during the regionally pronounced rainy season.

The equator passes a few degrees north of the main course of the Amazon (Fig. 1), and seasonal rainfall patterns produce rainy and dry seasons. In the Central Amazon Basin, precipitation varies between 2,000 and 3,000 mm per year, but the dry season does not last long enough to hinder the development of evergreen tropical rain forest. Further to the north and south of the equator, rainfall periodicity is more pronounced. The southern part of the Central Amazonian Shield, the northern part of the Guyana Shield, and the Orinoco basin all have distinct dry seasons and support mainly savanna vegetation (Beard 1953, Donselaar 1965, Sarmiento 1983). Annual precipitation in the savanna areas varies between 1,400 and 1,800 mm. On the eastern slopes of the Andes, local precipitation may be more than 5,000 mm year. In contrast, the central Andes and its western slopes are rather arid (Walter *et al.* 1975, Salati and Marques 1984).

Change in terrain elevation results in temperature ranges from very hot in the lowlands to below freezing in the high Andes. Along the elevation gradient, there is a distinct zonation, with different types of evergreen lowland rain forests occurring at lower elevations and different types of montane forests, cloud forests, and finally the Altiplano Paramo and Puna grasslands extending to the branches of Andean glaciers at about 5,000 m altitude.

The seasonality of the precipitation produces periodic excesses and shortages of water. Consequently, there are huge water level fluctuations in all rivers. Large rivers display monomodal flooding pattern that is synchronized with the dry and rainy seasons. Minor streams react more to local climatic events and often show a polymodal flood pattern with flooding frequency greatest during the rainy season. One exception is the Magdalena River, which has a bimodal flood pattern.

In spite of the fact that there is no clear cut border between streams with monomodal, bimodal, and polymodal flood patterns, this type of differentiation is important in wetland areas because each produces fundamentally different ecophysiological responses. The impact of one or two long lasting flood periods per year, each of which is several meters in amplitude, would result in a very different set of conditions in a wetland than the same period of flooding occurring as many short-term inundations.

Most rivers have fringing floodplains which may extend over many thousand of square kilometers along large rivers (Fig. 1). Floodplains with a monomodal flood pattern have a distinct type of seasonal pattern in an environment where other climatic parameters show little or no seasonal fluctuation. The flood induced seasonality changes terrestrial habitats into aquatic ones and produces enormous seasonal variations in the size of the different habitats, which affects both mobile and sessile aquatic and terrestrial organisms (Junk 1984a, Junk *et al.* 1989). Along minor streams, hydrological seasonality also occurs, but conditions are quite different due to the frequency and sometimes the unpredictability of the flooding.

Since large areas in the region are very flat and poorly drained, a pronounced rainy season also results in periodic sheet flooding that covers thousands of square kilometers (Fig. 1). This flood water originates exclusively as rain, but it may vary in depth from a few decimeters to more than a meter. At the end of the rainy season, these areas become dry due to evaporation and drainage, and they have soil water deficits that may last for several months. These periodic wetlands associated with a pronounced dry period are much more abundant than permanent wetlands in tropical South America. Many plant communities already described by phytosociologists, geographers, or forestry officers as terrestrial communities are, at least temporarily, wetland systems, and the wet period is often critically important in determining how the systems function. To understand their specific character, studies must deal with both aquatic and terrestrial aspects as both phases are closely linked with each other and interact extensively. Such wide-ranging aproaches are, however, very difficult and have seldom been employed (Bayley 1981, Junk 1980, Junk *et al.* 1989, Junk 1990).

Almost all wetlands in the region are periodically connected to rivers and this connection is essential to an understanding of their bio-geochemical and biological conditions. During the rise and fall of the flood waters, there is an exchange of dissolved and suspended inorganic and organic materials. Organisms drift or migrate from wetlands to the rivers, and vice versa. This mechanism provides the opportunity for genetic information to be exchanged over very long distances and even from one river system to another. Many aquatic plant and animal species therefore have very broad distributions in the region. Some wetland habitats are ephemeral as a result of periodic desiccation, but they are recolonized from adjacent rivers during the flooding season and thus maintain a high species diversity despite periodic disappearance.

Wetland soils of the region are difficult to classify due to both their heterogeneity and the lack of information about them. The majority of lowland soils are strongly acidic and extremely poor in dissolved minerals and nutrients. Most soils have a high sand or caolinitic clay content and thus have a very low ion exchange capacity. Often an impermeable layer of iron oxides or humates separates the surface layer of the soil from deeper layers. Periodic flooding with rainwater leads to a continuous loss of nutrients, and the soils become even more impoverished. Exceptions are recent alluvial soils of Andean origin which have higher nutrient levels and a greater ion exchange capacity due to a higher percentage of feldspars and which get periodic nutrient inputs from flooding. Soils may have a high salt content in semi-arid regions and in coastal areas that are influenced by marine tides.

Organic material accumulations are often slight in wetland soils since the large seasonal change in water level, in combination with high temperature, results in high decomposition rates. Therefore, there are no large peat deposits in the lowlands (Nikonov and Sluka 1964). When the water levels are stable and dissolved oxygen is absent, layers of organic material may accumulate, in such places as the swamps along the coast or isolated areas on the floodplains, where water stands for long periods of time (Junk 1983).

In the high Andes, low temperature favours the accumulation of organic material in *Sphagnum* bogs, reed swamps, and flush and cushion bogs. *Sphagnum* bogs are oligotrophic and strongly acidic, while flush and cushion bogs are mesotrophic and moderately acidic. Reed swamps are mesotrophic to eutrophic. In the dry Puna area, salinization of the soils occurs, and in salt pans (solares), accumulations of many meters of salt have been described.

Description of wetlands

Floodplains of large central Amazonian rivers

The Amazon, the mightiest river on earth, has an average discharge of about 175,000 m^3 $second^{-1}$, which corresponds to between 1/5 and 1/6 of all of the freshwater that the world's rivers transport to the oceans (Richey *et al.* 1989, Sioli 1984b). The main river and its large tributaries, the Tapajós, Xingú, Madeira, Purús, Juruá, Japurá, and Negro Rivers, show a distinct monomodal flood pattern. All have fringing floodplains which cover a total area of about 300,000 km^2 (Fig. 1). Melack (1984) indicates about 8,000 lakes in the floodplain of the Amazon.

Lowland floodplains are post-glacial in origin and were formed by a sea level rise, which began about 12,000 years ago and slowed down about 5,000 years ago (Milliman and Emery 1968, Schackleton and Opdyke 1973). In the Amazon basin, the influence of the postglacial sea level rise reached to the slopes of the Andes and, in the North and South, to the slopes of the archaic shields of the Guyanas and Central Brazil (Irion 1984).

There are considerable geomorphological differences between floodplains of different rivers or even different stretches of the same river (Sioli 1984b). The Amazon and the Japurá show little tendency to meander, whereas the Purús and Juruá Rivers have strongly meandering beds (Fig. 2a–d). Consequently, oxbow lakes are absent on the Amazon and Japurá floodplain but common on the floodplains of the Purús and Juruá.

Due to the lower sediment load, the floodplains of black and clear water rivers are less developed than those of white water rivers. A specific case is the floodplain of the Negro River downstream from its confluence with the Branco River (Fig. 2c), where there is an internal delta of hydrochemical origin. A change in pH due to the mixture of water from the two rivers results in the flocculation and subsequent sedimentation of fine suspended material, which formed the Anavilhanas Archipelago (Leenheer and Santos 1980). Since 1981, 3,500 km^2 of this area have been given the status of a protected "ecological station".

The nutrient content of floodplain substrates and associated water bodies is controlled primarily by the mineralogical composition of the transported sediments and the dissolved nutrient load of the rivers (Irion 1982, 1984, Victoria *et al.* 1989). Many rivers draining the Andes are "white water

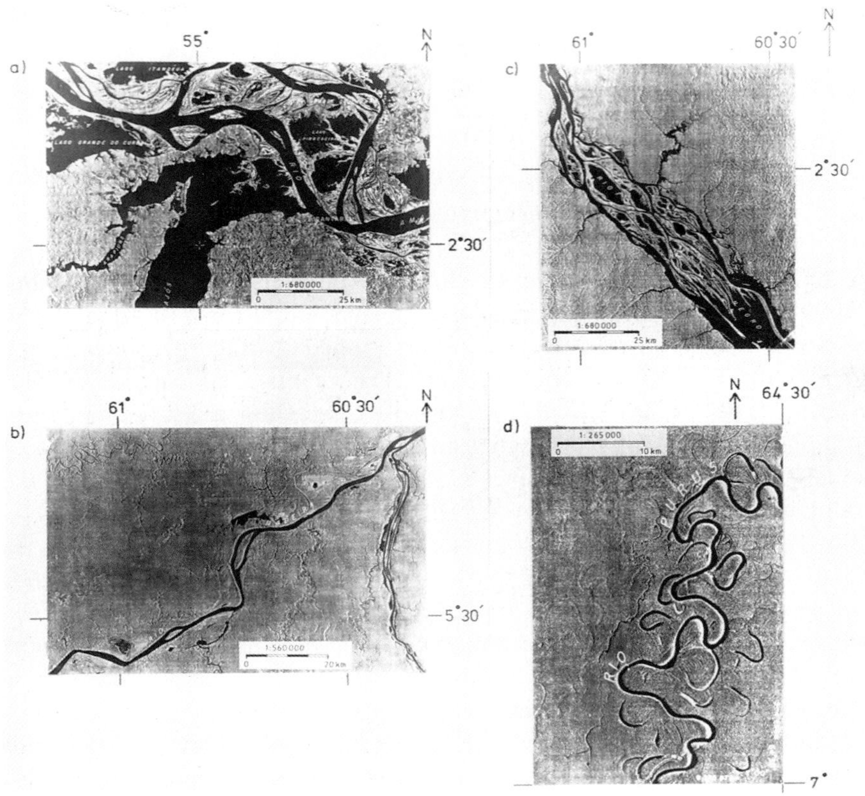

Figure 2. Aerial view of different types of floodplains in the Amazon basin: a) The Amazon River at Santarém (2°30′S, 55°W), b) The Madeira River, a white water river with a small floodplain due to geomorphological peculiarities(5°30′S, 61°W), c) The archipelago of Anavilhanas in the floodplain of the Negro River near Manaus (2°30′S, 61°W), d) The Purús River near its confluence with the Amazon River (7°S, 64°30′W). All photos from RADAM, Brasil.

rivers''. Among these are the Amazon itself and its tributaries, the Madeira, Purús, Juruá, and Japurá. They have rather fertile floodplains, locally called Várzea, and the waters are rich in suspended matter and electrolytes with a neutral pH (Sioli 1950, 1965, 1967, Gibbs 1967, Martinelli *et al.* 1989). In tributaries from the Guyana and Central Brazilian Shields, the quantity and fertility of the sediments is less, and dissolved minerals in the water are less abundant than in the ''clear water rivers'' Tapajós and Xingú. At the other end of the range are ''black water rivers'', such as the Rio Negro, which are nutrient poor, have an extremely light sediment load, acidic water, and infertile floodplains (Furch 1976, Furch and Klinge 1978). On the floodplains, the electrolyte content of the river water can be strongly modified by abiotic and biotic processes (Furch *et al.* 1983, Furch, K. 1984).

Figure 3. The floodplain forest on the Rio Negro (igapó) at high water.

The higher parts of most Central Amazonian floodplains are covered by more or less closed forests consisting of species highly adapted to inundation (Junk 1989b, Kubitzki 1989). Some tree and shrub species, such as *Eugenia inundata*, *Ruprechtia ternifolia*, *Coccoloba ovata*, *Myrcia* sp., can resist floods up to 10 m deep for an average flood period of 280 days (Fig. 3).

The main stress factor is probably oxygen deficiency in the flooded soils. During the high water period, many floodplain trees shed their leaves and reduce their growth rates, forming distinct annual rings as do trees in temperate regions (Worbes 1985, 1989, Worbes and Junk 1989). Some species, such as *Symmeria paniculata* and *Astrocaryum jauari*, retain their leaves even when inundated beneath several meters of water for many months (Furch, B. 1984). Fruiting occurs mainly during the high water period, and there is an adaptation for seed dispersal through the water and by fishes (Gottsberger 1978, Goulding 1981, Goulding *et al.* 1988).

Flood stress reduces the diversity of tree species on the floodplains. Near Manaus about 60 to 120 species per hectare have been reported (Revilla 1981, Worbes 1983, Revilla cited in Junk 1989b). This is about 50% of the number of species per hectare occurring in non-flooded Amazonian rain forests (Prance *et al.* 1976). Increased erosion and sedimentation rates can become additional stress factors. However, in comparison with other regions of the world, the number of flood resistant tree species is very high. About

Figure 4. Herbaceous plants in the floodplain of the Amazon River (várzea) at low water, dominated by *Alternanthera* spp. and *Oryza perennis*.

100 tree species have been reported from bottomland forests in the United States, including those with a flood tolerance of only a few days.

In addition to flooding, sedimentation, and erosion, the nutrient content of the water and sediments influence the species composition of floodplain forests. Prance (1979) distinguishes between the igapó forest on black and clear water rivers and the várzea forest on white water rivers. Other authors support the classification of Prance using invertebrates (Irmler 1977), herbaceous plants (Junk 1983), and productivity data (Junk and Furch 1984).

Aquatic and terrestrial herbaceous plants are abundant mainly on the floodplains of nutrient-rich white water rivers, where they colonize areas not covered by forest. Junk (1990) records 387 species belonging to 182 genera and 64 families along the várzea of the middle Amazon near Manaus. Of these, 330 were terrestrial, 34 aquatic, 20 palustrine, 17 aquatic with a pronounced terrestrial phase, and 2 terrestrial with a pronounced aquatic phase. However, only 17 species form large populations. Only *Echinochloa polystachya*, *Paspalum repens*, and *Oryza perennis* were dominant during the aquatic phase, while *Paspalum fasciculatum* and *Cynodon dactylon* predominated during the terrestrial phase (Figs. 4 and 5). In floodplains of black water rivers, herbaceous species occur in low abundance and aquatic species, except for *Utricularia foliosa*, are often totally absent. Floodplains of clear

Figure 5. Aquatic macrophytes in the floodplain of the Amazon River (várzea) at high water, dominated by *Paspalum repens* and *Echinochloa polystachya*, interspersed with a few *Eichhornia crassipes*.

water rivers are intermediate in respect to diversity and abundance of herbaceous species. Podostemonaceae are common in rapids of clear water rivers.

Maximum productivity by herbaceous species can reach up to 100 t dry weight ha^{-1} yr^{-1}, as shown for *Echinochloa polystachya* stands (Piedade 1988). Phytoplankton production varies from 60 kg ha^{-1} yr^{-1} in black water floodplain lakes to 6 t ha^{-1} yr^{-1} in white water floodplain lakes (Schmidt 1973, 1976, Rai and Hill 1984, Fisher and Parsley 1979, Forsberg *et al.* 1988). Litter fall in the igapó forest is about 6 t ha^{-1} yr^{-1} and about 10 t ha^{-1} yr^{-1} in the várzea forest (Adis *et al.* 1979).

According to Melack and Fisher (1990) the productivity of phytoplankton in Lago Calado, a floodplain lake of the Amazon near Manaus, was limited by phosphorus during rising and high water levels and by nitrogen during falling and low water levels. Inflowing Amazon water supplied most of the phosphorous to the lake; the majority of nitrogen was supplied by local rain and runoff. Small epilimnetic heterotroph organisms such as bacteria, fungi, protozoans, and rotifers were much more important than the macrozooplankton in the production of ammonium. Furch *et al.* (1989) stressed the importance of leaf litter from the floodplain forest as an additional source for bioelements to the aquatic system. They also point out that the input of nitrogen, phosphorous and potassium released from decomposing leaves may be as great or even greater than the input by the inflowing river water. These findings are in accord with Melack and Fisher (1990) who stress the net heterotrophic nature of the open water of Amazonian floodplain lakes.

The chemical composition of the leaves, bark, and wood of the floodplain trees reflects, to a certain extent, the availability of the elements in the water and soils. The várzea forests which are rich in nutrient elements, grow on floodplains that contain large amounts of most elements, especially Ca and Mg. The igapó forests grow on extremely impoverished substrates (Klinge *et al*. 1983, 1984).

Because of high primary production and quick decomposition, the Amazon river-floodplain system has to be considered a "hot spot" in the carbon cycle. Most of long living carbon compounds transported by the main river derive from terrestrial vegetation of the watershed, whereas the quickly decomposing carbon load derives mainly from the floodplain (Junk 1985, Ertel *et al*. 1986, Hedges *et al*. 1986, Richey *et al*. 1988, 1990). The Amazon floodplain is considered an important source of methane to the troposphere (Devol *et al*. 1990, Wassmann *et al*. 1992).

Like plants, animals show many anatomical, morphological, physiological, or ethological adaptations to flooding and drought (Irmler 1981). Different kinds of resting stages are known among lower animals. Horizontal and vertical migration of terrestrial arthropods has been described (Beck 1976, Adis 1984, in press). Large scale fish migrations occur during the period of flooding in all large rivers of tropical South America (Goulding 1981, Ribeiro 1983, Lowe-McConnell 1975, Bonetto and Castello 1985).

An additional stress factor for aquatic animals on the floodplain is oxygen deficiency, which does not occur in the rivers themselves (Melack and Fisher 1983). Oxygen deficiency is correlated with productivity, since highly productive floodplains contribute more organic matter to the water than poorly productive ones. Zoobenthos is restricted by the oxygen content (Irmler 1975, Reiss 1976a, b). Many fish species have morphological, anatomical, and physiological adaptations to oxygen stress (Kramer *et al*. 1978, Junk *et al*. 1983, Saint-Paul and Soares 1987, 1988).

The fairly regular flood pattern imposes a distinct seasonal pattern on the life cycles of aquatic as well as terrestrial animals. Most fish species have a distinct spawning season while the water is rising and when floodplains are becoming available as feeding grounds for the offspring. The mechanism triggering gonad development and spawning is not yet fully understood in many species. Terrestrial invertebrates colonizing floodplains begin reproduction when the flood recedes (Irmler 1976, 1979a,b, Adis 1984). Increasing daily temperatures may be one of the factors triggering gonad development and migration by some species (Irmler 1985).

As a result of the drastic change in environmental conditions, the majority of species tend to have short life cycles and high reproductive rates (r-strategy). The success of r-selected species, seems to be related to the nutrient supply within the system. On nutrient-poor floodplains, K-selected species may be favoured. For example, the decapod crustacean, *Macrobrachium amazonicum*, colonizes nutrient-rich white water floodplains and produces many small eggs, which develop into planktonic larval stages. In contrast, *Euryrhynchus burchelli* occurs on the nutrient-poor black water floodplains, produces few large eggs, and has no planktonic larval stages (Magalhães and Walker 1984). The spawning migrations of many fish species from nutrient-poor tributaries into the nutrient rich Amazon may be interpreted in the same way.

Aquatic and terrestrial floodplain phases are closely linked, as best shown by food webs. When water levels rise, fish move onto the floodplain to feed on abundant fruit, seeds, detritus, and terrestrial invertebrates (Goulding 1981). Aquatic food items are of minor importance for the fish fauna (Soares *et al.* 1986) except during very early life stages. When the water level is low, food becomes scarce and many fish starve. This is shown by distinct changes in the body fat content (Junk 1986).

Inland fisheries associated with floodplains are of major importance for supplying the protein to humans (Petrere Jr. 1978a, b, Smith 1979, Bayley 1981, 1982). Actual yields in the Amazon basin are estimated at about 200,000 t yr^{-1}, and there is a potential for 350,000 to 900,000 t yr^{-1} within the whole basin, if existing resources are properly used (Junk 1984b, Bayley and Petrere 1989). Since fish yields depend upon the size and fertility of the floodplains (Welcomme 1979), large scale modifications may have strong negative impacts on the fishery.

The Amazon floodplain and its lakes near Manaus have been studied for several years by the Instituto Nacional de Pesquisas da Amazonia (INPA) in collaboration with foreign institutions such as the Max-Planck-Institut für Limnologie, Plön, Germany, the Universities of Washington and California at Santa Barbara, USA, and the Office de la Recherche Scientifique e Technique Outre-Mer, France. A thorough survey of the extensive literature is provided by Sioli (1984a).

Floodplains of small central Amazonian streams and rivers

All papers dealing with Central Amazonian habitats distinguish between areas that are seasonally flooded by large rivers, such as the várzea and igapó, and the land not subject to flooding, the terra firme. A considerable percentage of the terra firme, however, includes water-logged and frequently flooded areas, since almost all small streams are accompanied by fringing wetlands of various sizes (Fig. 6).

A detailed soil map depicting several hundred km^2 north of Manaus shows a stream density of 2 km of stream channel per km^2. About 40% of the soils are considered hydromorphic (Falesi *et al.* 1971). The water levels in the

Figure 6. A small forest creek in Central Amazonia with *Mauritia flexuosa* growing in the frequently flooded, waterlogged valley.

streams show a sharp response to rainfall (Nortcliff and Thornes 1981). Stream density is greatest in the areas of abundant precipitation covered by rain forests. Therefore, most of this wetland type is concentrated in the Central Amazon basin. The total area covered by wetlands along small rivers or streams with a polymodal flood pattern may amount to about 1,000,000 km^2, which surpasses by far the total floodplain area of large rivers. In the savannas with extended dry seasons, these wetlands are often occupied by gallery forests. In the rain forest, the distinction is less clear because dense forests along streams merge into the unflooded rain forest. The species composition and ecology of the frequently flooded areas is not well known, but *Mauritia flexuosa* is one of the characteristic tree species (Fig. 6).

There is little limnological information available about Central Amazonian streams. Hydrochemical conditions vary considerably, depending upon the geology of the catchment area. Most streams are extremely poor in electrolytes and very acidic (Schmidt 1972, Furch and Junk 1980, Furch 1986). Due to shading by the dense forest, light becomes the limiting factor for aquatic primary production and the main carbon source for the aquatic food chain is allochthonous material (Knöppel 1970, Fittkau 1964, 1967, 1976). In savanna areas, rooted aquatic macrophytes may be abundant, even in streams with extremely small amounts of dissolved nutrients e.g. *Sagittaria rhombifolia*, *Eichhornia pauciflora*, *Elodea granatensis*, *Cabomba pihauhyensis*, *Nym-*

phaea rudgeana, *Nymphoides humboldtiana*, *Mayaca fluviatilis*, and *M. kunthii* (Furch and Junk 1980).

About 2,000 fish species are described from the Amazon catchment area (Geisler *et al.* 1971). Some have a rather restricted distribution, whereas others occur over very large areas and even in neighbouring river basins. Geisler *et al.* (1971) pointed out the importance of water chemistry differences among different types of water as chemical barriers to species occurrence and distribution. Fittkau (1973, 1982) stressed the comparatively low species diversity of aquatic invertebrates in Central Amazonian streams, in contrast to their highly diverse fish fauna. Similar observations had already been made by Patrick *et al.* (1966) in streams of the Andean foothills, as discussed in the next chapter.

The rivers and streams of the Andean foothills

The first comprehensive limnological study on the head waters of the Amazon, in the transition area between the vast Central Amazonian Basin and the high Andes, was undertaken by the Catherwood Foundation Peruvian Amazon Expedition. The purpose of the expedition was "to compare the pattern of aquatic life in rivers in the tropical zone with the pattern of aquatic life in rivers in the temperate zone in eastern and southern United States" (Patrick *et al.* 1966). Studies were conducted near Tingo Maria in steep gradient tributaries of the Huallaga River, a major headwater tributary of the Amazon River, and near Iquitos in areas with gradual slopes. Besides important taxonomic contributions, the most interesting discovery was that the total number of species of aquatic organisms of small body size, such as algae, protozoa and insects, was similar or sometimes a little lower than those in similar areas in the temperate zones. Fish abundance increased with increasing stream order. Considering the enormous species diversity in terrestrial habitats of the humid tropics, these findings were striking.

Patrick *et al.* (1966) related species diversity to the size of the river and the number of ecological niches available. Since small organisms have smaller, more densely packed ecological niches than larger ones, including fish, the number of niches and consequently the number of species in small tropical headwaters are comparable to those in headwaters and main streams in the temperate zones. Similar observations with respect to species diversity have been made by Fittkau (1973, 1982) in Central Amazonian forest streams. His explanation for the comparatively low species diversity, however, is somewhat different. He stresses a greater number of factors, such as low nutrient supplies and historical, environmental, and synecological influences on Amazonian lowland streams. Recent studies on species diversity and co-evolution in the tropics have shown that space is not necessarily the major limiting factor and that the niche concept must be seen as more differentiated than previously believed.

Additional detailed studies on rivers in the Andean foothills have been conducted by the University of Turku, Finland (Salo *et al.* 1986). They revealed some distinct differences from the Central Amazonian Rivers. River floodplains cover 12% of the 500,000 km^2 of the Peruvian headwaters analysed from Landsat images. Erosion and deposition processes are much more dynamic than in Central Amazonia. Rivers in the Andean foothills change their floodplains completely in a few decades. Erosion and sediment deposition are additional stress factors for floodplain forest growth. Such dynamics are considered very important for species diversity and tree speciation in Amazonian lowland forests (Kalliola *et al.* 1987).

Another striking feature of that region is the existence of many isolated oxbow lakes and swampy depressions in older, abandoned floodplains. The older floodplains cover 14.6% of the 0.5 million km^2 in the area analysed. According to Salo *et al.* (1986), this is indicative of major channel changes in the recent history of the western Amazon River system. Since isolated lakes are very rare features in Amazonia, they are of specific interest because they may be sites of increased speciation of aquatic organisms due to isolation. In addition, they may yield highly valuable palaeoclimatic information. The undisturbed accumulation of sediments over a considerable period of time should allow more accurate dating than does the use of sediments in recent floodplain lakes which are subjected to strong riverine influence.

Seidenschwarz (1986) studied phytosociological aspects of herbaceous plant-communities in the floodplains of the Yuyapichis and Pachitea Rivers in Peru and compared them with the pioneer vegetation along the roads and in crop plantations on nearby land not subject to flooding. He reports 245 herbaceous plant species, much fewer than the 387 species listed by Junk (1990) for the floodplain of the middle Amazon. This is explained by the fact that the slowly and regularly pulsing system of the middle Amazon provides better living conditions and more habitats for herbaceous plants than the irregularly pulsing systems of the headwaters. In both systems, there is obviously an invasion of r-selected, flood-tolerant weeds from Africa, Europe, and Asia.

Further studies on the headwaters of Amazonian rivers in the Andes are proceeding along the Pachytea-River through German Peruvian Collaboration at Panguana Station and at the recently established Peruvian Amazon Research Institute (IIAP) at Iquitos.

The Bananal savannas

The Araguaia River, the main tributary of the Tocantins River, passes through a vast sedimentation basin. Part of it forms Bananal Island, which is one of the largest (20,000 km^2) river islands in the world. The soils are sandy clays with an underlying lateritic layer that is up to several meters thick. Small, flat, granitic hills interrupt the monotony of the plain in some places.

In 1959, the first National Park in the Amazon basin was created on Bananal Island, protecting 5,623 km^2 of periodically flooded savanna, shallow

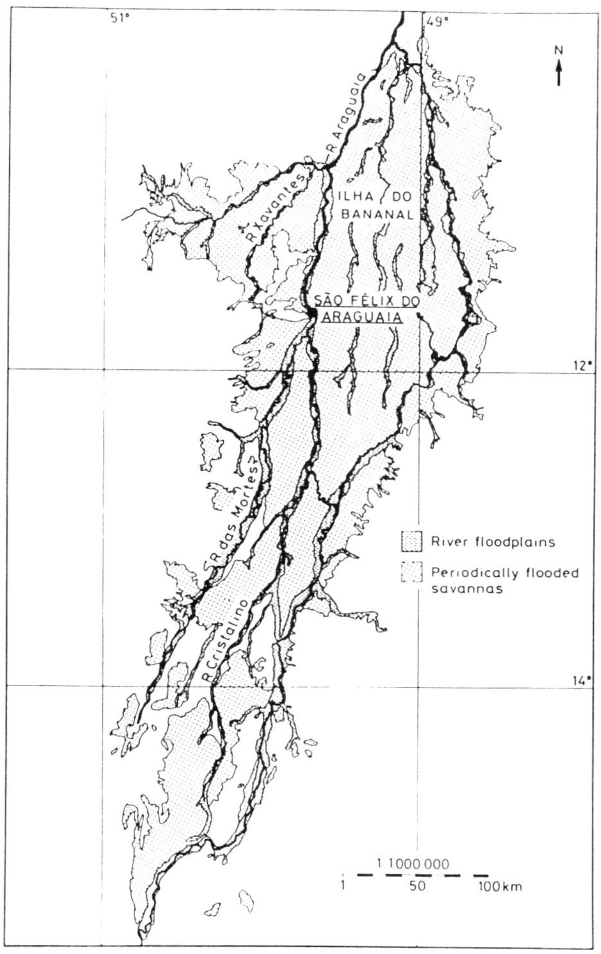

Figure 7. The distribution of the flooded savannas along the Tocantins River at Bananal island (RADAM/Brasil).

Figure 8. Aerial view of the Bananal savannas at low water. In the middle, a macrophyte-covered lake.

lakes and hills, which are not subject to flooding (Carvalho 1984). The national park is famous for its abundant aquatic bird fauna, which is similar to that of the Pantanal on the Paraná River. However, very few data are available on the ecology of the area.

The main channels of the Araguaia River and its large tributaries cut deeply into the plain and are rather well defined, probably due to the erosion-resistant material in the subsoil. Its water is transparent and contains a small suspended load during the dry period. The bed consists of sand, which also forms the beaches and islands.

The average precipitation varies between 1,600 and 1,800 mm per year. The main dry season occurs from June to September, and the main rainy season from December to April. Maximum flooding occurs in March and April. Flood and rain water, more than 1 m deep, cover as much as 65,000 km^2 (Fig. 7) at times (RADAM/Brasil 1981a,b). When the floods recede, hundreds of shallow temporary and permanent lakes remain. The plains are covered by grassy savanna vegetation with regularly spaced, drought adapted trees, such as *Curatella americana*. To escape flooding, termites build clay mounds around the bases of trees that are between 50 and 100 cm high. The mounds are ideal sites for the establishment of new seedlings when old trees die. The distance between the trees and their termite mounds may indicate the home ranges of specific termite populations (Fig. 8). Gallery forests occur in moist depressions around lakes and along rivers.

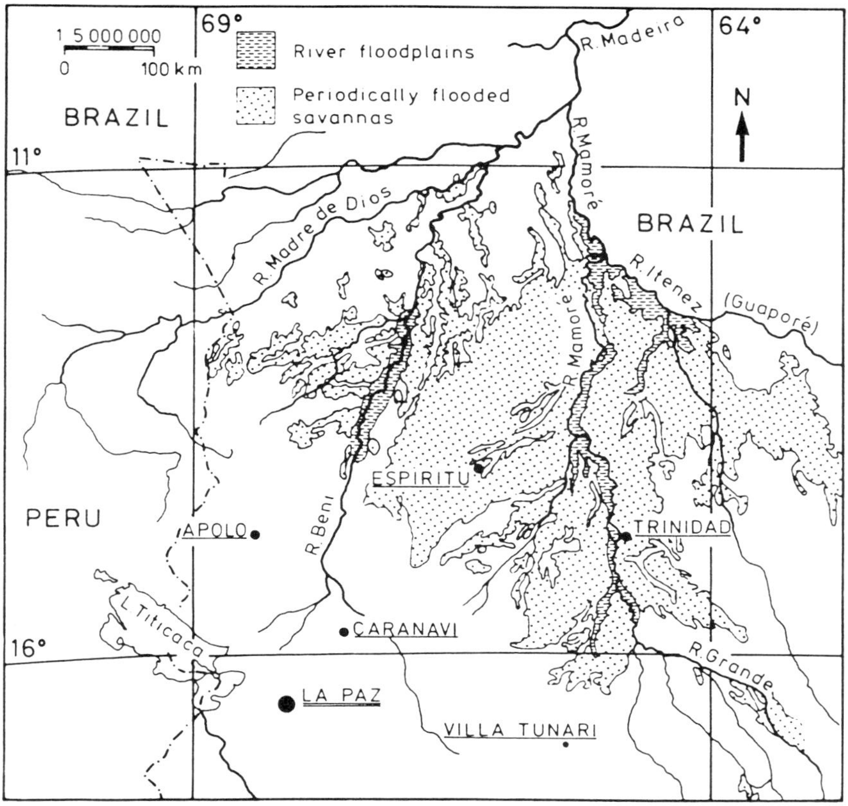

Figure 9. The distribution of the floodable savannas in the Bolivian lowlands (ERTS 1978).

The forests may have both drought and flood adapted species, *Astrocaryum jauari* and *Curatella americana*. Lakes are bordered and partially covered by aquatic macrophytes, but due to the low nutrient level of the water (Santos 1983), macrophyte growth is not as luxurious as in the várzea of the Amazon River, although the species composition is similar.

The active floodplain of the Araguaia River is covered by a rather species-rich floodplain forest. On sandy beaches and islands, shrubs, such as *Eugenia* spp., *Psidium* spp., *Alchornia castaneifolia*, *Sapium* spp., *Symmeria paniculata* and *Couepia paraensis*, are frequent. Their seeds are important food items for fishes during high water.

The floodable savannas of north east Bolivia

Up to 150,000 km^2 of the north eastern Bolivian lowlands between the Rivers Madre de Dios, Beni, Mamoré and Guaporé are flooded every year (Fig. 9). About 110,000 km^2 are covered by savannas containing more than 1,000 lakes. Drainage of the basin is hindered by the many granitic ridges of the Central Brazilian Shield, which emerge and form cascades in the beds of the rivers draining the region. According to Church (1898), the whole area was formerly a large lake. In the Quaternary, it was filled with fine fluvial and lacustrine sediments, varying in extent, but reaching a maximum depth of 5,500 m in the western part of the basin.

According to Beek and Bramão (1968) and Cochrane (1973), soils of the region include pale yellow latosols, groundwater laterite soils, and red-yellow podzolic soils. Flooding affects the pH of the soils, which varies from acidic in the low lying areas to alkaline in the uplands. Calcium carbonate is concentrated in the subsoil, and NaCl is enriched in the upper soil layers of former levees.

Precipitation varies between 1,600 mm (Trinidad) and 2,300 mm (Rurrenabaque) on the edges of the savanna and is about 2,000 mm at its centre. Year to year fluctuations may be extreme. In Espíritu for example, 1,322 mm were measured in 1974 and 2,454 mm in 1981. A two month dry season from mid-June to mid-August interrupts seven months of rainy weather (Beck 1983).

The high water period in the rivers occurs with a lag of almost three months after the start of the rainy season and lasts from the end of December to June. Maximum flooding occurs between January and March. Water level fluctuations may reach 11 m, as in the Mamoré River near Trinidad. Since the rivers cut deeply into the savanna, at low water level, their water level is 5 to 7 m below the savanna surface. Flooding varies with topographic situations but, on average, reaches only 60 cm over large areas of the savanna. Depending upon the total rainfall, between 60% and 90% of the savanna area floods every year.

Water chemistry data are scarce. The large rivers draining the Andes and their foothill slopes transport sediments and dissolved nutrients into areas adjacent to the main river courses. These areas are more fertile than the more distant ones, which are flooded only by rainwater (Haase 1990b).

At a first glance, the savannas are rather monotonous (Fig. 10). On more detailed analysis, a highly differentiated, patchy distribution of different grassland communities alternating with small bush and forest communities is revealed. The grasslands are occasionally interrupted by swampy depressions, lakes, and streams (Haase and Beck 1989). Beck (1983) distinguished more than 30 plant associations along the Yacuma River, a tributary of the Beni River. Haase (1989, 1990a) described 37 plant communities west of the Beni River. As in the flooded savannas of the Bananal area along the Araguaia River and in the Roraima-Rupununi region, flood tolerant species, such as *Bactris aff. glaucens* and *Pithecolobium aff. cauliflorus*, grow along with drought adapted and fire-resistant species, including *Curatella americana*, *Tabebuia caraiba* and *Cereus* sp.

Figure 10. The floodable savannas in the Bolivian lowlands at low water. In the upper foreground, *Andropogon laterale*, only slightly submerged at high water. In the lower background, *Paratheria prostrata* (Photo: Haase).

It is presumed that natural vegetation patterns have been dramatically altered by man as early as in the Pre-Columbian period. Large areas on the western bank of the Mamoré River, totalling 45,000 km^2, show extensive ridges and channels of anthropogenic origin (Plafker 1964, Denevan 1966). Economic exploitation is now restricted mainly to large scale cattle farming, which entails the use of fire for weed control during dry periods. Studies on the fauna of the area are still in their early stages. Capybaras (*Hydrochoerus hydrochaeris*) and many waterfowl are common around the permanent water bodies. The fish fauna seems to show similar migratory, feeding, and spawning behaviour as those on other Amazonian floodplains. Ecological studies in the floodplain of Rio Yacuma have been undertaken by Bolivian and German scientists from the universities of La Paz and Göttingen.

The pantanal of Mato Grosso

The Pantanal is a large floodplain covering about 140,000 km^2 in the center of South America. It covers the Alto Paraguai depression between 16° and 22°S and from 55° to 58°W, and has a terrain elevation 70 to 140 m above msl. (Fig. 11). This depression is situated between the old cristallin shield of Central Brazil and the young, uplifting Andes Mountains separating Brazil from Paraguay and Bolivia. The largest part of it belongs to Brazil and is divided between the States of Mato Grosso and Mato Grosso do Sul.

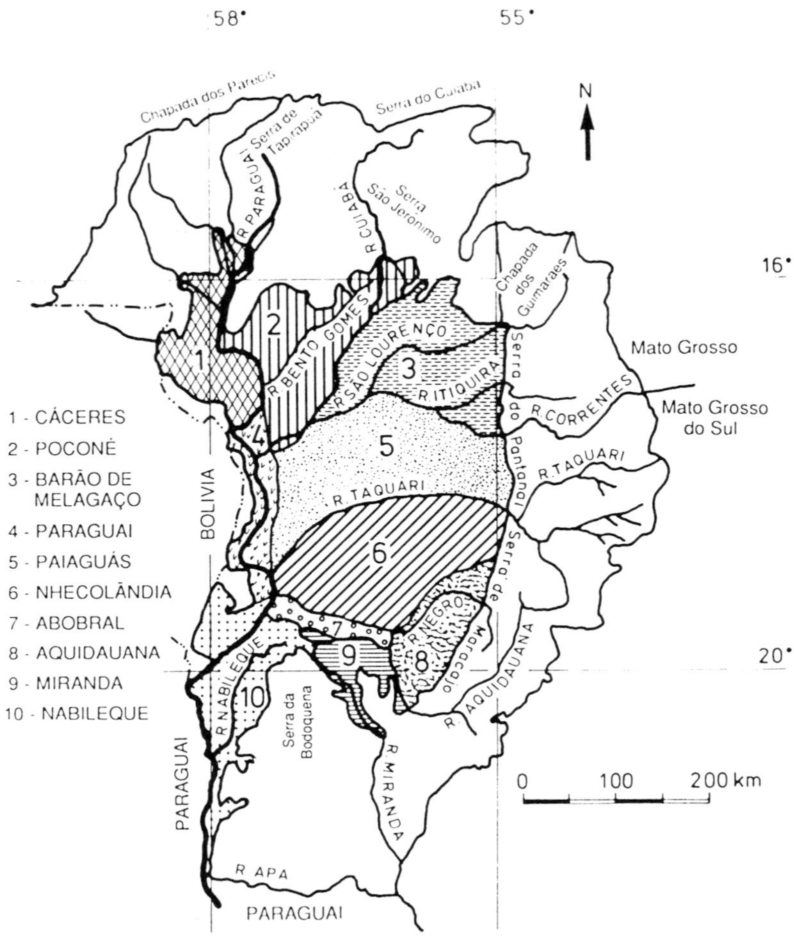

Figure 11. The Pantanal of Mato Grosso and its subunits according to Adamoli (1981).

The complex geological history and geomorphology of the area is described by Wilhelmy (1958), Adamoli (1981), Barros *et al.* (1982), Del'Arco *et al.* (1982), Alvarenga *et al.* (1984), Godoi Filho (1986), Ab'Saber (1988), and others. According to these authors, during the Cretaceous, the area of the Pantanal was an elevation which separated the Andean zone from the sedimentation basin of the Alto Paraguay. In post-Cretaceous periods, erosive processes excavated the depression of the Pantanal, in which, the upper Paraguay River and its tributaries deposit their sediments since Quarternary times.

Soils of the Pantanal are sandy and clayey sediments of fluvial and lacustrine origin, sometimes consolidated and partly or totally lateritic (Amaral

Filho 1986). Adamoli (1981) distinguished 10 different geomorphological subunits and Alvarenga *et al.* (1984) recognized 12 subunits, indicating a great complexity of the area (Fig. 11). Therefore, Brazilian scientists frequently use the term Pantanal in the plural form: os Pantanais.

The climate of the Pantanal is tropical with a pronounced seasonality in temperature and humidity. The hottest month is December with a mean temperature of 27.4°C; the coldest month is July with 21.4°C. In winter, the temperature may drop to O°C due to ingression of polar air masses, influencing the South American continent up to the equator (friagem).

The total annual precipitation varies between 1200 to 1400 mm. A pronounced dry season from May to September alternates with a rainy season from October to April. During the dry season, minimum precipitation values are only 5 mm in Cuiabá (July). In Corumbá, the corresponding value is 20 mm (August). During the rainy season, maximum values in January are 213 mm and 212 mm, respectively (Tarifa 1986).

The hydrology of the Pantanal is determined by the Paraguay River. The inclination of the area from east to west is 25 cm km^{-1}, but from north to south it is only 3 cm (Carvalho 1984). The rapid east-west drainage results in a large water supply during the rainy season and the subsequent flooding of the area due to the slow drainage to the south by the Paraguay River. While the small rivers entering the Pantanal show flood peaks in the middle of the rainy season from January to March, the highest water levels at Corumbá at the lower end of the Pantanal are reached at the beginning of the dry season from April to June. The large size of the area and the retardation of the drainage are responsible for the complex pattern of locally different hydrographs and changing water currents throughout the year.

In addition to the annual hydrological cycle, the Pantanal is subject to cycles lasting several years, which are sometimes responsible for several years of consecutive extended floods or droughts (Adamoli 1986). These events may be related to large-scale climatic changes in the Southern Hemisphere, such as the El Niño phenomenon.

With respect to the flood cycles, the Pantanal can be roughly divided into three subregions according to Adamoli (1986). The Alto Pantanal in the north is the upland section which is only slightly flooded. Floods last two to three month per year and cover about 20% of the area with 30 to 40 cm of water. The Baixo Pantanal in the south is a low-lying, very flat region, which is nearly totally flooded during the high water period. Both areas are connected by the Médio Pantanal, an area of intermediate elevations, which is flooded three to four months every year.

The chemical characteristics of the water bodies in the Pantanal vary strongly according to the source of the inflowing water and the local geomorphological conditions. The concentration of ions is generally low and varies normally from 10 $\mu S\,cm^{-1}$, that is, nearly distilled water, in areas frequently leached by rainwater, to about 100 $\mu S\,cm^{-1}$ in areas strongly influenced by the water of the tributaries. From the results of a limnological study in the upper Pantanal near Cuiaba, Silva (1990) reported an electric conductivity between 17.8 and 53.4 $\mu S\,cm^{-1}$ and pH-values from 5.6 to 8.1.

Electrolyte concentrations observed during flood periods were about twice as high as during low water, mainly due to increased levels of K, Ca, and Mg, probably derived from the surrounding catchment area. According to Furch and Junk (1980) tributaries that drain the geological formations of the Araras Group which contain Cambrian limestones and dolomites, and that drain the Alto Paraguay Strata of early Ordovician sandstones, clay shists, and sandy limestone banks, transport alkaline water (pH 8.05 to 8.4) with an electric conductivity up to 340 $\mu S\,cm^{-1}$.

Increased levels of nitrogen with total N reaching 4.3 $mg\,l^{-1}$ and phosphorous with total P reaching 0.7 $mg\,l^{-1}$ were recorded most often during low water periods. The enrichment is attributed to the excrement of many animals, which congregate in the remaining pools, an increased release from the sediments, which are stirred up by wind and animals, and the decomposition of the abundant aquatic macrophytes (Silva 1990).

Slightly brackish water has been reported to be present in some isolated lakes where evaporation exceeds precipitation over long periods. Mourão (1989) reported that Salina do Meio (Njecolandia, Mato Grosso do Sul) contains 1,222 $mg\,l^{-1}$ dissolved solids.

The oxygen concentration in the open water is normally sufficient due to the shallowness of the lakes, but near the bottom it is reduced during the high water period. In dense stands of aquatic macrophytes, the oxygen concentration may become greatly reduced, and H_2S may appear (Silva 1990).

Frequent, small-scale changes in topography result in a complex mosaic of habitats, as indicated by many different vegetation units and many ecotones, occurring side by side. A difference in elevation of only a few meters may separate a permanently dry habitat with a calcareous soil and a xeric vegetation including cacti, such as *Cereus peruvianus*, *C. bonplandii*, *Opuntia stenarthra* and *Parescia saccharosa* from permanently flooded habitats with an acid soil, covered by aquatic macrophytes. According to Loureiro *et al.* (1982), there are four main vegetation types in the Pantanal: sabana (Cerrado), steppe (vegetação chaquenha), semideciduous forest, and deciduous forest. In addition, many different sub-units and ecotones are found, depending upon specific local hydrological, geomorphological, edaphic, and climatic conditions.

The component species of the semideciduous and deciduous forests originate from among the northern Periamazonian vegetation, whereas floristic components of the Chaco are found in the extreme south of the Pantanal (Ab'Saber 1988). Recent phytogeographical papers by Adamoli (1981) and Prance and Schaller (1982) discuss the relationship between the Pantanal and the Cerrado. According to these authors, most of the plant species have a wide distribution. The number of endemic species is low.

Part of the Pantanal is covered by a very flood-resistant, poorly diversified forest. Common species are *Pithecolobium unifoliatum*, *Eugenia punicifolia*, *Myrcia multiflora*, and *Salacia laevigata*. Large low-lying areas are colonized by flood-tolerant or semi-aquatic grasses and sedges such as *Cyperus haspan*, *Rhynchospora corymbosa*, *Oryza subulata* and *Eleocharis* spp. Swampy areas are occupied by *Cyperus giganteus*, *Thalia geniculata*, *Echinodorus* spp., *Reussia subovata*, and other species. During the high water period free floating aquatic macrophytes dominate hundreds of km^2. These species include *Eichhornia crassipes*, *E. azurea*, *Paspalum repens* and *Scirpus cubensis*.

Figure 12. Herons, egrets, and storks in the Pantanal.

Figure 13. Caimans in pools at low water in the Pantanal.

Victoria amazonica is also common (Hoehne 1923, 1936, Prance and Schaller 1982, Silva 1990). A detailed analysis of growth, nutrient content, and distribution of the aquatic macrophytes in some floodplain lakes in the northern Pantanal was conducted by Silva (1990).

Like the flora of the Pantanal, its fauna is characterized by relatively few endemic species (Brown Jr. 1986). Species originate from the Amazon forest, the Cerrado, and the Chaco. Most of them have a wide distribution.

The Pantanal is famous for its abundant and diversified avifauna, which includes, among others, 13 herons and egrets, such as species of *Casmerodius*, *Egretta*, and *Pilherodius*; 3 storks (*Mycteria*, *Euxenura*, *Jabiru mycteria*); 6 ibises and spoonbills; and 5 kingfishers (Fig. 12). In addition, the Pantanal is used by dozens of migrating bird species as a wintering and stopover area. These include *Pluvialis dominica*, *Falco peregrinus*, and *Dolichonyx oryzivorus*. Waterfowl can be observed in enormous numbers, mainly during the dry period, when they concentrate in the remaining pools. The same is true for caimans, *Caiman yacare* (Fig. 13), and capybaras, *Hydrochoerus hydrochaeris*.

Human impact on the Pantanal was formerly small and confined to low-density cattle ranching. This has changed in the last decade. Agriculture, mainly soybean and some sugar cane crops, has become more and more important in the catchment area. This has resulted in an increased input of sediments, agro-industrial wastes, and agro-chemicals into the Pantanal.

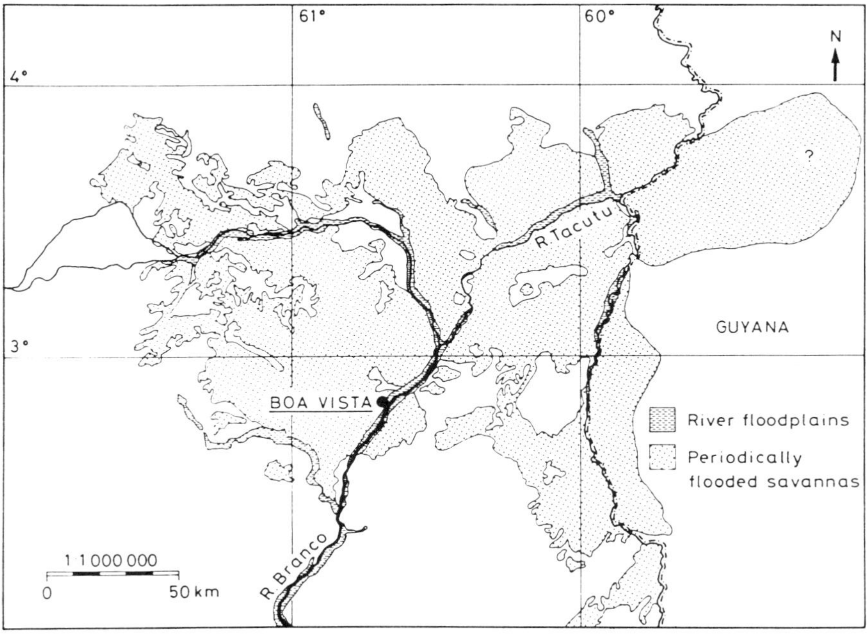

Figure 14. The distribution of the floodable savannas of Roraima and the Rupununi district (RADAM/Brasil).

Agriculture is beginning to extend into the Pantanal itself. Deforestation, illegal hunting, inadequate fishing, and the illegal trade with living animals are modifying plant and animal communities. Further serious threats are the input of mercury due to gold mining and the construction of flood-control measures.

One of the economically viable alternatives to the destructive use of natural resources in the Pantanal is tourism (Alho *et al.* 1988, Silva 1990). An increasing number of Brasilians and foreigners come to see the fascinating landscape and wildlife. However, tourism has begun to cause problems for the environment as well, because the infrastructure is insufficient to handle large numbers of tourists.

Savannas subject to flooding in the Roraima and Rupununi districts

After leaving the Pacaraima mountains, the Branco River and its tributaries pass through a vast, slightly undulating plain, about 100 to 130 m above sea-level. In Roraima Territory, about 20,000 km^2 is covered by savanna vegetation dominated by grasses (Fig. 14), such as *Trachypogon plumosus*, *Aristida setifolia*, *Axonopus aureus*, *Andropogon angustatus*, *Mesosetum lilii-*

Figure 15. Aerial view of the floodable savannas of Roraima at low water.

forme, *Paspalum stellatum* and *Paspalum melanospermum* (RADAM/Brasil 1975). The main tree species is the palm, *Mauritia flexuosa*, which is sometimes associated with *Virola* sp. and *Symphonia* sp. The topsoil is a sandy-clay that is underlaid in some areas by a lateritic layer up to 3 m deep. In some places, the laterite is exposed on the surface (Guerra 1957). Many hundreds of small, shallow depressions of a rather regular roundish shape are dispersed over the plain (Fig. 15). Ruellan (1957) attributed the formation of the deeper depressions to subterranean dissolution of gypsum. This explanation, however, has not been confirmed because the presence of gypsum deposits has not been proved. Brehmer (1972) attributed the depressions to subterranean transport of dissolved solids and fine particles.

Mean precipitation near Boa Vista is about 1,600 mm per year and it is concentrated between May and July when as much as 300 mm per month have been recorded. The heavy rainfall results in sheet flooding of the savanna. Depressions are then filled to form innumerable shallow lakes. Some lakes dry out every year, some only in years of extremely low precipitation, and others are permanent but shrink considerably in size during the dry season.

Most lakes have transparent water which is very poor in dissolved minerals (conductivity of 10 $\mu S_{20}\,cm^{-1}$) and slightly acidic with pH values of 5.5 to 6.5 (Reiss 1973, Furch, K. pers. comm.). These values correspond to those of Amazonian rain water (Anonymous 1972).

The permanent lakes, despite their oligotrophic character, are colonized by a rather well developed macrophyte community, including *Eichhornia crassipes*, *Thalia geniculata*, *Montrichardia arborescens*, *Cyperus articulatus*, *Eleocharis geniculata*, and several species of Alismataceae. They form detritus which is the food source for many aquatic animals. Reiss (1973) listed 2,400 individuals per m^2 of two *Tanytarsus* species in the benthos in sandy sediments of the central part of the lakes. In comparison, the aquatic invertebrate fauna of the seasonally flooded areas is not as abundant or diverse. In lakes supplied mostly by rain water, the fish fauna is poor in species due to pronounced seasonal changes in water level and to the complete disappearance of the lakes in extremely dry years. Recolonization occurs during extremely wet years when inundation covers almost the entire area and allows an exchange with other waterbodies.

Caimans (*Caiman crocodilus*) and capybaras are abundant in permanent lakes. There is also a remarkably rich and diverse avifauna. It includes the giant jabirú stork (*Jabiru mycteria*), which avoids densely forested Central Amazonian floodplains but is found in savanna floodplains in all parts of the lowlands.

The savannas of Roraima extend into Guyana to the east, where they are called Rupununi savannas (Eden 1970). They are part of a watershed between the Amazon and Essequibo River (Fig. 14) systems and cover an area of 13,000 km^2. According to Lowe-McConnell (1964), extensive areas of the Rupununi savannas are flooded to a depth of 1 to 2 m during the rainy season. Flooding is important for the region's fish fauna and has the same impact on their behaviour as flooding of the river floodplains of the Amazon and its tributaries. Fish disperse from the rivers over the flooded savanna to spawn and feed. They migrate back to the main rivers at the end of the rainy season. Many of them become stranded in the drying pools or are trapped in isolated ponds, where they await the next floods. From 60 to 70 of the 150 recorded species in the rivers of the Rupununi savannas were collected in isolated pools during the dry period (Lowe-McConnell 1964). Many species found in the area occur throughout the Amazon basin. This is not surprising because the Essequibo and Amazon systems are linked when the Rupununi savannas are flooded in the rainy season.

Seasonality in food availability for fishes in the Rupununi savannas leads to the conclusion that specialization of fishes for specific food resources develops chiefly during the rainy season, when a great variety of food items are available. Fishes are restricted to the remaining pools and food resources are limited in the dry season. These findings contradict those of Zaret and Rand (1971), who showed that there is specialization for specific food items during low water periods, when resources are scarce, but there is overlap during the high water seasons, when resources are abundant. Knöppel (1970), who analyzed food and feeding habits of fishes in Central Amazonian streams, found only a few distinct specialization of fishes to specific food items.

Studying the relationship between fishes and the floodplain forest in the Madeira River system, Goulding (1981) and Lowe-McConnell (1964) came to similar conclusions. However, Goulding stressed resource-sharing by closely related species when the water level is high and the importance of fat storage in many species as an adaptation to a low water starvation period. According to Junk (1976), Bayley (1982), and Soares *et al.* (1986), fish communities in tropical floodplains may not be food-limited during high water periods. The widely fluctuating hydrological conditions may control fish populations by means others than food supply. For example, survival rates may be lower during low water periods and recruitment may only successfully occur while the water level is rising. Predator-prey relationships may vary at times (Welcomme 1979).

Llanos of the Orinoco basin

The Orinoco River drains a catchment area of about 880,000 km^2. Annual precipitation may exceed 5,000 mm on the slopes of the Andes and in the Cassiquiare area, which is covered by tropical rain forest. The western part of the catchment area is a large sedimentary basin (Llanos) delimited to the west by the Andes, to the north by the Caribbean Cordilleras, and to the south by the Guyana Shield. In the Llanos, average annual precipitation is between 1,400 and 1,500 mm (Walter and Medina 1971). Considerable variation occurs from year to year. For instance, in San Fernando de Apure a maximum of 2,037 mm was reported in 1954. A minimum of only 840 mm was recorded in 1932.

Precipitation has a distinct seasonal pattern within the catchment area. About 60% falls between June and August and the least amount from January to March. The water level of the Orinoco and its headwater tributaries is influenced by the rains with a one to two month lag period. Levels are highest between June and November and lowest between December and May. The difference between the high and low water levels in the Orinoco River can be as much as 10 m.

The water of the Orinoco is turbid, low in dissolved minerals (conductivity 20–40 $\mu S_{20}\,cm^{-1}$) and slightly acidic (pH 6–7). Its tributaries may vary considerably in ionic contents depending upon the geology of the catchment area (Lewis and Weibezahn 1976, Paolini *et al.* 1983).

The Llanos is a huge (450,000 km^2), flat sedimentary basin in Venezuela and Columbia, which is between 50 m and 250 m above sea level. Northwest of the Orinoco, it extends from the Andes to the delta area and is divided into four subregions: piedmont, high plains, alluvial overflow plains, and aeolian plains (Ramia 1967, Sarmiento 1983). Permanent or seasonal wetlands occur along the rivers and in shallow depressions. The largest seasonal wetlands are located in the alluvial overflow plains in the central part of the Llanos (Fig. 1). Vila (1955) estimates that 60% of the State Apure (about 50,000 km^2) in Venezuela is flooded every year. Sarmiento *et al.* (1971) indicate that there are about 4,000 km^2 of periodically flooded savannas south of Barinas and the FAO (1966) reported that another 1,000 km^2 are located near Villavincencio Colombia, belonging to the same savanna complex in the piedmont region of the Andes. The most recent map of the vegetation in Venezuela (Huber and Alarconó 1988) indicates the occurrence of vegetation exposed to periodic flooding all over the Llanos and in the Venezuelan part of Amazonia. However, Huber and Alarconó's publication does not contain enough geographic data to produce a map, nor does it permit the calculation of the total area of the flooded region. In summary, a realistic estimate of savannas that are subjected to seasonal flooding seems to be about 80,000 km^2.

At the beginning of the rainy season, depressions (esteros) in the terrain fill with water to a depth of 1 m or more. Later, sheet flooding occurs over large, poorly drained areas. Near the large tributaries, including the Apure, Aranca, Capanaparo, Sinaruca and Meta, flooding is due to lateral overflow. Sandy, natural levees bordering streams constitute the highest parts of the plains. Areas covered by river water receive additional dissolved nutrients and fertile sediments, whereas the areas covered mainly by rain water are rather poor in nutrients.

When the water recedes, the Llanos begin to dry. In February and March, at the end of the dry season, only small moist areas remain along with shallow lakes (lagunas) in the centre. The lagunas are important grazing areas for herbivorous animals during the dry season and are thus of great importance to the whole Llanos system.

The system of the Llanos is dominated by savanna grasslands. Trees indicate waterlogged soils and include the palms *Mauritia minor*, *Mauritia flexuosa*, *Caraipa llanorum*, *Sabal mauritiaeformis*, and *Copernicia tectorum*. Levees along the rivers are often covered by gallery forest. Detailed descriptions of the vegetation are provided by Ramia (1959, 1974), FAO (1966), Blydenstein (1967), Sarmiento *et al.* (1971), Medina and Sarmiento (1979), and Medina (1980).

Productivity in rainwater-filled esteros is mostly due to the growth of *Leersia hexandra* and *Hymenachne amplexicaulis*, which are important grasses due to their great food value and frequent occurrence. In esteros filled with river water, *Paspalum fasciculatum* is the major species and above ground biomass is about 25.4 t ha^{-1} (Escobar 1977, Gonzáles-Jiménez 1979).

This value is lower than those recorded for *Paspalum fasciculatum* stands on the banks of the Amazon (Junk and Howard-Williams 1984, Junk 1990), but the nutrient content of the sediments in the Orinoco basin is probably lower and drought stress is more pronounced. An additional stress factor may be fire, but it is not known how important fire is in controlling the vegetation around the esteros.

Limnological data from the Llanos are rather scarce (Gessner 1955, 1965, Lewis and Weibezahn 1976, Vasquez 1984, Vasquez and Sanchez 1984, Blanco and Sanchez 1984). More data are available on fish and fishery resources (Mago 1970, Novoa 1982, 1989, Novoa *et al.* 1984, Perez 1984). A total of 494 species have been recorded (Lowe-McConnell 1975), and Bonetto and Castello (1985) have estimated fish yields in the delta area of 40 kg ha^{-1} yr^{-1} as compared to 12 kg ha^{-1} yr^{-1} in the Llanos.

The mammals of the Llanos are well known (Ojasti 1983). Detailed studies exist on the biology and ecology of the capybara (*Hydrochoerus hydrochaeris*) (Ojasti 1973). The capybara occurs in all lowland wetlands of tropical South America and only avoids small streams in densely forested areas. It is well adapted to aquatic life. In savanna areas with a pronounced dry period, including the Llanos, growth and reproduction occurs mainly in the rainy season when aquatic grasses of considerable nutritional value are abundant. At the end of the dry season, mortality is high due to lack of food and scarcity of aquatic habitats for shelter. Fat accumulation during the rainy season facilitates survival during the dry season (Ojasti 1978). The animals attain sexual maturity in eighteen months and have an average litter size of four. Two breeding cycles per year can be expected in optimal habitats (Ojasti 1973). This enables the capybara to quickly replace heavy population losses. It is thus an r-strategist.

Maximum population density of capybaras is estimated at 0.6 individuals ha^{-1}. Considering habitat availability during the dry season, the density increases to 2 animals ha^{-1}. At 1.6 animals ha^{-1} a steady state with respect to the plant biomass was maintained, whereas 3 animals ha^{-1} caused net destruction of the vegetation. At a stocking rate of 1.6 ha^{-1} the animals consumed 500 kg dry weight ha^{-1} yr^{-1} plant material. This is estimated to represent 3.5% of the annual primary production (Ojasti 1978). Due to their great numbers, capybaras are among the most effective herbivorous mammals in tropical South American wetlands.

Deer occur in much smaller numbers and are considered of secondary importance as grazers. Hofmann *et al.* (1976) estimated that there are 0.5 to 0.7 swamp deer (*Blastocerus dichotomus*) per km^2 in undisturbed habitats, whereas Brokx (1972) considers 4 to 8 white-tailed deer (*Odocoileus virginianus*) per km^2 to be the optimal stocking rate in the Llanos.

Experiments are being conducted to determine how rain water can be retained in predetermined areas (modulos) of the Venezuelan Llanos by the construction of dikes. It is intended to increase the carrying capacity of the savannas by increasing the primary production of aquatic grasses. If existing plans are carried out, 10,000 to 20,000 km^2 in the Apure would be affected (Salinas 1975). The first experimental results revealed an increased primary production and positive responses of water birds, fishes, crocodiles, capybaras and other animals (Pinowski and Morales 1981, Ramos *et al.* 1981). There is, however, a considerable concern about negative side effects, especially the spread of aquatic weeds and waterborne diseases. A final conclusion about the ecological impact of this type of management is not yet possible, since the experimental areas are still in the stage of transition.

Other periodically inundated savannas

Other periodically inundated savannas with permanently moist depressions occur all over tropical South America due to the flat topography, poorly drained soils, and pronounced seasonality of the precipitation. However, little is known about them. According to Janssen (1986), 40% of the savannas of Humaitá, which cover an area of 630 km^2 about 600 km south of Manaus, have to be considered as wetlands. This value compares well with the one given earlier about the percentage of waterlogged soils and periodically inundated areas in the region of the Central Amazonian rainforest. Fifteen percent of the Humaitá savannas are dominated by *Schizachyrium brevifolium*, 10% by *Cyperus haspan*, 5% by *Rhynchanthera grandiflora* and 10% by *Montrichardia arborescens/Mauritia armata* palm tree swamps. A similar flora is to be expected in the wetlands of the 1,200,000 km^2 savanna area of the Central Brazilian Cerrado. There is, however, no detailed information available.

The transition zone between rain forest and Cerrado in the northeastern part of Brazil is characterized by large stands of the hydrophylous Babaçú palm tree, *Orbignya martiana* (Hueck, 1966). These "Babaçuais" cover more than 100,000 km^2 of the Brazilian states of Maranhão and Piauí. About 80% of the annual precipitation of 1,500 to 2,200 mm falls during a six month rainy season, resulting in the periodic flooding of the depressions. The wetlands of those areas have not been studied in detail.

On the Pacific Coast, periodically inundated savannas (tembladeiras) were reported to occur in Colombia (Acosta-Solís 1966, Harling 1979). One such area of about 10,000 km^2 is situated north of Guayaquil and a smaller one of about 1,000 km^2 to the south. There are no details available about wetlands in those two areas.

The Bana woodlands and the Amazon Caatingas

In Southwest Venezuela, between the Cassiquiare and the Rio Negro, there are large areas of white sands, some of which are classified as Spodosols. These have an impermeable B horizon cemented by humates at a depth of 1 to 2 m. Water and soils in this area are extremely poor in nutrients and acidic. Two main vegetation types have been distinguished: the Bana woodlands and the Amazon Caatinga. The Banas develop on circular mounds a few hectares in area and a few meters in height. The Caatingas occur in depressions between the Banas (Klinge *et al.* 1977, Klinge and Herrera 1978, 1983, Bongers and Engelen 1982).

Differentiation of the Bana vegetation into the categories of Tall Bana, Low Bana, and Open Bana are related to the water supply and nutrient status (Bongers *et al.* 1985). A mean annual precipitation of 3,600 mm brings about waterlogging or even flooding of the Banas during periods of heavy rainfall and drought during periods of little rainfall. The Caatingas are periodically flooded due to insufficient drainage, but they are not subject to as many periodic droughts as the Banas. In addition to the hydrological gradient, there is probably also a nutrient gradient (Paolini 1978, Herrera 1979, Herrera *et al.* in press.).

Since periodic flooding or waterlogging affects the species composition and physionomy of the vegetation, the Banas, Amazonian Caatingas, and probably some other vegetation types on white sands in tropical South America have to be considered as wetlands. There is, however, no detailed information about their distribution, but areas such as these cover several thousand square kilometers in Venezuela (Huber 1982a, b, Huber and Alarcon 1988).

Acording to RADAM/Brasil (1975, 1978), there are large areas of poorly drained, waterlogged and periodically submerged soils with numerous small, shallow lakes in the region south of the Roraima savannas (Fig. 16). They are mostly covered by forest or shrub vegetation, sometimes with palms being dominant. There is, however, no information on these wetlands, which cover an area of 50,000 km^2.

The Magdalena river

The Magdalena River catchment area (Fig. 17) covers about 266,000 km^2. The uppermost sections of the rivers in the catchment area cut deeply into the Andes. They have rather small floodplains in spite of water level fluctuations of about 10 m. The flooding is irregular, but a distinct bimodal pattern appears when monthly values are averaged over several years (Dister and Garcia 1984, Lozano and Dister 1990). Maximum water levels are reached

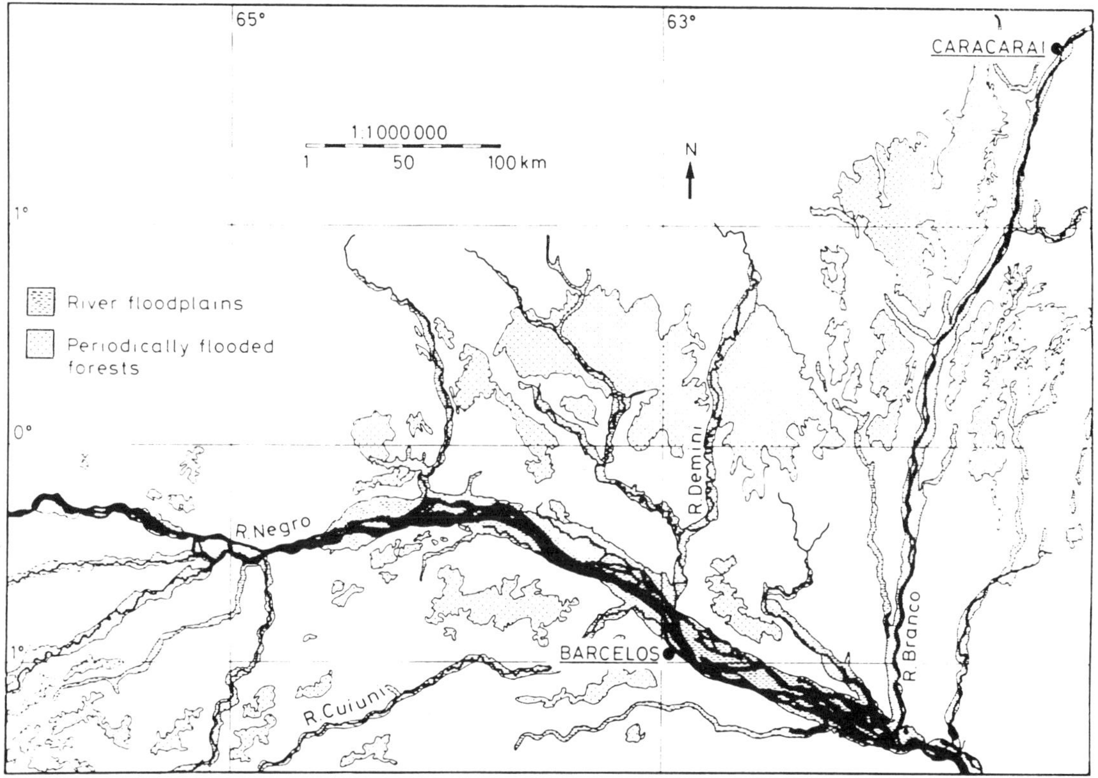

Figure 16. The distribution of the forested wetlands in the Rio Negro area (RADAM/Brasil 1975, 1978).

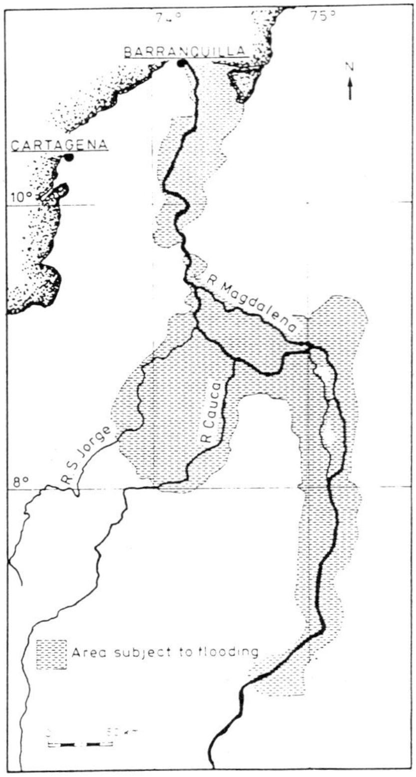

Figure 17. The distribution of the floodplains along the lower Magdalena River (according to Lonzano and Dister 1990).

in May and November and the minimum occur in January and August. In topographically lower areas, the river valleys widen, the gradient flattens, and large floodplain areas are formed. The flooded areas cover about 36,000 km^2 (about 5,000 km^2 in the delta) and the flood amplitude is about 4.5 m. On the floodplain and in the delta areas, there are many temporary and permanent floodplain lakes and backwaters (ciénagas) that are separated from the main river by levees. Water exchange between the depressions and the main channels occurs along narrow channels called caños.

The chemical composition of the water in the tributaries of the Magdalena River is quite variable. Conductivity varies from 45 $\mu S_{20}\ cm^{-1}$ in the Rio Sumapaz to 625 $\mu S_{20}\ cm^{-1}$ in the Rio Negro. The pH is 7 but tributaries may vary from slightly acidic to slightly alkaline. For example, the pH in the Rio Sumapaz is 5.9 and that in the Rio Guarino 8.4. The oxygen concentration in the water of the ciénagas is relatively high during the daytime but an oxygen deficiency may occur near the bottom at night (Dister and Lozano 1984, Lozano and Dister 1990).

Biological information is available on more than 150 fish species (Gery 1969), which are an important source of animal protein (Arias 1975); as 50,000 to 65,000 t ha^{-1} yr^{-1} are caught (Bonetto and Castello 1985, Barco and Villarreal 1989). The bimodal flood pattern of the Magdalena River induces two fish migration cycles.

The Columbian government was planning to construct about 10 reservoirs on a 500 km stretch of the river between Girardot and San Pablo to improve ship traffic and to generate hydroelectricity. This would markedly affect both the fishery and ecological conditions on the downstream floodplains and delta area. According to Dister (1983), modification of the hydraulic geometry of the Magdalena River may result in an increase in extreme floods. This would have a drastic effect on the plant and animal communities in the brackish water of the delta area. Such floods already occur at 6 to 7 years with catastrophic effects on the culture of oysters, *Crassostrea rhizophorae* (Kaufmann and Hevert 1973). At present, construction plans have been suspended.

High Andean wetlands

High Andean wetlands include the shallow littoral zones of deep lakes, such as Lake Titicaca (Fig. 1) and Lake Yunin, as well as shallow lakes, such as Lake Poopó and Lake Yanayacu, salt pans in the dry Puna region, *Sphagnum* bogs, cushion bogs, and reed swamps in the wet Paramo region.

Most information on these wetlands is available from the region of Lake Titicaca, which is situated in a 19,000 km^2 endorheic basin on the Planalto of Peru and Bolivia. According to Moon (1939) and Newell (1949), it was raised during the Miocene from a few hundred meters above sea level to about 4,000 m and formed as part of an internal drainage system between two mountain ranges. During a Pleistocene interglacial period, a large lake, Balliván, existed, which included the present-day Lakes Titicaca and Poopó. Since then, the climate has become drier and the lake levels are falling. Evaporation losses in the Titicaca area are about 1,500 mm yr^{-1} (Monheim 1956) compared to precipitation, which ranges from 1,800 mm yr^{-1} in the north of the basin to 300 mm in the south. The greater overall E/P ratio has lead to increasing salinity and, mainly in the southern part of the basin, to the formation of salt deposits.

Lake Titicaca has 2,810 km^2 of shallow littoral zones which constitute 36% of the entire lake. Large "totora" swamps, occupied by *Scirpus totora*, form in protected bays; the largest stands, occupying 300 km^2, are located in Puno bay. According to Gilson (1964), the macrophyte community is composed of only a few species. *Scirpus totora*, *Zanichellia palustris*, *Potamogeton strictus* and *Lilaeopsis* sp. occur to a depth of about 1 m. *Myriophyllum elatinoides* and *Elodea potamogeton* are found to depths above 8 m, *Chara* sp. to about 12 m, and the moss *Sciaronium* as deep as 29 m. The luxuriant aquatic macrophytes near the shores are used as food by cattle.

Most phytoplankton and many zooplankton species there are cosmopolitan. According to Löffler (1968), the lakes and wetlands of the high Andes are used by many Antarctic birds as resting places on their northerly migration route. They transport small aquatic organisms in their gut and on their bodies, distributing them along the Altiplano as far as Central America. Examples of this type of distribution are the crustacean genera *Boeckella* and *Pseudoboeckella*.

Larger aquatic animals are less easily transported between the high Andean water bodies and a number of endemic species occur in Lake Titicaca. The subfamily Orestiinae (Fam. Cyprinodontidae) provides an example of intralacustrine speciation (Villwock 1963, Kosswig and Villwock 1965). Fourteen of the 20 species of the genus *Orestias* are endemic to Lake Titicaca. According to Villwock (1972), the endemic fish fauna is endangered by foreign species, especially *Salmo gairdneri*, which have been introduced as early as 1937. With the introduction of rainbow trout, a sporozoon was also introduced, which has subsequently negatively affected the native fish species. Between 1950 and 1960, *Basilichthys bonariensis* (Atherinidae), *Salmo trutta*, *Salvelinus fontinalis*, and *S. namaycush* were introduced with less success, probably because of strong competition by the established rainbow trout. Other examples of endemism are molluscs, especially the genera *Taphius* (Planorbidae) and *Littoridina* (Hydrobiidae), of which many of the 24 species are endemic to Lake Titicaca.

The 320 km Desaguador River connects Lake Titicaca to Lake Poopó, a shallow 26,000 km^2 lake that is only 3 to 5 m deep. Lake Poopó is highly saline, even though there are periods when it overflows in the south west toward the Salar de Caipasa. Large salt depositions occur in the south, including the Salar del Huasco (Tricart 1969).

There is very little limnological information about other lakes in the high tropical Andes. Hegewald *et al.* (1976) reported some data on water chemistry and phytoplankton in 43 Peruvian lakes, including 13 which lie above 4,000 m and 10 which occur between 2,000 and 4,000 m. Gessner and Hammer (1967) mentioned 397 small lakes in the high Andes of Venezuela and provided some limnological information on two of them.

Detailed phytosociological information is available on the bogs, reed swamps, and mires (Gutte 1980). Cleef (1981) distinguished more than 40 associations of herbaceous plants and shrubs in the Paramo wetlands of the Colombian Cordillera Oriental. *Sphagnum* bogs are widely distributed at elevations between 2,800 and 3,800 m. The main species there are *Sphagnum sanctojosephense*, *S. magellanicum*, and *S. cuspidatum*. A hummock-hollow relief is characteristic for all *Sphagnum* bogs. The peat layer is generally 1 to 4 m thick and moderately acidic (pH 4.6–5.3). According to van der Hammen and Gonzalez (1960, 1965), the Colombian Paramo is between 3,000 and 5,000 years old. Cleef (1981) distinguished four main types of *Sphagnum* bogs in the Colombian Cordilleira: *Sphagnum* bogs with *Espeletia* and *Blechnum loxense*, those with *Swallenochloa*, those with giant *Puya*, and *Xyris-Sphagnum* bogs. *Sphagnum* bogs may develop into climax shrub communities dominated by *Aragoetum abientinae* and *Diplostephietum revoluti*. At higher elevations between 3,400 to 5,200 m, *Sphagnum* bogs are gradually replaced by flush and cushion bog vegetation, which occurs in small, steep valleys and on glacial valley floors, including former lake beds. Characteristic species are *Werneria pygmaea*, *Gentiana sedifolia*, *Distichia muscoides*, *Plantago rigida*, *Oritrophium peruvianum*, and *Oreobulus obtusangulus*. The flush and cushion bogs are mesotrophic with a pH of 5 to 6. Reed swamps and mires are found on wet slopes, flat marshy valley floors, lake shores, and along streams. Characteristic species are *Marchantaria plicata*, *Epilobium denticulatum*, and *E. meridense*. They occur in nearly all the *Carex* and *Cyperus* reed swamps and in *Calamagrostis ligularis* mire communities in the Paramos. The pH values in the root zone vary between 4.5 and 7.5. The layer of organic matter can be more than 8 m thick and it is mesotrophic to eutrophic.

The saline lakes and their shores in the dry Puna highlands support few higher plant species. Ruthsatz (1977) differentiated between ephemeral shallow lakes with different plant community zones, according to duration and depth of flooding and salinity. *Gamochaeta deserticola* is characteristic of higher areas, *Muhlenbergia fastigiata* in the middle parts, and *Marsilea punae* for the lowest-lying areas. Large populations of flamingos (*Phoenicopterus chilensis*, *Phoenicoparrus jamesi*, *P. andinus*) are reported to occur in those areas but other types of ecological information are not available.

Coastal wetlands

The coastal wetlands of tropical South America consist of mangrove (mangal) swamps and smaller areas of salt marshes. Large freshwater wetlands subject to tidal influence occur in the delta areas. Brackish and freshwater swamps occur on alluvial deposits along the coasts (Fig. 1). The estuary of the Amazon is estimated to cover 25,000 km^2 (Gourou 1950, Lima 1965) but this figure is questionable as recent maps indicate the existence of about 50,000 km^2 of periodically flooded alluvial soils in the Amazon estuary (Fig. 18). The Orinoco delta covers about 22,000 km^2 and the delta of Magdalena River covers about 5,000 km^2. Coastal wetlands in Surinam cover about

18,000 km^2 (Lindemann 1953), and in Guyana, about 15,000 km^2. For mangroves, the following data on coverage are provided (IUCN 1983): Brazil: 25,000 km^2, French Guyana: 55 km^2, Surinam: 1,150 km^2, Guyana: 1,500 km^2, Venezuela: 6,736 km^2, Colombia: 4,400 km^2, Ecuador: 1,601 km^2, and Peru: 280 km^2. The total area covered by marine and freshwater coastal wetlands thus may be as much as 120,000 km^2.

Whereas the world-wide literature on coastal wetlands is extensive, ecological studies on those of tropical South America are scarce (Chapman 1975). Considerable data exist on Lake Maracaibo, Venezuela (Rodriguez 1973). On the Pacific coast, mangroves are chiefly found in the Gulf of Guayaquil in Ecuador. They are restricted to the Tumbes River Delta in the south. It is generally assumed that the cold water of the Humboldt Current limits mangrove development along the Pacific coast. Some authors stress that other factors, such as a lack of adequate sheltered habitats and river deltas with fine sediments, may also be important (West 1956, 1977, Pannier and Pannier 1977, Godt 1985). The principal mangrove species are *Rhizophora harrisonii*, *Avicennia germinans*, and *Pelliciera rhizophorae*. *Avicennia tonduzzii* has been reported from some places in Ecuador.

On the Atlantic coast, mangal vegetation is found at the mouth of the Grajaú River near São Luis (Brazil), at the mouth of the Amazon River, and intermittently along the Guyanan and Venezuelan coastlines as far as the Magdalena River Delta. The principal species are *Rhizophora mangle*, *Avicennia germinans*, *A. schaueriana*, *Laguncularia racemosa*, and *Conocarpus erectus*. *Rhizophora racemosa* occurs from the mouth of the Amazon to the mouth of the Orinoco. *Rhizophora harrisonii* occurs from the Orinoco Delta along the Guyana Coast (West 1977).

Spartina brasiliensis salt marshes are known to develop on the mud flats on the seawardside of the mangroves. These have been described from the Guyanan and Brazilian coasts (Freyburg 1930, Martyn 1934). *Spartina* is presumed to accelerate sediment deposition around its extensive root system. *Spartina* wetlands are replaced by mangroves when the mud banks are sufficiently high to allow seedling establishment. Zonation is a characteristic feature of mangrove swamps. It has often been described as a sequence of successional stages (Lugo 1980). Several factors are considered important, including level gradients, salinity gradients, and differences in soil structure. According to West (1977), the classical description of the zonation in Atlantic tropical mangrove areas treats *Rhizophora mangle* as a pioneer species. Its stands are followed by belts of *Avicennia* spp., *Laguncularia racemosa*, and *Conocarpus erectus* as one moves inland to areas beyond tidal influence. The data available on the structure of mangrove communities along the coast of tropical South America, however, reveals a somewhat different picture; *Avicennia* is often the pioneer plant (Martyn 1934, Fanshawe 1952, Lindeman

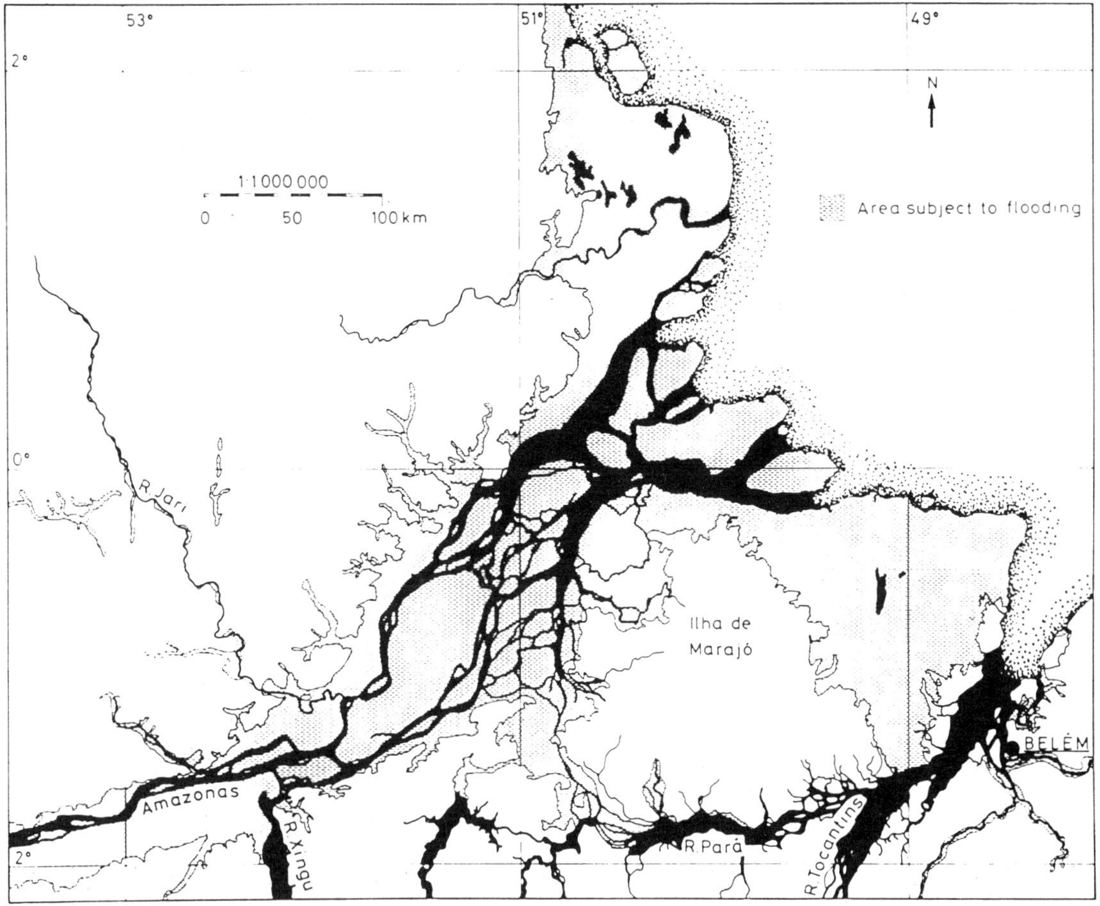

Figure 18. The wetlands in the Amazon River Delta area (RADAM/Brasil 1974a,b).

1953, Vann 1959, 1969, Boyé 1962, Gessner 1967). In delta areas, small changes in soil structure, salinity, and relief may result in a rather complex pattern of plant distribution. Salinity is often highest near the inland borders of mangrove swamps due to increased evapotranspiration. This results in increased osmotic stress on mangroves.

The detailed description of coastal vegetation of Surinam (Lanjouw 1936, Geijskes 1945, Lindemann 1953) is probably valid, with minor modifications, for most mangrove areas along the Atlantic coast north of the mouth of the Amazon River (Fanshawe 1952, Benoist 1924, Williams 1940, 1941, Pires 1966). According to Lindemann (1953), an area of 18,000 km^2 along the coast of Surinam is covered by alluvial sands and clays. In the eastern part, the sands are deposited by the sea in a series of ridges of considerable length. The oldest ones are found 15 to 20 km landward and rise to a height of 10 m above sea level. Westward the ridges decrease in number and extent, dominating immense swamps. Alluvial clays covered with water up to a few meters in depth form the bottom of the depressions between the ridges. While the seaward depressions are influenced by the tides, the landward ones are influenced only by rainwater, and their water level may rise a few meters above sea level. A marked salinity gradient is detectable. Plant communities are influenced by the moisture, soils, and salinity.

The saline coastal belt is colonized by mangal vegetation (*Avicennia* and *Laguncularia*) with the addition of the shrubs, *Hibiscus tiliaceus* and *Stenotaphrum secundatum*, the halophilous fern *Acrostichum aureum*, and other herbaceous plants, including *Ipomoea pes-caprae*, *Canavalia maritima*, and *Sporobolus virginicus* on the ridges. Other salt tolerant shrubs on the following ridges are *Rosenbergiodendron formosum*, *Dalbergia castophyllum*, and *Machaerium lunatum*. On the inland ridges trees including a *Cereus* sp., grow taller and reach a height of 10 m.

Hypersaline depressions are covered by *Eleocharis mutata* and a few other species. Brackish water depressions are colonized mainly by *Typha angustifolia*, *Cyperus articulatus*, and *Leersia hexandra*. Depressions with slightly brackish or fresh water are dominated by *Cyperus giganteus*, *Typha angustifolia*, *Scleria eggersiana*, *Thalia geniculata*, and *Montrichardia arborescens*. Old oligotrophic swamps are dominated by *Rhynchospora corymbosa*, *Montrichardia arborescens*, and the ferns *Blechnum indicum*, *Dryopteris gongylodes*, *Nephrolepis biserrata*, and *Pityrogramma calomelanus*. On deep peat layers, *Lagenocarpus guianensis* may form nearly monospecific stands up to 2 m in height together with some *Montrichardia* and *Rhynchospora gigantea*. In small flooded areas, different types of swamp forests develop; *Triplaris surinamensis*, *Virola surinamensis*, *Tabebuia insignis*, *Annona glabra*, *Ilex guianensis*, *Andira inermis*, *Mauritia flexuosa*, and the shrub, *Chrysobalanus icaco* are the dominant species.

Along the rivers, *Rhizophora mangle*, *Laguncularia racemosa* and *Machaerium lunatum* are predominant. Upriver, they are replaced by *Montrichardia arborescens*, *Bombax aquaticum*, and *Pterocarpus officinalis*. In the western half of Surinam and Guiana, *Mora excelsa* dominates large areas of the freshwater tidal forest. In French Guiana, *Rhizophora* and *Laguncularia* are restricted to the areas near the coast, whereas *Avicennia* and *Conocarpus* are also found further inland together with *Bombax aquaticum* and *Macrolobium bifolium*. In Guiana and in the Orinoco Delta, the mangal zone is found interspersed with *Caraipa guianensis*, *Pterocarpus officinalis*, *Euterpe edulis*, and *Manicaria saccifera*. Further upriver, *Bombax aquaticum* and *Macrolobium vampa* become abundant. In the Amazon delta, an increased number of palm species, such as *Euterpe oleracea*, *Mauritia flexuosa*, *Raphia taedigera*, *Manicaria saccifera*, *Iriartea exorhiza*, and *Astrocaryum murumuru*, are important floristic elements in the freshwater tidal zone. Furthermore, the giant Araceae *Montrichardia arborescens* occurs in large stands (Huber 1903, Pires, 1966).

Man-made wetlands

Man-made lakes

In the future, very large areas of man-made wetlands will be created by the construction of large hydroelectric power plants in tropical South America. To date only a few reservoirs have been constructed, including Brokopondo Reservoir on the Suriname River in Surinam (1,560 km^2), Guri Reservoir on the Caroni River in Venezuela (1,500 km^2), Tucuruí Reservoir on the Tocantins River (2,430 km^2), Balbina Reservoir on Uatumã River (2,360 km^2) and Samuel Reservoir on the Jamarí River (645 km^2) in Brazil. However, the number will increase substantially.

The hydroelectic potential of the Brazilian Amazon basin is estimated at 100,000 MW. Eighty resevoirs that would flood about 100,000 km^2, mostly covered with forest, were to be constructed. Because of water level fluctuations in the reservoirs, about 15% of the area would become periodically dry and form large man-made floodplains. Some rivers would be transformed into a series of reservoirs. On the Tocantins River and its tributaries, for examples, 27 reservoirs are planned with a hydroelectric potential of 25,300 MW (Goodland 1978, Junk and Mello 1986, Fig. 19).

The overall impact of so many large reservoirs is complex and difficult to evaluate. Data are available mainly from the Brokopondo Reservoir (Leentvaar 1966, 1973, 1979, van der Heide 1982). Futher data exist from Curua-Una Reservoir, a small reservoir covering about 80 km^2 in the Amazon basin (Junk *et al*. 1981, Junk 1982b). Negative effects on the floodplains of all rivers affected, due to more or less drastic changes in the hydraulic

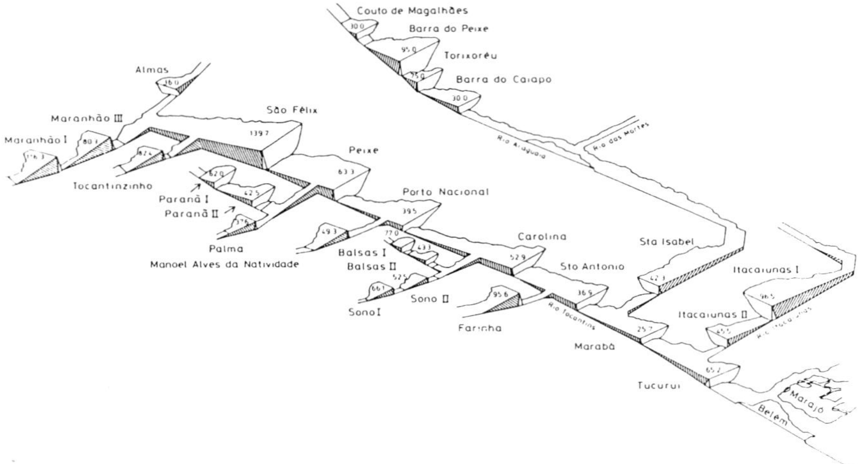

Figure 19. The hydroelectric power projects planned in the Tocantins-Araguaia basin (ELECTROBRAS).

geometry, are to be expected. In addition, interruption of the migration routes of aquatic animals, mainly fishes, will probably modify the faunistic and floristic composition of the reservoirs as well as areas downstream and upstream (Junk and Mello 1986). As shown for the Brokopondo Reservoir, the gain in new habitats in the reservoirs will not compensate for the loss of terrestrial and aquatic habitats in the flooded areas (Leentvaar 1984).

Paddy-fields

The number of plantations of wet rice in tropical South America is still small. In the Amazon basin, rice cultivation was initiated about 40 years ago (Camargo 1949) and currently small plantations exist near the coast. The largest plantation was established by D. Ludwig as part of the Jarí Project on the lower Amazon at the confluence of the Jarí and Amazon Rivers. The information available on the project indicates it was a highly mechanized system that used flood control and the large scale application of fertilizers and pesticides.

The Brazilian government started a project (PROVARZEA) to provide financial aid to stimulate the use of wetlands for agriculture and animal husbandry some years ago. Most of the funding has been awarded outside the Amazon basin, but some pilot projects with wet rice plantations have been started near Manaus. Other wet rice plantations exist on the Orinoco and along the coast; however, they are rather small. On the Orinoco, there is some concern about the spread of schistosomiasis due to the increasing area of the wet rice plantations.

Other man-made wetlands
The construction of ponds for aquaculture is still limited in tropical South America, since the potential of river fisheries has not yet been fully exploited. Shrimp farming is increasing in the saline coastal wetlands, mainly in Ecuador, where more than 400 km^2 of aquaculture projects have been reported by IUCN (1983). On the Atlantic coast, shrimp farms are increasing as well in number.

Road building has created a large number of small to medium sized wetlands, which cover many hundreds of square kilometers all over the lowlands of tropical South America. Roads often dam streams and create artificial lakes and swamps on the upstream side. In flat areas, roads are slightly elevated to minimize flooding during the rainy season; material for road construction is taken from areas adjacent to the roads. This method creates many ponds and channels along the roads, which are filled with water during the rainy season. Often, the roads act as dams, hinder runoff, and allow transport and migration of aquatic organisms across the borderlines from one catchment area to the next. In the future, such areas may become important as breeding grounds for *Anopheles* spp. and increase the spread of malaria in the area.

Man's impact on the wetlands of tropical South America

Man's impact on wetlands occurs at the species, community, and ecosystem levels with influence increasing from one level to the next. A discussion of the impacts at the species level would go beyond the scope of this chapter. Therefore, I will concentrate on major impacts at the ecosystem level. The most important alterations occur as side effects from the construction of large hydroelectric power projects, large scale deforestation, land reclamation for agricultural purposes and farming, pollution by industrial wastes and agrochemicals, and mining various minerals.

As already mentioned, many large rivers within the region will be impounded in the next decades. It may be argued that the building of reservoirs will increase the total area of aquatic habitats. On the other hand, natural wetlands will be eliminated or modified on a large scale in the area covered by the reservoirs and in the floodplains upstream and downstream due to the modification of the hydraulic geometry of the rivers and the interruption of the migration routes of aquatic organisms.

Replacement of the rain forest ecosystems in shallow parts of the reservoirs by flood tolerant tree species will take many decades, as shown at Lake Brokopondo, the first large reservoir in the South American humid tropics. Consequently, the reservoirs will become rather monotonous, uniform habitats for very long periods. According to Leentvaar (1984), the gain in new habitats in the reservoirs themselves will not compensate for the loss of terrestrial and aquatic habitats in the flooded areas.

Large constructions for ship traffic improvement may have negative side-effects on the wetlands as well. Actually, a multinational project is being discussed to build a channel ("hidrovia") through the Pantanal by cutting through the meanders and widening and deepening the main channel of Paraguay River to allow the passage of large ships to the Atlantic Ocean throughout the year. Such a project would modify the hydrology of large areas of the Pantanal and result in major changes within the whole system, including the floodplains along the lower Paraguay river.

Large-scale deforestation to permit agriculture and cattle raising threatens vast forested floodplain areas along most large Amazonian rivers as well as small wetlands along the small streams in the rain forest area. On the floodplains of large rivers, the highly complex and well adapted floodplain forest is replaced by grasslands, which are annually burned for weed control. The biocenosis is not adapted to fire stress; productivity and species diversity decrease as weeds of low food value become dominant (Junk 1989a). Agriculture is also beginning to advance into the floodable savannas (e.g. the Pantanal) using, in part, flood control measures which completely modify the floodplain characteristics of the system and result in an explosive development of weeds (Conceição and Paula 1986). *Ipomoea fistulosa*, a shrub-like, flood-resistant member of the family Convolvulaceae is spreading over large areas and is often combated with herbicides.

Clear cutting of forested areas affects the innumerable small wetlands along the streams by changing the hydrological regime, the sediment load, the light regime, and the food webs. There is, however, very little information about the undisturbed systems and nearly none about the impacts of disturbances on them. Furthermore, comparative studies about the level of similarity among these small wetlands are lacking. Therefore, the overall consequences of large scale deforestation on the reduction of aquatic plant and animal species and on losses of specific wetlands are currently hard to evaluate.

Increased grazing by cattle is noticeable on seasonally flooded savannas all over the continent. In contrast to forested floodplains, fire in the flooded savannas is a natural stress factor, but frequent burning and increased cattle grazing tends to stimulate the transformation of the swamp forest and the floodplain forest into species poor communities of grasses and herbaceous plants together with an increasing number of weeds (Junk 1989a, 1990).

Agroindustrial wastes from alcohol distilleries are becoming a serious threat for many rivers and connected wetlands (e.g. in the Pantanal area). There are nine alcohol distilleries planned or functioning in the State of Mato Grosso. These will produce 1,320 m^3 of distillate and 79,300 m^3 of residual water daily. Two distilleries operating in Mato Grosso do Sul have been linked to several fish kills (Alho *et al.* 1988).

Pesticides may become a major problem in floodplains that are intensively used for agricultural purposes. Natural selection of plants and animals in fertile floodplains selects r-strategists such as weeds, pests, and parasites, which then require the frequent use of agrochemicals for control. During floods, the remaining pesticides are transferred into the aquatic system, resulting in fish kills, as already demonstrated in the Pantanal, (Alho *et al.* 1988).

Pesticide application will likely increase in rural areas where there is growing conflict of interest between wetland protection and the control of malaria and schistosomiasis. Proposals for the control of the diseases include the application of toxic substances as well as the modification or even complete elimination of small wetlands to destroy the habitats of the vectors.

Illegal hunting and the export of animal skins have considerably reduced the numbers of many animal species. For the Brazilian Pantanal, Alho *et al.* (1988) cited the following species as endangered: *Dusicyon thous* (crab-eating fox), *Chrysocyon brachyurus* (maned wolf), *Speothus venaticus* (bush dog), *Felis pardalis* (ocelot), *Pantera onça* (jaguar), *Pteronura brasiliensis* (giant otter),*Lutra longicaudis* (neotropical river otter), *Myrmecophaga tridactyla* (giant anteater), *Priodontes giganteus* (giant armadillo), *Tolypuetes tricinctus* (threebanded armadillo), *Ozotocerus bezoarticus* (pampas deer), *Blastocerus dichotomus* (swamp deer). Furthermore, thousands of skins from caimans, *Caiman yacare*, and snakes, *Constrictor constrictor*, are exported to the USA and Europe, mainly Germany.

Furthermore, there is a considerable export of live animals, but data on live animal trade are not available or the volume is greatly underestimated. Brazilian authorities have indicated that there is a problem controlling smuggling of animal skins and live animals across the border between Brazil, Paraguay, and Bolivia. There is an urgent need for a major allocation of funds and human resources for better control and protection as well as improved protective legislation in the importing countries to eliminate the black market for skins and endangered animals.

In the past, large scale pollution by domestic and industrial wastes has been of minor importance since population densities were low, except around large cities. This situation is now changing due to the rapidly growing exploitation of mineral resources such as petroleum, coal, gold, iron, aluminium, and tin. As a by-product of gold mining, many tons of mercury are released every year due to the primitive methods of extraction into the atmosphere and the water in the Amazon basin and the Pantanal.

There is little information available about the pathways of metallic mercury within tropical ecosystems. However, accumulation of mercury in aquatic and terrestrial organisms is locally observed in various parts of Brasil (e.g. in the Madeira River and in the Pantanal). Alho *et al.* (1988) mentioned that kingfishers and raptors in the Pantanal near Poconé were affected. Recent studies of Nogueira (pers. comm.) in the same area confirm the findings for other organisms. Lacerda *et al.* (1988, 1990) give data about increased Hg-concentrations in fishes of various Brazilian rivers.

In addition to causing the direct toxic effects of mercury, gold mining also releases huge amounts of sediments into the rivers. The same is true for other large mining projects, for example, the aluminium and tin production. Heavy rains favour erosion and the transport of sediment into the aquatic systems. This was demonstrated on May 1, 1987 in a cassiterite mine north of Manaus. During a torrential rain storm, 96 mm of precipitation fell during 12 hours. This great amount of water destroyed the dams of the settling ponds, liberating thousands of cubic meters of sediments into a branch of the Alalaú River, which is located in the Rio Negro catchment area.

Further increases in the sediment load result from large scale agricultural activities, such as soy bean farming in Mato Grosso, Brasil, because erosion control measures are not effectively employed. The ecological and economic impacts of increased sediment load on riparian wetlands in tropical South America are still largely unknown but are likely to create very serious problems throughout the region. In Amazonian lowland streams, an increase in the amount of sediments results in a destruction of the riparian vegetation, destabilizes the stream bed, and leads to increased erosion during heavy rainstorms. The flooding destroys roads and bridges as observed in Mato Grosso, Brasil (Junk, unpublished). This affects the lotic ecosystem and other downstream wetlands. In the Pantanal of Mato Grosso, destruction of the floodplain forest due to increased sediment input from tributaries has been observed locally. Junk and Mello (1986) suppose that the useful life expectancy of many reservoirs for the generation of electric energy will be much shorter than previously calculated due to increased sediment transport of the tributaries.

Coastal wetlands are endangered by various human activities such as mining, diversion of fresh water, forest exploitation, development of agriculture and aquaculture, coastal development, and solid and liquid waste disposal. IUCN (1983) specifically mentioned the opening up of 5,000 km^2 of mangrove forest in the Orinoco Delta due to the clearing of timber and the establishment of more than 400 km^2 of aquaculture projects for penaeid shrimps. Large areas of coastal wetlands of the Guyanas have been reclaimed for agricultural purposes.

High Andean wetlands are suffering from increasing land reclamation for agriculture due to the rising population density. There are, however, no quantitative data about the impact of the changes. A major technological project may threaten the wetlands of the Titicaca Basin. As early as 1906, Guarini indicated the possibility of utilizing water from Lake Titicaca for energy production and for irrigation of the water-deficient areas along the Pacific coast. The project was again discussed in 1954 at the world-energy conference in Petropolis, Brasil (Forti 1953, Harnecker *et al.* 1954). Another alternative planned in 1929 proposed the drainage of surplus water into the Amazon basin (Monheim 1956). These projects take into account a lowering of the water level of Lake Titicaca from 5 to 7 m. This would reduce the lake surface by about 1,500 to 1,700 km^2 and would affect the productive Totora swamps. The export of water would affect the hydrology and salinity of the adjacent Lake Poopó and would probably affect the climate and the wetlands of the whole endorheic Titicaca basin. Currently, the project is not in an active stage of planning due to political and economic problems.

Recommendations

General recommendations concerning the protection of the wetlands in tropical South America are difficult to make since the wetlands suffer many different problems of different origin in the various countries of the subcontinent. Two general recommendations, however, can be made:

1) There is an urgent need for more basic information about most of the regions' wetlands. It is very hard for ecologists to argue with politicians and decision makers because of an insufficient data base.
2) There is an urgent need to compile a set of plans for the ecologically, sociologically, economically, technically, and politically viable alternatives to development schemes which include the destruction of wetlands. Any proposal based upon purely scientific or esthetic arguments has little or no chance of success over the long term.

The future of the wetlands of tropical South America will, to a considerable extent, depend upon the effectiveness and the speed by which these recommendations can be realized.

References

Ab'Saber, A. N. (1988) O Pantanal Mato-Grossense e a teoria dos refúgios. Revista Brasileira de Geografia 50: 9–57.

Acosta Solís, M. (1966) Las divisiones fitogeográficas y las formaciones geobotánicas del Ecuador. Revista Academia Colombiana 12: 401–447.

Adamoli, J. (1981) O Pantanal e suas relações fitogeográficas com os cerrados. Discussão sobre o conceito "Complexo do Pantanal". Congresso Nacional de Botânica, Teresina, 1981. Sociedade Brasileira de Botânica 32: 109–119.

Adamoli, J. (1986) A dinâmica das inundações no Pantanal. Anais do Primeiro Simpósio sobre Recursos Naturais e Sócio-econômicos do Pantanal. Corumbá 1984. Departamento de Difusão de Technologia, Documento No. 5, EMBRAPA, Brasília, Brazil: 51–61.

Adis, J. (1984) "Seasonal Igapó"-forest of Central Amazonian blackwater rivers and their terrestrial arthropod fauna. p. 246–268. *In*: H. Sioli (ed.). The Amazon: Limnology and Landscape Ecology of a Mighty Tropical River and its Basin. Dr. W. Junk Publisher, Dordrecht, The Netherlands.

Adis, J. (in press) Überlebensstrategien terrestrischer Invertebraten in Überschwemmungswäldern Zentralamazoniens. Habilitationsschrift, Universität Kiel, Germany.

Adis, J., Furch, K. and Irmler, U. (1979) Litter production of a central amazonian blackwater inundation forest. Tropical Ecology 20: 236–245.

Alho, C. J. R., Lacher, T. E. and Gonçalves, H. C. (1988) Environmental degradation in the Pantanal ecosystem. Bioscience 38: 164–171.

Alvarenga, S. M., Brasil, A. E., Pinheira, R. and Kux, H. J. H. (1984) Estudo geomorfológico aplicado à Bacia do Alto Rio Paraguai e Pantanais Matogrossenses. Boletim Técnico Projeto RADAM/BRASIL. Série Geomorfologia, Salvador, 187: 89–183.

Amaral Filho, Z. P. (1986) Solos do Pantanal Matogrossense. Anais do Primeiro Simposio sobre Recursos Naturais e Socio-econômicos do Pantanal. Corumbá 1984. Departamento de Difusao de Tecnologia, Documento No. 5, EMBRAPA, Brasilia, Brazil: 29–42.

Anonymous (1972) Regenwasseranalysen aus Zentralamazonien, ausgeführt in Manaus, Amazonas von Dr. Harald Ungemach. Amazoniana 3: 186–198.

Arias, P. A. (1975) Contribucion al conocimiento limnológico de Guarinocito y su relacion con el rio Magdalena. Thesis, Bogotá, Fundación University de Bogotá, Columbia.

Barco, M. V. and Villarreal, M. Z. (1989) Some ecological aspects and present state of the fishery of the Magdalena river basin, Columbia, South America. p. 409–421. *In*: D. P. Dodge (ed.). Proceedings of the International Large River Symposium (LARS). Canadian Special Publication of Fisheries and Aquatic Science 106.

Barros, A. M., Silva, R. H., Cardoso, O. R. F. A., Freire, F. A., Souza Jr., J. J., Rivetti, M., Luz, D. S., Palmeira, R. C. B. and Tassinari, C. C. G. (1982) Geologia. p. 25–175. *In*: Brasil. Ministério das Minas e Energia. Projeto RADAM/BRASIL. Folha SD, 21 Cuiabá. Levantamento de Recursos Naturais 26, Rio de Janeiro, Brasil.

Bayley, P. B. (1981) Fish yield from the Amazon in Brazil. Comparison with African river yields and management possibilities. Transactions of the American Fisheries Society 100: 351–359.

Bayley, P. B. (1982) Central Amazon fish populations: Biomass production and some dynamic characteristics. Ph. D. thesis, Dalhousie University, Halifax, Canada. 308 pp.

Bayley, P. B. and Petrere Jr., M. (1989) Amazon fisheries: Assessment methods, current status, and management options. p. 385–398. *In*: D. P. Dodge (ed.). Proceedings of the International Large River Symposium (LARS). Canadian Special Publication of Fisheries and Aquatic Science 106: 385–398.

Beard, J. S. (1953) The savanna vegetation of northern tropical America. Ecological Monographs 23: 149–215.

Beck, L. (1976) Zum Massenwechsel der Makro-Arthropodenfauna des Bodens in Überschwemmungswäldern des zentralen Amazonasgebietes. Amazoniana 6: 1–20.

Beck, S. (1983) Vegetationsökologische Grundlagen der Viehwirtschaft in den Überschwemmungs-Savannen des Rio Yacuma (Departamento Beni, Bolivien). Dissertationes Botanicae 80, Ganter Verlag, Vaduz, Schweiz. 214 pp.

Beek, K. J. and Bramão, D. L. (1968) Nature and geography of South American soils. p. 82–112. *In*: E. J. Fittkau, J. Illies, H. Klinge, G. H. Schwabe, and H. Sioli (eds.). Biogeography and Ecology in South America. W. Junk, The Hague, The Netherlands.

Benoist, R. (1924) La végétation de la Guyane Française. Bulletin Société Botanique de France 71: 1169–1177.

Blanco, L. J. and Sanchez, L. (1984) Contribución al estudio taxonómico de las diatomeas del Orinoco medio, bajo Caroni y algunas lagunas de inundación (Venezuela). Mémoires Société Sciences Naturelles La Salle, Caracas: 35–94.

Blydenstein, J. (1967) Tropical savanna vegetation of the Llanos of Colombia. Ecology 48: 1–15.

Bonetto, A. A. and Castello, H. P. (1985) Pesca y piscicultura en aguas continentales de America Latina. Secretaria General de los Estados Americanos Programa Regional de Desarollo Cientifico y Technológico, Washington, DC, USA. 118 pp.

Bongers, F. and Engelen, D. (1982) Some aspects of the structure and ecology of an Amazon Caatinga forest in southern Venezuela. Report Department Geobotany, University Nijmegen, The Netherlands. 58 pp.

Bongers, F., Engelen, D. and Klinge, H. (1985) Phytomass structure of natural communities on spodosols in southern Venezuela: the Bana woodland. Vegetatio 63: 13–34.

Boyé, M. (1962) Les paletuviers de littoral de la Guyane Francaise: ressources et problèmes d'exploitation. Cahiers Outre-mer 15: 271–290.

Bremer, H. (1972) Flußarbeit, Flächen- und Stufenbildung in den feuchten Tropen. Zeitschrift für Geomorphologie N. F. Supplement Bd. 14: 21–38.

Brokx, P. A. (1972) A Study of the Biology of the Venezuelan Whitetailed Deer (*Odocoileus virginianus gymnotis* Wiegmann 1833) with a Hypothesis on the Origin of American Cervids. Ph. D. thesis, University Waterloo, Canada. 355 pp.

Brown Jr., K. Z. (1986) Zoogeografia da região do Pantanal Matogrossense. Anais do Primeiro Simpósio sobre Recursos Naturais e Sócio-econômicos do Pantanal. Corumbá 1984. Departamento de Difusão de Technologia, Documento No. 5, EMBRAPA, Brasília, Brazil: 137–178.

Camargo, F. (1949) Reclamation of the Amazonian flood-lands near Belém. UNSCCUR Proceedings, Land resources: 598–602.

Carvalho, J. C. de M. (1984) The conservation of nature in the Brazilian Amazonia. p. 707–736. *In*: H. Sioli (ed.). The Amazon: Limnology and Landscape Ecology of a Mighty Tropical River and its Basin. Dr. W. Junk Publisher, Dordrecht, The Netherlands.

Chapman, V. J. (1975) Mangrove Vegetation. Cramer, Lehre. 425 pp.

Church, G. (1898) Argentine geography and the ancient Pampean sea. Geography Journal 12: 386–401.

Cleef, A. M. (1981) The vegetation of the Paramos of the Colombian Cordillera Oriental. Dissertationes Botanicae 61, Ganter Verlag, Vaduz, Switzerland. 320 pp.

Cochrane, T. T. (1973) El potential agrícola del uso de la tierra en Bolivia. Un mapa de sistema de tierras. Don Bosco, La Paz, Bolivia. 826 pp.

Conceição, C. A. and Paula, J. E. (1986) Contribuição para o conhecimento da flora do Pantanal Matogrossense e sua relação com a fauna e o homen. Anais do Primeiro Simpósio sobre Recursos Naturais e Sócio-econômicos do Pantanal. Corumbá 1984. Departamento de Difusão de Technologia, Documento No. 5, EMBRAPA, Brasília, Brazil: 107–130.

Del'Arco, J. D., Silva, R. H., Rarapanoff, I., Freire, F. A., Pereira, L. G. M., Souza, S. L., Luz, D. S., Palmeira, R. C. B. and Tassinari, C. C. G. (1982) Geologia. p. 25–175. *In*: Brasil. Ministério das Minas e Energia. Projeto RADAM/BRASIL. Folha SE, 21 Corumbá e parte da folha SE 20. Levantamento de Recursos Naturais 27, Rio de Janeiro, Brasil.

Denevan, W. M. (1966) The aboriginal cultural geography of the Llanos de Mojos of Bolivia. Ibero Amerincana (Berkeley) 48. 160 pp.

Devol, A. H., Richey, B. R., Forsberg, B. R. and Martinelli, L. A. (1990) Seasonal dynamics in methane emissions from the Amazon river to the troposphere. Journal of Geophysical Research 95: 16, 417–16,426.

Dister, E. (1983) Anthropogene Wasserstandsänderungen in Flußauen und ihre ökologischen Folgen: Beispiele vom Oberrhein und vom Rio Magdalena (Kolumbien). Verhandlungen der Gesellschaft für Ökologie 11: 89–100.

Dister, E. and Lozano, L. C. G. (1984) Ökologische Aspekte beim Ausbau des Rio Magdalena/Kolumbien. Biogeographica 19: 41–56.

Donselaar, J. van (1965) An ecological and phytogeographical study of north Suriname savannas. Wentia 14: 1–163.

Eden, M. J. (1970) Savanna vegetation in the Northern Rupununi, Guayana. Journal of Tropical Geography 30: 17–28.

Ertel, J. R., Hedges, J. I., Devol, A. H. and Richey, J. E. (1986) Dissolved humic substances of the Amazon river system. Limnology and Oceanography 31: 739–754.

ERTS (1978) Mapa de cobertura y uso actual de la tierra, Bolivia. Memoria explicativa. GEOBOL (Serie sensores remotos, La Paz) 2. La Paz, Bolivia. 116 pp.

Escobar, A. (1977) Estudio de las sabanas inundables de *Paspalum fasciculatum*. M. Sc. thesis, IVIC.

Falesi, I. C., Rodrigues, T. E., Morikawa, I. K. and Reis, R. S. (1971) Solos do distrito agropecuário da SUFRAMA (Trecho km 30–km 79, Rod. BR-174). Instituto de Pesquisa e Experimentação Agropecuária da Amazônia Ocidental (IPEAAOc), Serie Solos I. 99 pp.

Fanshawe, D. B. (1952) The Vegetation of British Guiana. Imperial Forestry Institute Paper No 29. Oxford University Press, London, England. 96 pp.

FAO (1966) Reconocimento edafologico de los Llanos Orientales, Colombia. III La vegetacion natural y la ganaderia en los Llanos Orientales. FAO, Rome, Italy. 233 pp.

Fisher Jr., T. J. and Parsley, P. E. (1979) Amazon lakes: water storage and nutrient stripping by algae. Limnology and Oceanography 24: 547–553.

Fittkau, E. J. (1964) Remarks on limnology of central Amazon rain-forest streams. Verhandlungen der Internationalen Vereinigung für Limnologie 15: 1092–1096.

Fittkau, E. J. (1967) On the ecology of Amazonian rain-forest streams. Atas do Simpósio sobre a Biota Amazônica 3 (Limnologia): 97–108.

Fittkau, E. J. (1973) Artenmannigfaltigkeit amazonischer Lebensräume aus ökologischer Sicht. Amazoniana 4: 321–340.

Fittkau, E. J. (1976) Kinal und Kinon, Lebensraum und Lebensgemeinschaft der Oberflächendrift am Beispiel amazonischer Fließgewässer. Biogeographica 7: 101–113.

Fittkau, E. J. (1982) Struktur, Funktion und Diversität zentralamazonischer Ökosysteme. Archiv für Hydrobiologie 95: 29–45.

Forsberg, B. R., Devol, A. H., Richey, J. E., Martinelli, L. A. and Santos, H. M. (1988) Factors controlling nutrient concentrations in Amazon floodplain lakes. Limnology and Oceanography 33: 41–56.

Forti, A. (1953) Un interesante caso hydrológico y un grandioso aprovechamiento sudamericano de fuerza motriz hidraulica. G. Kraft, Buenos Aires, Argentina. 152 pp.

Freyburg, G. V. (1930) Zerstörung und Sedimentation an der Mangroveküste Brasiliens. Leopoldina 6: 69–117.

Furch, B. (1984) Untersuchungen zur Überschwemmungstoleranz von Bäumen der Várzea und des Igapó. Blattpigmente. Biogeographica 19: 77–83.

Furch, K. (1976) Haupt- und Spurenmetallgehalte zentralamazonischer Gewässertypen. Biogeographica 7: 27–43.

Furch, K. (1984) Seasonal variations of the major cation content of the várzea-lake Lago Camaleão, middle Amazon, Brasil in 1981 and 1982. Verhandlungen der Internationalen Vereinigung für Limnologie 22: 1288–1293.

Furch, K. (1986) Hydrogeochemie von Fließgewässern im Bereich der Transamazonica (Nordbrasilien). Amazoniana 9: 371–409.

Furch, K. and Junk, W. J. (1980) Water chemistry and macrophytes of creeks and rivers in Southern Amazonia and the Central Brazilian Shield. p. 771–796. *In*: J. I. Furtado, (ed.). Tropical Ecology and Development. Proceedings of the Vth International Symposium of Tropical Ecology, 16–21 April 1979, Kuala Lumpur, Malaysia. Kuala Lumpur.

Furch, K. and Junk, W. J. (1980) Water chemistry and macrophytes of creeks and rivers in southern Amazonia and the Central Brasilian Shield. Tropical Ecology and Development: 171–196.

Furch, K. and Klinge, H. (1978) Towards a regional characterization of the biogeochemistry of alkali- and alkali-earth metals in northern South America. Acta Cientifica Venezolana 29: 434–444.

Furch, K., Junk, W. J., Dieterich, J. and Kochert, N. (1983) Seasonal variation in the major cation (Na, K, Mg and Ca) content of the water of Lago Camaleão, an Amazonian floodplain-lake near Manaus, Brazil. Amazoniana 8: 75–89.

Furch, K., Junk, W. J. and Campos, Z. E. S. (1989) Nutrient dynamics of decomposing leaves from Amazonian floodplain forest species in water. Amazoniana 11: 91–116.

Geijskes, D. C. (1945) Plants of the Surinam coastland. Journal New York Botanical Garden 46: 229–236.

Geisler, R., Knöppel, H. A. and Sioli, H. (1971) Ökologie der Süßwasserfische Amazoniens. Stand und Zukunftsaufgaben der Forschung. Naturwissenschaften 58: 303–311.

Géry, J. (1969) The fresh-water fishes of South America. p. 828–848. *In*: E. J. Fittkau, J. Illies, H. Klinge, G. H. Schwabe, and H. Sioli (eds.). Biogeography and Ecology in South America. Dr. W. Junk, Publisher, La Haya, The Netherlands.

Gessner, F. (1955) Die limnologischen Verhältnisse in den Seen und Flüssen von Venezuela. Verhandlungen der Internationalen Vereinigung für Limnologie 12: 284–295.

Gessner, F. (1965) Zur Limnologie des unteren Orinoco. Internationale Revue der gesamten Hydrobiologie 50: 305–333.

Gessner, F. (1967) Untersuchungen an der Mangrove in Ost-Venezuela. Internationale Revue der gesamten Hydrobiologie 52: 769–781.

Gessner, F. and Hammer, L. (1967) Limnologische Untersuchungen an Seen der venezolanischen Hochanden. Internationale Revue gesamten Hydrobiologie 52: 301–320.

Gibbs, R. J. (1967) The geochemistry of the Amazon river system. I. The factors that control the salinity and the composition and concentration of the suspended solids. Geological Society of America, Bulletin 78: 1203–1232.

Gilson, H. C. (1964) Lake Titicaca. Verhandlungen der Internationalen Vereinigung für Limnologie: 112–127.

Godoi Filho, J. D. (1986) Aspectos geológicos do Pantanal Matogrossense e de sua área de influência. Anais do Primeiro Simpósio sobre Recursos Naturais e Sócio econômicos do Pantanal. Corumbá 1984. p. 63–76. *In*: Departamento de Difusão de Tecnologia, Documento No. 5, EMBRAPA, Brasília, Brazil.

Godt, M. C. (1985) Ökologische Untersuchungen zur Verbreitung der Mangrove in Peru. M. Sc. thesis University Bielefeld, Germany. 145 pp.

Gonzáles-Jiménez, E. (1979) Primary and secondary productivity in flooded savannas. p. 620–625. *In*: UNESCO/UNEP/FAO Tropical Grazing Land Ecosystems. Paris, France.

Goodland, R. (1978) Environmental assessment of the Tucurui hydroproject Rio Tocantins, Amazonia, Brazil. Eletronorte, S.A. Brasilia, D.F., Brazil. 168 pp.

Gottsberger, G. (1978) Seed dispersal by fish in the inundated regions of Humaitá, Amazonia. Biotropica 10: 170–183.

Goulding, M. (1981) The Fishes and the Forest. Explorations in Amazonian Natural History. University of California Press, Berkeley, California, USA. 280 pp.

Goulding, M., Carvalho, M. L. and Ferreira, E. G. (1988) Rio Negro, Rich Life in Poor Water. SPB Academic Publishing, The Hague, The Netherlands. 200 pp.

Gourou, P. (1950) Observações geográficas na Amazonia. Revista Brasileira de Geografia 11: 355–408.

Guerra, A. T. (1957) Estudo geográfico do Territorio do Rio Branco. Biblioteca Geográfica Brasileira 13 (Serie A), I.B.G.E. Rio de Janeiro, Brasil. 252 pp.

Gutte, P. (1980) Beitrag zur Kenntnis zentralperuanischer Pflanzengesellschaften II. Die Hochandinen Moore und ihre Kontaktgesellschaften. Feddes Repertorium 91: 327–336.

Haase, R. (1989) Plant communities of a savanna in northern Bolivia I. Seasonally flooded grassland and gallery forest. Phytocoenologia 18: 55–81.

Haase, R. (1990a) Plant communities of a savanna in northern Bolivia II. Palm swamps, dry grassland, and shrubs. Phytocoenologia 18: 343–370.

Haase, R. (1990b) Community composition and soil properties in northern Bolivian savanna vegetation. Journal of Vegetation Science 1: 345–352.

Haase, R. and Beck, S. G. (1989) Structure and composition of savanna vegetation in northern Bolivia: A preliminary report. Brittonia 41: 80–100.

Harling, G. (1979) The vegetation types of Ecuador – a brief survey. p. 165–174. *In*: K. Larsen, and L. B. Holm-Nilsen (eds.). Tropical Botany. Academic Press, New York, New York, USA.

Harnecker, R., Salazar, R. and Santa Maria, D. (1954) Los principios técnicos y económicos do aprovechamiento internacional de las aguas del lago Titicaca. Panorama Económico, Santiago de Chile: 702–708.

Hedges, J. I., Clark, W. A., Quay, P. D., Richey, J. E., Devol, A. H. and Santos, H. M. (1986) Composition and fluxes of particulate organic material in the Amazon river. Limnology and Oceanography 31: 717–738.

Hegewald, E., Aldave, A. and Hakulit, T. (1976) Investigations of the lakes of Peru and their phytoplankton 1. Review of literature, description of the investigated waters and chemical data. Archiv für Hydrobiologie 78: 494–506.

Herrera, R. (1979) Nutrient Distribution and Cycling in an Amazon Caatinga Forest on Spodosols in Southern Venezuela. Ph. D. thesis, University Reading, England. 241 pp.

Harnecker, R., Salazar, R. and Santa Maria, d. (1954) Los Principios Técnicos y Económicos des Aprovechamiento Internacional de los Aguas del Lago Titicaca. Panorama Económico, Santiago de Chile: 702–708.

Herrera, R., Medina, E., Klinge, H., Jordan, C. F. and Uhl, C. (in press) Nutrient retention mechanisms in tropical forests: The Amazon Caatinga, San Carlos Pilot Project, Venezuela. *In*: F. di Castri, F. W. G. Baker, and M. Hadley (eds.). Ecology in Practice. Establishing a Scientific Basis for Land Management. Tycooly, Dublin, Ireland.

Hoehne, F. C. (1923) Phytophysionomia do Estado de Mato-Grosso. Companía Melhoramentos, São Paulo, Brazil: 53–58.

Hoehne, F. C. (1936) O grande Pantanal de Mato Grosso. Bolivian Agriculture, São Paulo, Brazil: 443–470.

Hofmann, R. K., Ponce del Prado, C. F. and Otte, K. C. (1976) Registrato de dos nuevas especies de mamiferos para el Peru *Odocoileus dichotomus* (Illiger 1811) y *Chrysocyon brachyurus* (Illiger 1811), con notas sobre su habitat. Revista Forestal Peru, 5: 61–81.

Huber, J. (1903) Contribuição á geographia physica dos Furos de Breves e da parte occidental de Marajó. Boletin do Museo Paraense de Historias Naturais 3: 447–498.

Huber, O. (1982a) Esbozo de los formaciones vegetales del Territorio Federal Amazonas, Venezuela. Ministerio del Ambiente y de los Recursos Naturales Renovables, Caracas. Serie Informe Técnico, DGSIIA/IT/103. 36 pp.

Huber, O. (1982b) Significance of savanna vegetation in the Amazon territory of Venezuela. p. 221–243. *In*: G. T. Prance, (ed.). Biological Diversification in the Tropics. Columbia University Press, New York, U.S.A.

Huber, O. and Alarcon, C. (1988) Mapa de vegetacion de Venezuela. Ministerio del Ambiente y de los recursos naturales renovables, Division de Vegetacion, Caracas, Venezuela.

Hueck, K. (1966) Die Wälder Südamerikas. Gustav Fischer Verlag, Stuttgart, Germany. 422 pp.

Irion, G. (1982) Mineralogical and geochemical contribution to climatic history in Central Amazonia during Quarternary time. Tropical Ecology 23: 76–85.

Irion, G. (1984) Sedimentation and sediments of Amazonian rivers and evolution of the Amazonian landscape since Pliocene times. p. 201–214. *In*: H. Sioli (ed.). The Amazon: Limnology and Landscape Ecology of a Mighty Tropical River and its Basin. Dr. W. Junk Publisher, Dordrecht, The Netherlands.

Irmler, U. (1975) Ecological studies of the aquatic soil invertebrates in three inundation forests of Central Amazonia. Amazoniana 5: 337–409.

Irmler, U. (1976) Zusammensetzung, Besiedlungsdichte und Biomasse der Makrofauna des Bodens in der emersen und submersen Phase zentralamazonischer Überschwemmungswälder. Biogeographica 7: 79–99.

Irmler, U. (1977) Inundation forests types in the vicinity of Manaus. Biogeographica 8: 17–29.

Irmler, U. (1979a) Considerations on structure and function of the "Central Amazonian inundation forest ecosystem" with particular emphasis on selected soil animals. Oecologia 43: 1–18.

Irmler, U. (1979b) Abundance fluctuations and habitat changes of soil beetles in Central Amazonian inundation forests (Coleoptera: Carabidae, Staphylinidae). Studies on Neotropical Fauna and Environment 14: 1–16.

Irmler, U. (1981) Überlebensstrategien von Tieren im saisonal überfluteten Überschwemmungswald. Zoologischer Anzeiger Jena 206: 26–38.

Irmler, U. (1985) Temperature dependant generationcycle for the cicindelid beetle *Pentacomia egregia* CHAUD. (Coleoptera, Carabidae, Cicindelinae) of the Amazon valley. Amazoniana 9: 431–439.

IUCN (1983) Global status of mangrove ecosystems. P. Saenger, E. J. Hegerl, and J. D. S. Davie (eds.). Commission on Ecology Paper 3. 88 pp.

Janssen, A. (1986) Flora und Vegetation der Savannen von Humaitá und ihre Standortbedingungen. Dissertationes Botanicae 93, J. Cramer, Berlin-Stuttgart, Germany. 325 pp.

Junk, W. J. (1976) Faunal ecological studies in inundated areas and the definition of habitats and ecological niches. Animal Research and Development 4: 47–54.

Junk, W. J. (1980) Areas inundáveis – um desafio para limnologia. Acta Amazonica 10: 775–795.

Junk, W. J. (1982a) Amazonian floodplains: Their ecology, present and potential use. Rev. Hydrobiol. Trop. 15: 285–301.

Junk, W. J. (1982b) Zur Entwicklung aquatischer Makrophyten in Curuá-Una, dem ersten Stausee in Zentralamazonien. Archiv für Hydrobiologie 95: 169–180.

Junk, W. J. (1983) Ecology of swamps on the Middle Amazon. p. 269–294. *In*: A. J. P. Gore (ed.). Ecosystems of the World. Mires, Swamp, Bog, Fen and Moor. 3 Regional studies. Elsevier Publication Company, Amsterdam, The Netherlands.

Junk, W. J. (1984a) Ecology of the várzea, floodplain of Amazonian whitewater rivers. p. 215–243. *In*: H. Sioli (ed.). The Amazon: Limnology and Landscape Ecology of a Mighty Tropical River and its Basin. Dr. W. Junk Publisher, Dordrecht, The Netherlands.

Junk, W. J. (1984b) Ecology, fisheries and fishculture in Amazonia. p. 443–476. *In*: H. Sioli (ed.). The Amazon: Limnology and Landscape Ecology of a Mighty Tropical River and its Basin. Dr. W. Junk Publisher, Dordrecht, The Netherlands.

Junk, W. J. (1985) The Amazon floodplain – a sink or source for organic carbon. p. 267–283. *In*: E. T. Degens, S. Kempe and R. Herrera (eds.). Transport of Carbon and Minerals in Major World Rivers. Mitteilungen aus dem Geologisch-Palaontologischen Institute der Universitat Hamburg, Heft 58 SCOPE/UNEP, Sonderband, Germany.

Junk, W. J. (1986) Temporary fat storage, an adapatation of some fish species to the waterlevel fluctuations and related environmental changes of the Amazon river. Amazoniana 9: 315–351.

Junk, W. J. (1989a) The use of Amazonian floodplains under ecological perspective. Interciencia 14: 317–322.

Junk, W. J. (1989b) Flood tolerance and tree distribution in central Amazonian floodplains. p. 41–64. *In*: L. B. Holm Nielsen, I. Nielsen, and H. Balslev (eds.). Tropical Forests: Botanical Dynamics, Speciation and Diversity. Academic Press, London, England.

Junk, W. J. (1990) Die Krautvegetation der Überschwemmungsgebiete des Amazonas (Várzea) bei Manaus und ihre Bedeutung für das Ökosystem. Habilitationsschrift, Universität Hamburg, Germany. 349 pp.

Junk, W. J. and Furch, K. (1984) The physical and chemical properties of Amazonian waters and their relationships with the biota. p. 3–17. *In*: G. T. Prance, and T. E. Lovejoy (eds.). Key Environments: Amazonia. Pergamon Press, Oxford, England.

Junk, W. J. and Howard-Williams, C. (1984) Ecology of aquatic macrophytes in Amazonia. p. 269–293. *In*: H. Sioli (ed). The Amazon: Limnology and Landscape Ecology of a Mighty Tropical River and its Basin. Dr. W. Junk, Publisher, The Hague, The Netherlands.

Junk, W. J. and Mello de, J. A. S. N. (1986) Impactos ecológicos das represas hidrelétricas na Amazonia brasileira. Tübinger Geographische Studien 95: 367–385.

Junk, W. J., Robertson, B. A., Darwich, A. J. and Vieira, J. (1981) Investigações limnológicas e ictiológicas em Curuá-Una, a primeira represa hidrelétrica na Amazônia central. Acta Amazonica 11: 687–716.

Junk, W. J., Soares, M. G. M. and Carvalho, F. M. (1983) Distribution of fish species in a lake of the Amazon river floodplain near Manaus (Lago Camaleão),with special reference to extreme oxygen conditions. Amazoniana 7: 397–431.

Junk, W. J., Bayley, P. B. and Sparks, R. E. (1989) The floodpulse concept in river-floodplain systems. p. 110–127. *In*: D. P. Dodge (ed.). Proceedings of the International Large River Symposium (LARS). Canadian Special Publication of Fisheries and Aquatic Science.

Kalliola, R., Salo, J. and Mäkinen, Y. (1987) Regeneración natural de selvas en la Amazonia peruana 1: Dinamica fluvial y sucesion riberena. Memorias 18B, Universidad Nacional Mayor de San Marcos, Lima, Peru.

Kaufmann, R. and Hevert, F. (1973) El regimen pluviométrico del Rio Magdalena y su importancia para la ciénaga Grande de Santa Marta. Mitt. Inst. Colombo/Alemán, Invest. Cient. 7: 121–137.

Klinge, H. and Herrera, R. (1978) Biomass studies in Amazon Caatinga forest in southern Venezuela. Tropical Ecology 19: 93–110.

Klinge, H. and Herrera, R. (1983) Phytomass structure of natural plant communities on spodosols in southern Venezuela: The tall Amazon Caatinga forest. Vegetatio 53: 65–84.

Klinge, H., Medina, E. and Herrera, R. (1977) Studies on the ecology of Amazon Caatinga forest in southern Venezuela. Acta Cientifica Venezulana 28: 270–276.

Klinge, H., Furch, K., Harms, E. and Revilla, J. (1983) Foliar nutrient levels of native tree species from Central Amazonia. I. Inundation forests. Amazoniana 8: 19–46.

Klinge, H., Furch, K. and Harms, E. (1984) Selected bioelements in bark and wood of native tree species from Central-Amazonian inundation forests. Amazoniana 9: 105–117.

Knöppel, H. A. (1970) Food of central Amazonian fishes. Amazoniana 2: 257–352.

Kosswig, C. and Villwock, W. (1965) Das Problem der intralakustrischen Speziation im Titicaca- und Llanaosee. Verhandlungen der Deutschen zoologischen Gessellschaft Kiel 1964 Zoologischer Anzeiger Supplement 28: 95–102.

Kramer, D. L., Lindsey, C. C., Moodie, G. E. E. and Stevens, E. D. (1978) The fishes and the aquatic environment of the central Amazon basin, with particular reference to respiratory pattern. Canadian Journal Zoology 56: 717–729.

Kubitzki, K. (1989) The ecogeographical differentiation of the Amazonian inundation forests. Plant Systematics and Evolution 162: 285–304.

Lacerda, L. D., Pfeiffer, W. C., Ott, A. T., Silveira, E. G. (1988) Mercury contaminations in the Madeira River, Amazon – Hg inputs to the environment. Biotropica 21: 91–93.

Lacerda, L. D., Pfeiffer, W. C., Malm, O., Souza, C. M. M., Silveira, E. G. and Bastos, W. R. (1990) Mercúrio no meio ambiente: risco potencial das áreas garimpeiras no Brasil. Acta Limnológica Brasileira 3: 969–977.

Lanjouw, J. (1936) Studies on the vegetation of the Suriname savannahs and swamps. Nederlandsch Kruidkundig Archief 46: 823–851.

Leenheer, J. A. and Santos, U. M. (1980) Considerações sobre os processos de sedimentação na água preta ácida do Rio Negro (Amazonia Central). Acta Amazônica 10: 343–355.

Leentvaar, P. (1966) The Brokopondo research project, Surinam. p. 33–42. *In*: R. H. Lowe-McConnell (ed.). Man-made Lakes. Academic Press, London, England.

Leentvaar, P. (1973) Further developments in Lake Brokopondo, Surinam. Amazoniana 4: 1–8.

Leentvaar, P. (1979) Additions and corrections to the Brokopondo study (Surinam). Amazoniana 6: 521–528.

Leentvaar, P. (1984) The Brokopondo barrage lake in Suriname, South America, and the planned Kabalebo project in West Suriname. p. 49–56. *In*: J. R. Cord and P. Leentvaar (eds.). Hydroenvironmental Indices: A Review and Evaluation of Their Use in the Assessment of the Environmental Impacts of Water Projects. IHP-D Project A 3.2, UNESCO, Paris, France.

Lewis, W. M. Jr. and Weibezahn, F. H. (1976) Chemistry, energy flow and community structure in some Venezuelan fresh waters. Archiv für Hydrobiologie (Supplement) 50: 145–207.

Lima, R. R. (1965) A agricultura nas várzeas do estuário do Amazonas. Boletim técnico do Instituto Agronômico del Norte 33: 1–164.

Lindemann, J. C. (1953) The vegetation of the coastal region of Suriname. *In*: J. A. de Hulster, J. Lanjouw and F. W. Ostedorf (eds.). The Vegetation of Suriname. Vol. 1 Part 1. Von Eedenfouds, Amsterdam, The Netherlands. 135 pp.

Löffler, H. (1968) Tropical high mountain lakes. Their distribution, ecology and zoogeographical importance. Colloquium Geographicum, Bonn, 9: 57–75.

Lozano, L. C. G. and Dister E. (1990) La planicie de inundacion del medio-bajo Magdalena: restauracion y conservacion de habitats.– Interciencia 15: 396–410.

Loureiro, R. L., Souza Lima, J. P. and Fonzar, B. C. (1982) Vegetação. p. 329–372. *In*: Brasil. Ministério das Minas e Energia. Projeto RADAM/BRASIL. Folha SE, 21 Corumbá e parte da folha SE 20. Levantamento de Recursos Naturais 27, Rio de Janeiro, Brasil.

Lowe-McConnell, R. H. (1964) The fishes of the Rupununi savanna district in British Guiana, South America. I. Ecological groupings of fish species and effects of the seasonal cycle on the fish. Journal of the Linnean Society Zoology 45: 103–144.

Lowe-McConnell, R. H. (1975) Fish Communities in Tropical Freshwater. Longman, London, United Kingdom. 337 pp.

Lugo, A.E. (1980) Mangrove ecosystems: successional or steady state? Biotropica 12, Supplement Tropical Succession: 65–72.

Magalhães, C. and Walter, J. (1984) Desenvolvimento larval e distribuição ecológica de seis espécies de camarões (Decapoda, Palaemonidae) da região amazônica. Resumos do 11 Congresso brasileiro de zoologia, Belém: 92–93.

Mago, F. M. L. (1970) Lista de los peces de Venezuela incluendo un estudio preliminar sobre la ictiogeografía del país. Ministerio de Agricultura y Cría, Oficina Nacional de Pesca, Caracas, Venezuela. 283 pp.

Martinelli, L. A., Victoria, R. L., Devol, A. H., Richey, J. E. and Forsberg, B. R. (1989) Suspended sediment load in the Amazon basin: an overview. Geo-Journal 19: 381–389.

Martyn, E. B. (1934) A note on the foreshore vegetation near Georgetown. Journal of Ecology 22: 292–298.

Medina, E. (1980) Ecology of tropical American savannas: an ecophysiological approach. p. 297–319. *In*: D. R. Harris (ed.). Human Ecology in Savanna Environments. Academic Press Incorporation, London, England.

Medina, E. and Sarmiento, G. (1979) Ecophysiological studies in the *Trachypogon* savanna (central Llanos). p. 612–619. *In*: UNESCO/UNEP/FAO (eds.). Tropical Grazing Land Ecosystems. Paris, France.

Melack, J. M. (1984) Amazon floodplain lakes: Shape, fetch, and stratification. Verhandlungen der Internationalen Vereinigung für Limnologie 22: 1278 1282.

Melack, J. M. and Fisher, T. R. (1983) Diel oxygen variations and their ecological implications in Amazon floodplain lakes. Archiv für Hydrobiologie 98: 422–442.

Melack, J. M. and Fisher, T. R. (1990) Comparative limnology of tropical floodplain lakes with an emphasis on the Central Amazon. Acta Limnologica Brasileira 3: 1–48.

Milliman, J. D. and Emery, K. O. (1968) Sealevels during the past 35,000 years. Science 162: 1121–1123.

Moon, H. P. (1939) The geology and physiography of the Altiplano of Peru and Bolivia. Transaction of the Linnean Society of London: 27–43.

Monheim, F. (1956) Beiträge zur Klimatologie und Hydrologie des Titicacabeckens. Heidelberger geographische Arbeiten 1: 1–152.

Mourão, G. M. (1989) Limnologia comparativa de três lagoas (duas "Baías" e uma "Salina") do Pantanal de Nhecolandia, M. S. Thesis, University São Carlos-SP, Brasil. 135 pp.

Newell, N. D. (1949) Geology of the Lake Titicaca region, Peru and Bolivia. Memoirs of the Geological Society of America 36: 1–111.

Nikonov, M. N. and V. P. Sluka (1964) The distribution of peatbogs. 8th International Congress of Soil Science. Bucharest: 1047–1055.

Nortcliff, S. and Thornes, J. B. (1981) Seasonal variations in the hydrology of a small forested catchment near Manaus, Amazonas, and the implications for its management. p. 31–57. *In*: R. Lal and E. W. Russel (eds.). Tropical Agricultural Hydrology. Watershed Management and Land Use. John Wiley and Sons, Chichester, England.

Novoa, D. F. (1982) Los recursos pesqueiros del rio Orinoco y su explotacion. Corp. Venez. de Guyana, Division de Desarrolo Agricola, Caracas, Venezuela. 386 pp.

Novoa, D. F. (1989) The multispecies fisheries of the Orinoco river: Development, present status, and management strategies. p. 422–428. In D.P. Dodge (ed.). Proceedings of the International Large River Symposium (LARS). Canadian Special Publication of Fisheries and Aquatic Science 106: 422–428.

Novoa, D. F., Ramos, F. and Cartaya, E. (1984) Las pesquerias artesanales del Rio Orinoco. Sector Cainara -Cabruta, Parte 1. Memorias de la Sociedad de Ciencias Naturales La Salle, Caracas: 163–215.

Ojasti, J. (1973) Estudio biologico del chiguire o capibara. Fondo Nacional de Investigaciones Agropecuarias, Caracas, Venezuela. 275 pp.

Ojasti, J. (1978) The Relation Between Population and Production of the Capybara (*Hydrochoerus hydrochaeris*). Ph. D. thesis University Georgia, Athens, Georgia, USA. 204 pp.

Ojasti, J. (1983) Ungulates and large rodents of South America. p. 427–439. *In*: F. Burliere (ed.). Ecosystems of the world 13: tropical savannas. Elsevier Scientific Publisher Company, Amsterdam, The Netherlands.

Paolini, J. (1978) Charakterisierung und Dynamik von Huminstoffsystemen in der amazonischen Caatinga bei San Carlos de Rio Negro, Venezuela. Ph. D. thesis, University Göttingen, Germany. 134 pp.

Paolini, J., Herrera, R. and Németh, A. (1983) Hydrochemistry of the Orinoco and Caroní rivers. Mitteilungen aus dem Geologisch-Paläontologischen Institut der Universität Hamburg. SCOPE/UNEP Sonderband Heft 55: 223–236.

Pannier, F. and Pannier, R. F. (1977) Interpretacion fisioecologica de la distribuicion de manglares en las costas del continente Suramericano. Interciencia, Caracas 2: 153–162.

Patrick, R., Aldrich, F. A., Cairns Jr., J., Drouet, F., Hohn, M. H., Roback, S. S., Skuja, H., Spangler, P. J., Swabey, Y. H. and Whitford, L. A. (1966) The Catherwood Foundation Peruvian-Amazon Expedition: Limnological and Systematic Studies. Monographs. Academy of Natural Sciences of Philadelphia. Philadelphia, Pennsylvania, USA. 495 pp.

Perez, L. E. (1984) Uso del habitat por la comunidad de peces de um rio tropical asociado a um bosque. Memorias de la Sociedad de Ciencias Naturales La Salle, Caracas: 143–162.

Petrere, Jr. M. (1978a) Pesca e esforço no Estado do Amazonas I. Esforço, e captura por unidade de esforço. Acta Amazonica 8: 439–454.

Petrere, Jr. M. (1978b) Pesca e esforço no Estado do Amazonas. II. Locais, aparelhos de captura e estatística de desembarque. Acta Amazonica 8, Supplement 2. 54 pp.

Piedade, M. T. F. (1988) Biomassa, produtividade e atividade fotossintética de *Echinochloa polystachya* Hitchkock (Graminea = Poaceae), capim semi-aquático da várzea Amazônica. Ph. D. Thesis, INPA/FUA, Manaus, Amazonas, Brasil. 148 pp.

Pinowski, J. and Morales, C. (1981) Aspectos ecologicos das aves de los modulos de Apure. Boletín de la Sociedad Venezolana de Ciencias Naturales 36: 67–78.

Pires, J. M. (1966) The estuaries of the Amazon and Oiapoque rivers and their floras. p. 211–218. *In*: UNESCO. Scientific problems of the humid tropical zone deltas and their implications. Proceedings Dacca Symposium.

Plafker, G. (1964) Oriented lakes and lineaments of Northeastern Bolivia. Geological Society of America, Bulletin 75: 503–522.

Prance, G. T. (1979) Notes on the vegetation of Amazonia III. The terminology of Amazonian forest types subject to inundation. Brittonia 31: 26–38.

Prance, G. T. and Schaller, G. B. (1982) Preliminary study of some vegetation types of the Pantanal, Mato Grosso, Brazil. Brittonia 34: 228–251.

Prance, G. T., Rodrigues, W. A. and Silva da, M. F. (1976) Inventário florestal de um hectare de mata de terra firme km 30 da estrada Manaus-Itacoatiara. Acta Amazonica 6: 9–35.

Putzer, H. (1984) The geological evolution of the Amazon basin and its mineral resources. p. 15–46. *In*: H. Sioli (ed.). The Amazon: Limnology and Landscape Ecology of a Mighty Tropical River and its Basin. Dr. W. Junk Publisher, Dordrecht, The Netherlands.

RADAM/Brasil (1974a): Levantamento de recursos naturais Vol. 5. (1974b): Levantamento de recursos naturais Vol. 6. (1975): Levantamento de recursos naturais Vol. 8. (1978): Levantamento de recursos naturais Vol. 18. (1981a): Levantamento de recursos naturais Vol.

22. (1981b): Levantamento de recursos naturais Vol. 25. Ministério das Minas e Energia, Departamento Nacional de Produção Mineral, Rio de Janeiro, Brazil.

Rai, H. and G. Hill (1984) Primary production ecosystem. p. 311–335. *In*: H. Sioli (ed.). The Amazon: Limnology and Landscape Ecology of a Mighty Tropical River and its Basin. Dr. W. Junk Publisher, Dordrecht, The Netherlands.

Ramia, M. (1959) Las savannas de Apure. Ministerio de Agricultura y Cría, Caracas, Venezuela. 134 pp.

Ramia, M. (1967) Tipos de sabanas en los Llanos de Venezuela. Boletín de la Sociedad Venezolana de Ciencias Naturales 27: 264–288.

Ramia, M. (1974) Plantas de las savannas llaneras. Monte Avila Editores, Caracas, Venezuela. 287 pp.

Ramos, S., Danielewski, S. and Colomine, G. (1981) Contribuicion a la ecologia de los vertebrados acuaticos en esteros y bajios de sabanas moduladas. Boletín de la Sociedad Venezolana de Ciencias Naturales 36: 79–103.

Reiss, F. (1973) Zur Hydrographie und Makrobenthosfauna tropischer Lagunen in den Savannen des Territorio de Roraima, Nordbrasilien. Amazoniana 4: 367–378.

Reiss, F. (1976a) Die Benthoszoozönosen zentralamazonischer Várzea-een und ihre Anpassungen an die jahresperiodischen Wasserstandsschwankungen. Biogeographica 7: 125–135.

Reiss, F. (1976b) Charakterisierung zentralamazonischer Seen aufgrund ihrer Makrobenthosfauna. Amazoniana 6: 123–134.

Revilla, J. D. (1981) Aspectos floristicos e fitosociológicos da floresta inundavel (Igapó) Praia Grande, Rio Negro, Amazonas, Brasil. M. Sc. thesis. CNPq, INPA, FUA, Manaus, Brazil.

Ribeiro, M. C. L. de B. (1983) As migrações dos jaraquis (Pisces, Prochilodontidae) no Rio Negro, Amazonas, Brasil. M. Sc. thesis, CNPq, INPA, FUA, Manaus, Brazil. 192 pp.

Richey, J. E., Devol, A. H., Wofsy, S. C., Victoria, R. and Ribeiro, M. N. G. (1988) Biogenic gases and the oxidation and reduction of carbon in Amazon river and floodplain lakes. Limnology and Oceanography 33: 551–561.

Richey, J. E., Nobre, C. and Deser, C. (1989) Amazon river discharge and climate variability – 1903 to 1985. Science 246: 101–103.

Richey, J. E., Hedges, J. I., Devol, A. H., Quay, P. D., Victoria, R., Martinelli, L. A. and Forsberg, B. R. (1990) Biogeochemistry of carbon in the Amazon river. Limnology and Oceanography 35: 352–371.

Rodriguez, G. (1973) El sistema de Maracaibo. Biologia y ambiente. IVIC, Caracas, Venezuela. 395 pp.

Ruellan, F. (1957) Espedições geomorfológicas no Territorio do Rio Branco. CNPq, INPA, Rio de Janeiro, Brazil. 170 pp.

Ruthsatz, B. (1977) Pflanzengesellschaften und ihre Lebensbedingungen in den andinen Halbwüsten Nordwest-Argentiniens. Dissertationes Botanicae 39, Ganter Verlag, Vaduz, Schweiz. 204 pp.

Saint-Paul, U. and Soares, M. G. M. (1987) Diurnal distribution and behavioral responses of fishes to extreme hypoxia in an Amazon floodplain lake. Environmental Biology of Fishes 20: 91–104.

Saint-Paul, U. and Soares, M. G. M. (1988) Ecomorphological adaptation to oxygen deficiency in Amazon floodplains by Serrasalmid fish of the genus *Mylossoma*. Journal Fisherie Biology 32: 231–236.

Salati, E. and Marques, J. (1984) Climatology of the Amazon region. p. 85–126. *In*: H. Sioli (ed.). The Amazon: Limnology and Landscape Ecology of a Mighty Tropical River and its Basin. Dr. W. Junk Publisher, Dordrecht, The Netherlands.

Salinas, P. J. (1975) Observaciones ecologicas sobre los modulos de Apure. Necesidad de una estacion de estudios ecologicos. Revista Forestal Venezolana, University de los Andes, Merida: 131–155.

Salo, Y., Kalliola, R., Häkkinen, I., Mäkinen, Y., Niemelä, P., Puhakka, M. and Coley, P. D. (1986) River dynamics and the diversity of Amazon lowland forest. Nature: 1005–1011.

Santos dos, A. (1983) Limnologia do sistema Tocantins-Araguaia: aspectos fisicos e biogeoquimicos. Ph. D. thesis, São Carlos, Brazil. 257 pp.

Sarmiento, G. (1983) The savannas of tropical America. p. 245–288. *In*: F. Burliere (ed.). Ecosystems of the World 13: Tropical Savannas. Elsevier Scientific Publication Company, Amsterdam, The Netherlands.

Sarmiento, G., Monasterio, M. and Silva, J. (1971) Reconocimento ecologico de los Llanos Occidentales. IV. El oeste del Estado Apure. Acta Cientifica Venezolana 22: 170–180.

Schmidt, G. W. (1972) Chemical properties of some waters in the tropical rain-forest region of Central Amazonia along the new road Manaus-Caracarai. Amazoniana 3: 199–207.

Schmidt, G. W. (1973) Primary production of phytoplankton in the three types of Amazonian waters. III. Primary productivity of phytoplankton in a tropical floodplain lake of Central Amazonia, Lago do Castanho, Amazonas, Brazil. Amazoniana 4: 379–404.

Schmidt, G. W. (1976) Primary production of phytoplankton in the three types of Amazonian waters. IV. On the primary productivity of phytoplankton in a bay of the lower Rio Negro (Amazonas, Brasil). Amazoniana 5: 517–528.

Scott, D. A. and Carbonell, M. (1986) A Directory of Neotropical Wetlands. IUCN Cambridge and IWRB Slimbridge, England. 665 pp.

Seidenschwarz, F. (1986) Pioniervegetation im Amazonasgebiet Perus. Ein pflanzensoziologischer Vergleich von vorandinem Flußufer und Kulturland. Monographs on Agriculture and Ecology of Warmer Climates. Vol. 3. Margraf, Tropical Scientific Books, Gaimersheim, West Germany. 226 pp.

Shackleton, N. J. and Opdyke, N. D. (1973) Oxygen isotope and paleomagnetic stratigraphy of equatorial Pacific core$_5$ V28–238: Oxygen isotope temperatures and ice volumes on a 10^5 and 10^6 year scale. Quarternary Research 3: 39–55.

Silva da, C. J. (1990) Influência da variação do nivel d'água sobre a estrutura e funcionamento de uma área alagável do Pantanal Matogrossense (Pantanal de Barão de Melgaço), Município de Santo Antonio de Leverger e Barão de Melgaço MT).– Ph.D. thesis, University São Carlos-SP, Brasil. 251 pp.

Sioli, H. (1950) Das Wasser im Amazonasgebiet. Forschungen und Fortschritte 26: 274–280.

Sioli, H. (1965) Bemerkungen zur Typologie amazonischer Flüsse. Amazoniana 1: 74–83.

Sioli, H. (1967) Studies in Amazonian waters. Atlas do Simpósio sôbre a biota Amazônica 3: 9–50.

Sioli, H. (ed.) (1984a) The Amazon: Limnology and Landscape Ecology of a Mighty Tropical River and its Basin. Dr. W. Junk Publisher, Dordrecht, The Netherlands. 763 pp.

Sioli, H. (1984b) The Amazon and its main affluents: Hydrography, morphology of the river courses, and river types. p. 127–165. *In*: H. Sioli (ed.). The Amazon: Limnology and Landscape Ecology of a Mighty Tropical River and its Basin. Dr. W. Junk Publisher, Dordrecht, The Netherlands.

Smith, N. J. H. (1979) A pesca no rio Amazonas. CNPq/INPA, Manaus, Brasil. 154 pp.

Soares, M. G. M., Almeida, R. G. and Junk, W. J. (1986) The trophic status of the fish fauna in Lago Camaleão, a floodplain lake in the middle Amazon. Amazoniana 9: 511–526.

Tarifa, J. R. (1986) O sistema climático do Pantanal. Da compreensão do sistema à definição de prioridade de pesquisa climatológica. Anais do Primeiro Simpósio sobre Recursos Naturais e Sócio-econômicos do Pantanal. Corumbá 1984. Departamento de Difusão de Technologia, Documento No. 5, EMBRAPA, Brasília, Brazil: 9–27.

Tricart, J. (1969) Le salar del Huasco. Revue de géomorphologie dynamique 19: 49–84.

Van der Hammen, T. and Gonzalez, E. (1960) Upper Pleistocene and Holocene climate and vegetation of the "sabana de Bogota" (Columbia, South America). Leidse Geologische medelingen 25: 261–31

Van der Hammen, T. and Gonzalez, E. (1965) A late-glacial and holocene pollen diagram from Cienaga del Visitador (Departamento Boyaca, Columbia). Leidse Geologische Nedelingen 32: 193–201.

Van der Heide, J. (1982) Lake Brokopondo: Filling Phase Limnology of a Man-made Lake in the Humid Tropics. Vrije Universiteit te Amsterdam, Amsterdam, The Netherlands. 428 pp.

Vann, J. H. (1959) The Physical Geography of the Lower Coastal Plain of the Guiana Coast. Technical Report no 1, Project NR 388–029 Contract 1575 (02), Geography Branch, Office of Naval Research, Louisiana State University Baton Rouge, Louisiana, USA. 91 pp.

Vann, J. H. (1969) Landforms, Vegetation and Sea Level Change Along the Guiana Coast of South America. Technicol Report no 3, Project NR 388–O28, Contract 4501 (00), Geography Branch, Office of Naval Research, State University College, Buffalo, New York, USA. 128 pp.

Vasquez, E. (1984) El zooplancton de la seccion baja de um rio de aguas negras (rio Caroni) y de um embalse hidroelectrico (Macagua I), Venezuela. Memorias de la Sociedad de Ciencias Naturales La Salle, Caracas: 109–129.

Vasquez, E. and Sanchez, L. (1984) Variacion estacional del plancton en dos sectores del rio Orinoco y uma laguna de inundacion adjacente. Memorias de la Sociedad de Ciencias Naturales La Salle, Caracas: 11–33.

Victoria, R. L., Martinelli, L. A., Richey, J. E., Devol, A. H., Forsberg, B. R. and Ribeiro, M. N. G. (1989) Spatial and temporal variations in soil chemistry on the Amazon floodplain. Geo-Journal 19: 45–52.

Vila, M. A. (1955) Aspectos geographicos del Estado Apure. Corp. Venezuela Fomento, Caracas, Venezuela. 263 pp.

Villwock, W. (1963) Die Gattung *Orestias* (Pisces Microcyprini) und die Frage der intralacustrischen Speziation im Titicaca Seengebiet. Verhandlungen der Deutschen zoologischen Gesellschaft Wien: 610–624.

Villwock, W. (1972) Gefahren für die endemische Fischfauna durch Einführungsversuche und Akklimatisation von Fremdfischen am Beispiel des Titicaca-Sees (Peru-Bolivien) und des Lanao-Sees (Mindanao/Philippinen). Verhandlungen der Internationalen Vereinigung für Limnologie 18: 1227–1234.

Walter, H. and Medina, E. (1971) Caracterizacion climatica de Venezuela sobre la base de climadiagramas de estaciones particulares. Boletín de la Sociedad Venezolana de Ciencias Naturales 29: 211–240.

Walter, H., Harnickell, E. and Mueller-Dombois, D. (1975) Klimadiagramm-Karten der einzelnen Kontinente und die ökologische Klimagliederung der Erde. Gustav Fischer Verlag, Stuttgart, Germany, Karte 2.

Wassmann, R., Thein, U. G., Whiticar, M. J., Rennenberg, H., Seiler, W.and Junk, W. J. (1992) Methane emissions from the Amazon floodplain: Characterization of production and transport. Global Biogeochemical Cycles 6: 3–13.

Welcomme, R. L. (1979) Fisheries Ecology of Floodplain Rivers. The Chancer Press Ltd., Bungay, Suffolk, United Kingdom. 317 pp.

West, R. C. (1956) Mangrove swamps of the Pacific coast of Colombia. American Association of American Geography 46: 98–121.

West, R. C. (1977) Tidal salt-marsh and mangal formations of Middle and South America. p. 193–213. *In*: V. J. Chapmann (ed.). Ecosystems of the World 1: Wet Coastal Ecosystems. Elsevier Scientific Publisher Company, Amsterdam, The Netherlands.

Wilhelmy, H. (1958) Das Große Pantanal. Die Umschau 18: 555–559.

Williams, L. (1940) Botanical exploration in the middle and lower Caura, Venezuela. Tropical Woods 62: 1–20.

Williams, L. (1941) Forests of the Venezuelan Guiana. Tropical Woods 68: 13–40.

Worbes, M. (1983) Vegetationskundliche Untersuchungen zweier Überschwemmungswälder in Zentralamazonien –vorläufige Ergebnisse. Amazoniana 8: 47–65.

Worbes, M. (1985) Lebensbedingungen und Artengefüge zentralamazonischer Überschwemmungswälder. Ph. D. thesis, Universität Göttingen, Germany. 166 pp.

Worbes, M. (1989) Growth rings, increment and age of trees in inundation forests, savannas and a mountain forest in the Neotropics. IAWA Bulletin n.s. 10: 109–122.

Worbes, M. and Junk, W.J. (1989) Dating tropical trees by means of ^{14}C from bomb tests. Ecology 70: 503–507.

Zaret, T. and Rand, A. S. (1971) Competition in tropical stream fishes: support for the competitive exclusion. Ecology 52: 336–342.

Subject index

Species index